Statistics for Business and Economics

FIRST EUROPEAN EDITION

Statistics for Business and Economics

FIRST EUROPEAN EDITION

Carlos Cortinhas and Ken Black

WILEY

John Wiley & Sons, Ltd.

Library of Congress Cataloging-in-Publication Data

Cortinhas, Carlos.
    Statistics for business and economics / Carlos Cortinhas and Ken Black. — 1st European ed.
      p. cm.
    Includes bibliographical references and index.
    ISBN 978-1-119-99366-7 (pbk.)
    1. Commercial statistics.   2. Economics—Statistical methods.   I. Black, Ken (Kenneth Urban)   II. Title.
  HF1017.C654   2012
  519.5—dc23

                                            2011051097

A catalogue record for this book is available from the British Library.

Set in 10/12pt MinionPro by MPS Limited, Chennai, India
Printed in Italy by Printer Trento Srl.

Critical Values from the _t_ Distribution

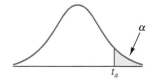

	Values of α for One-Tailed Test and α/2 for Two-Tailed Test					
df	$t_{.100}$	$t_{.050}$	$t_{.025}$	$t_{.010}$	$t_{.005}$	$t_{.001}$
1	3.078	6.314	12.706	31.821	63.656	318.289
2	1.886	2.920	4.303	6.965	9.925	22.328
3	1.638	2.353	3.182	4.541	5.841	10.214
4	1.533	2.132	2.776	3.747	4.604	7.173
5	1.476	2.015	2.571	3.365	4.032	5.894
6	1.440	1.943	2.447	3.143	3.707	5.208
7	1.415	1.895	2.365	2.998	3.499	4.785
8	1.397	1.860	2.306	2.896	3.355	4.501
9	1.383	1.833	2.262	2.821	3.250	4.297
10	1.372	1.812	2.228	2.764	3.169	4.144
11	1.363	1.796	2.201	2.718	3.106	4.025
12	1.356	1.782	2.179	2.681	3.055	3.930
13	1.350	1.771	2.160	2.650	3.012	3.852
14	1.345	1.761	2.145	2.624	2.977	3.787
15	1.341	1.753	2.131	2.602	2.947	3.733
16	1.337	1.746	2.120	2.583	2.921	3.686
17	1.333	1.740	2.110	2.567	2.898	3.646
18	1.330	1.734	2.101	2.552	2.878	3.610
19	1.328	1.729	2.093	2.539	2.861	3.579
20	1.325	1.725	2.086	2.528	2.845	3.552
21	1.323	1.721	2.080	2.518	2.831	3.527
22	1.321	1.717	2.074	2.508	2.819	3.505
23	1.319	1.714	2.069	2.500	2.807	3.485
24	1.318	1.711	2.064	2.492	2.797	3.467
25	1.316	1.708	2.060	2.485	2.787	3.450
26	1.315	1.706	2.056	2.479	2.779	3.435
27	1.314	1.703	2.052	2.473	2.771	3.421
28	1.313	1.701	2.048	2.467	2.763	3.408
29	1.311	1.699	2.045	2.462	2.756	3.396
30	1.310	1.697	2.042	2.457	2.750	3.385
40	1.303	1.684	2.021	2.423	2.704	3.307
50	1.299	1.676	2.009	2.403	2.678	3.261
60	1.296	1.671	2.000	2.390	2.660	3.232
70	1.294	1.667	1.994	2.381	2.648	3.211
80	1.292	1.664	1.990	2.374	2.639	3.195
90	1.291	1.662	1.987	2.368	2.632	3.183
100	1.290	1.660	1.984	2.364	2.626	3.174
150	1.287	1.655	1.976	2.351	2.609	3.145
200	1.286	1.653	1.972	2.345	2.601	3.131
∞	1.282	1.645	1.960	2.326	2.576	3.090

Tree Diagram Taxonomy of Inferential Techniques

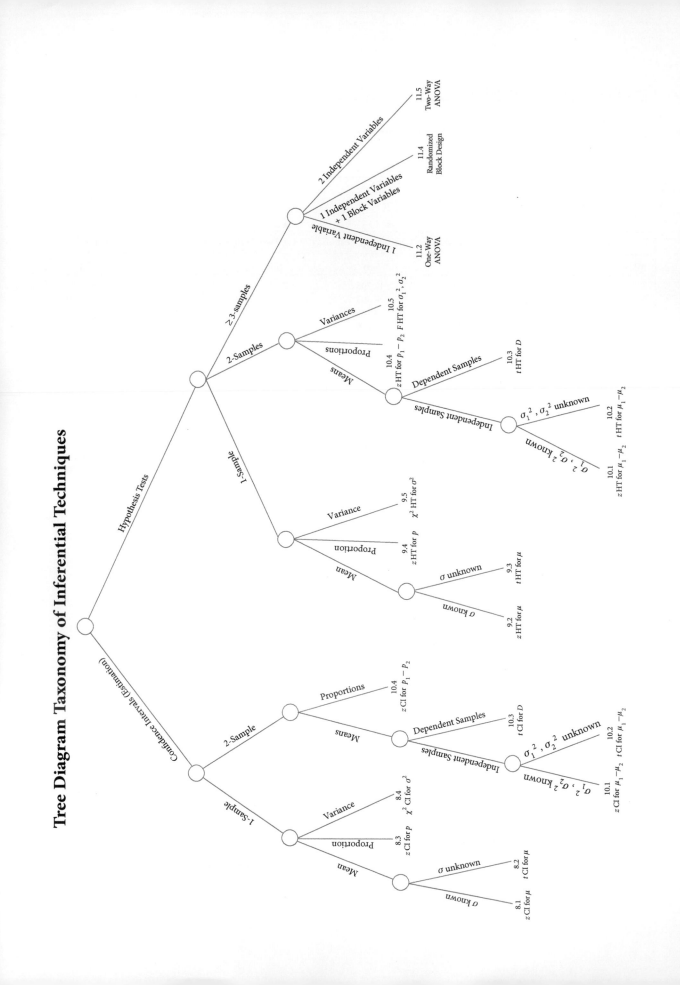

BRIEF CONTENTS

CONTENTS

UNIT V

CATEGORICAL DATA AND NON-PARAMETRIC STATISTICS

APPENDICES

WEBSITE MATERIALS

The first European edition of *Statistics for Business and Economics* is a timely adaptation of the best-selling *Business Statistics for Contemporary Decision Making* by Ken Black. It has been specifically created for European and international audiences and continues the successful tradition of using clear and complete student-friendly pedagogy to present and explain business statistics topics. The rich diversity of problems, examples and Decision Dilemmas presented in this edition are up-to-date and almost entirely based on European and international business and economics data and case studies. With the first European edition, the authors and Wiley have continued to expand the vast ancillary resources available through WileyPLUS, These resources are an important complement to the text, helping instructors to effectively deliver this subject matter and assisting students in their learning.

This edition is written and designed for a two-semester introductory undergraduate business statistics course or an MBA-level introductory course. In addition, with 19 chapters, the first European edition lends itself nicely to adaptation for a one-semester introductory business statistics course. The text is written with the assumption that the student has a basic algebra mathematical background. No calculus is used in the presentation of material in the text.

An underlying philosophical approach to the text is that every statistical tool presented in the book has some business application. While the text contains statistical rigor, it is written so that the student can readily see that the proper application of statistics in the business world goes hand-in-hand with good decision making. In this edition, statistics are presented as a means for converting data into useful information that can be used to assist the business decision maker in making more thoughtful, information-based decisions. Thus, the text presents business statistics as "value added" tools in this conversion process.

CHANGES TO THE FIRST EUROPEAN EDITION

Decision Dilemma and the Decision Dilemma Solved

The popular Decision Dilemma and Decision Dilemma Solved feature included in previous editions of the text has been retained in the first European edition. The Decision Dilemmas are real business vignettes that open each chapter and set the tone by presenting a business dilemma and asking a number of managerial or statistical questions, the solutions to which require the use of techniques presented in the chapter. The Decision Dilemma Solved feature discusses and answers the managerial and statistical questions posed in the Decision Dilemma using techniques from the chapter, thus bringing the chapter to a close. In the first European edition, seven new decision dilemmas have been introduced and a further seven have been updated and revised. Solutions given in the Decision Dilemma Solved feature have been revised for new data and for new versions of computer output. Below is an example of one of the new Decision Dilemmas in this edition:

What is the human resource cost of absenteeism? According to a recent survey by PwC Consultants, UK workers take 10 unauthorised days off from work each year. This number is similar to the average level of absenteeism in Western Europe (9.7 days), but is significantly higher than the 4.5 days average in Asia or the US (5.5 days). Why are the absenteeism figures in Europe so much worse than in Asia or the US? The reasons are not totally clear. As pointed out by the PwC study, one might assume the perceived US work culture of long hours and short holidays could lead to higher stress and sick rates. Our data suggests otherwise, or perhaps demonstrates that strong employee engagement and commitment can override workplace pressures.

Statistics in Business Today

The first European edition includes one or two Statistics in Business Today features in every chapter. This feature presents a real-life example of how the statistics presented in that chapter apply in the business world today. There are thirteen new Statistics in Business Today features in this edition, which have been added for timeliness and relevance to today's students, and others have been revised and updated. The thirteen new Statistics in Business Today features are "Where Are Carbonated Soft Drinks Sold in Europe?", "Recycling Statistics", "Green Tea or Coffee? Daily Drinking Habits in Japan", "Warehousing", "Teleworking in the European Union", "Sampling British Businesses", "What is the World's Favourite Food?", "Online Shopping Habits in Greater China", "Coffee Consumption in the United Kingdom", "Testing Hypotheses about Commuting", "Predicting the Price of a Car", "Inflation Targeting in the Euro Zone" and "Job Search for the Class of 2011". As an example, from "Recycling Statistics,"

In Britain, seven million trees are cut down every year just to make disposable nappies. Disposable nappies take 500 years to decompose. Each year, one person gets through 90 drink cans, 70 food cans, 107 bottles and jars and 45 kg of plastic in the United Kingdom. Each tonne of recycled paper saves about 17 trees. Over 20,000 tonnes of batteries are sent to landfill sites in the UK each year. It takes 50 times more energy to make a battery than it gives during its life. Every year, each British person throws into the dustbin about 50% of the organic garbage that could be composted.

New Problems

Every problem from the previous editios has been examined for timeliness, appropriateness, and logic before inclusion in the first European edition. Those that fell short were replaced or rewritten. While the total number of problems in the text is 950, a concerted effort has been made to include only problems that make a significant contribution to the learning process and that are relevant for a European audience. A large number of new problems have been added to the first European edition, replacing problems that have become less effective or relevant. All demonstration problems and example problems were thoroughly reviewed and edited for effectiveness. A demonstration problem is an extra example containing both a problem and its solution and is used as an additional pedagogical tool to supplement explanations and examples in the chapters. Virtually all example and demonstration problems in the first European edition are business oriented and contain the most current data available.

As with the previous editions, problems are located at the end of most sections in the chapters. A significant number of additional problems are provided at the end of each chapter in the Supplementary Problems section. The Supplementary Problems are "scrambled"—problems using the various techniques in the chapter are mixed—so that students can test themselves on their ability to discriminate and differentiate ideas and concepts.

New Databases

There are nine databases accompanying the first European edition, four of which are new to this edition. One new database is the New Vehicle Registrations in Europe, which includes data on new vehicle registrations for a period of 21 years for a group of sixteen European countries. A second new database is the Household Final Consumption database, which contains 36 annual entries for household consumption variables for the United Kingdom from 1975 to 2010. The third new database is the Energy database that contains price data on five energy variables over a period of 33 years for the group of countries in the Organisation for Economic Co-operation and Development (OECD). The fourth new database is the International Labour database and contains data on the unemployment rates from ten countries presented yearly over a 38-year period.

New Cases

A total of eleven new cases were added to the first European edition. These cases are end-of-chapter case studies, which are based on real data and real companies and give the student an opportunity to use statistical concepts and techniques presented in the chapter to solve a business dilemma. The new cases are "Caffè Nero: The Italian Coffee Co.", "Kodak Transitions Well Into The Digital Camera Market But Faces Big Challenges", "Does Money Make You Happy?", "Artnet AG", "European Banks Face Increasingly Complex Channel Challenges", "Manutan International Mail Order European Success", "The Tomoe Valve Company", "Audience Demand for UK Films", "Mobile Phone Quality and Price", "Nando's" and "Raleigh".

VIDEOTAPE TUTORIALS BY KEN BLACK

An exciting feature of the European edition package that will impact the effectiveness of student learning in business statistics and significantly enhance the presentation of course material is the series of videotape tutorials by Ken Black. With the advent of online business statistics courses, increasingly large class sizes, and the number of commuter students who have very limited access to educational resources on business statistics, it is often difficult for students to get the learning assistance that they need to bridge the gap between theory and application on their own. There are now 22 videotaped tutorial sessions on key difficult topics in business statistics delivered by Ken Black and available for all adopters on WileyPLUS. In addition, these tutorials can easily be uploaded for classroom usage to augment lectures and enrich classroom presentations. Because there is at least one video for each of the first 12 chapters, the instructor has the option to include at least one video in the template of each chapter's plan for most, if not all, of the course. While the video tutorials vary in length, a typical video is about 10 minutes in length. The 22 video tutorials are:

1. Chapter 1: Levels of Data Measurement
2. Chapter 2: Stem-and-Leaf Plot
3. Chapter 3: Computing Variance and Standard Deviation
4. Chapter 3: Understanding and Using the Empirical Rule
5. Chapter 4: Constructing and Solving Joint Probability Tables
6. Chapter 4: Solving Probability Word Problems
7. Chapter 5: Solving Binomial Distribution Problems, Part I
8. Chapter 5: Solving Binomial Distribution Problems, Part II
9. Chapter 6: Solving Problems Using the Normal Curve
10. Chapter 7: Solving for Probabilities of Sample Means Using the z Statistic
11. Chapter 8: Confidence Intervals
12. Chapter 8: Determining Which Inferential Technique to Use, Part I,

Confidence Intervals

13. Chapter 9: Hypothesis Testing Using the z Statistic
14. Chapter 9: Establishing Hypotheses
15. Chapter 9: Understanding p-Values
16. Chapter 9: Type I and Type II errors
17. Chapter 9: Two-Tailed Tests
18. Chapter 9: Determining Which Inferential Technique to Use, Part II,

Hypothesis Tests

19. Chapter 10: Hypothesis Tests of the Difference in Means of Two Independent

Populations Using the t Statistic

20. Chapter 11: Computing and Interpreting a One-Way ANOVA
21. Chapter 12: Testing the Regression Model I—Predicted Values, Residuals, and

Sum of Squares of Error

22. Chapter 12: Testing the Regression Model II—Standard Error of the Estimate and r^2

Each chapter of the first European edition contains the following sections: Learning Objectives, a Decision Dilemma, Demonstration Problems, Section Problems, Statistics in Business Today, Decision Dilemma Solved, Chapter Summary, Key Terms, Formulas, Ethical Considerations, Supplementary Problems, Analyzing the Databases, Case, Using the Computer, and Computer Output from both Excel and Minitab. In addition, Tree Diagrams of Inferential Techniques are presented at the beginning of a number of chapters to help students distinguish between the large plethora of alternative inferential techniques available.

- **Learning Objectives.** Each chapter begins with a statement of the chapter's main learning objectives. This statement gives the reader a list of key topics that will be discussed and the goals to be achieved from studying the chapter.

- **Decision Dilemma.** At the beginning of each chapter, a short case describes a real company or business situation in which managerial and statistical questions are raised. In most Decision Dilemmas, actual data are given and the student is asked to consider how the data can be analyzed to answer the questions.

- **Demonstration Problems.** Virtually every section of every chapter in the first European edition contains demonstration problems. A demonstration problem contains both an example problem and its solution, and is used as an additional pedagogical tool to supplement explanations and examples.

- **Section Problems.** There are over 950 problems in the text. Problems for practice are found at the end of almost every section of the text. Most problems utilize real data gathered from a plethora of sources. Included here are a few brief excerpts from some of the real-life problems in the text: "A survey of 3,219 consumers by GfK NOP for the Food Standards Agency showed that 49% of respondents correctly identified the use by date as the best indicator of whether food is safe to eat or not. Overall respondents were more likely to take heed of the use by/best before dates when using meat, dairy and egg products compared to bread and breakfast cereals. Around 55% said they would not cook and eat raw meat that was past its use by date compared to 27% of respondents when asked about bread and 26% when asked about breakfast cereals." "A survey conducted by the European Agency for Health and Safety at Work revealed that 79% of European managers were concerned about the levels of stress in the workplace. Despite high levels of concern, only 26 percent of EU organisations have procedures in place to deal with stress." "In a study by Syed Sultana recently published in the Global Journal of Finance and Management, it was determined that 20% of all stock investors in India are women. In addition, 22% of all Indian investors are not married."

- **Statistics in Business Today.** Every chapter in the first European edition contains at least one Statistics in Business Today feature. These focus boxes contain an interesting application of how techniques of that particular chapter are used in the business world today. They are usually based on real companies, surveys, or published research.

- **Decision Dilemma Solved.** Situated at the end of the chapter, the Decision Dilemma Solved feature addresses the managerial and statistical questions raised in the Decision Dilemma. Data given in the Decision Dilemma are analyzed computationally and by computer using techniques presented in the chapter. Answers to the managerial and statistical questions raised in the Decision Dilemma are arrived at by applying chapter concepts, thus bringing closure to the chapter.

- **Chapter Summary.** Each chapter concludes with a summary of the important concepts, ideas, and techniques of the chapter. This feature can serve as a preview of the chapter as well as a chapter review.

- **Key Terms.** Important terms are bolded and their definitions italicized throughout the text as they are discussed. At the end of the chapter, a list of the key terms from the chapter is presented. In addition, these terms appear with their definitions in an end-of-book glossary.

- **Formulas.** Important formulas in the text are highlighted to make it easy for a reader to locate them. At the end of the chapter, most of the chapter's formulas are listed together as a handy reference.

- **Ethical Considerations.** Each chapter contains an Ethical Considerations feature that is very timely, given the serious breach of ethics and lack of moral leadership of some business executives in recent years. With the abundance of statistical data and analysis, there is considerable potential for the misuse of statistics in business dealings.

The important Ethical Considerations feature underscores this potential misuse by discussing such topics as lying with statistics, failing to meet statistical assumptions, and failing to include pertinent information for decision makers. Through this feature, instructors can begin to integrate the topic of ethics with applications of business statistics. Here are a few excerpts from Ethical Considerations features: "It is unprofessional and unethical to draw cause-and-effect conclusions just because two variables are correlated." "The business researcher needs to conduct the experiment in an environment such that as many concomitant variables are controlled as possible. To the extent that this is not done, the researcher has an ethical responsibility to report that fact in the findings." "The reader is warned that the value lambda is assumed to be constant in a Poisson distribution experiment. Business researchers may produce spurious results if the value of lambda is used throughout a study; but because the study is conducted during different time periods, the value of lambda is actually changing." "In describing a body of data to an audience, it is best to use whatever statistical measures it takes to present a 'full' picture of the data. By limiting the descriptive measures used, the business researcher may give the audience only part of the picture and skew the way the receiver understands the data."

- **Supplementary Problems.** At the end of each chapter is an extensive set of additional problems. The Supplementary Problems are divided into three groups: Calculating the Statistics, which are strictly computational problems; Testing Your Understanding, which are problems for application and understanding; and Interpreting the Output, which are problems that require the interpretation and analysis of software output.

- **Analyzing the Databases.** There are nine major databases located on the student companion Web site that accompanies the first European edition. The end-of-chapter Analyzing the Databases section contains several questions/problems that require the application of techniques from the chapter to data in the variables of the databases. It is assumed that most of these questions/problems will be solved using a computer.

- **Case.** Each end-of-chapter case is based on a real company. These cases give the student an opportunity to use statistical concepts and techniques presented in the chapter to solve a business dilemma. Some cases feature very large companies—such as Mercedes, Coca-Cola, or Colgate Palmolive. Others pertain to small businesses—such as Caffè Nero, Nando's, or Raleigh—that have overcome obstacles to survive and thrive. Most cases include raw data for analysis and questions that encourage the student to use several of the techniques presented in the chapter. In many cases, the student must analyze software output in order to reach conclusions or make decisions.

- **Using the Computer.** The Using the Computer section contains directions for producing the Excel and Minitab software output presented in the chapter. It is assumed that students have a general understanding of a Microsoft Windows environment. Directions include specifics about menu bars, drop-down menus, and dialog boxes. Not every detail of every dialog box is discussed; the intent is to provide enough information for students to produce the same statistical output analyzed and discussed in the chapter. The sixth edition has a strong focus on both Excel and Minitab software packages. More than 250 Excel or Minitab computer-generated outputs are displayed.

- **Tree Diagrams of Inferential Techniques**. To assist the student in sorting out the plethora of confidence intervals and hypothesis testing techniques presented in the text, tree diagrams are presented at the beginning of Unit III and Chapters 8, 9, 10, and 17. The tree diagram at the beginning of Unit III displays virtually all of the inferential techniques presented in Chapters 8–10 so that the student can construct a view of the "forest from the trees" and determine how each technique plugs into the whole. At the beginning of Chapters 8–10, an additional tree diagram is presented to display the branch of the tree that applies to techniques in that particular chapter. Chapter 17 includes a tree diagram for just the nonparametric statistics presented in that chapter. In determining the appropriate inferential technique for a particular analysis, there are several key questions that the students should ask themselves, such as: Does the problem call for estimation (using a confidence interval) or testing (using a hypothesis test)? How many samples are being analyzed? Are you analyzing means, proportions, or variances? If means are being analyzed, is (are) the variance(s) known or not? If means from two samples are being analyzed, are the samples independent or related? If three or more samples are being analyzed, are there one or two independent variables and is there a blocking variable?

WileyPLUS is a powerful online tool that provides instructors and students with an integrated suite of teaching and learning resources, including an online version of the text, in one easy-to-use website. To learn more about WileyPLUS, request a instructor test drive or view a demo, please visit **www.wileyplus.com.**

WileyPLUS Tools for Instructors

WileyPLUS enables you to:

- Assign automatically graded homework, practice, and quizzes from the end of chapter and test bank.
- Track your students' progress in an instructor's grade book.
- Access all teaching and learning resources, including an online version of the text, and student and instructor supplements, in one easy-to-use website. These include full colour PowerPoint slides, an Excel manual for students, list of key formula and all answers to the problems in the book.
- Create class presentations using Wiley-provided resources, with the ability to customize and add your own materials.

WileyPLUS Resources for Students within WileyPLUS

In WileyPLUS, students will find various helpful tools, such as an eBook, the student study manual, videos with tutorials by the author, interactive learning activities, demonstration problems, databases, problem data and case data in Excel and Minitab, and an Excel manual guiding students through the use of Excel in the book.

- **Ebook.** The complete text is available on WileyPLUS with learning links to various features and tools to assist students in their learning.
- **Videos.** There are 22 videos of the author explaining concepts and demonstrating how to work problems for some of the more difficult topics.
- **Interactive Activities.** Interactive Activities are available, affording students the opportunity to learn concepts by iteratively experimenting with various values of statistics and parameters and observing the outcomes.
- **Learning Activities.** There are numerous learning activities to help the student better understand concepts and key terms. These activities have been developed to make learning fun, enjoyable, and challenging.
- **Data Sets.** Virtually all problems in the text along with the case problems and the databases are available to students in both Excel and Minitab format.
- **Student Study Guide.** Complete answers to all odd-numbered questions.
- **Demo Problems.** Step-by-step solved problems for each chapter.
- **Student's Guide to Using Excel.** A comprehensive guide that walks students through each problem in the book using Microsoft Excel.

ANCILLARY TEACHING AND LEARNING MATERIALS

Students' Companion Site:

www.wiley.com/college/cortinhas

The students' companion Web site contains:

- Student study guide containing answers to all the odd-numbered problems in the text.
- All databases in both Excel and Minitab formats for easy access and use.
- Excel and Minitab files of data from all text problems and all cases. Instructors and students now have the option of analyzing any of the data sets using the computer.
- Full and complete version of Chapter 18, Quality, and 19, Decision Analysis, in PDF format. This allows an instructor the option of covering the material in these chapters in the normal manner, while keeping the text manageable in size and length.
- Students' Guide to using Excel.

Instructor's Resource Kit

All instructor ancillaries are provided on the Instructor Resource Site. Included in this convenient format are:

- **Instructor's Manual.** Prepared by the authors, this manual contains the worked out solutions to virtually all problems in the text. In addition, this manual contains chapter objectives, chapter outlines, chapter teaching strategies, and solutions to the cases.
- **PowerPoint Presentation Slides.** The presentation slides contain graphics to help instructors create stimulating lectures. The PowerPoint slides may be adapted using PowerPoint software to facilitate classroom use.
- **Test Bank.** The Test Bank includes multiple-choice questions for each chapter. The Test Bank is provided in Microsoft Word format and in Respondus.
- **List of Key Formulae.**

ACKNOWLEDGEMENTS

For the First European edition:

List of reviewers

John Wiley & Sons and Carlos Cortinhas would like to thank the reviewers who took the time to provide us with their excellent insights and advice, which was then used to improve the first European edition. The reviewers include Suzanne McCallum, University of Glasgow; Dr Adrian Boucher, University of Birmingham; Assistant Professor Kim Christensen, Aarhus University; Luc Lawers, Katholieke Universiteit Leuven; Andrew Whalley, Royal Holloway, University of London; Dr James Freeman, University of Manchester; Dr Jan Kristian Woike, Université de Lausanne, Martyn Jarvis, University of Glamorgan.

I personally would like to thank the whole team at John Wiley & Sons publishing group for their enthusiasm and assistance on this project. I would especially like to thank Steve Hardman, Sarah Booth, Nicole Burnett and Jennifer Edgecombe.

I would also like to thank Rebecca Harwood for her invaluable contribution in proofreading and editing the early versions of the chapters, Alex Janes from the University of Exeter for his input in some of the case studies and Andrew Gourlay from the University of Exeter for his assistance in creating some of the support materials.

Finally, I would like to thank my wife, Becky, whose love, support, help and encouragement have enabled me to keep going and to complete this project.

—Carlos Cortinhas

For previous editions:

List of reviewers

John Wiley & Sons and I would like to thank the reviewers and advisors who cared enough and took the time to provide us with their excellent insights and advice. These colleagues include: Lihui Bai, Valparaiso University; Pam Boger, Ohio University; Parag Dhumal, Winona State University; Bruce Ketler, Grove City College; Peter Lenk, University of Michigan—Ann Arbor; Robert Montague, Southern Adventist University; Robert Patterson, Penn State University—Behrend; Victor Prybutok, University of North Texas; Nikolai Pulchritudoff, California State University—Los Angeles; Ahmad Saranjam, Northeastern University; Vijay Shah, West Virginia University; Daniel Shimshak, University of Massachusetts—Boston; Cheryl Staley, Lake Land College—Mattoon; Debbie Stiver, University of Nevada—Reno; Minghe Sun, University of Texas—San Antonio

As always, I wish to recognize my colleagues at the University of Houston–Clear Lake for their continued interest and support of this project. In particular, I want to thank William Staples, president; Carl Stockton, provost; and Ted Cummings, dean of the School of Business for their personal interest in the book and their administrative support.

There are several people within the John Wiley & Sons publishing group whom I would like to thank for their invaluable assistance on this project. These include: Franny Kelly, Maria Guarascio, Allie Morris, Lisé Johnson, and Diane Mars.

I want to express a special appreciation to my wife of 41 years, Carolyn, who is the love of my life and continues to provide both professional and personal support in my writing. Thanks also to my daughters, Wendi and Caycee, for their patience, love, and support.

—Ken Black

Carlos Cortinhas is currently a Senior Lecturer at the Department of Economics of the University of Exeter Business School, the United Kingdom. He earned a bachelor's degree in International Economic Relations at the University of Minho in Portugal, a Masters degree in International Economics at the University of Kobe, Japan, and a Ph.D. in International Economics at the University of Exeter in the United Kingdom.

In his academic career, Carlos has taught a large number of statistics courses at the undergraduate level and a number of International Economics and Macroeconomic courses at both undergraduate and postgraduate level. Aside from academia, Carlos has previously worked in banking (for the Industrial Bank of Japan, Tokyo) and in government institutions (for the Ministry for the Economy and Innovation, Portugal) where he worked on applied economics.

Carlos has published a number of journal articles, working papers and a book on applied economics. He is a Senior Fellow of the Higher Education Academy and a member of the Scientific Council of The Economic Policy Research Centre (NIPE).

Carlos Cortinhas and his wife, Becky, have three sons, Samuel, Gabriel and Benjamin.

Ken Black is currently professor of decision sciences in the School of Business at the University of Houston–Clear Lake. Born in Cambridge, Massachusetts, and raised in Missouri, he earned a bachelor's degree in mathematics from Graceland University, a master's degree in math education from the University of Texas at El Paso, a Ph.D. in business administration in management science, and a Ph.D. in educational research from the University of North Texas.

Since joining the faculty of UHCL in 1979, Professor Black has taught all levels of statistics courses, forecasting, management science, market research, and production/ operations management. In 2005, he was awarded the President's Distinguished Teaching Award for the university. He has published over 20 journal articles and 20 professional papers, as well as two textbooks: *Business Statistics: An Introductory Course* and *Business Statistics for Contemporary Decision Making*. Black has consulted for many different companies, including Aetna, the city of Houston, NYLCare, AT&T, Johnson Space Center, Southwest Information Resources, Connect Corporation, and Eagle Engineering.

Ken Black and his wife, Carolyn, have two daughters, Caycee and Wendi. His hobbies include playing the guitar, reading, travelling, and running.

Introduction

THE STUDY OF BUSINESS STATISTICS IS IMPORTANT, valuable, and interesting. However, because it involves a new language of terms, symbols, logic, and application of mathematics, it can be at times confounding. For many students, this text is their first and only introduction to business statistics, which instructors often teach as a 'survey course'. That is, the student is presented with an overview of the subject, including a waterfront of techniques, concepts, and formulas. It can be overwhelming! One of the main difficulties in studying business statistics in this way is to be able to see the forest for the trees: that is, sorting out the myriad topics so they make sense. With this in mind, the 17 chapters of this text have been organized into five units with each unit containing chapters that tend to present similar material. At the beginning of each unit, there is an introduction presenting the overlying themes to those chapters.

Unit I is titled Introduction because the four chapters (1–4) contained therein 'introduce' the study of business statistics. In Chapter 1, students will learn what statistics are, the concepts of descriptive and inferential statistics, and levels of data measurement. In Chapter 2, students will see how raw data can be organized using various graphical and tabular techniques to facilitate their use in making better business decisions. Chapter 3 introduces some essential and basic statistics that will be used to both summarize data and as tools for techniques introduced later in the text. There will also be discussion of distribution shapes. In Chapter 4, the basic laws of probability are presented. The notion of probability underlies virtually every business statistics topic, distribution, and technique, thereby making it important to acquire an appreciation and understanding of probability. In Unit I, the first four chapters, we are developing 'building blocks' that will enable students to understand and apply statistical concepts to analyse data that can assist present and future business managers in making better decisions.

CHAPTER 1

Introduction to Statistics

LEARNING OBJECTIVES

The primary objective of Chapter 1 is to introduce you to the world of statistics, thereby enabling you to:

1. List quantitative and graphical examples of statistics within a business context
2. Define important statistical terms, including population, sample, and parameter, as they relate to descriptive and inferential statistics
3. Compare the four different levels of data: nominal, ordinal, interval, and ratio

Statistics Describe the State of Business in India's Countryside

India is the second largest country in the world, with more than a billion people. Nearly three-quarters of the people live in rural areas scattered about the countryside in six million villages. In fact, it may be said that one in every 10 people in the world live in rural India. Presently, the population in rural India can be described as poor and semi-illiterate. With an annual per capita income of less than US$1 per day, rural India accounts for only about one-third of total national product sales. Less than 50% of households in rural India have electricity, and many of the roads are not paved. The annual per capita consumption for toothpaste is only 30 grams per person in rural India compared with 160 grams in urban India and 400 grams in the United States.

However, in addition to the impressive size of the population, there are other compelling reasons for companies to market their goods and services to rural India. The market of rural India has been growing at five times the rate of the urban India market. There is increasing agricultural productivity, leading to growth in disposable income, and there is a reduction in the gap between the tastes of urban and rural customers. The literacy level is increasing, and people are becoming more conscious of their lifestyles and opportunities for a better life.

Nearly two-thirds of all middle-income households in India are in rural areas, with the number of middle- and high-income households in rural India expected to grow from 80 million to 111 million over the next three years. More than one-third of all rural households now have a main source of income other than farming. Virtually every home has a radio, almost 20% have a television, and more than 30% have at least one bank account.

In the early 1990s, toothpaste consumption in rural India doubled, and the consumption of shampoo increased fourfold. Recently, other products have done well in rural India, accounting for nearly one-half of all of the country's sales of televisions, fans, bicycles, bath soap, and other products. According to MART, a New Delhi-based research organization, rural India buys 46% of all soft drinks and 49% of motorcycles sold in India.

In one year alone, the market for Coca-Cola in rural India grew by 37%, accounting for 80% of new Coke drinkers in India. Because of such factors, many US and Indian firms, such as Microsoft, General Electric, Kellogg's, Colgate-Palmolive, Hindustan Lever, Godrej, Nirma Chemical Works, and Mahotra Marketing, have entered the rural Indian market with enthusiasm. Marketing to rural customers often involves building categories by persuading them to try and to adopt products that they may not have used before. Rural India is a huge, relatively untapped market for businesses. However, entering such a market is not without risks and obstacles. The dilemma facing companies is whether to enter this marketplace and, if so, to what extent and how.

Managerial and Statistical Questions

1. Are the statistics presented in this report exact figures or estimates?
2. How and where could the researchers have gathered such data?
3. In measuring the potential of the rural India marketplace, what other statistics could have been gathered?
4. What levels of data measurement are represented by data on rural India?
5. How can managers use these and other statistics to make better decisions about entering this marketplace?

Source: Adapted from Raja Ramachandran, 'Understanding the Market Environment of India', *Business Horizons*, January 2000; P. Balakrishna and B. Sidharth, 'Selling in Rural India', *The Hindu Business Line* – Internet Edition, 16 February 2004; Rohit Bansal and Srividya Easwaran, 'Creative Marketing for Rural India', research paper, available at www.indiainfoline.com; Alex Steffen, 'Rural India Ain't What It Used to Be', *WorldChanging*, available at www.worldchanging.com/archives/001235.html; 'Corporates Turn to Rural India for Growth', BS Corporate Bureau in New Delhi, 21 August 2003, available at www.rediff.com/money/2003/aug/21rural.htm; Rajesh Jain, 'Tech Talk: The Discovery of India: Rural India', 20 June 2003, www.emergic.org/archives/indi/005721.php; 'Marketing to Rural India: Making the Ends Meet', 8 March 2007, in *India Knowledge@Wharton*, http://knowledge.wharton.upenn.edu/india/article.cfm?articleid=4172.

Every minute of the working day, decisions are made by businesses around the world that determine whether companies will be profitable and growing or whether they will stagnate and die. Most of these decisions are made with the assistance of information gathered about the marketplace, the economic and financial environment, the workforce, the competition, and other factors. Such information usually comes in the form of data or is accompanied by data. Business statistics provides the tool through which such data are collected, analysed, summarized, and presented to facilitate the decision-making process, and business statistics plays an important role in the ongoing saga of decision making within the dynamic world of business.

1.1 STATISTICS IN BUSINESS

Virtually every area of business uses statistics in decision making. Here are some recent examples:

- In a survey of some 1,000 Internet users, J.P. Morgan found that 40% of iPad owners also own a Kindle and another 23% intend to buy one in the next 12 months.
- A survey of 1,465 workers by Hotjobs reports that 55% of workers believe that the quality of their work is perceived the same when they work remotely as when they are physically in the office.
- A survey of 2,700 Chinese households by Credit Suisse showed that household income of the bottom 20% has risen by 50% from 2004 to 2009, while the top 10% has grown 255%. The savings rate has dropped from 26% to 12% during the same period.
- A recent Household Economic Survey by Statistics New Zealand determined that the average weekly household net expenditure in New Zealand was $956 and that households in the Wellington region averaged $120 weekly on recreation and culture. In addition, 75% of all households were satisfied or very satisfied with their material standard of living.
- Survey results published by Harris Interactive suggest that adult Internet users are now spending an average of 13 hours a week online. About 14% spends 24 or more hours a week online, while 20% of adult Internet users are online for only two hours or less a week.

- A Deloitte Retail 'Green' survey of 1,080 adults revealed that 54% agreed that plastic, non-compostable shopping bags should be banned.
- A 2010 Europe-wide survey on graduate employability conducted by Eurobarometer showed that, when it comes to graduate recruitment, 'soft' skills are just as valued as sector-specific and computer skills. Significant numbers of employers questioned said that the ability to work well in a team (98%), to adapt to new situations (97%), communication skills (96%), and knowledge of foreign languages (67%) were important when recruiting for their companies.

You can see from these few examples that there is a wide variety of uses and applications of statistics in business. Note that in most of these examples, business researchers have conducted a study and provided us rich and interesting information.

In this text we will examine several types of graphs for depicting data as we study ways to arrange or structure data into forms that are both meaningful and useful to decision makers. We will learn about techniques for sampling from a population that allow studies of the business world to be conducted more inexpensively and in a more timely manner. We will explore various ways to forecast future values and examine techniques for predicting trends. This text also includes many statistical tools for testing hypotheses and for estimating population values. These and many other exciting statistics and statistical techniques await us on this journey through business statistics. Let us begin.

1.2 BASIC STATISTICAL CONCEPTS

Business statistics, like many areas of study, has its own language. It is important to begin our study with an introduction of some basic concepts in order to understand and communicate about the subject. We begin with a discussion of the word *statistics*. The word *statistics* has many different meanings in our culture. *Webster's Third New International Dictionary* gives a comprehensive definition of **statistics** as *a science dealing with the collection, analysis, interpretation, and presentation of numerical data*. Viewed from this perspective, statistics includes all the topics presented in this text.

The study of statistics can be organized in a variety of ways. One of the main ways is to subdivide statistics into two branches: descriptive statistics and inferential statistics. To understand the difference between descriptive and inferential statistics, definitions of *population* and *sample* are helpful. *Webster's Third New International Dictionary* defines

population as *a collection of persons, objects, or items of interest.* The population can be a widely defined category, such as 'all automobiles', or it can be narrowly defined, such as 'all Volkswagen Polo cars produced from 2002 to 2005'. A population can be a group of people, such as 'all workers presently employed by Microsoft', or it can be a set of objects, such as 'all dishwashers produced on 3 February 2007 by the Bosch Group at the Dilligen plant'. The researcher defines the population to be whatever he or she is studying. When researchers *gather data from the whole population for a given measurement of interest,* they call it a **census**. Most people are familiar with the population Census. Every 10 years, most governments attempt to measure all persons living in the country.

A **sample** is *a portion of the whole* and, if properly taken, is representative of the whole. For various reasons (explained in Chapter 7), researchers often prefer to work with a sample of the population instead of the entire population. For example, in conducting quality-control experiments to determine the average life of light bulbs, a light bulb manufacturer might randomly sample only 75 light bulbs during a production run. Because of time and money limitations, a human resources manager might take a random sample of 40 employees instead of using a census to measure company morale.

If a business analyst is *using data gathered on a group to describe or reach conclusions about that same group,* the statistics are called **descriptive statistics**. For example, if an instructor produces statistics to summarize a class's examination effort and uses those statistics to reach conclusions about that class only, the statistics are descriptive.

Many of the statistical data generated by businesses are descriptive. They might include number of employees on vacation during June, average salary at the London office, corporate sales for 2011, average managerial satisfaction score on a company-wide census of employee attitudes, and average return on investment for Adecco for the years 2000 through 2010.

Another type of statistics is called **inferential statistics**. If a researcher *gathers data from a sample and uses the statistics generated to reach conclusions about the population from which the sample was taken,* the statistics are inferential statistics. The data gathered from the sample are used to infer something about a larger group. Inferential statistics are sometimes referred to as *inductive statistics*. The use and importance of inferential statistics continue to grow.

One application of inferential statistics is in pharmaceutical research. Some new drugs are expensive to produce, and therefore tests must be limited to small samples of patients. Utilizing inferential statistics, researchers can design experiments with small randomly selected samples of patients and attempt to reach conclusions and make inferences about the population.

Market researchers use inferential statistics to study the impact of advertising on various market segments. Suppose a soft drink company creates an advertisement depicting a dispensing machine that talks to the buyer, and market researchers want to measure the impact of the new advertisement on various age groups. The researcher could stratify the population into age categories ranging from young to old, randomly sample each stratum, and use inferential statistics to determine the effectiveness of the advertisement for the various age groups in the population. The advantage of using inferential statistics is that they enable the researcher to study effectively a wide range of phenomena without having to conduct a census. Most of the topics discussed in this text pertain to inferential statistics.

A *descriptive measure of the population* is called a **parameter**. Parameters are usually denoted by Greek letters. Examples of parameters are population mean (μ), population variance (σ^2), and population standard deviation (σ). A *descriptive measure of a sample* is called a **statistic**. Statistics are usually denoted by Roman letters. Examples of statistics are sample mean ($\bar{x}$), sample variance (s^2), and sample standard deviation (s).

Differentiation between the terms *parameter* and *statistic* is important only in the use of inferential statistics. A business researcher often wants to estimate the value of a parameter or conduct tests about the parameter. However, the calculation of parameters is usually either impossible or unfeasible because of the amount of time and money required to take a census. In such cases, the business researcher can take a random sample of the population, calculate a statistic on the sample, and infer by estimation the value of the parameter. The basis for inferential statistics, then, is the ability to make decisions about parameters without having to complete a census of the population.

For example, a manufacturer of washing machines would probably want to determine the average number of loads that a new machine can wash before it needs repairs. The parameter is the population mean or average number of washes per machine before repair. A company researcher takes a sample of machines, computes the number of washes before repair for each machine, averages the numbers, and estimates the population value or parameter by using the statistic, which in this case is the sample average. Figure 1.1 demonstrates the inferential process.

Inferences about parameters are made under uncertainty. Unless parameters are computed directly from the population, the statistician never knows with certainty whether the estimates or inferences made from samples are true. In an effort to estimate the level of confidence in the result of the process, statisticians use probability statements. For this and other reasons, part of this text is devoted to probability (Chapter 4).

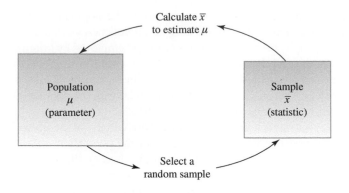

FIGURE 1.1

Process of Inferential Statistics to Estimate a Population Mean (μ)

Millions of numerical data are gathered in businesses every day, representing myriad items. For example, numbers represent euro costs of items produced, geographical locations of retail outlets, weights of shipments, and rankings of subordinates at yearly reviews. All such data should not be analysed the same way statistically because the entities represented by the numbers are different. For this reason, the business researcher needs to know the *level of data measurement* represented by the numbers being analysed.

The disparate use of numbers can be illustrated by the numbers 10 and 20, which could represent the weights of two objects being shipped, the ratings received on a consumer test by two different products, or football shirt numbers of a defender and a striker. Although 20 kilograms is twice as much as 10 kilograms, the striker is probably not twice as big as the defender! Averaging the two weights seems reasonable, but averaging the football shirt numbers makes no sense. The appropriateness of the data analysis depends on the level of measurement of the data gathered. The phenomenon represented by the numbers determines the level of data measurement. Four common levels of data measurement follow.

1. Nominal
2. Ordinal
3. Interval
4. Ratio

Nominal Level

The *lowest level of data measurement* is the **nominal level**. Numbers representing nominal-level data (the word *level* often is omitted) can be *used only to classify or categorise*. Employee identification numbers are an example of nominal data. The numbers are used only to differentiate employees and not to make a value statement about them. Many demographic questions in surveys result in data that are nominal because the questions are used for classification only. The following is an example of such a question that would result in nominal data: Which of the following employment classifications best describes your area of work?

1. Educator
2. Construction worker
3. Manufacturing worker
4. Lawyer
5. Doctor
6. Other

Suppose that, for computing purposes, an educator is assigned a 1, a construction worker is assigned a 2, a manufacturing worker is assigned a 3, and so on. These numbers should be used only to classify respondents. The number 1 does not denote the top classification. It is used only to differentiate an educator (1) from a lawyer (4).

Some other types of variables that often produce nominal-level data are sex, religion, ethnicity, geographic location, and place of birth. Social Security numbers, telephone numbers, employee ID numbers, and ZIP code numbers are further examples of nominal data. Statistical techniques that are appropriate for analysing nominal data are limited. However, some of the more widely used statistics, such as the chi-square statistic, can be applied to nominal data, often producing useful information.

Ordinal Level

Ordinal-level data measurement is higher than the nominal level. In addition to the nominal-level capabilities, ordinal-level measurement can be used to rank or order objects. For example, using ordinal data, a supervisor can evaluate three employees by ranking their productivity with the numbers 1 through 3. The supervisor could identify one employee as the most productive, one as the least productive, and one as somewhere between by using ordinal data. However, the supervisor could not use ordinal data to establish that the intervals between the employees ranked 1 and 2 and between the employees ranked 2 and 3 are equal; that is, she could not say that the differences in the amount of productivity between workers ranked 1, 2, and 3 are necessarily the same. With ordinal data, the distances or spacing represented by consecutive numbers are not always equal.

Some questionnaire Likert-type scales are considered by many researchers to be ordinal in level. The following is an example of one such scale:

This computer tutorial is	___	___	___	___	___
	not helpful	somewhat helpful	moderately helpful	very helpful	extremely helpful
	1	2	3	4	5

When this survey question is coded for the computer, only the numbers 1 through 5 will remain, not the adjectives. Virtually everyone would agree that a 5 is higher than a 4 on this scale and that ranking responses is possible. However, most respondents would not consider the differences between not helpful, somewhat helpful, moderately helpful, very helpful, and extremely helpful to be equal.

Mutual funds as investments are sometimes rated in terms of risk by using measures of default risk, currency risk, and interest rate risk. These three measures are applied to investments by rating them as having high, medium, and low risk. Suppose high risk is assigned a 3, medium risk a 2, and low risk a 1. If a fund is awarded a 3 rather than a 2, it carries more risk, and so on. However, the differences in risk between categories 1, 2, and 3 are not necessarily equal. Thus, these measurements of risk are only ordinal-level measurements. Another example of the use of ordinal numbers in business is the ranking of the top 50 most admired companies in *Fortune* magazine. The numbers ranking the companies are only ordinal in measurement. Certain statistical techniques are specifically suited to ordinal data, but many other techniques are not appropriate for use on ordinal data. For example, it does not make sense to say that the average of 'moderately helpful' and 'very helpful' is 'moderately helpful and a half'.

Because nominal and ordinal data are often derived from imprecise measurements such as demographic questions, the categorization of people or objects, or the ranking of items, *nominal and ordinal data* are **non-metric data** and are sometimes referred to as *qualitative data*.

Interval Level

Interval-level data measurement is the *next to the highest level of data in which the distances between consecutive numbers have meaning and the data are always numerical*. The distances represented by the differences between consecutive numbers are equal; that is, interval data have equal intervals. An example of interval measurement is degree Celsius temperature scale. With Celsius temperature numbers, the temperatures can be ranked, and the amounts of heat between consecutive readings, such as 20°, 21°, and 22°, are the same.

In addition, with interval-level data, the zero point is a matter of convention or convenience and not a natural or fixed zero point. Zero is just another point on the scale and does not mean the absence of the phenomenon. For example, zero degrees Celsius is not the lowest possible temperature. Some other examples of interval level data are the percentage change in employment, the percentage return on a stock, and the euro change in stock price.

Ratio Level

Ratio-level data measurement is *the highest level of data measurement.* Ratio data *have the same properties as interval data,* but ratio data have an *absolute zero,* and *the ratio of two numbers is meaningful.* The notion of absolute zero means that zero is fixed, and *the zero value in the data represents the absence of the characteristic being studied.* The value of zero cannot be arbitrarily assigned because it represents a fixed point. This definition enables the statistician to create *ratios* with the data.

Examples of ratio data are height, weight, time, volume, and Kelvin temperature. With ratio data, a researcher can state that a weight of 180 pounds is twice as much as 90 pounds or, in other words, make a ratio of 180:90. Many of the data gathered by machines in industry are ratio data.

Other examples in the business world that are ratio level in measurement are production cycle time, work measurement time, passenger miles, number of trucks sold, complaints per 10,000 fliers, and number of employees. With ratio-level data, no b factor is required in converting units from one measurement to another: that is, $y = ax$ (e.g., yards $= 3 \times$ feet).

Because interval- and ratio-level data are usually gathered by precise instruments often used in production and engineering processes, in national standardized testing, or in standardized accounting procedures, they are called **metric data** and are sometimes referred to as *quantitative* data.

Comparison of the Four Levels of Data

Figure 1.2 shows the relationships of the usage potential among the four levels of data measurement. The concentric squares denote that each higher level of data can be analysed by any of the techniques used on lower levels of data but, in addition, can be used in other statistical techniques. Therefore, ratio data can be analysed by any statistical technique applicable to the other three levels of data plus some others.

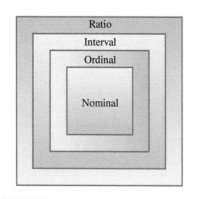

Nominal data are the most limited data in terms of the types of statistical analysis that can be used with them. Ordinal data allow the researcher to perform any analysis that can be done with nominal data and some additional analyses. With ratio data, a statistician can make ratio comparisons and appropriately do any analysis that can be performed on nominal, ordinal, or interval data. Some statistical techniques require ratio data and cannot be used to analyse other levels of data.

Statistical techniques can be separated into two categories: parametric statistics and non-parametric statistics. **Parametric statistics** require that data be interval or ratio. If the data are nominal or ordinal, **non-parametric statistics** must be used. Non-parametric statistics can also be used to analyse interval or ratio data. This text focuses largely on parametric statistics, with the exception of Chapter 16 and Chapter 17, which contain non-parametric techniques. Thus much of the material in this text requires that data be interval or ratio data.

FIGURE 1.2

Usage Potential of Various Levels of Data

DEMONSTRATION PROBLEM 1.1

Many changes continue to occur in the healthcare industry. Because of increased competition for patients among providers and the need to determine how providers can better serve their clientele, hospital administrators sometimes administer a quality satisfaction survey to their patients after the patient is released. The following types of questions are sometimes asked on such a survey. These questions will result in what level of data measurement?

1. How long ago were you released from the hospital?
2. Which type of unit were you in for most of your stay?

- Coronary care
- Intensive care
- Maternity care
- Medical unit
- Paediatric/children's unit
- Surgical unit

3. In choosing a hospital, how important was the hospital's location?

(circle one)

Very Important	Somewhat Important	Not Very Important	Not at All Important

4. How serious was your condition when you were first admitted to the hospital?

    __Critical   __Serious   __Moderate   __Minor

5. Rate the skill of your doctor:

    __Excellent   __Very Good   __Good   __Fair   __Poor

Solution

Question 1 is a time measurement with an absolute zero and is therefore ratio-level measurement. A person who has been out of the hospital for two weeks has been out twice as long as someone who has been out of the hospital for one week.

Question 2 yields nominal data because the patient is asked only to categorize the type of unit he or she was in. This question does not require a hierarchy or ranking of the type of unit. Questions 3, 4, and 5 are likely to result in ordinal-level data. Suppose a number is assigned the descriptors in each of these three questions. For question 3, 'very important' might be assigned a 4, 'somewhat important' a 3, 'not very important' a 2, and 'not at all important' a 1. Certainly, the higher the number, the more important is the hospital's location. Thus, these responses can be ranked by selection. However, the increases in importance from 1 to 2 to 3 to 4 are not necessarily equal. This same logic applies to the numeric values assigned in questions 4 and 5.

STATISTICS IN BUSINESS TODAY

Mobile Phone Use in Japan

The Communications and Information Network Association of Japan (CIAJ) conducts an annual study of mobile phone use in Japan. A recent survey was taken as part of this study using a sample of 600 mobile phone users, split evenly between men and women and almost equally distributed over six age brackets. The survey was administered in the greater Tokyo and Osaka metropolitan areas. The study produced several interesting findings. It was determined that 62.2% had replaced their handsets in the previous 10 months. A little more than 6% owned a second mobile phone. Of these, the objective of about two-thirds was to own one for business use and a second one for personal use. Of all those surveyed, 18.2% used their handsets to view videos, and another 17.3% were not currently using their handsets to view videos but were interested in doing so. Some of the everyday uses of mobile phones included e-mailing (91.7% of respondents), camera functions (77.7%), Internet searching (46.7%), and watching TV (28.0%). In the future, respondents hoped there would be mobile phones with high-speed data transmission that could be used to send and receive PC files (47.7%), for video services such as YouTube (46.9%), for downloading music albums (45.3%) and music videos (40.8%), and for downloading long videos such as movies (39.2%).

Statistical Analysis Using the Computer: Excel and Minitab

The advent of the modern computer opened many new opportunities for statistical analysis. The computer allows for storage, retrieval, and transfer of large data sets. Furthermore, computer software has been developed to analyse data by means of sophisticated statistical techniques. Some widely used statistical techniques, such as multiple regression, are so tedious and cumbersome to compute manually that they were of little practical use to researchers before computers were developed.

Business statisticians use many popular statistical software packages, including Minitab, SAS, and SPSS. Many computer spreadsheet software packages also have the capability of analysing data statistically. In this text, the computer statistical output presented is from both the Minitab and the Microsoft Excel software.

Decision Dilemma SOLVED

Statistics Describe the State of Business in India's Countryside

Several statistics were reported in the Decision Dilemma about rural India, including the average annual consumption of toothpaste per person, the percentage of households having electricity, and the percentage of households that have at least one bank account. The authors of the sources from which the Decision Dilemma was drawn never stated whether the reported statistics were based on actual data drawn from a census of rural India households or were based on estimates taken from a sample of rural households. If the data came from a census, then the totals, averages, and percentages presented in the Decision Dilemma are parameters. If, on the other hand, the data were gathered from samples, then they are statistics. Although governments, especially, do conduct censuses and at least some of the reported numbers could be parameters, more often than not, such data are gathered from samples of people or items. For example, in rural India, the government, academics, or business researchers could have taken random samples of households, gathering consumer statistics that are then used to estimate population parameters, such as percentage of households with televisions, and so forth.

In conducting research on a topic like consumer consumption in rural India, there is potential for a wide variety of statistics to be gathered that represent several levels of data. For example, ratio-level measurements on items such as income, number of children, age of household heads, number of livestock, and grams of toothpaste consumed per year might be obtained. On the other hand, if researchers use a Likert scale (1-to-5 measurements) to gather responses about the interests, likes, and preferences of rural India consumers, an ordinal-level measurement would be obtained, as would the ranking of products or brands in market research studies. Other variables, such as geographic location, sex, occupation, or religion, are usually measured with nominal data.

The decision to enter the rural India market is not just a marketing decision. It involves production capacity and schedule issues, transportation challenges, financial commitments, managerial growth or reassignment, accounting issues (accounting for rural India may differ from techniques used in traditional markets), information systems, and other related areas. With so much on the line, company decision makers need as much relevant information available as possible. In this Decision Dilemma, it is obvious to the decision maker that rural India is still quite poor and illiterate. Its capacity as a market is great. The statistics on the increasing sales of a few personal-care products look promising. What are the future forecasts for the earning power of people in rural India? Will major cultural issues block the adoption of the types of products that companies want to sell there? The answers to these and many other interesting and useful questions can be obtained by the appropriate use of statistics. The 750 million people living in rural India represent the second largest group of people in the world. It certainly is a market segment worth studying further.

ETHICAL CONSIDERATIONS

With the abundance and proliferation of statistical data, potential misuse of statistics in business dealings is a concern. It is, in effect, unethical business behaviour to use statistics out of context. Unethical businesspeople might use only selective data from studies to underscore their point, omitting statistics from the same studies that argue against their case. The results of statistical studies can be misstated or overstated to gain favour.

This chapter noted that if data are nominal or ordinal, then only non-parametric statistics are appropriate for analysis. The use of parametric statistics to analyse nominal and/or ordinal data is wrong and could be considered under some circumstances to be unethical.

In this text, each chapter contains a section on ethics that discusses how businesses can misuse the techniques presented in the chapter in an unethical manner. As both users and producers, business students need to be aware of the potential ethical pitfalls that can occur with statistics.

SUMMARY

Statistics is an important decision-making tool in business and is used in virtually every area of business. In this course, the word *statistics* is defined as the science of gathering, analysing, interpreting, and presenting data.

The study of statistics can be subdivided into two main areas: *descriptive statistics* and *inferential statistics*. Descriptive statistics result from gathering data from a body, group, or population and reaching conclusions only about that group. Inferential statistics are generated from the process of gathering sample data from a group, body, or population and reaching conclusions about the larger group from which the sample was drawn.

The appropriate type of statistical analysis depends on the level of data measurement, which can be (1) *nominal*, (2) *ordinal*, (3) *interval*, or (4) *ratio*. Nominal is the lowest level, representing classification of only such data as geographic location, sex, or Social Security number. The next level is ordinal, which provides rank ordering measurements in which the intervals between consecutive numbers do not necessarily represent equal distances. Interval is the next to highest level of data measurement in which the distances represented by consecutive numbers are equal. The highest level of data measurement is ratio, which has all the qualities of interval measurement, but ratio data contain an absolute zero and ratios between numbers are meaningful. Interval and ratio data sometimes are called *metric* or *quantitative* data. Nominal and ordinal data sometimes are called *non-metric* or *qualitative* data.

Two major types of inferential statistics are (1) *parametric statistics* and (2) *non-parametric statistics*. Use of parametric statistics requires interval or ratio data and certain assumptions about the distribution of the data. The techniques presented in this text are largely parametric. If data are only nominal or ordinal in level, non-parametric statistics must be used.

KEY TERMS

census
descriptive statistics
inferential statistics
interval-level data
metric data
nominal-level data
non-metric data
non-parametric statistics

ordinal-level data
parameter
parametric statistics
population
ratio-level data
sample
statistic
statistics

GO ONLINE TO DISCOVER THE EXTRA FEATURES FOR THIS CHAPTER

The Student Study Guide containing solutions to the odd-numbered questions, additional Quizzes and Concept Review Activities, Excel and Minitab databases, additional data files in Excel and Minitab, and more worked examples.
www.wiley.com/college/cortinhas

SUPPLEMENTARY PROBLEMS

1.1 Give a specific example of data that might be gathered from each of the following business disciplines: accounting, finance, human resources, marketing, information systems, production, and management. An example in the marketing area might be 'number of sales per month by each salesperson'.

1.2 State examples of data that can be gathered for decision-making purposes from each of the following industries: manufacturing, insurance, travel, retailing, communications, computing, agriculture, banking, and healthcare. An example in the travel industry might be the cost of business travel per day in various European cities.

1.3 Give an example of *descriptive* statistics in the recorded music industry. Give an example of how *inferential* statistics could be used in the recorded music industry. Compare the two examples. What makes them different?

1.4 Suppose you are an operations manager for a plant that manufactures batteries. Give an example of how you could use *descriptive* statistics to make better managerial decisions. Give an example of how you could use *inferential* statistics to make better managerial decisions.

1.5 Classify each of the following as nominal, ordinal, interval, or ratio data:
  a. The time required to produce each tyre on an assembly line
  b. The number of bottles of milk a family drinks in a month
  c. The ranking of four machines in your plant after they have been designated as excellent, good, satisfactory, and poor
  d. The telephone area code of clients in the United Kingdom
  e. The age of each of your employees
  f. The euro sales at the local pizza shop each month
  g. An employee's identification number
  h. The response time of an emergency unit

1.6 Classify each of the following as nominal, ordinal, interval, or ratio data:
  a. The ranking of a company by *Fortune* 500
  b. The number of tickets sold at a cinema on any given night
  c. The identification number on a questionnaire
  d. Per capita income
  e. The trade balance in pounds sterling
  f. Profit/loss in euros
  g. A company's tax identification
  h. The Standard & Poor's bond ratings of cities based on the following scales:

Rating	Grade
Highest quality	AAA
High quality	AA
Upper medium quality	A
Medium quality	BBB
Somewhat speculative	BB
Low quality, speculative	B
Low grade, default possible	CCC
Low grade, partial recovery possible	CC
Default, recovery unlikely	C

1.7 The Rathburn Manufacturing Company makes electric wiring, which it sells to contractors in the construction industry. Approximately 900 electric contractors purchase wire from Rathburn annually. Rathburn's director of marketing wants to determine electric contractors' satisfaction with Rathburn's wire. He developed a questionnaire that yields a satisfaction score between 10 and 50 for participant responses. A random sample of 35 of the 900 contractors is asked to complete a satisfaction survey. The satisfaction scores for the 35 participants are averaged to produce a mean satisfaction score.
  a. What is the population for this study?
  b. What is the sample for this study?
  c. What is the statistic for this study?
  d. What would be a parameter for this study?

ANALYZING THE DATABASES

DATABASE

Nine databases are available with this text, providing additional opportunities to apply the statistics presented in this course. These databases are located in WileyPLUS, and each is available in either Minitab or Excel format for your convenience. These nine databases represent a wide variety of business areas, such as agribusiness, consumer spending, energy, finance, healthcare, international labour, new vehicle registrations, manufacturing, and the stock market. Altogether, these databases contain 93 variables and 7,885 observations. The data are gathered from such reliable sources as the UK Office of National Statistics, the European Automobile Manufacturer's Association, the US Department of Agriculture, the American Hospital Association, the International Energy Agency, and *Moody's Handbook of Common Stocks*. Six of the nine databases contain time-series data. These databases are:

NEW VEHICLE REGISTRATIONS IN EUROPE

The new vehicle registrations in Europe database contains data on new vehicle registrations for a period of 21 years for a group of 16 European countries, including Austria, Belgium, Denmark, Finland, France, Germany, Greece, Ireland, Italy, the Netherlands, Portugal, Spain, Sweden, the United Kingdom, Norway, and Switzerland. The data refers to passenger cars and is derived from the European Automobile Manufacturer's Association (ACEA) statistics.

MANUFACTURING DATABASE

This database contains eight variables taken from 20 industries and 140 sub-industries. Some of the industries are food products, textile mill products, furniture, chemicals, rubber products, primary metals, industrial machinery, and transportation equipment. The eight variables are Number of Employees, Number of Production Workers, Value Added by Manufacture, Cost of Materials, Value of Industry Shipments, New Capital Expenditures, End-of-Year Inventories, and Industry Group. Two variables, Number of Employees and Number of Production Workers, are in units of 1,000. Four variables, Value Added by Manufacture, Cost of Materials, New Capital Expenditures, and End-of-Year Inventories, are in million-pound units. The Industry Group variable consists of numbers from 1 to 20 to denote the industry group to which the particular sub-industry belongs. Value of Industry Shipments has been recoded to the following 1-to-4 scale.

1 = £0 to £4.9 billion
2 = £5 billion to £13.9 billion
3 = £14 billion to £28.9 billion
4 = £29 billion or more

HOUSEHOLD FINAL CONSUMPTION EXPENDITURE DATABASE

The household final consumption expenditure database contains 12 variables: expenditure on Food and Non-Alcoholic Beverages, Alcoholic Beverages and Tobacco, Clothing and Footwear, Housing, Furnishings, Health, Transport, Communication, Recreation and Culture, Education, Restaurants and Hotels, and Miscellaneous Goods and Services. There are 36 annual entries for each variable in this database representing details of household final consumption expenditure for the United Kingdom from 1975 to 2010. The data are published by the UK Office of National Statistics.

INTERNATIONAL LABOUR DATABASE

This time-series database contains the civilian unemployment rates in per cent from 10 countries presented yearly over a 38-year period. The data are published by the Bureau of Labor Statistics of the US Department of Labor. The countries are the United States, Canada, Australia, Japan, France, Germany, Italy, the Netherlands, Sweden, and United Kingdom.

FINANCIAL DATABASE

The financial database contains observations on eight variables for 100 companies. The variables are Type of Industry, Total Revenues (€ millions), Total Assets (€ millions), Return on Equity (%), Earnings per Share (€), Average Yield (%), Dividends per Share (€), and Average Price per Earnings (P/E) ratio. The companies represent seven different types of industries. The variable "Type" displays a company's industry type as:

1 = clothing
2 = chemical
3 = electric power
4 = food
5 = healthcare products
6 = insurance
7 = petroleum

ENERGY DATABASE

The time-series energy database consists of price data on five energy variables over a period of 33 years for the group of countries in the Organisation for Economic Co-operation and Development (OECD). The five variables are price of Automotive Diesel, price of Premium Unleaded 95, price of Natural Gas, Steam Coal (per tonne), and price of Electricity. The prices of automotive Diesel and Premium Unleaded 95 are per litre, the prices of Natural Gas are per tens of millions kcal on a GCV basis, the prices of Steam Coal are per tonne and the prices of Electricity are per Kilowatt hour. The data are in USD and are published by the International Energy Agency.

INTERNATIONAL STOCK MARKET DATABASE

This database contains data on seven stock market indexes around the world with data representing monthly averages of each index over a period of five years resulting in 60 data points per variable. The stock market indexes include the American Dow Jones Industrial Average, the NASDAQ, Standard & Poor's 500, the Japanese Nikkei 225, the Hong Kong Hang Seng, the UK FTSE 100, and the Mexican IPC.

HOSPITAL DATABASE

This database contains observations for 11 variables on U.S. hospitals. These variables include Geographic Region, Control, Service, Number of Beds, Number of Admissions, Census, Number of Outpatients, Number of Births, Total Expenditures, Payroll Expenditures, and Personnel.

The region variable is coded from 1 to 7, and the numbers represent the following regions:

1 = South
2 = Northeast
3 = Midwest
4 = Southwest
5 = Rocky Mountain
6 = California
7 = Northwest

Control is a type of ownership. Four categories of control are included in the database:

1 = government, non-federal
2 = non-government, not-for-profit
3 = for-profit
4 = federal government

Service is the type of hospital. The two types of hospitals used in this database are:

1 = general medical
2 = psychiatric

The total expenditures and payroll variables are in units of $1,000.

AGRIBUSINESS TIME-SERIES DATABASE

The agribusiness time-series database contains the monthly weight (in 1,000 lbs.) of cold storage holdings for six different vegetables and for total frozen vegetables over a 14-year period. Each of the seven variables represents 168 months of data. The six vegetables are green beans, broccoli, carrots, sweet corn, onions, and green peas. The data are published by the National Agricultural Statistics Service of the U.S. Department of Agriculture.

ASSIGNMENT

Use the databases to answer the following questions.

1. In the manufacturing database, what is the level of data for each of the following variables?
    a. Number of Production Workers
    b. Cost of Materials
    c. Value of Industry Shipments
    d. Industry Group

2. In the hospital database, what is the level of data for each of the following variables?
    a. Region
    b. Control
    c. Number of Beds
    d. Personnel

3. In the financial database, what is the level of data for each of the following variables?
    a. Type of Industry
    b. Total Assets
    c. P/E Ratio

CASE

CAFFÈ NERO: THE ITALIAN COFFEE CO.

Caffè Nero is a privately held independent coffee retailer based in the United Kingdom that was founded in 1987 by Gerry Ford. His aim was to bring a premium, authentically Italian café to Great Britain. In 1997, Ford bought five retail sites in London and began to create his vision. It took him about a year to get the brand right. In 1998, three more coffee houses were opened in London, bringing the total to eight. During 1999, the company moved outside of London for the first time, opening in Manchester. By the end of 2000, Caffè Nero owned 31 stores and had become a national brand.

In March 2001, Caffè Nero joined the London Stock Exchange and subsequently became the largest publicly listed coffee house company in the UK. By the end of the year, Caffè Nero had 80 coffee houses in 24 cities and towns. In 2002, 29 Aroma stores were acquired, which established the firm as the largest independent coffee house chain in the UK, with 106 outlets. In 2004 another 12 sites were acquired from Coffee Republic and integrated into the group.

In October 2005, Caffè Nero was named the 20th fastest-growing company in Europe by *Business Week* magazine. That same year, Caffè Nero retained the number one position in Allegra's UK coffee rankings, marking its fifth consecutive win, and Gerry was named 'UK Entrepreneur of the Year' by the *Financial Times* and the London Stock Exchange. In early 2007, Gerry took Caffè Nero private, purchasing a majority equity position and removing Caffè Nero from the London Stock Exchange. At the time Caffè Nero was the largest privately owned, independent coffee house business in the UK and was valued at around £250 million. In 2007, the first international expansion was undertaken, with an outlet established in Turkey. At the same time a brand new Nero Express concept was launched within train stations in the UK. Since then, Caffè Nero has expanded to Dubai and the UAE.

Today, Caffè Nero has over 500 stores globally with more than 4,000 employees. Caffè Nero has been part of the coffee culture revolution and has helped to raise the bar on the coffee quality available to consumers through the introduction of hand-crafted artisan coffee in hundreds of towns and cities. Through this, Caffè Nero has become one of the best recognized and most admired coffee-house brands in the world.

Caffè Nero's meteoric success is certainly partly due to keeping the customer at the centre of its strategy and to its continuing market research. Caffè Nero endeavours to take its customers' opinions into account when making decisions on a whole array of areas – the taste of the coffee, new product ideas, the music played, and the atmosphere of the stores. For that purpose, Caffè Nero developed a permanent online 'customer panel' where registered members are regularly asked to complete a number of short surveys (around four a year) to help the company understand its customers' preferences.

DISCUSSION

Think about the market research that is conducted by Caffè Nero.

1. What are some of the populations that Caffè Nero might be interested in measuring for these studies? Does Caffè Nero actually attempt to contact entire populations? What samples are taken? In light of these two questions, how is the inferential process used by Caffè Nero in its market research? Can you think of any descriptive statistics that might be used by Caffè Nero in its decision-making process?

2. In the various market research efforts made by Caffè Nero, some of the possible measurements appear in the list below. Categorize these by level of data. Think of some other measurements that Caffè Nero researchers might have made to help them in this research effort, and categorize them by level of data.

   a. Number of coffees consumed per week per person
   b. Age of coffee purchaser in Caffè Nero
   c. Postal code of the survey respondent
   d. Pounds spent per month on coffee per person
   e. Time in between purchases of coffee
   f. Taste rating of a given type of coffee on a scale from 1 to 10, where 1 is very poor tasting and 10 is excellent taste
   g. Ranking of the taste of four types of coffee on a taste test
   h. Number representing the geographic location of the survey respondent
   i. Quality rating of a type of coffee as excellent, good, average, below average, poor
   j. Number representing the type of coffee being evaluated
   k. Gender of survey respondent

Source: Adapted from 'Caffè Nero: Our Story', available at www.caffenero.com/story/History_01.aspx (last accessed in September 2011); 'Caffè Nero: Customer Panel', available at www.caffenero.com/community/panel/default.aspx (last accessed in September 2011).

Charts and Graphs

LEARNING OBJECTIVES

The overall objective of Chapter 2 is for you to master several techniques for summarizing and depicting data, thereby enabling you to:

1. Construct a frequency distribution from a set of data
2. Construct different types of quantitative data graphs, including histograms, frequency polygons, ogives, dot plots, and stem-and-leaf plots, in order to interpret the data being graphed
3. Construct different types of qualitative data graphs, including pie charts, bar graphs, and Pareto charts, in order to interpret the data being graphed
4. Recognize basic trends in two-variable scatter plots of numerical data

Decision Dilemma

Energy Consumption Around the World

As most people suspect, the United States is the number one consumer of oil in the world, followed by China, Japan, India, Russia, Germany, South Korea, and Canada. China, however, is the world's largest consumer of coal, with the United States coming in second, followed by India, Japan, and Russia. The annual oil and coal consumption figures for eight of the top total energy-consuming nations in the world, according to figures released by the *BP Statistical Review of World Energy* for a recent year, are as follows:

Country	Oil Consumption (million tons)	Coal Consumption (million tons oil equivalent)
United States	943.1	573.7
China	368.0	1,311.4
Japan	228.9	125.3
India	128.5	208.0
Russia	125.9	94.5
Germany	112.5	86.0

(*continued*)

Country	Oil Consumption (million tons)	Coal Consumption (million tons oil equivalent)
South Korea	107.6	59.7
Canada	102.3	30.4

Managerial and Statistical Questions

Suppose you are an energy industry analyst and you are asked to prepare a brief report showing the leading energy-consumption countries in both oil and coal.

1. What is the best way to display the energy consumption data in a report? Are the raw data enough? Can you effectively display the data graphically?
2. Is there a way to graphically display oil and coal figures together so that readers can visually compare countries on their consumptions of the two different energy sources?

Source: 'BP Statistical Review of World Energy,' June 2008, available at www .bp.com/liveassets/bp_internet/globalbp/globalbp_uk_english/reports_ and_publications/statistical_energy_review_2008/STAGING/local_assets/ downloads/pdf/statistical_review_of_world_energy_full_review_2008.pdf.

In Chapters 2 and 3 many techniques are presented for reformatting or reducing data so that the data are more manageable and can be used to assist decision makers more effectively. Two techniques for grouping data are the frequency distribution and the stem-and-leaf plot presented in this chapter. In addition, this chapter discusses and displays several graphical tools for summarizing and presenting data, including histogram, frequency polygon, ogive, dot plot, bar chart, pie chart, and Pareto chart for one-variable data and the scatter plot for two-variable numerical data.

Raw data, or *data that have not been summarized in any way,* are sometimes referred to as **ungrouped data**. Table 2.1 contains 60 years of raw data of the unemployment rates for Canada. *Data that have been organized into a frequency distribution* are called **grouped data**. Table 2.2 presents a frequency distribution for the data displayed

TABLE 2.1

60 Years of Canadian Unemployment Rates (ungrouped data)

2.3	7.0	6.3	11.3	9.6
2.8	7.1	5.6	10.6	9.1
3.6	5.9	5.4	9.7	8.3
2.4	5.5	7.1	8.8	7.6
2.9	4.7	7.1	7.8	6.8
3.0	3.9	8.0	7.5	7.2
4.6	3.6	8.4	8.1	7.7
4.4	4.1	7.5	10.3	7.6
3.4	4.8	7.5	11.2	7.2
4.6	4.7	7.6	11.4	6.8
6.9	5.9	11.0	10.4	6.3
6.0	6.4	12.0	9.5	6.0

TABLE 2.2

Frequency Distribution of 60 Years of Unemployment Data for Canada (grouped data)

Class Interval	Frequency
1–under 3	4
3–under 5	12
5–under 7	13
7–under 9	19
9–under 11	7
11–under 13	5

in Table 2.1. The distinction between ungrouped and grouped data is important because the calculation of statistics differs between the two types of data. This chapter focuses on organizing ungrouped data into grouped data and displaying them graphically.

2.1 FREQUENCY DISTRIBUTIONS

One particularly useful tool for grouping data is the **frequency distribution**, which is *a summary of data presented in the form of class intervals and frequencies.* How is a frequency distribution constructed from raw data? That is, how are frequency distributions like the one displayed in Table 2.2 constructed from raw data like those presented in Table 2.1? Frequency distributions are relatively easy to construct. Although some guidelines and rules of thumb help in their construction, frequency distributions vary in final shape and design, even when the original raw data are identical. In a sense, frequency distributions are constructed according to individual business researchers' taste.

When constructing a frequency distribution, the business researcher should first determine the range of the raw data. The **range** often is defined as *the difference between the largest and smallest numbers.* The range for the data in Table 2.1 is 9.7 (12.0–2.3).

The second step in constructing a frequency distribution is to determine how many classes it will contain. One rule of thumb is to select between *5 and 15 classes.* If the frequency distribution contains too few classes, the data summary may be too general to be useful. Too many classes may result in a frequency distribution that does not aggregate the data enough to be helpful. The final number of classes is arbitrary. The business researcher arrives at a number by examining the range and determining a number of classes that will span the range adequately and also be meaningful to the user. The data in Table 2.1 were grouped into six classes for Table 2.2.

After selecting the number of classes, the business researcher must determine the width of the class interval. An approximation of the class width can be calculated by dividing the range by the number of classes. For the data in Table 2.1, this approximation would be 9.7/6 = 1.62. Normally, the number is rounded up to the next whole number, which in this case is 2. The frequency distribution must start at a value equal to or lower than the lowest number of the ungrouped data and end at a value equal to or higher than the highest number. The lowest unemployment rate is 2.3 and the highest is 12.0, so the business researcher starts the frequency distribution at 1 and ends it at 13. Table 2.2 contains the completed frequency distribution for the data in Table 2.1. Class endpoints are selected so that no value of the data can fit into more than one class. The class interval expression 'under' in the distribution of Table 2.2 avoids such a problem.

Class Midpoint

The *midpoint of each class interval* is called the **class midpoint** and is sometimes referred to as the **class mark**. It is *the value halfway across the class interval* and can be calculated as *the average of the two class endpoints.* For example, in the distribution of Table 2.2, the midpoint of the class interval '3–under 5' is 4, or (3 + 5)/2.

The class midpoint is important, because it becomes the representative value for each class in most group statistics calculations. The third column in Table 2.3 contains the class midpoints for all classes of the data from Table 2.2.

TABLE 2.3

Class Midpoints, Relative Frequencies, and Cumulative Frequencies for Unemployment Data

Interval	Frequency	Class Midpoint	Relative Frequency	Cumulative Frequency
1–under 3	4	2	.0667	4
3–under 5	12	4	.2000	16
5–under 7	13	6	.2167	29
7–under 9	19	8	.3167	48
9–under 11	7	10	.1167	55
11–under 13	5	12	.0833	60
Total	60			

Relative Frequency

Relative frequency is *the proportion of the total frequency that is in any given class interval in a frequency distribution.* Relative frequency is the individual class frequency divided by the total frequency. For example, from Table 2.3, the relative frequency for the class interval 5–under 7 is 13/60 = 0.2167. Consideration of the relative frequency is preparatory to the study of probability in Chapter 4. Indeed, if values were selected randomly from the data in Table 2.1, the probability of drawing a number that is '5–under 7' would be 0.2167, the relative frequency for that class interval. The fourth column of Table 2.3 lists the relative frequencies for the frequency distribution of Table 2.2.

Cumulative Frequency

The **cumulative frequency** is *a running total of frequencies through the classes of a frequency distribution.* The cumulative frequency for each class interval is the frequency for that class interval added to the preceding cumulative total. In Table 2.3 the cumulative frequency for the first class is the same as the class frequency: 4. The cumulative frequency for the second class interval is the frequency of that interval (12) plus the frequency of the first interval (4), which yields a new cumulative frequency of 16. This process continues through the last interval, at which point the cumulative total equals the sum of the frequencies (60). The concept of cumulative frequency is used in many areas, including sales cumulated over a fiscal year, sports scores during a contest (cumulated points), years of service, points earned in a course, and costs of doing business over a period of time. Table 2.3 gives cumulative frequencies for the data in Table 2.2.

DEMONSTRATION PROBLEM 2.1

The following data are the average weekly mortgage interest rates for a 40-week period:

7.29	7.23	7.11	6.78	7.47
6.69	6.77	6.57	6.80	6.88
6.98	7.16	7.30	7.24	7.16
7.03	6.90	7.16	7.40	7.05
7.28	7.31	6.87	7.68	7.03
7.17	6.78	7.08	7.12	7.31
7.40	6.35	6.96	7.29	7.16
6.97	6.96	7.02	7.13	6.84

Construct a frequency distribution for these data. Calculate and display the class midpoints, relative frequencies, and cumulative frequencies for this frequency distribution.

Solution

How many classes should this frequency distribution contain? The range of the data is 1.33 (7.68 − 6.35). If 7 classes are used, each class width is approximately:

$$\text{Class Width} = \frac{\text{Range}}{\text{Number of Classes}} = \frac{1.33}{7} = 0.19$$

If a class width of 0.20 is used, a frequency distribution can be constructed with endpoints that are more uniform looking and allow presentation of the information in categories more familiar to mortgage interest rate users.

The first class endpoint must be 6.35 or lower to include the smallest value; the last endpoint must be 7.68 or higher to include the largest value. In this case the frequency distribution begins at 6.30 and ends at 7.70. The resulting frequency distribution, class midpoints, relative frequencies, and cumulative frequencies are listed in the following table:

Interval	Frequency	Class Midpoint	Relative Frequency	Cumulative Frequency
6.30–under 6.50	1	6.40	.025	1
6.50–under 6.70	2	6.60	.050	3
6.70–under 6.90	7	6.80	.175	10
6.90–under 7.10	10	7.00	.250	20
7.10–under 7.30	13	7.20	.325	33
7.30–under 7.50	6	7.40	.150	39
7.50–under 7.70	1	7.60	.025	40
Total	40			

The frequencies and relative frequencies of these data reveal the mortgage interest rate classes that are likely to occur during the period. Most of the mortgage interest rates (36 of the 40) are in the classes starting with 6.70–under 6.90 and going through 7.30–under 7.50. The rates with the greatest frequency, 13, are in the 7.10–under 7.30 class.

2.1   PROBLEMS

2.1   The following data represent the afternoon high temperatures in Celsius for 50 construction days during a year in Barcelona.

10	26	22	13	23	25	27	8	13	13
17	33	4	12	12	16	21	13	22	33
8	9	32	7	6	11	10	9	11	8
8	31	6	7	12	22	28	16	16	20
16	8	16	10	32	5	21	11	15	5

a. Construct a frequency distribution for the data using five class intervals.
b. Construct a frequency distribution for the data using 10 class intervals.
c. Examine the results of (a) and (b) and comment on the usefulness of the frequency distribution in terms of temperature summarization capability.

2.2   A packaging process is supposed to fill small boxes of raisins with approximately 50 raisins so that each box will weigh the same. However, the number of raisins in each box will vary. Suppose 100 boxes of raisins are randomly sampled, the raisins counted, and the following data are obtained.

57	51	53	52	50	60	51	51	52	52
44	53	45	57	39	53	58	47	51	48
49	49	44	54	46	52	55	54	47	53
49	52	49	54	57	52	52	53	49	47
51	48	55	53	55	47	53	43	48	46
54	46	51	48	53	56	48	47	49	57
55	53	50	47	57	49	43	58	52	44
46	59	57	47	61	60	49	53	41	48
59	53	45	45	56	40	46	49	50	57
47	52	48	50	45	56	47	47	48	46

Construct a frequency distribution for these data. What does the frequency distribution reveal about the box fills?

2.3   The owner of a fast-food restaurant ascertains the ages of a sample of customers. From these data, the owner constructs the frequency distribution shown. For each class interval of the frequency distribution, determine the class midpoint, the relative frequency, and the cumulative frequency.

Class Interval	Frequency
0–under 5	6
5–under 10	8
10–under 15	17
15–under 20	23
20–under 25	18
25–under 30	10
30–under 35	4

What does the relative frequency tell the fast-food restaurant owner about customer ages?

2.4   The human resources manager for a large company commissions a study in which the employment records of 500 company employees are examined for absenteeism during the past year. The business researcher conducting the study organizes the data into a frequency distribution to assist the human resources manager in analysing the data. The frequency distribution is shown. For each class of the frequency distribution, determine the class midpoint, the relative frequency, and the cumulative frequency.

Class Interval	Frequency
0–under 2	218
2–under 4	207
4–under 6	56
6–under 8	11
8–under 10	8

2.5   List three specific uses of cumulative frequencies in business.

2.2   QUANTITATIVE DATA GRAPHS

One of the most effective mechanisms for presenting data in a form meaningful to decision makers is graphical depiction. Through graphs and charts, the decision maker can often get an overall picture of the data and reach some useful conclusions merely by studying the chart or graph. Converting data to graphics can be creative and artful. Often the most difficult step in this process is to reduce important and sometimes expensive data to a graphic picture that is both clear and concise and yet consistent with the message of the original data. One of the most important uses of graphical depiction in statistics is to help the researcher determine the shape of a distribution. Data graphs can generally be classified as quantitative or qualitative. Quantitative data graphs are plotted along a numerical scale, and qualitative graphs are plotted using non-numerical categories. In this section, we will examine five types of quantitative data graphs: (1) histogram, (2) frequency polygon, (3) ogive, (4) dot plot, and (5) stem-and-leaf plot.

Histograms

One of the more widely used types of graphs for quantitative data is the **histogram**. A histogram is a series of contiguous bars or rectangles that represent the frequency of data in given class intervals. If the class intervals used along the horizontal axis are equal, then the height of the bars represent the frequency of values in a given class interval. If the class intervals are unequal, then the areas of the bars (rectangles) can be used for relative comparisons of class frequencies. Construction of a histogram involves labelling the x-axis (abscissa) with the class endpoints and the y-axis (ordinate) with the frequencies, drawing a horizontal line segment from class endpoint to class endpoint at each frequency value, and connecting each line segment vertically from the frequency value to the x-axis to form a series of rectangles (bars). Figure 2.1 is a histogram of the frequency distribution in Table 2.2 produced by using the software package Minitab.

A histogram is a useful tool for differentiating the frequencies of class intervals. A quick glance at a histogram reveals which class intervals produce the highest frequency totals. Figure 2.1 clearly shows that the class interval 7–under 9 yields by far the highest frequency count (19). Examination of the histogram reveals where large increases or decreases occur between classes, such as from the 1–under 3 class to the 3–under 5 class, an increase of 8, and from the 7–under 9 class to the 9–under 11 class, a decrease of 12.

Note that the scales used along the x-and y-axes for the histogram in Figure 2.1 are almost identical. However, because ranges of meaningful numbers for the two variables being graphed often differ considerably, the graph may have different scales on the two axes. Figure 2.2 shows what the histogram of unemployment rates would look like if the scale on the y-axis were more compressed than that on the x-axis. Notice that less difference in the length of the rectangles appears to represent the frequencies in Figure 2.2. It is important that the user of the graph clearly understands the scales used for the axes of a histogram. Otherwise, a graph's creator can 'lie with statistics' by stretching or compressing a graph to make a point.*

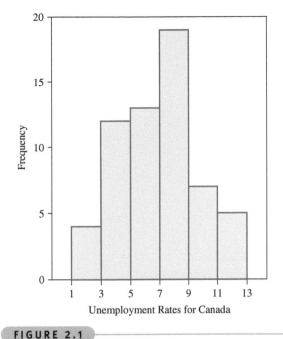

FIGURE 2.1

Minitab Histogram of Canadian Unemployment Data

Using Histograms to Get an Initial Overview of the Data

Because of the widespread availability of computers and statistical software packages to business researchers and decision makers, the histogram continues to grow in importance in yielding information about the shape of the distribution of a large database, the variability of the data, the central location of the data, and outlier data. Although most of these

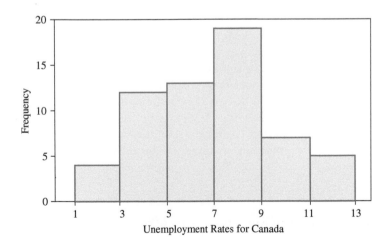

FIGURE 2.2

Minitab Histogram of Canadian Unemployment Data (y-axis compressed)

---

*It should be pointed out that the software package Excel uses the term *histogram* to refer to a frequency distribution. However, by checking Chart Output in the Excel histogram dialog box, a graphical histogram is also created.

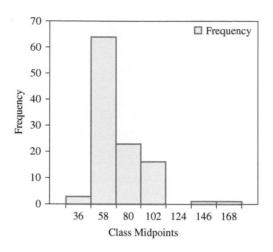

FIGURE 2.3

Histogram of Stock Volumes

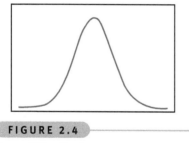

FIGURE 2.4

Normal Distribution

concepts are presented in Chapter 3, the notion of histogram as an initial tool to access these data characteristics is presented here.

A business researcher measured the volume of derivatives contracts traded on the NYSE Liffe derivatives exchange from January 2000 to May 2010 resulting in a database of 125 observations. Suppose a financial decision maker wants to use these data to reach some conclusions about the derivatives market. Figure 2.3 shows a Minitab-produced histogram of these data. What can we learn from this histogram? The great majority of the derivatives market volumes fall between 30 and 110 million contracts. The distribution takes on a shape that is high on the left end and tapered to the right. In Chapter 3 we will learn that the shape of this distribution is skewed towards the right end. In statistics, it is often useful to determine whether data are approximately normally distributed (bell-shaped curve) as shown in Figure 2.4. We can see by examining the histogram in Figure 2.3 that the derivatives market volume data are not normally distributed. Although the centre of the histogram is located near 70 million derivates, a large portion of derivatives volume observations falls in the lower end of the data somewhere between 40 million and 70 million derivatives. In addition, the histogram shows some outliers in the upper end of the distribution. Outliers are data points that appear outside of the main body of observations and may represent phenomena that differ from those represented by other data points. By observing the histogram, we notice a few data observations near 140 and 180 million. One could conclude that on a few trading days an unusually large volume of derivatives contracts are traded. These and other insights can be gleaned by examining the histogram and show that histograms play an important role in the initial analysis of data.

Frequency Polygons

A **frequency polygon**, like the histogram, is a graphical display of class frequencies. However, instead of using bars or rectangles like a histogram, in a frequency polygon each class frequency is plotted as a dot at the class midpoint, and the dots are connected by a series of line segments. Construction of a frequency polygon begins by scaling class midpoints along the horizontal axis and the frequency scale along the vertical axis. A dot is plotted for the associated frequency value at each class midpoint. Connecting these midpoint dots completes the graph. Figure 2.5 shows a frequency polygon of the distribution data from Table 2.2 produced by using the software package Excel. The information gleaned from frequency polygons and histograms is similar. As with the histogram, changing the scales of the axes can compress or stretch a frequency polygon, which affects the user's impression of what the graph represents.

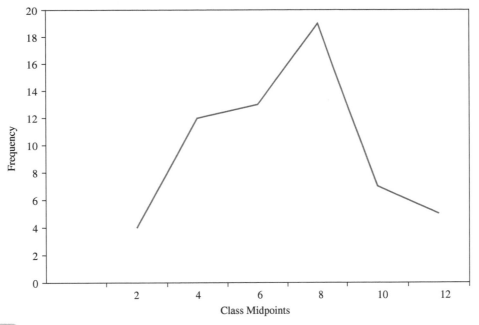

FIGURE 2.5

Excel-Produced Frequency Polygon of the Unemployment Data

Ogives

An **ogive** (o-jive) is *a cumulative frequency polygon*. Construction begins by labelling the x-axis with the class endpoints and the y-axis with the frequencies. However, the use of cumulative frequency values requires that the scale along the y-axis be great enough to include the frequency total. A dot of zero frequency is plotted at the beginning of the first class, and construction proceeds by marking a dot at the *end* of each class interval for the cumulative value. Connecting the dots then completes the ogive. Figure 2.6 presents an ogive produced by using Excel for the data in Table 2.2.

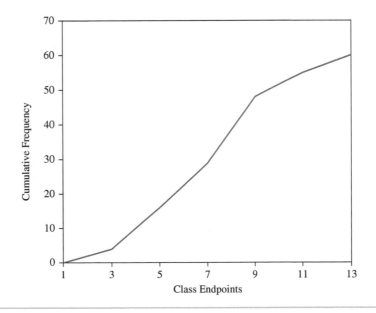

FIGURE 2.6

Excel-Produced Ogive of the Unemployment Data

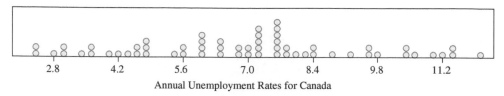

FIGURE 2.7

A Minitab-Produced Dot Plot of the Canadian Unemployment Data

Ogives are most useful when the decision maker wants to see *running totals*. For example, if a comptroller is interested in controlling costs, an ogive could depict cumulative costs over a fiscal year.

Steep slopes in an ogive can be used to identify sharp increases in frequencies. In Figure 2.6, a particularly steep slope occurs in the 7–under 9 class, signifying a large jump in class frequency totals.

Dot Plots

A relatively simple statistical chart that is generally used to display continuous, quantitative data is the **dot plot**. In a dot plot, each data value is plotted along the horizontal axis and is represented on the chart by a dot. If multiple data points have the same values, the dots will stack up vertically. If there are a large number of close points, it may not be possible to display all of the data values along the horizontal axis. Dot plots can be especially useful for observing the overall shape of the distribution of data points along with identifying data values or intervals for which there are groupings and gaps in the data. Figure 2.7 displays a Minitab-produced dot plot for the Canadian unemployment data shown in Table 2.1. Note that the distribution is relatively balanced with a peak near the centre. There are a few gaps to note, such as from 4.9 to 5.3, from 9.9 to 10.2, and from 11.5 to 11.9. In addition, there are groupings around 6.0, 7.1, and 7.5.

Stem-and-Leaf Plots

Another way to organize raw data into groups besides using a frequency distribution is a **stem-and-leaf plot**. This technique is simple and provides a unique view of the data. A stem-and-leaf plot is constructed by separating the digits for each number of the data into two groups, *a stem and a leaf*. The leftmost digits are the stem and consist of the higher-valued digits. The rightmost digits are the leaves and contain the lower values. If a set of data has only two digits, the stem is the value on the left and the leaf is the value on the right. For example, if 34 is one of the numbers, the stem is 3 and the leaf is 4. For numbers with more than two digits, division of stem and leaf is a matter of researcher preference.

Table 2.4 contains scores from an examination on plant safety policy and rules given to a group of 35 job trainees. A stem-and-leaf plot of these data is displayed in Table 2.5. One advantage of such a distribution is that the instructor can readily see whether the scores are in the upper or lower end of each bracket and also determine the spread of the scores. A second advantage of stem-and-leaf plots is that the values of the original raw data are retained (whereas most frequency distributions and graphic depictions use the class midpoint to represent the values in a class).

TABLE 2.4

Safety Examination Scores for Plant Trainees

86	77	91	60	55
76	92	47	88	67
23	59	72	75	83
77	68	82	97	89
81	75	74	39	67
79	83	70	78	91
68	49	56	94	81

TABLE 2.5

Stem-and-Leaf Plot for Plant Safety Examination Data

Stem	Leaf								
2	3								
3	9								
4	7	9							
5	5	6	9						
6	0	7	7	8	8				
7	0	2	4	5	5	6	7	7	8	9
8	1	1	2	3	3	6	8	9	
9	1	1	2	4	7				

DEMONSTRATION PROBLEM 2.2

The following data represent the costs (in euros) of a sample of 30 postal mailings by a company.

3.67	2.75	9.15	5.11	3.32	2.09
1.83	10.94	1.93	3.89	7.20	2.78
6.72	7.80	5.47	4.15	3.55	3.53
3.34	4.95	5.42	8.64	4.84	4.10
5.10	6.45	4.65	1.97	2.84	3.21

Using euros as a stem and cents as a leaf, construct a stem-and-leaf plot of the data.

Solution

Stem	Leaf					
1	83	93	97			
2	09	75	78	84		
3	21	32	34	53	55	67	89
4	10	15	65	84	95	
5	10	11	42	47		
6	45	72				
7	20	80				
8	64					
9	15					
10	94					

2.2 PROBLEMS

2.6 Construct a histogram and a frequency polygon for the following data:

Class Interval	Frequency
30–under 32	5
32–under 34	7
34–under 36	15
36–under 38	21
38–under 40	34
40–under 42	24
42–under 44	17
44–under 46	8

2.7 Construct a histogram and a frequency polygon for the following data:

Class Interval	Frequency
10–under 20	9
20–under 30	7
30–under 40	10
40–under 50	6
50–under 60	13
60–under 70	18
70–under 80	15

2.8   Construct an ogive for the following data:

Class Interval	Frequency
3–under 6	2
6–under 9	5
9–under 12	10
12–under 15	11
15–under 18	17
18–under 21	5

2.9   Construct a stem-and-leaf plot using two digits for the stem.

212	239	240	218	222	249	265	224
257	271	266	234	239	219	255	260
243	261	249	230	246	263	235	229
218	238	254	249	250	263	229	221
253	227	270	257	261	238	240	239
273	220	226	239	258	259	230	262
255	226						

2.10   The following data represent the number of passengers per flight in a sample of 50 flights from London to Amsterdam.

23	46	66	67	13	58	19	17	65	17
25	20	47	28	16	38	44	29	48	29
69	34	35	60	37	52	80	59	51	33
48	46	23	38	52	50	17	57	41	77
45	47	49	19	32	64	27	61	70	19

a.   Construct a dot plot for these data.

b.   Construct a stem-and-leaf plot for these data. What does the stem-and-leaf plot tell you about the number of passengers per flight?

2.3   QUALITATIVE DATA GRAPHS

In contrast to quantitative data graphs that are plotted along a numerical scale, qualitative graphs are plotted using non-numerical categories. In this section, we will examine three types of qualitative data graphs: (1) pie charts, (2) bar charts, and (3) Pareto charts.

Pie Charts

A **pie chart** is *a circular depiction of data where the area of the whole pie represents 100% of the data and slices of the pie represent a percentage breakdown of the sublevels.* Pie charts show the relative magnitudes of the parts to the whole. They are widely used in business, particularly to depict such things as budget categories, market share, and time/resource allocations. However, the use of pie charts is minimized in the sciences and technology because pie charts can lead to less accurate judgements than are possible with other types of graphs.* Generally, it is more difficult for the viewer to interpret the relative size of angles in a pie chart than to judge the length of rectangles in a bar chart. In the feature Statistics in Business Today 'Where Are Carbonated Soft Drinks Sold in Europe?' on page 34 graphical depictions of the percentage of sales by country are displayed by both a pie chart and a vertical bar chart.

Construction of the pie chart begins by determining the proportion of the subunit to the whole. Table 2.6 contains the leading branded oil refineries' operable capacity in barrels per day in a recent year. To construct a pie chart from these data, first convert the raw production capacity figures to proportions by dividing each production capacity figure

---

*William S. Cleveland, *The Elements of Graphing Data*. Monterey, CA: Wadsworth Advanced Books and Software, 1985.

2 CHARTS AND GRAPHS 31

TABLE 2.6

Leading Petroleum Refining Companies

Company	Operable Capacity (barrels per day)	Proportion	Degrees
Exxon Mobil	347,000	.1839	66.20
Petroplus	325,000	.1722	62.00
Royal Dutch Shell	246,000	.1304	46.93
Total	223,000	.1182	42.54
ConocoPhillips	221,000	.1171	42.16
Chevron	220,000	.1166	41.97
Ineos	205,000	.1086	39.11
Murco	100,000	.0530	19.08
Totals	1,887,000	1.0000	360.00

by the total production capacity figure. This proportion is analogous to relative frequency computed for frequency distributions. Because a circle contains 360°, each proportion is then multiplied by 360 to obtain the correct number of degrees to represent each item. For example, Exxon Mobil's capacity of 347,000 barrels represents a 0.1839 proportion of the total given $\left(\dfrac{347,000}{1,887,000} = 0.1839 \right)$. Multiplying this value by 360° results in an angle of 66.20°. The pie chart is then completed by determining each of the other angles and using a compass to lay out the slices. The pie chart in Figure 2.8, constructed by using Minitab, depicts the data from Table 2.6.

Bar Graphs

Another widely used qualitative data graphing technique is the **bar graph** or **bar chart**. A bar graph or chart contains two or more categories along one axis and a series of bars, one for each category, along the other axis. Typically, the length of the bar represents the magnitude of the measure (amount, frequency, money, percentage, etc.) for each category. The bar graph is qualitative because the categories are non-numerical, and it may be either horizontal or vertical. In Excel,

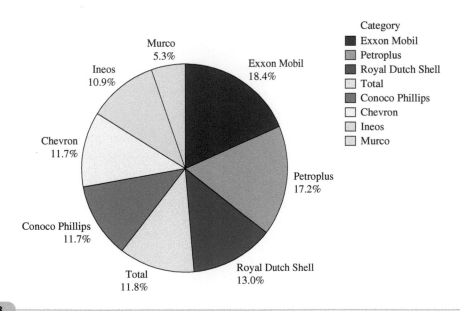

FIGURE 2.8

Minitab Pie Chart of Petroleum Refining Capacity by Brand

TABLE 2.7

Living Expenses of the Average Student in the UK

Category	Amount Spend (£)
Accommodation	3,700
Meals	1,450
Transport	490
Books and stationery	360
Clothes	360
Other general living expenses	1,340

horizontal bar graphs are referred to as 'bar charts' and vertical bar graphs are referred to as 'column charts'. A bar graph generally is constructed from the same type of data that is used to produce a pie chart. However, an advantage of using a bar graph over a pie chart for a given set of data is that for categories that are close in value, it is considered easier to see the difference in the bars of bar graph than discriminating between pie slices.

As an example, consider the data in Table 2.7 regarding how much the average university student spends during an academic year at an English university. Constructing a bar graph from these data, the categories are Accommodation, Meals, Transport, Books and stationery, Clothes, and Other general living expenses. Bars for each of these categories are made using the figures given in the table. The resulting bar graph is shown in Figure 2.9 produced by Excel.

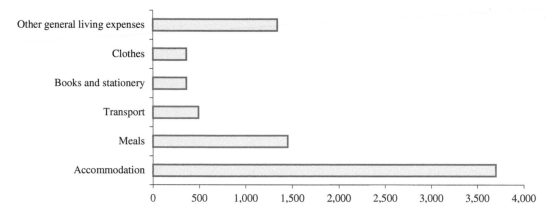

FIGURE 2.9

Bar Graph of Living Expenses of the Average Student

DEMONSTRATION PROBLEM 2.3

According to the Centre for Retail Research, the main sources of inventory shrinkage are customer theft, employee theft, supplier fraud, card fraud, robbery/burglary, criminal damage, and administrative error. The estimated annual pound sterling amount in shrinkage (£ millions) associated with each of these sources follows:

Customer theft	1,886
Employee theft	1,624
Supplier fraud	154
Card fraud	89
Robbery/burglary	71
Criminal damage	39
Administrative error	753
Total	4,616

Construct a pie chart and a bar chart to depict these data.

Solution

To produce a pie chart, convert each raw sterling amount to a proportion by dividing each individual amount by the total.

Customer theft	1,886/4,616 = .409
Employee theft	1,624/4,616 = .352
Supplier fraud	154/4,616 = .033
Card fraud	89/4,616 = .019
Robbery/burglary	71/4,616 = .015
Criminal damage	39/4,616 = .008
Administrative error	753/4,616 = .163
Total	1.000

Convert each proportion to degrees by multiplying each proportion by 360°.

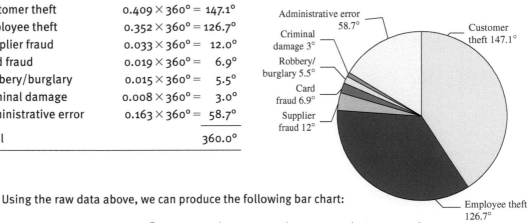

Customer theft	$0.409 \times 360° = 147.1°$
Employee theft	$0.352 \times 360° = 126.7°$
Supplier fraud	$0.033 \times 360° = 12.0°$
Card fraud	$0.019 \times 360° = 6.9°$
Robbery/burglary	$0.015 \times 360° = 5.5°$
Criminal damage	$0.008 \times 360° = 3.0°$
Administrative error	$0.163 \times 360° = 58.7°$
Total	360.0°

Using the raw data above, we can produce the following bar chart:

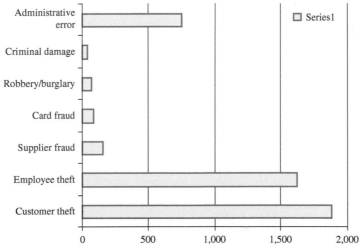

Pareto Charts

A third type of qualitative data graph is a Pareto chart, which could be viewed as a particular application of the bar graph. An important concept and movement in business is total quality management (see Chapter 18). One of the important aspects of total quality management is the constant search for causes of problems in products and processes. A graphical technique for displaying problem causes is Pareto analysis. Pareto analysis is a quantitative tallying of the number and types of defects that occur with a product or service. Analysts use this tally to produce *a vertical bar chart that displays the most common types of defects, ranked in order of occurrence from left to right*. The bar chart is called a **Pareto chart**.

Pareto charts were named after an Italian economist, Vilfredo Pareto, who observed more than 100 years ago that most of Italy's wealth was controlled by a few families who were the major drivers behind the Italian economy. Quality expert J. M. Juran applied this notion to the quality field by observing that poor quality can often be addressed by attacking a few major causes that result in most of the problems. A Pareto chart enables quality-management decision makers to separate the most important defects from trivial defects, which helps them to set priorities for needed quality improvement work.

Suppose the number of electric motors being rejected by inspectors for a company has been increasing. Company officials examine the records of several hundred of the motors in which at least one defect was found to determine which defects occurred more frequently. They find that 40% of the defects involved poor wiring, 30% involved a short in the coil, 25% involved a defective plug, and 5% involved cessation of bearings. Figure 2.10 is a Pareto chart constructed from this information. It shows that the main three problems with defective motors – poor wiring, a short in the coil, and a defective plug – account for 95% of the problems. From the Pareto chart, decision makers can formulate a logical plan for reducing the number of defects.

Company officials and workers would probably begin to improve quality by examining the segments of the production process that involve the wiring. Next, they would study the construction of the coil, then examine the plugs used and the plug-supplier process.

Figure 2.11 is a Minitab rendering of this Pareto chart. In addition to the bar chart analysis, the Minitab Pareto analysis contains a cumulative percentage line

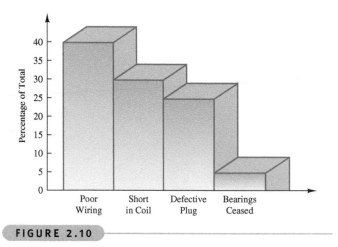

FIGURE 2.10

Pareto Chart for Electric Motor Problems

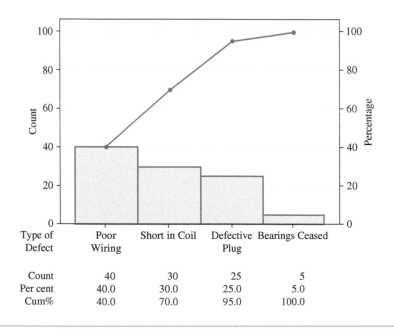

Type of Defect	Poor Wiring	Short in Coil	Defective Plug	Bearings Ceased
Count	40	30	25	5
Per cent	40.0	30.0	25.0	5.0
Cum%	40.0	70.0	95.0	100.0

FIGURE 2.11

Minitab Pareto Chart for Electric Motor Problems

graph. Observe the slopes on the line graph. The steepest slopes represent the more frequently occurring problems. As the slopes level off, the problems occur less frequently. The line graph gives the decision maker another tool for determining which problems to solve first.

Where Are Carbonated Soft Drinks Sold in Europe?

The carbonated soft drink market is an extremely large and growing market in the Europe and worldwide. In a recent year, 51 billion litres of carbonated soft drinks were sold in Europe alone. Where are soft drinks sold? The following data from Datamonitor indicate that the four leading markets for carbonated soft drink in Europe are Germany, the United Kingdom, Italy, and France.

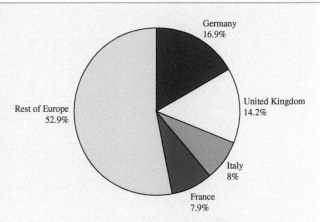

Category	Percentage Share
Germany	16.9
United Kingdom	14.2
Italy	8.0
France	7.9
Rest of Europe	52.9

These data can be displayed graphically several ways. Displayed here are an Excel pie chart and a Minitab bar chart of the data. Some statisticians prefer the histogram or the bar chart over the pie chart because they believe it is easier to compare categories that are similar in size.

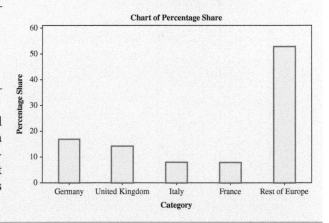

2.3 PROBLEMS

2.11 Shown here is a list of the top five banking groups in the United Kingdom, along with their assets (£ billions). Construct a pie chart and a bar graph to represent these data, and label the slices with the appropriate percentages. Comment on the effectiveness of using a pie chart to display the assets of these top banking companies.

Firm	Assets (£ billion)
HSBC	1,736
Lloyds Banking Group	1,195
Royal Bank of Scotland Group	2,508
Standard Chartered	299
Barclays	2,320

2.12 According to T-100 Domestic Market, the top seven airlines in the United States by domestic boardings in a recent year were Southwest Airlines with 81.1 million, Delta Airlines with 79.4 million, American Airlines with 72.6 million, United Airlines with 56.3 million, Northwest Airlines with 43.3 million, US Airways with 37.8 million, and Continental Airlines with 31.5 million. Construct a pie chart and a bar graph to depict this information.

2.13 The following list shows the top six pharmaceutical companies in the world and their total revenue figures ($ billions) for a recent year. Use this information to construct a pie chart and a bar graph to represent these six companies and their sales.

Pharmaceutical Company	Total Revenues ($ billions)
Johnson & Johnson	61.90
Pfizer	50.01
Roche	47.35
GlaxoSmithKline	45.83
Novartis	44.27
Sanofi-Aventis	41.99

2.14 An airline company uses a central telephone bank and a semi-automated telephone process to take reservations. It has been receiving an unusually high number of customer complaints about its reservation system. The company conducted a survey of customers, asking them whether they had encountered any of the following problems in making reservations: busy signal, disconnection, poor connection, too long a wait to talk to someone, could not get through to an agent, connected with the wrong person. Suppose a survey of 744 complaining customers resulted in the following frequency tally.

Number of Complaints	Complaint
184	Too long a wait
10	Transferred to the wrong person
85	Could not get through to an agent
37	Got disconnected
420	Busy signal
8	Poor connection

Construct a Pareto diagram from this information to display the various problems encountered in making reservations.

2.4 GRAPHICAL DEPICTION OF TWO-VARIABLE NUMERICAL DATA: SCATTER PLOTS

Many times in business research it is important to explore the relationship between two numerical variables. A more detailed statistical approach is given in Chapter 12, but here we present a graphical mechanism for examining the relationship between two numerical variables – the scatter plot (or scatter diagram). A **scatter plot** is *a two-dimensional graph plot of pairs of points from two numerical variables.*

As an example of two numerical variables, consider the data in Table 2.8. Displayed are the values of total life expectancy (in years, male and female) and GDP per capita in Sweden for the past 40 years. Do these two numerical variables exhibit any relationship? The Minitab scatter plot of these data displayed in Figure 2.12 clearly suggests that this is the case for most years in our sample as a clear upward trend can be seen on the graph. The scatter plot also shows, however, that in some years the level of income per person and life expectancy moved in opposite directions.

TABLE 2.8

Life Expectancy and GDP per capita in Sweden over the past 40 Years

Source: World Development Indicators, December 2010 edition.

Life Expectancy, Total (years)	GDP per capita (constant 2000 US$)	Life Expectancy, Total (years)	GDP per capita (constant 2000 US$)
74.08	15,731.60	77.73	23,436.45
74.65	16,526.55	77.54	23,488.28
74.62	16,562.03	77.67	23,072.48
74.72	16,892.89	78.00	22,662.55
74.87	17,536.22	78.06	22,065.15
74.98	18,044.16	78.65	22,788.22
74.98	18,440.99	78.74	23,550.83
74.97	18,578.08	78.96	23,898.05
75.38	18,214.38	79.20	24,527.46
75.47	18,488.46	79.34	25,552.00
75.52	19,149.38	79.43	26,725.78
75.74	19,437.42	79.65	27,879.15
76.03	19,375.23	79.75	28,151.71
76.33	19,595.96	79.85	28,753.85
76.55	19,934.32	80.11	29,320.32
76.82	20,776.53	80.50	30,439.66
76.67	21,200.55	80.64	31,290.27
76.93	21,763.25	80.87	32,431.94
77.09	22,447.19	81.01	33,259.26
76.98	22,952.15	81.24	32,866.26

(continued)

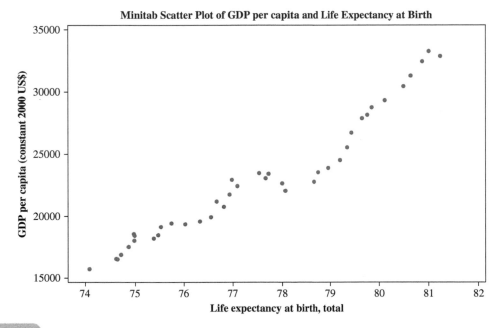

Minitab Scatter Plot of GDP per capita and Life Expectancy at Birth

FIGURE 2.12

Minitab Scatter Plot Life Expectancy and GDP per capita in Sweden over the past 40 Years

2.4 PROBLEMS

2.15 The US National Oceanic and Atmospheric Administration, National Marine Fisheries Service, publishes data on the quantity and value of domestic fishing in the United States. The quantity (in millions of pounds) of fish caught and used for human food and for industrial products (oil, bait, animal food, etc.) over a decade follows. Is a relationship evident between the quantity used for human food and the quantity used for industrial products for a given year? Construct a scatter plot of the data. Examine the plot and discuss the strength of the relationship of the two variables.

Human Food	Industrial Product
3,654	2,828
3,547	2,430
3,285	3,082
3,238	3,201
3,320	3,118
3,294	2,964
3,393	2,638
3,946	2,950
4,588	2,604
6,204	2,259

2.16 Are the advertising amounts spent by a company related to total sales revenue? The data below represent the advertising euros and the sales revenues for various companies in a given industry during a recent year. Construct a scatter plot of the data from the two variables and discuss the relationship between the two variables.

Advertising (in € millions)	Sales (in € millions)
4.2	155.7
1.6	87.3
6.3	135.6
2.7	99.0
10.4	168.2
7.1	136.9
5.5	101.4
8.3	158.2

Decision Dilemma SOLVED

Energy Consumption Around the World

The raw values as shown in the table in the Decision Dilemma are relatively easy to read and interpret. However, these numbers could also be displayed graphically in different ways to create interest and discussion among readers and to allow for more ease of comparisons. For example, shown below are side-by-side Excel pie charts displaying both oil and coal energy consumption figures by country. With such charts, the reader can visually see which countries are dominating consumption of each energy source and then can compare consumption segments across sources.

Pie Charts for World Oil and Coal Consumption (Top Eight Nations)

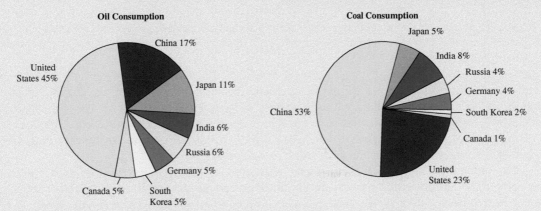

Sometimes it is difficult for the reader to discern the relative sizes of pie slices that are close in magnitude. For that reason, a bar chart might be a better way to display the data. Shown below is a Minitab-produced histogram of the oil consumption data. It is easy to see that the United States dominates world oil consumption.

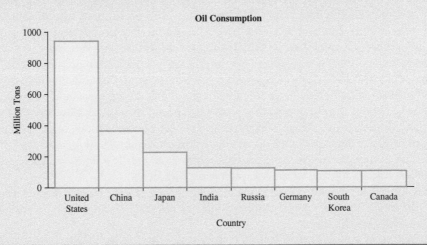

SUMMARY

The two types of data are grouped and ungrouped. Grouped data are data organized into a frequency distribution. Differentiating between grouped and ungrouped data is important, because statistical operations on the two types are computed differently.

Constructing a frequency distribution involves several steps. The first step is to determine the range of the data, which is the difference between the largest value and the smallest value. Next, the number of classes is determined, which is an arbitrary choice of the researcher. However, too few classes over-aggregate the data into meaningless categories, and too many classes do not summarize the data enough to be useful. The third step in constructing the frequency distribution is to determine the width of the class interval. Dividing the range of values by the number of classes yields the approximate width of the class interval.

The class midpoint is the midpoint of a class interval. It is the average of the class endpoints and represents the halfway point of the class interval. Relative frequency is a value computed by dividing an individual frequency by the sum of the frequencies. Relative frequency represents the proportion of total values that is in a given class interval. The cumulative frequency is a running total frequency tally that starts with the first frequency value and adds each ensuing frequency to the total.

Two types of graphical depictions are quantitative data graphs and qualitative data graphs. Quantitative data graphs presented in this chapter are histogram, frequency polygon, ogive, dot plot, and stem-and-leaf plot. Qualitative data graphs presented are pie chart, bar chart, and Pareto chart. In addition, two-dimensional scatter plots are presented. A histogram is a vertical bar chart in which a line segment connects class endpoints at the value of the frequency. Two vertical lines connect this line segment down to the x-axis, forming a rectangle. A frequency polygon is constructed by plotting a dot at the midpoint of each class interval for the value of each frequency and then connecting the dots. Ogives are cumulative frequency polygons. Points on an ogive are plotted at the class endpoints. A dot plot is a graph that displays frequency counts for various data points as dots graphed above the data point. Dot plots are especially useful for observing the overall shape of the distribution and determining both gaps in the data and high concentrations of data. Stem-and-leaf plots are another way to organize data. The numbers are divided into two parts, a stem and a leaf. The stems are the leftmost digits of the numbers and the leaves are the rightmost digits. The stems are listed individually, with all leaf values corresponding to each stem displayed beside that stem.

A pie chart is a circular depiction of data. The amount of each category is represented as a slice of the pie proportionate to the total. The researcher is cautioned in using pie charts because it is sometimes difficult to differentiate the relative sizes of the slices.

The bar chart or bar graph uses bars to represent the frequencies of various categories. The bar chart can be displayed horizontally or vertically.

A Pareto chart is a vertical bar chart that is used in total quality management to graphically display the causes of problems. The Pareto chart presents problem causes in descending order to assist the decision maker in prioritizing problem causes. The scatter plot is a two-dimensional plot of pairs of points from two numerical variables. It is used to graphically determine whether any apparent relationship exists between the two variables.

KEY TERMS

bar graph
class mark
class midpoint
cumulative frequency
dot plot
frequency distribution
frequency polygon
group data
histogram

ogive
Pareto chart
pie chart
range
relative frequency
scatter plot
stem-and-leaf plot
ungrouped data

GO ONLINE TO DISCOVER THE EXTRA FEATURES FOR THIS CHAPTER

The Student Study Guide containing solutions to the odd-numbered questions, additional Quizzes and Concept Review Activities, Excel and Minitab databases, additional data files in Excel and Minitab, and more worked examples.
www.wiley.com/college/cortinhas

SUPPLEMENTARY PROBLEMS

CALCULATING THE STATISTICS

2.17 For the following data, construct a frequency distribution with six classes:

57	23	35	18	21
26	51	47	29	21
46	43	29	23	39
50	41	19	36	28
31	42	52	29	18
28	46	33	28	20

2.18 For each class interval of the frequency distribution given, determine the class midpoint, the relative frequency, and the cumulative frequency.

Class Interval	Frequency
20–under 25	17
25–under 30	20
30–under 35	16
35–under 40	15
40–under 45	8
45–under 50	6

2.19 Construct a histogram, a frequency polygon, and an ogive for the following frequency distribution:

Class Interval	Frequency
50–under 60	13
60–under 70	27
70–under 80	43
80–under 90	31
90–under 100	9

2.20 Construct a dot plot from the following data:

16	15	17	15	15
15	14	9	16	15
13	10	8	18	20
17	17	17	18	23
7	15	20	10	14

2.21 Construct a stem-and-leaf plot for the data below. Let the leaf contain one digit.

312	324	289	335	298
314	309	294	326	317
290	311	317	301	316
306	286	308	284	324

2.22 Construct a pie chart from the following data:

Label	Value
A	55
B	121
C	83
D	46

2.23 Construct a bar graph from the following data:

Category	Frequency
A	7
B	12
C	14
D	5
E	19

2.24 An examination of rejects shows at least seven problems. A frequency tally of the problems follows. Construct a Pareto chart for these data.

Problem	Frequency
1	673
2	29
3	108
4	202
5	73
6	564
7	402

2.25 Construct a scatter plot for the following two numerical variables.

x	y
12	5
17	3
9	10
6	15
10	8
14	9
8	8

TESTING YOUR UNDERSTANDING

2.26 The Whitcomb Company manufactures a metal ring for industrial engines that usually weighs about 50 grams. A random sample of 50 of these metal rings produced the following weights (in grams).

51	53	56	50	44	47
53	53	42	57	46	55
41	44	52	56	50	57
44	46	41	52	69	53
57	51	54	63	42	47
47	52	53	46	36	58
51	38	49	50	62	39
44	55	43	52	43	42
57	49				

Construct a frequency distribution for these data using eight classes. What can you observe about the data from the frequency distribution?

2.27 An European distribution company surveyed 53 of its mid-level managers. The survey obtained the ages of these managers, which later were organized into the frequency distribution shown. Determine the class midpoint, relative frequency, and cumulative frequency for these data.

Class Interval	Frequency
20–under 25	8
25–under 30	6
30–under 35	5
35–under 40	12
40–under 45	15
45–under 50	7

2.28 Use the data from Problem 2.27 to
  a. Construct a histogram and a frequency polygon.
  b. Construct an ogive.

2.29 The following data are shaped roughly like a normal distribution (discussed in Chapter 6).

61.4	27.3	26.4	37.4	30.4	47.5
63.9	46.8	67.9	19.1	81.6	47.9
73.4	54.6	65.1	53.3	71.6	58.6
57.3	87.8	71.1	74.1	48.9	60.2
54.8	60.5	32.5	61.7	55.1	48.2
56.8	60.1	52.9	60.5	55.6	38.1
76.4	46.8	19.9	27.3	77.4	58.1
32.1	54.9	32.7	40.1	52.7	32.5
35.3	39.1				

Construct a frequency distribution starting with 10 as the lowest class beginning point and use a class width of 10. Construct a histogram and a frequency polygon for this frequency distribution and observe the shape of a normal distribution. On the basis of your results from these graphs, what does a normal distribution look like?

2.30 In a medium-sized southern European city, 86 houses are for sale, each having about 185 square metres of floor space. The asking prices vary. The frequency distribution shown contains the price categories for the 86 houses. Construct a histogram, a frequency polygon, and an ogive from these data.

Asking Price (€)	Frequency
100,000–under 120,000	21
120,000–under 140,000	27
140,000–under 160,000	18
160,000–under 180,000	11
180,000–under 200,000	6
200,000–under 220,000	3

2.31 Good, relatively inexpensive prenatal care can often prevent a lifetime of expense owing to complications resulting from a baby's low birth weight. A survey of a random sample of 57 new mothers asked them to estimate how much they spent on prenatal care. The researcher tallied the results and presented them in the frequency distribution shown. Use these data to construct a histogram, a frequency polygon, and an ogive.

Amount Spent on Prenatal Care (€)	Frequency of New Mothers
0–under 100	3
100–under 200	6
200–under 300	12
300–under 400	19
400–under 500	11
500–under 600	6

2.32 A consumer group surveyed food prices at 87 stores in the south-west of the United Kingdom. Among the food prices being measured was that of sugar. From the data collected, the group constructed the frequency distribution of the prices of 5 pounds of sugar in the stores surveyed. Compute a histogram, a frequency polygon, and an ogive for the following data.

Price (£)	Frequency
1.75–under 1.90	9
1.90–under 2.05	14
2.05–under 2.20	17
2.20–under 2.35	16
2.35–under 2.50	18
2.50–under 2.65	8
2.65–under 2.80	5

2.33 The top music genres according to SoundScan for a recent year are R&B, Alternative (Rock), Rap, and Country. These and other music genres along with the number of albums sold in each (in millions) are shown.

Genre	Albums Sold
R&B	146.4
Alternative	102.6
Rap	73.7
Country	64.5
Soundtrack	56.4
Metal	26.6
Classical	14.8
Latin	14.5

Construct a pie chart for these data displaying the percentage of the whole that each of these genres represents. Construct a bar chart for these data.

2.34 The following figures for US imports of agricultural products and manufactured goods were taken from selected years over a 30-year period (in \$ billions). The source of the data is the US International Trade Administration. Construct a scatter plot for these data and determine whether any relationship is apparent between the US imports of agricultural products and the US imports of manufactured goods during this time period.

Agricultural Products	Manufactured Goods
5.8	27.3
9.5	54.0
17.4	133.0
19.5	257.5
22.3	388.8
29.3	629.7

2.35 Shown here is a list of the verified CO_2 emissions (in tonnes) of a number of European Union countries, according to the European Commission. Construct a pie chart and a bar chart to depict this information.

Country	Total Release (in tonnes of CO_2)
Germany	487,004,055
United Kingdom	256,581,160
Italy	226,368,773
Poland	209,601,993
Spain	186,495,894
France	126,634,806
Czech Republic	87,834,758
Netherlands	79,874,658
Greece	72,717,006
Belgium	52,795,318

2.36 A manufacturing company produces plastic bottles for the dairy industry. Some of the bottles are rejected because of poor quality. Causes of poor-quality bottles include faulty plastic, incorrect labelling, discoloration, incorrect thickness, broken handle, and others. The following data for 500 plastic bottles that were rejected include the problems and the frequency of the problems. Use these data to construct a Pareto chart. Discuss the implications of the chart.

Problem	Number
Discoloration	32
Thickness	117
Broken handle	86
Fault in plastic	221
Labelling	44

2.37 A research organization selected 50 German towns with Census 2000 populations between 4,000 and 6,000 as a sample to represent small towns for survey purposes. The populations of these towns follow.

4,420	5,221	4,299	5,831	5,750
5,049	5,556	4,361	5,737	4,654
4,653	5,338	4,512	4,388	5,923
4,730	4,963	5,090	4,822	4,304
4,758	5,366	5,431	5,291	5,254
4,866	5,858	4,346	4,734	5,919
4,216	4,328	4,459	5,832	5,873
5,257	5,048	4,232	4,878	5,166
5,366	4,212	5,669	4,224	4,440
4,299	5,263	4,339	4,834	5,478

Construct a stem-and-leaf plot for the data, letting each leaf contain two digits.

INTERPRETING THE OUTPUT

2.38 Suppose 150 shoppers in an upmarket shopping centre are interviewed and one of the questions asked is the household income. Study the Minitab histogram of the following data and discuss what can be learned about the shoppers.

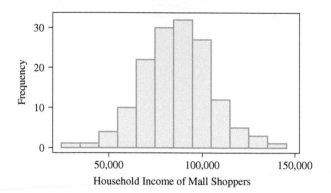

2.39 Study the following dot plot and comment on the general shape of the distribution. Discuss any gaps or heavy concentrations in the data.

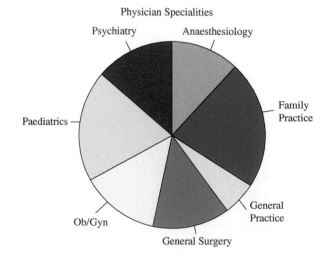

2.41 Suppose 100 Accounting firms are surveyed to determine how many audits they perform over a certain time. The data are summarized using the Minitab stem-and-leaf plot shown below. What can you learn about the number of audits being performed by these firms from this plot?

```
Stem-and-Leaf Display: Audits

Stem-and-leaf of Audits        N = 100
Leaf Unit = 1.0
   9        1     222333333
  16        1     4445555
  26        1     6666667777
  35        1     888899999
  39        2     0001
  44        2     22333
  49        2     55555
  (9)       2     677777777
  42        2     8888899
  35        3     000111
  29        3     223333
  23        3     44455555
  15        3     67777
  10        3     889
   7        4     0011
   3        4     222
```

2.40 Shown here is an Excel-produced pie chart representing medical specialities. What does the chart tell you about the various specialities?

2.42 The following Excel ogive shows toy sales by a company over a 12-month period. What conclusions can you reach about toy sales at this company?

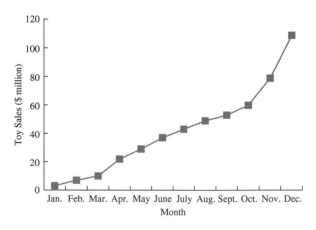

ANALYSING THE DATABASES

1. Using the Manufacturer database, construct a frequency distribution for the variable Number of Production Workers. What does the frequency distribution reveal about the number of production workers?

2. Using the New Vehicle Registrations database, construct a histogram for the variable Switzerland. How is the histogram shaped? Is it high in the middle or high near one or both ends of the data? Is it relatively constant in size across the class (uniform), or does it appear to have no shape? Does it appear to be nearly "normally" distributed?

3. Construct an ogive for the variable Type in the Financial database. The 100 companies in this database are each categorized into one of seven types of companies. These types are listed at the end of Chapter 1. Construct a pie chart of these types and discuss the output. For example, which type is most prevalent in the database and which is the least?

4. Using the international Labour database, construct a stem-and-leaf plot for Italy. What does the plot show about unemployment for Italy over the past 38 years? What does the plot fail to show?

CASE

SOAP COMPANIES DO BATTLE

Procter & Gamble has been the leading soap manufacturer in the United States since 1879, when it introduced Ivory soap. However, late in 1991, its major rival, Lever Bros. (Unilever), overtook it by grabbing 31.5% of the $1.6 billion personal soap market, of which Procter & Gamble had a 30.5% share. Lever Bros. had trailed Procter & Gamble since it entered the soap market with Lifebuoy in 1895. In 1990, Lever Bros. introduced a new soap, Lever 2000, into its product mix as a soap for the entire family. A niche for such a soap had been created because of the segmentation of the soap market into speciality soaps for children, women, and men. Lever Bros. felt that it could sell a soap for everyone in the family. Consumer response was strong; Lever 2000 rolled up $113 million in sales in 1991, putting Lever Bros. ahead of Procter & Gamble for the first time in the personal-soap revenue contest. Procter & Gamble still sells more soap, but Lever's brands cost more, thereby resulting in greater overall sales.

Needless to say, Procter & Gamble was quick to search for a response to the success of Lever 2000. Procter & Gamble looked at several possible strategies, including repositioning Safeguard, which has been seen as a male soap. Ultimately, Procter & Gamble responded to the challenge by introducing its Oil of Olay Moisturizing Bath Bar. In its first year of national distribution, this product was backed by a $24 million media effort. The new bath bar was quite successful and helped Procter & Gamble regain market share.

These two major companies continue to battle it out for domination in the personal soap market, along with the Dial Corporation and Colgate-Palmolive.

Shown below are sales figures in a recent year for personal soaps in the United States. Each of these soaps is produced by one of four soap manufacturers: Unilever, Procter & Gamble, Dial, and Colgate-Palmolive.

Soap	Manufacturer	Sales ($ millions)
Dove	Unilever	271
Dial	Dial	193
Lever 2000	Unilever	138
Irish Spring	Colgate-Palmolive	121
Zest	Procter & Gamble	115
Ivory	Procter & Gamble	94
Caress	Unilever	93
Olay	Procter & Gamble	69
Safeguard	Procter & Gamble	48
Coast	Dial	44

In 1983, the market shares for soap were Procter & Gamble with 37.1%, Lever Bros. (Unilever) with 24%, Dial with 15%, Colgate-Palmolive with 6.5%, and all others with 17.4%. By 1991, the market shares for soap were Lever Bros. (Unilever) with 31.5%, Procter & Gamble with 30.5%, Dial with 19%, Colgate-Palmolive with 8%, and all others with 11%.

DISCUSSION

1. Suppose you are making a report for Procter & Gamble displaying its share of the market along with the share of other companies for the years 1983, 1991, and the latest figures. Using either Excel or Minitab, produce graphs for the market shares of personal soap for each of these years. For the latest figures data, assume that the 'all others' total is about $119 million. What do you observe about the market shares of the various companies by studying the graphs? In particular, how is Procter & Gamble doing relative to previous years?

2. Suppose Procter & Gamble sells about 20 million bars of soap per week, but the demand is not constant and production management would like to get a better handle on how sales are distributed over the year. Let the following sales figures given in units of million bars represent the sales of bars per week over one year. Construct a histogram to represent these data. What do you see in the graph that might be helpful to the production (and sales) people?

17.1	19.6	15.4	17.4	15.0	18.5	20.6	18.4
20.0	20.9	19.3	18.2	14.7	17.1	12.2	19.9
18.7	20.4	20.3	15.5	16.8	19.1	20.4	15.4
20.3	17.5	17.0	18.3	13.6	39.8	20.7	21.3
22.5	21.4	23.4	23.1	22.8	21.4	24.0	25.2
26.3	23.9	30.6	25.2	26.2	26.9	32.8	26.3
26.6	24.3	26.2	23.8				

Construct a stem-and-leaf plot using the whole numbers as the stems. What advantages does the stem-and-leaf plot of these sales figures offer over the histogram? What are some disadvantages? Which would you use in discussions with production people, and why?

3. A random sample of finished soap bars in their packaging is tested for quality. All defective bars are examined for problem causes. Among the problems found were improper packaging, poor labelling, bad seal, shape of bar wrong, bar surface marred, wrong colour in bar, wrong bar fragrance, wrong soap consistency, and others. Some of the leading problem causes and the number of each are given here. Use a Pareto chart to analyse these problem causes. Based on your findings, what would you recommend to the company?

Problem Cause	Frequency
Bar surface	89
Colour	17
Fragrance	2
Label	32
Shape	8
Seal	47
Labelling	5
Soap consistency	3

Source: Adapted from Valerie Reitman, 'Buoyant Sales of Lever 2000 Soap Bring Sinking Sensation to Procter & Gamble', *The Wall Street Journal*, 19 March 1992, p. B1. Reprinted by permission of *The Wall Street Journal* © 1992, Dow Jones & Company, Inc. All rights reserved worldwide; Pam Weisz, '$40M Extends Lever 2000 Family', *Brandweek*, vol. 36, no. 32 (21 August 1995), p. 6; Laurie Freeman, 'P&G Pushes Back Against Unilever in Soap', *Advertising Age*, vol. 65, no. 41 (28 September 1994), p. 21; Jeanne Whalen and Pat Sloan, 'Intros Help Boost Soap Coupons', *Advertising Age*, vol. 65, no. 19 (2 May 1994), p. 30; and 'P&G Places Coast Soap up for Sale', *The Post*, World Wide Web Edition of *The Cincinnati Post*, 2 February 1999, www.cincypost.com.business/pg022599.html.

USING THE COMPUTER

EXCEL

- Excel offers the capability of producing many of the charts and graphs presented in this chapter. Most of these can be accessed by clicking on the **Insert** tab found along the top of an Excel worksheet (second tab from the left next to **Home**). In addition, Excel can generate frequency distributions and histograms using the **Data Analysis** feature.

- Many of the statistical techniques presented in this text can be performed in Excel using a tool called **Data Analysis**. To access this feature, select the **Data** tab along the top of an Excel worksheet. The **Data** tab is the fifth tab over from the left. If the **Data Analysis** feature has been uploaded into your Excel package, it will be found in the **Analysis** section at the top right of the **Data** tab page on the far right. If **Data Analysis** does not appear in the **Analysis** section, it must be added in. To add in **Data Analysis**: (1) Click on the Microsoft Office logo button located in the very topmost left of the Excel 2007 page (looks like an Office icon and is called **Office Button**); (2) now click on **Excel Options** located at the bottom of the pulldown menu; (3) from the menu of the left panel of the **Excel Options** dialog box, click on **Add-Ins**. From the resulting menu shown on the right side of the dialog box, highlight **Analysis ToolPak**. Click on **Go. . .** at the bottom of the page. An **Add-Ins** dialog box will appear with a menu. Check **Analysis ToolPak** and click on **OK**. Your **Data Analysis** feature is now uploaded onto your computer, and you need not add it in again. Now you can bring up the **Analysis ToolPak** feature at any time by going to the **Data** tab at the top of the Excel worksheet and clicking on **Data Analysis**.

- In Excel, frequency distributions are referred to as histograms, and the classes of a frequency distribution are referred to as bins. If you do not specify bins (classes), Excel will automatically determine the number of bins and assign class endpoints based on a formula. If you want to specify bins, load the class endpoints that you want to use into a column. To construct a frequency distribution, select the **Data** tab in the Excel worksheet and then select the **Data Analysis** feature (upper right). If this feature does not appear, you may need to add it (see above). Clicking on **Data Analysis**, the dialog box features a pulldown menu of many of the statistical analysis tools presented and used in this text. From this list, select **Histogram**. In the **Histogram** dialog box, place the location of the raw data values in the space beside **Input Range**. Place the location of the class endpoints (optional) in the space beside **Bin Range**. Leave this blank if you want Excel to determine the bins (classes). If you have labels, check **Labels**. If you want a histogram graph, check **Chart Output**. If you want an ogive, select **Cumulative Percentage** along with **Chart Output**. If you opt for this, Excel will yield a histogram graph with an ogive overlaid on it.

- Excel has excellent capability of constructing many different types of charts, including column charts, line charts, pie charts, bar charts, area charts, and XY (scatter) charts. To begin the process of producing these charts, select the **Insert** tab from the top of the Excel 2007 worksheet. In the **Charts** section, which is the middle section shown at the top of the **Insert** worksheet, there are icons for column, line, pie, bar, area, scatter, and other charts. Click on the icon representing the desired chart to begin construction. Each of these types of charts allow for several versions of the chart shown in the dropdown menu. For example, the pie chart menu contains four types of two-dimensional pie charts and two types of three-dimensional pie charts. To select a particular version of a type of chart, click on the type of chart and then the version of that chart that is desired.

- To construct a pie chart, enter the categories in one column and the data values of each category in another column in the Excel worksheet. Categories and data values could also be entered in rows instead of columns. Click and drag over the data for which the pie chart is to be constructed. From the **Insert** tab, select **Pie** from the **Charts** section and then select the type of pie chart to be constructed. The result is a pie chart from the data. Once the chart has been constructed, a set of three new tabs appears at the top of the worksheet under the general area of **Chart Tools** (see top upper right corner of worksheet). The three new tabs are **Design**, **Layout**, and **Format**. There are many options available for changing the design of the pie chart that can be accessed by clicking on the up and down arrow on the right end of the **Design** tab in the section called **Chart Styles**. On the far right end of the **Design** menu bar is a feature called **Move Chart Location**, which can be used to move the chart to another location or to a new sheet. On the far left end of the **Design** menu bar, there is a **Change Chart Type** feature that allows for changing the type of chart that has been constructed. The second group of features from the left at the top of the **Design** menu bar makes it possible to switch to another set of data (**Select Data**) or switch rows and columns (**Switch Row/Column**). There is a useful feature in the middle of the Design menu bar called

Quick Layout that offers several different layouts for the given chart type. For example, for pie charts, there are seven different possible layouts using titles, labels, and legends in different ways. Right-clicking while on the pie chart brings up a menu that includes **Format Data Labels** and **Format Data Series**. Clicking on **Format Data Labels** brings up another menu (shown on the left) that allows you to modify or edit various features of your graph, including **Label Options**, **Number**, **Fill**, **Border Color**, **Border Styles**, **Shadow**, **3-D Format**, and **Alignment**. Under **Label Options**, there are several different types of labels for pie charts and there are other various chart options available, such as **Series Name**, **Category Name**, **Value**, **Percentage**, and **Show Leader Lines**. In addition, it offers various options for the label location, such as **Center**, **Inside End**, **Outside End**, and **Best Fit**. It also offers the opportunity to include the legend key in the label. The **Number** option under **Format Data Labels. . .** allows for the usual Excel options in using numbers. The **Fill** option allows you to determine what type of fill you want to have for the chart. Options include **No fill**, **Solid fill**, **Gradient fill**, **Picture or texture fill**, and **Automatic**. Other options under **Format Data Labels. . .** allow you to manipulate the border colours and styles, shadow, 3-D format, and text alignment or layout. The **Layout** tab at the top of the worksheet page has a **Labels** panel located at the top of the worksheet page just to the left of the centre. In this section, you can further specify the location of the chart title by selecting **Chart Title**, the location of the legend by selecting **Legend**, or the location of the labels by selecting **Data Labels**. The **Format** tab at the top of the worksheet page contains a **Shape Styles** panel just to the left of the centre at the top of the worksheet. This panel contains options for visual styles of the graph (for more options, use the up and down arrow) and options for **Shape Fill**, **Shape Outline**, and **Shape Effects**. Other formatting options are available through the use of the **Format Selection** option on the far upper left of the **Current Selection** panel on the **Format** tab page.

- Frequency polygons can be constructed in Excel 2007 by using the **Histogram** feature. Follow the directions shown above to construct a histogram. Once the histogram is constructed, right-click on one of the 'bars' of the histogram. From the dropdown menu, select **Change Series Chart Type**. Next select a line chart type. The result will be a frequency polygon.

- An ogive can be constructed at least two ways. One way is to cumulate the data manually. Enter the cumulated data in one column and the class endpoints in another column. Click and drag over both columns. Go to the **Insert** tab at the top of the Excel worksheet. Select **Scatter** as the type of chart. Under the **Scatter** options, select the option with the solid lines. The result is an ogive. A second way is to construct a frequency distribution first using the **Histogram** feature in the **Data Analysis** tool. In the **Histogram** dialog box, enter the location of the data and enter the location of the class endpoints as bin numbers. Check **Cumulative Percentage** and **Chart Output** in the **Histogram** dialog box. Once the chart is constructed, right-click on one of the bars and select the **Delete** option. The result will be an ogive chart with just the ogive line graph (and bars eliminated).

- Bar charts and column charts are constructed in a manner similar to that of a pie chart. Begin by entering the categories in one column and the data values of each category in another column in the Excel worksheet. Categories and data values could also be entered in rows instead of columns. Click and drag over the data and categories for which the chart is to be constructed. Go to the **Insert** tab at the top of the worksheet. Select **Column** or **Bar** from the **Charts** section and then select the version of the chart to be constructed. The result is a chart from the data. Once the bar chart or column chart has been constructed, there are many options available to you. By right-clicking on the bars or columns, a menu appears that allows you, among other things, to label the columns or bars. This command is **Add Data Labels**. Once data labels are added, clicking on the bars or columns will allow you to modify the labels and the characteristics of the bars or columns by selecting **Format Data Labels. . .** or **Format Data Series. . . .** Usage of these commands is the same as when constructing or modifying pie charts (see above). Various options are also available under **Chart Tools** (see pie charts above).

- Pareto charts, as presented in the text, have categories and numbers of defects. As such, Pareto charts can be constructed as **Column** charts in Excel using the same commands (see above). However, the user will first need to place the categories and their associated frequencies in descending order. In addition, in constructing a histogram in Excel (see above), there is an option in the **Histogram** dialog box called **Pareto (sorted histogram)** in which Excel takes histogram data and presents the data with categories organized from highest frequency to lowest.

- Scatter diagrams can be constructed in Excel. Begin by entering the data for the two variables to be graphed in two separate rows or columns. You may either use a label for each variable or not. Click and drag over the data (and labels). Go to the **Insert** tab. From the **Charts** panel (upper middle), select **Scatter**. From the ensuing pulldown menu of scatter plot options, select one of the versions from the five presented. The result is the scatter chart. By right-clicking on the chart, various other chart options are available including, **Format Plot Area. . . .** The resulting menu associated with this command offers the usual chart options regarding fill, border colour, border styles, shadow, and 3-D format (see pie charts above). In addition, if you want to fit a line or curve to the data, right-click on one of the chart points. A menu pops up containing, among other options, **Add Trendline. . . .** From the **Trendline Options**, select the type of line or curve that you want to fit to the data. The result is a line or curve shown on the scatter plot attempting to fit to the points. Various other options are available regarding the line colour, style, and shadow.

MINITAB

- Minitab has the capability of constructing histograms, dot plots, stem-and-leaf charts, pie charts, bar charts, and Pareto charts. With the exception of Pareto charts, which are accessed through **Stat**, all of these other charts and graphs are accessed by selecting **Graph** on the menu bar.

- To construct a histogram, select **Graph** on the Minitab menu bar, and then select **Histogram**. The first dialog box offers four histogram options: **Simple, With Fit, With Outline and Groups**, and **With Fit and Groups.** Select the Simple option, which is also the default option. In the dialog box that follows beside **Graph variables**, insert the column location (or columns) containing the data for

which you want to create a histogram. There are several options from which to modify the histogram. Select **Scale** to adjust or modify the axes, ticks, and gridlines. Select **Labels** to title the graph and label the axes. Select **Data view** for optional ways to present the data, including bars, symbols, project lines, and areas in addition to presenting other options such as smoothing the data. Select **Multiple graphs** to create multiple separate graphs or to combine several graphs on one. Select **Data** options for several options in grouping data.

- To construct a dot plot, select **Graph** on the Minitab menu bar, and then select **Dotplot**. The first dialog box offers seven different ways to configure the plot. Select **Simple** to produce a dot plot like those shown in this chapter. In the dialog box that follows, insert the column location(s) containing the data for which you want to create the dot plot in **Graph variables**. There are several options available. Select **Scale** to adjust or modify the axes and ticks. Select **Labels** to title the graph and add footnotes. Select **Multiple graphs** to create multiple separate graphs or to combine several graphs on one. Select **Data options** for options in grouping data, frequencies, and subsets.

- To construct a stem-and-leaf chart, select **Stem-and-Leaf** from the **Graph** pulldown menu. In the **Stem-and-Leaf** dialog box, place the name of the column(s) containing the data in the **Graph variables** space. Click **OK** and the stem-and-leaf plot is generated. If you have a grouping variable in another column and want to use it, enter the location or name of the column into the **By variable** space. You can trim outliers from the plot by checking **Trim outliers.**

- To construct a pie chart, select **Graph** on the Minitab menu bar, and then select **Pie Chart** on the **Graph** pulldown menu. In the **Pie** Chart dialog box, there are two options to consider: **Chart counts of unique values** and **Chart values** from a table. If you have raw data and you want Minitab to group them for you into a pie chart, select **Chart counts of unique values.** You can also use this command even if you have text data. On the other hand, if your data are in one column and your categories are in another column, select **Chart values from a table**. The dialog box will ask for the name of the

Categorical variable and the name of the **Summary variables.** Several options are available to modify the pie chart, including **Labels**, **Multiple graphs**, and **Data options.** Several **Pie** Options are available, including how the pie slices are ordered, the starting angle, and the **option** of combining small categories.

- To construct a bar chart, select **Graph** on the Minitab menu bar, then select **Bar Chart**. In the **Bar Chart** dialog box, there are three options available. To construct a bar chart like those presented in the chapter, select **Simple**. In the dialog box that follows, enter the column(s) containing the data in **Categorical variables**. Several options are available to modify the bar chart, including **Chart Options**, **Labels**, **Scale**, **Data View**, **Multiple Graphs**, and **Data Options.**

- To construct a Pareto chart, select **Stat** from the menu bar, and then from the pulldown menu that appears, select **Quality Tools**. From the **Quality Tools** pulldown menu, select **Pareto Chart**. From the **Pareto Chart** dialog box, select **Chart defects table** if you have a summary of the defects with the reasons (**Labels in**) in one column and the frequency of occurrence (**Frequencies in**) in another column. Enter the location of the reasons in **Labels in** and the location of the frequencies in **Frequencies in**. If you have unsummarized data, you can select **Chart defects data in**. In the space provided, give the location of the column with all the defects that occurred. It is possible to have the defects either by name or with some code. If you want to have the labels in one column and the defects in another, then select **By variable in** and place the location of the labels there.

- To construct a scatter plot, select **Graph**, then select **Scatterplot**. In the **Scatterplot** dialog box, select the type of scatter plot you want from **Simple**, **With Groups**, **With Regression**, **With Regression and Groups**, **With Connect Line**, and **With Connect and Groups**. In the second dialog box, enter the x and y variable names/locations. There are several options from which to modify the scatter plot. Select **Scale** to adjust or modify the axes, ticks, and gridlines. Select **Labels** to title the graph and label the axes. Select **Data view** for optional ways to present the data, including bars, symbols, project lines, and areas in addition to presenting other options such as smoothing the data. Select **Multiple graphs** to create multiple separate graphs or to combine several graphs on one. Select **Data options** for several options in grouping data.

Descriptive Statistics

LEARNING OBJECTIVES

The focus of Chapter 3 is the use of statistical techniques to describe data, thereby enabling you to:

1. Apply various measures of central tendency – including the mean, median, and mode – to a set of ungrouped data
2. Apply various measures of variability – including the range, interquartile range, mean absolute deviation, variance, and standard deviation (using the empirical rule and Chebyshev's theorem) – to a set of ungrouped data
3. Compute the mean, median, mode, standard deviation, and variance of grouped data
4. Describe a data distribution statistically and graphically using skewness, kurtosis, and box-and-whisker plots
5. Use computer packages to compute various measures of central tendency, variation, and shape on a set of data, as well as to describe the data distribution graphically

Decision Dilemma

Choosing the Right Recycling Collection System

Most people seem to agree that recycling saves energy, reduces raw material extraction, and combats climate change. However, choosing the right recycling collection system continues to cause debate. In the UK, local authorities have to decide between two main scheme types: 'kerbside sort' schemes where recyclables are sorted into their respective materials on the lorry at the kerbside and 'co-mingled' collections where all recyclables are put into one compartment on the lorry before being taken to a Materials Recovery Facility (MRF), where automated sorting separates recoverable items by size, weight, and type. The Waste & Resources Action Programme reviewed alternative collection systems for municipal waste collection and concluded in June 2009 that kerbside sort systems have lower net costs and lead to much lower MRF rejection rates, which in turn minimizes the amount of waste sent to landfill. Conversely, waste and environmental consultants WYG undertook a review of recycling collection systems in May 2010 and found that in many cases co-mingled collection is more cost effective and leads to higher recycling rates. Similarly, one of the UK's largest waste management companies, Biffa, recently quoted a trial with an East London Local Authority involving 12,000 households that showed that a move to co-mingled collections increased

the average weekly tonnage of recycled materials collected by 58%.

Managerial and Statistical Questions

Virtually all of the statistics cited here are gleaned from studies or surveys.

1. Suppose a study of a co-mingled sort scheme is done in 50 households in a given council of the United Kingdom. The amount of recycled materials in kilograms collected over a month per household are presented below. Summarize the data so that study findings can be reported.

50	50	43	47	39	59	56	46	39	56
59	54	56	59	48	61	55	52	43	56
39	47	41	43	52	39	61	49	54	53
45	49	44	50	50	42	47	44	59	55
49	48	60	39	57	44	57	47	60	58

2. The average amount of waste per head in a year in the UK is about 592 kilograms. Suppose the standard deviation of the amount of waste per head per year in the UK is 83 kilograms. Within what range do most British households' amount of waste per year fall?

Chapter 2 presented graphical techniques for organizing and displaying data. Even though such graphical techniques allow the researcher to make some general observations about the shape and spread of the data, a more complete understanding of the data can be attained by summarizing the data using statistics. This chapter presents such statistical measures, including measures of central tendency, measures of variability, and measures of shape. The computation of these measures is different for ungrouped and grouped data. Hence we present some measures for both ungrouped and grouped data.

3.1 MEASURES OF CENTRAL TENDENCY: UNGROUPED DATA

One type of measure that is used to describe a set of data is the **measure of central tendency**. Measures of central tendency *yield information about the centre, or middle part, of a group of numbers*. Table 3.1 displays the wealth in billions of US dollars of the world's richest people in 2011 according to *Forbes* magazine. For these data, measures of central tendency can yield such information as the average wealth, the middle wealth, and the most frequently occurring wealth level. Measures of central tendency do not focus on the span of the data set or how far values are from the middle numbers. The measures of central tendency presented here for ungrouped data are the mode, the median, the mean, percentiles, and quartiles.

Mode

The **mode** is *the most frequently occurring value in a set of data.* For the data in Table 3.1 the mode is $22 billion because the figure that recurred the most times (two) was $22 billion. Sorting the data into an ordered array (an ordering of the numbers from smallest to largest) helps to locate the mode. The following is an ordered array of the values from Table 3.1:

21.3	22	22	23	23.3	23.5	24	24.5	25.5	26
26.5	27	30	31	31.1	39.5	41	50	56	74

This grouping makes it easier to see that 22 is the most frequently occurring number.

In the case of a tie for the most frequently occurring value, two modes are listed. Then the data are said to be **bimodal**. If a set of data is not exactly bimodal but contains two values that are more dominant than others, some researchers take the liberty of referring to the data set as bimodal even without an exact tie for the mode. Data sets with more than two modes are referred to as **multimodal**.

In the world of business, the concept of mode is often used in determining sizes. As an example, manufacturers who produce cheap rubber flip-flops that are sold for as little as €1.00 around the world might only produce them in one size in order to save on machine setup costs. In determining the one size to produce, the manufacturer would most likely produce flip-flops in the modal size. The mode is an appropriate measure of central tendency for nominal-level data.

Median

The **median** is *the middle value in an ordered array of numbers.* For an array with an odd number of terms, the median is the middle number. For an array with an even number of terms, the median is the average of the two middle numbers. The following steps are used to determine the median:

STEP 1: Arrange the observations in an ordered data array.

STEP 2: For an odd number of terms, find the middle term of the ordered array. It is the median.

STEP 3: For an even number of terms, find the average of the middle two terms. This average is the median.

Suppose a business researcher wants to determine the median for the following numbers:

15	11	14	3	21	17	22	16	19	16	5	7	19	8	9	20	4

The researcher arranges the numbers in an ordered array.

3	4	5	7	8	9	11	14	15	16	16	17	19	19	20	21	22

Because the array contains 17 terms (an odd number of terms), the median is the middle number, or 15.
If the number 22 is eliminated from the list, the array would contain only 16 terms.

3	4	5	7	8	9	11	14	15	16	16	17	19	19	20	21

Now, for an even number of terms, the statistician determines the median by averaging the two middle values, 14 and 15. The resulting median value is 14.5.

Another way to locate the median is by finding the $(n + 1)/2$ term in an ordered array. For example, if a data set contains 77 terms, the median is the 39th term. That is,

$$\frac{n+1}{2} = \frac{77+1}{2} = \frac{78}{2} = 39\text{th term}$$

This formula is helpful when a large number of terms must be manipulated.

TABLE 3.1

Wealth of World's Richest Billionaires (US$ billions)

74	31.1	26	23.3
56	31	25.5	23
50	30	24.5	22
41	27	24	22
39.5	26.5	23.5	21.3

Consider the offer price data in Table 3.1. Because this data set contains 20 values, or $n = 20$, the median for these data is located at the $(20 + 1)/2$ term, or the 10.5th term. This equation indicates that the median is located halfway between the 10th and 11th terms or the average of $26 billion and $26.5 billion. Thus, the median offer price for the 20 wealthiest people in the world is $26.25 billion.

The median is unaffected by the magnitude of extreme values. This characteristic is an advantage, because large and small values do not inordinately influence the median. For this reason, the median is often the best measure of location to use in the analysis of variables such as house costs, income, and age. Suppose, for example, that a real estate broker wants to determine the median selling price of 10 houses listed at the following prices (€).

67,000	105,000	148,000	5,250,000
91,000	116,000	167,000	
95,000	122,000	189,000	

The median is the average of the two middle terms, €116,000 and €122,000, or €119,000. This price is a reasonable representation of the prices of the 10 houses. Note that the house priced at €5,250,000 did not enter into the analysis other than to count as one of the 10 houses. If the price of the 10th house were €200,000, the results would be the same. However, if all the house prices were averaged, the resulting average price of the original 10 houses would be €635,000, higher than nine of the 10 individual prices.

A disadvantage of the median is that not all the information from the numbers is used. For example, information about the specific asking price of the most expensive house does not really enter into the computation of the median. The level of data measurement must be at least ordinal for a median to be meaningful.

Mean

The **arithmetic mean** is *the average of a group of numbers* and is computed by summing all numbers and dividing by the number of numbers. Because the arithmetic mean is so widely used, most statisticians refer to it simply as the *mean*.

The population mean is represented by the Greek letter mu (μ). The sample mean is represented by $\bar{x}$. The formulas for computing the population mean and the sample mean are given in the boxes that follow.

POPULATION MEAN
$$\mu = \frac{\Sigma x}{N} = \frac{x_1 + x_2 + x_3 + \cdots + x_N}{N}$$

SAMPLE MEAN
$$\bar{x} = \frac{\Sigma x}{n} = \frac{x_1 + x_2 + x_3 + \cdots + x_n}{n}$$

The capital Greek letter sigma (Σ) is commonly used in mathematics to represent a summation of all the numbers in a grouping. Also, N is the number of terms in the population, and n is the number of terms in the sample. The algorithm for computing a mean is to sum all the numbers in the population or sample and divide by the number of terms. It is inappropriate to use the mean to analyse data that are not at least interval level in measurement.

Suppose a company has five departments with 24, 13, 19, 26, and 11 workers each. The *population mean* number of workers in each department is 18.6 workers. The computations follow.

$$\begin{array}{r} 24 \\ 13 \\ 19 \\ 26 \\ \underline{11} \\ \Sigma x = 93 \end{array}$$

and

$$\mu = \frac{\Sigma x}{N} = \frac{93}{5} = 18.6$$

The calculation of a sample mean uses the same algorithm as for a population mean and will produce the same answer if computed on the same data. However, it is inappropriate to compute a sample mean for a population or a population mean for a sample. Because both populations and samples are important in statistics, a separate symbol is necessary for the population mean and for the sample mean.

DEMONSTRATION PROBLEM 3.1

The top 20 companies in the world in revenue in a recent year according to *Fortune* magazine's *Global 500* ranking follows.

Company	Annual Revenue (US$ billions)
Wal-Mart Stores	408
Royal Dutch Shell	285
Exxon Mobil	285
BP	246
Toyota Motor	204
Japan Post Holdings	202
Sinopec	188
State Grid	184
AXA	175
China National Petroleum	165
Chevron	164
ING Group	163
General Electric	157
Total	156
Bank of America Corp.	150
Volkswagen	146
ConocoPhillips	140
BNP Paribas	131
Assicurazioni Generali	126
Allianz	126

Compute the mode, the median, and the mean.

Solution

Mode: 285 and 126 (bimodal dataset)

Median: With 20 different companies in this group, $N = 20$. As this is an even number of terms, the median is located at the $(20 + 1)/2 = 10.5$th term, that is, between the 10th and 11th term. Since the 10th and 11th terms are 165 and 164, respectively, the median value is 164.5.

Mean: The sum of the revenue of the top 20 companies in the world is $3,801bn $= \Sigma x$

$$\mu = \frac{\Sigma x}{N} = \frac{3801}{20} = 190.05$$

The mean is affected by each and every value, which is an advantage. The mean uses all the data, and each data item influences the mean. It is also a disadvantage because extremely large or small values can cause the mean to be pulled towards the extreme value. Recall the preceding discussion of the 10 house prices. If the mean is computed for the 10 houses, the mean price is higher than the prices of nine of the houses because the €5,250,000 house is included in the calculation. The total price of the 10 houses is €6,350,000, and the mean price is €635,000.

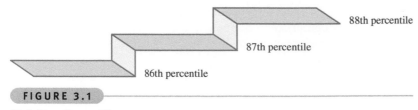

88th percentile

87th percentile

86th percentile

FIGURE 3.1

Stair-Step Percentiles

The mean is the most commonly used measure of central tendency because it uses each data item in its computation. It is a familiar measure, and it has mathematical properties that make it attractive to use in inferential statistics analysis.

Percentiles

Percentiles are *measures of central tendency that divide a group of data into 100 parts.* There are 99 percentiles because it takes 99 dividers to separate a group of data into 100 parts. The nth percentile is the value such that at least n per cent of the data are below that value and at most $(100 - n)$ per cent are above that value. Specifically, the 87th percentile is a value such that at least 87% of the data are below the value and no more than 13% are above the value. Percentiles are 'stair-step' values, as shown in Figure 3.1, because the 87th percentile and the 88th percentile have no percentile between. If a plant operator takes a safety examination and 87.6% of the safety exam scores are below that person's score, he or she still scores at only the 87th percentile, even though more than 87% of the scores are lower.

Percentiles are widely used in reporting test results. Almost all college or university students have taken some form of examination. In most cases, the results for these examinations are reported in percentile form and also as raw scores. Shown next is a summary of the steps used in determining the location of a percentile.

Steps in Determining the Location of a Percentile

1. Organize the numbers into an ascending-order array.
2. Calculate the percentile location (i) by:

$$i = \frac{P}{100}(N)$$

   where:
   - P = the percentile of interest
   - i = percentile location
   - N = number in the data set

3. Determine the location by either (a) or (b).
   a. If i is a whole number, the Pth percentile is the average of the value at the ith location and the value at the $(i+1)$st location.
   b. If i is not a whole number, the Pth percentile value is located at the whole number part of $i + 1$.

For example, suppose you want to determine the 80th percentile of 1,240 numbers. P is 80 and N is 1,240. First, order the numbers from lowest to highest. Next, calculate the location of the 80th percentile.

$$i = \frac{80}{100}(1240) = 992$$

Because $i = 992$ is a whole number, follow the directions in step 3(a). The 80th percentile is the average of the 992nd number and the 993rd number.

$$P_{80} = \frac{(992\text{nd number} + 993\text{rd number})}{2}$$

DEMONSTRATION PROBLEM 3.2

Determine the 30th percentile of the following eight numbers: 14, 12, 19, 23, 5, 13, 28, 17.

Solution

For these eight numbers, we want to find the value of the 30th percentile, so $N = 8$ and $P = 30$.
First, organize the data into an ascending-order array.

| 5 | 12 | 13 | 14 | 17 | 19 | 23 | 28 |

Next, compute the value of i.

$$i = \frac{30}{100}(8) = 2.4$$

Because i is not a whole number, step 3(b) is used. The value of $i + 1$ is 2.4 + 1, or 3.4. The whole-number part of 3.4 is 3. The 30th percentile is located at the third value. The third value is 13, so 13 is the 30th percentile. Note that a percentile may or may not be one of the data values.

Quartiles

Quartiles are *measures of central tendency that divide a group of data into four subgroups or parts.* The three quartiles are denoted as Q_1, Q_2, and Q_3. The first quartile, Q_1, separates the first, or lowest, one-fourth of the data from the upper three-fourths and is equal to the 25th percentile. The second quartile, Q_2, separates the second quarter of the data from the third quarter. Q_2 is located at the 50th percentile and equals the median of the data. The third quartile, Q_3, divides the first three-quarters of the data from the last quarter and is equal to the value of the 75th percentile. These three quartiles are shown in Figure 3.2.

Suppose we want to determine the values of Q_1, Q_2, and Q_3 for the following numbers:

| 106 | 109 | 114 | 116 | 121 | 122 | 125 | 129 |

The value of Q_1 is found at the 25th percentile, P_{25}, by:

$$\text{For } N = 8, \; i = \frac{25}{100}(8) = 2$$

Because i is a whole number, P_{25} is found as the average of the second and third numbers.

$$P_{25} = \frac{(109 + 114)}{2} = 111.5$$

The value of Q_1 is P_{25} is = 111.5. Notice that one-fourth, or two, of the values (106 and 109) are less than 111.5.

The value of Q_2 is equal to the median. Because the array contains an even number of terms, the median is the average of the two middle terms.

$$Q_2 = \text{median} = \frac{(116 + 121)}{2} = 118.5$$

Notice that exactly half of the terms are less than Q_2 and half are greater than Q_2.

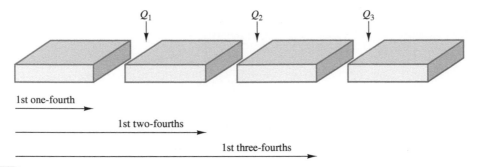

FIGURE 3.2

Quartiles

The value of Q_3 is determined by P_{75} as follows:

$$i = \frac{75}{100}(8) = 6$$

Because i is a whole number, P_{75} is the average of the sixth and the seventh numbers.

$$P_{75} = \frac{(122 + 125)}{2} = 123.5$$

The value of Q_3 is $P_{75} = 123.5$. Notice that three-fourths, or six, of the values are less than 123.5 and two of the values are greater than 123.5.

DEMONSTRATION PROBLEM 3.3

See below for the top 16 largest airlines in Europe in terms of passengers carried in a recent year according to *Wikipedia*. Determine the first, the second, and the third quartiles for these data.

Airline	Millions of passengers
Lufthansa Group	90.17
Ryanair	72.72
Air France-KLM	70.75
IAG	57.3
EasyJet	49.72
Air Berlin	31.77
Turkish Airlines	29.1
SAS Group	25.23
Alitalia	23.4
Swiss International Air Lines	14.17
Norwegian	13.03
Aeroflot	11.29
Vueling Airlines	11.03
Austrian Airlines Group	10.89
Aer Lingus	9.71
Wizz Air	9.6

Solution

For 16 airline companies, $N = 16$. $Q_1 = P_{25}$ is found by

$$i = \frac{25}{100}(16) = 4$$

Because i is a whole number, Q_1 is found to be the average of the fourth and fifth values from the bottom.

$$Q_1 = \frac{11.03 + 11.29}{2} = 11.16$$

$Q_2 = P_{50} = $ median; with 16 terms, the median is the average of the eighth and ninth terms.

$$Q_2 = \frac{23.4 + 25.23}{2} = 24.315$$

$Q_3 = P_{75}$ is solved by

$$i = \frac{75}{100}(16) = 12$$

Q_3 is found by averaging the 12th and 13th terms.

$$Q_3 = \frac{49.72 + 57.3}{2} = 53.51$$

3.1 PROBLEMS

3.1 Determine the mode for the following numbers:

   2   4   8   4   6   2   7   8   4   3   8   9   4   3   5

3.2 Determine the median for the numbers in Problem 3.1:

3.3 Determine the median for the following numbers:

   213   345   609   073   167   243   444   524   199   682

3.4 Compute the mean for the following numbers:

   17.3   44.5   31.6   40.0   52.8   38.8   30.1   78.5

3.5 Compute the mean for the following numbers:

   7   −2   5   9   0   −3   −6   −7   −4   −5   2   −8

3.6 Compute the 35th percentile, the 55th percentile, Q_1, Q_2, and Q_3 for the following data:

   16   28   29   13   17   20   11   34   32   27   25   30   19   18   33

3.7 Compute P_{20}, P_{47}, P_{83}, Q_1, Q_2, and Q_3 for the following data:

120	138	97	118	172	144
138	107	94	119	139	145
162	127	112	150	143	80
105	116	142	128	116	171

3.8   The list below shows the 15 largest banks in the world by assets compiled from *Forbes's* Global 2010 ranking of the world's largest companies. Compute the median and the mean assets from this group. Which of these two measures do you think is most appropriate for summarizing these data, and why? What is the value of Q_2? Determine the 63rd percentile for the data. Determine the 29th percentile for the data.

Bank	Country	Assets ($ billions)
BNP Paribas	France	2,952.22
HSBC Holdings	United Kingdom	2,355.83
Crédit Agricole	France	2,227.22
Bank of America	United States	2,223.30
Barclays	United Kingdom	2,223.04
Deutsche Bank	Germany	2,150.60
Lloyds Banking Group	United Kingdom	1,650.78
Société Générale Group	France	1,468.72
UniCredit Group	Italy	1,438.91
Banco Santander	Spain	1,438.68
ICBC	China	1,428.46
Wells Fargo	United States	1,243.65
China Construction Bank	China	1,106.20
Bank of China	China	1,016.31
Intesa Sanpaolo	Italy	877.66

3.9   Below are listed the 10 largest vehicle makers in the world and the number of vehicles produced by each in a recent year. Compute the median, Q_3, P_{20}, P_{60}, P_{80}, and P_{93} on these data.

Manufacturer	Production (millions)
Toyota Motor Corp.	9.37
General Motors	8.90
Volkswagen AG	6.19
Ford Motor Co.	5.96
Hyundai-Kia Automotive Group	3.96
Honda Motor Co. Ltd.	3.83
Nissan Motor Co.	3.68
PSA/Peugeot-Citreon SA	3.43
Chrysler LLC	2.68
Fiat S.p.A.	2.62

3.10   Below are listed the number of fatal accidents suffered by scheduled commercial airlines over a 17-year period according to the Air Transport Association of America. Using these data, compute the mean, median, and mode. What is the value of the third quartile? Determine P_{11}, P_{35}, P_{58}, and P_{67}.

4   4   4   1   4   2   4   3   8   6   4   4   1   4   2   3   3

3.2   MEASURES OF VARIABILITY: UNGROUPED DATA

Measures of central tendency yield information about the centre or middle part of a data set. However, business researchers can use another group of analytic tools, **measures of variability**, to *describe the spread or the dispersion of a set of data*. Using measures of variability in conjunction with measures of central tendency makes possible a more complete numerical description of the data.

For example, a company has 25 salespeople in the field, and the median annual sales figure for these people is £1.2 million. Are the salespeople being successful as a group or not? The median provides information about the sales of the person in the middle, but what about the other salespeople? Are all of them selling £1.2 million annually, or do the sales figures vary widely, with one person selling £5 million annually and another selling only £150,000 annually? Measures of variability provide the additional information necessary to answer that question.

Figure 3.3 shows three distributions in which the mean of each distribution is the same ($\mu = 50$) but the variabilities differ. Observation of these distributions shows that a measure of variability is necessary to complement the mean value in describing the data. Methods of computing measures of variability differ for ungrouped data and grouped data. This section focuses on seven measures of variability for ungrouped data: range, interquartile range, mean absolute deviation, variance, standard deviation, z scores, and coefficient of variation.

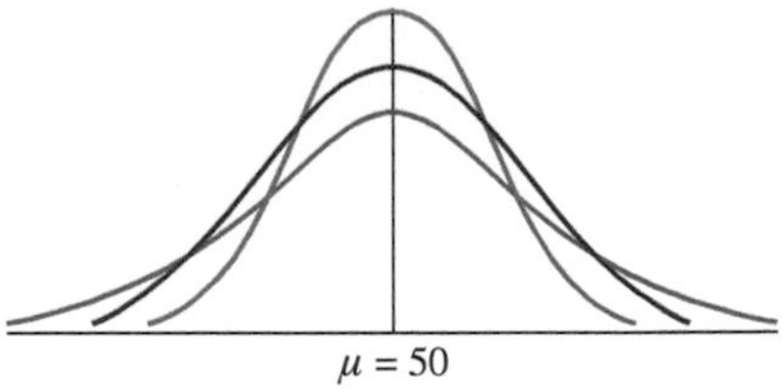

$\mu = 50$

FIGURE 3.3

Three Distributions with the Same Mean but Different Dispersions

Range

The **range** is *the difference between the largest value of a data set and the smallest value of a set.* Although it is usually a single numeric value, some business researchers define the range of data as the ordered pair of smallest and largest numbers (smallest, largest). It is a crude measure of variability, describing the distance to the outer bounds of the data set. It reflects those extreme values because it is constructed from them. An advantage of the range is its ease of computation. One important use of the range is in quality assurance, where the range is used to construct control charts. A disadvantage of the range is that, because it is computed with the values that are on the extremes of the data, it is affected by extreme values, and its application as a measure of variability is limited.

The data presented earlier in Table 3.1 represent the accumulated wealth of the world's richest people in a recent year. The lowest figure was $21.30 billion and the wealthiest person has a wealth of $74 billion. The range of the wealth of the top 20 richest individuals can be computed as the difference between the highest and lowest values:

$$\text{Range} = \text{Highest} - \text{Lowest} = \$74.00 \text{ billion} - \$21.3 \text{ billion} = \$52.7 \text{ billion}$$

Interquartile Range

Another measure of variability is the **interquartile range**. The interquartile range is *the range of values between the first and third quartile.* Essentially, it is the range of the middle 50% of the data and is determined by computing the value of $Q_3 - Q_1$. The interquartile range is especially useful in situations where data users are more interested in values towards the middle and less interested in extremes. In describing a real estate housing market, real estate agents might use the interquartile range as a measure of housing prices when describing the middle half of the market for buyers who are interested in houses in the midrange. In addition, the interquartile range is used in the construction of box-and-whisker plots.

INTERQUARTILE RANGE	$Q_3 - Q_1$

The following data indicate the top 15 exporters in the world in 2010 according to the CIA's *The World Factbook*:

Country	Exports (US$ billions)
China	1,506
Germany ·	1,337
United States	1,270
Japan	765.2
France	508.7
Korea, South	466.3
Italy	458.4
Netherlands	451.3
Canada	406.8
United Kingdom	405.6
Hong Kong	382.6
Russia	376.7
Singapore	351.2
Mexico	303.0
Belgium	279.2

What is the interquartile range for these data? The process begins by computing the first and third quartiles as follows:
Solving for $Q_1 = P_{25}$ when $N = 15$:

$$i = \frac{25}{100}(15) = 3.75$$

Because i is not a whole number, P_{25} is found as the fourth term from the bottom.

$$Q_1 = P_{25} = 376.7$$

Solving for $Q_3 = P_{75}$:

$$i = \frac{75}{100}(15) = 11.25$$

Because i is not a whole number, P_{75} is found as the 12th term from the bottom.

$$Q_3 = P_{75} = 765.2$$

The interquartile range is:

$$Q_3 - Q_1 = 765.2 - 376.7 = 388.5$$

The middle 50% of the volume of exports by the top 15 exporters in the world spans a range of 388.5 ($ billion).

Mean Absolute Deviation, Variance, and Standard Deviation

Three other measures of variability are the variance, the standard deviation, and the mean absolute deviation. They are obtained through similar processes and are, therefore, presented together. These measures are not meaningful unless the data are at least interval-level data. The variance and standard deviation are widely used in statistics. Although the standard deviation has some stand-alone potential, the importance of variance and standard deviation lies mainly in their role as tools used in conjunction with other statistical devices.

Suppose a small company started a production line to build computers. During the first five weeks of production, the output is 5, 9, 16, 17, and 18 computers, respectively. Which descriptive statistics could the owner use to measure the early progress of production? In an attempt to summarize these figures, the owner could compute a mean.

$$\begin{array}{c} x \\ \hline 5 \\ 9 \\ 16 \\ 17 \\ 18 \end{array}$$

$$\Sigma x = 65 \qquad \mu = \frac{\Sigma x}{N} = \frac{65}{5} = 13$$

What is the variability in these five weeks of data? One way for the owner to begin to look at the spread of the data is to subtract the mean from each data value. *Subtracting the mean from each value of data* yields the **deviation from the mean** $(x - \mu)$. Table 3.2 shows these deviations for the computer company production. Note that some deviations from the mean are positive and some are negative. Figure 3.4 shows that geometrically the negative deviations represent values that are below (to the left of) the mean and positive deviations represent values that are above (to the right of) the mean.

An examination of deviations from the mean can reveal information about the variability of data. However, the deviations are used mostly as a tool to compute other measures of variability. Note that in both Table 3.2 and Figure 3.4 these deviations total zero. This phenomenon applies to all cases. For a given set of data, the sum of all deviations from the arithmetic mean is always zero.

SUM OF DEVIATIONS FROM THE ARITHMETIC MEAN IS ALWAYS ZERO	$\Sigma(x - \mu) = 0$

This property requires considering alternative ways to obtain measures of variability.

One obvious way to force the sum of deviations to have a non-zero total is to take the absolute value of each deviation around the mean. Utilizing the absolute value of the deviations about the mean makes solving for the mean absolute deviation possible.

Mean Absolute Deviation

The **mean absolute deviation (MAD)** is *the average of the absolute values of the deviations around the mean for a set of numbers.*

MEAN ABSOLUTE DEVIATION	$\text{MAD} = \dfrac{\Sigma \lvert x - \mu \rvert}{N}$

TABLE 3.2

Deviations from the Mean for Computer Production

Number (x)	Deviations from the Mean ($x - \mu$)
5	$5 - 13 = -8$
9	$9 - 13 = -4$
16	$16 - 13 = +3$
17	$17 - 13 = +4$
18	$18 - 13 = +5$
$\Sigma x = 65$	$\Sigma(x - \mu) = 0$

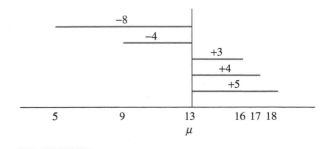

FIGURE 3.4

Geometric Distances from the Mean (from Table 3.2)

TABLE 3.3

MAD for Computer Production Data

x	$x - \mu$	$\|x - \mu\|$
5	-8	$+8$
9	-4	$+4$
16	$+3$	$+3$
17	$+4$	$+4$
18	$+5$	$+5$
$\Sigma x = 65$	$\Sigma(x - \mu) = 0$	$\Sigma\|x - \mu\| = 24$

$$\text{MAD} = \frac{\Sigma|x - \mu|}{N} = \frac{24}{5} = 4.8$$

Using the data from Table 3.2, the computer company owner can compute a mean absolute deviation by taking the absolute values of the deviations and averaging them, as shown in Table 3.3. The mean absolute deviation for the computer production data is 4.8.

Because it is computed by using absolute values, the mean absolute deviation is less useful in statistics than other measures of dispersion. However, in the field of forecasting, it is used occasionally as a measure of error.

Variance

Because absolute values are not conducive to easy manipulation, mathematicians developed an alternative mechanism for overcoming the zero-sum property of deviations from the mean. This approach utilizes the square of the deviations from the mean. The result is the variance, an important measure of variability.

The **variance** is *the average of the squared deviations about the arithmetic mean for a set of numbers.* The population variance is denoted by σ^2.

POPULATION VARIANCE	$\sigma^2 = \dfrac{\Sigma(x - \mu)^2}{N}$

Table 3.4 shows the original production numbers for the computer company, the deviations from the mean, and the squared deviations from the mean.

The sum of the squared deviations about the mean of a set of values – called the **sum of squares of x** and sometimes abbreviated as SS_x – is used throughout statistics. For the computer company, this value is 130. Dividing it by the number of data values (five weeks) yields the variance for computer production.

$$\sigma^2 = \frac{130}{5} = 26.0$$

TABLE 3.4

Computing a Variance and a Standard Deviation from the Computer Production Data

x	$x - \mu$	$(x - \mu)^2$
5	-8	64
9	-4	16
16	$+3$	9
17	$+4$	16
18	$+5$	25
$\Sigma x = 65$	$\Sigma(x - \mu) = 0$	$\Sigma(x - \mu)^2 = 130$

$$SS_x = \Sigma(x - \mu)^2 = 130$$

$$\text{Variance} = \sigma^2 = \frac{SS_x}{N} = \frac{\Sigma(x - \mu)^2}{N} = \frac{130}{5} = 26.0$$

$$\text{Standard Deviation} = \sigma = \sqrt{\frac{\Sigma(x - \mu)^2}{N}} = \sqrt{\frac{130}{5}} = 5.1$$

Because the variance is computed from squared deviations, the final result is expressed in terms of squared units of measurement. Statistics measured in squared units are problematic to interpret. Consider, for example, Lego attempting to interpret production costs in terms of squared Danish krone or Black & Decker measuring production output variation in terms of squared lawn mowers. Therefore, when used as a descriptive measure, variance can be considered as an intermediate calculation in the process of obtaining the standard deviation.

Standard Deviation

The standard deviation is a popular measure of variability. It is used both as a separate entity and as a part of other analyses, such as computing confidence intervals and in hypothesis testing (see Chapters 8, 9, and 10).

POPULATION STANDARD DEVIATION	$\sigma = \sqrt{\dfrac{\Sigma(x-\mu)^2}{N}}$

The **standard deviation** is *the square root of the variance.* The population standard deviation is denoted by σ.

Like the variance, the standard deviation utilizes the sum of the squared deviations about the mean (SS_x). It is computed by averaging these squared deviations (SS_x/N) and taking the square root of that average. One feature of the standard deviation that distinguishes it from a variance is that the standard deviation is expressed in the same units as the raw data, whereas the variance is expressed in those units squared. Table 3.4 shows the standard deviation for the computer production company: $\sqrt{26}$, or 5.1.

What does a standard deviation of 5.1 mean? The meaning of standard deviation is more readily understood from its use, which is explored in the next section. Although the standard deviation and the variance are closely related and can be computed from each other, differentiating between them is important, because both are widely used in statistics.

Meaning of Standard Deviation

What is a standard deviation? What does it do, and what does it mean? The most precise way to define standard deviation is by reciting the formula used to compute it. However, insight into the concept of standard deviation can be gleaned by viewing the manner in which it is applied. Two ways of applying the standard deviation are the **empirical rule** and **Chebyshev's theorem**.

Empirical Rule

The empirical rule is an important rule of thumb that *is used to state the approximate percentage of values that lie within a given number of standard deviations from the mean of a set of data if the data are normally distributed.*

The empirical rule is used only for three numbers of standard deviations: 1σ, 2σ, and 3σ. More detailed analysis of other numbers of σ values is presented in Chapter 6. Also discussed in further detail in Chapter 6 is the normal distribution, a unimodal, symmetrical distribution that is bell (or mound) shaped. The requirement that the data be normally distributed contains some tolerance, and the empirical rule generally applies as long as the data are approximately mound shaped.

	Distance from the Mean	Values Within Distance
EMPIRICAL RULE*	$\mu \pm 1\sigma$	68%
	$\mu \pm 2\sigma$	95%
	$\mu \pm 3\sigma$	99.7%

*Based on the assumption that the data are approximately normally distributed.

If a set of data is normally distributed, or bell shaped, approximately 68% of the data values are within one standard deviation of the mean, 95% are within two standard deviations, and almost 100% are within three standard deviations.

Suppose a recent report states that for Belgium, the average price of a litre of regular petrol is €1.704. Suppose regular petrol prices vary across the country with a standard deviation of €0.08 and are normally distributed. According to the empirical rule, approximately 68% of the prices should fall within $\mu \pm 1\sigma$, or €1.704 ± 1(€0.08). Approximately 68% of the prices should be between €1.624 and €1.784, as shown in Figure 3.5a. Approximately 95% should fall within $\mu \pm 2\sigma$, or €1.704 ± 2(€0.08); that is, between €1.544 and €1.864, as shown in Figure 3.5b. Nearly all regular petrol prices (99.7%) should fall between €1.464 and €1.944 ($\mu \pm 3\sigma$).

Note that with 68% of the petrol prices falling within one standard deviation of the mean, approximately 32% are outside this range. Because the normal distribution is symmetrical, the 32% can be split in half such that 16% lie in each tail of the distribution. Thus, approximately 16% of the petrol prices should be less than €1.624 and approximately 16% of the prices should be greater than €1.784.

Many phenomena are distributed approximately in a bell shape, including most human characteristics such as height and weight; therefore the empirical rule applies in many situations and is widely used.

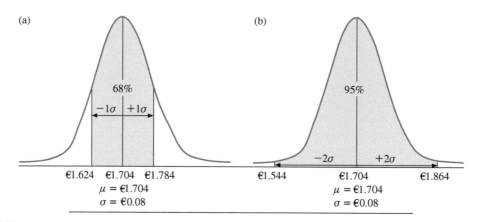

FIGURE 3.5

Empirical Rule for One and Two Standard Deviations of Petrol Prices

DEMONSTRATION PROBLEM 3.4

A company produces a lightweight valve that is specified to weigh 1,365 grams. Unfortunately, because of imperfections in the manufacturing process not all of the valves produced weigh exactly 1,365 grams. In fact, the weights of the valves produced are normally distributed with a mean weight of 1,365 grams and a standard deviation of 294 grams. Within what range of weights would approximately 95% of the valve weights fall? Approximately 16% of the weights would be more than what value? Approximately 0.15% of the weights would be less than what value?

Solution

Because the valve weights are normally distributed, the empirical rule applies. According to the empirical rule, approximately 95% of the weights should fall within $\mu \pm 2\sigma = 1,365 \pm 2(294) = 1,365 \pm 588$. Thus, approximately 95% should fall between 777 and 1,953. Approximately 68% of the weights should fall within $\mu \pm 1\sigma$, and 32% should fall outside this interval. Because the normal distribution is symmetrical, approximately 16% should lie above $\mu \pm 1\sigma = 1,365 + 294 = 1,659$. Approximately 99.7% of the weights should fall within $\mu \pm 3\sigma$, and 0.3% should fall outside this interval. Half of these, or 0.15%, should lie below $\mu - 3\sigma = 1,365 - 3(294) = 1,365 - 882 = 483$.

Chebyshev's Theorem

The empirical rule applies only when data are known to be approximately normally distributed. What do researchers use when data are not normally distributed or when the shape of the distribution is unknown? Chebyshev's theorem applies to all distributions regardless of their shape and thus can be used whenever the data distribution shape is unknown or is non-normal. Even though Chebyshev's theorem can in theory be applied to data that are normally distributed, the empirical rule is more widely known and is preferred whenever appropriate. Chebyshev's theorem is not a rule of thumb, as is the empirical rule, but rather it is presented in formula format and therefore can be more widely applied. Chebyshev's theorem states that *at least $1 - 1/k^2$ values will fall within $\pm k$ standard deviations of the mean regardless of the shape of the distribution.*

(Figure showing a distribution with -2σ and $+2\sigma$ marked, 75% of the area, and μ at the center)

FIGURE 3.6

Application of Chebyshev's Theorem for Two Standard Deviations

CHEBYSHEV'S THEOREM

Within k standard deviations of the mean, $\mu \pm k\sigma$, lie at least

$$1 - \frac{1}{k^2}$$

proportion of the values.
Assumption: $k > 1$

Specifically, Chebyshev's theorem says that at least 75% of all values are within $\pm 2\sigma$ of the mean regardless of the shape of a distribution because if $k = 2$, then $1 - 1/k^2 = 1 - 1/2^2 = 3/4 = 0.75$. Figure 3.6 provides a graphic illustration. In contrast, the empirical rule states that if the data are normally distributed 95% of all values are within $\mu \pm 2\sigma$. According to Chebyshev's theorem, the percentage of values within three standard deviations of the mean is at least 89%, in contrast to 99.7% for the empirical rule. Because a formula is used to compute proportions with Chebyshev's theorem, any value of k greater than 1 ($k > 1$) can be used. For example, if $k = 2.5$, at least 0.84 of all values are within $\mu \pm 2.5\sigma$, because $1 - 1/k^2 = 1 - 1/(2.5)^2 = 0.84$.

DEMONSTRATION PROBLEM 3.5

In the computing industry the average age of professional employees tends to be younger than in many other business professions. Suppose the average age of a professional employed by a particular computer firm is 28 with a standard deviation of six years. A histogram of professional employee ages within this firm reveals that the data are not normally distributed but rather are amassed in the 20s and that few workers are over 40. Apply Chebyshev's theorem to determine within what range of ages would at least 80% of the workers' ages fall.

Solution

Because the ages are not normally distributed, it is not appropriate to apply the empirical rule; and therefore Chebyshev's theorem must be applied to answer the question.

Chebyshev's theorem states that at least $1 - 1/k^2$ proportion of the values are within $\mu \pm k\sigma$. Because 80% of the values are within this range, let

$$1 - \frac{1}{k^2} = 0.80$$

Solving for k yields

$$0.20 = \frac{1}{k^2}$$
$$k^2 = 5.000$$
$$k = 2.24$$

Chebyshev's theorem says that at least 0.80 of the values are within ± 2.24 of the mean.

For $\mu = 28$ and $\sigma = 6$, at least 0.80, or 80%, of the values are within $28 \pm 2.24(6) = 28 \pm 13.4$ years of age or between 14.6 and 41.4 years old.

Population versus Sample Variance and Standard Deviation

The sample variance is denoted by s^2 and the sample standard deviation by s. The main use for sample variances and standard deviations is as estimators of population variances and standard deviations. Because of this, computation of the sample variance and standard deviation differs slightly from computation of the population variance and standard deviation. Both the sample variance and sample standard deviation use $n-1$ in the denominator instead of n because using n in the denominator of a sample variance results in a statistic that tends to underestimate the population variance. While discussion of the properties of *good estimators* is beyond the scope of this text, one of the properties of a good estimator is being *unbiased*. Whereas using n in the denominator of the sample variance makes it a *biased* estimator, using $n-1$ allows it to be an *unbiased* estimator, which is a desirable property in inferential statistics.

SAMPLE VARIANCE $\qquad\qquad s^2 = \dfrac{\Sigma(x - \bar{x})^2}{n-1}$

SAMPLE STANDARD DEVIATION $\qquad\qquad s = \sqrt{\dfrac{\Sigma(x - \bar{x})^2}{n-1}}$

Shown here is a sample of seven of the largest accounting firms in the United Kingdom and the number of UK partners associated with each firm as reported by *Accountancy Age*.

Name of firm	Number of UK partners
PricewaterhouseCoopers	858
Deloitte	681
KPMG	569
Ernst & Young	515
Grant Thornton UK	225
BDO Stoy Hayward	201
RSM Tenon Group	253

The sample variance and sample standard deviation can be computed by:

x	$(x - \bar{x})^2$
858	149,216.70
681	43,800.51
569	9,464.51
515	1,873.65

x	$(x - \bar{x})^2$
225	60,867.94
201	73,286.22
253	47,835.94
$\Sigma x = 3{,}302$	$\Sigma(x - \bar{x})^2 = 386{,}345.4$

$$\bar{x} = \frac{3302}{7} = 471.7$$

$$s^2 = \frac{\Sigma(x - \bar{x})^2}{n - 1} = \frac{386{,}345.4}{6} = 64{,}390.9$$

$$s = \sqrt{s^2} = \sqrt{64{,}390.9} = 253.75$$

The sample variance is 64,390.9, and the sample standard deviation is 253.75.

Computational Formulas for Variance and Standard Deviation

An alternative method of computing variance and standard deviation, sometimes referred to as the computational method or shortcut method, is available. Algebraically,

$$\Sigma(x - \mu)^2 = \Sigma x^2 - \frac{(\Sigma x)^2}{N}$$

and

$$\Sigma(x - \bar{x})^2 = \Sigma x^2 - \frac{(\Sigma x)^2}{n}$$

Substituting these equivalent expressions into the original formulas for variance and standard deviation yields the following computational formulas:

COMPUTATIONAL FORMULA FOR POPULATION VARIANCE AND STANDARD DEVIATION	$$\sigma^2 = \frac{\Sigma x^2 - \dfrac{(\Sigma x)^2}{N}}{N}$$ $$\sigma = \sqrt{\sigma^2}$$

COMPUTATIONAL FORMULA FOR SAMPLE VARIANCE AND STANDARD DEVIATION	$$s^2 = \frac{\Sigma x^2 - \dfrac{(\Sigma x)^2}{n}}{n - 1}$$ $$s = \sqrt{s^2}$$

These computational formulas utilize the sum of the x values and the sum of the x^2 values instead of the difference between the mean and each value and computed deviations. In the pre-calculator/computer era, this method usually was faster and easier than using the original formulas.

TABLE 3.5

Computational Formula Calculations of Variance and Standard Deviation for Computer Production Data

x	x^2
5	25
9	81
16	256
17	289
18	324
$\Sigma x = 65$	$\Sigma x^2 = 975$

$$\sigma^2 = \frac{975 - \frac{(65)^2}{5}}{5} = \frac{975 - 845}{5} = \frac{130}{5} = 26$$

$$\sigma = \sqrt{26} = 5.1$$

For situations in which the mean is already computed or is given, alternative forms of these formulas are

$$\sigma^2 = \frac{\Sigma x^2 - N\mu^2}{N}$$

$$s^2 = \frac{\Sigma x^2 - n(\bar{x})^2}{n-1}$$

Using the computational method, the owner of the start-up computer production company can compute a population variance and standard deviation for the production data, as shown in Table 3.5. (Compare these results with those in Table 3.4.)

DEMONSTRATION PROBLEM 3.6

The effectiveness of crown prosecutors can be measured by several variables, including the number of convictions per month, the number of cases handled per month, and the total number of years of conviction per month. A researcher uses a sample of five crown prosecutors in a city and determines the total number of years of conviction that each prosecutor won against defendants during the past month, as reported in the first column in the tabulations below. Compute the mean absolute deviation, the variance, and the standard deviation for these figures.

Solution

The researcher computes the mean absolute deviation, the variance, and the standard deviation for these data in the following manner.

x	$\lvert x - \bar{x} \rvert$	$(x - \bar{x})^2$
55	41	1,681
100	4	16
125	29	841
140	44	1,936
60	36	1,296
$\Sigma x = 480$	$\Sigma \lvert x - \bar{x} \rvert = 154$	$\Sigma (x - \bar{x})^2 = 5,770$

$$\bar{x} = \frac{\Sigma x}{n} = \frac{480}{5} = 96$$

$$MAD = \frac{154}{5} = 30.8$$

$$s^2 = \frac{5,770}{4} = 1,442.5 \text{ and } s = \sqrt{s^2} = 37.98$$

Computational formulas are then used to solve for s^2 and s and the results compared.

x	**x^2**
55	3,025
100	10,000
125	15,625
140	19,600
60	3,600
$\Sigma x = 480$	$\Sigma x^2 = 51,850$

$$s^2 = \frac{51,850 - \frac{(480)^2}{5}}{4} = \frac{51,850 - 46,080}{4} = \frac{5,770}{4} = 1,442.5$$

$$s = \sqrt{1,442.5} = 37.98$$

The results are the same. The sample standard deviation obtained by both methods is 37.98, or 38 years.

z Scores

A **z score** represents the number of standard deviations a value (x) is above or below the mean of a set of numbers when the data are normally distributed. Using z scores allows translation of a value's raw distance from the mean into units of standard deviations.

z SCORE

$$z = \frac{x - \mu}{\sigma}$$

For samples,

$$z = \frac{x - \bar{x}}{s}$$

If a z score is negative, the raw value (x) is below the mean. If the z score is positive, the raw value (x) is above the mean.

For example, for a data set that is normally distributed with a mean of 50 and a standard deviation of 10, suppose a statistician wants to determine the z score for a value of 70. This value ($x = 70$) is 20 units above the mean, so the z value is

$$z = \frac{70 - 50}{10} = +2.00$$

This z score signifies that the raw score of 70 is two standard deviations above the mean. How is this z score interpreted? The empirical rule states that 95% of all values are within two standard deviations of the mean if the data are approximately normally distributed. Figure 3.7 shows that because the value of 70 is two standard deviations above the mean ($z = +2.00$), 95% of the values are between 70 and the value ($x = 30$), that is two standard deviations below the mean, or $z = (30 - 50)/10 = -2.00$. Because 5% of the values are outside the range of two standard deviations from the mean and the normal distribution is symmetrical, 2½% (half of the 5%) are below the value of 30. Thus 97½% of the values are below

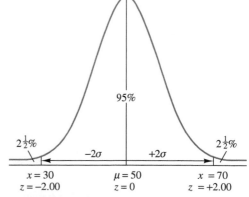

FIGURE 3.7

Percentage Breakdown of Scores Two Standard Deviations from the Mean the value of 70. Because a z score is the number of standard deviations an individual data value is from the mean, the empirical rule can be restated in terms of z scores.

Between $z = -1.00$ and $z = +1.00$ are approximately 68% of the values.
Between $z = -2.00$ and $z = +2.00$ are approximately 95% of the values.
Between $z = -3.00$ and $z = +3.00$ are approximately 99.7% of the values.

The topic of z scores is discussed more extensively in Chapter 6.

Coefficient of Variation

The **coefficient of variation** is a statistic that *is the ratio of the standard deviation to the mean expressed in percentage* and is denoted CV.

COEFFICIENT OF VARIATION

$$CV = \frac{\sigma}{\mu}(100)$$

The coefficient of variation essentially is a relative comparison of a standard deviation to its mean. The coefficient of variation can be useful in comparing standard deviations that have been computed from data with different means.

Suppose five weeks of average prices for stock A are 57, 68, 64, 71, and 62. To compute a coefficient of variation for these prices, first determine the mean and standard deviation: $\mu = 64.40$ and $\sigma = 4.84$. The coefficient of variation is

$$CV_A = \frac{\sigma_A}{\mu_A}(100) = \frac{4.84}{64.40}(100) = 0.075 = 7.5\%$$

The standard deviation is 7.5% of the mean.

Sometimes financial investors use the coefficient of variation or the standard deviation or both as measures of risk. Imagine a stock with a price that never changes. An investor bears no risk of losing money from the price going down because no variability occurs in the price. Suppose, in contrast, that the price of the stock fluctuates wildly. An investor who buys at a low price and sells for a high price can make a nice profit. However, if the price drops below what the investor buys it for, the stock owner is subject to a potential loss. The greater the variability is, the more the potential for loss. Hence, investors use measures of variability such as standard deviation or coefficient of variation to determine the risk of a stock. What does the coefficient of variation tell us about the risk of a stock that the standard deviation does not?

Suppose the average prices for a second stock, B, over these same five weeks are 12, 17, 8, 15, and 13. The mean for stock B is 13.00 with a standard deviation of 3.03. The coefficient of variation can be computed for stock B as

$$CV_B = \frac{\sigma_B}{\mu_B}(100) = \frac{3.03}{13}(100) = 0.233 = 23.3\%$$

The standard deviation for stock B is 23.3% of the mean.

With the standard deviation as the measure of risk, stock A is more risky over this period of time because it has a larger standard deviation. However, the average price of stock A is almost five times as much as that of stock B. Relative to the amount invested in stock A, the standard deviation of 4.84 may not represent as much risk as the standard deviation of 3.03 for stock B, which has an average price of only 13.00. The coefficient of variation reveals the risk of a stock in terms of the size of standard deviation relative to the size of the mean (in percentage). Stock B has a coefficient of variation that is nearly three times as much as the coefficient of variation for stock A. Using coefficient of variation as a measure of risk indicates that stock B is riskier.

The choice of whether to use a coefficient of variation or raw standard deviations to compare multiple standard deviations is a matter of preference. The coefficient of variation also provides an optional method of interpreting the value of a standard deviation.

3.2 PROBLEMS

3.11 A (population) data set contains the following seven values:

6   2   4   9   1   3   5

a. Find the range.
b. Find the mean absolute deviation.
c. Find the population variance.
d. Find the population standard deviation.
e. Find the interquartile range.
f. Find the z score for each value.

3.12 A (sample) data set contains the following eight values:

4   3   0   5   2   9   4   5

a. Find the range.
b. Find the mean absolute deviation.
c. Find the sample variance.
d. Find the sample standard deviation.
e. Find the interquartile range.

3.13 A data set contains the following six values:

12   23   19   26   24   23

a. Find the population standard deviation using the formula containing the mean (the original formula).
b. Find the population standard deviation using the computational formula.
c. Compare the results. Which formula was faster to use? Which formula do you prefer? Why do you think the computational formula is sometimes referred to as the 'shortcut' formula?

3.14 Use your calculator or computer to find the sample variance and sample standard deviation for the following data:

57	88	68	43	93
63	51	37	77	83
66	60	38	52	28
34	52	60	57	29
92	37	38	17	67

3.15 Use your calculator or computer to find the population variance and population standard deviation for the following data:

123	090	546	378
392	280	179	601
572	953	749	075
303	468	531	646

3.16 Determine the interquartile range on the following data:

44	18	39	40	59
46	59	37	15	73
23	19	90	58	35
82	14	38	27	24
71	25	39	84	70

3.17 According to Chebyshev's theorem, at least what proportion of the data will be within $\mu \pm k\sigma$ for each value of k?

a. $k = 2$

b. $k = 2.5$

c. $k = 1.6$

d. $k = 3.2$

3.18 Compare the variability of the following two sets of data by using both the population standard deviation and the population coefficient of variation:

Data Set 1	Data Set 2
49	159
82	121
77	138
54	152

3.19 A sample of 12 small accounting firms reveals the following numbers of professionals per office:

7	10	9	14	11	8
5	12	8	3	13	6

a. Determine the mean absolute deviation.

b. Determine the variance.

c. Determine the standard deviation.

d. Determine the interquartile range.

e. What is the z score for the firm that has six professionals?

f. What is the coefficient of variation for this sample?

3.20 Shown below are the 15 largest companies in China in a recent year according to *Fortune* magazine.

Company	Revenues ($ billions)
Sinopec	187.5
State Grid	184.5
China National Petroleum	165.5
China Mobile Communications	71.7
Industrial & Commercial Bank of China	69.2
China Construction Bank	58.3
China Life Insurance	57
China Railway Construction	52
China Railway Group	50.7
Agricultural Bank of China	49.8

(*continued*)

Company	Revenues ($ billions)
Bank of China	49.7
China Southern Power Grid	45.7
Dongfeng Motor	39.4
China State Construction Engineering	38.1
Sinochem Group	35.6

Assume that the data represent a population.

a. Find the range.

b. Find the mean absolute deviation.

c. Find the population variance.

d. Find the population standard deviation.

e. Find the interquartile range.

f. Find the z score for State Grid.

g. Find the coefficient of variation.

3.21 A distribution of numbers is approximately bell shaped. If the mean of the numbers is 125 and the standard deviation is 12, between what two numbers would approximately 68% of the values fall? Between what two numbers would 95% of the values fall? Between what two values would 99.7% of the values fall?

3.22 Some numbers are not normally distributed. If the mean of the numbers is 38 and the standard deviation is 6, what proportion of values would fall between 26 and 50? What proportion of values would fall between 14 and 62? Between what two values would 89% of the values fall?

3.23 According to Chebyshev's theorem, how many standard deviations from the mean would include at least 80% of the values?

3.24 The time needed to assemble a particular piece of furniture with experience is normally distributed with a mean time of 43 minutes. If 68% of the assembly times are between 40 and 46 minutes, what is the value of the standard deviation? Suppose 99.7% of the assembly times are between 35 and 51 minutes and the mean is still 43 minutes. What would the value of the standard deviation be now? Suppose the time needed to assemble another piece of furniture is not normally distributed and that the mean assembly time is 28 minutes. What is the value of the standard deviation if at least 77% of the assembly times are between 24 and 32 minutes?

3.25 Environmentalists are concerned about emissions of sulphur dioxide into the air. The average number of days per year in which sulphur dioxide levels exceed 150 milligrams per cubic metre in Milan, Italy, is 29. The number of days per year in which emission limits are exceeded is normally distributed with a standard deviation of 4.0 days. What percentage of the years would average between 21 and 37 days of excess emissions of sulphur dioxide? What percentage of the years would exceed 37 days? What percentage of the years would exceed 41 days? In what percentage of the years would there be fewer than 25 days with excess sulphur dioxide emissions?

3.26 Shown below are the per diem business travel expenses listed by Runzheimer International for 11 selected cities around the world. Use this list to calculate the z scores for Moscow, Beijing, Rio de Janeiro, and London. Treat the list as a sample.

City	Per Diem Expense ($)
Beijing	282
Hong Kong	361
London	430
Los Angeles	259
Mexico City	302
Moscow	376
New York (Manhattan)	457
Paris	305
Rio de Janeiro	343
Rome	297
Sydney	188

3.3   MEASURES OF CENTRAL TENDENCY AND VARIABILITY: GROUPED DATA

Grouped data do not provide information about individual values. Hence, measures of central tendency and variability for grouped data must be computed differently from those for ungrouped or raw data.

Measures of Central Tendency

Three measures of central tendency are presented here for grouped data: the mean, the median, and the mode.

Mean

For ungrouped data, the mean is computed by summing the data values and dividing by the number of values. With grouped data, the specific values are unknown. What can be used to represent the data values? The midpoint of each class interval is used to represent all the values in a class interval. This midpoint is weighted by the frequency of values in that class interval. The mean for grouped data is then computed by summing the products of the class midpoint and the class frequency for each class and dividing that sum by the total number of frequencies. The formula for the mean of grouped data follows.

MEAN OF GROUPED DATA

$$\mu_{grouped} = \frac{\Sigma fM}{N} = \frac{\Sigma fM}{\Sigma f} = \frac{f_1 M_1 + f_2 M_2 + \cdots + f_i M_i}{f_1 + f_2 + \cdots + f_i}$$

where:

i = the number of classes
f = class frequency
N = total frequencies

Table 3.6 gives the frequency distribution of the unemployment rates of Canada from Table 2.2. To find the mean of these data, we need Σf and ΣfM. The value of Σf can be determined by summing the values in the frequency column. To calculate ΣfM, we must first determine the values of M, or the class midpoints. Next we multiply each of these class midpoints by the frequency in that class interval, f, resulting in fM. Summing these values of fM yields the value of ΣfM.

Table 3.7 contains the calculations needed to determine the group mean. The group mean for the unemployment data is 6.93. Remember that because each class interval was represented by its class midpoint rather than by actual values, the group mean is only approximate.

TABLE 3.6

Frequency Distribution of 60 Years of Unemployment Data for Canada (Grouped Data)

Class Interval	Frequency	Cumulative Frequency
1–under 3	4	4
3–under 5	12	16
5–under 7	13	29
7–under 9	19	48
9–under 11	7	55
11–under 13	5	60

TABLE 3.7

Calculation of Grouped Mean

Class Interval	Frequency (f)	Class Midpoint (M)	fM
1–under 3	4	2	8
3–under 5	12	4	48
5–under 7	13	6	78
7–under 9	19	8	152
9–under 11	7	10	70
11–under 13	5	12	60
	$\Sigma f = N = 60$		$\Sigma fM = 416$

$$\mu = \frac{\Sigma fM}{\Sigma f} = \frac{416}{60} = 6.93$$

Median

The median for ungrouped or raw data is the middle value of an ordered array of numbers. For grouped data, solving for the median is considerably more complicated. The calculation of the median for grouped data is done by using the following formula.

MEDIAN OF GROUPED DATA

$$Median = L + \frac{\frac{N}{2} - cf_p}{f_{med}}(W)$$

where

L = the lower limit of the median class interval
cf_p = a cumulative total of the frequencies up to but not including the frequency of the median class
f_{med} = the frequency of the median class
W = the width of the median class interval
N = total number of frequencies

The first step in calculating a grouped median is to determine the value of $N/2$, which is the location of the median term. Suppose we want to calculate the median for the frequency distribution data in Table 3.6. Since there are 60 values (N), the value of $N/2$ is $60/2 = 30$. The median is the 30th term. The question to ask is where does the 30th term fall? This can be answered by determining the cumulative frequencies for the data, as shown in Table 3.6.

An examination of these cumulative frequencies reveals that the 30th term falls in the fourth class interval because there are only 29 values in the first three class intervals. Thus, the median value is in the fourth class interval somewhere between 7 and 9. The class interval containing the median value is referred to as the *median class interval*.

Since the 30th value is between 7 and 9, the value of the median must be at least 7. How much more than 7 is the median? The difference between the location of the median value, $N/2 = 30$, and the cumulative frequencies up to but not including the median class interval, $cf_p = 29$, tells how many values into the median class interval lies the value of the median. This is determined by solving for $N/2 - cf_p = 30 - 29 = 1$. The median value is located one value into the median class interval. However, there are 19 values in the median interval (denoted in the formula as f_{med}). The median value is 1/19 of the way through this interval.

$$\frac{\frac{N}{2} - cf_p}{f_{med}} = \frac{30 - 29}{19} = \frac{1}{19}$$

Thus, the median value is at least 7 – the value of L – and is 1/19 of the way across the median interval. How far is it across the median interval? Each class interval is 2 units wide (w). Taking 1/19 of this distance tells us how far the median value is into the class interval.

$$\frac{\frac{N}{2} - cf_p}{f_{med}}(W) = \frac{\frac{60}{2} - 29}{19}(2) = \frac{1}{19}(2) = 0.105$$

Adding this distance to the lower endpoint of the median class interval yields the value of the median.

$$\text{Median} = 7 + \frac{\frac{60}{2} - 29}{19}(2) = 7 + \frac{1}{19}(2) = 7 + 0.105 = 7.105$$

The median value of unemployment rates for Canada is 7.105. Keep in mind that like the grouped mean, this median value is merely approximate. The assumption made in these calculations is that the actual values fall uniformly across the median class interval – which may or may not be the case.

Mode

The *mode* for grouped data is *the class midpoint of the modal class. The modal class is the class interval with the greatest frequency.* Using the data from Table 3.7, the 7–under 9 class interval contains the greatest frequency, 19. Thus, the modal class is 7–under 9. The class midpoint of this modal class is 8. Therefore, the mode for the frequency distribution shown in Table 3.7 is 8. The modal unemployment rate is 8%.

Measures of Variability

Two measures of variability for grouped data are presented here: the variance and the standard deviation. Again, the standard deviation is the square root of the variance. Both measures have original and computational formulas.

	Original Formula	**Computational Version**
FORMULAS FOR POPULATION VARIANCE AND STANDARD DEVIATION OF GROUPED DATA	$\sigma^2 = \dfrac{\Sigma f(M - \mu)^2}{N}$ where: f = frequency M = class midpoint $N = \Sigma f$, or total of the frequencies of the sample μ = grouped mean for the population	$\sigma^2 = \dfrac{\Sigma fM^2 - \dfrac{(\Sigma fM)^2}{N}}{N}$

	Original Formula	**Computational Version**
FORMULAS FOR SAMPLE VARIANCE AND STANDARD DEVIATION OF GROUPED DATA	$s^2 = \dfrac{\Sigma f(M - \bar{x})^2}{n-1}$ $s = \sqrt{s^2}$ where: f = frequency M = class midpoint $N = \Sigma f$, or total of the frequencies of the sample $\bar{x}$ = grouped mean for the sample	$s^2 = \dfrac{\Sigma fM^2 - \dfrac{(\Sigma fM)^2}{n}}{n-1}$

Calculating Grouped Variance and Standard Deviation with the Original Formula

Class Interval	f	M	fM	$(M - \mu)$	$(M - \mu)^2$	$f(M - \mu)^2$
1–under 3	4	2	8	−4.93	24.305	97.220
3–under 5	12	4	48	−2.93	8.585	103.020
5–under 7	13	6	78	−0.93	0.865	11.245
7–under 9	19	8	152	1.07	1.145	21.755
9–under 11	7	10	70	3.07	9.425	65.975
11–under 13	5	12	60	5.07	25.705	128.525
	$\Sigma f = N = 60$		$\Sigma fM = 416$			$\Sigma f(M - \mu)^2 = 427.740$

$$\mu = \frac{\Sigma fM}{\Sigma f} = \frac{416}{60} = 6.93$$

$$\sigma^2 = \frac{\Sigma f(M - \mu)^2}{N} = \frac{427.74}{60} = 7.129$$

$$\sigma = \sqrt{7.129} = 2.670$$

For example, let us calculate the variance and standard deviation of the Canadian unemployment data grouped as a frequency distribution in Table 3.6. If the data are treated as a population, the computations are as follows.

For the original formula, the computations are given in Table 3.8. The method of determining σ^2 and σ by using the computational formula is shown in Table 3.9. In either case, the variance of the unemployment data is 7.129 (squared per cent), and the standard deviation is 2.67%. As with the computation of the grouped mean, the class midpoint is used to represent all values in a class interval. This approach may or may not be appropriate, depending on whether the average value in a class is at the midpoint. If this situation does not occur, then the variance and the standard deviation are only approximations. Because grouped statistics are usually computed without knowledge of the actual data, the statistics computed potentially may be only approximations.

Calculating Grouped Variance and Standard Deviation with the Computational Formula

Class Interval	f	M	fM	fM^2
1–under 3	4	2	8	16
3–under 5	12	4	48	192
5–under 7	13	6	78	468
7–under 9	19	8	152	1216
9–under 11	7	10	70	700
11–under 13	5	12	60	720
	$\Sigma f = N = 60$		$\Sigma fM = 416$	$\Sigma fM^2 = 3312$

$$\sigma^2 = \frac{\Sigma fM^2 - \frac{(\Sigma fM)^2}{N}}{N} = \frac{3312 - \frac{416^2}{60}}{60} = \frac{3312 - 2884.27}{60} = \frac{427.73}{60} = 7.129$$

$$\sigma = \sqrt{7.129} = 2.670$$

DEMONSTRATION PROBLEM 3.7

Compute the mean, median, mode, variance, and standard deviation on the following sample data:

Class Interval	Frequency	Cumulative Frequency
10–under 15	6	6
15–under 20	22	28
20–under 25	35	63
25–under 30	29	92
30–under 35	16	108
35–under 40	8	116
40–under 45	4	120
45–under 50	2	122

Solution

The mean is computed as follows:

Class	F	M	fM
10–under 15	6	12.5	75.0
15–under 20	22	17.5	385.0
20–under 25	35	22.5	787.5
25–under 30	29	27.5	797.5
30–under 35	16	32.5	520.0
35–under 40	8	37.5	300.0
40–under 45	4	42.5	170.0
45–under 50	2	47.5	95.0
	$\Sigma f = n = 122$		$\Sigma fM = 3{,}130.0$

$$\bar{x} = \frac{\Sigma fM}{\Sigma f} = \frac{3130}{122} = 25.66$$

The grouped mean is 25.66.

The grouped median is located at the 61st value (122/2). Observing the cumulative frequencies, the 61st value falls in the 20-under 25 class, making it the median class interval; and thus, the grouped median is at least 20. Since there are 28 cumulative values before the median class interval, 33 more (61 − 28) are needed to reach the grouped median. However, there are 35 values in the median class. The grouped median is located 33/35 of the way across the class interval, which has a width of 5. The grouped median is $20 + \frac{33}{35}(5) = 20 + 4.71 = 24.71$.

The grouped mode can be determined by finding the class midpoint of the class interval with the greatest frequency. The class with the greatest frequency is 20–under 25 with a frequency of 35. The midpoint of this class is 22.5, which is the grouped mode.

The variance and standard deviation can be found as shown next. First, use the original formula.

Class	f	M	$M - \bar{x}$	$(M - \bar{x})^2$	$f(M - \bar{x})^2$
10–under 15	6	12.5	−13.16	173.19	1,039.14
15–under 20	22	17.5	−8.16	66.59	1,464.98
20–under 25	35	22.5	−3.16	9.99	349.65

Class	f	M	$M - \bar{x}$	$(M - \bar{x})^2$	$f(M - \bar{x})^2$
25–under 30	29	27.5	1.84	3.39	98.31
30–under 35	16	32.5	6.84	46.79	748.64
35–under 40	8	37.5	11.84	140.19	1,121.52
40–under 45	4	42.5	16.84	283.59	1,134.36
45–under 50	2	47.5	21.84	476.99	953.98
	$\Sigma f = n = 122$				$\Sigma f(M - \bar{x})^2 = 6910.58$

$$s^2 = \frac{\Sigma f(M - \bar{x})^2}{n - 1} = \frac{6910.58}{121} = 57.11$$

$$s = \sqrt{57.11} = 7.56$$

Next, use the computational formula.

Class	f	M	fM	fM^2
10–under 15	6	12.5	75.0	937.50
15–under 20	22	17.5	385.0	6,737.50
20–under 25	35	22.5	787.5	17,718.75
25–under 30	29	27.5	797.5	21,931.25
30–under 35	16	32.5	520.0	16,900.00
35–under 40	8	37.5	300.0	11,250.00
40–under 45	4	42.5	170.0	7,225.00
45–under 50	2	47.5	95.0	4,512.50
	$\Sigma f = n = 122$		$\Sigma fM = 3,130.0$	$\Sigma fM^2 = 87,212.50$

$$s^2 = \frac{\Sigma fM^2 - \frac{(\Sigma fM)^2}{n}}{n - 1} = \frac{87,212.5 - \frac{(3,130)^2}{122}}{121} = \frac{6,910.04}{121} = 57.11$$

$$s = \sqrt{57.11} = 7.56$$

The sample variance is 57.11 and the standard deviation is 7.56.

3.3 PROBLEMS

3.27 Compute the mean, the median, and the mode for the following data:

Class	f
0–under 2	39
2–under 4	27
4–under 6	16
6–under 8	15
8–under 10	10
10–under 12	8
12–under 14	6

3.28 Compute the mean, the median, and the mode for the following data:

Class	f
1.2–under 1.6	220
1.6–under 2.0	150
2.0–under 2.4	90
2.4–under 2.8	110
2.8–under 3.2	280

3.29 Determine the population variance and standard deviation for the following data by using the original formula:

Class	f
20–under 30	7
30–under 40	11
40–under 50	18
50–under 60	13
60–under 70	6
70–under 80	4

3.30 Determine the sample variance and standard deviation for the following data by using the computational formula:

Class	f
5–under 9	20
9–under 13	18
13–under 17	8
17–under 21	6
21–under 25	2

3.31 A random sample of voters in Uppsala, Sweden, is classified by age group, as shown by the following data:

Age Group	Frequency
18–under 24	17
24–under 30	22
30–under 36	26
36–under 42	35
42–under 48	33
48–under 54	30
54–under 60	32
60–under 66	21
66–under 72	15

a. Calculate the mean of the data.
b. Calculate the mode.
c. Calculate the median.
d. Calculate the variance.
e. Calculate the standard deviation.

COEFFICIENT OF SKEWNESS

$$S_k = \frac{3(\mu - M_d)}{\sigma}$$

where:

S_k = coefficient of skewness
M_d = median

Suppose, for example, that a distribution has a mean of 29, a median of 26, and a standard deviation of 12.3. The coefficient of skewness is computed as

$$S_k = \frac{3(29 - 26)}{12.3} = +0.73$$

Because the value of S_k is positive, the distribution is positively skewed. If the value of S_k is negative, the distribution is negatively skewed. The greater the magnitude of S_k, the more skewed is the distribution.

Kurtosis

Kurtosis *describes the amount of peakedness of a distribution*. Distributions that are high and thin are referred to as **leptokurtic** distributions. Distributions that are flat and spread out are referred to as **platykurtic** distributions. Between these two types are distributions that are more 'normal' in shape, referred to as **mesokurtic** distributions. These three types of kurtosis are illustrated in Figure 3.12.

Box-and-Whisker Plots

Another way to describe a distribution of data is by using a box-and-whisker plot. A **box-and-whisker plot**, sometimes called a *box plot*, is *a diagram that utilizes the upper and lower quartiles along with the median and the two most extreme values to depict a distribution graphically*. The plot is constructed by using a box to enclose the median. This *box* is extended outward from the median along a continuum to the lower and upper quartiles, enclosing not only the median but also the middle 50% of the data. From the lower and upper quartiles, lines referred to as *whiskers* are extended out from the box towards the outermost data values. The box-and-whisker plot is determined from five specific numbers.

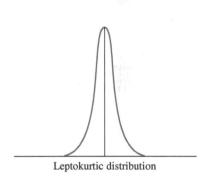

Leptokurtic distribution

1. The median (Q_2)
2. The lower quartile (Q_1)
3. The upper quartile (Q_3)
4. The smallest value in the distribution
5. The largest value in the distribution

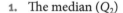

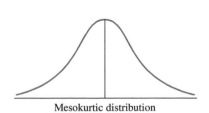

Platykurtic distribution

The box of the plot is determined by locating the median and the lower and upper quartiles on a continuum. A box is drawn around the median with the lower and upper quartiles (Q_1 and Q_3) as the box endpoints. These box endpoints (Q_1 and Q_3) are referred to as the *hinges* of the box.

Next the value of the interquartile range (IQR) is computed by $Q_3 - Q_1$. The interquartile range includes the middle 50% of the data and should equal the length of the box. However, here the interquartile range is used outside of the box also. At a distance of $1.5 \cdot$ IQR outward from the lower and upper quartiles are what are referred to as *inner fences*. A *whisker*, a line segment, is drawn from the lower hinge of the

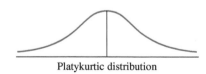

Mesokurtic distribution

FIGURE 3.12

Types of Kurtosis

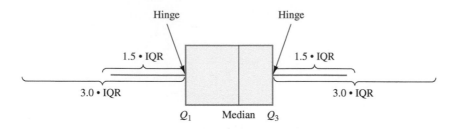

FIGURE 3.13

Box-and-Whisker Plot

box outward to the smallest data value. A second whisker is drawn from the upper hinge of the box outward to the largest data value. The inner fences are established as follows:

$$Q_1 - 1.5 \cdot IQR$$
$$Q_3 + 1.5 \cdot IQR$$

If data fall beyond the inner fences, then *outer* fences can be constructed:

$$Q_1 - 3.0 \cdot IQR$$
$$Q_3 + 3.0 \cdot IQR$$

Figure 3.13 shows the features of a box-and-whisker plot.

Data values outside the mainstream of values in a distribution are viewed as *outliers*. Outliers can be merely the more extreme values of a data set. However, sometimes outliers occur due to measurement or recording errors. Other times they are values so unlike the other values that they should not be considered in the same analysis as the rest of the distribution. Values in the data distribution that are outside the inner fences but within the outer fences are referred to as *mild outliers*. Values that are outside the outer fences are called *extreme outliers*. Thus, one of the main uses of a box-and-whisker plot is to identify outliers. In some computer-produced box-and-whisker plots (such as in Minitab), the whiskers are drawn to the largest and smallest data values within the inner fences. An asterisk is then printed for each data value located between the inner and outer fences to indicate a mild outlier. Values outside the outer fences are indicated by a zero on the graph. These values are extreme outliers.

Another use of box-and-whisker plots is to determine whether a distribution is skewed. The location of the median in the box can relate information about the skewness of the middle 50% of the data. If the median is located on the right side of the box, then the middle 50% are skewed to the left. If the median is located on the left side of the box, then the middle 50% are skewed to the right. By examining the length of the whiskers on each side of the box, a business researcher can make a judgement about the skewness of the outer values. If the longest whisker is to the right of the box, then the outer data are skewed to the right and vice versa. We shall use the data given in Table 3.10 to construct a box-and-whisker plot.

After organizing the data into an ordered array, as shown in Table 3.11, it is relatively easy to determine the values of the lower quartile (Q_1), the median, and the upper quartile (Q_3). From these, the value of the interquartile range can be computed.

The hinges of the box are located at the lower and upper quartiles, 69 and 80.5. The median is located within the box at distances of 4 from the lower quartile and 7.5 from the upper quartile. The distribution of the middle 50% of the data is skewed right, because the median is nearer to the lower or left hinge. The inner fence is constructed by

$$Q_1 - 1.5 \cdot IQR = 69 - 1.5(11.5) = 69 - 17.25 = 51.75$$

and

$$Q_3 + 1.5 \cdot IQR = 80.5 + 1.5(11.5) = 80.5 + 17.25 = 97.75$$

TABLE 3.10

Data for Box-and-Whisker Plot

71	87	82	64	72	75	81	69
76	79	65	68	80	73	85	71
70	79	63	62	81	84	77	73
82	74	74	73	84	72	81	65
74	62	64	68	73	82	69	71

TABLE 3.11

Data in Ordered Array with Quartiles and Median

87	85	84	84	82	82	82	81	81	81
80	79	79	77	76	75	74	74	74	73
73	73	73	72	72	71	71	71	70	69
69	68	68	65	65	64	64	63	62	62

$$Q_1 = 69$$
$$Q_2 = \text{median} = 73$$
$$Q_3 = 80.5$$
$$\text{IQR} = Q_3 - Q_1 = 80.5 - 69 = 11.5$$

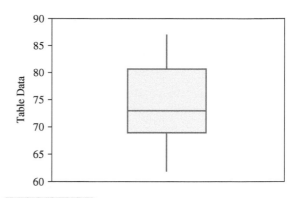

FIGURE 3.14

Minitab Box-and-Whisker Plot

The whiskers are constructed by drawing a line segment from the lower hinge outward to the smallest data value and a line segment from the upper hinge outward to the largest data value. An examination of the data reveals that no data values in this set of numbers are outside the inner fence. The whiskers are constructed outward to the lowest value, which is 62, and to the highest value, which is 87.

To construct an outer fence, we calculate $Q_1 - 3 \cdot \text{IQR}$ and $Q_3 + 3 \cdot \text{IQR}$, as follows.

$$Q_1 - 3 \cdot \text{IQR} = 69 - 3(11.5) = 69 - 34.5 = 34.5$$
$$Q_3 + 3 \cdot \text{IQR} = 80.5 + 3(11.5) = 80.5 + 34.5 = 115.0$$

Figure 3.14 is the Minitab computer printout for this box-and-whisker plot.

3.4 PROBLEMS

3.35 On a certain day the average closing price of a group of stocks on the Euronext Stock Exchange is €35 (to the nearest euro). If the median value is €33 and the mode is €21, is the distribution of these stock prices skewed? If so, how?

3.36 A local hotel offers ballroom dancing on Friday nights. A researcher observes the customers and estimates their ages. Discuss the skewness of the distribution of ages if the mean age is 51, the median age is 54, and the modal age is 59.

3.37 The sales volumes for the top real estate brokerage firms in the United States for a recent year were analysed using descriptive statistics. The mean annual dollar volume for these firms was $5.51 billion, the median was $3.19 billion, and the standard deviation was $9.59 billion. Compute the value of the Pearsonian coefficient of skewness and discuss the meaning of it. Is the distribution skewed? If so, to what extent?

3.38 Suppose the following data are the ages of Internet users obtained from a sample. Use these data to compute a Pearsonian coefficient of skewness. What is the meaning of the coefficient?

41	15	31	25	24
23	21	22	22	18
30	20	19	19	16
23	27	38	34	24
19	20	29	17	23

3.39 Construct a box-and-whisker plot on the data below. Do the data contain any outliers? Is the distribution of data skewed?

540	690	503	558	490	609
379	601	559	495	562	580
510	623	477	574	588	497
527	570	495	590	602	541

3.40   Suppose a consumer group asked 18 consumers to keep a yearly log of their shopping practices and that the data below represent the number of coupons used by each consumer over the yearly period. Use the data to construct a box-and-whisker plot. List the median, Q_1, Q_3, the endpoints for the inner fences, and the endpoints for the outer fences. Discuss the skewness of the distribution of these data and point out any outliers.

81	68	70	100	94	47	66	70	82
110	105	60	21	70	66	90	78	85

3.5   DESCRIPTIVE STATISTICS ON THE COMPUTER

Both Minitab and Excel yield extensive descriptive statistics. Even though each computer package can compute individual statistics such as a mean or a standard deviation, they can also produce multiple descriptive statistics at one time. Figure 3.15 displays a Minitab output for the descriptive statistics associated with the computer production data presented earlier in this section. The Minitab output contains, among other things, the mean, the median, the sample standard deviation, the minimum and maximum (which can then be used to compute the range), and Q_1 and Q_3 (from which the interquartile range can be computed). Excel's descriptive statistics output for the same computer production data is displayed in Figure 3.16. The Excel output contains the mean, the median, the mode, the sample standard deviation, the sample variance, and the range. The descriptive statistics feature on either of these computer packages yields a lot of useful information about a data set.

DESCRIPTIVE STATISTICS

Variable	N	N*	Mean	SE Mean	StDev	Minimum	Q_1
Computers Produced	5	0	13.00	2.55	5.70	5.00	7.00

Variable	Median	Q_3	Maximum
Computers Produced	16.00	17.50	18.00

FIGURE 3.15

Minitab Output for the Computer Production Problem

COMPUTER PRODUCTION DATA

Mean	13
Standard error	2.54951
Median	16
Mode	#N/A
Standard deviation	5.700877
Sample variance	32.5
Kurtosis	−1.71124
Skewness	−0.80959
Range	13
Minimum	5
Maximum	18
Sum	65
Count	5

FIGURE 3.16

Excel Output for the Computer Production Problem

Decision Dilemma SOLVED

Choosing the Right Recycling Collection System

The descriptive statistics presented in this chapter are excellent for summarizing and presenting data sets in more concise formats. For example, question 1 of the managerial and statistical questions in the Decision Dilemma reports recycling collection measurements for 50 UK households. Using Excel and/or Minitab, many of the descriptive statistics presented in this chapter can be applied to these data. The results are shown in Figures 3.17 and 3.18.

These computer outputs show that the average recycling materials collected usage is 50.2 kilograms with a standard deviation of about 6.794 kilograms. The median is 50 kilograms with a range of 22 kilograms (39 to 61). The first quartile is 44 kilograms and the third quartile is 56 kilograms. The mode is 39 kilograms. The Minitab graph and the skewness measures show that the data are slightly skewed to the left. Applying Chebyshev's theorem to the mean and standard deviation shows that at least 88.9% of the measurements should fall between 29.818 kilograms and 70.582 kilograms. An examination of the data and the minimum and maximum reveals that 100% of data actually fall within these limits.

According to the Decision Dilemma, the mean amount of waste per head is 592 kilograms with a standard deviation of 83 kilograms. If the amount of waste per head in the UK is approximately normally distributed, we can apply the empirical rule. According to the empirical rule, 68% of the times would fall within 509 and 675 kilograms, 95% of the times would fall within 426 and 758 kilograms, and 99.7% of the amount of waste per head would fall within 343 and 841 kilograms. If the data are not normally distributed, Chebyshev's theorem reveals that at least 75% of the times should fall between 426 and 758 kilograms and 88.9% should fall between 343 and 841 kilograms.

KILOGRAMS OF WASTE	
Mean	50.2
Standard Error	0.960867
Median	50
Mode	39
Standard Deviation	6.794355
Sample Variance	46.16327
Kurtosis	−1.15691
Skewness	−0.08036
Range	22
Minimum	39
Maximum	61
Sum	2510
Count	50

FIGURE 3.17

Excel Descriptive Statistics

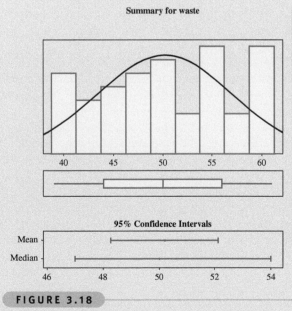

FIGURE 3.18

Minitab Descriptive Statistics

ETHICAL CONSIDERATIONS

In describing a body of data to an audience, it is best to use whatever measures it takes to present a 'full' picture of the data. By limiting the descriptive measures used, the business researcher may give the audience only part of the picture and can skew the way the receiver understands the data. For example, if a researcher presents only the mean, the audience will have no insight into the variability of the data; in addition, the mean might be inordinately large or small because of extreme values. Likewise, the choice of the median precludes a picture that includes the extreme values. Using the mode can cause the receiver of the information to focus only on values that occur often.

At least one measure of variability is usually needed with at least one measure of central tendency for the audience to begin to understand what the data look like. Unethical researchers might be tempted to present only the descriptive measure that will convey the picture of the data that they want the audience to see. Ethical researchers will instead use any and all methods that will present the fullest, most informative picture possible from the data.

Former US governor of Colorado, Richard Lamm, has been quoted as having said that 'Demographers are academics who can statistically prove that the average person in Miami is born Cuban and dies Jewish . . .'.* People are more likely to reach this type of conclusion if incomplete or misleading descriptive statistics are provided by researchers.

*Alan L. Otten, 'People Patterns/Odds and Ends', *The Wall Street Journal*, 29 June 1992, p. B1. Reprinted by permission of *The Wall Street Journal* © 1992, Dow Jones & Company, Inc. All Rights Reserved Worldwide.

SUMMARY

Statistical descriptive measures include measures of central tendency, measures of variability, and measures of shape. Measures of central tendency and measures of variability are computed differently for ungrouped and grouped data. Measures of central tendency are useful in describing data because they communicate information about the more central portions of the data. The most common measures of central tendency are the three Ms': mode, median, and mean. In addition, percentiles and quartiles are measures of central tendency.

The mode is the most frequently occurring value in a set of data. Among other things, the mode is used in business for determining sizes.

The median is the middle term in an ordered array of numbers containing an odd number of terms. For an array with an even number of terms, the median is the average of the two middle terms. A median is unaffected by the magnitude of extreme values. This characteristic makes the median a most useful and appropriate measure of location in reporting such things as income, age, and house prices.

The arithmetic mean is widely used and is usually what researchers are referring to when they use the word *mean*. The arithmetic mean is the average. The population mean and the sample mean are computed in the same way but are denoted by different symbols. The arithmetic mean is affected by every value and can be inordinately influenced by extreme values.

Percentiles divide a set of data into 100 groups, which means 99 percentiles are needed. Quartiles divide data into four groups. The three quartiles are Q_1, which is the lower quartile; Q_2, which is the middle quartile and equals the median; and Q_3, which is the upper quartile.

Measures of variability are statistical tools used in combination with measures of central tendency to describe data. Measures of variability provide information about the spread of the data values. These measures include the range, mean absolute deviation, variance, standard deviation, interquartile range, z scores, and coefficient of variation for ungrouped data.

One of the most elementary measures of variability is the range. It is the difference between the largest and smallest values. Although the range is easy to compute, it has limited usefulness. The interquartile range is the difference between the third and first quartile. It equals the range of the middle 50% of the data.

The mean absolute deviation (MAD) is computed by averaging the absolute values of the deviations from the mean. The mean absolute deviation provides the magnitude of the average deviation but without specifying its direction. The mean absolute deviation has limited usage in statistics, but interest is growing for the use of MAD in the field of forecasting.

Variance is widely used as a tool in statistics but is used little as a stand-alone measure of variability. The variance is the average of the squared deviations about the mean.

The square root of the variance is the standard deviation. It also is a widely used tool in statistics, but it is used more often than the variance as a stand-alone measure. The standard deviation is best understood by examining its applications in determining where data are in relation to the mean. The empirical rule and Chebyshev's theorem are statements about the proportions of data values that are within various numbers of standard deviations from the mean.

The empirical rule reveals the percentage of values that are within one, two, or three standard deviations of the mean for a set of data. The empirical rule applies only if the data are in a bell-shaped distribution.

Chebyshev's theorem also delineates the proportion of values that are within a given number of standard deviations from the mean. However, it applies to any distribution. The z score represents the number of standard deviations a value is from the mean for normally distributed data.

The coefficient of variation is a ratio of a standard deviation to its mean, given as a percentage. It is especially useful in comparing standard deviations or variances that represent data with different means.

Some measures of central tendency and some measures of variability are presented for grouped data. These measures include mean, median, mode, variance, and standard deviation. Generally, these measures are only approximate for grouped data because the values of the actual raw data are unknown.

Two measures of shape are skewness and kurtosis. Skewness is the lack of symmetry in a distribution. If a distribution is skewed, it is stretched in one direction or the other. The skewed part of a graph is its long, thin portion. One measure of skewness is the Pearsonian coefficient of skewness.

Kurtosis is the degree of peakedness of a distribution. A tall, thin distribution is referred to as leptokurtic. A flat distribution is platykurtic, and a distribution with a more normal peakedness is said to be mesokurtic.

A box-and-whisker plot is a graphical depiction of a distribution. The plot is constructed by using the median, the lower quartile, and the upper quartile. It can yield information about skewness and outliers.

KEY TERMS

arithmetic mean
bimodal
box-and-whisker plot
Chebyshev's theorem
coefficient of skewness
coefficient of variation (CV)
deviation from the mean
empirical rule
interquartile range
kurtosis
leptokurtic
mean
mean absolute deviation (MAD)
measures of central tendency
measures of shape

measures of variability
median
mesokurtic
mode
multimodal
percentiles
platykurtic
quartiles
range
skewness
standard deviation
sum of squares of x
variance
z score

FORMULAS

Population mean (ungrouped)

$$\mu = \frac{\Sigma x}{N}$$

Sample mean (ungrouped)

$$\bar{x} = \frac{\Sigma x}{n}$$

Mean absolute deviation

$$MAD = \frac{\Sigma |x - \mu|}{N}$$

Population variance (ungrouped)

$$\sigma^2 = \frac{\Sigma (x - \mu)^2}{N}$$

$$\sigma^2 = \frac{\Sigma x^2 - \frac{(\Sigma x)^2}{N}}{N}$$

$$\sigma^2 = \frac{\Sigma x^2 - N\mu^2}{N}$$

Population standard deviation (ungrouped)

$$\sigma = \sqrt{\sigma^2}$$

$$\sigma = \sqrt{\frac{\Sigma (x - \mu)^2}{N}}$$

$$\sigma = \sqrt{\frac{\Sigma x^2 - \frac{(\Sigma x)^2}{N}}{N}}$$

$$\sigma = \sqrt{\frac{\Sigma x^2 - N\mu^2}{N}}$$

Grouped mean

$$\mu_{\text{grouped}} = \frac{\Sigma fM}{N}$$

Grouped median

$$Median = L + \frac{\frac{N}{2} - cf_p}{f_{med}}(W)$$

Population variance (grouped)

$$\sigma^2 = \frac{\Sigma f(M - \mu)^2}{N} = \frac{\Sigma fM^2 - \frac{(\Sigma fM)^2}{N}}{N}$$

Population standard deviation (grouped)

$$\sigma = \sqrt{\frac{\Sigma f(M - \mu)^2}{N}} = \sqrt{\frac{\Sigma fM^2 - \frac{(\Sigma fM)^2}{N}}{N}}$$

Sample variance

$$s^2 = \frac{\Sigma (x - \bar{x})^2}{n - 1}$$

$$s^2 = \frac{\Sigma x^2 - \frac{(\Sigma x)^2}{n}}{n - 1}$$

$$s^2 = \frac{\Sigma x^2 - n(\bar{x})^2}{n - 1}$$

Sample standard deviation

$$s = \sqrt{s^2}$$

$$s = \sqrt{\frac{\Sigma(x - \bar{x})^2}{n-1}}$$

$$s = \sqrt{\frac{\Sigma x^2 - \frac{(\Sigma x)^2}{n}}{n-1}}$$

$$s = \sqrt{\frac{\Sigma x^2 - n(\bar{x})^2}{n-1}}$$

Chebyshev's theorem

$$1 - \frac{1}{k^2}$$

z score

$$z = \frac{x - \mu}{\sigma}$$

Coefficient of variation

$$CV = \frac{\sigma}{\mu}(100)$$

Interquartile range

$$IQR = Q_3 - Q_1$$

Sample variance (grouped)

$$s^2 = \frac{\Sigma f(M - \bar{x})^2}{n-1} = \frac{\Sigma fM^2 - \frac{(\Sigma fM)^2}{n}}{n-1}$$

Sample standard deviation (grouped)

$$s = \sqrt{\frac{\Sigma f(M - \bar{x})^2}{n-1}} = \sqrt{\frac{\Sigma fM^2 - \frac{(\Sigma fM)^2}{n}}{n-1}}$$

Pearsonian coefficient of skewness

$$S_k = \frac{3(\mu - M_d)}{\sigma}$$

SUPPLEMENTARY PROBLEMS

CALCULATING THE STATISTICS

3.41 The 2011 UK Census asked every household to report information on each person living there. Suppose for a sample of 30 households selected, the number of persons living in each was reported as follows:

2	3	1	2	6	4	2	1	5	3	2	3	1	2	2
1	3	1	2	2	4	2	1	2	8	3	2	1	1	3

Compute the mean, median, mode, range, lower and upper quartiles, and interquartile range for these data.

3.42 The 2011 UK Census also asked for each person's age. Suppose that a sample of 40 households taken from the census data showed the age of the

first person recorded on the census form to be as follows:

42	29	31	38	55	27	28
33	49	70	25	21	38	47
63	22	38	52	50	41	19
22	29	81	52	26	35	38
29	31	48	26	33	42	58
40	32	24	34	25		

Compute P_{10}, P_{80}, Q_1, Q_3, the interquartile range, and the range for these data.

3.43 Shown below are the top 20 companies in the computer industry by sales according to netvalley.com in a recent year. Compute the mean, median, P_{30}, P_{60}, P_{90}, Q_1, Q_3, range, and interquartile range on these data.

Company	Sales ($ millions)
IBM	91,134
Hewlett Packard	86,696
Verizon Communications	75,112
Dell	49,205
Microsoft	39,788
Intel	38,826
Motorola	36,843
Sprint	34,680
Canon	34,222
Ingram Micro	28,808
Cisco Systems	24,801
EDS	19,757
Xerox	15,701
Computer Sciences	14,059
Apple	13,931
Texas Instruments	13,392
Oracle	11,799
Sanmina-SCI	11,735
Arrow Electronics	11,164
Sun Microsystems	11,070

3.44 Shown below is the list of Dutch cities with more than 100,000 inhabitants according to the Netherland's Central Bureau of Statistics in a recent year. Use this population data to compute a mean and a standard deviation for these 23 cities.

City	Population
Amsterdam	778,825
Rotterdam	611,000
The Hague	488,553
Utrecht	307,081
Eindhoven	213,809
Tilburg	203,492
Almere	188,160
Groningen	187,298
Breda	173,299
Enschede	157,012
Apeldoorn	155,726
Nijmegen	151,030
Haarlem	150,611
Arnhem	147,018
Amersfoort	126,750
Dordrecht	118,540
Zoetermeer	118,020
Zwolle	118,000
Leiden	117,480

(continued)

City	Population
Maastricht	116,200
Ede	107,476
's-Hertogenbosch	102,220
Venlo	100,271

3.45 Shown here is a list of the largest oil exporting countries, ranked in terms of barrels per day according to the US Energy Information Administration. Use these as population data and answer the questions.

Exporting Nation	Thousands of barrels per day
Saudi Arabia	8,651
Russia	6,565
Norway	2,542
Iran	2,519
United Arab Emirates	2,515
Venezuela	2,203
Kuwait	2,150
Nigeria	2,146
Algeria	1,847
Mexico	1,676
Libya	1,525
Iraq	1,438
Angola	1,363
Kazakhstan	1,114
Canada	1,071

a. What are the values of the mean and the median? Compare the answers and state which you prefer as a measure of location for these data and why.

b. What are the values of the range and interquartile range? How do they differ?

c. What are the values of variance and standard deviation for these data?

d. What is the z score for Angola? What is the z score for Russia? Interpret these z scores.

e. Calculate the Pearsonian coefficient of skewness and comment on the skewness of this distribution.

3.46 The Carbon Dioxide Information Analysis Centre collected data for the United Nations on carbon dioxide emissions due to human activity. Following are the top 20 sovereign states in terms of annual CO_2 emissions in a recent year.

Country	Annual CO_2 emissions
China	6,538,367
United States	5,838,381
India	1,612,362
Russia	1,537,357
Japan	1,254,543
Germany	787,936
Canada	557,340
United Kingdom	539,617
South Korea	503,321
Iran	495,987
Mexico	471,459
Italy	456,428
South Africa	433,527
Saudi Arabia	402,450
Indonesia	397,143
Australia	374,045
France	371,757
Brazil	368,317
Spain	359,260
Ukraine	317,537

a. Calculate the mean, median, and mode.

b. Calculate the range, interquartile range, mean absolute deviation, sample variance, and sample standard deviation.

c. Compute the Pearsonian coefficient of skewness for these data.

d. Sketch a box-and-whisker plot.

3.47 The radio music listener market is diverse. Listener formats might include adult contemporary, album rock, top 40, oldies, rap, country and western, classical, and jazz. In targeting audiences, market researchers need to be concerned about the ages of the listeners attracted to particular formats. Suppose a market researcher surveyed a sample of 170 listeners of country music radio stations and obtained the following age distribution:

Age	Frequency
15–under 20	9
20–under 25	16
25–under 30	27
30–under 35	44
35–under 40	42
40–under 45	23
45–under 50	7
50–under 55	2

a. What are the mean and modal ages of country music listeners?

b. What are the variance and standard deviation of the ages of country music listeners?

3.48 A research agency administers a demographic survey to 90 telemarketing companies to determine the size of their operations. When asked to report how many employees now work in their telemarketing operation, the companies gave responses ranging from 1 to 100. The agency's analyst organizes the figures into a frequency distribution.

Number of Employees Working in Telemarketing	Number of Companies
0–under 20	32
20–under 40	16
40–under 60	13
60–under 80	10
80–under 100	19

a. Compute the mean, median, and mode for this distribution.

b. Compute the sample standard deviation for these data.

TESTING YOUR UNDERSTANDING

3.49 Financial analysts like to use the standard deviation as a measure of risk for a stock. The greater the deviation in a stock price over time, the more risky it is to invest in the stock. However, the average prices of some stocks are considerably higher than the average price of others, allowing for the potential of a greater standard deviation of price. For example, a standard deviation of £5.00 on a £10.00 stock is considerably different from a £5.00 standard deviation on a £40.00 stock. In this situation, a coefficient of variation might provide insight into risk. Suppose stock X costs an average of £32.00 per share and showed a standard deviation of £3.45 for the past 60 days. Suppose stock Y costs an average of £84.00 per share and showed a standard deviation of £5.40 for the past 60 days. Use the coefficient of variation to determine the variability for each stock.

3.50 The European Automobile Manufacturers' Association reported that the average age of a car on European Union roads in a recent year was

8.5 years. Suppose the distribution of ages of cars on EU roads is approximately bell-shaped. If 99.7% of the ages are between 1 year and 16 years, what is the standard deviation of car age? Suppose the standard deviation is 1.7 years and the mean is 8.5 years. Between what two values would 95% of the car ages fall?

3.51 According to a *Human Resources* report, a worker in the industrial countries spends on average 419 minutes a day on the job. Suppose the standard deviation of time spent on the job is 27 minutes.

   a. If the distribution of time spent on the job is approximately bell shaped, between what two times would 68% of the figures be? 95%? 99.7%?

   b. If the shape of the distribution of times is unknown, approximately what percentage of the times would be between 359 and 479 minutes?

   c. Suppose a worker spent 400 minutes on the job. What would that worker's z score be, and what would it tell the researcher?

3.52 Below are the 2009 per capita GDP figures for Central and Eastern European countries according to the United Nation's World Development Indicators.

Country	GDP per capita, PPP (constant 2005 $)
Albania	7,451
Armenia	4,794
Azerbaijan	8,752
Belarus	11,841
Bosnia and Herzegovina	7,267
Bulgaria	11,458
Croatia	16,225
Czech Republic	22,098
Georgia	4,335
Hungary	16,896
Kosovo	n.a.
Macedonia, FYR	8,742
Moldova	2,592
Montenegro	10,024
Poland	16,705
Romania	10,796
Russian Federation	13,611
Serbia	9,968
Slovak Republic	19,202
Slovenia	24,807
Ukraine	5,737

   a. Compute the mean and standard deviation for Albania, Armenia, Bulgaria, Croatia, and Czech Republic.

   b. Compute the mean and standard deviation for Hungary, Poland, Romania, Bosnia/Herzegovina, and Slovenia

   c. Use a coefficient of variation to compare the two standard deviations. Treat the data as population data.

3.53 The average daily car traffic in the main avenue of a medium-sized European city is 35,748. Suppose the median is 31,369 and the mode is 29,500. Is the distribution of daily traffic skewed? If so, how and why? Which of these measures of central tendency would you use to describe these data? Why?

3.54 The data presented below represents full-time average annual gross wages and salaries in the entire economy of selected Organization for Economic Cooperation and Development (OECD) member countries.

Country	Average full-time wages ($)
United States	51,493
Luxembourg	50,610
Switzerland	47,810
Australia	45,385
Netherlands	45,161
Ireland	45,160
United Kingdom	43,607
Norway	43,250
Denmark	42,173
Belgium	41,923
Austria	41,421
Germany	37,544
France	37,269
Sweden	35,582
Finland	34,903
Spain	32,957
Japan	32,816
South Korea	32,638
Italy	32,121
Greece	27,460
Portugal	22,666
Czech Republic	19,618
Hungary	18,220
Poland	17,812

a. Construct a box-and-whisker plot for these data.

b. Discuss the shape of the distribution from the plot.

c. Are there outliers?

d. What are they and why do you think they are outliers?

3.55 *Runzheimer International* publishes data on overseas business travel costs. They report that the average per diem total for a business traveller in Paris, France, is $349. Suppose the shape of the distribution of the per diem costs of a business traveller to Paris is unknown, but that 53% of the per diem figures are between $317 and $381. What is the value of the standard deviation? The average per diem total for a business traveller in Moscow is $415. If the shape of the distribution of per diem costs of a business traveller in Moscow is unknown and if 83% of the per diem costs in Moscow lie between $371 and $459, what is the standard deviation?

INTERPRETING THE OUTPUT

3.56 Bankersalmanac.com recently compiled a list of the top 50 banks in the World according to total assets. Leading the list was BNP Paribas of France, followed by The Royal Bank of Scotland Group plc from the UK. Following is an Excel analysis of total assets ($ billions) of these banks using the descriptive statistics feature. Study the output and describe in your own words what you can learn about the assets of these top 50 banks.

Top 50 Banks Worldwide	
Mean	1099.62
Standard Error	84.31088
Median	946.5
Mode	1162
Standard Deviation	596.1679
Sample Variance	355416.2
Kurtosis	1.452122
Skewness	1.292794
Range	2466
Minimum	486
Maximum	2952
Sum	54981
Count	50

3.57 *Hispanic Business* magazine publishes a list of the top 50 advertisers in the Hispanic market. The advertising spending for each of these 50 advertisers (in $ millions) was entered into a Minitab spreadsheet and the data were analysed using Minitab's Graphical Summary. Study the output from this analysis and describe the advertising expenditures of these top Hispanic market advertisers.

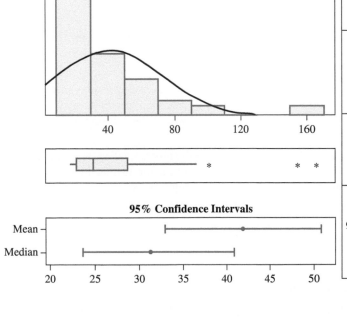

Top 50 Advertisers in the Hispanic Market

Anderson-Darling Normality Test
A-Squared   4.12
P-Value <   0.005

Mean   41.853
St Dev   31.415
Variance   986.892
Skewness   2.34450
Kurtosis   6.25496
N   50

Minimum   18.079
1st Quartile   21.617
Median   31.167
3rd Quartile   51.154
Maximum   162.695

95% Confidence Interval for Mean
32.925   50.781
95% Confidence Interval for Median
23.672   40.810
95% Confidence Interval for St Dev
26.242   39.147

95% Confidence Intervals

Mean

Median

3.58 Excel's descriptive statistics feature was used to analyse the number of employees for the top 50 employers extracted from the Fortune Global 500. Summarize what you have learned about the number of employees for these companies by studying the output.

Top 50 Employers in the World	
Mean	438,068.8
Standard Error	50,255.13
Median	325,425
Mode	# n.a.
Standard Deviation	355,357.4
Sample Variance	1.26e + 11
Kurtosis	12.80329
Skewness	3.554087
Range	1,850,722
Minimum	249,278
Maximum	2,100,000
Sum	21,903,439
Count	50

3.59 The Nielsen Company compiled a list of the top 25 advertisers in African American media. Shown below are a Minitab descriptive statistics analysis of the annual advertising spending (in $ millions) by these companies in African American media and a box plot of these data. Study this output and summarize the expenditures of these top 25 advertisers in your own words.

Variable	N	N*	Mean	SE Mean	StDev	Minimum	Q1
Advertisers	25	0	27.24	2.84	14.19	16.20	21.25
Variable	Median	Q3	Maximum				
Advertisers	24.00	27.50	89.70				

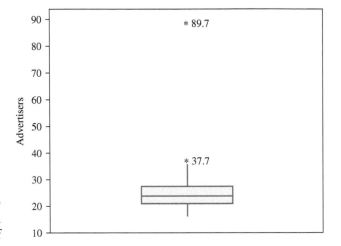

ANALYSING THE DATABASES

1. What are the mean and the median amounts of new capital expenditures for industries in the Manufacturing database? Comparing the mean and the median for these data, what do these statistics tell you about the data?

2. Analyse green beans cold storage holdings in the agribusiness time-series database using a descriptive statistics feature of either Excel or Minitab. By examining such statistics as the mean, median, mode, range, standard deviation, and a measure of skewness, describe green beans cold storage holdings over this period of time.

3. Using measures of central tendency including the mean, median, and mode, describe transport spending and household final consumption expenditure for British households in the Household Final Consumption database. Compare the two results by determining approximately what per cent of household final consumption expenditure is due to transport spending.

4. Using the Financial database, study earnings per share for Type 2 and Type 7 companies (chemical and petrochemical) using statistics. Compute a coefficient of variation for both Type 2 and Type 7. Compare the two coefficients and comment on them.

5. Using the Hospital database, construct a box-and-whisker plot for births. Thinking about hospitals and birthing facilities, comment on why the box-and-whisker plot looks like it does.

COCA-COLA GOES SMALL IN RUSSIA

The Coca-Cola Company is the number-one seller of soft drinks in the world. Every day an average of more than 1.5 billion servings of Coca-Cola, Diet Coke, Sprite, Fanta, and other products of Coca-Cola are enjoyed around the world. The company has the world's largest production and distribution system for soft drinks and sells more than twice as many soft drinks as its nearest competitor. Coca-Cola products are sold in more than 200 countries around the globe.

For several reasons, the company believes it will continue to grow internationally. One reason is that disposable income is rising. Another is that outside the United States and Europe, the world is getting younger. In addition, reaching world markets is becoming easier as political barriers fall and transportation difficulties are overcome. Still another reason is that the sharing of ideas, cultures, and news around the world creates market opportunities. Part of the company mission is for Coca-Cola to maintain the world's most powerful trademark and effectively utilize the world's most effective and pervasive distribution system.

In June 1999, Coca-Cola Russia introduced a 200 millilitre (about 6.8 ounce) Coke bottle in Volgograd, Russia, in a campaign to market Coke to its poorest customers. This strategy was successful for Coca-Cola in other countries, such as India. The bottle sells for 12 cents, making it affordable to almost everyone. In 2001, Coca-Cola enjoyed a 25% volume growth in Russia, including an 18% increase in unit case sales of Coca-Cola.

Today, Coca-Cola beverages are produced and sold in Russia by the company's authorized local bottling partner, Coca-Cola HBC Russia, based in Moscow. The Coca-Cola business system directly employs approximately 4,000 people in Russia, and more than 70% of all supplies required by the company are sourced locally.

DISCUSSION

1. Because of the variability of bottling machinery, it is likely that every 200 millilitre bottle of Coca-Cola does not contain exactly 200 millilitre of fluid. Some bottles may contain more fluid and others less. Because 200 millilitre bottle fills are somewhat unusual, a production engineer wants to test some of the bottles from the first production runs to determine how close they are to the 200 millilitre specification. Suppose the data below are the fill measurements from a random sample of 50 bottles. Use the techniques presented in this chapter to describe the sample. Consider measures of central tendency, variability, and skewness. Based on this analysis, how is the bottling process working?

200.1	199.9	200.2	200.2	200.0
200.1	200.9	200.1	200.3	200.5
199.7	200.4	200.3	199.8	199.3
200.1	199.4	199.6	199.2	200.2
200.4	199.8	199.9	200.2	199.6
199.6	200.4	200.4	200.6	200.6
200.1	200.8	199.9	200.0	199.9
200.3	200.5	199.9	201.1	199.7
200.2	200.5	200.2	199.7	200.9
200.2	199.5	200.6	200.3	199.8

2. Suppose that at another plant Coca-Cola is filling bottles with the more traditional 20 ounces of fluid. A lab randomly samples 150 bottles and tests the bottles for fill volume. The descriptive statistics are given in both Minitab and Excel computer output. Write a brief report to supervisors summarizing what this output is saying about the process.

Minitab Output

```
Descriptive Statistics: Bottle Fills

Variable      Total Count    Mean  SE Mean  StDev  Variance  CoefVar  Minimum       Q₁
Bottle Fills          150  20.008  0.00828  0.101    0.0103     0.51   19.706  19.940
Variable           Median      Q₃  Maximum  Range       IQR
Bottle Fills       19.997  20.079   20.263  0.557     0.139
```

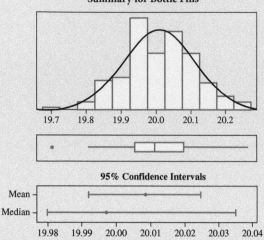

Excel Output

Bottle Fills

Mean	20.00817
Standard error	0.008278
Median	19.99697
Mode	# n.a.
Standard deviation	0.101388
Sample variance	0.010279
Kurtosis	−0.11422
Skewness	0.080425
Range	0.557666
Minimum	19.70555
Maximum	20.26322
Sum	3001.225
Count	150

Summary for Bottle Fills

95% Confidence Intervals

Anderson-Darling Normality Test	
A-Squared	0.32
P-Value	0.531
Mean	20.008
St Dev	0.101
Variance	0.010
Skewness	0.080479
Kurtosis	−0.116220
N	150
Minimum	19.706
1st Quartile	19.940
Median	19.997
3rd Quartile	20.079
Maximum	20.263

95% Confidence Interval for Mean
19.992	20.025

95% Confidence Interval for Median
19.980	20.035

95% Confidence Interval for St Dev
0.091	0.114

Source: Adapted from 'Coke, Avis Adjust in Russia', *Advertising Age*, 5 July 1999, p. 25; Coca-Cola website at www.coca-cola.com/home.html. Coca-Cola Company's 2001 annual report is found at www2.coca-cola.com/investors/annualreport/2001/index.html. Coca-Cola's website for information on Russia is located at www2.cocacola.com/ourcompany/cfs/cfs_include/cfs_russia_include.html. View Coca-Cola Company's 2007 annual report at: www.thecoca-colacompany.com/investors/pdfs/2007_annual_review.pdf.

USING THE COMPUTER

EXCEL

- While Excel has the capability of producing many of the statistics in this chapter piecemeal, there is one Excel feature, **Descriptive Statistics**, that produces many of these statistics in one output.

- To use the **Descriptive Statistics** feature, begin by selecting the **Data** tab on the Excel worksheet. From the **Analysis** panel at the right top of the **Data** tab worksheet, click on **Data Analysis**. If your Excel worksheet does not show the **Data Analysis** option, then you can load it as an add-in following directions given in Chapter 2. From the **Data Analysis** pulldown menu, select **Descriptive Statistics**. In the **Descriptive Statistics** dialog box, enter the location of the data to be analysed in **Input Range**. Check **Labels in the First Row** if your data contains a label in the first row (cell). Check the box beside **Summary statistics**. The **Summary statistics** feature computes a wide variety of descriptive statistics.

The output includes the mean, the median, the mode, the standard deviation, the sample variance, a measure of kurtosis, a measure of skewness, the range, the minimum, the maximum, the sum, and the count.

- The **Rank and Percentile** feature of the **Data Analysis** tool of Excel has the capability of ordering the data, assigning ranks to the data, and yielding the percentiles of the data. To access this command, click on **Data Analysis** (see above) and select **Rank and Percentile** from the menu. In the **Rank and Percentile** dialog box, enter the location of the data to be analysed in **Input Range**. Check **Labels in the First Row** if your data contains a label in the first row (cell).

- Many of the individual statistics presented in this chapter can be computed using the **Insert Function** (f_x) of Excel. To access the **Insert Function**, go to the **Formulas** tab on an Excel worksheet (top centre tab). The **Insert Function** is on the far left of the menu bar. In the **Insert Function** dialog box at the top, there is a pulldown menu where it says **Or select a category**. From the pulldown menu associated with this command, select **Statistical**. There are 83 different statistics that can be computed using one of these commands. Select the one that you want to compute and enter the location of the data. Some of the more useful commands in this menu are **AVERAGE, MEDIAN, MODE, SKEW, STDEV**, and **VAR**.

MINITAB

- Minitab is capable of performing many of the tasks presented in this chapter, including descriptive statistics and box plots. To begin Descriptive Statistics, select **Stat** on the menu bar, and then from the pulldown menu select **Basic Statistics**. From the **Basic Statistics** menu, select either **Display Descriptive Statistics** or **Graphical Summary**. If you select **DisplayDescriptiveStatistics** in the dialog box that appears, input the column(s) to be analysed in the box labelled **Variables**. If you click **OK**, then your output will include the sample size, mean, median, standard deviation, minimum, the first quartile, and the third quartile. If in the **Display Descriptive Statistics** dialog box you select the option **Graphs**, you will have several other output options that are relatively self-explanatory. The options include: **Histogram of data; Histogram of data, with normal curve; Individual value plot**; and **Boxplot of data**. If in the **Display Descriptive Statistics** dialog box you select the option **Statistics**, you have the option of selecting any of 24 statistics offered to appear in your output.

- On the other hand, you may opt to use the **Graphical Summary** option under **Basic Statistics**. If you use this option, in the dialog box that appears input the column(s) to be analysed in the box labelled **Variables**. If you click **OK**, then your output will include a histogram graph of the data with the normal curve superimposed, a box plot of the data, the same descriptive statistics listed above for the output from **Display Descriptive Statistics**, along with skewness and kurtosis statistics and other output that pertain to Chapter 8 topics.

- A variety of descriptive statistics can be obtained through the use of **Column Statistics** or **Row Statistics**. To begin, select **Calc** from the menu bar. From the pulldown menu, select either **Column Statistics** or **Row Statistics**, depending on where the data are located. For **Column Statistics**, in the space below **Input variable**, enter the column to be analysed. For **Row Statistics**, in the space below **Input variables**, enter the rows to be analysed. Check which **Statistic** you want to compute from **Sum, Mean, Standard deviation, Minimum, Maximum, Range, Median**, and **N total**. Minitab will only allow you to select one statistic at a time.

- Minitab can produce box-and-whisker plots. To begin, select **Graph** from the menu bar, and then select **Boxplot** from the pulldown menu. In the Boxplot dialog box, there are four options – **One Y Simple, One Y With Groups, Multiple Y's Simple**, and **Multiple Y's With Groups** – and you must select one.

- After you select one of these types of box plots, another dialog box appears with several options to select, including: **Scale . . ., Labels . . ., Data View . . ., Multiple Graphs . . .**, and **Data Options . . .**

- Enter the location of the variables to be graphed in the top box labelled **Graph variables**, and click **OK**.

CHAPTER 4

Probability

LEARNING OBJECTIVES

The main objective of Chapter 4 is to help you understand the basic principles of probability, thereby enabling you to:

1. Describe what probability is and when one would use it
2. Differentiate among three methods of assigning probabilities: the classical method, relative frequency of occurrence, and subjective probability
3. Deconstruct the elements of probability by defining experiments, sample spaces, and events, classifying events as mutually exclusive, collectively exhaustive, complementary, or independent, and counting possibilities
4. Compare marginal, union, joint, and conditional probabilities by defining each one
5. Calculate probabilities using the general law of addition, along with a probability matrix, the complement of a union, or the special law of addition if necessary
6. Calculate joint probabilities of both independent and dependent events using the general and special laws of multiplication
7. Calculate conditional probabilities with various forms of the law of conditional probability, and use them to determine if two events are independent
8. Calculate conditional probabilities using Bayes' rule

Decision Dilemma

Equality of the Sexes in the Workplace

A large body of European legislative texts is dedicated to the issue of equality between women and men. These texts mainly comprise various treaty provisions and directives concerning access to employment, equal pay, maternity protection, parental leave, social security and occupational social security, the burden of proof in discrimination cases and self-employment. In addition, each member state has to implement the EU law into its own legislation, so local variations in the hiring procedures can occur across the European Union. Company hiring procedures in the UK, for example, must comply with the 2010 Equality Act and must be carried out in a way that promotes equal opportunities to ensure that no unlawful discrimination occurs. The act allows companies to pursue *positive action* when hiring so as to take under-representation of female workers (and other 'particular groups') into account when selecting between two equally qualified candidates for recruitment or promotion. Today, company hiring procedures must simultaneously be within the framework of European and national guidelines.

How does a company defend its hiring practices or know when they are within acceptable bounds? How can individuals or groups who feel they have been the victims of illegal hiring practices 'prove' their case? How can a group demonstrate that they have been 'adversely impacted' by a company's discriminatory hiring practices?

Statistics are widely used in employment discrimination actions and by companies in attempting to meet both the EU and members states guidelines. Substantial quantities of human resources data are logged and analysed on a daily basis.

Managerial and Statistical Questions

Assume that a small portion of the human resources data was gathered on a client company.

1. Suppose some legal concern has been expressed that a disproportionate number of managerial people at the client company are men. If a worker is randomly selected from the client company, what is the probability that the worker is a woman? If a managerial person is randomly selected, what is the probability that the person is a woman? What factors might enter into the apparent discrepancy between probabilities?

2. Suppose a special bonus is being given to one person in the technical area this year. If the bonus is randomly awarded, what is the probability that it will go to a woman, given that worker is in the technical area? Is this discrimination against male technical workers? What factors might enter into the awarding of the bonus other than random selection?

3. Suppose that at the annual party the name of an employee of the client company will be drawn randomly to win a trip to Hawaii. What is the probability that a professional person will be the winner?

4. What is the probability that the winner will be either a man or a clerical worker? What is the probability that the winner will be a woman and in management? Suppose the winner is a man. What is the probability that the winner is from the technical group, given that the winner is a man?

Client Company Human Resource Data by Sex

Type of Position	Sex		Total
	Male	Female	
Managerial	8	3	11
Professional	31	13	44
Technical	52	17	69
Clerical	9	22	31
Total	100	55	155

Source: Adapted from www.equalities.gov.uk/equality_bill.aspx and http://europa.eu/legislation_summaries/employment_and_social_policy/equality_between_men_and_women/index_en.htm. Accessed January 2012.

In business, most decision making involves uncertainty. For example, an operations manager does not know definitely whether a valve in the plant is going to malfunction or continue to function – or, if it continues, for how long. When should it be replaced? What is the chance that the valve will malfunction within the next week? In the banking industry, what are the new vice CEO's prospects for successfully turning a department around? The answers to these questions are uncertain.

In the case of a multistorey building, what are the chances that a fire-extinguishing system will work when needed if redundancies are built in? Businesspeople must address these and thousands of similar questions daily. Because such questions do not have definite answers, the decision making is based on uncertainty. In many of these situations, a probability can be assigned to the likelihood of an outcome. This chapter is about learning how to determine or assign probabilities.

4.1 INTRODUCTION TO PROBABILITY

Chapter 1 discussed the difference between descriptive and inferential statistics. Much statistical analysis is inferential, and probability is the basis for inferential statistics. Recall that inferential statistics involves taking a sample from a population, computing a statistic on the sample, and inferring from the statistic the value of the corresponding parameter of the population. The reason for doing so is that the value of the parameter is unknown. Because it is unknown, the analyst conducts the inferential process under uncertainty. However, by applying rules and laws, the analyst can often assign a probability of obtaining the results. Figure 4.1 depicts this process.

Suppose a quality control inspector selects a random sample of 40 light bulbs from a population of brand X bulbs and computes the average number of hours of luminance for the sample bulbs. By using techniques discussed later in this text, the specialist estimates the average number of hours of luminance for the *population* of brand X light bulbs from this sample information. Because the light bulbs being analysed are only a sample of the population, the average number of hours of luminance for the 40 bulbs may or may not accurately estimate the average for all bulbs in the population. The results are uncertain. By applying the laws presented in this chapter, the inspector can assign a value of probability to this estimate.

In addition, probabilities are used directly in certain industries and industry applications. For example, the insurance industry uses probabilities in actuarial tables to determine the likelihood of certain outcomes in order to set specific rates and coverages. The gaming industry uses probability values to establish charges and payoffs. One way to determine whether a company's hiring practices meet EU and member states' guidelines mentioned in the Decision Dilemma is to compare various proportional breakdowns of their employees (by ethnicity, gender, age, etc.) to the proportions in the general population from which the employees are hired. In comparing the company figures with those of the general population, the courts could study the probabilities of a company randomly hiring a certain profile of employees from a given population. In other industries, such as manufacturing and aerospace, it is important to know the life of a mechanized part and the probability that it will malfunction at any given length of time in order to protect the firm from major breakdowns.

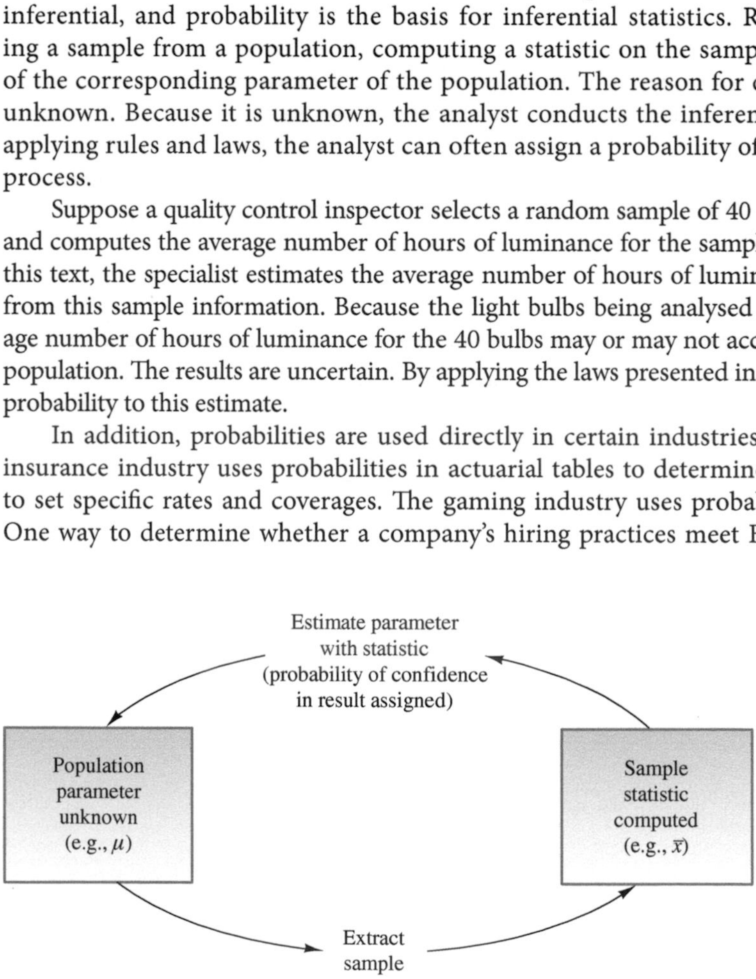

FIGURE 4.1

Probability in the Process of Inferential Statistics

4.2 METHODS OF ASSIGNING PROBABILITIES

The three general methods of assigning probabilities are (1) the classical method, (2) the relative frequency of occurrence method, and (3) subjective probabilities.

Classical Method of Assigning Probabilities

When probabilities are assigned based on laws and rules, the method is referred to as the **classical method of assigning probabilities**. This method involves an experiment, which is *a process that produces outcomes*, and an event, *which is an outcome of an experiment.*

When we assign probabilities using the classical method, the probability of an individual event occurring is determined as the ratio of the number of items in a population containing the event (n_e) to the total number of items in the population (N). That is, $P(E) = n_e/N$. For example, if a company has 200 workers and 70 are female, the probability of randomly selecting a female from this company is $70/200 = 0.35$.

<table>
<tr>
<td>

CLASSICAL METHOD OF
ASSIGNING PROBABILITIES

</td>
<td>

$$P(E) = \frac{n_e}{N}$$

where:
    N = total possible number of outcomes of an experiment
    n_e = the number of outcomes in which the event occurs out of
        N outcomes

</td>
</tr>
</table>

Suppose, in a particular plant, three machines make a given product. Machine A always produces 40% of the total number of this product. Ten per cent of the items produced by machine A are defective. If the finished products are well mixed with regard to which machine produced them and if one of these products is randomly selected, the classical method of assigning probabilities tells us that the probability that the part was produced by machine A and is defective is 0.04. This probability can be determined even before the part is sampled because with the classical method, the probabilities can be determined **a priori;** that is, *they can be determined prior to the experiment.*

Because n_e can never be greater than N (no more than N outcomes in the population could possibly have attribute e), the highest value of any probability is 1. If the probability of an outcome occurring is 1, the event is certain to occur. The smallest possible probability is 0. If none of the outcomes of the N possibilities has the desired characteristic, e, the probability is $0/N = 0$, and the event is certain not to occur.

<table>
<tr>
<td>RANGE OF POSSIBLE PROBABILITIES</td>
<td>$0 \leqslant P(E) \leqslant 1$</td>
</tr>
</table>

Thus, probabilities are non-negative proper fractions or non-negative decimal values greater than or equal to 0 and less than or equal to 1.

Probability values can be converted to percentages by multiplying by 100. Meteorologists often report weather probabilities in percentage form. For example, when they forecast a 60% chance of rain for tomorrow, they are saying that the probability of rain tomorrow is 0.60.

Relative Frequency of Occurrence

The **relative frequency of occurrence method** of assigning probabilities is based on cumulated historical data. With this method, *the probability of an event occurring is equal to the number of times the event has occurred in the past divided by the total number of opportunities for the event to have occurred.*

PROBABILITY BY RELATIVE FREQUENCY OF OCCURRENCE	$\dfrac{\text{Number of Times an Event Occurred}}{\text{Total Number of Opportunities for the Event to Occur}}$

Relative frequency of occurrence is not based on rules or laws but on what has occurred in the past. For example, a company wants to determine the probability that its inspectors are going to reject the next batch of raw materials from a supplier. Data gathered from company record books show that the supplier sent the company 90 batches in the past, and inspectors rejected 10 of them. By the method of relative frequency of occurrence, the probability of the inspectors rejecting the next batch is 10/90, or 0.11. If the next batch is rejected, the relative frequency of occurrence probability for the subsequent shipment would change to $11/91 = 0.12$.

Subjective Probability

The *subjective method of assigning probability is based on the feelings or insights of the person determining the probability*. **Subjective probability** comes from the person's intuition or reasoning. Although not a scientific approach to probability, the subjective method often is based on the accumulation of knowledge, understanding, and experience stored and processed in the human mind. At times it is merely a guess. At other times, subjective probability can potentially yield accurate probabilities. Subjective probability can be used to capitalize on the background of experienced workers and managers in decision making.

Suppose a director of transportation for an oil company is asked the probability of getting a shipment of oil out of Saudi Arabia to Germany within three weeks. A director who has scheduled many such shipments, has a knowledge of Saudi politics, and has an awareness of current climatological and economic conditions may be able to give an accurate probability that the shipment can be made on time.

Subjective probability also can be a potentially useful way of tapping a person's experience, knowledge, and insight and using them to forecast the occurrence of some event. An experienced airline mechanic can usually assign a meaningful probability that a particular plane will have a certain type of mechanical difficulty. Physicians sometimes assign subjective probabilities to the life expectancy of people who have cancer.

4.3   STRUCTURE OF PROBABILITY

In the study of probability, developing a language of terms and symbols is helpful. The structure of probability provides a common framework within which the topics of probability can be explored.

Experiment

As previously stated, an **experiment** is *a process that produces outcomes*. Examples of business-oriented experiments with outcomes that can be statistically analysed might include the following:

- Interviewing 20 randomly selected consumers and asking them which brand of appliance they prefer
- Sampling every 200th bottle of ketchup from an assembly line and weighing the contents
- Testing new pharmaceutical drugs on samples of cancer patients and measuring the patients' improvement
- Auditing every 10th account to detect any errors
- Recording the Euro Stoxx 50 stock index on the first Monday of every month for 10 years

Event

Because an **event** is *an outcome of an experiment*, the experiment defines the possibilities of the event. If the experiment is to sample five bottles coming off a production line, an event could be to get one defective and four good bottles. In an experiment to roll a die, one event could be to roll an even number and another event could be to roll a number greater than two. Events are denoted by uppercase letters: that is, italic capital letters (e.g., A and E_1, E_2, . . .) represent the general or abstract case, and roman capital letters (e.g., H and T for heads and tails) denote specific things and people.

Elementary Events

Events that cannot be decomposed or broken down into other events are called **elementary events**. Elementary events are denoted by lowercase letters (e.g., e_1, e_2, e_3, . . .). Suppose the experiment is to roll a die. The elementary events for this experiment are to roll a 1 or roll a 2 or roll a 3, and so on. Rolling an even number is an event, but it is not an elementary event because the even number can be broken down further into events 2, 4, and 6.

In the experiment of rolling a die, there are six elementary events {1, 2, 3, 4, 5, 6}. Rolling a pair of dice results in 36 possible elementary events (outcomes). For each of the six elementary events possible on the roll of one die, there are six possible elementary events on the roll of the second die, as depicted in the tree diagram in Figure 4.2. Table 4.1 contains a list of these 36 outcomes.

In the experiment of rolling a pair of dice, other events could include outcomes such as two even numbers, a sum of 10, a sum greater than five, and others. However, none of these events is an elementary event because each can be broken down into several of the elementary events displayed in Table 4.1.

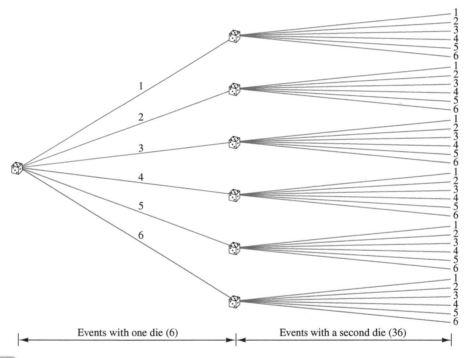

Events with one die (6)   Events with a second die (36)

FIGURE 4.2

Possible Outcomes for the Roll of a Pair of Dice

TABLE 4.1

All Possible Elementary Events in the Roll of a Pair of Dice (Sample Space)

(1,1)	(2,1)	(3,1)	(4,1)	(5,1)	(6,1)
(1,2)	(2,2)	(3,2)	(4,2)	(5,2)	(6,2)
(1,3)	(2,3)	(3,3)	(4,3)	(5,3)	(6,3)
(1,4)	(2,4)	(3,4)	(4,4)	(5,4)	(6,4)
(1,5)	(2,5)	(3,5)	(4,5)	(5,5)	(6,5)
(1,6)	(2,6)	(3,6)	(4,6)	(5,6)	(6,6)

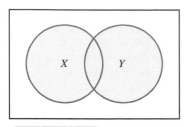

FIGURE 4.3

A Union

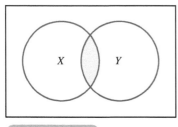

FIGURE 4.4

An Intersection

Sample Space

A **sample space** is *a complete roster or listing of all elementary events for an experiment*. Table 4.1 is the sample space for the roll of a pair of dice. The sample space for the roll of a single die is {1, 2, 3, 4, 5, 6}.

Sample space can aid in finding probabilities. Suppose an experiment is to roll a pair of dice. What is the probability that the dice will sum to 7? An examination of the sample space shown in Table 4.1 reveals that there are six outcomes in which the dice sum to 7 – {(1,6), (2,5), (3,4), (4,3), (5,2), (6,1)} – in the total possible 36 elementary events in the sample space. Using this information, we can conclude that the probability of rolling a pair of dice that sum to 7 is 6/36, or 0.1667. However, using the sample space to determine probabilities is unwieldy and cumbersome when the sample space is large. Hence, statisticians usually use other more effective methods of determining probability.

Unions and Intersections

Set notation, the use of braces to group numbers, is used as *a symbolic tool for unions and intersections* in this chapter. The **union** of X, Y is *formed by combining elements from each of the sets* and is denoted $X \cup Y$. An element qualifies for the union of X, Y if it is in either X or Y or in both X and Y. The union expression $X \cup Y$ can be translated to 'X or Y.' For example, if

$$X = \{1, 4, 7, 9\} \quad \text{and} \quad Y = \{2, 3, 4, 5, 6\}$$

$$X \cup Y = \{1, 2, 3, 4, 5, 6, 7, 9\}$$

Note that all the values of X and all the values of Y qualify for the union. However, none of the values is listed more than once in the union. In Figure 4.3, the shaded region of the Venn diagram denotes the union.

An intersection is denoted $X \cap Y$. To qualify for intersection, an element must be in both X and Y. The **intersection** *contains the elements common to both sets*. Thus the intersection symbol, $\cap$, is often read as *and*. The intersection of X, Y is referred to as X and Y. For example, if

$$X = \{1, 4, 7, 9\} \quad \text{and} \quad Y = \{2, 3, 4, 5, 6\}$$

$$X \cap Y = \{4\}$$

Note that only the value 4 is common to both sets X and Y. The intersection is more exclusive than and hence equal to or (usually) smaller than the union. Elements must be characteristic of both X and Y to qualify. In Figure 4.4, the shaded region denotes the intersection.

Mutually Exclusive Events

Two or more events are **mutually exclusive events** if *the occurrence of one event precludes the occurrence of the other event(s)*. This characteristic means that mutually exclusive events cannot occur simultaneously and therefore can have no intersection.

A manufactured part is either defective or non-defective: the part cannot be both defective and non-defective at the same time because 'non-defective' and 'defective' are mutually exclusive categories. In a sample of the manufactured products, the event of selecting a defective part is mutually exclusive with the event of selecting a non-defective part. Suppose an office building is for sale and two different potential buyers have placed bids on the building. It is not possible for both buyers to purchase the building; therefore, the event of buyer A purchasing the building is mutually exclusive with the event of buyer B purchasing the building. In the toss of a single coin, heads and tails are mutually exclusive events. The person tossing the coin gets either a head or a tail but never both.

The probability of two mutually exclusive events occurring at the same time is zero.

| MUTUALLY EXCLUSIVE EVENTS X AND Y | $P(X \cap Y) = 0$ |

Independent Events

Two or more events are **independent events** if *the occurrence or non-occurrence of one of the events does not affect the occurrence or non-occurrence of the other event(s)*. Certain experiments, such as rolling dice, yield independent events; each die is independent of the other. Whether a 6 is rolled on the first die has no influence on whether a 6 is rolled on the second die. Coin tosses always are independent of each other. The event of getting a head on the first toss of a coin is independent of getting a head on the second toss. It is generally believed that certain human characteristics are independent of other events. For example, left-handedness is probably independent of the possession of a credit card. Whether a person wears glasses or not is probably independent of the brand of milk preferred.

Many experiments using random selection can produce either an independent or non-independent event, depending on how the experiment is conducted. In these experiments, the outcomes are independent if sampling is done with replacement; that is, after each item is selected and the outcome is determined, the item is restored to the population and the population is shuffled. This way, each draw becomes independent of the previous draw. Suppose an inspector is randomly selecting bolts from a bin that contains 5% defects. If the inspector samples a defective bolt and returns it to the bin, on the second draw there are still 5% defects in the bin regardless of the fact that the first outcome was a defect. If the inspector does not replace the first draw, the second draw is not independent of the first; in this case, fewer than 5% defects remain in the population. Thus the probability of the second outcome is dependent on the first outcome.

If X and Y are independent, the following symbolic notation is used.

INDEPENDENT EVENTS X AND Y	$P(X|Y) = P(X)$ and $P(Y|X) = P(Y)$

$P(X|Y)$ denotes the probability of X occurring given that Y has occurred. If X and Y are independent, then the probability of X occurring given that Y has occurred is just the probability of X occurring. Knowledge that Y has occurred does not impact the probability of X occurring because X and Y are independent. For example, P (prefers Pepsi|person is right-handed $= P$ (prefers Pepsi) because a person's handedness is independent of brand preference.

Collectively Exhaustive Events

A list of **collectively exhaustive events** contains *all possible elementary events for an experiment*. Thus, all sample spaces are collectively exhaustive lists. The list of possible outcomes for the tossing of a pair of dice contained in Table 4.1 is a collectively exhaustive list. The sample space for an experiment can be described as a list of events that are mutually exclusive and collectively exhaustive. Sample space events do not overlap or intersect, and the list is complete.

Complementary Events

The **complement** of event A is denoted A', pronounced 'not A'. All *the elementary events of an experiment not in A comprise its complement.* For example, if in rolling one die, event A is getting an even number, the complement of A is getting an odd number. If event A is getting a 5 on the roll of a die, the complement of A is getting a 1, 2, 3, 4, or 6. The complement of event A contains whatever portion of the sample space that event A does not contain, as the Venn diagram in Figure 4.5 shows.

Using the complement of an event sometimes can be helpful in solving for probabilities because of the following rule.

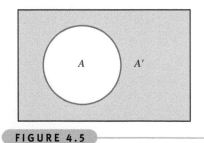

FIGURE 4.5

The Complement of Event A

PROBABILITY OF THE COMPLEMENT OF A	$P(A') = 1 - P(A)$

Suppose 32% of the employees of a company have a university degree. If an employee is randomly selected from the company, the probability that the person does not have a university degree is $1 - 0.32 = 0.68$. Suppose 42% of all parts produced in a plant are moulded by machine A and 31% are moulded by machine B. If a part is randomly selected, the probability that it was moulded by neither machine A nor machine B is $1 - 0.73 = 0.27$. (Assume that a part is only moulded on one machine.)

Counting the Possibilities

In statistics, a collection of techniques and rules for counting the number of outcomes that can occur for a particular experiment can be used. Some of these rules and techniques can delineate the size of the sample space. Presented here are three of these counting methods.

The *mn* Counting Rule

Suppose a customer decides to buy a certain brand of new car. Options for the car include two different engines, five different paint colours, and three interior packages. If each of these options is available with each of the others, how many different cars could the customer choose from? To determine this number, we can use the **mn counting rule**.

THE *mn* COUNTING RULE	For an operation that can be done *m* ways and a second operation that can be done *n* ways, the two operations then can occur, in order, in *mn* ways. This rule can be extended to cases with three or more operations.

Using the *mn* counting rule, we can determine that the automobile customer has $(2)(5)(3) = 30$ different car combinations of engines, paint colours, and interiors available.

Suppose a scientist wants to set up a research design to study the effects of gender (M, F), marital status (single never married, divorced, married), and economic class (lower, middle, and upper) on the frequency of airline ticket purchases per year. The researcher would set up a design in which 18 different samples are taken to represent all possible groups generated from these customer characteristics.

$$\text{Number of Groups} = (\text{Sex})(\text{Marital Status})(\text{Economic Class})$$

$$= (2)(3)(3) = 18 \text{ Groups}$$

Sampling from a Population With Replacement

In the second counting method, sampling *n* items from a population of size *N* *with replacement* would provide

$$(N)^n \text{ possibilities}$$

where *N* is the population size and *n* is the sample size.

For example, each time a die, which has six sides, is rolled, the outcomes are independent (with replacement) of the previous roll. If a die is rolled three times in succession, how many different outcomes can occur? That is, what is the size of the sample space for this experiment? The size of the population, *N*, is 6, the six sides of the die. We are sampling three rolls, $n = 3$. The sample space is

$$(N)^n = (6)^3 = 216$$

Suppose in a lottery six numbers are drawn from the digits 0 through 9, with replacement (digits can be reused). How many different groupings of six numbers can be drawn? *N* is the population of 10 numbers (0 to 9) and *n* is the sample size, six numbers.

$$(N)^n = (10)^6 = 1,000,000$$

That is, a million six-digit numbers are available!

Combinations: Sampling from a Population Without Replacement

The third counting method uses combinations, sampling *n* items from a population of size *N* without replacement provides

$$_N C_n = \binom{N}{n} = \frac{N!}{n!(N-n)!}$$

possibilities.

For example, suppose a small law firm has 16 employees and three are to be selected randomly to represent the company at the annual conference of the International Bar Association. How many different combinations of lawyers could be sent to the conference? This situation does not allow sampling with replacement because three *different* lawyers will be selected to go. This problem is solved by using combinations. $N = 16$ and $n = 3$, so

$$_N C_n = {}_{16}C_3 = \frac{16!}{3!13!} = 560$$

A total of 560 combinations of three lawyers could be chosen to represent the firm.

4.3   PROBLEMS

4.1   A supplier shipped a lot of six parts to a company. The lot contained three defective parts. Suppose the customer decided to randomly select two parts and test them for defects. How large a sample space is the customer potentially working with? List the sample space. Using the sample space list, determine the probability that the customer will select a sample with exactly one defect.

4.2   Given $X = \{1, 3, 5, 7, 8, 9\}$, $Y = \{2, 4, 7, 9\}$, and $Z = \{1, 2, 3, 4, 7\}$, solve the following:

a. $X \cup Z = $ _____
b. $X \cap Y = $ _____
c. $X \cap Z = $ _____
d. $X \cup Y \cup Z = $ _____
e. $X \cap Y \cap Z = $ _____

f. $(X \cup Y) \cap Z = $ _____
g. $(Y \cap Z) \cup (X \cap Y) = $ _____
h. X or $Y = $ _____
i. Y and $X = $ _____

4.3   If a population consists of the positive even numbers through to 30 and if $A = \{2, 6, 12, 24\}$, what is A'?

4.4   A company's customer service toll-free telephone system is set up so that the caller has six options. Each of these six options leads to a menu with four options. For each of these four options, three more options are available. For each of these three options, another three options are presented. If a person calls the toll-free number for assistance, how many total options are possible?

4.5   A bin contains six parts. Two of the parts are defective and four are acceptable. If three of the six parts are selected from the bin, how large is the sample space? Which counting rule did you use, and why? For this sample space, what is the probability that exactly one of the three sampled parts is defective?

4.6   A company places a seven-digit serial number on each part that is made. Each digit of the serial number can be any number from 0 to 9. Digits can be repeated in the serial number. How many different serial numbers are possible?

4.7   A small company has 20 employees. Six of these employees will be selected randomly to be interviewed as part of an employee satisfaction programme. How many different groups of six can be selected?

4.4   MARGINAL, UNION, JOINT, AND CONDITIONAL PROBABILITIES

Four particular types of probability are presented in this chapter. The first type is **marginal probability**. Marginal probability is denoted $P(E)$, where E is some event. A marginal probability is usually *computed by dividing some subtotal by the whole*. An example of marginal probability is the probability that a person owns a Ford car. This probability is computed by dividing the number of Ford owners by the total number of car owners. The probability of a person wearing glasses is also a marginal probability. This probability is computed by dividing the number of people wearing glasses by the total number of people.

A second type of probability is the union of two events. Union probability is denoted $P(E_1 \cup E_2)$, where E_1 and E_2 are two events. $P(E_1 \cup E_2)$ is the probability that E_1 will occur or that E_2 will occur or that both E_1 and E_2 will occur. An example of union probability is the probability that a person owns a Ford or a Volkswagen. To qualify for the union, the person only has to have at least one of these cars. Another example is the probability of a person wearing glasses or having red hair. All people wearing glasses are included in the union, along with all redheads and all redheads who wear glasses. In a company, the probability that a person is male or a clerical worker is a union probability. A person qualifies for the union by being male or by being a clerical worker or by being both (a male clerical worker).

A third type of probability is the intersection of two events, or joint probability. The joint probability of events E_1 and E_2 occurring is denoted $P(E_1 \cap E_2)$. Sometimes $P(E_1 \cap E_2)$ is read as the probability of E_1 and E_2. To qualify for the intersection, both events must occur. An example of joint probability is the probability of a person owning both a Ford and a Volkswagen. Owning one type of car is not sufficient. A second example of joint probability is the probability that a person is a redhead and wears glasses.

The fourth type is conditional probability, which is denoted $P(E_1|E_2)$. This expression is read: the probability that E_1 will occur given that E_2 is known to have occurred. Conditional probabilities involve knowledge of some prior information. The information that is known or given is written to the right of the vertical line in the probability statement. An example of conditional probability is the probability that a person owns a Volkswagen given that she owns a Ford. This conditional probability is only a measure of the proportion of Ford owners who have a Volkswagen – not the proportion of total car owners who own a Volkswagen. Conditional probabilities are computed by determining the number of items that have an outcome out of some subtotal of the population. In the car owner example, the possibilities are reduced to Ford owners, and then the number of Volkswagen owners out of those Ford owners is determined. Another example of a conditional probability is the probability that a worker in a company is a professional given that he is male. Of the four probability types, only conditional probability does not have the population total as its denominator. Conditional probabilities have a population subtotal in the denominator. Figure 4.6 summarizes these four types of probability.

Marginal	Union	Joint	Conditional
$P(X)$	$P(X \cup Y)$	$P(X \cap Y)$	$P(X|Y)$
The probability of X occurring	The probability of X or Y occurring	The probability of X and Y occurring	The probability of X occurring given that Y has occurred
Uses total possible outcomes in denominator	Uses total possible outcomes in denominator	Uses total possible outcomes in denominator	Uses subtotal of the possible outcomes in denominator

FIGURE 4.6

Marginal, Union, Joint, and Conditional Probabilities

Green Tea or Coffee? Daily Drinking Habits in Japan

According to a survey by Japan-Guide.com, the most popular beverage in Japan is green tea with 54% of respondents drinking green tea on a daily basis. The second most frequently drunk beverage is coffee at 52%, followed by water (46%), milk (37%), English tea (25%), Chinese tea (22%) and fruit juice (14%). Converting these percentages to proportions, each could be considered to be a marginal probability. For example, if a person is randomly selected in Japan, there is a 54% probability that he/she drinks green tea on a daily basis, $P(G) = 0.54$.

Suppose further analysis shows that 35% of people that drink green tea daily also drink coffee on a daily basis. Converting this figure to probability results in the joint probability: $P(G \cap C) = 0.35$. Subtracting this value from the 0.54 that drink green tea daily, we can determine that 19% of green tea drinkers do not drink coffee on a daily basis.

Suppose 59% of those who drink green tea on a daily basis are female. This can be restated as a conditional probability: $P(F|G) = 0.59$.

Each of the four types of probabilities discussed in this chapter can be applied to the data on daily drinking habits in Japan. Further breakdowns of these statistics using probabilities can offer insights into how to better serve non-alcoholic drinks customers and how to better market beverages in Japan.

4.5 ADDITION LAWS

Several tools are available for use in solving probability problems. These tools include sample space, tree diagrams, the laws of probability, probability matrices, and insight. Because of the individuality and variety of probability problems, some techniques apply more readily in certain situations than in others. No best method is available for solving all probability problems. In some instances, the probability matrix lays out a problem in a readily solvable manner. In other cases, setting up the probability matrix is more difficult than solving the problem in another way. The probability laws almost always can be used to solve probability problems.

Four laws of probability are presented in this chapter: the addition laws, conditional probability, the multiplication laws, and Bayes' rule. The addition laws and the multiplication laws each have a general law and a special law.

The general law of addition is used to find the probability of the union of two events, $P(X \cup Y)$. The expression $P(X \cup Y)$ denotes the probability of X occurring or Y occurring or both X and Y occurring.

GENERAL LAW OF ADDITION
$$P(X \cup Y) = P(X) + P(Y) - P(X \cap Y)$$
where X, Y are events and $(X \cap Y)$ is the intersection of X and Y.

The Work Foundation recently commissioned a survey of over 1,000 workers on the degree of job satisfaction in the UK. Of the respondents, 78% said they found their work 'stimulating and challenging' while 69% said their work was a 'source of personal fulfilment'. If one of the survey respondents was randomly selected and asked about her job satisfaction, what is the probability that this person would select 'stimulating and challenging' or a 'source of personal fulfilment'?

Let C represent the event 'stimulating and challenging'. Let S represent the event 'source of personal fulfilment'. The probability of a person responding with C or S can be symbolized statistically as a union probability by using the law of addition.

$$P(C \cup S)$$

To successfully satisfy the search for a person who finds their work stimulating and challenging or a source of personal fulfilment, we need only find someone who wants at least one of those two events. Because 78% of the surveyed people responded that they found their work stimulating and challenging, then $P(C) = 0.78$. In addition, because 69%

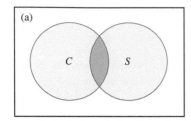

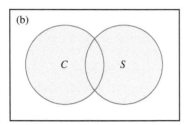

FIGURE 4.7

Solving for the Union in the Job
Satisfaction Problem

responded that they found their work a source of personal fulfilment, $P(S) = 0.69$. Either of these would satisfy the requirement of the union. Thus, the solution to the problem seems to be

$$P(C \cup S) = P(C) + P(S) = 0.78 + 0.69 = 1.47$$

However, we have already established that probabilities cannot be more than 1. What is the problem here? Notice that all people who responded that they found work to be *both* stimulating and challenging *and* a source of personal fulfilment are included in *each* of the marginal probabilities $P(C)$ and $P(S)$. Certainly a respondent who finds her work to be both stimulating and challenging and a source of personal fulfilment should be included as agreeing with at least one of the statements. However, because they are included in the $P(C)$ *and* the $P(S)$, the people who find their work both stimulating and challenging and a source of personal fulfilment are *double counted*. For that reason, the general law of addition subtracts the intersection probability, $P(C \cap S)$.

In Figure 4.7, Venn diagrams illustrate this discussion. Notice that the intersection area of C and S is double shaded in diagram A, indicating that it has been counted twice. In diagram B, the shading is consistent throughout C and S because the intersection area has been subtracted out. Thus diagram B illustrates the proper application of the general law of addition.

So what is the answer to The Work Foundation's union probability question? Suppose 56% of all respondents to the survey had said that they found their work to be *both* stimulating and challenging *and* a source of personal fulfilment: $P(C \cap S) = 0.56$. Then we could use the general law of addition to solve for the probability that a person responds that they find their work to be *either* stimulating and challenging *or* a source of personal fulfilment.

$$P(C \cup S) = P(C) + P(S) - P(C \cap S) = 0.78 + 0.69 - 0.56 = 0.91$$

Hence, 91% of the workers surveyed responded that their jobs were *either* stimulating and challenging *or* a source of personal fulfilment.

Probability Matrices

In addition to the formulas, another useful tool in solving probability problems is a probability matrix. A **probability matrix** *displays the marginal probabilities and the intersection probabilities of a given problem.* Union probabilities and conditional probabilities must be computed from the matrix. Generally, a probability matrix is constructed as a two-dimensional table with one variable on each side of the table. For example, in the job satisfaction problem, source of personal fulfilment would be on one side of the table and stimulating and challenging on the other. In this problem, a Yes row and a No row would be created for one variable and a Yes column and a No column would be created for the other variable, as shown in Table 4.2.

Once the matrix is created, we can enter the marginal probabilities. $P(C) = 0.78$ is the marginal probability that a person responds yes to finding their jobs stimulating and challenging. This value is placed in the 'margin' in the row of Yes to stimulating and challenging, as shown in Table 4.3. If $P(C) = 0.78$, then 22% of the people surveyed did not find their jobs stimulating and challenging. Thus, $P(\text{not } C) = 1 - 0.78 = 0.22$. This value, also a marginal probability, goes in the row indicated by No under stimulating and challenging. In the column under Yes for source of personal

TABLE 4.2

Probability Matrix for the Job Satisfaction Problem

		Source of Personal Fulfilment	
		Yes	No
Stimulating and Challenging	Yes		
	No		

TABLE 4.3

Probability Matrix for the Job Satisfaction Problem

		Source of Personal Fulfilment		
		Yes	No	
Stimulating and Challenging	Yes	.56	.22	.78
	No	.13	.09	.22
		.69	.31	1.00

fulfilment, the marginal probability $P(S) = 0.69$ is recorded. Finally, the marginal probability of No for source of personal fulfilment, $P(\text{not } S) = 1 - 0.69 = 0.31$ is placed in the No column.

In this probability matrix, all four marginal probabilities are given or can be computed simply by using the probability of a complement rule, $P(\text{not } S) = 1 - P(S)$. The intersection of stimulating and challenging and source of personal fulfilment is given as $P(C \cap S) = 0.56$. This value is entered into the probability matrix in the cell under Yes Yes, as shown in Table 4.3. The rest of the matrix can be determined by subtracting the cell values from the marginal probabilities. For example, subtracting 0.56 from 0.78 and getting 0.22 yields the value for the cell under Yes for stimulating and challenging and No for source of personal fulfilment. In other words, 22% of all respondents said that they found their jobs to be stimulating and challenging but not a source of personal fulfilment. Filling out the rest of the matrix results in the probabilities shown in Table 4.3.

Now we can solve the union probability, $P(C \cup S)$ in at least two different ways using the probability matrix. The focus is on the Yes row for stimulating and challenging and the Yes column for source of personal fulfilment, as displayed in Table 4.4. The probability of a person finding their job to be stimulating and challenging *or* a source of personal fulfilment, $P(C \cup S)$ can be determined from the probability matrix by adding the marginal probabilities of Yes for stimulating and challenging and Yes for source of personal fulfilment and then subtracting the Yes Yes cell, following the pattern of the general law of probabilities.

$$P(C \cup S) = 0.78 \text{ (from Yes row)} + 0.69 \text{ (from Yes column)} - 0.56 \text{ (from Yes Yes cell)} = 0.91$$

Another way to solve for the union probability from the information displayed in the probability matrix is to sum all cells in any of the Yes rows or columns. Observe the following from Table 4.4:

$P(C \cup S) = 0.56$ (from Yes Yes cell)

$\qquad + 0.22$ (from Yes on stimulating and challenging and No on source of personal fulfilment)

$\qquad + 0.13$ (from No on stimulating and challenging and Yes on source of personal fulfilment)

$\qquad = 0.91$

TABLE 4.4

Yes Row and Yes Column for Probability Matrix of the Job Satisfaction Problem

		Source of Personal Fulfilment		
		Yes	*No*	
Stimulating and Challenging	*Yes*	.56	.22	.78
	No	.13		
		.69		

DEMONSTRATION PROBLEM 4.1

The client company data from the Decision Dilemma reveal that 155 employees worked one of four types of positions. Shown here again is the raw values matrix (also called a contingency table) with the frequency counts for each category and for subtotals and totals containing a breakdown of these employees by type of position and by sex. If an employee of the company is selected randomly, what is the probability that the employee is female or a professional worker?

Company Human Resource Data

		Sex		
		Male	*Female*	
Type of Position	*Managerial*	8	3	11
	Professional	31	13	44
	Technical	52	17	69
	Clerical	9	22	31
		100	55	155

Solution

Let F denote the event of female and P denote the event of professional worker. The question is

$$P(F \cup P) = ?$$

By the general law of addition,

$$P(F \cup P) = P(F) + P(P) - P(F \cap P)$$

Of the 155 employees, 55 are women. Therefore, $P(F) = 55/155 = 0.355$. The 155 employees include 44 professionals. Therefore, $P(P) = 44/155 = 0.284$. Because 13 employees are both female and professional, $P(F \cap P) = 13/155 = 0.084$. The union probability is solved as

$$P(F \cup P) = 0.355 + 0.284 - 0.084 = 0.555$$

To solve this probability using a matrix, you can either use the raw values matrix shown previously or convert the raw values matrix to a probability matrix by dividing every value in the matrix by the value of N, 155. The raw value matrix is used in a manner similar to that of the probability matrix. To compute the union probability of selecting a person who is either female or a professional worker from the raw value matrix, add the number of people in the Female column (55) to the number of people in the Professional row (44), then subtract the number of people in the intersection cell of Female and Professional (13). This step yields the value $55 + 44 - 13 = 86$. Dividing this value (86) by the value of N (155) produces the union probability.

$$P(F \cup P) = 86/155 = 0.555$$

A second way to produce the answer from the raw value matrix is to add all the cells one time that are in either the Female column or the Professional row

$$3 + 13 + 17 + 22 + 31 = 86$$

and then divide by the total number of employees, $N = 155$, which gives

$$P(F \cup P) = 86/155 = 0.555$$

DEMONSTRATION PROBLEM 4.2

Shown here are the raw values matrix and corresponding probability matrix for the results of a national survey of 200 executives who were asked to identify the geographic locale of their company and their company's industry type. The executives were only allowed to select one locale and one industry type.

Raw Values Matrix

		North D	*South E*	*East F*	*West G*	
		\multicolumn{4}{c}{*Geographic Location*}		
	Finance A	24	10	8	14	56
Industry Type	*Manufacturing B*	30	6	22	12	70
	Communications C	28	18	12	16	74
		82	34	42	42	200

By dividing every value of the raw values matrix by the total (200), the corresponding probability matrix (shown below) can be constructed.

Probability Matrix

		Geographic Location				
		North D	South E	East F	West G	
Industry Type	Finance A	.12	.05	.04	.07	.28
	Manufacturing B	.15	.03	.11	.06	.35
	Communications C	.14	.09	.06	.08	.37
		.41	.17	.21	.21	1.00

Suppose a respondent is selected randomly from these data.

   a. What is the probability that the respondent is from the East (F)?

   b. What is the probability that the respondent is from the communications industry (C) or from the North (D)?

   c. What is the probability that the respondent is from the South (E) or from the finance industry (A)?

Solution

   a. $P(\text{East}) = P(F) = 0.21$

   b. $P(C \cup D) = P(C) + P(D) - P(C \cap D) = 0.37 + 0.41 - 0.14 = 0.64$

   c. $P(E \cup A) = P(E) + P(A) - P(E \cap A) = 0.17 + 0.28 - 0.05 = 0.40$

In computing the union by using the general law of addition, the intersection probability is subtracted because it is already included in both marginal probabilities. This adjusted probability leaves a union probability that properly includes both marginal values and the intersection value. If the intersection probability is subtracted out a second time, the intersection is removed, leaving the probability of X *or* Y but not *both*.

$$P(X \text{ or } Y \text{ but not both}) = P(X) + P(Y) - P(X \cap Y) - P(X \cap Y)$$
$$= P(X \cup Y) - P(X \cap Y)$$

Figure 4.8 is the Venn diagram for this probability.

Complement of a Union

The probability of the union of two events X and Y represents the probability that the outcome is *either X or* it is Y or it is *both X and Y*. The union includes everything except the possibility that it is neither (X or Y). Another way to state it is as *neither X nor Y*, which can symbolically be represented as $P(\text{not } X \cap \text{not } Y)$. Because it is the only possible case other than the union of X or Y, it is the **complement of a union**. Stated more formally,

$$P(\text{neither } X \text{ nor } Y) = P(\text{not } X \cap \text{not } Y) = 1 - P(X \cup Y).$$

Examine the Venn diagram in Figure 4.9. Note that the complement of the union of X, Y is the shaded area outside the circles. This area represents the neither X nor Y region.

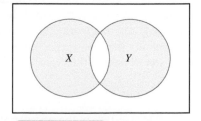

FIGURE 4.8

The *X* or *Y* but Not Both Case

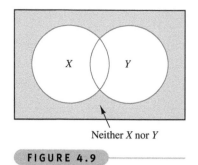

Neither X nor Y

FIGURE 4.9

The Complement of a Union: The Neither/Nor Region

In the survey about job satisfaction discussed earlier, the probability that a randomly selected worker would find their job to be stimulating and challenging *or a* source of personal fulfilment was determined to be

$$P(C \cup S) = P(C) + P(S) - P(C \cap S) = 0.78 + 0.69 - 0.56 = 0.91$$

The probability that a worker would find their job to be *neither* stimulating and challenging *nor a* source of personal fulfilment is calculated as the complement of this union.

$$P(\text{neither C nor S}) = P(\text{not C} \cap \text{not S}) = 1 - P(C \cup S) = 1 - 0.91 = 0.09$$

Thus 9% of the workers found their jobs to be neither stimulating and challenging nor *a* source of personal fulfilment. In Table 4.3, this *neither/nor* probability is found in the No No cell of the matrix, 0.09.

Special Law of Addition

If two events are mutually exclusive, the probability of the union of the two events is the probability of the first event plus the probability of the second event. Because mutually exclusive events do not intersect, nothing has to be subtracted.

SPECIAL LAW OF ADDITION	If X, Y are mutually exclusive, $P(X \cup Y) = P(X) + P(Y)$

The special law of addition is a special case of the general law of addition. In a sense, the general law fits all cases. However, when the events are mutually exclusive, a zero is inserted into the general law formula for the intersection, resulting in the special law formula.

In the survey job satisfaction, the respondents were allowed to choose more than one possible sentence that best described their jobs. Therefore, it is most likely that virtually none of the job satisfaction choices were mutually exclusive, and the special law of addition would not apply to that example.

In another survey, however, respondents were allowed to select only one option for their answer, which made the possible options mutually exclusive. In this survey, conducted by Jobsite.co.uk workers were asked what most contributes to satisfaction at work and were given only the following selections from which to choose only one answer:

- Salary
- Working environment
- Challenging work
- Career progression
- Interaction with colleagues
- Training
- Fun

Salary was cited by the most workers (21%), followed by working environment (20%), challenging work (18%), career progression (15%), interaction with colleagues (12%), training (8%), and fun (6%). If a worker who responded to this survey is selected (or if the survey actually reflects the views of the working public and a worker in general is selected) and that worker is asked which of the given selections most contributes to his/her job satisfaction, what is the probability that the worker will respond that it is either challenging work or career progression?

Let M denote the event 'challenging work' and I denote the event 'career progression'. The question is:

$$P(M \cup I) = ?$$

Because 18% of the survey respondents said 'challenging work',

$$P(M) = 0.18$$

Because 15% of the survey respondents said 'career progression',

$$P(I) = 0.15$$

Because it was not possible to select more than one answer,

$$P(M \cap I) = 0.0000$$

Implementing the special law of addition gives

$$P(M \cup I) = P(M) + P(I) = 0.18 + 0.15 = 0.33$$

DEMONSTRATION PROBLEM 4.3

If a worker is randomly selected from the company described in Demonstration Problem 4.1, what is the probability that the worker is either technical or clerical? What is the probability that the worker is either a professional or a clerical?

Solution

Examine the raw value matrix of the company's human resources data shown in Demonstration Problem 4.1. In many raw value and probability matrices like this one, the rows are non-overlapping or mutually exclusive, as are the columns. In this matrix, a worker can be classified as being in only one type of position and as either male or female but not both. Thus, the categories of type of position are mutually exclusive, as are the categories of sex, and the special law of addition can be applied to the human resource data to determine the union probabilities.

Let T denote technical, C denote clerical, and P denote professional. The probability that a worker is either technical or clerical is

$$P(T \cup C) = P(T) + P(C) = \frac{69}{155} + \frac{31}{155} = \frac{100}{155} = 0.645$$

The probability that a worker is either professional or clerical is

$$P(P \cup C) = P(P) + P(C) = \frac{44}{155} + \frac{31}{155} = \frac{75}{155} = 0.484$$

DEMONSTRATION PROBLEM 4.4

Use the data from the matrices in Demonstration Problem 4.2. What is the probability that a randomly selected respondent is from the South or the West?

$$P(E \cup G) = ?$$

Solution

Because geographic location is mutually exclusive (the work location is either in the South or in the West but not in both),

$$P(E \cup G) = P(E) + P(G) = 0.17 + 0.21 = 0.38$$

4.5   PROBLEMS

4.8   Given $P(A) = 0.10$, $P(B) = 0.12$, $P(C) = 0.21$, $P(A \cap C) = 0.05$, and $P(B \cap C) = 0.03$, solve the following:

   a. $P(A \cup C) = $ _____

   b. $P(B \cup C) = $ _____

   c.   If A and B are mutually exclusive, $P(A \cup B) = $ _____

4.9   Use the values in the matrix to solve the equations given.

	D	E	F
A	5	8	12
B	10	6	4
C	8	2	5

   a. $P(A \cup D) = $ _____

   b. $P(E \cup B) = $ _____

   c. $P(D \cup E) = $ _____

   d. $P(C \cup F) = $ _____

4.10   Use the values in the matrix to solve the equations given.

	E	F
A	.10	.03
B	.04	.12
C	.27	.06
D	.31	.07

   a. $P(A \cup F) = $ _____

   b. $P(E \cup B) = $ _____

   c. $P(B \cup C) = $ _____

   d. $P(E \cup F) = $ _____

4.11   Suppose that 47% of all Chinese people have flown in an airplane at least once and that 28% of all Chinese people have ridden on a train at least once. What is the probability that a randomly selected Chinese person has either ridden on a train or flown in an airplane? Can this problem be solved? Under what conditions can it be solved? If the problem cannot be solved, what information is needed to make it solvable?

4.12   According to the Organization for Economic Cooperation and Development, 64% of women aged 15 to 64 years of age participate in the labour force. Suppose 68% of the women in that age group are married. Suppose also that 55% of all women 15 to 64 years of age are married and are participating in the labour force.

   a.   What is the probability that a randomly selected woman in that age group is married or is participating in the labour force?

   b.   What is the probability that a randomly selected woman in that age group is married or is participating in the labour force but not both?

   c.   What is the probability that a randomly selected woman in that age group is neither married nor participating in the labour force?

4.13   According to Nielsen Media Research, approximately 67% of all US households with television have cable TV. Seventy-four per cent of all US households with television have two or more TV sets. Suppose 55% of all US households with television have cable TV and two or more TV sets. A US household with television is randomly selected.

   a.   What is the probability that the household has cable TV or two or more TV sets?

   b.   What is the probability that the household has cable TV or two or more TV sets but not both?

   c.   What is the probability that the household has neither cable TV nor two or more TV sets?

   d.   Why does the special law of addition not apply to this problem?

4.14 According to a recent survey of 5,828 job applications in the financial sector by Powerchex, 15% of CVs of applicants to the financial sector contained some form of discrepancies, including hidden criminal records, lies about academic records or professional qualifications, amongst others. The percentage of CVs with discrepancies varies greatly depending on the type of financial service company to which the candidates were applying for. The percentage of CVs with discrepancies was 20% for stockbrokerage firms, 15% for insurance companies, and 10% for banks. Assume that these percentages are true for the population of financial companies in the United Kingdom and that 7% of all CVs with discrepancies were from candidates that applied for both insurance companies and banks with the same CV that was found to have discrepancies in both cases.

   a. What is the probability that a randomly selected CV with discrepancies comes from a candidate that applied to either insurance companies or banks?

   b. What is the probability that a randomly selected CV with discrepancies comes from a candidate that applied to either insurance companies or banks but not both?

   c. What is the probability that a randomly selected CV with discrepancies comes from a candidate that did not apply to either insurance companies or banks?

   d. Construct a probability matrix for this problem and indicate the locations of your answers for parts (a), (b), and (c) on the matrix.

4.6 MULTIPLICATION LAWS

General Law of Multiplication

As stated in Section 4.4, the probability of the intersection of two events $(X \cap Y)$ is called the joint probability. The general law of multiplication is used to find the joint probability.

GENERAL LAW OF MULTIPLICATION	$P(X \cap Y) = P(X) \cdot P(Y|X) = P(Y) \cdot P(X|Y)$

The notation $X \cap Y$ means that both X *and* Y must *happen*. The general law of multiplication gives the probability that *both* event X and event Y will occur at the same time.

According to the Office for National Statistics, 46% of the UK labour force is female. In addition, 42% of the women in the labour force work part-time. What is the probability that a randomly selected member of the UK labour force is a woman *and* works part-time? This question is one of joint probability, and the general law of multiplication can be applied to answer it.

Let W denote the event that the member of the labour force is a woman. Let T denote the event that the member is a part-time worker. The question is:

$$P(W \cap T) = ?$$

According to the general law of multiplication, this problem can be solved by

$$P(W \cap T) = P(W) \cdot P(T|W)$$

Since 46% of the labour force is women, $P(W) = 0.46 P(T|W)$ is a conditional probability that can be stated as the probability that a worker is a part-time worker given that the worker is a woman. This condition is what was given in the statement that 42% *of the women in the labour force* work part-time. Hence, $P(T|W) = 0.42$. From there it follows that

$$P(W \cap T) = P(W) \cdot P(T|W) = (0.46)(0.42) = 0.193$$

It can be stated that 19.3% of the UK labour force are women *and* work part-time. The Venn diagram in Figure 4.10 shows these relationships and the joint probability.

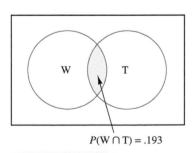

$P(W \cap T) = .193$

FIGURE 4.10

Joint Probability that a Woman Is in the Labour Force and Is a Part-Time Worker

TABLE 4.5

Probability Matrix of Company Human Resource Data

		Male	Female	
		____**Sex**____		
Type of Position	Managerial	.052	.019	.071
	Professional	.200	.084	.284
	Technical	.335	.110	.445
	Clerical	.058	.142	.200
		.645	.355	1.000

Determining joint probabilities from raw value or probability matrices is easy because every cell of these matrices is a joint probability. In fact, some statisticians refer to a probability matrix as a *joint probability table*.

For example, suppose the raw value matrix of the client company data from Demonstration Problem 4.1 and the Decision Dilemma is converted to a probability matrix by dividing by the total number of employees ($N = 155$), resulting in Table 4.5. Each value in the cell of Table 4.5 is an intersection, and the table contains all possible intersections (joint probabilities) for the events of sex and type of position. For example, the probability that a randomly selected worker is male *and* a technical worker, $P(M \cap T)$, is 0.335. The probability that a randomly selected worker is female *and* a professional worker, $P(F \cap T)$, is 0.084. Once a probability matrix is constructed for a problem, usually the easiest way to solve for the joint probability is to find the appropriate cell in the matrix and select the answer. However, sometimes because of what is given in a problem, using the formula is easier than constructing the matrix.

DEMONSTRATION PROBLEM 4.5

A company has 140 employees, of which 30 are supervisors. Eighty of the employees are married, and 20% of the married employees are supervisors. If a company employee is randomly selected, what is the probability that the employee is married and is a supervisor?

Solution

Let M denote married and S denote supervisor. The question is:

$$P(M \cap S) = ?$$

First, calculate the marginal probability.

$$P(M) = \frac{80}{140} = 0.5714$$

Then, note that 20% of the married employees are supervisors, which is the conditional probability, $P(S|M) = 0.20$. Finally, applying the general law of multiplication gives

$$P(M \cap S) = P(M) \cdot P(S|M) = (0.5714)(0.20) = 0.1143$$

Hence, 11.43% of the 140 employees are married and are supervisors.

DEMONSTRATION PROBLEM 4.6

From the data obtained from the interviews of 200 executives in Demonstration Problem 4.2, find

a. $P(B \cap E)$

b. $P(G \cap A)$

c. $P(B \cap C)$

Probability Matrix

	Geographic Location				
	North D	South E	East F	West G	
Finance A	.12	.05	.04	.07	.28
Industry Type Manufacturing B	.15	.03	.11	.06	.35
Communications C	.14	.09	.06	.08	.37
	.41	.17	.21	.21	1.00

Solution

a. From the cell of the probability matrix, $P(B \cap E) = 0.03$. To solve by the formula, $P(B \cap E) = P(B) \cdot P(E|B)$, first find $P(B)$:

$$P(B) = 0.35$$

The probability of E occurring given that B has occurred, $P(E|B)$, can be determined from the probability matrix as $P(E|B) = 0.03/0.35$. Therefore,

$$P(B \cap E) = P(B) \cdot P(E|B) = (0.35)\left(\frac{0.03}{0.35}\right) = 0.03$$

Although the formula works, finding the joint probability in the cell of the probability matrix is faster than using the formula.

An alternative formula is $P(B \cap E) = P(E) \cdot P(B|E)$, but $P(E) = 0.17$. Then $P(B|E)$ means the probability of B if E is given. There are 0.17 Es in the probability matrix and 0.03 Bs in these Es. Hence,

$$P(B|E) = \frac{0.03}{0.17} \text{ and } P(B \cap E) = P(E) \cdot P(B|E) = (0.17)\left(\frac{0.03}{0.17}\right) = 0.03$$

b. To obtain $P(G \cap A)$, find the intersecting cell of G and A in the probability matrix, 0.07, or use one of the following formulas:

$$P(G \cap A) = P(G) \cdot P(A|G) = (0.21)\left(\frac{0.07}{0.21}\right) = 0.07$$

or

$$P(G \cap A) = P(A) \cdot P(G|A) = (0.28)\left(\frac{0.07}{0.28}\right) = 0.07$$

c. The probability $P(B \cap C)$ means that one respondent would have to work both in the manufacturing industry and the communications industry. The survey used to gather data from the 200 executives, however, requested that each respondent specify only one industry type for his or her company. The matrix shows no intersection for these two events. Thus B and C are mutually exclusive. None of the respondents is in both manufacturing and communications. Hence,

$$P(B \cap C) = 0.0$$

Special Law of Multiplication

If events X and Y are independent, a special law of multiplication can be used to find the intersection of X and Y. This special law utilizes the fact that when two events X, Y are independent, $P(X|Y) = P(X)$ and $P(Y|X) = P(Y)$. Thus, the general law of multiplication, $P(X \cap Y) = P(X) \cdot P(Y|X)$, becomes $P(X \cap Y) = P(X) \cdot P(Y)$ when X and Y are independent.

SPECIAL LAW OF MULTIPLICATION	If X, Y are independent, $P(X \cap Y) = P(X) \cdot P(Y)$

A study recently released by the Office for National Statistics found that 56% of all adults in the UK regularly donated to charitable causes in the previous 12 months. Another study by the European Foundation for the Improvement of Living and Working Conditions reported that 53% of employees in the UK have undergone training paid for or provided by their employer in the previous 12 months. Is the number of adults that regularly donate to charity independent of the number of employees that underwent paid training in the last 12 months? If they are independent, then the probability of a person being randomly selected who regularly donated to charity *and* that underwent paid training is found as follows. Let D denote donations to charity and T denote paid training.

$$P(D) = 0.56$$

$$P(T) = 0.53$$

$$P(D \cap T) = (0.56)(0.53) = 0.2968$$

Therefore, 29.68% of the population in the UK regularly donated to charity *and* underwent paid training in the past 12 months.

DEMONSTRATION PROBLEM 4.7

A manufacturing firm produces pads of bound paper. Three per cent of all paper pads produced are improperly bound. An inspector randomly samples two pads of paper, one at a time. Because a large number of pads are being produced during the inspection, the sampling being done, in essence, is with replacement. What is the probability that the two pads selected are both improperly bound?

Solution

Let I denote improperly bound. The problem is to determine

$$P(I_1 \cap I_2) = ?$$

The probability of I = 0.03, or 3% are improperly bound. Because the sampling is done with replacement, the two events are independent. Hence,

$$P(I_1 \cap I_2) = P(I_1) \cdot P(I_2) = (0.03)(0.03) = 0.0009$$

TABLE 4.6

Contingency Table of Data from Independent Events

	D	E	
A	8	12	20
B	20	30	50
C	6	9	15
	34	51	85

Most probability matrices contain variables that are not independent. If a probability matrix contains independent events, the special law of multiplication can be applied. If not, the special law cannot be used. In Section 4.7 we explore a technique for determining whether events are independent. Table 4.6 contains data from independent events.

DEMONSTRATION PROBLEM 4.8

Use the data from Table 4.6 and the special law of multiplication to find $P(B \cap D)$.

Solution

$$P(B \cap D) = P(B) \cdot P(D) = \frac{50}{85} \cdot \frac{34}{85} = 0.2353$$

This approach works *only* for contingency tables and probability matrices in which the variable along one side of the matrix is *independent* of the variable along the other side of the matrix. Note that the answer obtained by using the formula is the same as the answer obtained by using the cell information from Table 4.6.

$$P(B \cap D) = \frac{20}{85} = 0.2353$$

4.6 PROBLEMS

4.15 Use the values in the contingency table to solve the equations given.

	C	D	E	F
A	5	11	16	8
B	2	3	5	7

a. $P(A \cap E) =$ _____

b. $P(D \cap B) =$ _____

c. $P(D \cap E) =$ _____

d. $P(A \cap B) =$ _____

4.16 Use the values in the probability matrix to solve the equations given.

	D	E	F
A	.12	.13	.08
B	.18	.09	.04
C	.06	.24	.06

a. $P(E \cap B) =$ _____

b. $P(C \cap F) =$ _____

c. $P(E \cap D) =$ _____

4.17 a. A batch of 50 parts contains six defects. If two parts are drawn randomly one at a time without replacement, what is the probability that both parts are defective?

b. If this experiment is repeated, with replacement, what is the probability that both parts are defective?

4.18 According to the non-profit group Zero Population Growth, 78% of the US population now lives in urban areas. Scientists at Princeton University and the University of Wisconsin report that about 15% of all American adults care for ill relatives. Suppose that 11% of adults living in urban areas care for ill relatives.

a. Use the general law of multiplication to determine the probability of randomly selecting an adult from the US population who lives in an urban area and is caring for an ill relative.

   b. What is the probability of randomly selecting an adult from the US population who lives in an urban area and does not care for an ill relative?

   c. Construct a probability matrix and show where the answer to this problem lies in the matrix.

   d. From the probability matrix, determine the probability that an adult lives in a non-urban area and cares for an ill relative.

4.19 A recent study by Syed Sultana, published in the *Global Journal of Finance and Management*, based on a sample of 150 investors, revealed that 37% of all Indian stockmarket investors are 35 years old or less. In addition, the study determined that 47% of all Indian stockholders have a postgraduate degree. Suppose 10% of all Indian adults aged 35 years old or less have a postgraduate degree. An Indian adult is randomly selected.

   a. What is the probability that the adult is more than 35 years old?

   b. What is the probability that the investor is less than 35 years old and has a postgraduate degree?

   c. What is the probability that the investor is less than 35 years old or has a postgraduate degree?

   d. What is the probability that the adult has neither a postgraduate degree nor is less than 35 years old?

   e. What is the probability that the adult is not less than 35 years old or has no postgraduate degree?

   f. What is the probability that the adult is less than 35 years old and has no postgraduate education?

4.20 According to the Office for National Statistics, 86% of all UK households have a satellite, digital, or cable receiver and 75% have a personal computer. Suppose 81% of all UK households having a satellite, digital, or cable receiver have a personal computer. A UK household is randomly selected.

   a. What is the probability that the household has a satellite, digital, or cable receiver and a personal computer?

   b. What is the probability that the household has a satellite, digital, or cable receiver or a personal computer?

   c. What is the probability that the household has a satellite, digital, or cable receiver and does not have a personal computer?

   d. What is the probability that the household has neither a satellite, digital, or cable receiver nor a personal computer?

   e. What is the probability that the household does not have a satellite, digital, or cable receiver and does have a personal computer?

4.21 A recent study by the European Commission found that out of the citizens who exercise regularly in Sweden, 51% said they exercise regularly in a park or in the outdoors and 14% reported exercising regularly in a fitness centre. Suppose 22% of the people who exercise regularly in a park or in the outdoors say that they also exercise regularly in a fitness centre.

   a. What is the probability of randomly selecting a Swede who exercises regularly, and finding out that they regularly exercise in a park or in the outdoors and do not regularly exercise in a fitness centre?

   b. What is the probability of randomly selecting a Swede who exercises regularly, and finding out that they neither regularly exercise in a park or in the outdoors nor regularly exercise in a fitness centre?

   c. What is the probability of randomly selecting a Swede who exercises regularly, and finding out that they do not regularly exercise in a park or in the outdoors and regularly exercise in a fitness centre?

4.22 Statistic Netherlands states that 84% of all Dutch households have a microwave. In addition, 47% of all Dutch households have a dishwasher. Suppose 45% of all Dutch households have both a microwave and dishwasher. A Dutch household is randomly selected.

   a. What is the probability that the household has a microwave or a dishwasher?

   b. What is the probability that the household has neither a microwave nor a dishwasher?

   c. What is the probability that the household does not have a microwave and does have a dishwasher?

   d. What is the probability that the household does have a microwave and does not have a dishwasher?

4.7 CONDITIONAL PROBABILITY

Conditional probabilities are computed based on the prior knowledge that a business researcher has on one of the two events being studied. If X, Y are two events, the conditional probability of X occurring given that Y is known or has occurred is expressed as $P(X|Y)$ and is given in the *law of conditional probability*.

LAW OF CONDITIONAL PROBABILITY	$P(X|Y) = \dfrac{P(X \cap Y)}{P(Y)} = \dfrac{P(X) \cdot P(Y|X)}{P(Y)}$

The conditional probability of $(X|Y)$ is the probability that X will occur given Y. The formula for conditional probability is derived by dividing both sides of the general law of multiplication by $P(Y)$.

In the study by the Work Foundation to determine the degree of job satisfaction in the UK, 78% of the respondents said they found work stimulating and challenging and 69% said their work was a source of personal fulfilment. In addition, suppose 56% of the respondents reported their work to be both stimulating and challenging and a source of personal fulfilment. A worker is selected randomly and asked his job satisfaction. This worker finds his work to be stimulating and challenging. What is the probability that this worker finds his work to be a source of personal fulfilment? That is, what is the probability that a randomly selected person finds his work to be a source of personal fulfilment *given that* he finds his work stimulating and challenging ? In symbols, the question is

$$P(S|C) = ?$$

Note that the given part of the information is listed to the right of the vertical line in the conditional probability. The formula solution is

$$P(S|C) = \frac{P(S \cap C)}{P(C)}$$

but

$$P(C) = 0.78 \text{ and } P(S \cap C) = 0.56$$

therefore

$$P(S|C) = \frac{P(S \cap C)}{P(C)} = \frac{0.56}{0.78} = 0.7179$$

Approximately 72% of workers who find work to be stimulating and challenging believe their work to be a source of personal fulfilment.

Note in Figure 4.11 that the area for C in the Venn diagram is completely shaded because it is given that the worker finds their work to be stimulating and challenging. Also notice that the intersection of C and S is more heavily shaded. This portion of stimulating and challenging includes a source of personal fulfilment. It is the only part of a source of personal fulfilment that is in stimulating and challenging, and because the person is known to find their work stimulating and challenging, it is the only area of interest that includes a source of personal fulfilment.

Examine the probability matrix in Table 4.7 for the office design problem. None of the probabilities given in the matrix are conditional probabilities. To reiterate what has been previously stated, a probability matrix contains only two types of probabilities, marginal and joint. The cell values are all joint probabilities and the subtotals in the margins are marginal probabilities. How

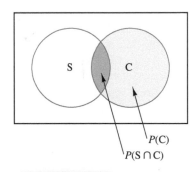

FIGURE 4.11

Conditional Probability of Stimulating and Challenging Given a Source of Personal Fulfilment

TABLE 4.7

Job Satisfaction Problem Probability Matrix

		Source of Personal Fulfilment		
		Yes	*No*	
Stimulating and	*Yes*	.56	.22	.78
Challenging	*No*	.13	.09	.22
		.69	.31	1.00

are conditional probabilities determined from a probability matrix? The law of conditional probabilities shows that a conditional probability is computed by dividing the joint probability by the marginal probability. Thus, the probability matrix has all the necessary information to solve for a conditional probability.

What is the probability that a randomly selected worker does not find his work stimulating and challenging given that the worker believes his work to be a source of personal fulfilment? That is,

$$P(\text{not } C|S) = ?$$

The law of conditional probability states that

$$P(\text{not } C|S) = \frac{P(\text{not } C \cap S)}{P(S)}$$

Notice that because S is given, we are interested only in the column that is shaded in Table 4.7, which is the Yes column for a source of personal fulfilment. The marginal probability, $P(S)$, is the total of this column and is found in the margin at the bottom of the table as 0.69. $P(\text{not } C|S)$ is found as the intersection of No for stimulating and challenging and Yes for a source of personal fulfilment. This value is 0.13. Hence, $P(\text{not } C|S)$ is 0.11. Therefore,

$$P(\text{not } C|S) = \frac{P(\text{not } C \cap S)}{P(S)} = \frac{0.13}{0.69} = 0.188$$

The second version of the conditional probability law formula is

$$P(X|Y) = \frac{P(X) \cdot P(Y \cdot X)}{P(Y)}$$

This version is more complex than the first version, $P(X \cap Y)/P(Y)$. However, sometimes the second version must be used because of the information given in the problem – for example, when solving for $P(X|Y)$ but $P(Y|X)$ is given. The second version of the formula is obtained from the first version by substituting the formula for $P(X \cap Y) = P(X) \cdot P(Y|X)$ into the first version.

As an example, in Section 4.6, data relating to women in the UK labour force were presented. Included in this information was the fact that 46% of the UK labour force is female and that 42% of the females in the UK labour force work part-time. Suppose that 27.4% of all British workers are known to be part-time workers. What is the probability that a randomly selected UK worker is a woman if that person is known to be a part-time worker? Let W denote the event of selecting a woman and T denote the event of selecting a part-time worker. In symbols, the question to be answered is

$$P(W|T) = ?$$

The first form of the law of conditional probabilities is

$$P(W|T) = \frac{P(W \cap T)}{P(T)}$$

Note that this version of the law of conditional probabilities requires knowledge of the joint probability, $P(W \cap T)$, which is not given here. We therefore try the second version of the law of conditional probabilities, which is

$$P(W|T) = \frac{P(W) \cdot P(T|W)}{P(T)}$$

For this version of the formula, everything is given in the problem.

$$P(W) = 0.46$$

$$P(T) = 0.274$$

$$P(T|W) = 0.42$$

The probability of a worker being a woman given that the person works part-time can now be computed.

$$P(W|T) = \frac{P(W)P(T|W)}{P(T)} = \frac{(0.46)(0.42)}{(0.274)} = 0.705$$

Hence, 70.5% of the part-time workers are women.

In general, this second version of the law of conditional probabilities is likely to be used for solving $P(X|Y)$ when $P(X \cap Y)$ is unknown but $P(Y|X)$ is known.

DEMONSTRATION PROBLEM 4.9

The data from the executive interviews given in Demonstration Problem 4.2 are repeated here. Use these data to find:

   a. $P(B|F)$

   b. $P(G|C)$

   c. $P(D|F)$

Probability Matrix

		Geographic Location				
		North D	South E	East F	West G	
	Finance A	.12	.05	.04	.07	.28
Industry Type	Manufacturing B	.15	.03	.11	.06	.35
	Communications C	.14	.09	.06	.08	.37
		.41	.17	.21	.21	1.00

Solution

  a. $P(B|F) = \dfrac{P(B \cap F)}{P(F)} = \dfrac{0.11}{0.21} = 0.524$

Determining conditional probabilities from a probability matrix by using the formula is a relatively painless process. In this case, the joint probability, $P(B \cap F)$, appears in a cell of the matrix (0.11); the marginal probability, $P(F)$, appears in a margin (0.21). Bringing these two probabilities together by formula produces the answer: $0.11/0.21 = 0.524$. This answer means that 52.4% of the East executives (the F values) are in manufacturing (the B values).

  b. $P(G|C) = \dfrac{P(G \cap C)}{P(C)} = \dfrac{0.08}{0.37} = 0.216$

This result means that 21.6% of the responding communications industry executives, (C) are from the West (G).

  c. $P(D|F) = \dfrac{P(D \cap F)}{P(F)} = \dfrac{0.00}{0.21} = 0.00$

Because D and F are mutually exclusive, $P(D \cap F)$ is zero and so is $P(D|F)$. The rationale behind $P(D|F) = 0$ is that, if F is given (the respondent is known to be located in the East), the respondent could not be located in D (the North).

Independent Events

| INDEPENDENT EVENTS X, Y | To test to determine if X and Y are independent events, the following must be true. $$P(X|Y)=P(X) \quad \text{and} \quad P(Y|X)=P(Y)$$ |

In each equation, it does not matter that X or Y is given because X and Y are *independent*. When X and Y are independent, the conditional probability is solved as a marginal probability.

Sometimes, it is important to test a contingency table of raw data to determine whether events are independent. If *any* combination of two events from the different sides of the matrix fail the test, $P(X|Y) = P(X)$, the matrix does not contain independent events.

DEMONSTRATION PROBLEM 4.10

Test the matrix for the 200 executive responses to determine whether industry type is independent of geographic location.

Raw Values Matrix

		Geographic Location				
		North D	South E	East F	West G	
Industry Type	Finance A	24	10	8	14	56
	Manufacturing B	30	6	22	12	70
	Communications C	28	18	12	16	74
		82	34	42	42	200

Solution

Select one industry and one geographic location (say, A – Finance and G – West). Does $P(A|G) = P(A)$?

$$P(A|G) = \frac{14}{42} \quad \text{and} \quad P(A) = \frac{56}{200}$$

Does $14/42 = 56/200$? No, $0.33 \neq 0.28$. Industry and geographic location are not independent because at least one exception to the test is present.

Newspaper Advertising Reading Habits of Canadians

A national survey by Ipsos Reid for the Canadian Newspaper Association reveals some interesting statistics about newspaper advertising reading habits of Canadians. Sixty-six per cent of Canadians say that they enjoy reading the page advertising and the product inserts that come with a newspaper. The percentage is higher for women (70%) than men (62%), but 73% of households with children enjoy doing so.

While the percentage of those over 55 years of age who enjoy reading such ads is 71%, the percentage is only 55% for those in the 18- to 34-year-old category. These percentages decrease with increases in education as revealed by the fact that while 70% of those with a high school education enjoy reading such ads, only 55% of those having a university degree do so. Canadians living in the Atlantic region lead the country in this regard with 74%, in contrast to those living in British Columbia (63%) and Quebec (62%).

These facts can be converted to probabilities: the probability that a Canadian enjoys reading such ads is 0.66. Many of the other statistics represent conditional probabilities. For example, the probability that a Canadian enjoys such ads given that the Canadian is a woman is 0.70; and the probability that a Canadian enjoys such ads given that the Canadian has a college degree is 0.55. About 13% of the Canadian population resides in British Columbia. From this and from the conditional probability that a Canadian enjoys such ads given that they live in British Columbia (0.63), one can compute the joint probability that a randomly selected Canadian enjoys such ads and lives in British Columbia $(0.13)(0.63) = 0.0819$. That is, 8.19% of all Canadians live in British Columbia and enjoy such ads.

DEMONSTRATION PROBLEM 4.11

Determine whether the contingency table shown as Table 4.6 and repeated here contains independent events.

	D	E	
A	8	12	20
B	20	30	50
C	6	9	15
	34	51	85

Solution

Check the first cell in the matrix to find whether $P(A|D) = P(A)$.

$$P(A|D) = \frac{8}{34} = 0.2353$$

$$P(A) = \frac{20}{85} = 0.2353$$

The checking process must continue until all the events are determined to be independent. In this matrix, all the possibilities check out. Thus, Table 4.6 contains independent events.

4.7   PROBLEMS

4.23   Use the values in the contingency table to solve the equations given.

	E	F	G
A	15	12	8
B	11	17	19
C	21	32	27
D	18	13	12

   a. $P(G|A) =$ _____
   b. $P(B|F) =$ _____
   c. $P(C|E) =$ _____
   d. $P(E|G) =$ _____

4.24 Use the values in the probability matrix to solve the equations given.

	C	D
A	.36	.44
B	.11	.09

   a. $P(C|A) =$ _____
   b. $P(B|D) =$ _____
   c. $P(A|B) =$ _____

4.25 The results of a survey asking, 'Do you have a calculator and/or a computer in your home?' follow:

		Calculator		
		Yes	No	
Computer	Yes	46	3	49
	No	11	15	26
		57	18	75

Is the variable 'calculator' independent of the variable 'computer'? Why or why not?

4.26 In 2010, business failures in the United Kingdom numbered 19,946, according to Experian. The construction industry accounted for 3,267 of these business failures. The north-west accounted for 2,382 of the business failures. Suppose that 458 of all business failures were construction businesses located in the north-west 0. A failed business is randomly selected from this list of business failures.

   a. What is the probability that the business is located in the north-west?
   b. What is the probability that the business is in the construction industry or located in the north-west?
   c. What is the probability that the business is in the construction industry if it is known that the business is located in the north-west?
   d. What is the probability that the business is located in the north-west if it is known that the business is a construction business?
   e. What is the probability that the business is not located in the north-west if it is known that the business is not a construction business?
   f. Given that the business is a construction business, what is the probability that the business is not located in the north-west?

4.27 Arthur Andersen Enterprise Group/National Small Business United, Washington, conducted a national survey of small-business owners to determine the challenges for growth for their businesses. The top challenge, selected by 46% of the small-business owners, was the economy. A close second was finding qualified workers (37%). Suppose 15% of the small-business owners selected both the economy and finding qualified workers as challenges for growth. A small-business owner is randomly selected.

   a. What is the probability that the owner believes the economy is a challenge for growth if the owner believes that finding qualified workers is a challenge for growth?
   b. What is the probability that the owner believes that finding qualified workers is a challenge for growth if the owner believes that the economy is a challenge for growth?
   c. Given that the owner does not select the economy as a challenge for growth, what is the probability that the owner believes that finding qualified workers is a challenge for growth?

   d. What is the probability that the owner believes neither that the economy is a challenge for growth nor that finding qualified workers is a challenge for growth?

4.28 According to a survey by Nielsen and Yahoo, 86% of all mobile phone owners use the Internet on their handset while watching TV. Suppose that 20% of those who use the Internet on their mobile phone while watching TV to access a social networking site. Assuming that these figures are true for all mobile phone owners, if a mobile phone owner is randomly selected, determine the following probabilities:

   a. The mobile phone owner uses the Internet on her handset while watching TV and accesses a social networking site.

   b. The mobile phone owner does not access a social networking site given that she uses the Internet on her handset while watching TV.

   c. The mobile phone owner does not access a social networking site and does use the Internet on her handset while watching TV.

4.29 According *to the Association of British Insurers,* 49% of UK residents have life insurance while 16% have a personal pension scheme. Suppose that 27% of UK residents with life insurance also have a personal pension scheme. If a UK resident is randomly selected, determine the following probabilities:

   a. The UK resident does not have a personal pension scheme given that he does have life insurance.

   b. The UK resident has a personal pension scheme given that he does not have life insurance.

   c. The UK resident does not have life insurance if it is known that he does have a personal pension scheme.

   d. The UK resident does not have life insurance if it is known that he does not have a personal pension scheme.

4.30 In a study undertaken by Catalyst, 43% of women senior executives agreed or strongly agreed that a lack of role models was a barrier to their career development. In addition, 46% agreed or strongly agreed that gender-based stereotypes were barriers to their career advancement. Suppose 77% of those who agreed or strongly agreed that gender-based stereotypes were barriers to their career advancement agreed or strongly agreed that the lack of role models was a barrier to their career development. If one of these female senior executives is randomly selected, determine the following probabilities:

   a. What is the probability that the senior executive does not agree or strongly agree that a lack of role models was a barrier to her career development given that she does agree or strongly agree that gender-based stereotypes were barriers to her career development?

   b. What is the probability that the senior executive does not agree or strongly agree that gender-based stereotypes were barriers to her career development given that she does agree or strongly agree that the lack of role models was a barrier to her career development?

   c. If it is known that the senior executive does not agree or strongly agree that gender-based stereotypes were barriers to her career development, what is the probability that she does not agree or strongly agree that the lack of role models was a barrier to her career development?

4.8 REVISION OF PROBABILITIES: BAYES' RULE

An extension to the conditional law of probabilities is Bayes' rule, which was developed by and named after Thomas Bayes (1702–1761). **Bayes' rule** is *a formula that extends the use of the law of conditional probabilities to allow revision of original probabilities with new information.*

| BAYES' RULE | $$P(X_i|Y) = \frac{P(X_i) \cdot P(Y|X_i)}{P(X_1) \cdot P(Y|X_1) + P(X_2) \cdot P(Y|X_2) + \cdots + P(X_n) \cdot P(Y|X_n)}$$ |

Recall that the law of conditional probability for

$$P(X_i|Y)$$

is

$$P(X_i|Y) = \frac{P(X_i) \cdot P(Y|X_i)}{P(Y)}$$

Compare Bayes' rule with this law of conditional probability. The numerators of Bayes' rule and the law of conditional probability are the same – the intersection of X_i and Y shown in the form of the general rule of multiplication. The new feature that Bayes' rule uses is found in the denominator of the rule:

$$P(X_1) \cdot P(Y|X_1) + P(X_2) \cdot P(Y|X_2) + \cdots + P(X_n) \cdot P(Y|X_n)$$

The denominator of Bayes' rule includes a product expression (intersection) for every partition in the sample space, Y, including the event (X_i) itself. The denominator is thus a collective exhaustive listing of mutually exclusive outcomes of Y. This denominator is sometimes referred to as the 'total probability formula'. It represents a weighted average of the conditional probabilities, with the weights being the prior probabilities of the corresponding event.

By expressing the law of conditional probabilities in this new way, Bayes' rule enables the statistician to make new and different applications using conditional probabilities. In particular, statisticians use Bayes' rule to 'revise' probabilities in light of new information.

A particular type of bar code scanner is produced by only two companies, Cilico Microelectronics Corporation from China and AUM Corporation from India. Suppose Cilico produces 65% of the bar code scanners and that AUM produces 35%. Eight per cent of the bar code scanners produced by Cilico are defective and 12% of the AUM bar code scanners are defective. A customer purchases a new bar code scanner. What is the probability that Cilico produced the bar code scanner? What is the probability that AUM produced the bar code scanner? The bar code scanner is tested, and it is defective. Now what is the probability that Cilico produced the bar code scanner? That AUM produced the bar code scanner?

The probability was 0.65 that the bar code scanner came from Cilico and 0.35 that it came from AUM. These are called *prior* probabilities because they are based on the original information.

The new information that the bar code scanner is defective changes the probabilities because one company produces a higher percentage of defective bar code scanners than the other company does. How can this information be used to update or revise the original probabilities? Bayes' rule allows such updating. One way to lay out a revision of a probabilities problem is to use a table. Table 4.8 shows the analysis for the bar code scanner problem.

The process begins with the prior probabilities: 0.65 Cilico and 0.35 AUM. These prior probabilities appear in the second column of Table 4.8. Because the product is found to be defective, the conditional probabilities, $P(\text{defective}|\text{Cilico})$ and $P(\text{defective}|\text{AUM})$ should be used. Eight per cent of Cilico's bar code scanners are defective: $P(\text{defective}|\text{Cilico}) = 0.08$. Twelve per cent of AUM's bar code scanners are defective: $P(\text{defective}|\text{AUM}) = 0.12$. These two conditional probabilities

TABLE 4.8

Bayesian Table for Revision of Bar Code Scanner Problem Probabilities

Event	Prior Probability $P(E_i)$	Conditional Probability $P(d|E_i)$	Joint Probability $P(E_i \cap d)$	Posterior or Revised Probability
Cilico	.65	.08	.052	$\frac{.052}{.094} = .553$
AUM	.35	.12	.042	$\frac{.042}{.094} = .447$
			$P(\text{defective}) = .094$	

appear in the third column. Eight per cent of Cilico's 65% of the bar code scanners are defective: (0.08)(0.65) = 0.052, or 5.2% of the total. This figure appears in the fourth column of Table 4.8; it is the joint probability of getting a bar code scanner that was made by Cilico and is defective. Because the purchased bar code scanner is defective, these are the only Cilico bar code scanners of interest. Twelve per cent of AUM's 35% of the bar code scanners are defective. Multiplying these two percentages yields the joint probability of getting a AUM bar code scanner that is defective. This figure also appears in the fourth column of Table 4.8: (0.12)(0.35) = 0.042; that is, 4.2% of all bar code scanners are made by AUM and are defective. This percentage includes the only AUM bar code scanners of interest because the bar code scanner purchased is defective.

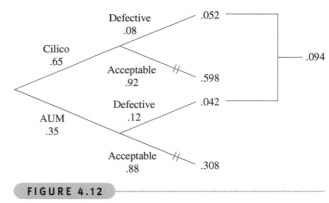

FIGURE 4.12

Tree Diagram for Bar Code Scanner Problem Probabilities

Column 4 is totalled to get 0.094, indicating that 9.4% of all bar code scanners are defective (Cilico and defective = 0.052 + AUM and defective = 0.042). The other 90.6% of the bar code scanners, which are acceptable, are not of interest because the bar code scanner purchased is defective. To compute the fifth column, the posterior or revised probabilities, involves dividing each value in column 4 by the total of column 4. For Cilico, 0.052 of the total bar code scanners are Cilico's and defective out of the total of 0.094 that are defective. Dividing 0.052 by 0.094 yields 0.553 as a revised probability that the purchased bar code scanner was made by Cilico. This probability is lower than the prior or original probability of 0.65 because fewer of Cilico's bar code scanners (as a percentage) are defective than those produced by AUM. The defective bar code scanner is now less likely to have come from Cilico than before the knowledge of the defective bar code scanner. AUM's probability is revised by dividing the 0.042 joint probability of the bar code scanner being made by AUM and defective by the total probability of the bar code scanner being defective (0.094). The result is 0.042/0.094 = 0.447. The probability that the defective bar code scanner is from AUM increased because a higher percentage of AUM bar code scanners are defective.

Tree diagrams are another common way to solve Bayes' rule problems. Figure 4.12 shows the solution for the bar code scanner problem. Note that the tree diagram contains all possibilities, including both defective and acceptable bar code scanners. When new information is given, only the pertinent branches are selected and used. The joint probability values at the end of the appropriate branches are used to revise and compute the posterior possibilities. Using the total number of defective bar code scanners, 0.052 + 0.042 = 0.094, the calculation is as follows:

$$\text{Revised Probability: Cilico} = \frac{0.052}{0.094} = 0.553$$

$$\text{Revised Probability: AUM} = \frac{0.042}{0.094} = 0.447$$

DEMONSTRATION PROBLEM 4.12

Machines A, B, and C all produce the same two parts, X and Y. Of all the parts produced, machine A produces 60%, machine B produces 30%, and machine C produces 10%. In addition,

40% of the parts made by machine A are part X.

50% of the parts made by machine B are part X.

70% of the parts made by machine C are part X.

A part produced by this company is randomly sampled and is determined to be an X part. With the knowledge that it is an X part, revise the probabilities that the part came from machine A, B, or C.

Solution

The prior probability of the part coming from machine A is 0.60, because machine A produces 60% of all parts. The prior probability is 0.30 that the part came from B and 0.10 that it came from C. These prior probabilities are more pertinent if nothing is known about the part. However, the part is known to be an X part. The conditional probabilities show that different machines produce different proportions of X parts. For example, 0.40 of the parts made by machine A are X parts, but 0.50 of the parts made by machine B and 0.70 of the parts made by machine C are X parts. It makes sense that the probability of the part coming from machine C would increase and that the probability that the part was made on machine A would decrease because the part is an X part.

The following table shows how the prior probabilities, conditional probabilities, joint probabilities, and marginal probability, $P(X)$, can be used to revise the prior probabilities to obtain posterior probabilities.

Event	Prior $P(E_i)$	Conditional $P(X|E_i)$	Joint $P(X \cap E_i)$	Posterior
A	.60	.40	$(.60)(.40) = .24$	$\dfrac{.24}{.46} = .52$
B	.30	.50	.15	$\dfrac{.15}{.46} = .33$
C	.10	.70	$\dfrac{.07}{P(X) = .46}$	$\dfrac{.07}{.46} = .15$

After the probabilities are revised, it is apparent that the probability of the part being made at machine A decreased and that the probabilities that the part was made at machines B and C increased. A tree diagram presents another view of this problem.

$$\text{Revised Probabilities: Machine A: } \frac{0.24}{0.46} = 0.52$$

$$\text{Machine B: } \frac{0.15}{0.46} = 0.33$$

$$\text{Machine C: } \frac{0.07}{0.46} = 0.15$$

4.8 PROBLEMS

4.31 In a manufacturing plant, machine A produces 10% of a certain product, machine B produces 40% of this product, and machine C produces 50% of this product. Five per cent of machine A products are defective, 12% of machine B products are defective, and 8% of machine C products are defective. The company inspector has just sampled a product from this plant and has found it to be defective. Determine the revised probabilities that the sampled product was produced by machine A, machine B, or machine C.

4.32 Alex, Alicia, and Yaoqi fill orders in a fast-food restaurant. Alex incorrectly fills 20% of the orders he takes. Alicia incorrectly fills 12% of the orders she takes. Yaoqi incorrectly fills 5% of the orders he takes. Alex fills 30% of all orders, Alicia fills 45% of all orders, and Yaoqi fills 25% of all orders. An order has just been filled.

a. What is the probability that Alicia filled the order?

b. If the order was filled by Yaoqi, what is the probability that it was filled correctly?

c. Who filled the order is unknown, but the order was filled incorrectly. What are the revised probabilities that Alex, Alicia, or Yaoqi filled the order?

d. Who filled the order is unknown, but the order was filled correctly. What are the revised probabilities that Alex, Alicia, or Yaoqi filled the order?

4.33 In a small town, two lawn companies fertilize lawns during the summer. GreenThumb Service has 72% of the market. Thirty per cent of the lawns fertilized by GreenThumb could be rated as very healthy one month after service. LawnMaster has the other 28% of the market. Twenty per cent of the lawns fertilized by LawnMaster could be rated as very healthy one month after service. A lawn that has been treated with fertilizer by one of these companies within the past month is selected randomly. If the lawn is rated as very healthy, what are the revised probabilities that GreenThumb or LawnMaster treated the lawn?

4.34 Suppose 70% of all companies are classified as small companies and the rest as large companies. Suppose further, 82% of large companies provide training to employees, but only 18% of small companies provide training. A company is randomly selected without knowing if it is a large or small company; however, it is determined that the company provides training to employees. What are the prior probabilities that the company is a large company or a small company? What are the revised probabilities that the company is large or small? Based on your analysis, what is the overall percentage of companies that offer training?

Decision Dilemma SOLVED

Equality of the Sexes in the Workplace

The client company data given in the Decision Dilemma are displayed in a raw values matrix form. Using the techniques presented in this chapter, it is possible to statistically answer the managerial questions. If a worker is randomly selected from the 155 employees, the probability that the worker is a woman, $P(W)$, is 55/155, or 0.355. This marginal probability indicates that roughly 35.5% of all employees of the client company are women. Given that the employee has a managerial position, the probability that the employee is a woman, $P(W|M)$ is 3/11, or 0.273. The proportion of managers at the company who are women is lower than the proportion of all workers at the company who are women. Several factors might be related to this discrepancy, some of which may be defensible by the company – including experience, education, and prior history of success – and some may not.

Suppose a technical employee is randomly selected for a bonus. What is the probability that a female would be selected given that the worker is a technical employee? That is, $P(F|T) = ?$ Applying the law of conditional probabilities to the raw values matrix given in the Decision Dilemma, $P(F|T) = 17/69 = 0.246$. Using the concept of complementary events, the probability that a man is selected given that the employee is a technical person is $1 - 0.246 = 0.754$. It is more than three times as likely that

a randomly selected technical person is a male. If a woman were the one chosen for the bonus, a man could argue discrimination based on the mere probabilities. However, the company decision makers could then present documentation of the choice criteria based on productivity, technical suggestions, quality measures, and others.

Suppose a client company employee is randomly chosen to win a trip to Greece. The marginal probability that the winner is a professional is $P(P) = 44/155 = 0.284$. The probability that the winner is either a male or is a clerical worker, a union probability, is:

$$P(M \cup C) = P(M) + P(C) - P(M \cap C)$$

$$= \frac{100}{155} + \frac{31}{155} - \frac{9}{155} = \frac{122}{155} = 0.787$$

The probability of a male or clerical employee at the client company winning the trip is 0.787. The probability that the winner is a woman *and* a manager, a joint probability, is

$$P(F \cap M) = 3/155 = 0.019$$

There is less than a 2% chance that a female manager will be selected randomly as the trip winner.

(*continued*)

What is the probability that the winner is from the technical group if it is known that the employee is a male? This conditional probability is as follows:

$$P(T|M) = 52/100 = 0.52.$$

Many other questions can be answered about the client company's human resource situation using probabilities.

The probability approach to a human resource pool is a factual, numerical approach to people selection taken without regard to individual talents, skills, and worth to the company. Of course, in most instances, many other considerations go into the hiring, promoting, and rewarding of workers besides the random draw of their name. However, company management should be aware that attacks on hiring, promotion, and reward practices are sometimes made using statistical analyses such as those presented here. It is not being argued here that management should base decisions merely on the probabilities within particular categories. Nevertheless, being aware of the probabilities, management can proceed to undergird their decisions with documented evidence of worker productivity and worth to the organization.

ETHICAL CONSIDERATIONS

One of the potential misuses of probability occurs when subjective probabilities are used. Most subjective probabilities are based on a person's feelings, intuition, or experience. Almost everyone has an opinion on something and is willing to share it. Although such probabilities are not strictly unethical to report, they can be misleading and disastrous to other decision makers. In addition, subjective probabilities leave the door open for unscrupulous people to overemphasize their point of view by manipulating the probability.

The decision maker should remember that the laws and rules of probability are for the 'long run'. If a coin is tossed, even though the probability of getting a head is 0.5, the result will be either a head or a tail. It isn't possible to get a half head. The probability of getting a head (0.5) will probably work out in the long run, but in the short run an experiment might produce 10 tails in a row. Suppose the probability of striking oil on a geological formation is 0.10. This probability means that, in the long run, if the company drills enough holes on this type of formation, it should strike oil in about 10% of the holes. However, if the company has only enough money to drill one hole, it will either strike oil or have a dry hole. The probability figure of 0.10 may mean something different to the company that can afford to drill only one hole than to the company that can drill many hundreds. Classical probabilities could be used unethically to lure a company or client into a potential short-run investment with the expectation of getting at least something in return, when in actuality the investor will either win or lose. The oil company that drills only one hole will not get 10% back from the hole. It will either win or lose on the hole. Thus, classical probabilities open the door for unsubstantiated expectations, particularly in the short run.

SUMMARY

The study of probability addresses ways of assigning probabilities, types of probabilities, and laws of probabilities. Probabilities support the notion of inferential statistics. Using sample data to estimate and test hypotheses about population parameters is done with uncertainty. If samples are taken at random, probabilities can be assigned to outcomes of the inferential process.

Three methods of assigning probabilities are (1) the classical method, (2) the relative frequency of occurrence method, and (3) subjective probabilities. The classical method can assign probabilities a priori, or before the experiment takes place. It relies on the laws and rules of probability. The relative frequency of occurrence method assigns probabilities based on historical data or empirically derived data. Subjective probabilities are based on the feelings, knowledge, and experience of the person determining the probability.

Certain special types of events necessitate amendments to some of the laws of probability: mutually exclusive events and independent events. Mutually exclusive events are events that cannot occur at the same time, so the probability of their intersection is zero. With independent events, the occurrence of one has no impact or influence on the occurrence of the other. Certain experiments, such as those involving coins or dice, naturally produce independent events. Other experiments produce independent events when the experiment is conducted with replacement. If events are independent, the joint probability is computed by multiplying the marginal probabilities, which is a special case of the law of multiplication.

Three techniques for counting the possibilities in an experiment are the *mn* counting rule, the N^n possibilities, and combinations. The *mn* counting rule is used to determine how many total possible ways an experiment can occur in a series of sequential operations. The N^n formula is applied when sampling is being done with replacement or events are independent. Combinations are used to determine the possibilities when sampling is being done without replacement.

Four types of probability are marginal probability, conditional probability, joint probability, and union probability. The general law of addition is used to compute the probability of a union. The general law of multiplication is used to compute joint probabilities. The conditional law is used to compute conditional probabilities.

Bayes' rule is a method that can be used to revise probabilities when new information becomes available; it is a variation of the conditional law. Bayes' rule takes prior probabilities of events occurring and adjusts or revises those probabilities on the basis of information about what subsequently occurs.

KEY TERMS

a priori
Bayes' rule
classical method of assigning probabilities
collectively exhaustive events
combinations
complement
complement of a union
conditional probability
elementary events
event
experiment
independent events

intersection
joint probability
marginal probability
mn counting rule
mutually exclusive events
probability matrix
relative frequency of occurrence
sample space
set notation
subjective probability
union
union probability

GO ONLINE TO DISCOVER THE EXTRA FEATURES FOR THIS CHAPTER

The Student Study Guide containing solutions to the odd-numbered questions, additional Quizzes and Concept Review Activities, Excel and Minitab databases, additional data files in Excel and Minitab, and more worked examples.
www.wiley.com/college/cortinhas

FORMULAS

Counting rule

$$mn$$

Sampling with replacement

$$N^n$$

Sampling without replacement

$$_NC_n$$

Combination formula

$$_NC_n = \binom{N}{n} = \frac{N!}{n!(N-n)!}$$

General law of addition

$$P(X \cup Y) = P(X) + P(Y) - P(X \cap Y)$$

Special law of addition

$$P(X \cup Y) = P(X) + P(Y)$$

General law of multiplication

$$P(X \cap Y) = P(X) \cdot P(Y|X) = P(Y) \cdot (X|Y)$$

Special law of multiplication

$$P(X \cap Y) = P(X) \cdot P(Y)$$

Law of conditional probability

$$P(X|Y) = \frac{P(X \cap Y)}{P(Y)} = \frac{P(X) \cdot P(Y|X)}{P(Y)}$$

Bayes' rule

$$P(X_i|Y) = \frac{P(X_i) \cdot P(Y|X_i)}{P(X_1) \cdot P(Y|X_1) + P(X_2) \cdot P(Y|X_2) + \cdots + P(X_n) \cdot P(Y|X_n)}$$

SUPPLEMENTARY PROBLEMS

CALCULATING THE STATISTICS

4.35 Use the values in the contingency table to solve the equations given.

		D	E
	A	10	20
Variable 2	B	15	5
	C	30	15

a. $P(E) = $ _____
b. $P(B \cup D) = $ _____
c. $P(A \cap E) = $ _____
d. $P(B|E) = $ _____
e. $P(A \cup B) = $ _____
f. $P(B \cap C) = $ _____
g. $P(D|C) = $ _____
h. $P(A|B) = $ _____
i. Are variables 1 and 2 independent? Why or why not?

4.36 Use the values in the contingency table to solve the equations given.

	D	E	F	G
A	3	9	7	12
B	8	4	6	4
C	10	5	3	7

a. $P(F \cap A) = $ _____
b. $P(A|B) = $ _____
c. $P(B) = $ _____
d. $P(E \cap F) = $ _____
e. $P(D|B) = $ _____
f. $P(B|D) = $ _____
g. $P(D \cup C) = $ _____
h. $P(F) = $ _____

4.37 The following probability matrix contains a breakdown on the age and gender of European Professional Performers in a recent year, as reported by the International Federation of Actors (FIA).

		Age (years)							
		<20	30–35	36–40	41–50	51–60	61–70	>71	
Gender	Male	0.076	0.073	0.064	0.103	0.097	0.039	0.010	0.462
	Female	0.152	0.112	0.071	0.105	0.065	0.028	0.004	0.538
		0.229	0.185	0.135	0.208	0.162	0.067	0.014	1.000

a. What is the probability that one randomly selected performer is 36–40 years old?

b. What is the probability that one randomly selected performer is both a woman and 41–50 years old?

c. What is the probability that one randomly selected performer is a man or is 36–40 years old?

d. What is the probability that one randomly selected performer is less than 20 years old or 51–60 years old?

e. What is the probability that one randomly selected performer is a woman if she is 41–50 years old?

f. What is the probability that a randomly selected performer is neither a woman nor 51–60 years old?

TESTING YOUR UNDERSTANDING

4.38 Purchasing Survey asked purchasing professionals what sales traits impressed them most in a sales representative. Seventy-eight per cent selected 'thoroughness'. Forty per cent responded 'knowledge of your own product'. The purchasing professionals were allowed to list more than one trait. Suppose 27% of the purchasing professionals listed both 'thoroughness' and 'knowledge of your own product' as sales traits that impressed them most. A purchasing professional is randomly sampled.

a. What is the probability that the professional selected 'thoroughness' or 'knowledge of your own product'?

b. What is the probability that the professional selected neither 'thoroughness' nor 'knowledge of your own product'?

c. If it is known that the professional selected 'thoroughness', what is the probability that the professional selected 'knowledge of your own product'?

d. What is the probability that the professional did not select 'thoroughness' and did select 'knowledge of your own product'?

4.39 The Office for National Statistics publishes data on e-Society, the use of information and communication technologies by UK households. In 2010, the most common activity performed by men on the Internet was sending/receiving email (9 in 10 adults) while listening to web radio/watching web television was performed by 52% of adult males in the UK. Suppose 50% of adult males used the Internet to send/receive emails and listen to web radio/watch web television.

If an adult male is randomly selected, determine the following probabilities:

a. The adult male sent/received email given that he listened to web radio/watched web television.

b. The adult male listened to web radio/watched web television given that he sent/received email.

c. The adult male listened to web radio/watched web television or sent/received email.

d. The adult male sent/received email and he did not listen to web radio/watch web television.

e. The adult male does not listen to web radio/ watch web television if he sends/receives email.

4.40 The following probability matrix contains the UK national distribution of blood groups as reported by the National Blood Service. A UK citizen is randomly selected.

		Blood type (ABO blood system)				
		O	A	B	AB	
RhD factor	Positive	.37	.35	.08	.03	.83
	Negative	.07	.07	.02	.01	.17
		.44	.42	.1	.04	1

a. What is the probability that the person's blood is type A and negative?

b. What is the probability that the person's blood is not type O and negative?

c. Suppose the chosen person is known to be negative. What is the probability that the person's blood is type A?

d. Suppose the chosen person's blood is known not to be negative. What is the probability that the person's blood is not type O?

e. Suppose the chosen person's blood is known to be negative. What is the probability that the person's blood is neither type A nor type O?

4.41 In a certain city, 30% of the families have a MasterCard, 20% have an American Express card, and 25% have a Visa card. Eight per cent of the families have both a MasterCard and an American Express card. Twelve per cent have both a Visa card and a MasterCard. Six per cent have both an American Express card and a Visa card.

a. What is the probability of selecting a family that has either a Visa card or an American Express card?

b. If a family has a MasterCard, what is the probability that it has a Visa card?

c. If a family has a Visa card, what is the probability that it has a MasterCard?

d. Is possession of a Visa card independent of possession of a MasterCard? Why or why not?

e. Is possession of an American Express card mutually exclusive of possession of a Visa card?

4.42 A few years ago, a survey commissioned by *The World Almanac* and *Maturity News* Service reported that 51% of the respondents did not believe the Social Security system will be secure in 20 years. Of the respondents who were age 45 or older, 70% believed the system will be secure in 20 years. Of the people surveyed, 57% were under age 45. One respondent is selected randomly.

a. What is the probability that the person is age 45 or older?

b. What is the probability that the person is younger than age 45 and believes that the Social Security system will be secure in 20 years?

c. If the person selected believes the Social Security system will be secure in 20 years, what is the probability that the person is 45 years old or older?

d. What is the probability that the person is younger than age 45 or believes the Social Security system will not be secure in 20 years?

4.43 A recent survey conducted by Rubicon Consulting found that 47% of iPhone users changed carriers just to use Apple's smartphone. The survey also found that 9.8% of iPhone users also use a BlackBerry. In addition, ChangeWave Research reported iPhone's market share to be 30% and BlackBerry's 40%. An iPhone user is selected randomly.

a. What is the probability that this person owns both types of smartphones?

b. What is the probability that this person owns an iPhone or a BlackBerry?

c. What is the probability that this person neither owns an iPhone nor a BlackBerry?

d. What is the probability that this person owns an iPhone and does not own a BlackBerry?

4.44 A survey by Tripadvisor studied the types of activities British people favour while on holidays in a recent year. Among other things, 64% of British people reported preferring a beach holiday and 62% reported getting in touch with work. Respondents to the study were allowed to select more than one activity. Suppose that of those who took a beach holiday, 78% checked in with work. One of these survey respondents is selected randomly.

a. What is the probability that while on holiday this respondent checked in with work and took a beach holiday?

b. What is the probability that while on holiday this respondent neither took a beach holiday nor checked in with work?

c. What is the probability that while on holiday this respondent took a beach holiday given that the respondent checked in with work?

d. What is the probability that while on holiday this respondent did not check in with work given that the respondent took a beach holiday?

e. What is the probability that while on holiday this respondent did not check in with work given that the respondent did not take a beach holiday?

f. Construct a probability matrix for this problem.

4.45 A study on ethics in the workplace by the Ethics Resource Center and Kronos, Inc., revealed that 35% of employees admit to keeping quiet when they see co-worker misconduct. Suppose 75% of employees who admit to keeping quiet when they see co-worker misconduct call in sick when they are well. In addition, suppose that 40% of the employees who call in sick when they are well admit to keeping quiet when they see co-worker misconduct. If an employee is randomly selected, determine the following probabilities:

a. The employee calls in sick when well and admits to keeping quiet when seeing co-worker misconduct.

b. The employee admits to keeping quiet when seeing co-worker misconduct or calls in sick when well.

c. Given that the employee calls in sick when well, he or she does not keep quiet when seeing co-worker misconduct.

d. The employee neither keeps quiet when seeing co-worker misconduct nor calls in sick when well.

e. The employee admits to keeping quiet when seeing co-worker misconduct and does not call in sick when well.

4.46 The British Broadcasting Corporation has recently published data on the number of complaints against police officers for a recent year from 43 out of 52 UK police forces. The number one complaint was incivility, impoliteness, and intolerance (18%), followed by other assault (17%). Several other complaints were noted, including other neglect/failure in duty (16%), unlawful/unnecessary arrest (6%), oppressive conduct (6%), and lack of fairness or impartiality (5%). These complaint categories are mutually exclusive. Assume that the results of this survey can be inferred to all citizens in the UK. If a UK citizen is randomly selected, determine the following probabilities:

a. The complaints about other neglect/failure in duty or oppressive conduct.

b. The complaints about incivility, impoliteness and intolerance, and other assault.

c. The complaints about lack of fairness or impartiality given that the citizen complained about other assault.

d. The citizen does not complain about other neglect/failure in duty nor does he complain about unlawful/unnecessary arrest.

4.47 Companies use employee training for various reasons, including employee loyalty, certification, quality, and process improvement. In a national survey of companies, BI Learning Systems reported that 56% of the responding companies named employee retention as a top reason for training. Suppose 36% of the companies replied that they use training for process improvement and for employee retention. In addition, suppose that of the companies that use training for process improvement, 90% use training for employee retention. A company that uses training is randomly selected.

a. What is the probability that the company uses training for employee retention and not for process improvement?

b. If it is known that the company uses training for employee retention, what is the probability that it uses training for process improvement?

c. What is the probability that the company uses training for process improvement?

d. What is the probability that the company uses training for employee retention or process improvement?

e. What is the probability that the company neither uses training for employee retention nor uses training for process improvement?

f. Suppose it is known that the company does not use training for process improvement. What is the probability that the company does use training for employee retention?

4.48 According to Eurostat, 6.22% of French nationals are aged 25–29 years old. The activity that people in this group spend more time on is work, followed by leisure. For 25- to 29-year-olds 19.3% of daily time use is dedicated to work. Suppose that 67.6% of the daily time of French nationals in the 25- to 29-year-old group is dedicated neither to work nor leisure. If one of these people is selected randomly and asked

about his time use in a randomly selected period of the day, determine the following probabilities:

a. The person used his time to work and not for leisure.

b. The person used his time for leisure.

c. The person used his time for leisure given that the person worked.

d. The person did not work given that the person did not use his time for leisure.

4.49 In a study of incentives used by companies to retain mature workers by The Conference Board, it was reported that 41% use flexible work arrangements. Suppose that of those companies that do not use flexible work arrangements, 10% give time off for volunteerism. In addition, suppose that of those companies that use flexible work arrangements, 60% give time off for volunteerism. If a company is randomly selected, determine the following probabilities:

a. The company uses flexible work arrangements or gives time off for volunteerism.

b. The company uses flexible work arrangements and does not give time off for volunteerism.

c. Given that the company does not give time off for volunteerism, the company uses flexible work arrangements.

d. The company does not use flexible work arrangements given that the company does give time off for volunteerism.

e. The company does not use flexible work arrangements or the company does not give time off for volunteerism.

4.50 A small independent physicians' practice has three doctors. Dr Sarabia sees 41% of the patients, Dr Tran sees 32%, and Dr Jackson sees the rest. Dr Sarabia requests blood tests on 5% of her patients, Dr Tran requests blood tests on 8% of his patients, and Dr Jackson requests blood tests on 6% of her patients. An auditor randomly selects a patient from the past week and discovers that the patient had a blood test as a result of the physician visit. Knowing this information, what is the probability that the patient saw Dr Sarabia? For what percentage of all patients at this practice are blood tests requested?

4.51 A survey by the Arthur Andersen Enterprise Group/ National Small Business United attempted to determine what the leading challenges are for the growth and survival of small businesses. Although the economy and finding qualified workers were the leading challenges, several others were listed in the results of the study, including regulations, listed by 30% of the companies, and the tax burden, listed by 35%. Suppose that 71% of the companies listing regulations as a challenge listed the tax burden as a challenge. Assume these percentages hold for all small businesses. If a small business is randomly selected, determine the following probabilities:

a. The small business lists both the tax burden and regulations as a challenge.

b. The small business lists either the tax burden or regulations as a challenge.

c. The small business lists either the tax burden or regulations but not both as a challenge.

d. The small business lists regulations as a challenge given that it lists the tax burden as a challenge.

e. The small business does not list regulations as a challenge given that it lists the tax burden as a challenge.

f. The small business does not list regulations as a challenge given that it does not list the tax burden as a challenge.

4.52 According to US Census Bureau figures, 35.3% of all Americans are in the 0–24 age category, 14.2% are age 25–34, 16.0% are in the 35–44 age category, and 34.5% are 45 or older. A study by Jupiter Media Metrix determined that Americans use their leisure time in different ways according to age. For example, of those who are 45 or older, 39% read a book or a magazine for more than 10 hours per week. Of those who are 24 or under, only 11% read a book or a magazine for more than 10 hours per week. The percentage figures for reading a book or a magazine for more than 10 hours per week are 24% for the 25–34 age bracket and 27% the 35–44 age bracket. Suppose an American is randomly selected and it is determined that he or she reads a book or a magazine for more than 10 hours per week. Revise the probabilities that he or she is in any given age category. Using these figures, what is the overall percentage of the US population that reads a book or a magazine for more than 10 hours per week?

4.53 A retail study by Deloitte revealed that 54% of adults surveyed believed that plastic, non-compostable shopping bags should be banned. Suppose 41% of adults regularly recycle aluminium cans and believe that plastic, non-compostable shopping bags should be banned. In addition, suppose that 60% of adults who do not believe that plastic, non-compostable shopping bags should be banned do recycle. If an adult is randomly selected:

a. What is the probability that the adult recycles and does not believe that plastic, non-compostable shopping bags should be banned?

b. What is the probability that the adult does recycle?

c. What is the probability that the adult does recycle or does believe that plastic, non-compostable shopping bags should be banned?

d. What is the probability that the adult does not recycle or does not believe that plastic, non-compostable shopping bags should be banned?

e. What is the probability that the adult does not believe that plastic, non-compostable shopping bags should be banned given that the adult does recycle?

ANALYSING THE DATABASES

1. In the Manufacturing database, what is the probability that a randomly selected SIC Code industry is in industry group 13? What is the probability that a randomly selected SIC Code industry has a value of industry shipments of 4 (see Chapter 1 for coding)? What is the probability that a randomly selected SIC Code industry is in industry group 13 and has a value of industry shipments of 2? What is the probability that a randomly selected SIC Code industry is in industry group 13 or has a value of industry shipments of 2? What is the probability that a randomly selected SIC Code industry neither is in industry group 13 nor has a value of industry shipments of 2?

2. Use the Hospital database. Construct a raw values matrix for region and for type of control. You should have a 7 × 4 matrix. Using this matrix, answer the following questions: (refer to Chapter 1 for category members). What is the probability that a randomly selected hospital is in the Midwest if the hospital is known to be for-profit? If the hospital is known to be in the South, what is the probability that it is a government, non-federal hospital? What is the probability that a hospital is in the Rocky Mountain region or a not-for-profit, non-government hospital? What is the probability that a hospital is a for-profit hospital located in California?

CASE

COLGATE-PALMOLIVE MAKES A 'TOTAL' EFFORT

In the mid-1990s, Colgate-Palmolive developed a new toothpaste for the US market, Colgate Total, with an antibacterial ingredient that was already being successfully sold overseas. However, the word *antibacterial* was not allowed for such products by the Food and Drug Administration rules. So Colgate-Palmolive had to come up with another way of marketing this and other features of their new toothpaste to US consumers. Market researchers told Colgate-Palmolive that consumers were weary of trying to discern among the different advantages of various toothpaste brands and wanted simplification in their shopping lives. In response, the name 'Total' was given to the product in the United States: the one word would convey that the toothpaste is the 'total' package of various benefits.

Young & Rubicam developed several commercials illustrating Total's benefits and tested the commercials with focus groups. One commercial touting Total's long-lasting benefits was particularly successful. Meanwhile, in 1997, Colgate-Palmolive received FDA approval for Total, five years after the company had applied for it. The product was launched in the United States in January of 1998 using commercials that were designed from the more successful ideas of the focus group tests. Total was introduced with a $100 million advertising campaign. Ten months later, 21% of all United States households had purchased Total for the first time. During this same time period, 43% of those who initially tried Total purchased it again. A year after its release, Total was the number one toothpaste in the United States. Total is advertised as not just a toothpaste but as a protective shield that protects you for a full range of oral health problems for up to 12 hours. Total is now offered in a variety of forms, including Colgate Total Advanced Whitening, Colgate Total Advanced Clean, Colgate Total Advanced Fresh Gel, Colgate Total Clean Mint Paste, Colgate Total Whitening Paste, Colgate Total Whitening Gel, Colgate Total Plus Whitening Liquid, and Colgate Total Mint Stripe Gel. In the United States, market share for Colgate Total toothpaste was 16.2% in the second quarter of 2008, which was its highest quarterly share ever.

DISCUSSION

1. What probabilities are given in this case? Use these probabilities and the probability laws to determine what percentage of US households purchased Total at least twice in the first 10 months of its release.

2. Is age category independent of willingness to try new products? According to the US Census Bureau, approximately 20% of all Americans are in the 45–64 age category. Suppose 24% of the consumers who purchased Total for the first time during the initial 10-month period were in the 45–64 age category. Use this information to determine whether age is independent of the initial purchase of Total during the introductory time period. Explain your answer.

3. Using the probabilities given in Question 2, calculate the probability that a randomly selected US consumer is either in the 45–64 age category or purchased Total during the initial 10-month period. What is the probability that a randomly selected person purchased Total in the first 10 months given that the person is in the 45–64 age category?

4. Suppose 32% of all toothpaste consumers in the United States saw the Total commercials. Of those who saw the commercials, 40% purchased Total at least once in the first 10 months of its introduction. Of those who did not see the commercials, 12.06% purchased Total at least once in the first 10 months of its introduction. Suppose a toothpaste consumer is randomly selected and it is learned that they purchased Total during the first 10 months of its introduction. Revise the probability that this person saw the Total commercials and the probability that the person did not see the Total commercials.

Source: Colgate-Palmolive's home page at www.colgate.com/app/Colgate/US/HomePage.cvsp, Total's homepage at www.colgate.com/app/ColgateTotal/US/EN/Products.cvsp, and at answers.com, available at www.answers.com/topic/colgate-palmolive-company, 2008.

Distributions and Sampling

UNIT II OF THIS TEXTBOOK INTRODUCES YOU TO THE CONCEPT of statistical distribution. In lay terms, a statistical distribution is a numerical or graphical depiction of frequency counts or probabilities for various values of a variable that can occur. Distributions are important because most of the analyses done in business statistics are based on the characteristics of a particular distribution. In Unit II, you will study eight distributions: six population distributions and two sampling distributions.

Six population distributions are presented in Chapters 5 and 6. These population distributions can be categorized as discrete distributions or continuous distributions. Discrete distributions are introduced in Chapter 5, and they include the binomial distribution, the Poisson distribution, and the hypergeometric distribution. Continuous distributions are presented in Chapter 6, and they include the uniform distribution, the normal distribution, and the exponential distribution. Information about sampling is discussed in Chapter 7 along with two sampling distributions, the sampling distribution of $\bar{x}$ and the sampling distribution of $\hat{p}$. Three more population distributions are introduced later in the text in Unit III. These include the t distribution and the chi-square distribution in Chapter 8 and the F distribution in Chapter 10.

CHAPTER 5

Discrete Distributions

LEARNING OBJECTIVES

The overall learning objective of Chapter 5 is to help you understand a category of probability distributions that produces only discrete outcomes, thereby enabling you to:

1. Define a random variable in order to differentiate between a discrete distribution and a continuous distribution
2. Determine the mean, variance, and standard deviation of a discrete distribution
3. Solve problems involving the binomial distribution using the binomial formula and the binomial table
4. Solve problems involving the Poisson distribution using the Poisson formula and the Poisson table
5. Solve problems involving the hypergeometric distribution using the hypergeometric formula

Decision Dilemma

Life with a Mobile Phone

As early as 1947, scientists understood the basic concept of a mobile phone as a type of two-way radio. Seeing the potential of crude mobile car phones, researchers understood that by using a small range of service areas (mobiles) with frequency reuse, they could increase the capacity for mobile phone usage significantly even though the technology was not then available. During that same year, AT&T proposed the allocation of a large number of radio-spectrum frequencies by the Federal Communications Commission (FCC) that would thereby make widespread mobile phone service feasible. At the same time, the FCC decided to limit the amount of frequency capacity available such that only 23 phone conversations could take place simultaneously. In 1968, the FCC reconsidered its position and freed the airwaves for more phones. About this time, AT&T and Bell Labs proposed to the FCC a system in which they would construct a series of many small, low-powered broadcast towers, each of which would broadcast to a 'mobile' covering a few miles. Taken as a whole, such 'cells' could be used to pass phone calls from mobile to mobile, thereby reaching a large area.

The first company to actually produce a mobile phone was Motorola, and Dr Martin Cooper, then of Motorola and considered the inventor of the first modern portable handset, made his first call on the portable mobile phone in 1973. By 1977, AT&T and Bell Labs had developed a prototype cellular phone system that was tested in Chicago by 2,000 trial customers. After the first commercial mobile phone system began operation in Japan by NTT in 1979, and Motorola and American Radio developed a second US cell system in 1981, the FCC authorized commercial cellular service in the United States in 1982. By 1987, mobile phone subscribers had exceeded 1 million customers in the United States, and as frequencies were getting crowded, the FCC authorized alternative cellular technologies, opening up new opportunities for development. Since that time, researchers have developed a number of advances that have increased capacity exponentially. Nowadays, the world's largest individual mobile operator is China Mobile with over 500 million mobile phone subscribers.

Today in the United States, over 14% of mobile phone owners use only cellular phones, and the trend is rising. According to a Harris Poll of 9132 surveyed adults, 89% of adults have a mobile phone. In an Associated Press/America Online Pew Poll of 1,200 mobile phone users, it was discovered that two-thirds of all mobile phone users said that it would be hard to give up their mobile phones, and 26% responded that they cannot imagine life without their mobile phones. In spite of Americans' growing dependence on their mobile phones, not everyone is happy about their usage. Almost 9 out of 10 mobile users encounter others using their phones in an annoying way. In addition, 28% claim that sometimes they do not drive as safely as they should because they are using mobile phones. Now, there are multiple uses for the mobile phone, including picture taking, text messaging, game playing, and others. According to the study, two-thirds of mobile phone owners in the 18 to 29 age bracket sent text messages using their mobile phones, 55% take pictures with their phones, 47% play games on the phones, and 28% use the Internet through their mobile phones.

Managerial and Statistical Questions

1. One study reports that 14% of mobile phone owners in the United States use only cellular phones (no land line). Suppose you randomly select 20 Americans, what is the probability that more than 7 of the sample use only mobile phones?

2. Another study reports that 9 out of 10 mobile users worldwide encounter others using their phones in an annoying way. Based on this, if you were to randomly select 25 mobile phone users, what is the probability that fewer than 20 report that they encounter others using their phones in an annoying way?

3. Suppose a survey of mobile phone users shows that, on average, a mobile phone user receives 3.6 calls per day. If this figure is true, what is the probability that a mobile phone user receives no calls in a day? What is the probability that a mobile phone user receives five or more calls in a day?

Sources: Mary Bellis, 'Selling the Mobile Phone, Part 1: History of Cellular Phones', in *About Business & Finance*, an America Online site, Selling the Mobile Phone – History of Cellular Phones, at http://inventors.about .com/library/weekly/aa070899.htm; *USA Today* Tech, 'For Many, Their Mobile Phone Has Become Their Only Phone', at www.usatoday.com/tech/ news/2003-03-24-mobile-phones x.htm; and Will Lester, 'A Love-Hate Relationship', *Houston Chronicle*, 4 April 2006, p. D4; www.harrisinteractive .com/harris_poll/index.asp?PID = 890.

In statistical experiments involving chance, outcomes occur randomly. As an example of such an experiment, a battery manufacturer randomly selects three batteries from a large batch of batteries to be tested for quality. Each selected battery is to be rated as good or defective. The batteries are numbered from 1 to 3, a defective battery is designated with a D, and a good battery is designated with a G. All possible outcomes are shown in Table 5.1. The expression $D_1 G_2 D_3$ denotes one particular outcome in which the first and third batteries are defective and the second battery is good. In this chapter, we examine the probabilities of events occurring in experiments that produce discrete distributions. In particular, we will study the binomial distribution, the Poisson distribution, and the hypergeometric distribution.

TABLE 5.1

All Possible Outcomes for the Battery Experiment

G_1	G_2	G_3
D_1	G_2	G_3
G_1	D_2	G_3
G_1	G_2	D_3
D_1	D_2	G_3
D_1	G_2	D_3
G_1	D_2	D_3
D_1	D_2	D_3

5.1   DISCRETE VERSUS CONTINUOUS DISTRIBUTIONS

A **random variable** is *a variable that contains the outcomes of a chance experiment.* For example, suppose an experiment is to measure the arrivals of automobiles at the London congestion charge zone during a 30-second period. The possible outcomes are: 0 cars, 1 car, 2 cars, ..., n cars. These numbers $(0, 1, 2, ..., n)$ are the values of a random variable. Suppose another experiment is to measure the time between the completion of two tasks in a production line. The values will range from 0 seconds to n seconds. These time measurements are the values of another random variable. The two categories of random variables are (1) discrete random variables and (2) continuous random variables.

A random variable is a **discrete random variable** *if the set of all possible values is at most a finite or a countable infinite number of possible values.* In most statistical situations, discrete random variables produce values that are nonnegative whole numbers. For example, if six people are randomly selected from a population and how many of the six are left-handed is to be determined, the random variable produced is discrete. The only possible numbers of left-handed people in the sample of six are 0, 1, 2, 3, 4, 5, and 6. There cannot be 2.75 left-handed people in a group of six people; obtaining non-whole number values is impossible. Other examples of experiments that yield discrete random variables include the following:

1. Randomly selecting 25 people who consume soft drinks and determining how many people prefer diet soft drinks
2. Determining the number of defects in a batch of 50 items
3. Counting the number of people who arrive at a store during a five-minute period
4. Sampling 100 registered voters and determining how many voted for the Prime Minister in the last election

The battery experiment described at the beginning of the chapter produces a distribution that has discrete outcomes. Any one trial of the experiment will contain 0, 1, 2, or 3 defective batteries. It is not possible to get 1.58 defective batteries. It could be said that discrete random variables are usually generated from experiments in which things are 'counted' not 'measured'.

Continuous random variables *take on values at every point over a given interval.* Thus continuous random variables have no gaps or unassumed values. It could be said that continuous random variables are generated from experiments in which things are 'measured' not 'counted'. For example, if a person is assembling a product component, the time it takes to accomplish that feat could be any value within a reasonable range such as 3 minutes 36.4218 seconds or 5 minutes 17.5169 seconds. A list of measures for which continuous random variables might be generated would include time, height, weight, and volume. Other examples of experiments that yield continuous random variables include the following:

1. Sampling the volume of liquid nitrogen in a storage tank
2. Measuring the time between customer arrivals at a retail outlet
3. Measuring the lengths of newly designed automobiles
4. Measuring the weight of grain in a grain elevator at different points of time

Once continuous data are measured and recorded, they become discrete data because the data are rounded off to a discrete number. Thus in actual practice, virtually all business data are discrete. However, for practical reasons, data analysis is facilitated greatly by using continuous distributions on data that were continuous originally.

The outcomes for random variables and their associated probabilities can be organized into distributions. The two types of distributions are **discrete distributions**, *constructed from discrete random variables,* and **continuous distributions**, *based on continuous random variables.*

In this chapter, three discrete distributions are presented:

1. binomial distribution
2. Poisson distribution
3. hypergeometric distribution

In addition, six continuous distributions are discussed later in this text:

1. uniform distribution
2. normal distribution
3. exponential distribution
4. t distribution
5. chi-square distribution
6. F distribution

Discrete

Continuous

5.2 DESCRIBING A DISCRETE DISTRIBUTION

TABLE 5.2

Discrete Distribution of Occurrence of Daily Crises

Number of Crises	Probability
0	.37
1	.31
2	.18
3	.09
4	.04
5	.01

How can we describe a discrete distribution? One way is to construct a graph of the distribution and study the graph. The histogram is probably the most common graphical way to depict a discrete distribution.

Observe the discrete distribution in Table 5.2. An executive is considering out-of-town business travel for a given Friday. She recognizes that at least one crisis could occur on the day that she is gone and she is concerned about that possibility. Table 5.2 shows a discrete distribution that contains the number of crises that could occur during the day that she is gone and the probability that each number will occur. For example, there is a 0.37 probability that no crisis will occur, a 0.31 probability of one crisis, and so on. The histogram in Figure 5.1 depicts the distribution given in Table 5.2. Notice that the x-axis of the histogram contains the possible outcomes of the experiment (number of crises that might occur) and that the y-axis contains the probabilities of these occurring.

It is readily apparent from studying the graph of Figure 5.1 that the most likely number of crises is 0 or 1. In addition, we can see that the distribution is discrete in that no probabilities are shown for values in between the whole-number crises.

Mean, Variance, and Standard Deviation of Discrete Distributions

What additional mechanisms can be used to describe discrete distributions besides depicting them graphically? The measures of central tendency and measures of variability discussed in Chapter 3 for grouped data can be applied to discrete distributions to compute a mean, a variance, and a standard deviation. Each of those three descriptive measures (mean, variance, and standard

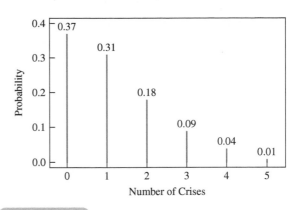

FIGURE 5.1

Minitab Histogram of Discrete Distribution of Crises Data

deviation) is computed on grouped data by using the class midpoint as the value to represent the data in the class interval. With discrete distributions, using the class midpoint is not necessary because the discrete value of an outcome (0, 1, 2, 3, . . .) is used to represent itself. Thus, instead of using the value of the class midpoint (M) in computing these descriptive measures for grouped data, the discrete experiment's outcomes (x) are used. In computing these descriptive measures on grouped data, the frequency of each class interval is used to weight the class midpoint. With discrete distribution analysis, the probability of each occurrence is used as the weight.

Mean or Expected Value

The **mean** or **expected value** of a discrete distribution is *the long-run average of occurrences*. We must realize that any one trial using a discrete random variable yields only one outcome. However, if the process is repeated long enough, the average of the outcomes are most likely to approach a long-run average, expected value, or mean value. This mean, or expected, value is computed as follows.

MEAN OR EXPECTED VALUE OF A DISCRETE DISTRIBUTION	$$\mu = E(x) = \Sigma[x \cdot P(x)]$$ where: $E(x) =$ long-run average $x =$ an outcome $P(x) =$ probability of that outcome

As an example, let's compute the mean or expected value of the distribution given in Table 5.2. See Table 5.3 for the resulting values. In the long run, the mean or expected number of crises on a given Friday for this executive is 1.15 crises. Of course, the executive will never have 1.15 crises.

Variance and Standard Deviation of a Discrete Distribution

The variance and standard deviation of a discrete distribution are solved for by using the outcomes (x) and probabilities of outcomes [$P(x)$] in a manner similar to that of computing a mean. In addition, the computations for

TABLE 5.3

Computing the Mean of the Crises Data

x	$P(x)$	$x \cdot P(x)$
0	.37	.00
1	.31	.31
2	.18	.36
3	.09	.27
4	.04	.16
5	.01	.05
	$\Sigma[x \cdot P(x)] = 1.15$	
	$\mu = 1.15$ crises	

TABLE 5.4

Calculation of Variance and Standard Deviation on Crises Data

x	$P(x)$	$(x - \mu)^2$	$(x - \mu)^2 \cdot P(x)$
0	.37	$(0-1.15)^2 = 1.32$	$(1.32)(.37) = .49$
1	.31	$(1-1.15)^2 = .02$	$(0.02)(.31) = .01$
2	.18	$(2-1.15)^2 = .72$	$(0.72)(.18) = .13$
3	.09	$(3-1.15)^2 = 3.42$	$(3.42)(.09) = .31$
4	.04	$(4-1.15)^2 = 8.12$	$(8.12)(.04) = .32$
5	.01	$(5-1.15)^2 = 14.82$	$(14.82)(.01) = .15$
			$\Sigma[(x - \mu)^2 \cdot P(x)] = 1.41$

The variance of $\sigma^2 = \Sigma[(x - \mu)^2 \cdot P(x)] = 1.41$

The standard deviation is $\sigma = \sqrt{1.41} = 1.19$ crises.

variance and standard deviations use the mean of the discrete distribution. The formula for computing the variance follows.

VARIANCE OF A DISCRETE DISTRIBUTION	$$\sigma^2 = \Sigma[(x - \mu)^2 \cdot P(x)]$$ where: x = an outcome $P(x)$ = probability of a given outcome μ = mean

The standard deviation is then computed by taking the square root of the variance.

STANDARD DEVIATION OF A DISCRETE DISTRIBUTION	$$\sigma = \sqrt{\Sigma[(x - \mu)^2 \cdot P(x)]}$$

The variance and standard deviation of the crisis data in Table 5.2 are calculated and shown in Table 5.4. The mean of the crisis data is 1.15 crises. The standard deviation is 1.19 crises, and the variance is 1.41.

DEMONSTRATION PROBLEM 5.1

Recently, the Loterie Nationale of Belgium launched a scratch card game called Quick Cash. In this game, total instant winnings of €3.16 million were available on 2.6 million €2 tickets, with ticket prizes ranging from €2 to €20,000. Shown here are the various prizes and the probability of winning each prize. Use these data to compute the expected value of the game, the variance of the game, and the standard deviation of the game.

Prize (x)	Probability P(x)
€20,000	.00000154
€2,000	.00000308
€200	.00000769
€20	.03846154
€4	.03461538
€2	.13461538
€0	.79229538

Solution

The mean is computed as follows.

Prize (x)	Probability P(x)	x · P(x)
€20,000	.00000154	.03077
€2,000	.00000308	.00615
€200	.00000769	.00154
€20	.03846154	.76923
€4	.03461538	.13846
€2	.13461538	.26923
€0	.79229538	.00000
		$\Sigma[x \cdot P(x)] = 1.21538$
	$\mu = E(x) = \Sigma[x \cdot P(x)] = 1.21538$	

The expected payoff for a €2 ticket in this game is 1.21 cents. If a person plays the game for a long time, he or she could expect to average about 1.20 cents in winnings. In the long run, the participant will lose about €2.00 − 1.21 = 0.79 or about 80 cents a game. Of course, an individual will never win €1.20 in any one game.

Using this mean, $\mu = 1.21538$ the variance and standard deviation can be computed as follows.

x	P(x)	$(x - \mu)^2$	$(x - \mu)^2 \cdot P(x)$
€20,000	.00000154	399951386.09254	615.30982
€2,000	.00000308	3995139.93870	12.29274
€200	.00000769	39515.32331	0.30396
€20	.03846154	352.86178	13.57161
€4	.03461538	7.75408	0.26841
€2	.13461538	0.61562	0.08287
€0	.79229538	1.47716	1.17035
			$\Sigma[(x - \mu)^2 \cdot P(x)] = 642.99976$

$$\sigma^2 = \Sigma[(x - \mu)^2 \cdot P(x)] = 642.99976$$

$$\sigma = \sqrt{\sigma^2} = \sqrt{\Sigma[(x - \sigma)^2 \cdot P(x)]} = \sqrt{642.9976} = 25.35744$$

The variance is 642.99976 (euro)^2 and the standard deviation is €25.36.

5.2   PROBLEMS

5.1 Determine the mean, the variance, and the standard deviation of the following discrete distribution:

x	P(x)
1	.238
2	.290
3	.177
4	.158
5	.137

5.2   Determine the mean, the variance, and the standard deviation of the following discrete distribution:

x	P(x)
0	.103
1	.118
2	.246
3	.229
4	.138
5	.094
6	.071
7	.001

5.3   The data below are the result of a historical study of the number of flaws found in a porcelain cup produced by a manufacturing firm. Use these data and the associated probabilities to compute the expected number of flaws and the standard deviation of flaws.

Flaws	Probability
0	.461
1	.285
2	.129
3	.087
4	.038

5.4   Suppose 20% of the people in a city prefer Pepsi-Cola as their soft drink of choice. If a random sample of six people is chosen, the number of Pepsi drinkers could range from zero to six. Shown here are the possible numbers of Pepsi drinkers in a sample of six people and the probability of that number of Pepsi drinkers occurring in the sample. Use the data to determine the mean number of Pepsi drinkers in a sample of six people in the city, and compute the standard deviation.

Number of Pepsi Drinkers	Probability
0	.262
1	.393
2	.246
3	.082
4	.015
5	.002
6	.000

5.3   BINOMIAL DISTRIBUTION

Perhaps the most widely known of all discrete distributions is the **binomial distribution**. The binomial distribution has been used for hundreds of years. Several assumptions underlie the use of the binomial distribution:

ASSUMPTIONS OF THE BINOMIAL DISTRIBUTION

- The experiment involves n identical trials.
- Each trial has only two possible outcomes denoted as success or as failure.
- Each trial is independent of the previous trials.
- The terms p and q remain constant throughout the experiment, where the term p is the probability of getting a success on any one trial and the term $q = (1-p)$ is the probability of getting a failure on any one trial.

As the word *binomial* indicates, any single trial of a binomial experiment contains only two possible outcomes. These two outcomes are labelled *success* or *failure*. Usually the outcome of interest to the researcher is labelled a success. For example, if a quality control analyst is looking for defective products, he would consider finding a defective product a success even though the company would not consider a defective product a success. If researchers are studying left-handedness, the outcome of getting a left-handed person in a trial of an experiment is a success. The other possible outcome of a trial in a binomial experiment is called a failure. The word *failure* is used only in opposition to success. In the preceding experiments, a failure could be to get an acceptable part (as opposed to a defective part) or to get a right-handed person (as opposed to a left-handed person). In a binomial distribution experiment, any one trial can have only two possible, mutually exclusive outcomes (right-handed/left-handed, defective/good, male/female, etc.).

The binomial distribution is a discrete distribution. In n trials, only x successes are possible, where x is a whole number between 0 and n. For example, if five parts are randomly selected from a batch of parts, only 0, 1, 2, 3, 4, or 5 defective parts are possible in that sample. In a sample of five parts, getting 2.714 defective parts is not possible, nor is getting eight defective parts possible.

In a binomial experiment, the trials must be independent. This constraint means that either the experiment is by nature one that produces independent trials (such as tossing coins or rolling dice) or the experiment is conducted with replacement. The effect of the independent trial requirement is that p, the probability of getting a success on one trial, remains constant from trial to trial. For example, suppose 5% of all parts in a bin are defective. The probability of drawing a defective part on the first draw is $p = 0.05$. If the first part drawn is not replaced, the second draw is not independent of the first, and the p value will change for the next draw. The binomial distribution does not allow for p to change from trial to trial within an experiment. However, if the population is large in comparison with the sample size, the effect of sampling without replacement is minimal, and the independence assumption essentially is met, that is, p remains relatively constant.

Generally, if the sample size, n, is less than 5% of the population, the independence assumption is not of great concern. Therefore the acceptable sample size for using the binomial distribution with samples taken *without* replacement is

$$n < 5\%N$$

where n is the sample size and N is the population size.

For example, suppose 10% of the population of the world is left-handed and that a sample of 20 people is selected randomly from the world's population. If the first person selected is left-handed – and the sampling is conducted without replacement – the value of $p = 0.10$ is virtually unaffected because the population of the world is so large. In addition, with many experiments the population is continually being replenished even as the sampling is being done. This condition often is the case with quality control sampling of products from large production runs. Some examples of binomial distribution problems follow.

1. Suppose a machine producing computer chips has a 6% defective rate. If a company purchases 30 of these chips, what is the probability that none is defective?
2. One ethics survey study suggested that 83% of publicly traded companies have an ethics code. From a random sample of 15 companies, what is the probability that at least 10 have an ethics code?
3. A survey found that nearly 67% of company buyers stated that their company had programmes for preferred buyers. If a random sample of 50 company buyers is taken, what is the probability that 40 or more have companies with programmes for preferred buyers?

Solving a Binomial Problem

A survey of relocation administrators by Runzheimer International revealed several reasons why workers reject relocation offers. Included in the list were family considerations, financial reasons, and others. Four per cent of the respondents said they rejected relocation offers because they received too little relocation help. Suppose five workers who just rejected relocation offers are randomly selected and interviewed. Assuming the 4% figure holds for all workers rejecting relocation, what is the probability that the first worker interviewed rejected the offer because of too little relocation help and the next four workers rejected the offer for other reasons?

Let T represent too little relocation help and R represent other reasons. The sequence of interviews for this problem is as follows:

$$T_1, R_2, R_3, R_4, R_5$$

The probability of getting this sequence of workers is calculated by using the special rule of multiplication for independent events (assuming the workers are independently selected from a large population of workers). If 4% of the workers rejecting relocation offers do so for too little relocation help, the probability of one person being randomly selected from workers rejecting relocation offers who does so for that reason is 0.04, which is the value of p. The other 96% of the workers who reject relocation offers do so for other reasons. Thus the probability of randomly selecting a worker from those who reject relocation offers who does so for other reasons is $1 - 0.04 = 0.96$, which is the value for q. The probability of obtaining this sequence of five workers who have rejected relocation offers is

$$P(T_1 \cap R_2 \cap R_3 \cap R_4 \cap R_5) = (0.04)(0.96)(0.96)(0.96)(0.96) = 0.03397$$

Obviously, in the random selection of workers who rejected relocation offers, the worker who did so because of too little relocation help could have been the second worker or the third or the fourth or the fifth. All the possible sequences of getting one worker who rejected relocation because of too little help and four workers who did so for other reasons follow.

$$T_1, R_2, R_3, R_4, R_5$$
$$R_1, T_2, R_3, R_4, R_5$$
$$R_1, R_2, T_3, R_4, R_5$$
$$R_1, R_2, R_3, T_4, R_5$$
$$R_1, R_2, R_3, R_4, T_5$$

The probability of each of these sequences occurring is calculated as follows:

$$(0.04)(0.96)(0.96)(0.96)(0.96) = 0.03397$$
$$(0.96)(0.04)(0.96)(0.96)(0.96) = 0.03397$$
$$(0.96)(0.96)(0.04)(0.96)(0.96) = 0.03397$$
$$(0.96)(0.96)(0.96)(0.04)(0.96) = 0.03397$$
$$(0.96)(0.96)(0.96)(0.96)(0.04) = 0.03397$$

Note that in each case the final probability is the same. Each of the five sequences contains the product of 0.04 and four 0.96s. The commutative property of multiplication allows for the reordering of the five individual probabilities in any one sequence. The probabilities in each of the five sequences may be reordered and summarized as $(0.04)^1(0.96)^4$. Each sequence contains the same five probabilities, which makes recomputing the probability of each sequence unnecessary. What *is* important is to determine how many different ways the sequences can be formed and multiply that figure by the probability of one sequence occurring. For the five sequences of this problem, the total probability of getting exactly one worker who rejected relocation because of too little relocation help in a random sample of five workers who rejected relocation offers is

$$5(0.04)^1(0.96)^4 = 0.16987$$

An easier way to determine the number of sequences than by listing all possibilities is to use *combinations* to calculate them. (The concept of combinations was introduced in Chapter 4.) Five workers are being sampled, so $n = 5$, and the problem is to get one worker who rejected a relocation offer because of too little relocation help, $x = 1$. Hence $_nC_x$ will yield the number of possible ways to get x successes in n trials. For this problem, $_5C_1$ tells the number of sequences of possibilities.

$$_5C_1 = \frac{5!}{1!(5-1)!} = 5$$

Weighting the probability of one sequence with the combination yields

$$_5C_1(0.04)^1(0.96)^4 = 0.16987$$

Using combinations simplifies the determination of how many sequences are possible for a given value of x in a binomial distribution.

As another example, suppose 70% of all Danish people believe cleaning up the environment is an important issue. What is the probability of randomly sampling four Danes and having exactly two of them say that they believe cleaning up the environment is an important issue? Let E represent the success of getting a person who believes cleaning up the environment is an important issue. For this example, $p = 0.70$. Let N represent the failure of not getting a person who believes cleaning up is an important issue (N denotes not important). The probability of getting one of these persons is $q = 0.30$.

The various sequences of getting two Es in a sample of four follow.

$$E_1, E_2, N_3, N_4$$
$$E_1, N_2, E_3, N_4$$
$$E_1, N_2, N_3, E_4$$
$$N_1, E_2, E_3, N_4$$
$$N_1, E_2, N_3, E_4$$
$$N_1, N_2, E_3, E_4$$

Two successes in a sample of four can occur six ways. Using combinations, the number of sequences is

$$_4C_2 = 6 \text{ ways}$$

The probability of selecting any individual sequence is

$$(0.70)^2(0.30)^2 = 0.0441$$

Thus the overall probability of getting exactly two people who believe cleaning up the environment is important out of four randomly selected people, when 70% of Danes believe cleaning up the environment is important, is

$$_4C_2(0.70)^2(0.30)^2 = 0.2646$$

Generalizing from these two examples yields the binomial formula, which can be used to solve binomial problems.

$$P(x) = {}_nC_x \cdot p^x \cdot q^{n-x} = \frac{n!}{x!(n-x)!} \cdot p^x \cdot q^{n-x}$$

BINOMIAL FORMULA

where:

$n =$ the number of trials (or the number being sampled)
$x =$ the number of successes desired
$p =$ the probability of getting a success in one trial
$q = 1 - p =$ the probability of getting a failure in one trial

The binomial formula summarizes the steps presented so far to solve binomial problems. The formula allows the solution of these problems quickly and efficiently.

DEMONSTRATION PROBLEM 5.2

A Gallup survey found that 65% of all financial consumers were very satisfied with their primary financial institution. Suppose that 25 financial consumers are sampled and if the Gallup survey result still holds true today, what is the probability that exactly 19 are very satisfied with their primary financial institution?

Solution

The value of p is 0.65 (very satisfied), the value of $q = 1 - p = 1 - 0.65 = 0.35$ (not very satisfied), $n = 25$, and $x = 19$. The binomial formula yields the final answer.

$$_{25}C_{19}(0.65)^{19}(0.35)^6 = (177,100)(0.00027884)(0.00183827) = 0.0908$$

If 65% of all financial consumers are very satisfied, about 9.08% of the time the researcher would get exactly 19 out of 25 financial consumers who are very satisfied with their financial institution. How many very satisfied consumers would one expect to get in 25 randomly selected financial consumers? If 65% of the financial consumers are very satisfied with their primary financial institution, one would expect to get about 65% of 25 or $(0.65)(25) = 16.25$ very satisfied financial consumers. While in any individual sample of 25 the number of financial consumers who are very satisfied cannot be 16.25, business researchers understand the x values near 16.25 are the most likely occurrences.

DEMONSTRATION PROBLEM 5.3

According to the Manchester City Council, approximately 6% of all workers in Manchester, UK, are unemployed. In conducting a random telephone survey in Manchester, what is the probability of getting two or fewer unemployed workers in a sample of 20?

Solution

This problem must be worked as the union of three problems: (1) zero unemployed, $x = 0$; (2) one unemployed, $x = 1$; and (3) two unemployed, $x = 2$. In each problem, $p = 0.06$, $q = 0.94$, and $n = 20$. The binomial formula gives the following result.

$$
\begin{array}{ccccc}
x=0 & & x=1 & & x=2 \\
{}_{20}C_0(0.06)^0(0.94)^{20} & + & {}_{20}C_1(0.06)^1(0.94)^{19} & + & {}_{20}C_2(0.06)^2(0.94)^{18} & = \\
0.2901 & + & 0.3703 & + & 0.2246 & = 0.8850
\end{array}
$$

If 6% of the workers in Manchester are unemployed, the telephone surveyor would get zero, one, or two unemployed workers 88.5% of the time in a random sample of 20 workers. The requirement of getting two or fewer is satisfied by getting zero, one, or two unemployed workers. Thus this problem is the union of three probabilities. Whenever the binomial formula is used to solve for cumulative success (not an exact number), the probability of each x value must be solved and the probabilities summed. If an actual survey produced such a result, it would serve to validate the census figures.

Using the Binomial Table

Anyone who works enough binomial problems will begin to recognize that the probability of getting $x = 5$ successes from a sample size of $n = 18$ when $p = 0.10$ is the same no matter whether the five successes are left-handed people, defective parts, brand X purchasers, or any other variable. Whether the sample involves people, parts, or products does not matter in terms of the final probabilities. The essence of the problem is the same: $n = 18$, $x = 5$, and $p = 0.10$. Recognizing this fact, mathematicians constructed a set of binomial tables containing pre-solved probabilities.

Two parameters, n and p, describe or characterize a binomial distribution. Binomial distributions actually are a family of distributions. Every different value of n and/or every different value of p gives a different binomial distribution, and tables are available for various combinations of n and p values. Because of space limitations, the binomial tables presented in this text are limited. Table A.2 in Appendix A contains binomial tables. Each table is headed by a value of n. Nine values of p are presented in each table of size n. In the column below each value of p is the binomial distribution for that combination of n and p. Table 5.5 contains a segment of Table A.2 with the binomial probabilities for $n = 20$.

TABLE 5.5

Excerpt from Table A.2, Appendix A

n = 20					Probability				
x	.1	.2	.3	.4	.5	.6	.7	.8	.9
0	.122	.012	.001	.000	.000	.000	.000	.000	.000
1	.270	.058	.007	.000	.000	.000	.000	.000	.000
2	.285	.137	.028	.003	.000	.000	.000	.000	.000
3	.190	.205	.072	.012	.001	.000	.000	.000	.000
4	.090	.218	.130	.035	.005	.000	.000	.000	.000
5	.032	.175	.179	.075	.015	.001	.000	.000	.000
6	.009	.109	.192	.124	.037	.005	.000	.000	.000
7	.002	.055	.164	.166	.074	.015	.001	.000	.000
8	.000	.022	.114	.180	.120	.035	.004	.000	.000
9	.000	.007	.065	.160	.160	.071	.012	.000	.000
10	.000	.002	.031	.117	.176	.117	.031	.002	.000
11	.000	.000	.012	.071	.160	.160	.065	.007	.000
12	.000	.000	.004	.035	.120	.180	.114	.022	.000
13	.000	.000	.001	.015	.074	.166	.164	.055	.002
14	.000	.000	.000	.005	.037	.124	.192	.109	.009
15	.000	.000	.000	.001	.015	.075	.179	.175	.032
16	.000	.000	.000	.000	.005	.035	.130	.218	.090
17	.000	.000	.000	.000	.001	.012	.072	.205	.190
18	.000	.000	.000	.000	.000	.003	.028	.137	.285
19	.000	.000	.000	.000	.000	.000	.007	.058	.270
20	.000	.000	.000	.000	.000	.000	.001	.012	.122

DEMONSTRATION PROBLEM 5.4

Solve the binomial probability for $n = 20$, $p = 0.40$, and $x = 10$ by using Table A.2, Appendix A.

Solution

To use Table A.2, first locate the value of n. Because $n = 20$ for this problem, the portion of the binomial tables containing values for $n = 20$ presented in Table 5.5 can be used. After locating the value of n, search horizontally across the top of the table for the appropriate value of p. In this problem, $p = 0.40$. The column under 0.40 contains the probabilities for the binomial distribution of $n = 20$ and $p = 0.40$. To get the probability of $x = 10$, find the value of x in the leftmost column and locate the probability in the table at the intersection of $p = 0.40$ and $x = 10$. The answer is 0.117. Working this problem by the binomial formula yields the same result.

$$_{20}C_{10}(0.40)^{10}(0.60)^{10} = 0.1171$$

DEMONSTRATION PROBLEM 5.5

According to Information Resources, which publishes data on market share for various products, Oreos control about 10% of the market for cookie brands. Suppose 20 purchasers of cookies are selected randomly from the population. What is the probability that fewer than four purchasers choose Oreos?

Solution

For this problem, $n = 20$, $p = 0.10$, and $x < 4$. Because $n = 20$, the portion of the binomial tables presented in Table 5.5 can be used to work this problem. Search along the row of p values for 0.10. Determining the probability of getting $x < 4$ involves summing the probabilities for $x = 0$, 1, 2, and 3. The values appear in the x column at the intersection of each x value and $p = 0.10$.

x Value	Probability
0	.122
1	.270
2	.285
3	.190
$(x < 4) = .867$	

If 10% of all cookie purchasers prefer Oreos and 20 cookie purchasers are randomly selected, about 86.7% of the time fewer than four of the 20 will select Oreos.

Using the Computer to Produce a Binomial Distribution

TABLE 5.6

Minitab Output for the Binomial Distribution of $n = 23$, $p = 0.64$

Probability Density Function	
Binomial with $n = 23$ and $p = 0.64$	
x	$P(X = x)$
0	.000000
1	.000000
2	.000000
3	.000001
4	.000006
5	.000037
6	.000199
7	.000858
8	.003051
9	.009040
10	.022500
11	.047273
12	.084041
13	.126420
14	.160533
15	.171236
16	.152209
17	.111421
18	.066027
19	.030890
20	.010983
21	.002789
22	.000451
23	.000035

Both Excel and Minitab can be used to produce the probabilities for virtually any binomial distribution. Such computer programs offer yet another option for solving binomial problems besides using the binomial formula or the binomial tables. Actually, the computer packages in effect print out what would be a column of the binomial table. The advantages of using statistical software packages for this purpose are convenience (if the binomial tables are not readily available and a computer is) and the potential for generating tables for many more values than those printed in the binomial tables.

For example, a study of bank customers stated that 64% of all financial consumers believe banks are more competitive today than they were five years ago. Suppose 23 financial consumers are selected randomly and we want to determine the probabilities of various x values occurring. Table A.2 in Appendix A could not be used because only nine different p values are included and $p = 0.64$ is not one of those values. In addition, $n = 23$ is not included in the table. Without the computer, we are left with the binomial formula as the only option for solving binomial problems for $n = 23$ and $p = 0.64$. Particularly if the cumulative probability questions are asked (for example, $x \leq 10$), the binomial formula can be a tedious way to solve the problem.

Shown in Table 5.6 is the Minitab output for the binomial distribution of $n = 23$ and $p = 0.64$. With this computer output, a researcher could obtain or calculate the probability of any occurrence within the binomial distribution of $n = 23$ and $p = 0.64$. Table 5.7 contains the Minitab output for the particular binomial problem, $P(x \leq 10)$ when $n = 23$ and $p = 0.64$, solved by using Minitab's cumulative probability capability.

TABLE 5.7

Minitab Output for the Binomial Problem, $P(x \leq 10)$, $n = 23$, and $p = 0.64$

Cumulative Distribution Function
Binomial with $n = 23$ and $p = 0.64$
$x\ P(X \Leftarrow x)$
10 0.0356916

TABLE 5.8

Excel Output for Demonstration Problem 5.3 and the Binomial Distribution of $n = 20$, $p = 0.06$

x	Prob(x)	
0	.2901	
1	.3703	
2	.2246	The probability $x \leq 2$ when $n = 20$ and $p = .06$ is .8850
3	.0860	
4	.0233	
5	.0048	
6	.0008	
7	.0001	
8	.0000	
9	.0000	

Shown in Table 5.8 is Excel output for all values of x that have probabilities greater than 0.000001 for the binomial distribution discussed in Demonstration Problem 5.3 ($n = 20$, $p = 0.06$) and the solution to the question posed in Demonstration Problem 5.3.

Mean and Standard Deviation of a Binomial Distribution

A binomial distribution has an expected value or a long-run average, which is denoted by μ. The value of μ is determined by $n \cdot p$. For example, if $n = 10$ and $p = 0.4$, then $\mu = n \cdot p = (10)(0.4) = 4$. The long-run average or expected value means that, if n items are sampled over and over for a long time and if p is the probability of getting a success on one trial, the average number of successes per sample is expected to be $n \cdot p$. If 40% of all graduate business students at a large university are women and if random samples of 10 graduate business students are selected many times, the expectation is that, on average, four of the 10 students would be women.

MEAN AND STANDARD DEVIATION OF A BINOMIAL DISTRIBUTION	$\mu = n \cdot p$
	$\sigma = \sqrt{n \cdot p \cdot q}$

Examining the mean of a binomial distribution gives an intuitive feeling about the likelihood of a given outcome.

According to one study, 64% of all financial consumers believe banks are more competitive today than they were five years ago. If 23 financial consumers are selected randomly, what is the expected number who believe banks are more competitive today than they were five years ago? This problem can be described by the binomial distribution of $n = 23$ and $p = 0.64$ given in Table 5.6. The mean of this binomial distribution yields the expected value for this problem.

In the long run, if 23 financial consumers are selected randomly over and over and if indeed 64% of all financial consumers believe banks are more competitive today, then the experiment should average 14.72 financial consumers out of 23 who believe banks are more competitive today. Realize that because the binomial distribution is a discrete distribution you will never actually get 14.72 people out of 23 who believe banks are more competitive today. The mean of the distribution does reveal the relative likelihood of any individual occurrence. Examine Table 5.6. Notice that the highest probabilities are those near $x = 14.72$: $P(x = 15) = 0.1712$, $P(x = 14) = 0.1605$, and $P(x = 16) = 0.1522$. All other probabilities for this distribution are less than these probabilities.

The standard deviation of a binomial distribution is denoted σ and is equal to $\sqrt{n \cdot p \cdot q}$. The standard deviation for the financial consumer problem described by the binomial distribution in Table 5.6 is

$$\sigma = \sqrt{n \cdot p \cdot q} = \sqrt{(23)(0.64)(0.36)} = 2.30$$

TABLE 5.9

Probabilities for Three Binomial
Distributions with $n = 8$

	Probabilities for		
x	$p = .20$	$p = .50$	$p = .80$
0	.1678	.0039	.0000
1	.3355	.0312	.0001
2	.2936	.1094	.0011
3	.1468	.2187	.0092
4	.0459	.2734	.0459
5	.0092	.2187	.1468
6	.0011	.1094	.2936
7	.0001	.0312	.3355
8	.0000	.0039	.1678

Chapter 6 shows that some binomial distributions are nearly bell shaped and can be approximated by using the normal curve. The mean and standard deviation of a binomial distribution are the tools used to convert these binomial problems to normal curve problems.

Graphing Binomial Distributions

The graph of a binomial distribution can be constructed by using all the possible x values of a distribution and their associated probabilities. The x values usually are graphed along the x-axis and the probabilities are graphed along the y-axis.

Table 5.9 lists the probabilities for three different binomial distributions: $n = 8$ and $p = 0.20$, $n = 8$ and $p = 0.50$, and $n = 8$ and $p = 0.80$. Figure 5.2 displays Excel graphs for each of these three binomial distributions. Observe how the shape of the distribution changes as the value of p increases. For $p = 0.50$, the distribution is symmetrical. For $p = 0.20$, the distribution is skewed right and for $p = 0.80$, the distribution is skewed left. This pattern makes sense because the mean of the binomial distribution $n = 8$ and $p = 0.50$ is 4, which is in the middle of the distribution. The mean of the distribution $n = 8$ and $p = 0.20$ is 1.6, which results in the highest probabilities being near $x = 2$ and $x = 1$. This graph peaks early and stretches towards the higher values of x. The mean of the distribution $n = 8$ and $p = 0.80$ is 6.4, which results in the highest probabilities being near $x = 6$ and $x = 7$. Thus the peak of the distribution is nearer to 8 than to 0 and the distribution stretches back towards $x = 0$.

$$\mu = n \cdot p = 23(0.64) = 14.72$$

In any binomial distribution the largest x value that can occur is n and the smallest value is zero. Thus the graph of any binomial distribution is constrained by zero and n. If the p value of the distribution is not 0.50, this constraint will result in the graph 'piling up' at one end and being skewed at the other end.

DEMONSTRATION PROBLEM 5.6

A manufacturing company produces 10,000 plastic mugs per week. This company supplies mugs to another company, which packages the mugs as part of picnic sets. The second company randomly samples 10 mugs sent from the supplier. If two or fewer of the sampled mugs are defective, the second company accepts the lot. What is the probability that the lot will be accepted if the mug manufacturing company actually is producing mugs that are 10% defective? 20% defective? 30% defective? 40% defective?

Solution

In this series of binomial problems, $n = 10$, $x \leq 2$, and p ranges from 0.10 to 0.40. From Table A.2 – and cumulating the values – we have the following probability of $x \leq 2$ for each p value and the expected value ($\mu = n \cdot p$).

p	Lot Accepted $P(x \leq 2)$	Expected Number of Defects (μ)
.10	.930	1.0
.20	.677	2.0
.30	.382	3.0
.40	.167	4.0

These values indicate that if the manufacturing company is producing 10% defective mugs, the probability is relatively high (0.930) that the lot will be accepted by chance. For higher values of p, the probability of lot acceptance by chance decreases. In addition, as p increases, the expected value moves away from the acceptable values, $x \leq 2$. This move reduces the chances of lot acceptance.

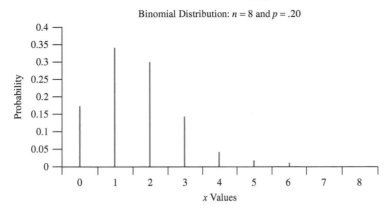

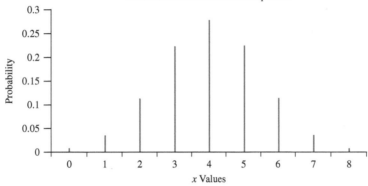

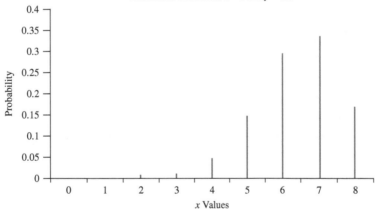

FIGURE 5.2

Excel Graphs of Three Binomial Distributions with $n = 8$

Plastic Bags vs. Bringing Your Own in Japan

In a move to protect and improve the environment, governments and companies around the world are making an effort to reduce the use of plastic bags by shoppers for transporting purchased food and goods. Specifically, in Yamagata City in northern Japan, the city concluded an agreement with seven local food supermarket chains to reduce plastic bag use in May of 2008 by having them agree to charge for the use of such bags. Before the agreement, in April of 2008, the average percentage of shoppers bringing their own shopping bags was about 35%. By the end of June, with some of the supermarket chains participating, the percentage had risen to almost 46%. However, by August, when 39 stores of the nine supermarket chains (two other chains joined the agreement) were charging for the use of plastic bags, the percentage rose to nearly 90%. It is estimated that the reduction of carbon dioxide emissions by this initiative is about 225 tonnes during July and August alone.

Source: www.japanfs.org/en/pages/028631.html.

5.3 PROBLEMS

5.5 Solve the following problems by using the binomial formula:
    a. If $n = 4$ and $p = 0.10$, find $P(x = 3)$.
    b. If $n = 7$ and $p = 0.80$, find $P(x = 4)$.
    c. If $n = 10$ and $p = 0.60$, find $P(x \geq 7)$.
    d. If $n = 12$ and $p = 0.45$, find $P(5 \leq x \leq 7)$.

5.6 Solve the following problems by using the binomial tables (Table A.2):
    a. If $n = 20$ and $p = 0.50$, find $P(x = 12)$.
    b. If $n = 20$ and $p = 0.30$, find $P(x > 8)$.
    c. If $n = 20$ and $p = 0.70$, find $P(x < 12)$.
    d. If $n = 20$ and $p = 0.90$, find $P(x \leq 16)$.
    e. If $n = 15$ and $p = 0.40$, find $P(4 \leq x \leq 9)$.
    f. If $n = 10$ and $p = 0.60$, find $P(x \geq 7)$.

5.7 Solve for the mean and standard deviation of the following binomial distributions:
    a. $n = 20$ and $p = 0.70$
    b. $n = 70$ and $p = 0.35$
    c. $n = 100$ and $p = 0.50$

5.8 Use the probability tables in Table A.2 and sketch the graph of each of the binomial distributions below. Note on the graph where the mean of the distribution falls.
    a. $n = 6$ and $p = 0.70$
    b. $n = 20$ and $p = 0.50$
    c. $n = 8$ and $p = 0.80$

5.9 What is the first big change that British motorists made due to higher petrol prices? According to a survey commissioned by Halfords, 63% said that they were driving slower. However, 11% said that they were using public transport more. If these figures are true for all British drivers, and if 15 such drivers are randomly sampled and asked what is the first big change they made due to higher petrol prices,
    a. What is the probability that exactly six said that it was using public transport more?
    b. What is the probability that none of them said that it driving slower?
    c. What is the probability that more than nine said that it was driving slower?

5.10 *The Wall Street Journal* reported some interesting statistics on the job market. One statistic is that 40% of all workers say they would change jobs for 'slightly higher pay'. In addition, 88% of companies say that there is a shortage of qualified job candidates. Suppose 16 workers are randomly selected and asked if they would change jobs for 'slightly higher pay'.

a. What is the probability that nine or more say yes?

b. What is the probability that three, four, five, or six say yes?

c. If 13 companies are contacted, what is the probability that exactly 10 say there is a shortage of qualified job candidates?

d. If 13 companies are contacted, what is the probability that all of the companies say there is a shortage of qualified job candidates?

e. If 13 companies are contacted, what is the expected number of companies that would say there is a shortage of qualified job candidates?

5.11 An increasing number of consumers of all ages are choosing to watch video content over the Internet. According to a survey conducted by Accenture, around 80% of consumers aged between 35 and 44 are now accessing video via desktops, laptops, Internet-connected TVs and mobile devices. Suppose a random sample of 20 consumers in this age group is contacted and interviewed about their video consumption habits.

a. What is the probability that 13 or more of these consumers are choosing to watch video content over the Internet via their televisions, PCs, smartphones, and other electronic devices?

b. What is the probability that more than 17 of these consumers are choosing to watch video content over the Internet via their televisions, PCs, smartphones, and other electronic devices?

c. What is the probability that fewer than 11 of these consumers are choosing to watch video content over the Internet via their televisions, PCs, smartphones, and other electronic devices?

5.12 According to the results from the 5th European Working Conditions Survey, conducted by the European Foundation for the Improvement of Living and Working Conditions in 2010, European workers say working conditions are slowly improving. The survey also found that for most workers (about 70%), working intensity remains at a high level and that 34% of workers received employer-provided training. Assuming these results are true for all European workers, then

a. If 16 European workers are randomly sampled, what is the probability that more than 10 of them find their work intensity remains high?

b. If 10 European workers are randomly selected, what is the probability that exactly six of them have had employer-provided training?

5.13 In the past few years, outsourcing overseas has become more frequently used than ever before by European companies, especially in the area of customer services. However, outsourcing is not without problems. A recent survey by *Purchasing* indicates that 20% of the companies that outsource overseas use a consultant. Suppose 15 companies that outsource overseas are randomly selected.

a. What is the probability that exactly five companies that outsource overseas use a consultant?

b. What is the probability that more than nine companies that outsource overseas use a consultant?

c. What is the probability that none of the companies that outsource overseas use a consultant?

d. What is the probability that between four and seven (inclusive) companies that outsource overseas use a consultant?

e. Construct a graph for this binomial distribution. In light of the graph and the expected value, explain why the probability results from parts (a) to (d) were obtained.

5.14 According to The Scottish Government's website, the average class size in publicly funded primary schools in Scotland had been declining over the years and is currently set at 23 students per class. Whilst most primary schools (52%) have classes of up to 25 students, a sizeable number of schools (16%) still have classes with 30 or more students. Suppose a complete list of class sizes in all Scottish publicly funded primary schools is available and 18 are randomly selected from that list.

a. What is the expected number of primary schools that have an average class size of up to 25 students? What is the expected number of schools with an average class size of 30 or more?

b. What is the probability that at least eight primary schools have an average class size of up to 25 students?

    c. What is the probability that two, three, or four primary schools have an average class size of up to 25 students?

    d. What is the probability that none of the public schools have an average class size of up to 25 students? What is the probability that none have an average class size of 30 or more? Which probability is higher and why?

5.4   POISSON DISTRIBUTION

The Poisson distribution is another discrete distribution. It is named after Simeon-Denis Poisson (1781–1840), a French mathematician, who published its essentials in a paper in 1837. The Poisson distribution and the binomial distribution have some similarities but also several differences. The binomial distribution describes a distribution of two possible outcomes designated as successes and failures from a given number of trials. The **Poisson distribution** *focuses only on the number of discrete occurrences over some interval or continuum.* A Poisson experiment does not have a given number of trials (*n*) as a binomial experiment does. For example, whereas a binomial experiment might be used to determine how many Indian-made cars are in a random sample of 20 cars, a Poisson experiment might focus on the number of cars randomly arriving at an automobile repair facility during a 10-minute interval.

The Poisson distribution describes the occurrence of *rare events*. In fact, the Poisson formula has been referred to as the *law of improbable events.* For example, serious accidents at a chemical plant are rare, and the number per month might be described by the Poisson distribution. The Poisson distribution often is used to describe the number of random arrivals per some time interval. If the number of arrivals per interval is too frequent, the time interval can be reduced enough so that a rare number of occurrences is expected. Another example of a Poisson distribution is the number of random customer arrivals per five-minute interval at a small boutique on weekday mornings.

The Poisson distribution also has an application in the field of management science. The models used in queuing theory (theory of waiting lines) usually are based on the assumption that the Poisson distribution is the proper distribution to describe random arrival rates over a period of time.

The Poisson distribution has the following characteristics:

- It is a discrete distribution.
- It describes rare events.
- Each occurrence is independent of the other occurrences.
- It describes discrete occurrences over a continuum or interval.
- The occurrences in each interval can range from zero to infinity.
- The expected number of occurrences must hold constant throughout the experiment.

    Examples of Poisson-type situations include the following:

1. Number of telephone calls per minute at a small business
2. Number of hazardous waste sites per province in Holland
3. Number of arrivals at an autoroute toll booth per minute between 3 A.M. and 4 A.M. in January on the A68 *autoroute* in France that connects Toulouse to Albi
4. Number of sewing flaws per pair of jeans during production
5. Number of times a tyre blows on a commercial aircraft per week

Each of these examples represents a rare occurrence of events for some interval. Note that, although time is a more common interval for the Poisson distribution, intervals can range from a county in the United States to a pair of jeans. Some of the intervals in these examples might have zero occurrences. Moreover, the average occurrence per interval for many of these examples is probably in the single digits (1–9).

If a Poisson-distributed phenomenon is studied over a long period of time, a *long-run average* can be determined. This average is denoted **lambda** (λ). Each Poisson problem contains a lambda value from which the probabilities of particular occurrences are determined. Although *n* and *p* are required to describe a binomial distribution, a Poisson distribution can be described by λ alone. The Poisson formula is used to compute the probability of occurrences over an interval for a given lambda value.

$$P(x) = \frac{\lambda^x e^{-\lambda}}{x!}$$

POISSON FORMULA

where:
$x = 0, 1, 2, 3, \ldots$
$\lambda = $ long-run average
$e = 2.718282$

Here, x is the number of occurrences per interval for which the probability is being computed and $e = 2.718282$ is the base of natural logarithms.

A word of caution about using the Poisson distribution to study various phenomena is necessary. The λ value must hold constant throughout a Poisson experiment. The researcher must be careful not to apply a given lambda to intervals for which lambda changes. For example, the average number of customers arriving at a Sears store during a one-minute interval will vary from hour to hour, day to day, and month to month. Different times of the day or week might produce different lambdas. The number of flaws per pair of jeans might vary from Monday to Friday. The researcher should be specific in describing the interval for which λ is being used.

Working Poisson Problems by Formula

Suppose bank customers arrive randomly on weekday afternoons at an average of 3.2 customers every four minutes. What is the probability of exactly five customers arriving in a four-minute interval on a weekday afternoon? The lambda for this problem is 3.2 customers per four minutes. The value of x is five customers per four minutes. The probability of five customers randomly arriving during a four-minute interval when the long-run average has been 3.2 customers per four-minute interval is

$$\frac{(3.2^5)(e^{-3.2})}{5!} = \frac{(335.54)(0.0408)}{120} = 0.1141$$

If a bank averages 3.2 customers every four minutes, the probability of five customers arriving during any one four-minute interval is 0.1141.

DEMONSTRATION PROBLEM 5.7

Bank customers arrive randomly on weekday afternoons at an average of 3.2 customers every four minutes. What is the probability of having more than seven customers in a four-minute interval on a weekday afternoon?

Solution

$$\lambda = 3.2 \text{ customers}/4 \text{ minutes}$$
$$x > 7 \text{ customers}/4 \text{ minutes}$$

In theory, the solution requires obtaining the values of $x = 8, 9, 10, 11, 12, 13, 14, \ldots \infty$. In actuality, each x value is determined until the values are so far away from $\lambda = 3.2$ that the probabilities approach zero. The exact probabilities are then summed to find $x > 7$.

$$P(x = 8 \mid \lambda = 3.2) = \frac{(3.2^8)(e^{-3.2})}{8!} = 0.0111$$

$$P(x=9 \mid \lambda=3.2) = \frac{(3.2^9)(e^{-3.2})}{9!} = 0.0040$$

$$P(x=10 \mid \lambda=3.2) = \frac{(3.2^{10})(e^{-3.2})}{10!} = 0.0013$$

$$P(x=11 \mid \lambda=3.2) = \frac{(3.2^{11})(e^{-3.2})}{11!} = 0.0004$$

$$P(x=12 \mid \lambda=3.2) = \frac{(3.2^{12})(e^{-3.2})}{12!} = 0.0001$$

$$P(x=13 \mid \lambda=3.2) = \frac{(3.2^{13})(e^{-3.2})}{13!} = 0.0000$$

$$P(x>7) = P(x \geq 8) = 0.0169$$

If the bank has been averaging 3.2 customers every four minutes on weekday afternoons, it is unlikely that more than seven people would randomly arrive in any one four-minute period. This answer indicates that more than seven people would randomly arrive in a four-minute period only 1.69% of the time. Bank officers could use these results to help them make staffing decisions.

DEMONSTRATION PROBLEM 5.8

A bank has an average random arrival rate of 3.2 customers every four minutes. What is the probability of getting exactly 10 customers during an eight-minute interval?

Solution

$$\lambda = 3.2 \text{ customers}/4 \text{ minutes}$$
$$x = 10 \text{ customers}/8 \text{ minutes}$$

This example is different from the first two Poisson examples in that the intervals for lambda and the sample are different. The intervals must be the same in order to use λ and x together in the probability formula. The right way to approach this dilemma is to adjust the interval for lambda so that it and x have the same interval. The interval for x is eight minutes, so lambda should be adjusted to an eight-minute interval. Logically, if the bank averages 3.2 customers every four minutes, it should average twice as many, or 6.4 customers, every eight minutes. If x were for a two-minute interval, the value of lambda would be halved from 3.2 to 1.6 customers per two-minute interval. The wrong approach to this dilemma is to equalize the intervals by changing the x value. Never adjust or change x in a problem. Just because 10 customers arrive in one eight-minute interval does not mean that there would necessarily have been five customers in a four-minute interval. There is no guarantee how the 10 customers are spread over the eight-minute interval. Always adjust the lambda value. After lambda has been adjusted for an eight-minute interval, the solution is

$$\lambda = 6.4 \text{ customers}/8 \text{ minutes}$$
$$x = 10 \text{ customers}/8 \text{ minutes}$$

$$\frac{(6.4)^{10} e^{-6.4}}{10!} = 0.0528$$

Using the Poisson Tables

Every value of lambda determines a different Poisson distribution. Regardless of the nature of the interval associated with a lambda, the Poisson distribution for a particular lambda is the same. Table A.3, Appendix A, contains the Poisson distributions for selected values of lambda. Probabilities are displayed in the table for each x value associated with a given lambda if the probability has a non-zero value to four decimal places. Table 5.10 presents a portion of Table A.3 that contains the probabilities of $x \leq 9$ if lambda is 1.6.

TABLE 5.10

Poisson Table for $\lambda = 1.6$

x	Probability	x	Probability
0	.2019	5	.0176
1	.3230	6	.0047
2	.2584	7	.0011
3	.1378	8	.0002
4	.0551	9	.0000

(continued)

DEMONSTRATION PROBLEM 5.9

If a real estate office sells 1.6 houses on an average weekday and sales of houses on weekdays are Poisson distributed, what is the probability of selling exactly four houses in one day? What is the probability of selling no houses in one day? What is the probability of selling more than five houses in a day? What is the probability of selling 10 or more houses in a day? What is the probability of selling exactly four houses in two days?

Solution

$$\lambda = 1.6 \text{ houses/day}$$
$$P(x = 4 \mid \lambda = 1.6) = ?$$

Table 5.10 gives the probabilities for $\lambda = 1.6$. The left column contains the x values. The line $x = 4$ yields the probability 0.0551. If a real estate firm has been averaging 1.6 houses sold per day, only 5.51% of the days would it sell exactly four houses and still maintain the lambda value. Line 1 of Table 5.10 shows the probability of selling no houses in a day (0.2019). That is, on 20.19% of the days, the firm would sell no houses if sales are Poisson distributed with $\lambda = 1.6$ houses per day. Table 5.10 is not cumulative. To determine $P(x > 5)$, more than 5 houses, find the probabilities of $x = 6, x = 7, x = 8, x = 9, \ldots x = ?$. However, at $x = 9$, the probability to four decimal places is zero, and Table 5.10 stops when an x value zeros out at four decimal places. The answer for $x > 5$ follows.

x	Probability
6	.0047
7	.0011
8	.0002
9	.0000
$x > 5 =$	.0060

What is the probability of selling 10 or more houses in one day? As the table zeros out at $x = 9$, the probability of $x \geq 10$ is essentially 0.0000 – that is, if the real estate office has been averaging only 1.6 houses sold per day, it is virtually impossible to sell 10 or more houses in a day. What is the probability of selling exactly four houses in two days? In this case, the interval has been changed from one day to two days. Lambda is for one day, so an adjustment must be made: a lambda of 1.6 for one day converts to a lambda of 3.2 for two days. Table 5.10 no longer applies, so Table A.3 must be used to solve this problem. The answer is found by looking up $\lambda = 3.2$ and $x = 4$ in Table A.3: the probability is 0.1781.

Mean and Standard Deviation of a Poisson Distribution

The mean or expected value of a Poisson distribution is λ. It is the long-run average of occurrences for an interval if many random samples are taken. Lambda usually is not a whole number, so most of the time actually observing lambda occurrences in an interval is impossible.

For example, suppose $\lambda = 6.5$/interval for some Poisson-distributed phenomenon. The resulting numbers of x occurrences in 20 different random samples from a Poisson distribution with $\lambda = 6.5$ might be as follows.

| 6 | 9 | 7 | 4 | 8 | 7 | 6 | 6 | 10 | 6 | 5 | 5 | 8 | 4 | 5 | 8 | 5 | 4 | 9 | 10 |

Computing the mean number of occurrences from this group of 20 intervals gives 6.6. In theory, for infinite sampling the long-run average is 6.5. Note from the samples that, when $\lambda = 6.5$, several 5s and 6s occur. Rarely would sample occurrences of 1, 2, 3, 11, 12, 13, . . . occur when $\lambda = 6.5$. Understanding the mean of a Poisson distribution gives a feel for the actual occurrences that are likely to happen.

STATISTICS IN BUSINESS TODAY

Air Passengers' Complaints

In recent years, airline passengers have expressed much more dissatisfaction with airline service than ever before. Complaints include flight delays, lost baggage, long runway delays with little or no onboard service, overbooked flights, cramped space due to fuller flights, cancelled flights, and grumpy airline employees. A majority of dissatisfied fliers merely grin and bear it. However, an increasing number of passengers log complaints with the US Department of Transportation. In the mid-1990s, the average number of complaints per 100,000 passengers boarded was 0.66. In ensuing years, the average rose to 0.74, 0.86, 1.08, and 1.21.

In a recent year, according to the US Department of Transportation, Southwest Airlines had the fewest average number of complaints per 100,000 with 0.27, followed by ExpressJet Airlines with 0.44, Alaska Airlines with 0.50, SkyWest Airlines with 0.53, and Frontier Airlines with 0.82. Within the top 10 largest US airlines, US Airways had the highest average number of complaints lodged against it – 2.11 complaints per 100,000 passengers.

Because these average numbers are relatively small, it appears that the actual number of complaints per 100,000 is rare and may follow a Poisson distribution. In this case, λ represents the average number of complaints and the interval is 100,000 passengers. For example, using $\lambda = 1.21$ complaints (average for all airlines), if 100,000 boarded passengers were contacted, the probability that exactly three of them logged a complaint to the Department of Transportation could be computed as

$$\frac{(1.21)^3 e^{-1.21}}{3!} = 0.0880$$

That is, if 100,000 boarded passengers were contacted over and over, 8.80% of the time exactly three would have logged complaints with the US Department of Transportation.

The variance of a Poisson distribution also is λ. The standard deviation is $\sqrt{\lambda}$. Combining the standard deviation with Chebyshev's theorem indicates the spread or dispersion of a Poisson distribution. For example, if $\lambda = 6.5$, the variance also is 6.5, and the standard deviation is 2.55. Chebyshev's theorem states that at least $1 - 1/k^2$ values are within k standard deviations of the mean. The interval $\mu \pm 2\sigma$ contains at least $1 - (1/2^2) = 0.75$ of the values. For $\mu = \lambda = 6.5$ and $\sigma = 2.55$, 75% of the values should be within the $6.5 \pm 2(2.55) = 6.5 \pm 5.1$ range. That is, the range from 1.4 to 11.6 should include at least 75% of all the values. An examination of the 20 values randomly generated for a Poisson distribution with $\lambda = 6.5$ shows that actually 100% of the values are within this range.

Graphing Poisson Distributions

The values in Table A.3, Appendix A, can be used to graph a Poisson distribution. The x values are on the x-axis and the probabilities are on the y-axis. Figure 5.3 is a Minitab graph for the distribution of values for $\lambda = 1.6$.

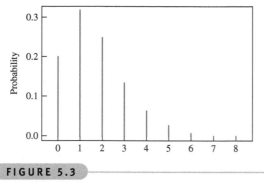

FIGURE 5.3

Minitab Graph of the Poisson Distribution for $\lambda = 1.6$

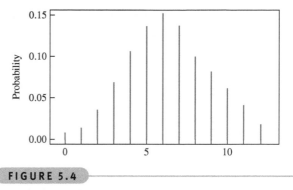

FIGURE 5.4

Minitab Graph of the Poisson Distribution for $\lambda = 6.5$

The graph reveals a Poisson distribution skewed to the right. With a mean of 1.6 and a possible range of x from zero to infinity, the values obviously will 'pile up' at 0 and 1. Consider, however, the Minitab graph of the Poisson distribution for $\lambda = 6.5$ in Figure 5.4. Note that with $\lambda = 6.5$, the probabilities are greatest for the values of 5, 6, 7, and 8. The graph has less skewness, because the probability of occurrence of values near zero is small, as are the probabilities of large values of x.

Using the Computer to Generate Poisson Distributions

Using the Poisson formula to compute probabilities can be tedious when one is working problems with cumulative probabilities. The Poisson tables in Table A.3, Appendix A, are faster to use than the Poisson formula. However, Poisson tables are limited by the amount of space available, and Table A.3 only includes probability values for Poisson distributions with lambda values to the tenths place in most cases. For researchers who want to use lambda values with more precision or who feel that the computer is more convenient than textbook tables, some statistical computer software packages are an attractive option.

Minitab will produce a Poisson distribution for virtually any value of lambda. For example, according to the Office for National Statistics, Great Britain has one of the lowest yearly road death rates in the world, at 5.4 per 100,000 population. If these cases are Poisson distributed, lambda is 5.4 per year. What does the Poisson probability distribution for this lambda look like? Table 5.11 contains the Minitab computer output for this distribution.

Excel can also generate probabilities of different values of x for any Poisson distribution. Table 5.12 displays the probabilities produced by Excel for the real estate problem from Demonstration Problem 5.9 using a lambda of 1.6.

Approximating Binomial Problems by the Poisson Distribution

Certain types of binomial distribution problems can be approximated by using the Poisson distribution. Binomial problems with large sample sizes and small values of p, which then generate rare events, are potential candidates for use of the Poisson distribution. As a rule of thumb, if $n > 20$ and $n \cdot p \leq 7$, the approximation is close enough to use the Poisson distribution for binomial problems.

If these conditions are met and the binomial problem is a candidate for this process, the procedure begins with computation of the mean of the binomial distribution, $\mu = n \cdot p$. Because μ is the expected value of the

TABLE 5.11

Minitab Output for the Poisson Distribution $\lambda = 5.4$

Probability Density Function	
Poisson with mean = 5.4	
x	$P(X=x)$
0	.004517
1	.024390
2	.065852
3	.118533
4	.160020
5	.172821
6	.155519
7	.119987
8	.080991
9	.048595
10	.026241
11	.012882
12	.005797
13	.002408
14	.000929
15	.000334

x	Probability
0	.2019
1	.3230
2	.2584
3	.1378
4	.0551
5	.0176
6	.0047
7	.0011
8	.0002
9	.0000

binomial, it translates to the expected value, λ, of the Poisson distribution. Using μ as the λ value and using the x value of the binomial problem allows approximation of the probability from a Poisson table or by the Poisson formula.

Large values of n and small values of p usually are not included in binomial distribution tables thereby precluding the use of binomial computational techniques. Using the Poisson distribution as an approximation to such a binomial problem in such cases is an attractive alternative; and indeed, when a computer is not available, it can be the only alternative.

As an example, the following binomial distribution problem can be worked by using the Poisson distribution: $n = 50$ and $p = 0.03$. What is the probability that $x = 4$? That is, $P(x = 4 | n = 50 \text{ and } p = 0.03) = ?$

To solve this equation, first determine lambda:

$$\lambda = \mu = n \cdot p = (50)(0.03) = 1.5$$

As $n > 20$ and $n \cdot p \leq 7$, this problem is a candidate for the Poisson approximation. For $x = 4$, Table A.3 yields a probability of 0.0471 for the Poisson approximation. For comparison, working the problem by using the binomial formula yields the following results:

$$_{50}C_4(0.03)^4(0.97)^{46} = 0.0459$$

The Poisson approximation is 0.0012 different from the result obtained by using the binomial formula to work the problem.

A Minitab graph of this binomial distribution follows.

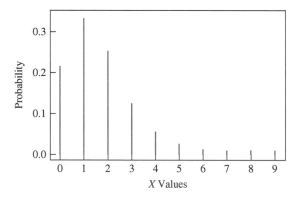

With $\lambda = 1.5$ the Poisson distribution can be generated. A Minitab graph of this Poisson distribution follows.

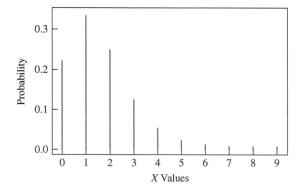

In comparing the two graphs, it is difficult to tell the difference between the binomial distribution and the Poisson distribution because the approximation of the binomial distribution by the Poisson distribution is close.

DEMONSTRATION PROBLEM 5.10

Suppose the probability of a bank making a mistake in processing a deposit is 0.0003. If 10,000 deposits (*n*) are audited, what is the probability that more than six mistakes were made in processing deposits?

Solution

$$\lambda = \mu = n \cdot p = (10{,}000)(0.0003) = 3.0$$

Because $n > 20$ and $n \cdot p \le 7$, the Poisson approximation is close enough to analyse $x > 6$. Table A.3 yields the following probabilities for $\lambda = 3.0$ and $x \ge 7$.

	$\lambda = 3.0$
x	Probability
7	.0216
8	.0081
9	.0027
10	.0008
11	.0002
12	.0001
$x > 6 = .0335$	

To work this problem by using the binomial formula requires starting with $x = 7$.

$$_{10{,}000}C_7(0.0003)^7(0.9997)^{9993}$$

This process would continue for *x* values of 8, 9, 10, 11, . . . , until the probabilities approach zero. Obviously, this process is impractical, making the Poisson approximation an attractive alternative.

5.4 PROBLEMS

5.15 Find the following values by using the Poisson formula:
  a. $P(x = 5|\lambda = 2.3)$
  b. $P(x = 2|\lambda = 3.9)$
  c. $P(x \le 3|\lambda = 4.1)$
  d. $P(x = 0|\lambda = 2.7)$
  e. $P(x = 1|\lambda = 5.4)$
  f. $P(4 \le x \le 5|\lambda = 4.4)$

5.16 Find the following values by using the Poisson tables in Appendix A:
  a. $P(x = 6|\lambda = 3.8)$
  b. $P(x > 7|\lambda = 2.9)$
  c. $P(3 \le x \le 9|\lambda = 4.2)$

    d. $P(x = 0 | \lambda = 1.9)$

    e. $P(x \leq 6 | \lambda = 2.9)$

    f. $P(5 < x \leq 8 | \lambda = 5.7)$

5.17 Sketch the graphs of the Poisson distributions below. Compute the mean and standard deviation for each distribution. Locate the mean on the graph. Note how the probabilities are graphed around the mean.

    a. $\lambda = 6.3$

    b. $\lambda = 1.3$

    c. $\lambda = 8.9$

    d. $\lambda = 0.6$

5.18 On Monday mornings, Santander Bank only has one teller window open for deposits and withdrawals. Experience has shown that the average number of arriving customers in a four-minute interval on Monday mornings is 2.8, and each teller can serve more than that number efficiently. These random arrivals at this bank on Monday mornings are Poisson distributed.

    a. What is the probability that on a Monday morning exactly six customers will arrive in a four-minute interval?

    b. What is the probability that no one will arrive at the bank to make a deposit or withdrawal during a four-minute interval?

    c. Suppose the teller can serve no more than four customers in any four-minute interval at this window on a Monday morning. What is the probability that, during any given four-minute interval, the teller will be unable to meet the demand? What is the probability that the teller will be able to meet the demand? When demand cannot be met during any given interval, a second window is opened. What percentage of the time will a second window have to be opened?

    d. What is the probability that exactly three people will arrive at the bank during a two-minute period on Monday mornings to make a deposit or a withdrawal? What is the probability that five or more customers will arrive during an eight-minute period?

5.19 A restaurant manager is interested in taking a more statistical approach to predicting customer load. She begins the process by gathering data. One of the restaurant hosts or hostesses is assigned to count customers every five minutes from 7 P.M. until 8 P.M. every Saturday night for three weeks. The data are shown here. After the data are gathered, the manager computes lambda using the data from all three weeks as one data set as a basis for probability analysis. What value of lambda did she find? Assume that these customers randomly arrive and that the arrivals are Poisson distributed. Use the value of lambda computed by the manager and help the manager calculate the probabilities in parts (a) through (e) for any given five-minute interval between 7 P.M. and 8 P.M. on a Saturday night.

Number of Arrivals		
Week 1	**Week 2**	**Week 3**
3	1	5
6	2	3
4	4	5
6	0	3
2	2	5
3	6	4
1	5	7
5	4	3
1	2	4
0	5	8
3	3	1
3	4	3

a. What is the probability that no customers arrive during any given five-minute interval?

b. What is the probability that six or more customers arrive during any given five-minute interval?

c. What is the probability that during a 10-minute interval fewer than four customers arrive?

d. What is the probability that between three and six (inclusive) customers arrive in any 10-minute interval?

e. What is the probability that exactly eight customers arrive in any 15-minute interval?

5.20 According to the United National Environmental Program and World Health Organization, in Mumbai, India, air pollution standards for particulate matter are exceeded an average of 5.6 days in every three-week period. Assume that the distribution of the number of days exceeding the standards per three-week period is Poisson distributed.

a. What is the probability that the standard is not exceeded on any day during a three-week period?

b. What is the probability that the standard is exceeded exactly six days of a three-week period?

c. What is the probability that the standard is exceeded 15 or more days during a three-week period? If this outcome actually occurred, what might you conclude?

5.21 The average number of annual holiday trips per family in Croatia is Poisson distributed, with a mean of 1.7 trips per year. What is the probability of randomly selecting a Croatian family and finding the following?

a. The family did not make a holiday trip last year.

b. The family took exactly one holiday trip last year.

c. The family took two or more holiday trips last year.

d. The family took three or fewer holiday trips over a three-year period.

e. The family took exactly four holiday trips during a six-year period.

5.22 Ship collisions in the English Channel are rare. Suppose the number of collisions are Poisson distributed, with a mean of 1.2 collisions every four months.

a. What is the probability of having no collisions occur over a four-month period?

b. What is the probability of having exactly two collisions in a two-month period?

c. What is the probability of having one or fewer collisions in a six-month period? If this outcome occurred, what might you conclude about ship channel conditions during this period? What might you conclude about ship channel safety awareness during this period? What might you conclude about weather conditions during this period? What might you conclude about lambda?

5.23 A pen company averages 1.2 defective pens per carton produced (200 pens). The number of defects per carton is Poisson distributed.

a. What is the probability of selecting a carton and finding no defective pens?

b. What is the probability of finding eight or more defective pens in a carton?

c. Suppose a purchaser of these pens will quit buying from the company if a carton contains more than three defective pens. What is the probability that a carton contains more than three defective pens?

5.24 A medical researcher estimates that 0.00004 of the population has a rare blood disorder. If the researcher randomly selects 100,000 people from the population,

a. What is the probability that seven or more people will have the rare blood disorder?

b. What is the probability that more than 10 people will have the rare blood disorder?

c. Suppose the researcher gets more than 10 people who have the rare blood disorder in the sample of 100,000 but that the sample was taken from a particular geographic region. What might the researcher conclude from the results?

5.25 A data firm records a large amount of data. Historically, 0.9% of the pages of data recorded by the firm contain errors. If 200 pages of data are randomly selected,

a. What is the probability that six or more pages contain errors?

b. What is the probability that more than 10 pages contain errors?

c. What is the probability that none of the pages contain errors?

d. What is the probability that fewer than five pages contain errors?

5.26 A high percentage of people who fracture or dislocate a bone see a doctor for that condition. Suppose the percentage is 99%. Consider a sample in which 300 people are randomly selected who have fractured or dislocated a bone.

    a. What is the probability that exactly five of them did not see a doctor?

    b. What is the probability that fewer than four of them did not see a doctor?

    c. What is the expected number of people who would not see a doctor?

5.5 HYPERGEOMETRIC DISTRIBUTION

Another discrete statistical distribution is the hypergeometric distribution. Statisticians often use the **hypergeometric distribution** to complement the types of analyses that can be made by using the binomial distribution. Recall that the binomial distribution applies, in theory, only to experiments in which the trials are done with replacement (independent events). The hypergeometric distribution applies only to experiments in which the trials are done without replacement.

The hypergeometric distribution, like the binomial distribution, consists of two possible outcomes: success and failure. However, the user must know the size of the population and the proportion of successes and failures in the population to apply the hypergeometric distribution. In other words, because the hypergeometric distribution is used when sampling is done without replacement, information about population makeup must be known in order to re-determine the probability of a success in each successive trial as the probability changes.

The hypergeometric distribution has the following characteristics:

- It is discrete distribution.
- Each outcome consists of either a success or a failure.
- Sampling is done without replacement.
- The population, N, is finite and known.
- The number of successes in the population, A, is known.

HYPERGEOMETRIC FORMULA

$$P(x) = \frac{{}_{A}C_x \cdot {}_{N-A}C_{n-x}}{{}_{N}C_n}$$

where:

N = size of the population
n = sample size
A = number of successes in the population
x = number of successes in the sample, sampling is done *without* replacement

A hypergeometric distribution is characterized or described by three parameters: N, A, and n. Because of the multitude of possible combinations of these three parameters, creating tables for the hypergeometric distribution is practically impossible. Hence, the researcher who selects the hypergeometric distribution for analysing data must use the hypergeometric formula to calculate each probability. Because this task can be tedious and time-consuming, most researchers use the hypergeometric distribution as a fallback position when working binomial problems without replacement. Even though the binomial distribution theoretically applies only when sampling is done with replacement and p stays constant, recall that, if the population is large enough in comparison with the sample size, the impact of sampling without replacement on p is minimal. Thus the binomial distribution can be used in some situations when sampling is done without replacement. Because of the tables available, using the binomial distribution instead of the hypergeometric distribution whenever possible is preferable. As a rule of thumb, if the sample size is less than 5% of the population, use of the binomial distribution rather than the hypergeometric distribution is acceptable when sampling is done without replacement. The hypergeometric distribution yields the exact probability, and the binomial distribution yields a good approximation of the probability in these situations.

In summary, the hypergeometric distribution should be used instead of the binomial distribution when the following conditions are present:

1. Sampling is being done without replacement.
2. $n \geq 5\% N$.

Hypergeometric probabilities are calculated under the assumption of equally likely sampling of the remaining elements of the sample space.

As an application of the hypergeometric distribution, consider the following problem: 24 people, of whom eight are women, apply for a job. If five of the applicants are sampled randomly, what is the probability that exactly three of those sampled are women?

This problem contains a small, finite population of 24, or $N = 24$. A sample of five applicants is taken, or $n = 5$. The sampling is being done without replacement, because the five applicants selected for the sample are five different people. The sample size is 21% of the population, which is greater than 5% of the population ($n/N = 5/24 = 0.21$). The hypergeometric distribution is the appropriate distribution to use. The population breakdown is $A = 8$ women (successes) and $N - A = 24 - 8 = 16$ men. The probability of getting $x = 3$ women in the sample of $n = 5$ is

$$\frac{{}_8C_3 \cdot {}_{16}C_2}{{}_{24}C_5} = \frac{(56)(120)}{42,504} = .1581$$

Conceptually, the combination in the denominator of the hypergeometric formula yields all the possible ways of getting n samples from a population, N, including the ones with the desired outcome. In this problem, there are 42,504 ways of selecting five people from 24 people. The numerator of the hypergeometric formula computes all the possible ways of getting x successes from the A successes available and $n - x$ failures from the $N - A$ available failures in the population. There are 56 ways of getting three women from a pool of eight, and there are 120 ways of getting two men from a pool of 16. The combinations of each are multiplied in the numerator because the joint probability of getting x successes and $n - x$ failures is being computed.

DEMONSTRATION PROBLEM 5.11

Suppose 18 major investment banks operate in the United Kingdom and that 12 are located in the City of London. If three investment banks are selected randomly from the entire list, what is the probability that one or more of the selected investment banks are located in the City of London?

Solution

$$N = 18, n = 3, A = 12, \text{ and } x \geq 1$$

This problem is actually three problems in one: $x = 1$, $x = 2$, and $x = 3$. Sampling is being done without replacement, and the sample size is 16.6% of the population. Hence this problem is a candidate for the hypergeometric distribution. The solution follows.

$$\begin{array}{ccc} x=1 & x=2 & x=3 \end{array}$$

$$\frac{{}_{12}C_1 \cdot {}_6C_2}{{}_{18}C_3} + \frac{{}_{12}C_2 \cdot {}_6C_1}{{}_{18}C_3} + \frac{{}_{12}C_3 \cdot {}_6C_0}{{}_{18}C_3} =$$

$$0.2206 + 0.4853 + 0.2896 = 0.9755$$

An alternative solution method using the law of complements would be one minus the probability that none of the investment banks is located in the City of London, or

$$1 - P(x = 0 | N = 18, n = 3, A = 12)$$

Thus,

$$1 - \frac{_{12}C_0 \cdot {_6}C_3}{_{18}C_3} = 1 - 0.0245 = 0.9755$$

Using the Computer to Solve for Hypergeometric Distribution Probabilities

Using Minitab or Excel, it is possible to solve for hypergeometric distribution probabilities on the computer. Both software packages require the input of N, A, n, and x. In either package, the resulting output is the exact probability for that particular value of x. The Minitab output for the example presented in this section, where $N = 24$ people of whom $A = 8$ are women, $n = 5$ are randomly selected, and $x = 3$ are women, is displayed in Table 5.13. Note that Minitab represents successes in the population as 'M'. The Excel output for this same problem is presented in Table 5.14.

TABLE 5.13

Minitab Output for Hypergeometric Problem

```
Probability Density Function

Hypergeometric with N=24, M=8 and n=24
   X              P(X=x)
   3              .158103
```

TABLE 5.14

Excel Output for a Hypergeometric Problem

The probability of $x = 3$ when $N = 24$, $n = 5$ and $A = 8$ is 0.158103

5.5 PROBLEMS

5.27 Compute the following probabilities by using the hypergeometric formula:
  a. The probability of $x = 3$ if $N = 11$, $A = 8$, and $n = 4$
  b. The probability of $x < 2$ if $N = 15$, $A = 5$, and $n = 6$
  c. The probability of $x = 0$ if $N = 9$, $A = 2$, and $n = 3$
  d. The probability of $x > 4$ if $N = 20$, $A = 5$, and $n = 7$

5.28 Shown here are the top 19 companies in the world in terms of oil refining capacity. Some of the companies are privately owned and others are state owned. Suppose six companies are randomly selected.
  a. What is the probability that exactly one company is privately owned?
  b. What is the probability that exactly four companies are privately owned?
  c. What is the probability that all six companies are privately owned?
  d. What is the probability that none of the companies is privately owned?

Company	Ownership Status
Exxon Mobil	Private
Royal Dutch/Shell	Private
British Petroleum	Private

(*continued*)

Company	Ownership Status
Sinopec	Private
Valero Energy	Private
Petroleos de Venezuela	State
Total	Private
ConocoPhillips	Private
China National	State
Saudi Arabian	State
Chevron	Private
Petroleo Brasilerio	State
Petroleos Mexicanos	State
National Iranian	State
OAO Yukos	Private
Nippon	Private
OAO Lukoil	Private
Repsol YPF	Private
Kuwait National	State

5.29 *The Hay Group* lists the top 20 Best Companies for Leadership annually. According to the latest report, General Electric is number 1, followed by Procter & Gamble and Intel Corporation. Suppose that of the 20 firms on the list, eight are in some type of computer-related business and that five firms are randomly selected.

a. What is the probability that none of the firms is in some type of computer-related business?

b. What is the probability that all five firms are in some type of computer-related business?

c. What is the probability that exactly three are in non-computer-related business?

5.30 W. Edwards Deming in his red bead experiment had a box of 4,000 beads, of which 800 were red and 3,200 were white.* Suppose a researcher were to conduct a modified version of the red bead experiment. In her experiment, she has a bag of 20 beads, of which four are red and 16 are white. This experiment requires a participant to reach into the bag and randomly select five beads without replacement.

a. What is the probability that the participant will select exactly four white beads?

b. What is the probability that the participant will select exactly four red beads?

c. What is the probability that the participant will select all red beads?

5.31 Shown here are the top 10 largest hotels in the world ranked by number of rooms according to *Wikipedia*.

Rank	Name	City	Number of Rooms
1	Izmailovo	Moscow	7,500
2	The Venetian/Palazzo	Las Vegas	7,117
3	First World Hotel Genting Highlands	Malaysia	6,118
4	MGM Grand Las Vegas	Las Vegas	5,044
5	Wynn Las Vegas/Encore Las Vegas	Las Vegas	4,750
6	Luxor Las Vegas	Las Vegas	4,408
7	Mandalay Bay/THEhotel	Las Vegas	4,332
8	Ambassador City Jomtien	Pattaya, Thailand	4,219
9	Excalibur Hotel and Casino	Las Vegas	4,008
10	Aria Resort & Casino	Las Vegas	4,004

Suppose four of these hotels are selected randomly.

a. What is the probability that exactly one hotel is in Moscow?

b. What is the probability that none of the hotels are in Pattaya?

c. What is the probability that exactly three of the hotels have more than 5,000 rooms?

*Mary Walton, 'Deming's Parable of Red Beads', *Across the Board* (February 1987): 43–48.

5.32 A company produces and ships 16 personal computers knowing that four of them have defective wiring. The company that purchased the computers is going to thoroughly test three of the computers. The purchasing company can detect the defective wiring. What is the probability that the purchasing company will find the following?

a. No defective computers

b. Exactly three defective computers

c. Two or more defective computers

d. One or no defective computers

5.33 A European police force has 18 police officers eligible for promotion. Eleven of the 18 are from an ethnic minority group. Suppose only five of the police officers are chosen for promotion and that one is from an ethnic minority group. If the officers chosen for promotion had been selected by chance alone, what is the probability that one or fewer of the five promoted officers would have been from an ethnic minority group? What might this result indicate?

Decision Dilemma SOLVED

Life with a Mobile Phone

Suppose that 14% of mobile phone owners in the United States use only cellular phones. If 20 Americans are randomly selected, what is the probability that more than seven use only mobile phones? Converting the 14% to a proportion, the value of p is 0.14, and this is a classic binomial distribution problem with $n = 20$ and $x > 7$. Because the binomial distribution probability tables (Appendix A, Table A.2) do not include $p = 0.14$, the problem will have to be solved using the binomial formula for each of $x = 8$, 9, 10, 11, ..., 20.

$$\text{For } x = 8: \quad {}_{20}C_8(0.14)^8(0.86)^{12} = 0.0030$$

Solving for $x = 9$, 10, and 11 in a similar manner results in probabilities of 0.0007, 0.0001 and 0.0000, respectively. Since the probabilities 'zero out' at $x = 11$, we need not proceed on to $x = 12, 13, 14, \ldots, 20$. Summing these four probabilities ($x = 8$, $x = 9$, $x = 10$, and $x = 11$) results in a total probability of 0.0038 as the answer to the posed question. To further understand these probabilities, we calculate the expected value of this distribution as:

$$\mu = n \cdot p = 20(0.14) = 2.8$$

In the long run, one would expect to average about 2.8 Americans out of every 20 who consider their mobile phone as their primary phone number. In light of this, there is a very small probability that more than seven Americans would do so.

Another study stated that 9 out of 10 mobile users encounter others using their phones in an annoying way. Converting this to $p = 0.90$ and using $n = 25$ and $x < 20$ this, too, is a binomial problem, but it can be solved by using the binomial tables obtaining the values shown below:

x	Probability
19	.024
18	.007
17	.002
16	.000

The total of these probabilities is 0.033. Probabilities for all other values ($x \leq 15$) are displayed as 0.000 in the binomial probability table and are not included here. If 90% of all mobile phone users encounter others using their phones in an annoying way, the probability is very small (0.033) that out of 25 randomly selected mobile phone users fewer than 20 encounter others using their phones in an annoying way. The expected number in any random sample of 25 is $(25)(0.90) = 22.5$.

Suppose, on average, mobile phone users receive 3.6 calls per day. Given that information, what is the probability that a mobile phone user receives no calls per day? Since random telephone calls are generally thought to be Poisson distributed, this problem can be solved by using either the Poisson probability formula or the Poisson tables (A.3, Appendix A). In this problem, $\lambda = 3.6$ and $x = 0$, and the probability associated with this is:

$$\frac{\lambda^x e^{-\lambda}}{x!} = \frac{(3.6)^0 e^{-3.6}}{0!} = 0.0273$$

x	Probability
5	.1377
6	.0826
7	.0425
8	.0191
9	.0076
10	.0028
11	.0009
12	.0003
13	.0001
14	.0000
total	.2936

What is the probability that a mobile phone user receives five or more calls in a day? Since this is a cumulative probability question ($x \geq 5$), the best option is to use the Poisson probability tables (Table A.3 in Appendix A) to obtain:

There is a 29.36% chance that a mobile phone user will receive five or more calls per day if, on average, such a mobile phone user averages 3.6 calls per day.

SUMMARY

Probability experiments produce random outcomes. A variable that contains the outcomes of a random experiment is called a random variable. Random variables such that the set of all possible values is at most a finite or countable infinite number of possible values are called discrete random variables. Random variables that take on values at all points over a given interval are called continuous random variables. Discrete distributions are constructed from discrete random variables. Continuous distributions are constructed from continuous random variables. Three discrete distributions are the binomial distribution, Poisson distribution, and hypergeometric distribution.

The binomial distribution fits experiments when only two mutually exclusive outcomes are possible. In theory, each trial in a binomial experiment must be independent of the other trials. However, if the population size is large enough in relation to the sample size ($n < 5\%N$), the binomial distribution can be used where applicable in cases where the trials are not independent. The probability of getting a desired outcome on any one trial is denoted as p, which is the probability of getting a success. The binomial formula is used to determine the probability of obtaining x outcomes in n trials. Binomial distribution problems can be solved more rapidly with the use of binomial tables than by formula. Table A.2 of Appendix A contains binomial tables for selected values of n and p.

The Poisson distribution usually is used to analyse phenomena that produce rare occurrences. The only information required to generate a Poisson distribution is the long-run average, which is denoted by lambda (λ). The Poisson distribution pertains to occurrences over some interval. The assumptions are that each occurrence is independent of other occurrences and that the value of lambda remains constant throughout the experiment. Poisson probabilities can be determined by either the Poisson formula or the Poisson tables in Table A.3 of Appendix A. The Poisson distribution can be used to approximate binomial distribution problems when n is large ($n > 20$), p is small, and $n \cdot p \leq 7$.

The hypergeometric distribution is a discrete distribution that is usually used for binomial-type experiments when the population is small and finite and sampling is done without replacement. Because using the hypergeometric distribution is a tedious process, using the binomial distribution whenever possible is generally more advantageous.

KEY TERMS

binomial distribution
continuous distributions
continuous random variables
discrete distributions
discrete random variables

hypergeometric distribution
lambda (λ)
mean/expected value
Poisson distribution
random variable

GO ONLINE TO DISCOVER THE EXTRA FEATURES FOR THIS CHAPTER

The Student Study Guide containing solutions to the odd-numbered questions, additional Quizzes and Concept Review Activities, Excel and Minitab databases, additional data files in Excel and Minitab, and more worked examples.
www.wiley.com/college/cortinhas

FORMULAS

Mean (expected) value of a discrete distribution

$$\mu = E(x) = \Sigma[x \cdot P(x)]$$

Variance of a discrete distribution

$$\sigma^2 = \Sigma[(x - \mu)^2 \cdot P(x)]$$

Standard deviation of a discrete distribution

$$\sigma = \sqrt{\Sigma[(x - \mu)^2 \cdot P(x)]}$$

Binomial formula

$$_nC_x \cdot p^x \cdot q^{n-x} = \frac{n!}{x!(n-x)!} \cdot p^x \cdot q^{n-x}$$

Mean of a binomial distribution

$$\mu = n \cdot p$$

Standard deviation of a binomial distribution

$$\sigma = \sqrt{n \cdot p \cdot q}$$

Poisson Formula

$$P(x) = \frac{\lambda^x e^{-\lambda}}{x!}$$

Hypergeometric formula

$$P(x) = \frac{_AC_x \cdot _{N-A}C_{n-x}}{_NC_n}$$

ETHICAL CONSIDERATIONS

Several points must be emphasized about the use of discrete distributions to analyse data. The independence and/or size assumptions must be met in using the binomial distribution in situations where sampling is done without replacement. Size and lambda assumptions must be satisfied in using the Poisson distribution to approximate binomial problems. In either case, failure to meet such assumptions can result in spurious conclusions.

As n increases, the use of binomial distributions to study exact x-value probabilities becomes questionable in decision making. Although the probabilities are mathematically correct, as n becomes larger, the probability of any particular x value becomes lower because there are more values among which to split the probabilities. For example, if $n = 100$ and $p = 0.50$, the probability of $x = 50$ is 0.0796. This probability of occurrence appears quite low, even though $x = 50$ is the expected value of this distribution and is also the value most likely to occur. It is more useful to decision makers and, in a sense, probably more ethical to present cumulative values for larger sizes of n. In this example, it is probably more useful to examine $P(x > 50)$ than $P(x = 50)$.

The reader is warned in the chapter that the value of λ is assumed to be constant in a Poisson distribution experiment. Researchers may produce spurious results because the λ value changes during a study. For example, suppose the value of λ is obtained for the number of customer arrivals at a toy store between 7 P.M. and 9 P.M. in the month of December. Because December is an active month in terms of traffic volume through a toy store, the use of such a λ to analyse arrivals at the same store between noon and 2 P.M. in February would be inappropriate and, in a sense, unethical.

Errors in judgement such as these are probably more a case of misuse than lack of ethics. However, it is important that statisticians and researchers adhere to assumptions and appropriate applications of these techniques. The inability or unwillingness to do so opens the way for unethical decision making.

SUPPLEMENTARY PROBLEMS

CALCULATING THE STATISTICS

5.34 Solve for the probabilities of the following binomial distribution problems by using the binomial formula:

  a. If $n = 11$ and $p = 0.23$, what is the probability that $x = 4$?

  b. If $n = 6$ and $p = 0.50$, what is the probability that $x \geq 1$?

  c. If $n = 9$ and $p = 0.85$, what is the probability that $x > 7$?

  d. If $n = 14$ and $p = 0.70$, what is the probability that $x \leq 3$?

5.35 Use Table A.2, Appendix A, to find the values of the following binomial distribution problems:

  a. $P(x = 14 | n = 20$ and $p = 0.60)$

  b. $P(x < 5 | n = 10$ and $p = 0.30)$

  c. $P(x \geq 12 | n = 15$ and $p = 0.60)$

  d. $P(x > 20 | n = 25$ and $p = 0.40)$

5.36 Use the Poisson formula to solve for the probabilities of the following Poisson distribution problems:

  a. If $\lambda = 1.25$, what is the probability that $x = 4$?

  b. If $\lambda = 6.37$, what is the probability that $x \leq 1$?

  c. If $\lambda = 2.4$ what is the probability that $x > 5$?

5.37 Use Table A.3, Appendix A, to find the following Poisson distribution values:

  a. $P(x = 3 | \lambda = 1.8)$

  b. $P(x < 5 | \lambda = 3.3)$

  c. $P(x \geq 3 | \lambda = 2.1)$

  d. $P(2 < x \leq 5 | \lambda = 4.2)$

5.38 Solve the following problems by using the hypergeometric formula:

  a. If $N = 6$, $n = 4$, and $A = 5$, what is the probability that $x = 3$?

  b. If $N = 10$, $n = 3$, and $A = 5$, what is the probability that $x \leq 1$?

  c. If $N = 13$, $n = 5$, and $A = 3$, what is the probability that $x \geq 2$?

TESTING YOUR UNDERSTANDING

5.39 In a study by Syed Sultana recently published in the *Global Journal of Finance and Management*, it was determined that 20% of all stock investors in India are women. In addition, 22% of all Indian investors are not married. Suppose a random sample of 25 stock investors is taken.

  a. What is the probability that exactly eight are women?

  b. What is the probability that 12 or more are women?

  c. How many women would you expect to find in a random sample of 25 stock investors?

d. Suppose a random sample of 20 Indian stock investors is taken. What is the probability that exactly eight persons are not married?

e. Suppose a random sample of 20 Indian stock investors is taken. What is the probability that fewer than six persons are not married?

f. Suppose a random sample of 20 Indian stock investors is taken. What is the probability that none of the persons is not married?

g. Suppose a random sample of 20 Indian stock investors is taken. What is the probability that 12 or more persons are not married?

h. For parts e–g, what exact number of adults would produce the highest probability? How does this compare with the expected number?

5.40 A service station has a pump that distributes diesel fuel to automobiles. The station owner estimates that only about 3.2 cars use the diesel pump every two hours. Assume the arrivals of diesel pump users are Poisson distributed.

a. What is the probability that three cars will arrive to use the diesel pump during a one-hour period?

b. Suppose the owner needs to shut down the diesel pump for half an hour to make repairs. However, the owner hates to lose any business. What is the probability that no cars will arrive to use the diesel pump during a half-hour period?

c. Suppose five cars arrive during a one-hour period to use the diesel pump. What is the probability of five or more cars arriving during a one-hour period to use the diesel pump? If this outcome actually occurred, what might you conclude?

5.41 In a particular manufacturing plant, two machines (A and B) produce a particular part. One machine (B) is newer and faster. In one five-minute period, a lot consisting of 32 parts is produced. Twenty-two are produced by machine B and the rest by machine A. Suppose an inspector randomly samples a dozen of the parts from this lot.

a. What is the probability that exactly three parts were produced by machine A?

b. What is the probability that half of the parts were produced by each machine?

c. What is the probability that all of the parts were produced by machine B?

d. What is the probability that seven, eight, or nine parts were produced by machine B?

5.42 Suppose that for every lot of 100 computer chips a company produces an average of 1.4 are defective. Another company buys many lots of these chips at a time, from which one lot is selected randomly and tested for defects. If the tested lot contains more than three defects, the buyer will reject all the lots sent in that batch. What is the probability that the buyer will accept the lots? Assume that the defects per lot are Poisson distributed.

5.43 The National Public Health Service for Wales reports that 33% of all Welsh adults have at least one chronic condition. Suppose you live in a county where the environment is conducive to good health and low stress and you believe the conditions in your county promote good health. To investigate this theory, you conduct a random telephone survey of 20 adults in your county.

a. On the basis of the figure from the National Public Health Service for Wales, what is the expected number of adults in your survey who have a chronic condition?

b. Suppose only one person in your survey has a chronic condition. What is the probability of getting one or fewer people with a chronic condition in a sample of 20 if 33% of the adult population has a health problem? What do you conclude about your state from the sample data?

5.44 A survey conducted by the European Agency for Health and Safety at Work revealed that 79% of European managers were concerned about the levels of stress in the workplace. Despite high levels of concern, only 26 per cent of EU organisations have procedures in place to deal with stress. The survey said work-related stress is 'very acute' in health and social work, where 91% of companies regard it as of some or major concern, and in education, where the figure is 84%.

a. Suppose a random sample of 10 European managers is selected. What is the probability that more than seven of them are concerned about stress at work? What is the expected number of managers who are concerned about stress at work?

b. Suppose a random sample of 15 European health and social companies is selected. What is the expected number of these sampled companies who regard work-related stress as of some or major concern? What is the probability that none of the companies regard work-related stress as of some or major concern?

c. Suppose a sample of seven companies from the education sector is selected randomly. What is the probability that all seven say they regard work-related stress as of some or major concern?

5.45 According to Padgett Business Services, 20% of all small-business owners say the most important advice for starting a business is to prepare for long hours and hard work. Twenty-five per cent say the most important advice is to have good financing ready. Nineteen per cent say having a good plan is the most important advice; 18% say studying the industry is the most important advice; and 18% list other advice. Suppose 12 small-business owners are contacted, and assume that the percentages hold for all small-business owners.

a. What is the probability that none of the owners would say preparing for long hours and hard work is the most important advice?

b. What is the probability that six or more owners would say preparing for long hours and hard work is the most important advice?

c. What is the probability that exactly five owners would say having good financing ready is the most important advice?

d. What is the expected number of owners who would say having a good plan is the most important advice?

5.46 According to a recent survey, the probability that a passenger files a complaint about a particular airline is 0.000014. Suppose 100,000 passengers who flew with this particular airline are randomly contacted.

a. What is the probability that exactly five passengers filed complaints?

b. What is the probability that none of the passengers filed complaints?

c. What is the probability that more than six passengers filed complaints?

5.47 A hair stylist has been in business one year. Sixty per cent of his customers are walk-in business. If he randomly samples eight of the people from last week's list of customers, what is the probability that three or fewer were walk-ins? If this outcome actually occurred, what would be some of the explanations for it?

5.48 A recent Barclaycard Business Travel survey showed that six out of 10 business travellers would not want in-flight mobile technology introduced even if there were no issues with the phones interfering with aircraft communications systems. If this information is correct and if a researcher randomly selects 25 business travellers,

a. What is the probability that exactly 12 oppose the use of mobile phones in flight?

b. What is the probability that more than 17 oppose the use of mobile phones in flight?

c. What is the probability that fewer than eight oppose the use of mobile phones in flight? If the researcher actually got fewer than eight, what might she conclude about the Barclaycard Business Travel survey?

5.49 A survey conducted by the Consumer Reports National Research Center reported, among other things, that women spend an average of 1.2 hours per week shopping online. Assume that hours per week shopping online are Poisson distributed. If this survey result is true for all women and if a woman is randomly selected,

a. What is the probability that she did not shop at all online over a one-week period?

b. What is the probability that a woman would shop three or more hours online during a one-week period?

c. What is the probability that a woman would shop fewer than five hours in a three-week period?

5.50 According to the World Association of Newspapers, the top 25 newspapers in the world ranked according to circulation are:

Rank	Title	Country
1	*Yomiuri Shimbun*	Japan
2	*The Asahi Shimbun*	Japan
3	*Mainichi Shimbun*	Japan
4	*Nihon Keizai Shimbun*	Japan
5	*Chunichi Shimbun*	Japan
6	*Bild*	Germany
7	*Sankei Shimbun*	Japan
8	*Canako Xiaoxi* (Beijing)	China
9	*People's Daily*	China
10	*Tokyo Sports*	Japan
11	*The Sun*	United Kingdom
12	*The Chosun Ilbo*	South Korea
13	*USA Today*	USA
14	*The Wall Street Journal*	USA
15	*Daily Mail*	UK
16	*The Joongang Ilbo*	South Korea
17	*The Dong-A Ilbo*	South Korea
18	*Nikkan Sports*	Japan
19	*Hokkaido Shimbun*	Japan

(continued)

Rank	Title	Country
20	*Dainik Jagran*	India
21	*Yangtse Evening Post*	China
22	*Sports Nippon*	Japan
23	*The Nikkan Gendai*	Japan
24	*Times of India*	India
25	*Guangzhou Daily*	China

Suppose a researcher wants to sample a portion of these newspapers and compare the sizes of the business sections of the Sunday papers. She randomly samples eight of these newspapers.

a. What is the probability that the sample contains exactly one newspaper located in Germany?

b. What is the probability that half of the newspapers are ranked in the top 10 by circulation?

c. What is the probability that none of the newspapers is located in South Korea?

d. What is the probability that exactly three of the newspapers are located in countries that begin with the letter *U* ?

5.51 An office in Lisbon has 24 workers including management. Eight of the workers commute to work from the south side of the Tagus River. Suppose six of the office workers are randomly selected.

a. What is the probability that all six workers commute from the south side of the Tagus river?

b. What is the probability that none of the workers commute from the south side of the Tagus river?

c. Which probability from parts (a) and (b) was greatest? Why do you think this is?

d. What is the probability that half of the workers do not commute from the south side of the Tagus river?

5.52 According to Virgin Vacations, 50% of the workers in Amsterdam commute to work by bicycle, making it 'The Bike Capital of the World'. If 25 Amsterdam workers are randomly selected, what is the expected number that commute to work by bicycle? Graph the binomial distribution for this sample. What are the mean and the standard deviation for this distribution? What is the probability that more than 12 of the selected workers commute to work by bicycle? Explain conceptually and from the graph why you would get this probability. Suppose you randomly sample 25 Amsterdam workers and actually get 14 who commute to work by bicycle. Is this outcome likely?

5.53 One of the earliest applications of the Poisson distribution was in analysing incoming calls to a telephone switchboard. Analysts generally believe that random phone calls are Poisson distributed. Suppose phone calls to a switchboard arrive at an average rate of 2.4 calls per minute.

a. If an operator wants to take a one-minute break, what is the probability that there will be no calls during a one-minute interval?

b. If an operator can handle at most five calls per minute, what is the probability that the operator will be unable to handle the calls in any one-minute period?

c. What is the probability that exactly three calls will arrive in a two-minute interval?

d. What is the probability that one or fewer calls will arrive in a 15-second interval?

5.54 A Guardian/ICM poll revealed that only 6% of Europeans say they have a great deal of trust in their government. Another researcher randomly selects 70 Europeans.

a. What is the expected number of these who would say they have a great deal of trust in their government?

b. What is the probability that eight or more say they have a great deal of trust in their government?

c. What is the probability that between three and six (inclusive) have a great deal of trust in their government?

5.55 Suppose that in the book-keeping operation of a large corporation the probability of a recording error on any one billing is 0.005. Suppose the probability of a recording error from one billing to the next is constant, and 1,000 billings are randomly sampled by an auditor.

a. What is the probability that fewer than four billings contain a recording error?

b. What is the probability that more than 10 billings contain a billing error?

c. What is the probability that all 1,000 billings contain no recording errors?

5.56 According to the Norwegian Ministry of Children and Equality, about 34% of all Norwegian doctors are women. Your company has just hired eight doctors and none is a woman. If a group of women doctors want to sue your company for discriminatory hiring practices, would they have a strong case based on these numbers? Use the binomial distribution to determine the probability of the company's hiring result occurring randomly, and comment on the potential justification for a lawsuit.

5.57 The table below lists the 25 largest universities in England according to enrolment figures from The Higher Education Statistics Agency.

University	Region of Institution	Number of Students
The Open University	SEAS	209,705
The University of Manchester	NWES	40,400
The Manchester Metropolitan University	NWES	35,520
Sheffield Hallam University	YORH	35,410
The University of Nottingham	EMID	34,120
The University of Leeds	YORH	33,585
University of the West of England, Bristol	SWES	32,840
The University of Central Lancashire	NWES	32,295
The University of Plymouth	SWES	32,200
The University of Northumbria at Newcastle	NEAS	31,425
The University of Birmingham	WMID	30,125
The University of Warwick	WMID	28,870
The University of Greenwich	LOND	28,805
The University of Teesside	NEAS	28,630
Leeds Metropolitan University	YORH	27,865
University of Hertfordshire	EAST	27,650
Kingston University	LOND	27,085
The University of East London	LOND	26,930
The Nottingham Trent University	EMID	26,455
The University of Sheffield	YORH	25,970
Liverpool John Moores University	NWES	25,860
Edge Hill University	NWES	25,320
London South Bank University	LOND	24,945
Birmingham City University	WMID	24,835
King's College London	LOND	24,500

a. If five different universities are selected randomly from the list, what is the probability that three of them have enrolments of 28,000 or more?

b. If eight different universities are selected randomly from the list, what is the probability that two or fewer are universities from the 'YORH' region?

c. Suppose universities are being selected randomly from this list with replacement. If five universities are sampled, what is the probability that the sample will contain exactly two universities in 'EAST' region?

5.58 In one European city, the government has 14 repossessed houses, which are evaluated to be worth about the same. Ten of the houses are on the north side of town and the rest are on the west side. A local contractor submitted a bid to purchase four of the houses. Which houses the contractor will get is subject to a random draw.

a. What is the probability that all four houses selected for the contractor will be on the north side of town?

b. What is the probability that all four houses selected for the contractor will be on the west side of town?

c. What is the probability that half of the houses selected for the contractor will be on the west side and half on the north side of town?

5.59 The Conseil National de l'Ordre des Médecins produces an annual atlas on the spread of doctors in France. The national average was 3.09 doctors per 1,000 habitants but there are large differences between regions. The region with the lowest number was Picardie, with 2.39 doctors per 1,000 habitants and the highest was Provence-Alpes-Côte d'Azur with 3.74. Assume that the numbers of doctors per 1,000 habitants in France is Poisson distributed.

a. What is the probability of randomly selecting 1,000 French habitants and finding no doctor?

b. What is the probability of randomly selecting 2,000 French habitants and finding six doctors?

c. What is the probability of randomly selecting 3,000 Picardie habitants and finding fewer than seven doctors?

INTERPRETING THE OUTPUT

5.60 Study the Minitab output. Discuss the type of distribution, the mean, standard deviation, and why the probabilities fall as they do.

```
Probability Density Function
Binomial with n=15 and n=0.36
   x                      P(X=x)

   0                      .001238
   1                      .010445
   2                      .041128
   3                      .100249
```

(continued)

x	P(X = x)
4	.169170
5	.209347
6	.196263
7	.141940
8	.079841
9	.034931
10	.011789
11	.003014
12	.000565
13	.000073
14	.000006
15	.000000

5.61 Study the Excel output. Explain the distribution in terms of shape and mean. Are these probabilities what you would expect? Why or why not?

x Values	Poisson Probabilities: $\lambda = 2.78$
0	.0620
1	.1725
2	.2397
3	.2221
4	.1544
5	.0858
6	.0398
7	.0158
8	.0055
9	.0017
10	.0005
11	.0001

5.62 Study the graphical output from Excel. Describe the distribution and explain why the graph takes the shape it does.

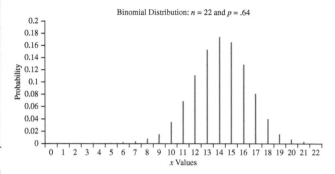

5.63 Study the Minitab graph. Discuss the distribution including type, shape, and probability outcomes.

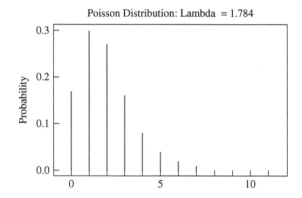

ANALYSING THE DATABASES

1. Use the Financial database. What proportion of the database companies are in the clothing industry? Use this as the value of p in a binomial distribution. If you were to randomly select 12 of these companies, what is the probability that fewer than three would be in the clothing industry? If you were to randomly select 25 of these companies, what is the probability that exactly eight would be in the clothing industry?

2. Use the Hospital database. What is the breakdown between hospitals that are general medical hospitals and those that are psychiatric hospitals in this database of 200 hospitals? (Hint: In Service, 1 = general medical and 2 = psychiatric). Using these figures and the hypergeometric distribution, determine the probability of randomly selecting 16 hospitals from the database and getting exactly nine that are psychiatric hospitals. Now, determine the number of hospitals in this database that are for-profit (Hint: In Control, 3 = for-profit). From this number, calculate p, the proportion of hospitals that are for-profit. Using this value of p and the binomial distribution, determine the probability of randomly selecting 30 hospitals and getting exactly 10 that are for-profit.

KODAK TRANSITIONS WELL INTO THE DIGITAL CAMERA MARKET BUT FACES BIG CHALLENGES

George Eastman was born in 1854 in Upstate New York. A high school dropout because of the untimely death of his father and the ensuing financial needs of his family, Eastman began his business career as a 14-year-old office boy in an insurance company, eventually taking charge of policy filing and even policy writing. In a few years, after studying accounting on the side, he became a bank clerk. When Eastman was 24, he was making plans for a trip to the Caribbean when a co-worker suggested that he record the trip. Eastman bought considerable photographic equipment and learned how to take pictures. He described the complete outfit of his photographic equipment as a 'pack-horse load'. Eastman never made the trip to the Caribbean, but he did spend the rest of his life in pursuit of simplifying the complicated photographic process, reducing the size of the load, and making picture taking available to everyone.

After experimenting for several years with various chemical procedures related to the photographic process at night while maintaining his job at the bank during the day, Eastman found a dry plate formula that worked, and he patented a machine for preparing large numbers of dry plates. In April 1880, he leased the third floor of a building in Rochester, New York, and began to manufacture dry plates for sale. Soon Eastman realized that what he was actually doing was attempting to make photography an everyday affair. He described it as trying 'to make the camera as convenient as the pencil'. In Eastman's experiments, he discovered how to use paper with a layer of plain, soluble gelatin mounted on a roll holder to take pictures rather than using dry plates. In 1885, he began mass advertising his products, and in 1888, he introduced the Kodak camera. By 1896, 100,000 Kodak cameras had been produced, and film and photographic paper was being made at a rate of about 400 miles a month. Today, the trademark, 'Kodak', coined by Eastman, himself, is known around the world for excellence in photographic products. Kodak has manufacturing operations in North America, South America, and Europe, and Kodak products are available in virtually every country in the world. Kodak's products include digital cameras, healthcare scanners, printing equipment, imaging, radiography, projectors, film, digital imaging products, and many more.

Kodak has recently faced aggressive competition in the digital camera arena, particularly from Canon, Fujifilm, Nikon, and Sony. As a result, it has seen its market share slowly eroded over the years. Kodak has been trying to win back market share by expanding its presence in the Asia-Pacific region and by forming a number of license agreements, including some with the mobile phone producers Nokia and LG. By the year 2010, the Eastman Kodak company had 18,800 employees and sales of over $7.2 billion.

DISCUSSION

Suppose you are a part of a Kodak team whose task it is to examine quality, customer satisfaction, and market issues. Using techniques presented in this chapter, analyse and discuss the following questions:

1. According to Digitimes Research, Kodak is number six in the sales of digital cameras in the world, with a market share of 7.6%. However, your team wants to confirm that this figure is constant for various geographic segments of the world. In an effort to study this issue, a random sample of 30 current purchasers of digital cameras is taken in each of the distribution networks in America, Europe, the Middle East and Africa, and Asia Pacific. If the 16% market share figure is constant across regions, how many of the 30 purchases of digital cameras would the company expect to be Kodak cameras in each region? If eight or more of the 30 purchases in America are Kodak, what might that tell the team? If fewer than three of the 30 purchases in Europe are Kodak, what does that mean? Suppose none of the 30 purchases in the Middle East and Africa is Kodak brand digital cameras. Is it still possible that Kodak holds 16% of the market share? Why or why not? If, indeed, Kodak holds a 16% share of the market in Asia Pacific, is it likely that in a sample of 30 purchases that 20 or more are Kodak? Explain to the team.

2. Digital cameras have quickly replaced film cameras in recent years. Companies that did not respond to the rapid market change were severely hurt and several went out of business. Kodak responded by embracing the new digital picture-taking platform,

while continuing its efforts in the production and marketing of film and film cameras, thereby trying to create a seamless shift into the digital market. Kodak finally stopped producing the famous Kodachrome film in 2009. Suppose the Kodak team wants to ascertain if people take more or fewer pictures with the digital format than they did with film cameras. Suppose in a previous study using film cameras, it was determined that, on average during daylight hours, families on holiday took 1.5 pictures per hour. Using this figure as a guide, if the Kodak team randomly samples families on holiday in various parts of the world who are using digital cameras to take pictures, what is the probability that a family takes four or more pictures in an hour? What is the probability that a family takes six or more pictures per hour? What is the probability that a family takes nine or more pictures per hour? What might the answers to these questions indicate about the usage of digital cameras versus film cameras?

3. According to a survey by J.D. Power and Associates, Kodak V series camera ranked second highest in customer satisfaction in the point and shoot segment in a recent year, after Fujifilm's Finepix S model. Suppose a consumer group conducts a study of 60 recent purchasers of digital cameras, of which 14 own a Kodak V series camera. In the study, camera owners are asked to rate their satisfaction with their cameras on a scale from 0 to 100. The top 10 satisfaction scores are taken, and 4 of the top 10 are from owners of Kodak V series camera. Is this about what is to be expected given the number of owners of this camera in the pool of 60 purchasers? If not, how can you explain the disparity? Suppose seven of the top 10 satisfaction scores were obtained from Kodak V series camera. What might this indicate?

Adapted from: Information found at Kodak's website: www.kodak.com and 'Kodak Tops USA Digital Camera Market' at website www.letsgo digital.org/en/news/articles/story_6315.html; www.macworld.com/article/55236/2007/02/cameras.html; www.hoovers.com/eastman-kodak/--ID_10500--/free-co-factsheet.xhtml; http://businesscenter.jdpower.com/news/pressrelease.aspx?ID=2008105; Eastman Kodak Company, Company profile, Datamonitor, 13 September 2010.

USING THE COMPUTER

EXCEL

- Excel can be used to compute exact or cumulative probabilities for particular values of discrete distributions including the binomial, Poisson, and hypergeometric distributions.

- Calculation of probabilities from each of these distributions begins with the **Insert Function** (f_x). To access the **Insert Function**, go to the **Formulas** tab on an Excel worksheet (top centre tab). The **Insert Function** is on the far left of the menu bar. In the **Insert Function** dialog box at the top, there is a pulldown menu where it says **Or select a category**. From the pulldown menu associated with this command, select **Statistical**.

- To compute probabilities from a binomial distribution, select **BINOMDIST** from the **Insert Function's Statistical** menu. In the **BINOMDIST** dialog box, there are four lines to which you must respond. On the first line, **Number_s**, enter the value of x, the number of successes. On the second line,

Trials, enter the number of trials (sample size, n). On the third line, **Probability_s**, enter the value of p. The fourth line, **Cumulative**, requires a logical response of either TRUE or FALSE. Place TRUE in the slot to get the cumulative probabilities for all values from 0 to x. Place FALSE in the slot to get the exact probability of getting x successes in n trials.

- To compute probabilities from a Poisson distribution, select **POISSON** from the **Insert Function's Statistical** menu. In the **POISSON** dialog box, there are three lines to which you must respond. On the first line, **X**, enter the value of x, the number of events. On the second line, **Mean**, enter the expected number, λ. The third line, **Cumulative**, requires a logical response of either TRUE or FALSE. Place TRUE in the slot to get the cumulative probabilities for all values from 0 to x. Place FALSE in the slot to get the exact probability of getting x successes when λ is the expected number.

- To compute probabilities from a hypergeometric distribution, select **HYPGEOMDIST** from the **Insert Function's Statistical** menu. In the **HYPGEOMDIST** dialog box, there are four lines to which you must respond. On the first line, **Sample_s**, enter the value of x, the number of successes in the sample. On the second line, **Number_ sample**, enter the size of the sample, n. On the third line, **Population_s**, enter the number of successes in the population. The fourth line, **Number_pop**, enter the size of the population, N.

MINITAB

- Probabilities can be computed using Minitab for the binomial distribution, the Poisson distribution, and the hypergeometric distribution.

- To begin binomial distribution probabilities, select **Calc** on the menu bar. Select **Probability Distributions** from the pulldown menu. From the long second pulldown menu, select **Binomial**. From the dialog box, check how you want the probabilities to be calculated from **Probability**, **Cumulative probability**, or **Inverse cumulative probability**. **Probability** yields the exact probability n, p, and x. **Cumulative probability** produces the cumulative probabilities for values less than or equal to x. **Inverse probability** yields the inverse of the cumulative probabilities. If you want to compute probabilities for several values of x, place them in a column, and list the column location in the **Input column**. If you want to compute the probability for a particular value of x, check **Input constant**, and enter the value of x.

- To begin Poisson distribution probabilities, select **Calc** on the menu bar. Select **Probability Distributions**

from the pulldown menu. From the long second pulldown menu, select **Poisson**. From the dialog box, check how you want the probabilities to be calculated from **Probability**, **Cumulative probability**, or **Inverse cumulative probability**. **Probability** yields the exact probability of a particular λ, and x. **Cumulative probability** produces the cumulative probabilities for values less than or equal to x. **Inverse probability** yields the inverse of the cumulative probabilities. If you want to compute probabilities for several values of x, place them in a column, and list the column location in the **Input column**. If you want to compute the probability for a particular value of x, check **Input constant**, and enter the value of x.

- To begin hypergeometric distribution probabilities, select **Calc** on the menu bar. Select **Probability Distributions** from the pulldown menu. From the long second pulldown menu, select **Hypergeometric**. From the dialog box, check how you want the probabilities to be calculated from **Probability**, **Cumulative probability**, or **Inverse cumulative probability**. **Probability** yields the exact probability of a particular combination of N, A, n, and x. Note that Minitab uses M for number of successes in the population instead of A. **Cumulative probability** produces the cumulative probabilities for values less than or equal to x. **Inverse probability** yields the inverse of the cumulative probabilities. If you want to compute probabilities for several values of x, place them in a column, and list the column location in the **Input column**. If you want to compute the probability for a particular value of x, check **Input constant**, and enter the value of x.

Continuous Distributions

LEARNING OBJECTIVES

The primary learning objective of Chapter 6 is to help you understand continuous distributions, thereby enabling you to:

1. Solve for probabilities in a continuous uniform distribution
2. Solve for probabilities in a normal distribution using z scores and for the mean, the standard deviation, or a value of x in a normal distribution when given information about the area under the normal curve
3. Solve problems from the discrete binomial distribution using the continuous normal distribution and correcting for continuity
4. Solve for probabilities in an exponential distribution and contrast the exponential distribution to the discrete Poisson distribution

Decision Dilemma

The Cost of Absenteeism in the Workplace

What is the human resource cost of absenteeism? According to a recent survey by PwC Consultants, UK workers take 10 unauthorised days off from work each year. This number is similar to the average level of absenteeism in Western Europe (9.7 days), but is significantly higher than the 4.5 days average in Asia or the 5.5 days in the US. Why are the absenteeism figures in Europe so much worse than in Asia or the US? The reasons are not totally clear. As pointed out by the PwC study, one might assume the perceived US work culture of long hours and short holidays could lead to higher stress and sick rates. The data suggests otherwise, or perhaps demonstrates that strong employee engagement and commitment can override workplace pressures. The study also suggests that labour laws favouring employers in the US and Asia also play a part, with a sense among workers that there is more at stake if they are not committed. There is also the suggestion that European employers might not be investing enough in the health and well-being of their workforce. US (and Asian) firms tend to take greater responsibility for staff well-being, whether providing gyms in the workplace or access to counsellors. The PwC survey estimates that absenteeism is costing British business approximately £32 billion per annum. Similarly, the US Bureau of Labor Statistics estimates that American businesses lose an average of 2.8 million workdays each year due to unplanned absences. These absences cost US employers more than $74 billion.

Managerial and Statistical Questions

1. A survey conducted by the Small Firms Association in Ireland reported that small firms with fewer than 50 employees are less likely to have workers absent on sick leave than larger organizations. While small firms have an average of five working days lost to absenteeism, that number is at least double for large companies. Suppose that number of days of absenteeism is uniformly distributed across all employees varying from 0 days to 15 days. What percentage of employees are absent between two and four days a year? What percentage of employees are absent for 10 days or more?

2. As the result of another survey, it was estimated that the average annual cost of unscheduled absenteeism is $660 per employee in the US. Suppose such costs are normally distributed with a standard deviation of $300. Based on these figures, what is the probability that a randomly selected employee gives rise to absenteeism costs more than $500? What percentage of employees has absenteeism costs less than $600?

3. An absenteeism survey determined that 35% of all unscheduled absenteeism is caused by personal illness. If this is true, what is the probability of randomly sampling 120 unscheduled absences and finding out that more than 50 were caused by personal illness?

Source: Date from websites: www.pwc.com, www.ibec.ie, and www.bls.gov.

Whereas Chapter 5 focused on the characteristics and applications of discrete distributions, this chapter concentrates on information about continuous distributions. Continuous distributions are constructed from continuous random variables in which values are taken on for every point over a given interval and are usually generated from experiments in which things are 'measured' as opposed to 'counted'. With continuous distributions, probabilities of outcomes occurring between particular points are determined by calculating the area under the curve between those points. In addition, the entire area under the whole curve is equal to 1. The many continuous distributions in statistics include the uniform distribution, the normal distribution, the exponential distribution, the t distribution, the chi-square distribution, and the F distribution. This chapter presents the uniform distribution, the normal distribution, and the exponential distribution.

6.1 THE UNIFORM DISTRIBUTION

The **uniform distribution**, sometimes referred to as the **rectangular distribution**, is *a relatively simple continuous distribution in which the same height, or f(x), is obtained over a range of values.* The following probability density function defines a uniform distribution.

PROBABILITY DENSITY FUNCTION OF A UNIFORM DISTRIBUTION	$f(x) = \begin{cases} \dfrac{1}{b-a} & \text{for } a \le x \le b \\ 0 & \text{for all other values} \end{cases}$

Figure 6.1 is an example of a uniform distribution. In a uniform, or rectangular, distribution, the total area under the curve is equal to the product of the length and the width of the rectangle and equals 1. Because the distribution lies, by definition, between the *x* values of *a* and *b*, the length of the rectangle is $(b - a)$. Combining this area calculation with the fact that the area equals 1, the height of the rectangle can be solved as follows.

$$\text{Area of Rectangle} = (\text{Length})(\text{Height}) = 1$$

But

$$\text{Length} = (b - a)$$

Therefore,

$$(b - a)(\text{Height}) = 1$$

and

$$\text{Height} = \frac{1}{(b-a)}$$

These calculations show why, between the *x* values of *a* and *b*, the distribution has a constant height of $1/(b-a)$.

The mean and standard deviation of a uniform distribution are given as follows.

MEAN AND STANDARD DEVIATION OF A UNIFORM DISTRIBUTION	$\mu = \dfrac{a+b}{2}$ $\sigma = \dfrac{b-a}{\sqrt{12}}$

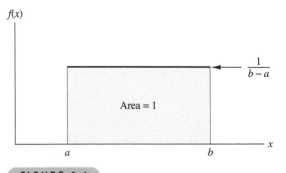

FIGURE 6.1

Uniform Distribution

Many possible situations arise in which data might be uniformly distributed. As an example, suppose a production line is set up to manufacture machine braces in lots of five per minute during a shift. When the lots are weighed, variation among the weights is detected, with lot weights ranging from 41 to 47 grams in a uniform distribution. The height of this distribution is

$$f(x) = \text{Height} = \frac{1}{(b-a)} = \frac{1}{(47-41)} = \frac{1}{6}$$

The mean and standard deviation of this distribution are

$$\text{Mean} = \frac{a+b}{2} = \frac{41+47}{2} = \frac{88}{2} = 44$$

$$\text{Standard Deviation} = \frac{b-a}{\sqrt{12}} = \frac{47-41}{\sqrt{12}} = \frac{6}{3.464} = 1.732$$

Figure 6.2 provides the uniform distribution for this example, with its mean, standard deviation, and the height of the distribution.

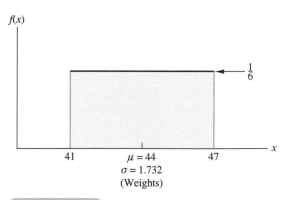

FIGURE 6.2

Distribution of Lot Weights

Determining Probabilities in a Uniform Distribution

With discrete distributions, the probability function yields the value of the probability. For continuous distributions, probabilities are calculated by determining the area over an interval of the function. With continuous distributions, there is no area under the curve for a single point. The following equation is used to determine the probabilities of x for a uniform distribution between a and b.

PROBABILITIES IN A UNIFORM DISTRIBUTION

$$P(x) = \frac{x_2 - x_1}{b-a}$$

where:

$$a \leq x_1 \leq x_2 \leq b$$

Remember that the area between a and b is equal to 1. The probability for any interval that includes a and b is 1. The probability of $x \geq b$ or of $x \leq a$ is zero because there is no area above b or below a.

Suppose that on the machine braces problem we want to determine the probability that a lot weighs between 42 and 45 grams. This probability is computed as follows:

$$P(x) = \frac{x_2 - x_1}{b-a} = \frac{45-42}{47-41} = \frac{3}{6} = 0.5000$$

Figure 6.3 displays this solution.

The probability that a lot weighs more than 48 grams is zero, because $x = 48$ is greater than the upper value, $x = 47$, of the uniform distribution. A similar argument gives the probability of a lot weighing less than 40 grams. Because 40 is less than the lowest value of the uniform distribution range, 41, the probability is zero.

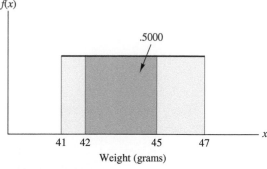

FIGURE 6.3

Solved Probability in a Uniform Distribution

DEMONSTRATION PROBLEM 6.1

Suppose the amount of time it takes to assemble a plastic module ranges from 27 to 39 seconds and that assembly times are uniformly distributed. Describe the distribution. What is the probability that a given assembly will take between 30 and 35 seconds? Fewer than 30 seconds?

Solution

$$f(x) = \frac{1}{39-27} = \frac{1}{12}$$

$$\mu = \frac{a+b}{2} = \frac{39+27}{2} = 33$$

$$\sigma = \frac{b-a}{\sqrt{12}} = \frac{39-27}{\sqrt{12}} = \frac{12}{\sqrt{12}} = 3.464$$

The height of the distribution is 1/12. The mean time is 33 seconds with a standard deviation of 3.464 seconds.

$$P(30 \le x \le 35) = \frac{35-30}{39-27} = \frac{5}{12} = 0.4167$$

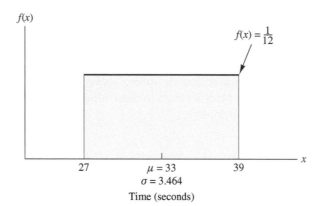

There is a 0.4167 probability that it will take between 30 and 35 seconds to assemble the module.

$$P(x < 30) = \frac{30-27}{39-27} = \frac{3}{12} = 0.2500$$

There is a 0.2500 probability that it will take less than 30 seconds to assemble the module. Because there is no area less than 27 seconds, $P(x < 30)$ is determined by using only the interval $27 \le x < 30$. In a continuous distribution, there is no area at any one point (only over an interval). Thus the probability $x < 30$ is the same as the probability of $x \le 30$.

DEMONSTRATION PROBLEM 6.2

According to the site Confused.com, the average annual cost for car insurance in the United Kingdom now stands at £835. Suppose automobile insurance costs are uniformly distributed in the United Kingdom with a range of from £200 to £1,182. What is the standard deviation of this uniform distribution? What is the height of the distribution? What is the probability that a person's annual cost for automobile insurance in the United Kingdom is between £410 and £825?

Solution

The mean is given as £691. The value of a is £200 and b is £1,182.

$$\sigma = \frac{b-a}{\sqrt{12}} = \frac{1,182-200}{\sqrt{12}} = 283.5$$

The height of the distribution is $1/(1,182-200) = 1/982 = 0.001$.

$$x_1 = 410 \text{ and } x_2 = 825$$

$$P(410 \leq x \leq 825) = \frac{825-410}{1,182-200} = \frac{415}{982} = 0.4226$$

The probability that a randomly selected person pays between £410 and £825 annually for car insurance in the United Kingdom is 0.4226. That is, about 42.26% of all people in the UK pay in that range.

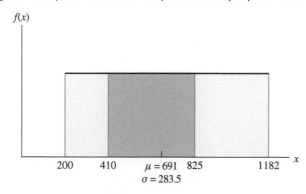

Using the Computer to Solve for Uniform Distribution Probabilities

Using the values of a, b, and x, Minitab has the capability of computing probabilities for the uniform distribution. The resulting computation is a cumulative probability from the left end of the distribution to each x value. As an example, the probability question, $P(410 \leq x \leq 825)$, from Demonstration Problem 6.2 can be worked using Minitab. Minitab computes the probability of $x \leq 825$ and the probability of $x \leq 410$, and these results are shown in Table 6.1. The final answer to the probability question from Demonstration Problem 6.2 is obtained by subtracting these two probabilities:

$$P(410 \leq x \leq 825) = 0.6365 - 0.2138 = 0.4227$$

Excel does not have the capability of directly computing probabilities for the uniform distribution.

TABLE 6.1

Minitab Output for Uniform Distribution

Cumulative Distribution Function	
Continuous uniform on 200 to 1182	
x	$P(X <= x)$
825	.636456
410	.213849

6.1   PROBLEMS

6.1   Values are uniformly distributed between 200 and 240.
- **a.**   What is the value of $f(x)$ for this distribution?
- **b.**   Determine the mean and standard deviation of this distribution.
- **c.**   Probability of $(x > 230) = ?$
- **d.**   Probability of $(205 \leqslant x \leqslant 220) = ?$
- **e.**   Probability of $(x \leqslant 225) = ?$

6.2 x is uniformly distributed over a range of values from 8 to 21.
- **a.**   What is the value of $f(x)$ for this distribution?
- **b.**   Determine the mean and standard deviation of this distribution.
- **c.**   Probability of $(10 \leqslant x < 17) = ?$
- **d.**   Probability of $(x < 22) = ?$
- **e.**   Probability of $(x \geqslant 7) = ?$

6.3   The retail price of a medium-sized box of a well-known brand of breakfast cereal ranges from €2.80 to €3.14. Assume these prices are uniformly distributed. What are the average price and standard deviation of prices in this distribution? If a price is randomly selected from this list, what is the probability that it will be between €3.00 and €3.10?

6.4   The average fill volume of a regular can of soft drink is 300 millilitres. Suppose the fill volume of these cans ranges from 295 to 305 millilitres and is uniformly distributed. What is the height of this distribution? What is the probability that a randomly selected can contains more than 302 millilitres? What is the probability that the fill volume is between 297 and 301 millilitres?

6.5   Suppose the average Greek household spends €2,100 a year on all types of insurance. Suppose the figures are uniformly distributed between the values of €400 and €3,800. What are the standard deviation and the height of this distribution? What proportion of households spends more than €3,000 a year on insurance? More than €4,000? Between €700 and €1,500?

6.2   NORMAL DISTRIBUTION

Probably the most widely known and used of all distributions is the **normal distribution**. It fits many human characteristics, such as height, weight, length, speed, IQ, scholastic achievement, and years of life expectancy, among others. Like their human counterparts, living things in nature, such as trees, animals, insects, and others, have many characteristics that are normally distributed.

Many variables in business and industry also are normally distributed. Some examples of variables that could produce normally distributed measurements include the annual cost of household insurance, the cost per square metres of renting warehouse space, and managers' satisfaction with support from ownership on a five-point scale. In addition, most items produced or filled by machines are normally distributed.

Because of its many applications, the normal distribution is an extremely important distribution. Besides the many variables mentioned that are normally distributed, the normal distribution and its associated probabilities are an integral part of statistical process control. When large enough sample sizes are taken, many statistics are normally distributed regardless of the shape of the underlying distribution from which they are drawn (as discussed in Chapter 7). Figure 6.4 is the graphic representation of the normal distribution: the normal curve.

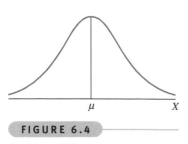

FIGURE 6.4

The Normal Curve

History of the Normal Distribution

Discovery of the normal curve of errors is generally credited to mathematician and astronomer Karl Gauss (1777–1855), who recognized that the errors of repeated measurement of objects are often normally distributed.* Thus the normal distribution is sometimes referred to as the *Gaussian distribution* or the *normal curve of error*. A modern-day analogy of Gauss's work might be the distribution of measurements of machine-produced parts, which often yield a normal curve of error around a mean specification.

To a lesser extent, some credit has been given to Pierre-Simon de Laplace (1749–1827) for discovering the normal distribution. However, many people now believe that Abraham de Moivre (1667–1754), a French mathematician, first understood the normal distribution. De Moivre determined that the binomial distribution approached the normal distribution as a limit. De Moivre worked with remarkable accuracy. His published table values for the normal curve are only a few ten-thousandths off the values of currently published tables.**

The normal distribution exhibits the following characteristics:

- It is a continuous distribution.
- It is a symmetrical distribution about its mean.
- It is asymptotic to the horizontal axis.
- It is unimodal.
- It is a family of curves.
- Area under the curve is 1.

The normal distribution is symmetrical. Each half of the distribution is a mirror image of the other half. Many normal distribution tables contain probability values for only one side of the distribution because probability values for the other side of the distribution are identical because of symmetry.

In theory, the normal distribution is asymptotic to the horizontal axis. That is, it does not touch the x-axis, and it goes forever in each direction. The reality is that most applications of the normal curve are experiments that have finite limits of potential outcomes. For example, even though GMAT scores are analysed by the normal distribution, the range of scores on each part of the GMAT is from 200 to 800.

The normal curve sometimes is referred to as the *bell-shaped curve*. It is unimodal in that values *mound up* in only one portion of the graph – the centre of the curve. The normal distribution actually is a family of curves. Every unique value of the mean and every unique value of the standard deviation result in a different normal curve. In addition, *the total area under any normal distribution is 1*. The area under the curve yields the probabilities, so the total of all probabilities for a normal distribution is 1. Because the distribution is symmetric, the area of the distribution on each side of the mean is 0.5.

Probability Density Function of the Normal Distribution

The normal distribution is described or characterized by two parameters: the mean, μ, and the standard deviation, σ. The values of μ and σ produce a normal distribution. The density function of the normal distribution is

$$f(x) = \frac{1}{\sigma\sqrt{2\pi}} e^{-1/2[(x-\mu)/\sigma]^2}$$

where:

μ = mean of x
σ = standard deviation of x
π = 3.14159..., and
e = 2.71828...

*John A. Ingram and Joseph G. Monks, *Statistics for Business and Economics*. San Diego: Harcourt Brace Jovanovich, 1989.
**Roger E. Kirk, *Statistical Issues: A Reader for the Behavioral Sciences*. Monterey, CA: Brooks/Cole, 1972.

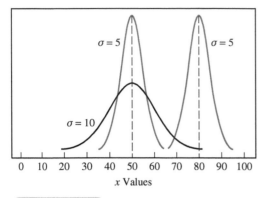

FIGURE 6.5

Normal Curves for Three Different Combinations of Means and Standard Deviations

Using Integral Calculus to determine areas under the normal curve from this function is difficult and time-consuming. Therefore, virtually all researchers use table values to analyse normal distribution problems rather than this formula.

Standardized Normal Distribution

Every unique pair of μ and σ values defines a different normal distribution. Figure 6.5 shows the Minitab graphs of normal distributions for the following three pairs of parameters.

1. $\mu = 50$ and $\sigma = 5$
2. $\mu = 80$ and $\sigma = 5$
3. $\mu = 50$ and $\sigma = 10$

Note that every change in a parameter (μ or σ) determines a different normal distribution. This characteristic of the normal curve (a family of curves) could make analysis by the normal distribution tedious because volumes of normal curve tables – one for each different combination of μ and σ – would be required. Fortunately, a mechanism was developed by which all normal distributions can be converted into a single distribution: the z distribution. This process yields the **standardized normal distribution** (or curve). The conversion formula for any x value of a given normal distribution follows.

z FORMULA

$$z = \frac{x - \mu}{\sigma}, \sigma \neq 0$$

A **z score** is *the number of standard deviations that a value, x, is above or below the mean*. If the value of x is less than the mean, the z score is negative; if the value of x is more than the mean, the z score is positive; and if the value of x equals the mean, the associated z score is zero. This formula allows conversion of the distance of any x value from its mean into standard deviation units. A standard z score table can be used to find probabilities for any normal curve problem that has been converted to z scores. The **z distribution** is *a normal distribution with a mean of 0 and a standard deviation of 1*. Any value of x at the mean of a normal curve is zero standard deviations from the mean. Any value of x that is one standard deviation above the mean has a z value of 1. The empirical rule, introduced in Chapter 3, is based on the normal distribution in which about 68% of all values are within one standard deviation of the mean regardless of the values of μ and σ. In a z distribution, about 68% of the z values are between $z = -1$ and $z = +1$.

The z distribution probability values are given in Table A.5. Because it is so frequently used, the z distribution is also printed inside the cover of this text. For discussion purposes, a list of z distribution values is presented in Table 6.2.

Table A.5 in Appendix A gives the total area under the z curve between 0 and any point on the positive z-axis. Since the curve is symmetric, the area under the curve between z and 0 is the same whether z is positive or negative (the sign on the z value designates whether the z score is above or below the mean). The table areas or probabilities are always positive.

Solving Normal Curve Problems

The mean and standard deviation of a normal distribution and the z formula and table enable a researcher to determine the probabilities for intervals of any particular values of a normal curve. One example is the many possible probability values of GMAT scores examined next.

The Graduate Management Aptitude Test (GMAT), produced by the Graduate Management Admission Council (GMAC), is widely used by graduate schools of business mainly in the United States, but also in other English-speaking countries as an entrance requirement. Assuming that the scores are normally distributed, probabilities of achieving scores over various ranges of the GMAT can be determined. In a recent year, the mean GMAT score was 494 and the standard deviation was about 100. What is the probability that a randomly selected score from this administration of the GMAT is between 600 and the mean? That is,

$$P(494 \leqslant x \leqslant 600 \mid \mu = 494 \text{ and } \sigma = 100) = ?$$

TABLE 6.2

z Distribution

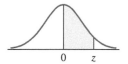

Second Decimal Place in *z*

z	0.00	0.01	0.02	0.03	0.04	0.05	0.06	0.07	0.08	0.09
0.0	.0000	.0040	.0080	.0120	.0160	.0199	.0239	.0279	.0319	.0359
0.1	.0398	.0438	.0478	.0517	.0557	.0596	.0636	.0675	.0714	.0753
0.2	.0793	.0832	.0871	.0910	.0948	.0987	.1026	.1064	.1103	.1141
0.3	.1179	.1217	.1255	.1293	.1331	.1368	.1406	.1443	.1480	.1517
0.4	.1554	.1591	.1628	.1664	.1700	.1736	.1772	.1808	.1844	.1879
0.5	.1915	.1950	.1985	.2019	.2054	.2088	.2123	.2157	.2190	.2224
0.6	.2257	.2291	.2324	.2357	.2389	.2422	.2454	.2486	.2517	.2549
0.7	.2580	.2611	.2642	.2673	.2704	.2734	.2764	.2794	.2823	.2852
0.8	.2881	.2910	.2939	.2967	.2995	.3023	.3051	.3078	.3106	.3133
0.9	.3159	.3186	.3212	.3238	.3264	.3289	.3315	.3340	.3365	.3389
1.0	.3413	.3438	.3461	.3485	.3508	.3531	.3554	.3577	.3599	.3621
1.1	.3643	.3665	.3686	.3708	.3729	.3749	.3770	.3790	.3810	.3830
1.2	.3849	.3869	.3888	.3907	.3925	.3944	.3962	.3980	.3997	.4015
1.3	.4032	.4049	.4066	.4082	.4099	.4115	.4131	.4147	.4162	.4177
1.4	.4192	.4207	.4222	.4236	.4251	.4265	.4279	.4292	.4306	.4319
1.5	.4332	.4345	.4357	.4370	.4382	.4394	.4406	.4418	.4429	.4441
1.6	.4452	.4463	.4474	.4484	.4495	.4505	.4515	.4525	.4535	.4545
1.7	.4554	.4564	.4573	.4582	.4591	.4599	.4608	.4616	.4625	.4633
1.8	.4641	.4649	.4656	.4664	.4671	.4678	.4686	.4693	.4699	.4706
1.9	.4713	.4719	.4726	.4732	.4738	.4744	.4750	.4756	.4761	.4767
2.0	.4772	.4778	.4783	.4788	.4793	.4798	.4803	.4808	.4812	.4817
2.1	.4821	.4826	.4830	.4834	.4838	.4842	.4846	.4850	.4854	.4857
2.2	.4861	.4864	.4868	.4871	.4875	.4878	.4881	.4884	.4887	.4890
2.3	.4893	.4896	.4898	.4901	.4904	.4906	.4909	.4911	.4913	.4916
2.4	.4918	.4920	.4922	.4925	.4927	.4929	.4931	.4932	.4934	.4936
2.5	.4938	.4940	.4941	.4943	.4945	.4946	.4948	.4949	.4951	.4952
2.6	.4953	.4955	.4956	.4957	.4959	.4960	.4961	.4962	.4963	.4964
2.7	.4965	.4966	.4967	.4968	.4969	.4970	.4971	.4972	.4973	.4974
2.8	.4974	.4975	.4976	.4977	.4977	.4978	.4979	.4979	.4980	.4981
2.9	.4981	.4982	.4982	.4983	.4984	.4984	.4985	.4985	.4986	.4986
3.0	.4987	.4987	.4987	.4988	.4988	.4989	.4989	.4989	.4990	.4990
3.1	.4990	.4991	.4991	.4991	.4992	.4992	.4992	.4992	.4993	.4993
3.2	.4993	.4993	.4994	.4994	.4994	.4994	.4994	.4995	.4995	.4995
3.3	.4995	.4995	.4995	.4996	.4996	.4996	.4996	.4996	.4996	.4997
3.4	.4997	.4997	.4997	.4997	.4997	.4997	.4997	.4997	.4997	.4998
3.5	.4998									
4.0	.49997									
4.5	.499997									
5.0	.4999997									
6.0	.499999999									

Figure 6.6 is a graphical representation of this problem.

The z formula yields the number of standard deviations that the x value, 600, is away from the mean.

$$z = \frac{x - \mu}{\sigma} = \frac{600 - 494}{100} = \frac{106}{100} = 1.06$$

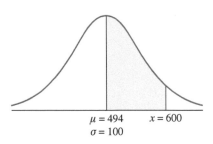

$\mu = 494$ $x = 600$
$\sigma = 100$

FIGURE 6.6

Graphical Depiction of the Area Between a Score of 600 and a Mean on a GMAT

The z value of 1.06 reveals that the GMAT score of 600 is 1.06 standard deviations more than the mean. The z distribution values in Table 6.2 give the probability of a value being between this value of x and the mean. The whole-number and tenths-place portion of the z score appear in the first column of Table 6.2 (the 1.0 portion of this z score). Across the top of the table are the values of the hundredths-place portion of the z score. For this z score, the hundredths-place value is 6. The probability value in Table 6.2 for z = 1.06 is 0.3554. The shaded portion of the curve at the top of the table indicates that the probability value given *always* is the probability or area between an x value and the mean. In this particular example, that is the desired area. Thus the answer is that 0.3554 of the scores on the GMAT are between a score of 600 and the mean of 494. Figure 6.7(a) depicts graphically the solution in terms of x values. Figure 6.7(b) shows the solution in terms of z values.

(a)   (b)

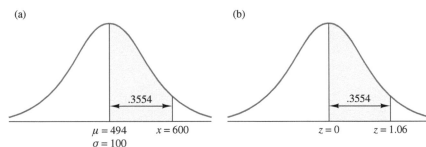

$\mu = 494$ $x = 600$ $z = 0$ $z = 1.06$
$\sigma = 100$

FIGURE 6.7

Graphical Solutions to the GMAT Problem

DEMONSTRATION PROBLEM 6.3

What is the probability of obtaining a score greater than 700 on a GMAT test that has a mean of 494 and a standard deviation of 100? Assume GMAT scores are normally distributed.

$$P(x > 700 \mid \mu = 494 \text{ and } \sigma = 100) = ?$$

Solution

Examine the following diagram.

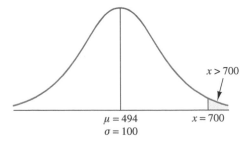

$x > 700$

$\mu = 494$ $x = 700$
$\sigma = 100$

This problem calls for determining the area of the upper tail of the distribution. The z score for this problem is

$$z = \frac{x - \mu}{\sigma} = \frac{700 - 494}{100} = \frac{206}{100} = 2.06$$

Table 6.2 gives a probability of 0.4803 for this z score. This value is the probability of randomly drawing a GMAT with a score between the mean and 700. Finding the probability of getting a score greater than 700, which is the tail of the distribution, requires subtracting the probability value of 0.4803 from 0.5000, because each half of the distribution contains 0.5000 of the area. The result is 0.0197. Note that an attempt to determine the area of $x \geq 700$ instead of $x > 700$ would have made no difference because, in continuous distributions, the area under an exact number such as $x = 700$ is zero. A line segment has no width and hence no area.

$$
\begin{array}{rl}
0.5000 & \text{(probability of } x \text{ greater than the mean)} \\
-0.4803 & \text{(probability of } x \text{ between 700 and the mean)} \\
\hline
0.0197 & \text{(probability of } x \text{ greater than 700)}
\end{array}
$$

The solution is depicted graphically in (a) for x values and in (b) for z values.

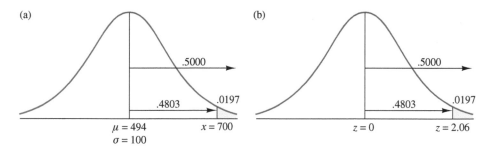

(a) $\mu = 494$, $\sigma = 100$, .5000, .4803, .0197, $x = 700$
(b) $z = 0$, .5000, .4803, .0197, $z = 2.06$

DEMONSTRATION PROBLEM 6.4

For the same GMAT examination, what is the probability of randomly drawing a score that is 550 or less?

$$P(x \leqslant 550 \mid \mu = 494 \text{ and } \sigma = 100) = ?$$

Solution

A sketch of this problem is shown here. Determine the area under the curve for all values less than or equal to 550.

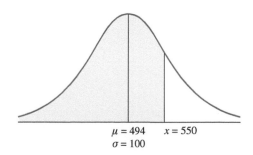

$\mu = 494$ $x = 550$
$\sigma = 100$

The *z* value for the area between 550 and the mean is equal to:

$$z = \frac{x - \mu}{\sigma} = \frac{550 - 494}{100} = \frac{56}{100} = 0.56$$

The area under the curve for $z = 0.56$ is 0.2123, which is the probability of getting a score between 550 and the mean. However, obtaining the probability for all values less than or equal to 550 also requires including the values less than the mean.

Because one-half or 0.5000 of the values are less than the mean, the probability of $x \leqslant 550$ is found as follows:

0.5000	(probability of values less than the mean)
+0.2123	(probability of values between 550 and the mean)
0.7123	(probability of values $\leqslant 550$)

This solution is depicted graphically in (a) for *x* values and in (b) for *z* values.

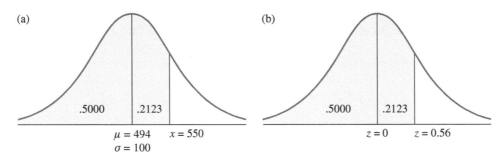

DEMONSTRATION PROBLEM 6.5

What is the probability of randomly obtaining a score between 300 and 600 on the GMAT exam?

$$P(300 < x < 600 \,|\, \mu = 494 \text{ and } \sigma = 100) = ?$$

Solution

The following sketch depicts the problem graphically: determine the area between $x = 300$ and $x = 600$, which spans the mean value. Because areas in the *z* distribution are given in relation to the mean, this problem must be worked as two separate problems and the results combined.

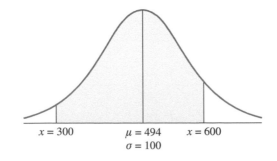

A *z* score is determined for each *x* value.

$$z = \frac{x - \mu}{\sigma} = \frac{600 - 494}{100} = \frac{106}{100} = 1.06$$

and

$$z = \frac{x - \mu}{\sigma} = \frac{300 - 494}{100} = \frac{-194}{100} = -1.94$$

Note that this *z* value ($z = -1.94$) is negative. A negative *z* value indicates that the *x* value is below the mean and the *z* value is on the left side of the distribution. None of the *z* values in Table 6.2 is negative. However, because the normal distribution is symmetric, probabilities for *z* values on the left side of the distribution are the same as the values on the right side of the distribution. The negative sign in the *z* value merely indicates that the area is on the left side of the distribution. The probability is always positive.

The probability for $z = 1.06$ is 0.3554; the probability for $z = -1.94$ is 0.4738. The solution of $P(300 < x < 600)$ is obtained by summing the probabilities.

0.3554	(probability of a value between the mean and 600)
+0.4738	(probability of a value between the mean and 300)
0.8292	(probability of a value between 300 and 600)

Graphically, the solution is shown in (a) for *x* values and in (b) for *z* values.

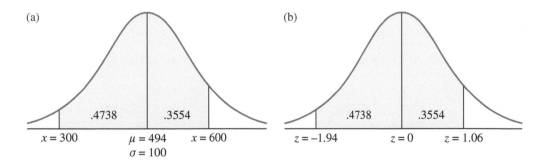

(a) .4738 .3554 $x = 300$ $\mu = 494$ $x = 600$ $\sigma = 100$

(b) .4738 .3554 $z = -1.94$ $z = 0$ $z = 1.06$

DEMONSTRATION PROBLEM 6.6

What is the probability of getting a score between 350 and 450 on the same GMAT exam?

$$P(350 < x < 450 \,|\, \mu = 494 \text{ and } \sigma = 100) = ?$$

Solution

The following sketch reveals that the solution to the problem involves determining the area of the shaded slice in the lower half of the curve.

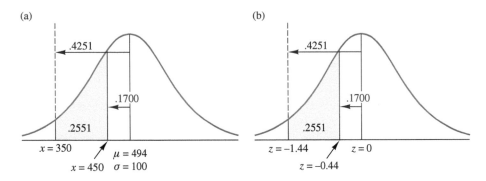

In this problem, the two x values are on the same side of the mean. The areas or probabilities of each x value must be determined and the final probability found by determining the difference between the two areas.

$$z = \frac{x - \mu}{\sigma} = \frac{350 - 494}{100} = \frac{-144}{100} = -1.44$$

and

$$z = \frac{x - \mu}{\sigma} = \frac{450 - 494}{100} = \frac{-44}{100} = -0.44$$

The probability associated with $z = -1.44$ is 0.4251.
The probability associated with $z = -0.44$ is 0.1700.

Subtracting gives the solution.

0.4251	(probability of a value between 350 and the mean)
−0.1700	(probability of a value between 450 and the mean)
0.2551	(probability of a value between 350 and 450)

Graphically, the solution is shown in (a) for x values and in (b) for z values.

DEMONSTRATION PROBLEM 6.7

Hotels.com publishes The Hotel Price Index (HPI), a regular survey of hotel prices in major city destinations across the world. The HPI is based on bookings made on Hotels.com, and prices shown are those actually paid by customers rather than advertised rates. If 86.65% of the prices of hotel rooms per night in Buenos Aires, Argentina, are less than £129.96 and if the standard deviation of hotel price per room is £36, what is the average hotel price per room in Buenos Aires? Assume that hotel prices per room per night are normally distributed.

Solution

In this problem, the standard deviation and an x value are given; the object is to determine the value of the mean. Examination of the z score formula reveals four variables: x, μ, σ, and z. In this problem, only two of the four variables are given. Because solving one equation with two unknowns is impossible, one of the other unknowns must be determined. The value of z can be determined from the normal distribution table (Table 6.2).

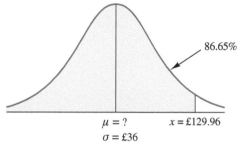

Because 86.65% of the values are less than $x = £129.96$, 36.65% of hotel room prices per night are between £129.96 and the mean. The other 50% of hotel room prices per night are in the lower half of the distribution. Converting the percentage to a proportion yields 0.3665 of the values between the x value and the mean. What z value is associated with this area? This area, or probability, of 0.3665 in Table 6.2 is associated with the z value of 1.11. This z value is positive, because it is in the upper half of the distribution. Using the z value of 1.11, the x value of £129.96, and the σ value of £36 allows solving for the mean algebraically.

$$z = \frac{x - \mu}{\sigma}$$

$$1.11 = \frac{£129.96 - \mu}{£36}$$

and

$$\mu = £129.96 - (£36)(1.11) = £129.96 - £39.96 = £90$$

The mean hotel room price per night in Buenos Aires is £90.

DEMONSTRATION PROBLEM 6.8

The OECD Environmental Data Compendium publishes figures on municipal waste generation in the OECD group of countries. In a recent year, the average amount of waste generated per person per year in Japan was 400 kilograms. Suppose the annual amount of municipal waste generated per person is normally distributed, with a standard deviation of 116 kilograms. Of the annual amounts of municipal waste generated per person in Japan, 67.72% would be greater than what amount?

Solution

The mean and standard deviation are given, but x and z are unknown. The problem is to solve for a specific x value when 0.6772 of the x values are greater than that value.

If 0.6772 of the values are greater than x, then 0.1772 are between x and the mean (0.6772 − 0.5000). Table 6.2 shows that the probability of 0.1772 is associated with a z value of 0.46. Because x is less than the mean, the z value actually is −0.46. Whenever an x value is less than the mean, its associated z value is negative and should be reported that way.

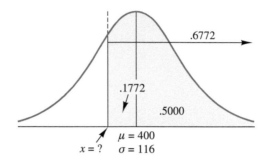

Solving the z equation yields

$$z = \frac{x - \mu}{\sigma}$$

$$-0.46 = \frac{x - 400}{116}$$

and

$$x = 400 + (-0.046)(116) = 346.64$$

Thus 67.72% of the annual average amount of municipal waste generated per person in Japan weighs more than 346 kilograms.

Warehousing

A recent study by J. Allen and M. Browne reported the warehousing land use patterns in urban areas in England and other western European countries. They identified three major trends that have taken place in recent decades: (i) de-industrialization (which led to a decline in demand for warehousing in urban areas), (ii) spatial centralization of stockholding (which increased the use of large distribution centres along the motorway network), and (iii) rapid rising urban land prices (which led to the suburbanization of warehousing). These trends have led to fewer but larger warehouses. The average warehouse size in England and Wales is about 772 square metres. To visualize such an 'average' warehouse, picture one that is a rectangle measuring 70 metres by 11 metres.

Suppose that the sizes of warehouses are normally distributed. Using the mean values already given and the standard deviations, techniques presented in this section could be used to determine, for example, the probability that a randomly selected warehouse is larger than 800 square metres.

Using the Computer to Solve for Normal Distribution Probabilities

Both Excel and Minitab can be used to solve for normal distribution probabilities. In each case, the computer package uses μ, σ, and the value of x to compute a cumulative probability from the left. Shown in Table 6.3 are Excel and Minitab output for the probability question addressed in Demonstration Problem 6.6: $P(350 < x < 450 \mid \mu = 494$ and $\sigma = 100)$. Since both computer packages yield probabilities cumulated from the left, this problem is solved manually with the computer output by finding the difference in $P(x < 450)$ and $P(x < 350)$.

TABLE 6.3

Excel and Minitab Normal Distribution Output for Demonstration Problem 6.6

Excel Output

	x Value	
	450	.3264
	350	.0735
		.2528

Minitab Output

```
CUMULATIVE DISTRIBUTION FUNCTION

Normal with mean = 494 and standard
deviation = 100
           x          P(X <= x)
         450           .329969
         350           .074934
Prob (350 < x < 450) = 0.255035
```

6.2 PROBLEMS

6.6 Determine the probabilities for the following normal distribution problems:

a. $\mu = 604$, $\sigma = 56.8$, $x \leqslant 635$

b. $\mu = 48$, $\sigma = 12$, $x < 20$

c. $\mu = 111$, $\sigma = 33.8$, $100 \leqslant x < 150$

d. $\mu = 264$, $\sigma = 10.9$, $250 < x < 255$

e. $\mu = 37$, $\sigma = 4.35$, $x > 35$

f. $\mu = 156$, $\sigma = 11.4$, $x \geqslant 170$

6.7 The study by J. Allen and M. Browne showed that the average warehouse size in London was 683 square metres. Suppose warehouse sizes are normally distributed and that the standard deviation is 100 metres. A warehouse in London is randomly selected.

a. What is the probability that the warehouse size is greater than 600 square metres?

b. What is the probability that the warehouse size is less than 400 square metres?

c. What is the probability that the warehouse size is between 700 and 800 square metres?

6.8 According to a study commissioned by the German water industry association (BGW) cited by *Wikipedia*, the average annual per capita water bill was €85 in France and Germany, €59 in Italy, and €95 in England and Wales. Suppose the average annual per capita water bills are normally distributed with a standard deviation of €11.35 in all of the above cases.

a. What is the probability that a randomly selected average annual water bill is more than €85 in France?

b. What is the probability that a randomly selected annual water bill is between €45 and €70 in Italy?

c. What is the probability that a randomly selected annual water bill is between €65 and €75 in England and Wales?

d. What is the probability that a randomly selected annual water bill is no more than €40 in Germany?

6.9 According to the Portuguese Ministry of Finance, income tax returns in Portugal averaged €729 in refunds for taxpayers in a recent year. One explanation of this figure is that taxpayers would rather have the government keep back too much money during the year than to owe it money at the end of the year. Suppose the average amount of tax at the end of a year is a refund of €729, with a standard deviation of €414. Assume that amounts owed or due on tax returns are normally distributed.

a. What proportion of tax returns show a refund greater than €1,000?

b. What proportion of the tax returns show that the taxpayer owes money to the government?

c. What proportion of the tax returns show a refund between €200 and €700?

6.10 According to CIPD 2010 Absence Survey, absence costs UK business on average £600 per employee per year. The same survey also determined that the average absence was higher in the public sector compared with private sector services, at 9.6 days per employee per year compared with 6.6 days. Suppose the absence costs are normally distributed, with a standard deviation of £500.

a. What proportion of the costs are between £300 and £900?

b. What proportion of the costs are greater than £1,700?

c. What proportion of the costs are between £200 and £400?

d. Suppose the standard deviation is unknown, but 90.82% of the costs are more than £320. What would be the value of the standard deviation?

e. Suppose the mean value is unknown, but the standard deviation is still £500. How much would the average cost be if 79.95% of the costs were less than £960?

6.11 Suppose you are working with a data set that is normally distributed, with a mean of 200 and a standard deviation of 47. Determine the value of x from the following information:

a. 60% of the values are greater than x.

b. x is less than 17% of the values.

c. 22% of the values are less than x.

d. x is greater than 55% of the values.

6.12 According to the recent *Global Humanitarian Assistance* report, Luxembourg is the largest donor of humanitarian aid on a per citizen basis, with an average donation of $114.4 per person. Suppose that donations per person are normally distributed with a standard deviation of $62.

a. What proportion of donations per person are greater than $300?

b. What proportion of citizens in Luxembourg do not donate any amount for humanitarian purposes?

6.13 Suppose the standard deviation for Problem 6.7 is unknown but the mean is still 683 square metres. If 72.24% of all warehouses in London have a size greater than 544 square metres, what is the standard deviation?

6.14 Suppose the mean size of warehouses in Paris is unknown but the standard deviation is known to be 175 square metres. What is the value of the mean warehouse size in Paris if 29% of warehouses in Paris have a size less than 400 square metres?

6.15 The average annual wind speed on Earth is approximately 56 kilometres per hour and the fastest daily average was recorded in Port Martin in Antarctica in 1951 at 174 kilometres/per hour. Suppose that the average wind speed in kilometres per hour for Moscow is 21.6. Suppose wind speed measurements are normally distributed for a given geographic location. If 22.45% of the time the wind speed measurements are more than 27 kilometres per hour, what is the standard deviation of wind speed in Moscow?

6.16 According to the Scottish Government website the average cumulated university student loan debt for a student leaving Scottish universities was £5,970 in 2009. Assume that the standard deviation of such student loan debt is £1,532. Thirty per cent of these students leaving universities owe more than what amount?

6.3   USING THE NORMAL CURVE TO APPROXIMATE BINOMIAL DISTRIBUTION PROBLEMS

For certain types of binomial distribution problems, the normal distribution can be used to approximate the probabilities. As sample sizes become large, binomial distributions approach the normal distribution in shape regardless of the value of p. This phenomenon occurs faster (for smaller values of n) when p is near 0.50. Figures 6.8 to 6.10

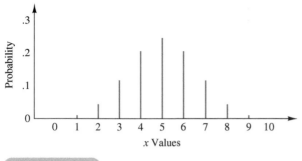

FIGURE 6.8

The Binomial Distribution for $n = 10$ and $p = 0.50$

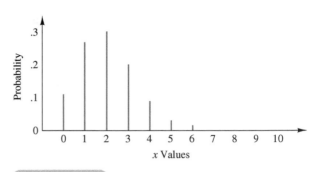

FIGURE 6.9

The Binomial Distribution for $n = 10$ and $p = 0.20$

show three binomial distributions. Note in Figure 6.8 that even though the sample size, n, is only 10, the binomial graph bears a strong resemblance to a normal curve.

The graph in Figure 6.9 ($n = 10$ and $p = 0.20$) is skewed to the right because of the low p value and the small size. For this distribution, the expected value is only 2 and the probabilities pile up at $x = 0$ and 1. However, when n becomes large enough, as in the binomial distribution ($n = 100$ and $p = 0.20$) presented in Figure 6.10, the graph is relatively symmetric around the mean ($\mu = n \cdot p = 20$) because enough possible outcome values to the left of $x = 20$ allow the curve to fall back to the x-axis.

For large n values, the binomial distribution is cumbersome to analyse without a computer. Table A.2 goes only to $n = 25$. The normal distribution is a good approximation for binomial distribution problems for large values of n.

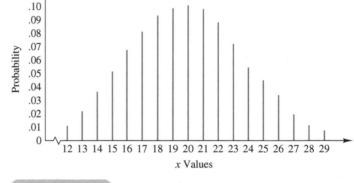

FIGURE 6.10

The Binomial Distribution for $n = 100$ and $p = 0.20$

To work a binomial problem by the normal curve requires a translation process. The first part of this process is to convert the two parameters of a binomial distribution, n and p, to the two parameters of the normal distribution, μ and σ. This process utilizes formulas from Chapter 5:

$$\mu = n \cdot p \text{ and } \sigma = \sqrt{n \cdot p \cdot q}$$

After completion of this, a test must be made to determine whether the normal distribution is a good enough approximation of the binomial distribution:

Does the interval $\mu \pm 3\sigma$ lie between 0 and n?

Recall that the empirical rule states that approximately 99.7%, or almost all, of the values of a normal curve are within three standard deviations of the mean. For a normal curve approximation of a binomial distribution problem to be acceptable, all possible x values should be between 0 and n, which are the lower and upper limits, respectively, of a binomial distribution. If $\mu \pm 3\sigma$ is not between 0 and n, do *not* use the normal distribution to work a binomial problem because the approximation is not good enough. Upon demonstration that the normal curve is a good approximation for a binomial problem, the procedure continues. Another rule of thumb for determining when to use the normal curve to approximate a binomial problem is that the approximation is good enough if both $n \cdot p > 5$ and $n \cdot q > 5$.

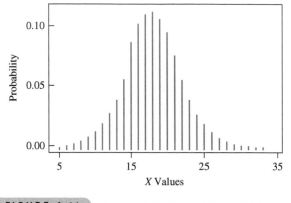

FIGURE 6.11

Graph of the Binomial Problem: $n = 60$ and $p = 0.30$

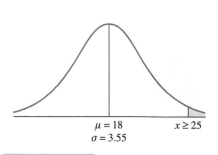

FIGURE 6.12

Graph of Apparent Solution of Binomial Problem Worked by the Normal Curve

The process can be illustrated in the solution of the binomial distribution problem.

$$P(x \geqslant 25 \mid n = 60 \text{ and } p = 0.30) = ?$$

Note that this binomial problem contains a relatively large sample size and that none of the binomial tables in Appendix A.2 can be used to solve the problem. This problem is a good candidate for use of the normal distribution. Translating from a binomial problem to a normal curve problem gives

$$\mu = n \cdot p = (60)(0.30) = 18 \text{ and } \sigma = \sqrt{n \cdot p \cdot q} = 3.55$$

The binomial problem becomes a normal curve problem.

$$P(x \geqslant 25 \mid \mu = 18 \text{ and } \sigma = 3.55) = ?$$

Next, the test is made to determine whether the normal curve sufficiently fits this binomial distribution to justify the use of the normal curve.

$$\mu \pm 3\sigma = 18 \pm 3(3.55) = 18 \pm 10.65$$

$$7.35 \leqslant \mu \pm 3\sigma \leqslant 28.65$$

This interval is between 0 and 60, so the approximation is sufficient to allow use of the normal curve. Figure 6.11 is a Minitab graph of this binomial distribution. Notice how closely it resembles the normal curve. Figure 6.12 is the apparent graph of the normal curve version of this problem.

Correcting for Continuity

The translation of a discrete distribution to a continuous distribution is not completely straightforward. A correction of +0.50 or –0.50 or ±0.50, depending on the problem, is required. This correction ensures that most of the binomial problem's information is correctly transferred to the normal curve analysis. This correction is called the **correction for continuity**, which is *made during conversion of a discrete distribution into a continuous distribution.*

Figure 6.13 is a portion of the graph of the binomial distribution, $n = 60$ and $p = 0.30$. Note that with a binomial distribution, all the probabilities are concentrated on the whole numbers. Thus, the answers for $x \geqslant 25$ are found by summing the probabilities for $x = 25, 26, 27, \ldots, 60$. There are no values between 24 and 25, 25 and 26, $\ldots$, 59, and 60. Yet, the normal distribution is continuous, and values are present all along the x-axis. A correction must be made for this discrepancy for the approximation to be as accurate as possible.

As an analogy, visualize the process of melting iron rods in a furnace. The iron rods are like the probability values on each whole number of a binomial distribution. Note that the binomial graph in Figure 6.13 looks like a series of iron rods in a line. When the rods are placed in a furnace, they melt down and spread out. Each rod melts and moves to fill the area between it and the adjacent rods. The result is a continuous sheet of solid iron (continuous iron) that looks

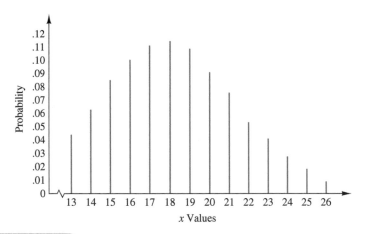

FIGURE 6.13

Graph of a Portion of the Binomial Problem: $n = 60$ and $p = 0.30$

TABLE 6.4

Rules of Thumb for the Correction for Continuity

Values Being Determined	Corrections
$x >$	$+.50$
$x \geq$	$-.50$
$x <$	$-.50$
$x \leq$	$+.50$
$\leq x \leq$	$-.50$ and $+.50$
$< x <$	$+.50$ and $-.50$
$x =$	$-.50$ and $+.50$

like the normal curve. The melting of the rods is analogous to spreading the binomial distribution to approximate the normal distribution.

How far does each rod spread towards the others? A good estimate is that each rod goes about halfway towards the adjacent rods. In other words, a rod that was concentrated at $x = 25$ spreads to cover the area from 24.5 to 25.5; $x = 26$ becomes continuous from 25.5 to 26.5; and so on. For the problem $P(x \geq 25 \mid n = 60$ and $p = 0.30)$, conversion to a continuous normal curve problem yields $P(x \geq 24.5 \mid \mu = 18$ and $\sigma = 3.55)$. The correction for continuity was -0.50 because the problem called for the inclusion of the value of 25 along with all greater values; the binomial value of $x = 25$ translates to the normal curve value of 24.5 to 25.5. Had the binomial problem been to analyse $P(x > 25)$, the correction would have been $+0.50$, resulting in a normal curve problem of $P(x \geq 25.5)$. The latter case would begin at more than 25 because the value of 25 would not be included.

The decision as to how to correct for continuity depends on the equality sign and the direction of the desired outcomes of the binomial distribution. Table 6.4 lists some rules of thumb that can help in the application of the correction for continuity.

For the binomial problem $P(x \geq 25 \mid n = 60$ and $p = 0.30)$, the normal curve becomes $P(x \geq 24.5 \mid \mu = 18$ and $\sigma = 3.55)$, as shown in Figure 6.14, and

$$z = \frac{x - \mu}{\sigma} = \frac{24.5 - 18}{3.55} = 1.83$$

The probability (Table 6.2) of this z value is 0.4664. The answer to this problem lies in the tail of the distribution, so the final answer is obtained by subtracting

$$
\begin{array}{r}
0.5000 \\
-0.4664 \\
\hline
0.0336
\end{array}
$$

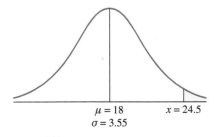

FIGURE 6.14

Graph of the Solution to the Binomial Problem Worked by the Normal Curve

TABLE 6.5

Probability Values for the Binomial Problem: $n = 60$ and $p = 0.30$, and $x > 25$

x Value	Probability
25	.0167
26	.0096
27	.0052
28	.0026
29	.0012
30	.0005
31	.0002
32	.0001
33	.0000
$x \geq 25$	.0361

Had this problem been worked by using the binomial formula, the solution would have been as shown in Table 6.5. The difference between the normal distribution approximation and the actual binomial values is only $0.0025(0.0361 - 0.0336)$.

DEMONSTRATION PROBLEM 6.9

Work the following binomial distribution problem by using the normal distribution:

$$P(x = 12) \mid n = 25 \text{ and } p = 0.40) = ?$$

Solution

Find μ and σ.

$$\mu = n \cdot p = (25)(0.40) = 10.0$$

$$\sigma = \sqrt{n \cdot p \cdot q} = \sqrt{(25)(0.40)(0.60)} = 2.45$$

$$\text{test}: \mu \pm 3\sigma = 10.0 \pm 3(2.45) = 2.65 \text{ to } 17.35$$

This range is between 0 and 25, so the approximation is close enough. Correct for continuity next. Because the problem is to determine the probability of x being exactly 12, the correction entails both −0.50 and +0.50. That is, a binomial probability at x = 12 translates to a continuous normal curve area that lies between 11.5 and 12.5. The graph of the problem follows:

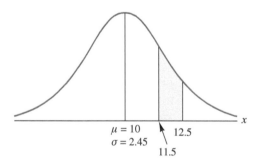

Then,

$$z = \frac{x - \mu}{\sigma} = \frac{12.5 - 10}{2.45} = 1.02$$

and

$$z = \frac{x - \mu}{\sigma} = \frac{11.5 - 10}{2.45} = 0.61$$

$z = 1.02$ produces a probability of 0.3461.

$z = 0.61$ produces a probability of 0.2291.

The difference in areas yields the following answer:

$$0.3461 - 0.2291 = 0.1170$$

Had this problem been worked by using the binomial tables, the resulting answer would have been 0.114. The difference between the normal curve approximation and the value obtained by using binomial tables is only 0.003.

DEMONSTRATION PROBLEM 6.10

Solve the following binomial distribution problem by using the normal distribution:

$$P(x < 27 \mid n = 100 \text{ and } p = 0.37) = ?$$

Solution

Because neither the sample size nor the p value is contained in Table A.2, working this problem by using binomial distribution techniques is impractical. It is a good candidate for the normal curve. Calculating μ and σ yields

$$\mu = n \cdot p = (100)(0.37) = 37.0$$
$$\sigma = \sqrt{n \cdot p \cdot q} = \sqrt{(100)(0.37)(0.63)} = 4.83$$

Testing to determine the closeness of the approximation gives

$$\mu \pm 3\sigma = 37 \pm 3(4.83) = 37 \pm 14.49$$

The range 22.51 to 51.49 is between 0 and 100. This problem satisfies the conditions of the test. Next, correct for continuity: $x < 27$ as a binomial problem translates to $x \leq 26.5$ as a normal distribution problem. The graph of the problem follows.

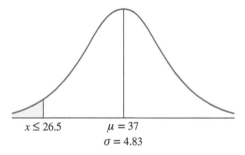

$$x \leq 26.5 \qquad \mu = 37$$
$$\sigma = 4.83$$

Then,

$$z = \frac{x - \mu}{\sigma} = \frac{26.5 - 37}{4.83} = -2.17$$

Table 6.2 shows a probability of 0.4850. Solving for the tail of the distribution gives

$$0.5000 - 0.4850 = 0.0150$$

which is the answer.

Had this problem been solved by using the binomial formula, the probabilities would have been the following:

x Value	Probability
26	.0059
25	.0035
24	.0019
23	.0010
22	.0005
21	.0002
20	.0001
$x < 27$	.0131

The answer obtained by using the normal curve approximation (0.0150) compares favourably to this exact binomial answer. The difference is only 0.0019.

STATISTICS IN BUSINESS TODAY

Teleworking in the European Union

A teleworker can be briefly defined as an employee that works with a PC away from the employer's premises on a regular basis. Telework is mostly viewed in positive terms by governments, employers, and trade unions alike. Governments, unions, and employer organizations tend to view telework as a means to enhance productivity and employment, while at the same time facilitating overall policy goals in terms of health, the environment, and family life. Telework is a growing phenomenon throughout the EU Member States. The average proportion of employees involved in telework in the EU27 countries increased from about 5% in 2000 to 7% in 2005. According to findings of the European Working Conditions Survey (EWCS), the highest proportion of telework is observed in the Czech Republic and Denmark, where about one out of seven employees is regularly involved in telework. Above average rates are also observed in the Benelux countries (Belgium, the Netherlands, and Luxembourg) and the Nordic states (including Norway). On the other end of the scale, Italy, Portugal, and Bulgaria have the lowest proportion of teleworkers with less than 2.3% of workers engaging in telework. Telework was also found to vary greatly across industries, occupations, and educational level. Telework is more prevalent among workers in the real estate, financial intermediation, and education sectors and for workers with higher levels of skills and qualifications. The survey also found significant different proportions of male and female teleworkers with about 8.1% of male employees engaging in telework versus only 5.8% of female employees using this form of work.

Source: Telework in the European Union, European Foundation for the Improvement of Living and Working Conditions, 2010.

6.3   PROBLEMS

6.17   Convert the binomial distribution problems below to normal distribution problems. Use the correction for continuity.

   a. $P(x \leq 16 \mid n = 30 \text{ and } p = 0.70)$

   b. $P(10 < x \leq 20 \mid n = 25 \text{ and } p = 0.50)$

   c. $P(x = 22 \mid n = 40 \text{ and } p = 0.60)$

   d. $P(x > 14 \mid n = 16 \text{ and } p = 0.45)$

6.18   Use the test $\mu \pm 3\sigma$ to determine whether the following binomial distributions can be approximated by using the normal distribution.

   a. $n = 8$ and $p = 0.50$

   b. $n = 18$ and $p = 0.80$

   c. $n = 12$ and $p = 0.30$

   d. $n = 30$ and $p = 0.75$

   e. $n = 14$ and $p = 0.50$

6.19   Where appropriate, work the following binomial distribution problems by using the normal curve. Also, use Table A.2 to find the answers by using the binomial distribution and compare the answers obtained by the two methods.

   a. $P(x = 8 \mid n = 25 \text{ and } p = 0.40) = ?$

   b. $P(x \geq 13 \mid n = 20 \text{ and } p = 0.60) = ?$

   c. $P(x = 7 \mid n = 15 \text{ and } p = 0.50) = ?$

   d. $P(x < 3 \mid n = 10 \text{ and } p = 0.70) = ?$

6.20   The Zimmerman Agency conducted a study for Residence Inn by Marriott of business travellers who take trips of five nights or more. According to this study, 37% of these travellers enjoy sightseeing more than any other activity that they do not get to do as much at home. Suppose 120 randomly selected business travellers who take trips of five nights or more are contacted. What is the probability that fewer than 40 enjoy sightseeing more than any other activity that they do not get to do as much at home?

6.21   One study on managers' satisfaction with management tools reveals that 59% of all managers use self-directed work teams as a management tool. Suppose 70 managers selected randomly in the European Union are interviewed. What is the probability that fewer than 35 use self-directed work teams as a management tool?

6.22   In a recent online poll conducted by the *Webcopyplus*, 93% of Internet users indicated they favour speed and readability over appearance when visiting websites. When Internet users were asked what was most likely to drive them away from a website, 51% indicated slow load times. Suppose 200 Internet users are randomly contacted.

    a.   What is the probability that more than 175 Internet users say they favour speed and readability over appearance when visiting websites?

    b.   What is the probability that between 185 and 190 (inclusive) Internet users indicate they favour speed and readability over appearance when visiting websites?

    c.   What is the probability that between 85 and 110 (inclusive) Internet users say that slow load times is the most likely reason to drive them away from a website?

    d.   What is the probability that fewer than 100 Internet users say that slow load times is the most likely reason to drive them away from a website?

6.23   Market researcher Gartner Dataquest reports that Hewlett-Packard holds 17.9% of the global PC market. Suppose a business researcher randomly selects 130 recent purchasers of a PC.

    a.   What is the probability that more than 29 PC purchasers bought a HP computer?

    b.   What is the probability that between 18 and 20 PC purchasers (inclusive) bought a HP computer?

    c.   What is the probability that fewer than 25 PC purchasers bought a HP computer?

    d.   What is the probability that exactly 23 PC purchasers bought a HP computer?

6.24   A study about strategies for competing in the global marketplace states that 52% of the respondents agreed that companies need to make direct investments in foreign countries. It also states that about 70% of those responding agree that it is attractive to have a joint venture to increase global competitiveness. Suppose CEOs of 95 manufacturing companies are randomly contacted about global strategies.

    a.   What is the probability that between 44 and 52 (inclusive) CEOs agree that companies should make direct investments in foreign countries?

    b.   What is the probability that more than 56 CEOs agree with that assertion?

    c.   What is the probability that fewer than 60 CEOs agree that it is attractive to have a joint venture to increase global competitiveness?

    d.   What is the probability that between 55 and 62 (inclusive) CEOs agree with that assertion?

6.4   EXPONENTIAL DISTRIBUTION

Another useful continuous distribution is the exponential distribution. It is closely related to the Poisson distribution. Whereas the Poisson distribution is discrete and describes random occurrences over some interval, the **exponential distribution** is *continuous and describes a probability distribution of the times between random occurrences*. The following are the characteristics of the exponential distribution.

- It is a continuous distribution.
- It is a family of distributions.
- It is skewed to the right.
- The x values range from zero to infinity.
- Its apex is always at $x = 0$.
- The curve steadily decreases as x gets larger.

The exponential probability distribution is determined by the following:

EXPONENTIAL PROBABILITY DENSITY FUNCTION	$$f(x) = \lambda e^{-\lambda x}$$ where: $x \geqslant 0$ $\lambda > 0$ and $e = 2.71828\ldots$

An exponential distribution can be characterized by the one parameter, λ. Each unique value of λ determines a different exponential distribution, resulting in a family of exponential distributions. Figure 6.15 shows graphs of exponential distributions for four values of λ. The points on the graph are determined by using λ and various values of x in the probability density formula. The mean of an exponential distribution is $\mu = 1/\lambda$, and the standard deviation of an exponential distribution is $\sigma = 1/\lambda$.

Probabilities of the Exponential Distribution

Probabilities are computed for the exponential distribution by determining the area under the curve between two points. Applying calculus to the exponential probability density function produces a formula that can be used to calculate the probabilities of an exponential distribution.

PROBABILITIES OF THE RIGHT TAIL OF THE EXPONENTIAL DISTRIBUTION	$$P(x \geqslant x_0) = e^{-\lambda x_0}$$ where: $x_0 \geqslant 0$

To use this formula requires finding values of e^{-x}. These values can be computed on most calculators or obtained from Table A.4 in Appendix A, which contains the values of e^{-x}, for selected values of x. x_0 is the fraction of the interval or the number of intervals between arrivals in the probability question and λ is the average arrival rate.

For example, arrivals at a bank are Poisson distributed with a λ of 1.2 customers every minute. What is the average time between arrivals and what is the probability that at least 2 minutes will elapse between one arrival and the next arrival? Since the interval for lambda is 1 minute and we want to know the probability that at least two minutes transpire between arrivals (twice the lambda interval), x_0 is 2.

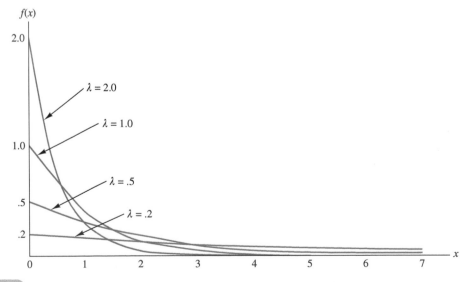

FIGURE 6.15

Graphs of Some Exponential Distributions

Interarrival times of random arrivals are exponentially distributed. The mean of this exponential distribution is $\mu = 1/\lambda = 1/1.2 = 0.833$ minute (50 seconds). On average, 0.833 minute, or 50 seconds, will elapse between arrivals at the bank. The probability of an interval of two minutes or more between arrivals can be calculated by

$$P(x \geqslant 2 \mid \lambda = 1.2) = e^{-1.2(2)} = 0.0907.$$

About 9.07% of the time when the rate of random arrivals is 1.2 per minute, two minutes or more will elapse between arrivals, as shown in Figure 6.16.

This problem underscores the potential of using the exponential distribution in conjunction with the Poisson distribution to solve problems. In operations research and management science, these two distributions are used together to solve queuing problems (theory of waiting lines). The Poisson distribution can be used to analyse the arrivals to the queue, and the exponential distribution can be used to analyse the interarrival time.

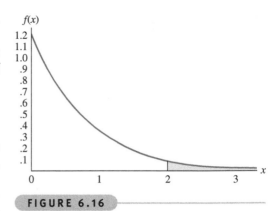

FIGURE 6.16

Exponential Distribution for $\lambda = 1.2$ and Solution for $x \geqslant 2$

DEMONSTRATION PROBLEM 6.11

A manufacturing firm has been involved in statistical quality control for several years. As part of the production process, parts are randomly selected and tested. From the records of these tests, it has been established that a defective part occurs in a pattern that is Poisson distributed on the average of 1.38 defects every 20 minutes during production runs. Use this information to determine the probability that less than 15 minutes will elapse between any two defects.

Solution

The value of λ is 1.38 defects per 20-minute interval. The value of μ can be determined by

$$\mu = \frac{1}{\lambda} = \frac{1}{1.38} = 0.7246$$

On the average, it is 0.7246 of the interval, or $(0.7246)(20 \text{ minutes}) = 14.49$ minutes, between defects. The value of x_0 represents the desired number of intervals between arrivals or occurrences for the probability question. In this problem, the probability question involves 15 minutes and the interval is 20 minutes. Thus x_0 is 15/20, or 0.75 of an interval. The question here is to determine the probability of there being less than 15 minutes between defects. The probability formula always yields the right tail of the distribution – in this case, the probability of there being 15 minutes or more between arrivals. By using the value of x_0 and the value of λ, the probability of there being 15 minutes or more between defects can be determined.

$$P(x \geqslant x_0) = P(x \geqslant 0.75) = e^{-\lambda x_0} = e^{(-1.38)(0.75)} = e^{-1.035} = 0.3552$$

The probability of 0.3552 is the probability that at least 15 minutes will elapse between defects. To determine the probability of there being less than 15 minutes between defects, compute $1 - P(x)$. In this case, $1 - 0.3552 = 0.6448$. There is a probability of 0.6448 that less than 15 minutes will elapse between two defects when there is an average of 1.38 defects per 20-minute interval or an average of 14.49 minutes between defects.

TABLE 6.6

Excel and Minitab Output for Exponential Distribution

Excel Output

x Value	Probability < x Value
.75	.6448

Minitab Output

```
CUMULATIVE DISTRIBUTION FUNCTION

Exponential with mean = 0.7246
  x    P(X <= x)
 .75    .644793
```

Using the Computer to Determine Exponential Distribution Probabilities

Both Excel and Minitab can be used to solve for exponential distribution probabilities. Excel uses the value of λ and x_0, but Minitab requires μ (equals $1/\lambda$) and x_0. In each case, the computer yields the cumulative probability from the left (the complement of what the probability formula shown in this section yields). Table 6.6 provides Excel and Minitab output for the probability question addressed in Demonstration Problem 6.11.

6.4   PROBLEMS

6.25 Use the probability density formula to sketch the graphs of the following exponential distributions:
   a. $\lambda = 0.1$
   b. $\lambda = 0.3$
   c. $\lambda = 0.8$
   d. $\lambda = 3.0$

6.26 Determine the mean and standard deviation of the following exponential distributions:
   a. $\lambda = 3.25$
   b. $\lambda = 0.7$
   c. $\lambda = 1.1$
   d. $\lambda = 6.0$

6.27 Determine the following exponential probabilities:
   a. $P(x \geq 5 \mid \lambda = 1.35)$
   b. $P(x < 3 \mid \lambda = 0.68)$
   c. $P(x > 4 \mid \lambda = 1.7)$
   d. $P(x < 6 \mid \lambda = 0.80)$

6.28 The average length of time between arrivals at an expressway tollbooth is 23 seconds. Assume that the time between arrivals at the tollbooth is exponentially distributed.
   a. What is the probability that one minute or more will elapse between arrivals?
   b. If a car has just passed through the tollbooth, what is the probability that no car will show up for at least three minutes?

6.29 A busy restaurant determined that between 6.30 P.M. and 9.00 P.M. on Friday nights, the arrivals of customers are Poisson distributed with an average arrival rate of 2.44 per minute.
   a. What is the probability that at least 10 minutes will elapse between arrivals?
   b. What is the probability that at least five minutes will elapse between arrivals?

  c. What is the probability that at least one minute will elapse between arrivals?

  d. What is the expected amount of time between arrivals?

6.30 During the summer at a small private airport in South East Asia, the unscheduled arrival of airplanes is Poisson distributed with an average arrival rate of 1.12 planes per hour.

  a. What is the average interarrival time between planes?

  b. What is the probability that at least 2 hours will elapse between plane arrivals?

  c. What is the probability of two planes arriving less than 10 minutes apart?

6.31 The exponential distribution can be used to solve Poisson-type problems in which the intervals are not time. The Air Transport Users Council (AUC) recently published statistics on the baggage handling performance of the member airlines of the Association of European Airlines. Turkish Airlines led the region in fewest occurrences of mishandled baggage, with a mean rate of 4.5 per 1,000 passengers. Assume mishandled baggage occurrences are Poisson distributed. Using the exponential distribution to analyse this problem, determine the average number of passengers between occurrences. Suppose baggage has just been mishandled.

  a. What is the probability that at least 500 passengers will have their baggage handled properly before the next mishandling occurs?

  b. What is the probability that the number will be fewer than 200 passengers?

6.32 The Foundation Corporation specializes in constructing the concrete foundations for new houses in the Ukraine. The company knows that because of soil types, moisture conditions, variable construction, and other factors, most foundations will eventually need major repair. On the basis of its records, the company's president believes that a new house foundation on average will not need major repair for 20 years. If she wants to guarantee the company's work against major repair but wants to have to honour no more than 10% of its guarantees, for how many years should the company guarantee its work? Assume that occurrences of major foundation repairs are Poisson distributed.

6.33 During the dry month of September, one European city has measurable rain on average only two days per month. If the arrival of rainy days is Poisson distributed in this city during the month of September, what is the average number of days that will pass between measurable rain? What is the standard deviation? What is the probability during this month that there will be a period of less than two days between rain?

Decision Dilemma SOLVED

The Cost of Absenteeism in the Workplace

The Small Firms Association in Ireland reported that, on average, a small-firm employee takes five days of unauthorized time off work per year. Suppose that number of days of absenteeism is uniformly distributed across all small firm employees varying from 0 days to 10 days. Using techniques presented in Section 6.1, this uniform distribution can be described by $a = 0$, $b = 10$, and $\mu = 5$. The probability that an employee takes between two and four days of unauthorized days off work can be determined by the following calculation assuming that $x_1 = 2$ and $x_2 = 4$:

$$P(x) = \frac{x_2 - x_1}{b - a} = \frac{4 - 2}{15 - 0} = \frac{2}{15} = 0.133$$

Thus, 13.3% of employees take between two and four days of unauthorized time off work.

  The probability of an employee being absent from work for 10 days or more can be calculated as:

$$P(x) = \frac{x_2 - x_1}{b - a} = \frac{15 - 10}{15 - 0} = \frac{5}{15} = 0.333$$

Around 33% of employees have 10 or more days of unauthorized absence from work. Note that here x_2 is 15 since 15 days is the upper end of the distribution.

  It is estimated by another study that the annual average cost of absenteeism is around €660 per employee. If such costs are normally distributed with a standard

(continued)

deviation of €300, the probability that the annual average cost of absenteeism is more than €700 can be calculated using techniques from Section 6.2 as:

$$z = \frac{x - \mu}{\sigma} = \frac{700 - 660}{300} = 0.13$$

The area associated with this z value is 0.0517 and the tail of the distribution is $0.5000 - 0.0517 = 0.4483$. That is, 44.83% of the time, absenteeism costs more than €700 per employee. The probability that absenteeism costs less than €600 can be determined in a similar manner:

$$z = \frac{x - \mu}{\sigma} = \frac{550 - 660}{300} = -0.36$$

The area associated with this z value is 0.1406 and the tail of the distribution is $0.5000 - 0.1406 = 0.3594$. That is, 35.94% of the time, the cost of absenteeism is less than €550 per employee.

Thirty-five per cent of all unscheduled absenteeism is caused by personal illness. Using techniques presented in Section 6.3, the probability that more than 50 of 120 randomly selected unscheduled absences were caused by personal illness can be determined. With $n = 120$, $p = 0.35$, and $x > 50$, this binomial distribution problem can be converted into a normal distribution problem by:

$$\mu = n \cdot p = (120)(0.35) = 42$$

and

$$\sigma = \sqrt{n \cdot p \cdot q} = \sqrt{1(120)(0.35)(0.65)} = 5.225$$

Since $42 \pm 3(5.225)$ is between 0 and 120, it is appropriate to use the normal distribution to approximate this binomial problem. Applying the correction for continuity, $x \geqslant 50.5$. The z value is calculated as:

$$z = \frac{x - \mu}{\sigma} = \frac{50.5 - 42}{5.225} = 1.63$$

The area associated with this z value is 0.4484 and the tail of the distribution is $0.500 - 0.4484 = 0.0516$. That is, 5.16% of the time, more than 50 out of 120 unscheduled absences are due to personal illness.

ETHICAL CONSIDERATIONS

Several points must be considered in working with continuous distributions. Is the population being studied the same population from which the parameters (mean, standard deviation, λ) were determined? If not, the results may not be valid for the analysis being done. Invalid or spurious results can be obtained by using the parameters from one population to analyse another population. For example, a market study in Norway may result in the conclusion that the amount of fish eaten per month by adults is normally distributed with the average of 1.3 kilograms of fish per month. A market researcher in Poland should not assume that these figures apply to her population. People in Poland probably have quite different fish-eating habits than people in Norway, and the application of Norway population parameters to Poland probably will result in questionable conclusions.

As was true with the Poisson distribution in Chapter 5, the use of λ in the exponential distribution should be judicious because a λ for one interval in a given time period or situation may not be the same as the λ for the same interval in a different time period or situation. For example, the number of arrivals per five-minute time period at a restaurant on Friday night is not likely to be the same as the number of arrivals in a five-minute time period at that same restaurant from 2 P.M. to 4 P.M. on weekdays. In using established parameters such as μ and λ, a researcher should be certain that the population from which the parameter was determined is, indeed, the same population being studied.

Sometimes a normal distribution is used to analyse data when, in fact, the data are not normal. Such an analysis can contain bias and produce false results. Certain techniques for testing a distribution of data can determine whether they are distributed a certain way. Some of the techniques are presented in Chapter 16. In general, the techniques in this chapter can be misused if the wrong type of distribution is applied to the data or if the distribution used for analysis is the right one but the parameters (μ, σ, λ) do not fit the data of the population being analysed.

SUMMARY

This chapter discussed three different continuous distributions: the uniform distribution, the normal distribution, and the exponential distribution. With continuous distributions, the value of the probability density function does not yield the probability but instead gives the height of the curve at any given point. In fact, with continuous distributions, the probability at any discrete point is 0.0000. Probabilities are determined over an interval. In each case, the probability is the area under the curve for the interval being considered. In each distribution, the probability or total area under the curve is 1.

Probably the simplest of these distributions is the uniform distribution, sometimes referred to as the rectangular distribution. The uniform distribution is determined from a probability density function that contains equal values along some interval between the points a and b. Basically, the height of the curve is the same everywhere between these two points. Probabilities are determined by calculating the portion of the rectangle between the two points a and b that is being considered.

The most widely used of all distributions is the normal distribution. Many phenomena are normally distributed, including characteristics of most machine-produced parts, many measurements of the biological and natural environment, and many human characteristics such as height, weight, IQ, and achievement test scores. The normal curve is continuous, symmetrical, unimodal, and asymptotic to the axis; actually, it is a family of curves.

The parameters necessary to describe a normal distribution are the mean and the standard deviation. For convenience, data that are being analysed by the normal curve should be standardized by using the mean and the standard deviation to compute z scores. A z score is the distance that an x value is from the mean, μ, in units of standard deviations. With the z score of an x value, the probability of that value occurring by chance from a given normal distribution can be determined by using a table of z scores and their associated probabilities.

The normal distribution can be used to work certain types of binomial distribution problems. Doing so requires converting the n and p values of the binomial distribution to μ and σ of the normal distribution. When worked by using the normal distribution, the binomial distribution solution is only an approximation. If the values of $\mu \pm 3\sigma$ are within a range from 0 to n, the approximation is reasonably accurate. Adjusting for the fact that a discrete distribution problem is being worked by using a continuous distribution requires a correction for continuity. The correction for continuity involves adding or subtracting 0.50 to the x value being analysed. This correction usually improves the normal curve approximation.

Another continuous distribution is the exponential distribution. It complements the discrete Poisson distribution. The exponential distribution is used to compute the probabilities of times between random occurrences. The exponential distribution is a family of distributions described by one parameter, λ. The distribution is skewed to the right and always has its highest value at $x = 0$.

KEY TERMS

correction for continuity
exponential distribution
normal distribution

rectangular distribution
standardized normal
   distribution

uniform distribution
z distribution
z score

GO ONLINE TO DISCOVER THE EXTRA FEATURES FOR THIS CHAPTER

The Student Study Guide containing solutions to the odd-numbered questions, additional Quizzes and Concept Review Activities, Excel and Minitab databases, additional data files in Excel and Minitab, and more worked examples.
www.wiley.com/college/cortinhas

FORMULAS

Probability density function of a uniform distribution

$$f(x)=\begin{cases} \dfrac{1}{b-a} & \text{for } a \leq x \leq b \\ 0 & \text{for all other values} \end{cases}$$

Mean and standard deviation of a uniform distribution

$$\mu = \frac{a+b}{2}$$

$$\sigma = \frac{b-a}{\sqrt{12}}$$

Probability density function of the normal distribution

$$f(x)=\frac{1}{\sigma\sqrt{2\pi}}e^{-(1/2)[(x-\mu)/\sigma]^2}$$

z formula

$$z=\frac{x-\mu}{\sigma}$$

Conversion of a binomial problem to the normal curve

$$\mu=n \cdot p \text{ and } \sigma=\sqrt{n \cdot p \cdot q}$$

Exponential probability density function

$$f(x) = \lambda e^{-\lambda x}$$

Probabilities of the right tail of the exponential distribution

$$P(x \geq x_0) = e^{-\lambda x_0}$$

SUPPLEMENTARY PROBLEMS

CALCULATING THE STATISTICS

6.34 Data are uniformly distributed between the values of 6 and 14. Determine the value of $f(x)$. What are the mean and standard deviation of this distribution? What is the probability of randomly selecting a value greater than 11? What is the probability of randomly selecting a value between 7 and 12?

6.35 Assume a normal distribution and find the following probabilities:
 a. $P(x<21 \,|\, \mu = 25 \text{ and } \sigma = 4)$
 b. $P(x \geq 77 \,|\, \mu = 50 \text{ and } \sigma = 9)$
 c. $P(x>47 \,|\, \mu = 50 \text{ and } \sigma = 6)$
 d. $P(13 <x< 29 \,|\, \mu = 23 \text{ and } \sigma = 4)$
 e. $P(x \geq 105 \,|\, \mu = 90 \text{ and } \sigma = 2.86)$

6.36 Work the following binomial distribution problems by using the normal distribution. Check your answers by using Table A.2 to solve for the probabilities.
 a. $P(x = 12 \,|\, n = 25 \text{ and } p = 0.60)$
 b. $P(x>5 \,|\, n = 15 \text{ and } p = 0.50)$
 c. $P(x \leq 3 \,|\, n = 10 \text{ and } p = 0.50)$
 d. $P(x \geq 8 \,|\, n = 15 \text{ and } p = 0.40)$

6.37 Find the probabilities for the following exponential distribution problems:
 a. $P(x \geq 3 \,|\, \lambda = 1.3)$
 b. $P(x<2 \,|\, \lambda = 2.0)$

 c. $P(1 \leq x \leq 3 \,|\, \lambda = 1.65)$
 d. $P(x > 2 \,|\, \lambda = 0.405)$

TESTING YOUR UNDERSTANDING

6.38 The OECD reports that of persons who usually work full-time, the average number of hours worked per week in South Korea is 43.4. Assume that the number of hours worked per week for those who usually work full-time is normally distributed. Suppose 12% of these workers work more than 48 hours. Based on this percentage, what is the standard deviation of number of hours worked per week for these workers?

6.39 A recent Scottish Household survey showed that one in four people aged over 50 years of age volunteered some of his or her time in the year before. If this figure holds for the entire population and if a random sample of 150 people 50 years of age or older is taken, what is the probability that more than 50 of those sampled do volunteer work?

6.40 An entrepreneur opened a small hardware store in a high street mall. During the first few weeks, business was slow, with the store averaging only one customer every 20 minutes in the morning. Assume that the random arrival of customers is Poisson distributed.

a. What is the probability that at least one hour would elapse between customers?
b. What is the probability that 10 to 30 minutes would elapse between customers?
c. What is the probability that less than five minutes would elapse between customers?

6.41 In a recent year, the average price of a Windows 7 Home Premium Upgrade was £84.99 according to a price comparison website. Assume that prices of the Windows 7 Home Premium Upgrade that year were normally distributed, with a standard deviation of £8.53. If a retailer of computer software was randomly selected that year:

a. What is the probability that the price of a Windows 7 Home Premium Upgrade was below £80?
b. What is the probability that the price was above £95?
c. What is the probability that the price was between £83 and £84?

6.42 According to Reuters, Toni's Freilandeier (Toni's free-range eggs) is the largest producer of free-range and organic free-range eggs in Austria, with an output of 80 million eggs per year. Suppose egg production per year in Austria is normally distributed, with a standard deviation of 83 million eggs. If during only 3% of the years Toni's Freilandeier produce more than 2,655 million eggs, what is the mean egg production by Toni's Freilandeier?

6.43 The US Bureau of Labor Statistics releases figures on the number of full-time wage and salary workers with flexible schedules. The numbers of full-time wage and salary workers in each age category are almost uniformly distributed by age, with ages ranging from 18 to 65 years. If a worker with a flexible schedule is randomly drawn from the US workforce, what is the probability that he or she will be between 25 and 50 years of age? What is the mean value for this distribution? What is the height of the distribution?

6.44 A business convention holds its registration on Wednesday morning from 9.00 A.M. until 12.00 noon. Past history has shown that registrant arrivals follow a Poisson distribution at an average rate of 1.8 every 15 seconds. Fortunately, several facilities are available to register convention members.

a. What is the average number of seconds between arrivals to the registration area for this conference based on past results?

b. What is the probability that 25 seconds or more would pass between registration arrivals?
c. What is the probability that less than five seconds will elapse between arrivals?
d. Suppose the registration computers went down for a one-minute period. Would this condition pose a problem? What is the probability that at least one minute will elapse between arrivals?

6.45 A recent UBS Prices and Earnings report lists the average monthly apartment rent price in some large cities around the world. According to their report, the average cost of renting an apartment in Frankfurt is €900. Suppose that the standard deviation of the cost of renting an apartment in Frankfurt is €96 and that apartment rents in Frankfurt are normally distributed. If a Frankfurt apartment is randomly selected, what is the probability that the price is:

a. €1,000 or more?
b. Between €900 and €1,100?
c. Between €825 and €925?
d. Less than €700?

6.46 According to *The Wirthlin Report*, 24% of all workers say that their job is very stressful. If 60 workers are randomly selected:

a. What is the probability that 17 or more say that their job is very stressful?
b. What is the probability that more than 22 say that their job is very stressful?
c. What is the probability that between eight and 12 (inclusive) say that their job is very stressful?

6.47 The 2010 Annual Survey of Hours and Earnings shows that the average weekly pay for full-time emp-loyees in the UK was £499. Suppose weekly wages in the UK are normally distributed with a standard deviation of £246. A UK worker is randomly selected.

a. What is the probability that the worker's weekly salary is more than £600?
b. What is the probability that the worker's weekly salary is less than £450?
c. What is the probability that the worker's weekly salary is more than £400?
d. What is the probability that the worker's weekly salary is between £440 and £520?

6.48 Suppose interarrival times at a hospital emergency room during a weekday are exponentially distributed, with an average interarrival time of nine minutes. If the arrivals are Poisson distributed, what would the average number of arrivals per hour be? What is the probability that less than five minutes will elapse between any two arrivals?

6.49 Suppose the average speeds of passenger trains travelling from Amsterdam in Holland to Geneva in Switzerland are normally distributed, with a mean average speed of 88 miles per hour and a standard deviation of 6.4 miles per hour.
 a. What is the probability that a train will average less than 70 miles per hour?
 b. What is the probability that a train will average more than 80 miles per hour?
 c. What is the probability that a train will average between 90 and 100 miles per hour?

6.50 The Conference Board published information on why companies expect to increase the number of part-time jobs and reduce full-time positions. Eighty-one per cent of the companies said the reason was to get a flexible workforce. Suppose 200 companies that expect to increase the number of part-time jobs and reduce full-time positions are identified and contacted. What is the expected number of these companies that would agree that the reason is to get a flexible workforce? What is the probability that between 150 and 155 (not including the 150 or the 155) would give that reason? What is the probability that more than 158 would give that reason? What is the probability that fewer than 144 would give that reason?

6.51 According to the US Bureau of the Census, about 75% of commuters in the United States drive to work alone. Suppose 150 US commuters are randomly sampled.
 a. What is the probability that fewer than 105 commuters drive to work alone?
 b. What is the probability that between 110 and 120 (inclusive) commuters drive to work alone?
 c. What is the probability that more than 95 commuters drive to work alone?

6.52 According to figures released by the UN Food and Agriculture Organization, the world production of wheat over the past 20 years has been approximately uniformly distributed. Suppose the mean production over this period was 600 million metric tons. If the height of this distribution is 0.005 million of metric tons, what are the values of a and b for this distribution?

6.53 The Luxembourg Income Study and the Organization for Economic Cooperation and Development publish internationally comparable data on household income for a large number of countries. The mean after-tax and after-social contributions family income was $39,688 (international dollars) for Luxembourg in a recent year. Suppose that 60% of the after-tax and after-social contributions family incomes in Luxembourg are between $30,088 and $49,288 and that these incomes are normally distributed. What is the standard deviation of after-tax and after-social contributions family incomes in Luxembourg?

6.54 According to the Polk Company, a survey of households using the Internet in buying or leasing cars reported that 81% were seeking information about prices. In addition, 44% were seeking information about products offered. Suppose 75 randomly selected households who are using the Internet in buying or leasing cars are contacted.
 a. What is the expected number of households who are seeking price information?
 b. What is the expected number of households who are seeking information about products offered?
 c. What is the probability that 67 or more households are seeking information about prices?
 d. What is the probability that fewer than 23 households are seeking information about products offered?

6.55 Nearly one-third of the world's tropical cyclones form within the western Pacific basin. Businesses along this area worry about the threat of hurricanes especially during the peak season, from June through October. Businesses in the South-East Asia area become especially nervous as the region just north-east of the Philippines is the most active place on earth for tropical cyclones. Suppose the arrival of tropical cyclones during this season is Poisson distributed, with an average of three tropical cyclones entering the South-East Asian region during the five-month season. If a tropical cyclone has just entered the South-East Asian region:
 a. What is the probability that at least one month will pass before the tropical cyclone enters the region again?
 b. What is the probability that another tropical cyclone will enter the region again in two weeks or less?
 c. What is the average amount of time between tropical cyclones entering the region?

6.56 With the growing emphasis on technology and the changing business environment, many employers are discovering that training such as re-education, skill development, and personal growth are of great assistance in staying ahead of the competition and in maintaining a motivated workforce. A recent European Foundation for the Improvement of Living and Working Conditions survey found that on-the-job training is becoming more prevalent in the European Union with 30% of employees having undergone training in the 12 months prior to the survey. If 50 EU workers are randomly sampled, what is the probability that fewer than 18 have experienced on-the-job training in the past 12 months? What is the expected number? What is the probability that between 12 and 17 (inclusive) have had on-the-job training in the 12 months prior to the survey?

6.57 According to a comparison website, the average rental cost of a Eurocopter EC120 Colibri helicopter is 1,000 per hour. Suppose the rental costs of a Eurocopter EC120 Colibri helicopter are normally distributed with a standard deviation of €175 per hour. At what rental cost would only 20% of the rental costs be less? At what rental cost would 65% of the rental costs be more? What rental cost would be more than 85% of rental costs?

6.58 Supermarkets usually become busy at about 5 P.M. on weekdays, because many workers stop by on the way home to shop. Suppose at that time arrivals at a supermarket's express checkout station are Poisson distributed, with an average of 0.8 person/minute. If the clerk has just checked out the last person in line, what is the probability that at least one minute will elapse before the next customer arrives? Suppose the clerk wants to go to the manager's office to ask a quick question and needs 2.5 minutes to do so. What is the probability that the clerk will get back before the next customer arrives?

6.59 In a recent year, the average daily circulation of the *Times of India* was 3,140,000. Suppose the standard deviation is 150,940. Assume the paper's daily circulation is normally distributed. On what percentage of days would circulation pass 3,300,000? Suppose the paper cannot support the fixed expenses of a full-production setup if the circulation drops below 2,600,000. If the probability of this even occurring is low, the production manager might try to keep the full crew in place and not disrupt operations. How often will this event happen, based on this historical information?

6.60 Incoming phone calls generally are thought to be Poisson distributed. If an operator averages 2.2 phone calls every 30 seconds, what is the expected (average) amount of time between calls? What is the probability that a minute or more would elapse between incoming calls? Two minutes?

INTERPRETING THE OUTPUT

6.61 Shown here is a Minitab output. Suppose the data represent the number of sales associates who are working in a department store in any given retail day. Describe the distribution including the mean and standard deviation. Interpret the shape of the distribution and the mean in light of the data being studied. What do the probability statements mean?

```
Cumulative Distribution Function

Continuous uniform on 11 to 32

     x            P(X<=x)
    28            0.80952
    34            1.00000
    16            0.23810
    21            0.47619
```

6.62 A manufacturing company produces a metal rod. Use the Excel output shown here to describe the weight of the rod. Interpret the probability values in terms of the manufacturing process.

Normal Distribution
Mean = 227 mg.
Standard Deviation = 2.3 mg.

x Value	Probability < x Value
220	0.0012
225	0.1923
227	0.5000
231	0.9590
238	1.0000

6.63 Suppose the Minitab output shown here represents the analysis of the length of home-use mobile phone calls in terms of minutes. Describe the distribution of mobile phone call lengths and interpret the meaning of the probability statements.

Cumulative Distribution Function	
Normal with mean = 2.35 and standard deviation = 0.11	
x	$P(X <= x)$
2.60	0.988479
2.45	0.818349
2.30	0.324718
2.00	0.000732

Exponential Distribution

x Value	Probability < x Value
0.1	0.3630
0.2	0.5942
0.5	0.8951
1.0	0.9890
2.4	1.0000

Cumulative Distribution Function	
Exponential with mean = 0.221729	
X	$P(X <= x)$
0.1	0.363010
0.2	0.594243
0.5	0.895127
1.0	0.989002
2.4	0.999980

6.64 A restaurant averages 4.51 customers per 10 minutes during the summer in the late afternoon. Shown here are Excel and Minitab output for this restaurant. Discuss the type of distribution used to analyse the data and the meaning of the probabilities.

ANALYSING THE DATABASES

1. The Financial database contains a variable, Earnings per Share, which represents the amount of earnings per each outstanding share of a company's stock. Calculate the mean and standard deviation for this variable that is approximately normally distributed in this database. Using the mean and standard deviation, calculate the probability that a randomly selected company reports earnings per share of more than €2.4. What is the probability that randomly selected company reports a earnings per share of less than €1.2? What is the probability that a randomly selected company reports an earnings per share between €2 and €3?

2. Select the Agribusiness time-series database. Create a histogram graph for onions and for broccoli. Each of these variables is approximately normally distributed. Compute the mean and the standard deviation for each distribution. The data in this database represent the monthly weight (in thousands of kilograms) of each vegetable. In terms of monthly weight, describe each vegetable (onions and broccoli). If a month were randomly selected from the onion distribution, what is the probability that the weight would be more than 50,000? What is the probability that the weight would be between 25,000 and 35,000? If a month were randomly selected from the broccoli distribution, what is the probability that the weight would be more than 100,000? What is the probability that the weight would be between 135,000 and 170,000?

3. From the Hospital database, it can be determined that some hospitals admit around 50 patients per day. Suppose we select a hospital that admits 50 patients per day. Assuming that admittance only occurs within a 12-hour time period each day and that admittance is Poisson distributed, what is the value of lambda for per hour for this hospital? What is the interarrival time for admittance based on this figure? Suppose a person was just admitted to the hospital. What is the probability that it would be more than 30 minutes before the next person was admitted? What is the probability that there would be less than 10 minutes before the next person was admitted?

MERCEDES GOES AFTER YOUNGER BUYERS

Mercedes and BMW have been competing head-to-head for market share in the luxury-car market for more than four decades. Back in 1959, BMW (Bayerische Motoren Werke) almost went bankrupt and nearly sold out to Daimler-Benz, the maker of Mercedes-Benz cars. BMW was able to recover to the point that in 1992 it passed Mercedes in worldwide sales. Among the reasons for BMW's success was its ability to sell models that were more luxurious than previous models but still focused on consumer quality and environmental responsibility. In particular, BMW targeted its sales pitch to the younger market, whereas Mercedes retained a more mature customer base.

In response to BMW's success, Mercedes has been trying to change its image by launching several products in an effort to attract younger buyers who are interested in sporty, performance-oriented cars. BMW, influenced by Mercedes, is pushing for more refinement and comfort. In fact, one automotive expert says that Mercedes wants to become BMW, and vice versa. However, according to one motoring expert, the focus is still on luxury and comfort for Mercedes while BMW focuses on performance and driving dynamics. Even though each company produces many different models, two relatively comparable coupe models are the BMW 3 Series Coupé 335i and the Mercedes CLK350 Coupé. In a recent year, the national German market price for the BMW 3 Series Coupé 335i was €43,000 and for the Mercedes CLK350 Coupé was €47,000. Petrol consumption for both of these cars is around 17 miles per gallon (mpg) in town and 25 mpg on the highway.

DISCUSSION

1. Suppose Mercedes is concerned that dealer prices of the CLK350 are not consistent and that even though the average price is €47,000, actual prices are normally distributed with a standard deviation of €2,981. Suppose also that Mercedes believes that, at €45,000, the CLK350 is priced out of the BMW 335i market. What percentage of the dealer prices for the Mercedes CLK350 is more than €45,000 and hence priced out of the BMW 335i market? The average price for a BMW 335i is €43,000. Suppose these prices are also normally distributed with a

standard deviation of €2,367. What percentage of BMW dealers are pricing the BMW 335i at more than the average price for a Mercedes CLK350? What might this mean to BMW if dealers were pricing the 335i at this level? What percentage of Mercedes dealers are pricing the CLK350 at less than the average price of a BMW 335i?

2. Suppose that highway fuel consumption rates for both of these cares are uniformly distributed over a range of from 20 to 30 mpg. What proportion of these cars would fall into the 22 to 27 mpg range? Compute the proportion of cars that get more than 28 mpg. What proportion of cars would get less than 23 mpg?

3. Suppose that in one dealership an average of 1.37 CLKs is sold every three hours (during a 12-hour showroom day) and that sales are Poisson distributed. The following Excel-produced probabilities indicate the occurrence of different between-sales times based on this information. Study the output and interpret it for the salespeople. For example, what is the probability that less than an hour will elapse between sales? What is the probability that more than a day (12-hour day) will pass before the next sale after a car has been sold? What can the dealership managers do with such information? How can it help in staffing? How can such information be used as a tracking device for the impact of advertising? Is there a chance that these probabilities would change during the year? If so, why?

Portion of 3-Hour Time Frame	Cumulative Exponential Probabilities from Left
.167	.2045
.333	.3663
.667	.5990
1	.7459
2	.9354
3	.9836
4	.9958
5	.9989

USING THE COMPUTER

EXCEL

- Excel can be used to compute cumulative probabilities for particular values of x from either a normal distribution or an exponential distribution.

- Calculation of probabilities from each of these distributions begins with the **Insert Function** (f_x). To access the **Insert Function**, go to the **Formulas** tab on an Excel worksheet (top centre tab). The **Insert Function** is on the far left of the menu bar. In the **Insert Function** dialog box at the top, there is a pulldown menu where it says **Or select a category**. From the pulldown menu associated with this command, select **Statistical**.

- To compute probabilities from a normal distribution, select **NORMDIST** from the **Insert Function's Statistical** menu. In the **NORMDIST** dialog box, there are four lines to which you must respond. On the first line, **X**, enter the value of x. On the second line, **Mean**, enter the value of the mean. On the third line, **Standard_dev**, enter the value of the standard deviation. The fourth line, **Cumulative**, requires a logical response of either TRUE or FALSE. Place TRUE in the slot to get the cumulative probabilities for all values up to x. Place FALSE in the slot to get the value of the probability density function for that combination of x, the mean, and the standard deviation. In this chapter, we are more interested in the cumulative probabilities and will enter TRUE most of the time.

- To compute probabilities from an exponential distribution, select **EXPONDIST** from the **Insert Function's Statistical** menu. In the **EXPONDIST** dialog box, there are three lines to which you must respond. On the first line, **X**, enter the value of x. On the second line, **Lambda**, enter the value of lambda. The third line, **Cumulative**, requires a logical response of either TRUE or FALSE. Place TRUE in the slot to get the cumulative probabilities for all values up to x. Place FALSE in the slot to get the value of the probability density function for that combination of x and lambda. In this chapter, we are more interested in the cumulative probabilities and will enter TRUE most of the time.

MINITAB

- Probabilities can be computed using Minitab for many different distributions, including the uniform distribution, the normal distribution, and the exponential distribution.

- To begin uniform distribution probabilities, select **C̲alc** on the menu bar. Select **Probability Distributions** from the pulldown menu. From the long second pulldown menu, select **U̲niform**. From the dialog box, check how you want the probabilities to be calculated from **Probability density**, **Cumulative probability**, or **Inverse probability**. **Probability density** yields the value of the probability density for a particular combination of a, b, and x. **Cumulative probability** produces the cumulative probabilities for values less than or equal to x. **Inverse probability** yields the inverse of the cumulative probabilities. Here we are mostly interested in **Cumulative probability**. On the line, **Lower endpoint:**, enter the value of a. On the line, **Upper endpoint:**, enter the value of b. If you want to compute probabilities for several values of x, place them in a column, list the column location in **Input column**. If you want to compute the probability for a particular value of x, check **Input constant**, and enter the value of x.

- To begin normal distribution probabilities, select **C̲alc** on the menu bar. Select **Probability Distri̲butions** from the pulldown menu. From the long second pulldown menu, select **N̲ormal**. From the dialog box, check how you want the probabilities to be calculated from **Probability density**, **Cumulative probability**, or **Inverse probability**. **Probability density** yields the value of the probability density for a particular combination of μ, σ, and x. **Cumulative probability** produces the cumulative probabilities for values less than or equal to x. **Inverse probability** yields the inverse of the cumulative probabilities. Here we are mostly interested in **Cumulative probability**. In the space beside **Mean**, enter the value of the mean. In the space beside **Standard deviation**, enter the value of the standard deviation. If you want to compute probabilities for several values of x, place them in a column, list the column location in **Input column**. If you want to compute the probability for a particular value of x, check **Input constant**, and enter the value of x.

- To begin exponential distribution probabilities, select **Calc** on the menu bar. Select **Probability Distributions** from the pulldown menu. From the long second pulldown menu, select **Exponential**. From the dialog box, check how you want the probabilities to be calculated from **Probability density, Cumulative probability,** or **Inverse probability**. **Probability density** yields the value of the probability density for a particular combination of x_0 and μ. **Cumulative probability** produces the cumulative probabilities for values less than or equal to x. **Inverse probability** yields the inverse of the cumulative probabilities. Here we are mostly interested in **Cumulative probability**. In the space beside **Scale**, enter a scale value to define the exponential distribution. The scale parameter equals the mean, when the threshold parameter equals 0. *Note*: Minitab uses the mean, $\mu = 1/\lambda$, not the value of λ. In the space beside **Threshold**, enter a threshold number to define the exponential distribution. If you want to compute probabilities for several values of x, place them in a column, list the column location in **Input column**. If you want to compute the probability for a particular value of x, check **Input constant**, and enter the value of x.

CHAPTER 7

Sampling and Sampling Distributions

LEARNING OBJECTIVES

The two main objectives for Chapter 7 are to give you an appreciation for the proper application of sampling techniques and an understanding of the sampling distributions of two statistics, thereby enabling you to:

1. Contrast sampling to census and differentiate among different methods of sampling, which include simple, stratified, systematic, and cluster random sampling; and convenience, judgement, quota, and snowball non-random sampling, by assessing the advantages associated with each

2. Describe the distribution of a sample's mean using the central limit theorem, correcting for a finite population if necessary

3. Describe the distribution of a sample's proportion using the z formula for sample proportions

Decision Dilemma

Private Health Insurance in India

The majority of the Indian population is unable to access high-quality healthcare provided by private players as a result of high costs. However, this is changing dramatically with the advent of health financing as a preferred tool for covering most healthcare expenditures. Health financing essentially involves paying a health service that has been arranged for under the financing contract. Health insurance coverage – both government sponsored and private – has been increasing very rapidly since the opening up of the market with the creation of a new regulator in 2000, the Insurance Regulatory and Development Authority (IRDA). The number of participants in the industry has gone from six insurers in 2000 to 48 insurers today, with 40 of those being private and eight from the public sector.

Currently only 2.6% of the 1.21 billion Indian population has private health insurance coverage. The Rashtriya Swasthya Bima Yojana (RSBY) health insurance scheme introduced in 2008, which provides hospitalization coverage for the below-poverty-line (BPL) population, has become the most widespread government scheme, covering a further 7.4% of the population. The other major social scheme is the Employees' State Insurance Scheme (ESIS), which provides coverage for employees in the 'organized' employment sector. A total of 4.8% of the population is covered by this scheme. There are also numerous of community schemes which offer limited coverage for small premiums (i.e. micro-insurance). Together these schemes cover less than 1% of the population. Despite the rapid increase in both social and private health insurance, almost three-quarters of India's population still lacks any type of coverage. There is, therefore, huge opportunity for insurance providers entering into the Indian healthcare market.

Both current and potential insurance providers would like to have a number of important questions answered to help define their strategy for the future. What are the characteristics of people with private insurance in India? What are the characteristics of people without any health insurance coverage? What makes people join a private insurance scheme and how much are they willing to pay for insurance premiums? How does a business researcher go about surveying the Indian population on Healthcare?

Managerial and Statistical Questions

Suppose researchers decide to survey the Indian population to ascertain the people's attitudes towards and expectations of private insurance schemes.

1. Should the researchers take a census of the entire Indian population or just a sample? What are the reasons for each?
2. If a sample is used, what type of sampling technique would gain the most valid information? How can the researchers be certain that the sample of people is representative of the population?
3. How can survey questions be analysed quantitatively?

Sources: Adapted from 'The Private Health Insurance Market in India', Report published by Business Insights on 13 May 2011; website of the Indian Insurance and Development Regulatory Authority (www.irda.gov.in/ADMINCMS/cms/frmGeneral_NoYearList.aspx?DF=AR&mid=11.1); and the *Wikipedia* entry on 'Healthcare in India' (http://en.wikipedia.org/wiki/Healthcare_in_India).

This chapter explores the process of sampling and the sampling distributions of some statistics. How do we obtain the data used in statistical analysis? Why do researchers often take a sample rather than conduct a census? What are the differences between random and non-random sampling? This chapter addresses these and other questions about sampling.

In addition to sampling theory, the distributions of two statistics: the sample mean and the sample proportion are presented. It has been determined that statistics such as these are approximately normally distributed under certain conditions. Knowledge of the uses of the sample mean and sample proportion is important in the study of statistics and is basic to much of statistical analysis.

7.1   SAMPLING

Sampling is widely used in business as a means of gathering useful information about a population. Data are gathered from samples and conclusions are drawn about the population as a part of the inferential statistics process. Often, a sample provides a reasonable means for gathering such useful decision-making information that might be otherwise unattainable and unaffordable.

Reasons for Sampling

Taking a sample instead of conducting a census offers several advantages:

1. The sample can save money.
2. The sample can save time.
3. For given resources, the sample can broaden the scope of the study.
4. Because the research process is sometimes destructive, the sample can save product.
5. If accessing the population is impossible, the sample is the only option.

A sample can be cheaper to obtain than a census for a given magnitude of questions. For example, if an eight-minute telephone interview is being undertaken, conducting the interviews with a sample of 100 customers rather than with a population of 100,000 customers obviously is less expensive. In addition to the cost savings, the significantly smaller number of interviews usually requires less total time. Thus, if obtaining the results is a matter of urgency, sampling can provide them more quickly. With the volatility of some markets and the constant barrage of new competition and new ideas, sampling has a strong advantage over a census in terms of research turnaround time.

If the resources allocated to a research project are fixed, more detailed information can be gathered by taking a sample than by conducting a census. With resources concentrated on fewer individuals or items, the study can be broadened in scope to allow for more specialized questions. One organization budgeted €100,000 for a study and opted to take a census instead of a sample by using a mail survey. The researchers mailed thousands of copies of a questionnaire on client satisfaction. The questionnaire contained 20 questions to which the respondent could answer Yes or No by punching out a perforated hole. The information retrieved amounted to the percentages of respondents who answered Yes and No on the 20 questions. For the same amount of money, the company could have taken a random sample from the population, held interactive one-on-one sessions with highly trained interviewers, and gathered detailed information about the process being studied. By using the money for a sample, the researchers could have spent significantly more time with each respondent and thus increased the potential for gathering useful information.

Some research processes are destructive to the product or item being studied. For example, if light bulbs are being tested to determine how long they burn or if chocolate bars are being taste tested to determine whether the taste is acceptable, the product is destroyed. If a census were conducted for this type of research, no product would be left to sell. Hence, taking a sample is the only realistic option for testing such products.

Sometimes a population is virtually impossible to access for research. For example, some people refuse to answer sensitive questions, and some telephone numbers are unlisted. Some items of interest (like a 1967 Porsche) are so scattered that locating all of them would be extremely difficult. When the population is inaccessible for these or other reasons, sampling is the only option.

STATISTICS IN BUSINESS TODAY

Sampling British Businesses

The Office for National Statistics, the United Kingdom's national statistical agency, administers a monthly survey of production and service businesses for the UK. This Monthly Business Survey (MBS) collects turnover, new orders, and employment information from UK businesses in production and service industries. The MBS data are used as indicators of the economic condition of production industries and services in the UK along with inputs for the UK's gross domestic product, economic studies, and econometric models. The sampling frame for the MBS is the Inter-Departmental Business Register (IDBR). The employment recorded on the Businesses' register is used as the basis for stratification with complete enumeration for businesses with employment exceeding an industry-specific threshold; a random sample is selected for each of the other strata. The target population consists of all statistical establishments on the IDBR, covering businesses in all parts of the economy except for some very small business and some non-profit organisations, thus representing nearly 99% of UK economic activity. The frame is further reduced by eliminating the smallest units of the survey population. As a result, there are approximately 32,000 establishments in the sample. Before the sample is taken, the sampling frame is stratified by stratified simple random sampling, but there is also100% coverage of businesses above an industry-specific employment threshold considered 'large'. Further stratification is then made within each cell by company size so that similar-sized companies are grouped together. Selected establishments are required to respond to the survey, and data are collected directly from survey respondents and extracted from administrative files. Sampled companies are contacted either by mail or telephone, whichever they prefer.

Source: Office for National Statistics website, at www.ons.gov.uk/about/surveys/a-z-of-surveys/monthly-business-survey--production-and-services-/index.html.

Reasons for Taking a Census

Sometimes it is preferable to conduct a census of the entire population rather than taking a sample. There are at least two reasons why a business researcher may opt to take a census rather than a sample, providing there is adequate time and money available to conduct such a census: (i) to eliminate the possibility that by chance a randomly selected sample may not be representative of the population and (ii) for the safety of the consumer.

Even when proper sampling techniques are implemented in a study, there is the possibility a sample could be selected by chance that does not represent the population. For example, if the population of interest is all Land Rover owners in the French region of Aquitaine, a random sample of Land Rover owners could yield mostly farmers when, in fact, many of the Land Rover owners in Aquitaine are urban dwellers. If the researcher or study sponsor cannot tolerate such a possibility, then taking a census may be the only option.

In addition, sometimes a census is taken to protect the safety of the consumer. For example, there are some products, such as airplanes or heart defibrillators, in which the performance of such is so critical to the consumer that 100% of the products are tested, and sampling is not a reasonable option.

Frame

Every research study has a target population that consists of the individuals, institutions, or entities that are the object of investigation. The sample is taken from a population *list, map, directory, or other source used to represent the population*. This list, map, or directory is called the **frame**, which can be school lists, trade association lists, or even lists sold by list brokers. Ideally, a one-to-one correspondence exists between the frame units and the population units. In reality, the frame and the target population are often different. For example, suppose the target population is all families living in Amsterdam. A feasible frame would be the residential pages of the Amsterdam telephone books. How would the frame differ from the target population? Some families have no telephone. Other families have unlisted numbers. Still other families might have moved and/or changed numbers since the directory was printed. Some families even have multiple listings under different names.

Frames that have *over-registration* contain the target population units plus some additional units. Frames that have *under-registration* contain fewer units than does the target population. Sampling is done from the frame, not the target population. In theory, the target population and the frame are the same. In reality, a business researcher's goal is to minimize the differences between the frame and the target population.

Random versus Non-random Sampling

The two main types of sampling are random and non-random. In **random sampling** *every unit of the population has the same probability of being selected into the sample*. Random sampling implies that chance enters into the process of selection. For example, most people would like to believe that winners of a magazine competition or the winning numbers for the EuroMillions lottery are selected by some random draw of numbers.

In **non-random sampling** *not every unit of the population has the same probability of being selected into the sample.* Members of non-random samples are not selected by chance. For example, they might be selected because they are at the right place at the right time or because they know the people conducting the research.

Sometimes random sampling is called *probability sampling* and non-random sampling is called *non-probability sampling.* Because every unit of the population is not equally likely to be selected, assigning a probability of occurrence in non-random sampling is impossible. The statistical methods presented and discussed in this text are based on the assumption that the data come from random samples. *Non-random sampling methods are not appropriate techniques for gathering data to be analysed by most of the statistical methods presented in this text.* However, several non-random sampling techniques are described in this section, primarily to alert you to their characteristics and limitations.

Random Sampling Techniques

The four basic random sampling techniques are simple random sampling, stratified random sampling, systematic random sampling, and cluster (or area) random sampling. Each technique offers advantages and disadvantages. Some techniques are simpler to use, some are less costly, and others show greater potential for reducing sampling error.

Simple Random Sampling

The most elementary random sampling technique is **simple random sampling**. Simple random sampling can be viewed as the basis for the other random sampling techniques. With simple random sampling, each unit of the frame is numbered from 1 to N (where N is the size of the population). Next, a table of random numbers or a random number generator is used to select n items into the sample. A random number generator is usually a computer program that allows computer-calculated output to yield random numbers. Table 7.1 contains a brief table of random numbers. Table A.1 in Appendix A contains a full table of random numbers. These numbers are random in all directions. The spaces in the table are there only for ease of reading the values. For each number, any of the 10 digits (0–9) is equally likely, so getting the same digit twice or more in a row is possible.

As an example, from the population frame of companies listed in Table 7.2, we will use simple random sampling to select a sample of six companies. First, we number every member of the population. We select as many digits for each unit sampled as there are in the largest number in the population. For example, if a population has 2,000 members, we select four-digit numbers. Because the population in Table 7.2 contains 30 members, only two digits need be selected for each number. The population is numbered from 01 to 30, as shown in Table 7.3.

The object is to sample six companies, so six different two-digit numbers must be selected from the table of random numbers. Because this population contains only 30 companies, all numbers greater than 30 (31–99) must be ignored. If, for example, the number 67 is selected, the process is continued until a value between 1 and 30 is obtained. If the same number occurs more than once, we proceed to another number. For ease of understanding, we start with the first pair of digits in Table 7.1 and proceed across the first row until $n = 6$ different values between 01 and 30 are selected. If additional numbers are needed, we proceed across the second row, and so on. Often a researcher will start at some randomly selected location in the table and proceed in a predetermined direction to select numbers.

In the first row of digits in Table 7.1, the first number is 91. This number is out of range so it is cast out. The next two digits are 56. Next is 74, followed by 25, which is the first usable number. From Table 7.3, we see that 25 is the

TABLE 7.1

A Brief Table of Random Numbers

91567	42595	27958	30134	04024	86385	29880	99730
46503	18584	18845	49618	02304	51038	20655	58727
34914	63974	88720	82765	34476	17032	87589	40836
57491	16703	23167	49323	45021	33132	12544	41035
30405	83946	23792	14422	15059	45799	22716	19792
09983	74353	68668	30429	70735	25499	16631	35006
85900	07119	97336	71048	08178	77233	13916	47564

TABLE 7.2

A Population Frame of 30 Companies

A.P. Moller–Maersk	Eni	Microsoft
Allianz	Exxon Mobil	Mitsubishi Corporation
Apple Inc.	Gazprom	Munich Re
Bayer	Hewlett-Packard	Novartis
Boeing	ING Group	PetroChina
China Mobile	Intel	Rio Tinto
CNP Assurances	Itausa	Sanofi-Aventis
Credit Suisse	JX Holdings	Temasek Holdings
Dai-ichi Life	LG Group	Toshiba
Dexia	Metro AG	UniCredit

TABLE 7.3

Numbered Population of 30 Companies

01	A.P. Moller–Maersk	11	Eni	21	Microsoft
02	Allianz	12	Exxon Mobil	22	Mitsubishi Corporation
03	Apple Inc.	13	Gazprom	23	Munich Re
04	Bayer	14	Hewlett-Packard	24	Novartis
05	Boeing	15	ING Group	25	PetroChina
06	China Mobile	16	Intel	26	Rio Tinto
07	CNP Assurances	17	Itausa	27	Sanofi-Aventis
08	Credit Suisse	18	JX Holdings	28	Temasek Holdings
09	Dai-ichi Life	19	LG Group	29	Toshiba
10	Dexia	20	Metro AG	30	UniCredit

number associated with PetroChina, so PetroChina is the first company selected into the sample. The next number is 95, unusable, followed by 27, which is usable. Twenty-seven is the number for Sanofi-Aventis, so this company is selected. Continuing the process, we pass over the numbers 95 and 83. The next usable number is 01, which is the value for A.P. Moller–Maersk. Thirty-four is next, followed by 04 and 02, both of which are usable. These numbers are associated with Bayer and Allianz, respectively. Continuing along the first row, the next usable number is 29, which is associated with Toshiba. Because this selection is the sixth, the sample is complete. The following companies constitute the final sample.

A.P. Moller–Maersk

Allianz

Bayer

PetroChina

Sanofi-Aventis

Toshiba

Simple random sampling is easier to perform on small than on large populations. The process of numbering all the members of the population and selecting items is cumbersome for large populations.

Stratified Random Sampling

A second type of random sampling is **stratified random sampling**, in which the population is divided into non-overlapping subpopulations called strata. The researcher then extracts a random sample from each of the

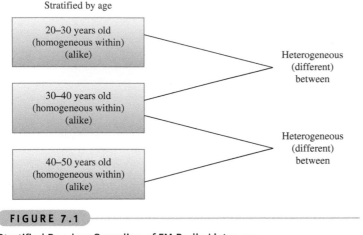

Stratified by age

FIGURE 7.1

Stratified Random Sampling of FM Radio Listeners

subpopulations. The main reason for using stratified random sampling is that it has the potential for reducing sampling error. Sampling error occurs when, by chance, the sample does not represent the population. With stratified random sampling, the potential to match the sample closely to the population is greater than it is with simple random sampling because portions of the total sample are taken from different population subgroups. However, stratified random sampling is generally more costly than simple random sampling because each unit of the population must be assigned to a stratum before the random selection process begins.

Strata selection is usually based on available information. Such information may have been gleaned from previous censuses or surveys. Stratification benefits increase as the strata differ more. Internally, a stratum should be relatively homogeneous; externally, strata should contrast with each other. Stratification is often done by using demographic variables, such as sex, socioeconomic class, geographic region, religion, and ethnicity. For example, if a parliamentary election poll is to be conducted by a market research firm, what important variables should be stratified? The sex of the respondent might make a difference because a gender gap in voter preference has been noted in past elections; that is, men and women tended to vote differently in national elections. Geographic region also provides an important variable in national elections because voters are influenced by local cultural values that differ from region to region.

In FM radio markets, age of listener is an important determinant of the type of programming used by a station. Figure 7.1 contains a stratification by age with three strata, based on the assumption that age makes a difference in preference of programming. This stratification implies that listeners 20 to 30 years of age tend to prefer the same type of programming, which is different from that preferred by listeners 30 to 40 and 40 to 50 years of age. Within each age subgroup (stratum), *homogeneity* or alikeness is present; between each pair of subgroups a difference, or *heterogeneity*, is present.

Stratified random sampling can be either proportionate or disproportionate. **Proportionate stratified random sampling** occurs *when the percentage of the sample taken from each stratum is proportionate to the percentage that each stratum is within the whole population.* For example, suppose voters are being surveyed in Bavaria, Germany and the sample is being stratified by religion as Catholic, Protestant, Muslim, and others. If Bavaria's population is 34% Catholic and if a sample of 1,000 voters is taken, the sample would require inclusion of 340 Catholics to achieve proportionate stratification. Any other number of Catholics would be disproportionate stratification. The sample proportion of other religions would also have to follow population percentages. Or consider the city of London, where the population is approximately 12.1% South Asian. If a researcher is conducting a city-wide poll in London and if stratification is by ethnicity, a proportionate stratified random sample should contain 12.1% South Asians. Hence, an ethnically proportionate stratified sample of 160 residents from London's 7,600,000 residents should contain approximately 19 South Asians. *Whenever the proportions of the strata in the sample are different from the proportions of the strata in the population,* **disproportionate stratified random sampling** occurs.

Systematic Sampling

Systematic sampling is a third random sampling technique. Unlike stratified random sampling, systematic sampling is not done in an attempt to reduce sampling error. Rather, systematic sampling is used because of its convenience and relative ease of administration. With **systematic sampling**, *every kth item is selected to produce a sample of size n from a population of size N.* The value of k, sometimes called the sampling cycle, can be determined by the following formula. If k is not an integer value, the whole-number value should be used.

DETERMINING THE
VALUE OF k

$$k = \frac{N}{n}$$

where:

n = sample size
N = population size
k = size of interval for selection

As an example of systematic sampling, a management information systems researcher wanted to sample the manufacturers in Barcelona. He had enough financial support to sample 1,000 companies (n). The *Directory of Barcelona Manufacturers* listed approximately 17,000 total manufacturers in Barcelona (N) in alphabetical order. The value of k was 17 (17,000/1,000) and the researcher selected every 17th company in the directory for his sample.

Did the researcher begin with the first company listed or the 17th or one somewhere between? In selecting every kth value, a simple random number table should be used to select a value between 1 and k inclusive as a starting point. The second element for the sample is the starting point plus k. In the example, $k = 17$, so the researcher would have gone to a table of random numbers to determine a starting point between 1 and 17. Suppose he selected the number 5. He would have started with the 5th company, then selected the 22nd (5 + 17), and then the 39th, and so on.

Besides convenience, systematic sampling has other advantages. Because systematic sampling is evenly distributed across the frame, a knowledgeable person can easily determine whether a sampling plan has been followed in a study. However, a problem with systematic sampling can occur if the data are subject to any periodicity, and the sampling interval is in syncopation with it. In such a case, the sampling would be non-random. For example, if a list of 150 university students is actually a merged list of five classes with 30 students in each class and if each of the lists of the five classes has been ordered with the names of top students first and bottom students last, then systematic sampling of every 30th student could cause selection of all top students, all bottom students, or all mediocre students; that is, the original list is subject to a cyclical or periodic organization. Systematic sampling methodology is based on the assumption that the source of population elements is random.

Cluster (or Area) Sampling

A fourth type of random sampling is cluster (or area) sampling, which involves dividing the population into non-overlapping areas, or clusters. However, in contrast to stratified random sampling where strata are homogeneous within, cluster sampling identifies clusters that tend to be internally heterogeneous. In theory, each cluster contains a wide variety of elements, and the cluster is a miniature, or microcosm, of the population. Examples of clusters are towns, companies, homes, universities, areas of a city, and geographic regions. Often clusters are naturally occurring groups of the population and are already identified, such as provinces or the Food and Agriculture Organization's 27 Major Fishing Areas. Although area sampling usually refers to clusters that are areas of the population, such as geographic regions and cities, the terms *cluster sampling* and *area sampling* are used interchangeably in this text.

After randomly selecting clusters from the population, the business researcher either selects all elements of the chosen clusters or randomly selects individual elements into the sample from the clusters. One example of business research that makes use of clustering is test marketing of new products. Often in test marketing, a country is divided into clusters of test market cities, and individual consumers within the test market cities are surveyed.

Sometimes the clusters are too large, and a second set of clusters is taken from each original cluster. This technique is called **two-stage sampling**. For example, a researcher could divide the United Kingdom into clusters of cities. She could then divide the cities into clusters of blocks and randomly select individual houses from the block clusters. The first stage is selecting the test cities and the second stage is selecting the blocks.

Cluster or area sampling offers several advantages. Two of the foremost advantages are convenience and cost. Clusters are usually convenient to obtain, and the cost of sampling from the entire population is reduced because the scope of the study is reduced to the clusters. The cost per element is usually lower in cluster or area sampling than in stratified sampling because of lower element listing or locating costs. The time and cost of contacting elements of the population can be reduced, especially if travel is involved, because clustering reduces the distance to the sampled elements. In addition, administration of the sample survey can be simplified. Sometimes cluster or area sampling is the only feasible approach because the sampling frames of the individual elements of the population are unavailable and therefore other random sampling techniques cannot be used.

Cluster or area sampling also has several disadvantages. If the elements of a cluster are similar, cluster sampling may be statistically less efficient than simple random sampling. In an extreme case – when the elements of a cluster are the same – sampling from the cluster may be no better than sampling a single unit from the cluster. Moreover, the costs and problems of statistical analysis are greater with cluster or area sampling than with simple random sampling.

Non-random Sampling

Sampling techniques used to select elements from the population by any mechanism that does not involve a random selection process are called **non-random sampling techniques**. Because chance is not used to select items from the samples, these techniques are non-probability techniques and are not desirable for use in gathering data to be analysed by the methods of inferential statistics presented in this text. Sampling error cannot be determined objectively for these sampling techniques. Four non-random sampling techniques are presented here: convenience sampling, judgement sampling, quota sampling, and snowball sampling.

Convenience Sampling

In **convenience sampling**, *elements for the sample are selected for the convenience of the researcher*. The researcher typically chooses elements that are readily available, nearby, or willing to participate. The sample tends to be less variable than the population because in many environments the extreme elements of the population are not readily available. The researcher will select more elements from the middle of the population. For example, a convenience sample of homes for door-to-door interviews might include houses where people are at home, houses with no dogs, houses near the street, first-floor apartments, and houses with friendly people. In contrast, a random sample would require the researcher to gather data only from houses and apartments that have been selected randomly, no matter how inconvenient or unfriendly the location. If a research firm is located in a mall, a convenience sample might be selected by interviewing only shoppers who pass the shop and look friendly.

Judgement Sampling

Judgement sampling occurs when *elements selected for the sample are chosen by the judgement of the researcher*. Researchers often believe they can obtain a representative sample by using sound judgement, which will result in saving time and money. Sometimes ethical, professional researchers might believe they can select a more representative sample than the random process will provide. They might be right! However, some studies show that random sampling methods outperform judgement sampling in estimating the population mean, even when the researcher who is administering the judgement sampling is trying to put together a representative sample. When sampling is done by judgement, calculating the probability that an element is going to be selected into the sample is not possible. The sampling error cannot be determined objectively because probabilities are based on *non-random* selection.

Other problems are associated with judgement sampling. The researcher tends to make errors of judgement in one direction. These systematic errors lead to what are called *biases*. The researcher also is unlikely to include extreme elements. Judgement sampling provides no objective method for determining whether one person's judgement is better than another's.

Quota Sampling

A third non-random sampling technique is **quota sampling**, which appears to be similar to stratified random sampling. Certain population subclasses, such as age group, gender, or geographic region, are used as strata. However, instead of randomly sampling from each stratum, the researcher uses a non-random sampling method to gather data from one stratum until the desired quota of samples is filled. Quotas are described by quota controls, which set the sizes of the samples to be obtained from the subgroups. Generally, a quota is based on the proportions of the subclasses in the population. In this case, the quota concept is similar to that of proportional stratified sampling.

Quotas often are filled by using available, recent, or applicable elements. For example, instead of randomly interviewing people to obtain a quota of Italian Australians, the researcher would go to the Italian area of Sydney and interview there until enough responses are obtained to fill the quota. In quota sampling, an interviewer would begin by asking a few filter questions; if the respondent represents a subclass whose quota has been filled, the interviewer would terminate the interview.

Quota sampling can be useful if no frame is available for the population. For example, suppose a researcher wants to stratify the population into owners of different types of cars but fails to find any lists of Toyota van owners. Through quota sampling, the researcher would proceed by interviewing all car owners and casting out non-Toyota van owners until the quota of Toyota van owners is filled.

Quota sampling is less expensive than most random sampling techniques because it essentially is a technique of convenience. However, cost may not be meaningful because the quality of non-random and random sampling techniques cannot be compared. Another advantage of quota sampling is the speed of data gathering. The researcher does not have to call back or send out a second questionnaire if he does not receive a response; he just moves on to the next element. Also, preparatory work for quota sampling is minimal.

The main problem with quota sampling is that, when all is said and done, it still is only a *non-random* sampling technique. Some researchers believe that if the quota is filled by *randomly* selecting elements and discarding those not from a stratum, quota sampling is essentially a version of stratified random sampling. However, most quota sampling is carried out by the researcher going where the quota can be filled quickly. The object is to gain the benefits of stratification without the high field costs of stratification. Ultimately, it remains a non-probability sampling method.

Snowball Sampling

Another non-random sampling technique is **snowball sampling**, in which *survey subjects are selected based on referral from other survey respondents*. The researcher identifies a person who fits the profile of subjects wanted for the study. The researcher then asks this person for the names and locations of others who would also fit the profile of subjects wanted for the study. Through these referrals, survey subjects can be identified cheaply and efficiently, which is particularly useful when survey subjects are difficult to locate. It is the main advantage of snowball sampling; its main disadvantage is that it is non-random.

STATISTICS IN BUSINESS TODAY

What is the World's Favourite Food?

A new global survey into the way the world eats today conducted by Globe Scan for Oxfam, an international charity organization, revealed that pasta is the world's favourite food, narrowly beating meat or rice dishes and pizza. The results of the independent poll across 17 countries show the spread of Western diets across the world. Although national dishes are still popular – such as paella in Spain, schnitzel in Germany and biryani in India – pizza and pasta are now the favourite foods of many, with more than half of the countries (nine out of 17) listing one or both in their top three foods. The research was carried out for Oxfam's recently launched Grow campaign, which is calling on governments and big businesses to fix the global food system to ensure that everyone has enough to eat. The survey also shows that the cost of food is a common worry worldwide. In fact, more than half of the people questioned globally said they are not eating the same food as they did two years ago, with two-fifths of those saying it was because food was becoming more expensive. In the UK, those citing the cost of food as their greatest concern rose to nearly eight out of 10 people (78%). This study demonstrates the usefulness of sampling. Since this was a global survey, inclusion of all the elements in the relevant population would be too time-consuming and expensive and therefore a sample had to be chosen instead. However, the fact that there is no practically accessible sampling frame implies that random sampling techniques could not be used and non-random methods such as quota and judgement sampling had to be used instead.

Sampling Error

Sampling error occurs *when the sample is not representative of the population*. When random sampling techniques are used to select elements for the sample, sampling error occurs by chance. Many times the statistic computed on the sample is not an accurate estimate of the population parameter because the sample was not representative of the population. This result is caused by sampling error. With random samples, sampling error can be computed and analysed.

Non-sampling Errors

All errors other than sampling errors are **non-sampling errors**. The many possible non-sampling errors include missing data, recording errors, input processing errors, and analysis errors. Other non-sampling errors result from the measurement instrument, such as errors of unclear definitions, defective questionnaires, and poorly conceived concepts. Improper definition of the frame is a non-sampling error. In many cases, finding a frame that perfectly fits the population is impossible. Insofar as it does not fit, a non-sampling error has been committed.

Response errors are also non-sampling errors. They occur when people do not know, will not say, or overstate. Virtually no statistical method is available to measure or control for non-sampling errors. The statistical techniques presented in this text are based on the assumption that none of these non-sampling errors were committed. The researcher must eliminate these errors through carefully planning and executing the research study.

7.1 PROBLEMS

7.1 Develop a frame for the population of each of the following research projects:

   a. Measuring the job satisfaction of all union employees in a company.

   b. Conducting a telephone survey in Rotterdam, Holland, to determine the level of interest in opening a new hunting and fishing speciality store in a shopping centre.

   c. Interviewing passengers of a major airline about its food service.

   d. Studying the quality-control programmes of boat manufacturers.

   e. Attempting to measure the corporate culture of cable television companies.

7.2 Make a list of 20 people you know. Include men and women, various ages, various educational levels, and so on. Number the list and then use the random number list in Table 7.1 to select six people randomly from your list. How representative of the population is the sample? Find the proportion of men in your population and in your sample. How do the proportions compare? Find the proportion of 20-year-olds in your sample and the proportion in the population. How do they compare?

7.3 Use the random numbers in Table A.1 of Appendix A to select 10 of the companies from the 30 companies listed in Table 7.2. Compare the types of companies in your sample with the types in the population. How representative of the population is your sample?

7.4 For each of the following research projects, list three variables for stratification of the sample.

   a. A nationwide study of motels and hotels is being conducted. An attempt will be made to determine the extent of the availability of online links for customers. A sample of motels and hotels will be taken.

   b. A consumer panel is to be formed by sampling people in Stuttgart, Germany. Members of the panel will be interviewed periodically in an effort to understand current consumer attitudes and behaviours.

   c. A large soft drinks company wants to study the characteristics of the European bottlers of its products, but the company does not want to conduct a census.

   d. The business research bureau of a large university is conducting a project in which the bureau will sample paper-manufacturing companies.

7.5 In each of the following cases, the variable represents one way that a sample can be stratified in a study. For each variable, list some strata into which the variable can be divided.

   a. Age of respondent (person).

   b. Size of company (sales volume).

   c. Size of retail outlet (square metres).

   d. Geographic location.

   e. Occupation of respondent (person).

   f. Type of business (company).

7.6 A city's telephone book lists 100,000 people. If the telephone book is the frame for a study, how large would the sample size be if systematic sampling were done on every 200th person?

7.7 If every 11th item is systematically sampled to produce a sample size of 75 items, approximately how large is the population?

7.8 If a company employs 3,500 people and if a random sample of 175 of these employees has been taken by systematic sampling, what is the value of k? The researcher would start the sample selection between what two values? Where could the researcher obtain a frame for this study?

7.9 For each of the following research projects, list at least one area or cluster that could be used in obtaining the sample:
a. A study of road conditions in Brazil.
b. A study of European offshore oil wells.
c. A study of the environmental impact of petrochemical plants in Asia.

7.10 Give an example of how judgement sampling could be used in a study to determine how university students feel about an increase in university fees.

7.11 Give an example of how convenience sampling could be used in a study of *Fortune 500* executives to measure corporate attitude towards paternity leave for employees.

7.12 Give an example of how quota sampling could be used to conduct sampling by a company test-marketing a new personal computer.

7.2 SAMPLING DISTRIBUTION OF $\bar{x}$

In the inferential statistics process, a researcher selects a random sample from the population, computes a statistic on the sample, and reaches conclusions about the population parameter from the statistic. In attempting to analyse the sample statistic, it is essential to know the distribution of the statistic. So far we have studied several distributions, including the binomial distribution, the Poisson distribution, the hypergeometric distribution, the uniform distribution, the normal distribution, and the exponential distribution.

In this section we explore the sample mean, $\bar{x}$, as the statistic. The sample mean is one of the more common statistics used in the inferential process. To compute and assign the probability of occurrence of a particular value of a sample mean, the researcher must know the distribution of the sample means. One way to examine the distribution possibilities is to take a population with a particular distribution, randomly select samples of a given size, compute the sample means, and attempt to determine how the means are distributed.

Suppose a small finite population consists of only $N = 8$ numbers:

$$54 \quad 55 \quad 59 \quad 63 \quad 64 \quad 68 \quad 69 \quad 70$$

Using the Excel-produced histogram below, we can see the shape of the distribution of this population of data.

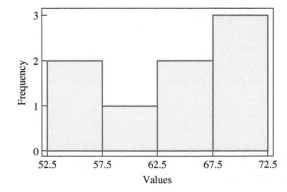

Suppose we take all possible samples of size $n = 2$ from this population with replacement. The result is the following pairs of data.

(54,54)	(55,54)	(59,54)	(63,54)
(54,55)	(55,55)	(59,55)	(63,55)
(54,59)	(55,59)	(59,59)	(63,59)
(54,63)	(55,63)	(59,63)	(63,63)
(54,64)	(55,64)	(59,64)	(63,64)
(54,68)	(55,68)	(59,68)	(63,68)
(54,69)	(55,69)	(59,69)	(63,69)
(54,70)	(55,70)	(59,70)	(63,70)
(64,54)	(68,54)	(69,54)	(70,54)
(64,55)	(68,55)	(69,55)	(70,55)
(64,59)	(68,59)	(69,59)	(70,59)
(64,63)	(68,63)	(69,63)	(70,63)
(64,64)	(68,64)	(69,64)	(70,64)
(64,68)	(68,68)	(69,68)	(70,68)
(64,69)	(68,69)	(69,69)	(70,69)
(64,70)	(68,70)	(69,70)	(70,70)

The means of each of these samples follow.

54	54.5	56.5	58.5	59	61	61.5	62
54.5	55	57	59	59.5	61.5	62	62.5
56.5	57	59	61	61.5	63.5	64	64.5
58.5	59	61	63	63.5	65.5	66	66.5
59	59.5	61.5	63.5	64	66	66.5	67
60	61.5	63.5	65.5	66	68	68.5	69
61.5	62	64	66	66.5	68.5	69	69.5
62	62.5	64.5	66.5	67	69	69.5	70

Again using an Excel-produced histogram, we can see the shape of the distribution of these sample means.

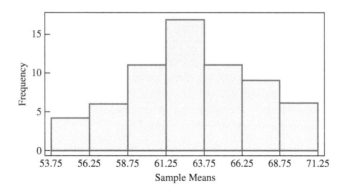

Notice that the shape of the histogram for sample means is quite unlike the shape of the histogram for the population. The sample means appear to 'pile up' towards the middle of the distribution and 'tail off' towards the extremes.

Figure 7.2 is a Minitab histogram of the data from a Poisson distribution of values with a population mean of 1.25. Note that the histogram is skewed to the right. Suppose 90 samples of size $n = 30$ are taken randomly from a Poisson distribution with $\lambda = 1.25$ and the means are computed on each sample. The resulting distribution of sample means is displayed in Figure 7.3. Notice that although the samples were drawn from a Poisson distribution, which is skewed to the right, the sample means form a distribution that approaches a symmetrical, nearly normal-curve-type distribution.

Suppose a population is uniformly distributed. If samples are selected randomly from a population with a uniform distribution, how are the sample means distributed? Figure 7.4 displays the Minitab histogram distributions of sample means from five different sample sizes. Each of these histograms represents the distribution of sample means from 90 samples generated randomly from a uniform distribution in which $a = 10$ and $b = 30$.

Observe the shape of the distributions. Notice that even for small sample sizes, the distributions of sample means for samples taken from the uniformly distributed population begin to 'pile up' in the middle. As sample sizes become much larger, the sample mean distributions begin to approach a normal distribution and the variation among the means decreases.

So far, we have examined three populations with different distributions. However, the sample means for samples taken from these populations appear to be approximately normally distributed, especially as the sample sizes become larger. What would happen to the distribution of sample means if we studied populations with differently shaped distributions? The answer to that question is given in the **central limit theorem**.

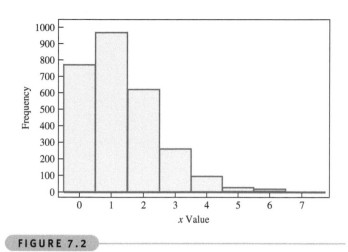

FIGURE 7.2

Minitab Histogram of a Poisson Distributed Population, $\lambda = 1.25$

CENTRAL LIMIT THEOREM

If samples of size n are drawn randomly from a population that has a mean of μ and a standard deviation of σ, the sample means, $\bar{x}$, are approximately normally distributed for sufficiently large sample sizes ($n \geq 30$) regardless of the shape of the population distribution. If the population is normally distributed, the sample means are normally distributed for any size sample.

From mathematical expectation* it can be shown that the mean of the sample means is the population mean

$$\mu_{\bar{x}} = \mu$$

and the standard deviation of the sample means (called the standard error of the mean) is the standard deviation of the population divided by the square root of the sample size

$$\sigma_{\bar{x}} = \frac{\sigma}{\sqrt{n}}$$

*The derivations are beyond the scope of this text and are not shown.

The central limit theorem creates the potential for applying the normal distribution to many problems when sample size is sufficiently large. Sample means that have been computed for random samples drawn from normally distributed populations are normally distributed. However, the real advantage of the central limit theorem comes when sample data drawn from populations not normally distributed or from populations of unknown shape also can be analysed by using the normal distribution because the sample means are normally distributed for sufficiently large sample sizes.* Column 1 of Figure 7.5 shows four different population distributions. Each succeeding column displays the shape of the distribution of the sample means for a particular sample size. Note in the bottom row

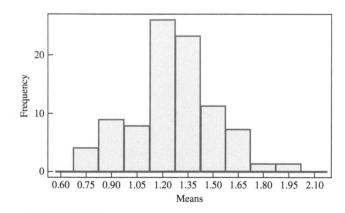

FIGURE 7.3

Minitab Histogram of Sample Means

*The actual form of the central limit theorem is a limit function of calculus. As the sample size increases to infinity, the distribution of sample means literally becomes normal in shape.

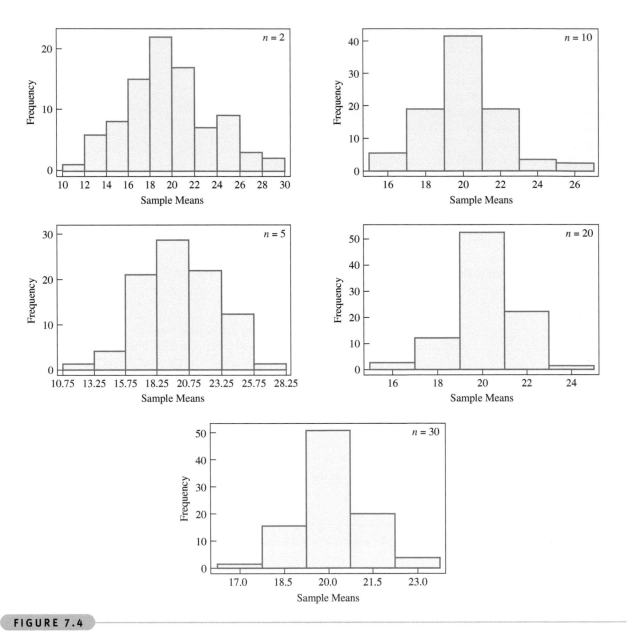

FIGURE 7.4

Minitab Outputs for Sample Means from 90 Samples Ranging in Size from $n = 2$ to $n = 30$ from a Uniformly Distributed Population with $a = 10$ and $b = 30$

for the normally distributed population that the sample means are normally distributed even for $n = 2$. Note also that with the other population distributions, the distribution of the sample means begins to approximate the normal curve as n becomes larger. For all four distributions, the distribution of sample means is approximately normal for $n = 30$.

How large must a sample be for the central limit theorem to apply? The sample size necessary varies according to the shape of the population. However, in this text (as in many others), a sample of *size 30 or larger* will suffice. Recall that if the population is normally distributed, the sample means are normally distributed for sample sizes as small as $n = 1$.

The shapes displayed in Figure 7.5 coincide with the results obtained empirically from the random sampling shown in Figures 7.3 and 7.4. As shown in Figure 7.5, and as indicated in Figure 7.4, as sample size increases, the distribution narrows, or becomes more leptokurtic. This trend makes sense because the standard deviation of the mean is $\sigma/\sqrt{n}$. This value will become smaller as the size of n increases.

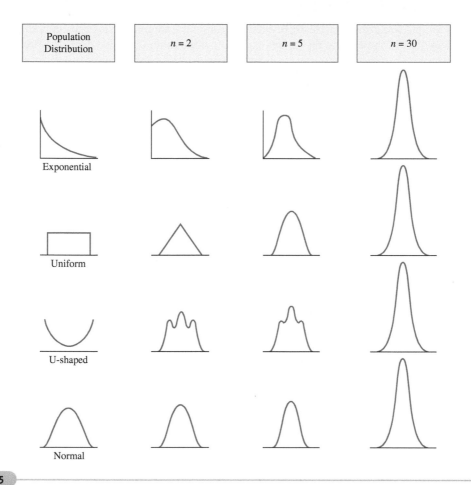

FIGURE 7.5

Shapes of the Distributions of Sample Means for Three Sample Sizes Drawn from Four Different Population Distributions

In Table 7.4, the means and standard deviations of the means are displayed for random samples of various sizes ($n = 2$ through $n = 30$) drawn from the uniform distribution of $a = 10$ and $b = 30$ shown in Figure 7.4. The population mean is 20, and the standard deviation of the population is 5.774. Note that the mean of the sample means for each sample size is approximately 20 and that the standard deviation of the sample means for each set of 90 samples is approximately equal to $\sigma/\sqrt{n}$. A small discrepancy occurs between the standard deviation of the sample means and $\sigma/\sqrt{n}$, because not all possible samples of a given size were taken from the population (only 90). In theory, if all possible

TABLE 7.4

$\mu_{\bar{x}}$ and $\sigma_{\bar{x}}$ of 90 Random Samples for Five Different Sizes[a]

Sample Size	Mean of Sample Means	Standard Deviation of Sample Means	μ	$\dfrac{\sigma}{\sqrt{n}}$
$n = 2$	19.92	3.87	20	4.08
$n = 5$	20.17	2.65	20	2.58
$n = 10$	20.04	1.96	20	1.83
$n = 20$	20.20	1.37	20	1.29
$n = 30$	20.25	0.99	20	1.05

[a] Randomly generated by using Minitab from a uniform distribution with $a = 10$, $b = 30$.

samples for a given sample size are taken exactly once, the mean of the sample means will equal the population mean and the standard deviation of the sample means will equal the population standard deviation divided by the square root of n.

The central limit theorem states that sample means are normally distributed regardless of the shape of the population for large samples and for any sample size with normally distributed populations. Thus, sample means can be analysed by using z scores. Recall from Chapters 3 and 6 the formula to determine z scores for individual values from a normal distribution:

$$z = \frac{x - \mu}{\sigma}$$

If sample means are normally distributed, the z score formula applied to sample means would be

$$z = \frac{\bar{x} - \mu}{\sigma_{\bar{x}}}$$

This result follows the general pattern of z scores: the difference between the statistic and its mean divided by the statistic's standard deviation. In this formula, the mean of the statistic of interest is $\mu_{\bar{x}}$, and *the standard deviation of the statistic of interest* is $\sigma_{\bar{x}}$, sometimes referred to as the **standard error of the mean**. To determine $\mu_{\bar{x}}$, the researcher would randomly draw out all possible samples of the given size from the population, compute the sample means, and average them. This task is virtually impossible to accomplish in any realistic period of time. Fortunately, $\mu_{\bar{x}}$ equals the population mean, μ, which is easier to access. Likewise, to determine directly the value of $\sigma_{\bar{x}}$, the researcher would take all possible samples of a given size from a population, compute the sample means, and determine the standard deviation of sample means. This task also is practically impossible. Fortunately, $\sigma_{\bar{x}}$ can be computed by using the population standard deviation divided by the square root of the sample size.

As sample size increases, the standard deviation of the sample means becomes smaller and smaller because the population standard deviation is being divided by larger and larger values of the square root of n. The ultimate benefit of the central limit theorem is a practical, useful version of the z formula for sample means.

z FORMULA FOR SAMPLE MEANS	$z = \dfrac{\bar{x} - \mu}{\dfrac{\sigma}{\sqrt{n}}}$

When the population is normally distributed and the sample size is 1, this formula for sample means becomes the z formula for individual values that we used in Chapter 6. The reason is that the mean of one value is that value, and when $n = 1$ the value of $\sigma/\sqrt{n} = \sigma$.

Suppose, for example, that the mean expenditure per customer at a tyre store is €85.00, with a standard deviation of €9.00. If a random sample of 40 customers is taken, what is the probability that the sample average expenditure per customer for this sample will be €87.00 or more? Because the sample size is greater than 30, the central limit theorem can be used, and the sample means are normally distributed. With $\mu = $ €85.00, $\sigma = $ €9.00 and the z formula for sample means, z is computed as

$$z = \frac{\bar{x} - \mu}{\dfrac{\sigma}{\sqrt{n}}} = \frac{€87.00 - €85.00}{\dfrac{€9.00}{\sqrt{40}}} = \frac{€2.00}{€1.42} = 1.41$$

From the z distribution (Table A.5), $z = 1.41$ produces a probability of 0.4207. This number is the probability of getting a sample mean between €87.00 and €85.00 (the population mean). Solving for the tail of the distribution yields

$$0.5000 - 0.4207 = 0.0793$$

which is the probability of $\bar{x} \geq $ €87.00 That is, 7.93% of the time, a random sample of 40 customers from this population will yield a sample mean expenditure of €87.00 or more. Figure 7.6 shows the problem and its solution.

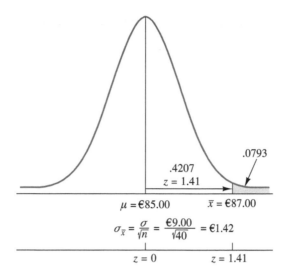

.0793

.4207
$z = 1.41$

$\mu = €85.00$ $\bar{x} = €87.00$

$$\sigma_{\bar{x}} = \frac{\sigma}{\sqrt{n}} = \frac{€9.00}{\sqrt{40}} = €1.42$$

$z = 0$ $z = 1.41$

Graphical Solution to the Tyre Store Example

DEMONSTRATION PROBLEM 7.1

Suppose that during any hour in a large department store, the average number of shoppers is 448, with a standard deviation of 21 shoppers. What is the probability that a random sample of 49 different shopping hours will yield a sample mean between 441 and 446 shoppers?

Solution

For this problem, $\mu = 448$, $\sigma = 21$, and $n = 49$. The problem is to determine $P(441 \leq \bar{x} \leq 446)$. The following diagram depicts the problem:

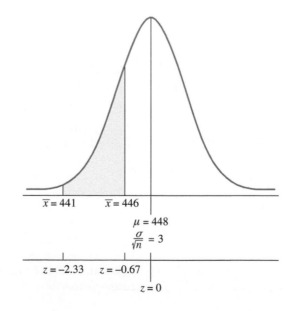

$\bar{x} = 441$ $\bar{x} = 446$

$\mu = 448$

$\dfrac{\sigma}{\sqrt{n}} = 3$

$z = -2.33$ $z = -0.67$

$z = 0$

Solve this problem by calculating the z scores and using Table A.5 to determine the probabilities:

$$z = \frac{441 - 448}{\frac{21}{\sqrt{49}}} = \frac{-7}{3} = -2.33$$

and

$$z = \frac{446 - 448}{\frac{21}{\sqrt{49}}} = \frac{-2}{3} = -0.67$$

z Value	Probability
-2.33	.4901
-0.67	$-.2486$
	.2415

The probability of a value being between $z = -2.33$ and -0.67 is 0.2415; that is, there is a 24.15% chance of randomly selecting 49 hourly periods for which the sample mean is between 441 and 446 shoppers.

Sampling from a Finite Population

The example shown in this section and Demonstration Problem 7.1 was based on the assumption that the population was infinitely or extremely large. In cases of a finite population, *a statistical adjustment can be made to the z formula for sample means*. The adjustment is called the **finite correction factor:** $\sqrt{\dfrac{N-n}{N-1}}$. It operates on the standard deviation of sample mean, σ_x. Following is the z formula for sample means when samples are drawn from finite populations.

z FORMULA FOR SAMPLE MEANS OF A FINITE POPULATION	$z = \dfrac{\bar{x} - \mu}{\dfrac{\sigma}{\sqrt{n}}\sqrt{\dfrac{N-n}{N-1}}}$

If a random sample of size 35 were taken from a finite population of only 500, the sample mean would be less likely to deviate from the population mean than would be the case if a sample of size 35 were taken from an infinite population. For a sample of size 35 taken from a finite population of size 500, the finite correction factor is

$$\sqrt{\frac{500 - 35}{500 - 1}} = \sqrt{\frac{465}{499}} = 0.965$$

TABLE 7.5

Finite Correction Factor for Some Sample Sizes

Population Size	Sample Size	Value of Correction Factor
2,000	30 ($<5\%N$)	.993
2,000	500	.866
500	30	.971
500	200	.775
200	30	.924
200	75	.793

Thus the standard deviation of the mean – sometimes referred to as the standard error of the mean – is adjusted downward by using 0.965. As the size of the finite population becomes larger in relation to sample size, the finite correction factor approaches 1. In theory, whenever researchers are working with a finite population, they can use the finite correction factor. A rough rule of thumb for many researchers is that, if the sample size is less than 5% of the finite population size or $n/N < 0.05$, the finite correction factor does not significantly modify the solution. Table 7.5 contains some illustrative finite correction factors.

DEMONSTRATION PROBLEM 7.2

A production company's 350 hourly employees average 37.6 years of age, with a standard deviation of 8.3 years. If a random sample of 45 hourly employees is taken, what is the probability that the sample will have an average age of less than 40 years?

Solution

The population mean is 37.6, with a population standard deviation of 8.3; that is, $\mu = 37.6$ and $\sigma = 8.3$. The sample size is 45, but it is being drawn from a finite population of 350; that is, $n = 45$ and $N = 350$. The sample mean under consideration is 40, or $\bar{x} = 40$. The following diagram depicts the problem on a normal curve:

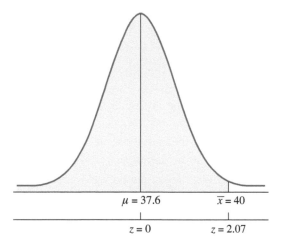

$$\mu = 37.6 \qquad \bar{x} = 40$$
$$z = 0 \qquad z = 2.07$$

Using the z formula with the finite correction factor gives

$$z = \frac{40 - 37.6}{\dfrac{8.3}{\sqrt{45}} \sqrt{\dfrac{350 - 45}{350 - 1}}} = \frac{2.4}{1.157} = 2.07$$

This z value yields a probability (Table A.5) of 0.4808. Therefore, the probability of getting a sample average age of less than 40 years is $0.4808 + 0.5000 = 0.9808$. Had the finite correction factor not been used, the z value would have been 1.94, and the final answer would have been 0.9738.

7.2 PROBLEMS

7.13 A population has a mean of 50 and a standard deviation of 10. If a random sample of 64 is taken, what is the probability that the sample mean is each of the following?
   a. Greater than 52.
   b. Less than 51.
   c. Less than 47.
   d. Between 48.5 and 52.4.
   e. Between 50.6 and 51.3.

7.14   A population is normally distributed, with a mean of 23.45 and a standard deviation of 3.8. What is the probability of each of the following?

   a.   Taking a sample of size 10 and obtaining a sample mean of 22 or more.
   b.   Taking a sample of size 4 and getting a sample mean of more than 26.

7.15   Suppose a random sample of size 36 is drawn from a population with a mean of 278. If 86% of the time the sample mean is less than 280, what is the population standard deviation?

7.16   A random sample of size 81 is drawn from a population with a standard deviation of 12. If only 18% of the time a sample mean greater than 300 is obtained, what is the mean of the population?

7.17   Find the probability in each case.

   a. $N = 1000$, $n = 60$, $\mu = 75$, and $\sigma = 6$; $P(\bar{x} < 76.5) = ?$
   b. $N = 90$, $n = 36$, $\mu = 108$, and $\sigma = 3.46$; $P(107 < \bar{x}\ 107.7) = ?$
   c. $N = 250$, $n = 100$, $\mu = 35.6$, and $\sigma = 4.89$; $P(\bar{x} \geq 36) = ?$
   d. $N = 5000$, $n = 60$, $\mu = 125$, and $\sigma = 13.4$; $P(\bar{x} \leq 123) = ?$

7.18   A study published in *Trends in Agricultural Economics*, recently reported that the average annual consumption of fresh fruit per person in Japan was 27.5 kilograms. The standard deviation of fresh fruit consumption is about 10.6 kilograms. Suppose a researcher took a random sample of 38 Japanese and had them keep a record of the fresh fruit they ate for one year.

   a.   What is the probability that the sample average would be less than 25 kilograms?
   b.   What is the probability that the sample average would be between 26 and 30 kilograms?
   c.   What is the probability that the sample average would be less than 32 kilograms?
   d.   What is the probability that the sample average would be between 22 and 24 kilograms?

7.19   Suppose a housing estate on the south side of Pori, Finland, contains 1,500 houses. The housing estate was built in 1993. A sample of 100 houses is selected randomly and evaluated by an appraiser. If the mean appraised value of a house in this housing estate for all houses is €177,000, with a standard deviation of €8,500, what is the probability that the sample average is greater than €185,000?

7.20   Suppose the average checkout bill at a large supermarket is £65.12, with a standard deviation of £21.45. Twenty-three per cent of the time when a random sample of 45 customer bills is examined, the sample average should exceed what value?

7.21   According to the Broadcasters' Audience Research Board, people in the United Kingdom watch an average of 30 hours of TV per week. Suppose the standard deviation is eight hours and a random sample of 42 UK residents is taken.

   a.   What is the probability that the sample average is more than 32 hours?
   b.   What is the probability that the sample average is less than 27.5 hours?
   c.   What is the probability that the sample average is less than 22 hours? If the sample average actually is less than 22 hours, what would it mean in terms of the Broadcasters' Audience Research Board figures?
   d.   Suppose the population standard deviation is unknown. If 71% of all sample means are greater than 28 hours and the population mean is still 30 hours, what is the value of the population standard deviation?

7.3   SAMPLING DISTRIBUTION OF $\hat{p}$

Sometimes in analysing a sample, a researcher will choose to use the sample proportion, denoted $\hat{p}$. If research produces *measurable* data such as weight, distance, time, and income, the sample mean is often the statistic of choice. However, if research results in *countable* items such as how many people in a sample choose Fanta as their soft drink or how many people in a sample have a flexible work schedule, the sample proportion is often the statistic of choice. Whereas the mean is computed by averaging a set of values, the **sample proportion** is *computed by dividing the frequency with which a given characteristic occurs in a sample by the number of items in the sample.*

SAMPLE PROPORTION

$$\hat{p} = \frac{x}{n}$$

where:

x = number of items in a sample that have the characteristic
n = number of items in the sample

For example, in a sample of 100 factory workers, 30 workers might belong to a union. The value of $\hat{p}$ for this characteristic, union membership, is $30/100 = 0.30$. In a sample of 500 businesses in shopping centres, if 10 are shoe stores, then the sample proportion of shoe stores is $10/500 = 0.02$. The sample proportion is a widely used statistic and is usually computed on questions involving Yes or No answers. For example, do you have at least a high school education? Are you predominantly right-handed? Are you female? Do you belong to the student accounting association?

How does a researcher use the sample proportion in analysis? The central limit theorem applies to sample proportions in that the normal distribution approximates the shape of the distribution of sample proportions if $n \cdot p > 5$ and $n \cdot q > 5$ (p is the population proportion and $q = 1 - p$). The mean of sample proportions for all samples of size n randomly drawn from a population is p (the population proportion) and the standard deviation of sample proportions is $\sqrt{\dfrac{p \cdot q}{n}}$ sometimes referred to as the **standard error of the proportion**. Sample proportions also have a z formula.

z FORMULA FOR SAMPLE PROPORTIONS FOR $n \cdot p > 5$ AND $n \cdot q > 5$

$$z = \frac{\hat{p} - p}{\sqrt{\dfrac{p \cdot q}{n}}}$$

where:

$\hat{p}$ = sample proportion
n = sample size
p = population proportion
$q = 1 - p$

Suppose 60% of the electrical contractors in a region use a particular brand of wire. What is the probability of taking a random sample of size 120 from these electrical contractors and finding that 0.50 or less use that brand of wire? For this problem,

$$p = 0.60 \quad \hat{p} = 0.50 \quad n = 120$$

The z formula yields

$$z = \frac{0.50 - 0.60}{\sqrt{\dfrac{(0.60)(0.40)}{120}}} = \frac{-0.10}{0.0447} = -2.24$$

From Table A.5 in Appendix A, the probability corresponding to $z = -2.24$ is 0.4875. For $z < -2.24$ (the tail of the distribution), the answer is $0.5000 - 0.4875 = 0.0125$. Figure 7.7 shows the problem and solution graphically.

This answer indicates that a researcher would have difficulty (probability of 0.0125) finding that 50% or less of a sample of 120 contractors use a given brand of wire if indeed the population market share for that wire is 0.60. If this sample result actually occurs, either it is a rare chance result, the 0.60 proportion does not hold for this population, or the sampling method may not have been random.

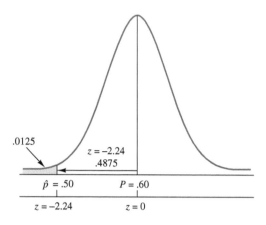

FIGURE 7.7

Graphical Solution to the Electrical Contractor Example

DEMONSTRATION PROBLEM 7.3

If 10% of a population of parts is defective, what is the probability of randomly selecting 80 parts and find-ing that 12 or more parts are defective?

Solution

Here, $p = 0.10$, $\hat{p} = 12/80 = 0.15$, and $n = 80$. Entering these values in the z formula yields

$$z = \frac{0.15 - 0.10}{\sqrt{\dfrac{(0.10)(0.90)}{80}}} = \frac{0.05}{0.0335} = 1.49$$

Table A.5 gives a probability of 0.4319 for a z value of 1.49, which is the area between the sample pro-portion, 0.15, and the population proportion, 0.10. The answer to the question is

$$P(\hat{p} \geq 0.15) = 0.5000 - 0.4319 = 0.0681$$

Thus, about 6.81% of the time, 12 or more defective parts would appear in a random sample of 80 parts when the population proportion is 0.10. If this result actually occurred, the 10% proportion for popu-lation defects would be open to question. The diagram shows the problem graphically.

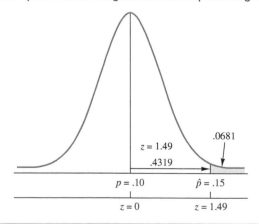

7.3   PROBLEMS

7.22 A given population proportion is 0.25. For the given value of n, what is the probability of getting each of the following sample proportions?

   a. $n = 110$ and $\hat{p} \leq 0.21$

   b. $n = 33$ and $\hat{p} > 0.24$

   c. $n = 59$ and $0.24 \leq \hat{p} < 0.27$

   d. $n = 80$ and $\hat{p} < 0.30$

   e. $n = 800$ and $\hat{p} < 0.30$

7.23 A population proportion is 0.58. Suppose a random sample of 660 items is sampled randomly from this population.

   a. What is the probability that the sample proportion is greater than 0.60?

   b. What is the probability that the sample proportion is between 0.55 and 0.65?

   c. What is the probability that the sample proportion is greater than 0.57?

   d. What is the probability that the sample proportion is between 0.53 and 0.56?

   e. What is the probability that the sample proportion is less than 0.48?

7.24 Suppose a population proportion is 0.40, and 80% of the time when you draw a random sample from this population you get a sample proportion of 0.35 or more. How large a sample were you taking?

7.25 If a population proportion is 0.28 and if the sample size is 140, 30% of the time the sample proportion will be less than what value if you are taking random samples?

7.26 According to a study by Decision Analyst, 21% of the people who have credit cards are very close to the total limit on the card(s). Suppose a random sample of 600 credit card users is taken. What is the probability that more than 150 credit card users are very close to the total limit on their card(s)?

7.27 According to a survey by Accountemps, 48% of executives believe that employees are most productive on Tuesdays. Suppose 200 executives are randomly surveyed.

   a. What is the probability that fewer than 90 of the executives believe employees are most productive on Tuesdays?

   b. What is the probability that more than 100 of the executives believe employees are most productive on Tuesdays?

   c. What is the probability that more than 80 of the executives believe employees are most productive on Tuesdays?

7.28 A Travel Weekly International Air Transport Association survey asked business travellers about the purpose of their most recent business trip. Nineteen per cent responded that it was for an internal company visit. Suppose 950 business travellers are randomly selected.

   a. What is the probability that more than 25% of the business travellers say that the reason for their most recent business trip was an internal company visit?

   b. What is the probability that between 15% and 20% of the business travellers say that the reason for their most recent business trip was an internal company visit?

   c. What is the probability that between 133 and 171 of the business travellers say that the reason for their most recent business trip was an internal company visit?

Decision Dilemma SOLVED

Private Health Insurance in India: Surveying the Population

Because of limited resources, limited time, and an extremely large population (1.21 billion people), most work on the attitudes towards and expectations of private insurance schemes of the Indian people would have to be accomplished through the use of random sampling. However, to ensure that the sample is representative of the population, it is necessary to ensure certain groups are included. To achieve this objective and also to reduce sampling error, a proportionate stratified sampling technique would be appropriate in this case. Such a sampling strategy could include as strata factors such things as geographic location of the person selected (metropolitan/regional/rural), age, gender, religion, household income, and other variables.

The latest population census (2011) could serve as a sampling frame for the study. If access is granted to use the census information, the researcher could then identify strata within the census and randomly sample people from these lists. To ensure that the sample is representative of the population, the sample characteristics (e.g. age, gender, household income) could be compared with those obtained from the census to determine whether they are appropriately similar.

Questions on the attitude towards and expectations of health insurance schemes are not easy to formulate in a way that produces valid data. Experts on the measurement of such things should be consulted. However, if questions are asked in a way that produces numerical responses that can be averaged, sample means can be computed. If sample sizes are large enough, the central limit theorem can be invoked, enabling the business researcher to analyse mean sample responses as though they came from normally distributed populations.

Some of the questions asked might require only a Yes or No response. For example, the question, 'Are you considering joining a private insurance scheme in the next 12 months?' requires only a Yes or No response. These responses, when tallied, can be used to compute sample proportions. If the sample sizes are large enough, the business researcher can assume from the central limit theorem that sample proportions come from a normal distribution, which can provide the basis for analysis.

ETHICAL CONSIDERATIONS

The art and science of sampling has potential for breeding unethical behaviour. Considerable research is reported under the guise of random sampling when, in fact, non-random sampling is used. Remember, if non-random sampling is used, probability statements about sampling error are not appropriate. Some researchers purport to be using stratified random sampling when they are actually using quota sampling. Others claim to be using systematic random sampling when they are actually using convenience or judgement sampling.

In the process of inferential statistics, researchers use sample results to make conclusions about a population. These conclusions are disseminated to the interested public. The public often assumes that sample results truly reflect the state of the population. If the sample does not reflect the population because questionable sampling practices were used, it could be argued that unethical research behaviour occurred. Valid, representative sampling is not an easy task. Researchers and statisticians should exercise extreme caution in taking samples to be sure the results obtained reflect the conditions in the population as nearly as possible.

The central limit theorem is based on large samples unless the population is normally distributed. In analysing small-sample data, it is an unethical practice to assume a sample mean is from a normal distribution unless the population can be shown with some confidence to be normally distributed. Using the normal distribution to analyse sample proportions is also unethical if sample sizes are smaller than those recommended by the experts.

SUMMARY

For much business research, successfully conducting a census is virtually impossible and the sample is a feasible alternative. Other reasons for sampling include cost reduction, potential for broadening the scope of the study, and loss reduction when the testing process destroys the product.

To take a sample, a population must be identified. Often the researcher cannot obtain an exact roster or list of the population and so must find some way to identify the population as closely as possible. The final list or directory used to represent the population and from which the sample is drawn is called the frame.

The two main types of sampling are random and non-random. Random sampling occurs when each unit of the population has the same probability of being selected for the sample. Non-random sampling is any sampling that is not random. The four main types of random sampling discussed are simple random sampling, stratified sampling, systematic sampling, and cluster, or area, sampling.

In simple random sampling, every unit of the population is numbered. A table of random numbers or a random number generator is used to select n units from the population for the sample.

Stratified random sampling uses the researcher's prior knowledge of the population to stratify the population into subgroups. Each subgroup is internally homogeneous but different from the others. Stratified random sampling is an attempt to reduce sampling error and ensure that at least some of each of the subgroups appears in the sample. After the strata are identified, units can be sampled randomly from each stratum. If the proportions of units selected from each subgroup for the sample are the same as the proportions of the subgroups in the population, the process is called proportionate stratified sampling. If not, it is called disproportionate stratified sampling.

With systematic sampling, every kth item of the population is sampled until n units have been selected. Systematic sampling is used because of its convenience and ease of administration.

Cluster or area sampling involves subdividing the population into non-overlapping clusters or areas. Each cluster or area is a microcosm of the population and is usually heterogeneous within. Individual units are then selected randomly from the clusters or areas to get the final sample. Cluster or area sampling is usually done to reduce costs. If a set of second clusters or areas is selected from the first set, the method is called two-stage sampling.

Four types of non-random sampling were discussed: convenience, judgement, quota, and snowball. In convenience sampling, the researcher selects units from the population to be in the sample for convenience. In judgement sampling, units are selected according to the judgement of the researcher. Quota sampling is similar to stratified sampling, with the researcher identifying subclasses or strata. However, the researcher selects units from each stratum by some non-random technique until a specified quota from each stratum is filled. With snowball sampling, the researcher obtains additional sample members by asking current sample members for referral information.

Sampling error occurs when the sample does not represent the population. With random sampling, sampling error occurs by chance. Non-sampling errors are all other research and analysis errors that occur in a study. They include recording errors, input errors, missing data, and incorrect definition of the frame.

According to the central limit theorem, if a population is normally distributed, the sample means for samples taken from that population also are normally distributed regardless of sample size. The central limit theorem also says that if the sample sizes are large ($n \geq 30$), the sample mean is approximately normally distributed regardless of the distribution shape of the population. This theorem is extremely useful because it enables researchers to analyse sample data by using the normal distribution for virtually any type of study in which means are an appropriate statistic, as long as the sample size is large enough. The central limit theorem states that sample proportions are normally distributed for large sample sizes.

KEY TERMS

central limit theorem

cluster (or area) sampling

convenience sampling

disproportionate stratified random sampling

finite correction factor

frame

judgement sampling

non-random sampling

non-random sampling
   techniques
non-sampling errors
proportionate stratified
quota sampling
random sampling
sample proportion
sampling error

simple random sampling
snowball sampling
standard error of the mean
standard error of the proportion
stratified random sampling
systematic sampling
two-stage sampling

GO ONLINE TO DISCOVER THE EXTRA FEATURES FOR THIS CHAPTER

The Student Study Guide containing solutions to the odd-numbered questions, additional Quizzes and Concept Review Activities, Excel and Minitab databases, additional data files in Excel and Minitab, and more worked examples. **www.wiley.com/college/cortinhas**

FORMULAS

Determining the value of k

$$k = \frac{N}{n}$$

z formula for sample means

$$z = \frac{\bar{x} - \mu}{\frac{\sigma}{\sqrt{n}}}$$

z formula for sample means when there is a finite population

$$z = \frac{\bar{x} - \mu}{\frac{\sigma}{\sqrt{n}} \sqrt{\frac{N-n}{N-1}}}$$

Sample proportion

$$\hat{p} = \frac{x}{n}$$

z formula for sample proportions

$$z = \frac{\hat{p} - p}{\sqrt{\frac{p \cdot q}{n}}}$$

SUPPLEMENTARY PROBLEMS

CALCULATING THE STATISTICS

7.29 The mean of a population is 76 and the standard deviation is 14. The shape of the population is unknown. Determine the probability of each of the following occurring from this population.

a. A random sample of size 35 yielding a sample mean of 79 or more.

b. A random sample of size 140 yielding a sample mean of between 74 and 77.

c. A random sample of size 219 yielding a sample mean of less than 76.5.

7.30 Forty-six per cent of a population possesses a particular characteristic. Random samples are taken from this population. Determine the probability of each of the following occurrences:

a. The sample size is 60 and the sample proportion is between 0.41 and 0.53.

b. The sample size is 458 and the sample proportion is less than 0.40.

c. The sample size is 1,350 and the sample proportion is greater than 0.49.

TESTING YOUR UNDERSTANDING

7.31 Suppose the age distribution in a city is as follows.

Under 18	22%
18–25	18%
26–50	36%
51–65	10%
Over 65	14%

A researcher is conducting proportionate stratified random sampling with a sample size of 250. Approximately how many people should he sample from each stratum?

7.32 Candidate Thomsen believes she will receive 0.55 of the total votes cast in her county. However, in an attempt to validate this figure, her pollster contacts a random sample of 600 registered voters in the county. The poll results show that 298 of the voters say they are committed to voting for her. If she actually has 0.55 of the total vote, what is the probability of getting a sample proportion this small or smaller? Do you think she actually has 55% of the vote? Why or why not?

7.33 Determine a possible frame for conducting random sampling in each of the following studies:

a. The average amount of overtime per week for production workers in a plastics company in Croatia.

b. The average number of employees in all LIDL supermarkets in Europe.

c. A survey of chartered accountants in England and Wales.

7.34 A particular car costs an average of €17,755 in Western Europe. The standard deviation of prices is €650. Suppose a random sample of 30 dealerships in Western Europe is taken, and their managers are asked what they charge for this automobile. What is the probability of getting a sample average cost of less than €17,500? Assume that only 120 dealerships in all of Western Europe sell this model.

7.35 A company has 1,250 employees, and you want to take a simple random sample of $n = 60$ employees. Explain how you would go about selecting this sample by using the table of random numbers. Are there numbers that you cannot use? Explain.

7.36 Suppose the average client charge per hour for out-of-court work by lawyers in the city of Berlin is €125. Suppose further that a random telephone sample of 32 lawyers in Berlin is taken and that the sample average charge per hour for out-of-court work is €110. If the population variance is €525, what is the probability of getting a sample mean of €110 or larger? What is the probability of getting a sample mean larger than €135 per hour? What is the probability of getting a sample mean of between €120 and €130 per hour?

7.37 A survey of 3,219 consumers by GfK NOP for the Food Standards Agency showed that 49% of respondents correctly identified the use by date as the best indicator of whether food is safe to eat or not. Overall respondents were more likely to take heed of the use by/best before dates when using meat, dairy and egg products compared to bread and breakfast cereals. Around 55% said they would not cook and eat raw meat that was past its use by date compared to 27% of respondents when asked about bread and 26% when asked about breakfast cereals. Suppose a random sample of 1,100 consumers is taken and each is asked a number of questions on food safety at home.

a. What is the probability that more than 550 consumers identify the use by date as the best indicator of whether food is safe to eat or not?

b. What is the probability that fewer than 650 consumers would not cook and eat raw meat that was past its use by date?

c. What is the probability that between 24% and 26% of consumers would not eat bread that was past its use by date?

7.38 Suppose you are sending out questionnaires to a randomly selected sample of 100 managers. The frame for this study is the membership list of the European Managers Association. The questionnaire contains demographic questions about the company and its top manager. In addition, it asks questions about the manager's leadership style. Research assistants are to score and enter the responses into the computer as soon as they are received. You are to conduct a statistical analysis of the data. Name and describe four non-sampling errors that could occur in this study.

7.39 A researcher is conducting a study of a *Fortune Global* 500 company that has factories, distribution centres, and retail outlets across a large number of countries. How can she use cluster or area sampling to take a random sample of employees of this firm?

7.40 A directory of computer shops in India contains 12,080 alphabetized entries. Explain how systematic sampling could be used to select a sample of 300 outlets.

7.41 In an effort to cut costs and improve profits, many companies have been turning to outsourcing. In fact, according to a European outsourcing survey by Ernst & Young, 70% of companies surveyed outsourced at least one function of their business, with 20% set to increase their level of outsourcing in the next two years. Suppose 565 of these companies are contacted.

a. What is the probability that 367 or more companies already outsource at least one function of their business?

b. What is the probability that 379 or more companies already outsource at least one function of their business?

c. What is the probability that 50% or less of these companies already outsource at least one function of their business?

7.42 The average cost of a one-bedroom apartment in a town is £550 per month. What is the probability of randomly selecting a sample of 50 one-bedroom apartments in this town and getting a sample mean of less than £530 if the population standard deviation is £100?

7.43 *RWE Npower Renewables* reports that the average UK household uses 4,799 kilowatt-hours (kWh) per year. A random sample of 51 UK households is monitored for one year to determine electricity usage. If the population standard deviation of annual usage is 1,039 kWh, what is the probability that the sample mean will be each of the following?

a. More than 5,000 kWh.

b. More than 4,900 kWh.

c. Between 4,700 and 4,800 kWh.

d. Less than 4,650 kWh.

e. Less than 4,500 kWh.

7.44 Use Table A.1 to select 20 three-digit random numbers. Did any of the numbers occur more than once? How is it possible for a number to occur more than once? Make a stem-and-leaf plot of the numbers with the stem being the left digit. Do the numbers seem to be equally distributed, or are they bunched together?

7.45 Direct marketing companies are turning to the Internet for new opportunities. A recent study by Gruppo, Levey, & Co. showed that 73% of all direct marketers conduct transactions on the Internet.

Suppose a random sample of 300 direct marketing companies is taken.

a. What is the probability that between 210 and 234 (inclusive) direct marketing companies are turning to the Internet for new opportunities?

b. What is the probability that 78% or more of direct marketing companies are turning to the Internet for new opportunities?

c. Suppose a random sample of 800 direct marketing companies is taken. Now what is the probability that 78% or more are turning to the Internet for new opportunities? How does this answer differ from the answer in part (b)? Why do the answers differ?

7.46 According to a survey by New Zealand's Ministry of Social Development, 33% of all people 15 years of age or older do volunteer work. No differences between genders were found when all respondents were taken into account. The survey also found that voluntary work was slightly more prevalent among older people, particularly for females. Females over 45 years old volunteered significantly more (36%) than females in the 15–24 age range (24%).

a. What is the probability of randomly sampling 140 people 15 years of age or older and getting 52 or more who do volunteer work?

b. What is the probability of getting 41 or fewer from this group?

c. Suppose a sample of 300 women 45 years of age or older is selected randomly from the New Zealand population. What is the probability that the sample proportion of those who do volunteer work is between 32% and 40%?

7.47 Suppose you work for a large Chinese firm that has 20,000 employees. The CEO calls you in and asks you to determine employee attitudes toward the company. She is willing to commit ¥100,000 to this project. What are the advantages of taking a sample versus conducting a census? What are the tradeoffs?

7.48 In a particular area of the North, an estimated 75% of the homes use heating oil as the principal heating fuel during the winter. A random telephone survey of 150 homes is taken in an attempt to determine whether this figure is correct. Suppose 120 of the 150 homes surveyed use heating oil as the principal heating fuel. What is the probability of getting a sample proportion this large or larger if the population estimate is true?

7.49 Eurostat compiles average monthly wage figures for various European countries by age group and sex. The most recent survey showed that the monthly gross earnings was €3,268 for Denmark, €2,856 for Sweden, and €2,571 for Finland. Assume that in all three countries, the standard deviation of monthly wage rates is €270.

a. Suppose 40 workers are selected randomly from across Denmark and asked what their monthly wage is. What is the probability that the sample average will be between €3,200 and €3,300?

b. Suppose 35 workers are selected randomly from across Sweden. What is the probability that the sample average will exceed €2,900?

c. Suppose 50 workers are selected randomly from across Finland. What is the probability that the sample average will be less than €2,500?

7.50 Give a variable that could be used to stratify the population for each of the following studies. List at least four subcategories for each variable.

a. A political party wants to conduct a poll prior to a parliamentary election.

b. A soft drink company wants to take a sample of soft drink purchases in an effort to estimate market share.

c. A retail outlet wants to interview customers over a one-week period.

d. An eyeglasses manufacturer and retailer wants to determine the demand for prescription eyeglasses in its marketing region.

7.51 According to the Society of Motor Manufacturers and Traders, black is the most popular car colour with 24.7% of all new cars being this colour. Next on the list of the most popular colours are silver with 21.4%, blue with 16.5%, grey with 14.4%, and red with 10.2%. The colours beige, yellow, and brown were the least chosen with a combined preference of less than 2% of all new cars. A survey of 65 randomly selected new car owners is taken. What is the probability that 17 or more chose black for their new car?

ANALYSING THE DATABASES

1. Let the Manufacturing database be the frame for a population of manufacturers to be studied. This database has 140 different SIC codes. Suppose you want to randomly sample six of these SIC codes from these using simple random sampling. Explain how would you go about doing this. Explain how you would take a systematic sample of size 10 from this frame. Examining the variables in the database, name two variables that could be used to stratify the population. Explain how these variables could be used in stratification and why each variable might produce important strata.

2. Consider the Financial database. Compute the mean and standard deviation for annual total revenues for this population. Now take a random sample of 30 of the companies in this database and compute the sample mean. Using techniques presented in this chapter, determine the probability of getting a mean this large or larger from the population. Work this problem both with and without the finite correction factor and compare the results by discussing the differences in answers.

3. Use the Hospital database and determine the proportion of hospitals that are under the control of non-government not-for-profit organizations (category 2). Assume that this proportion represents the entire population of all hospitals. If you randomly selected 500 hospitals from across the United States, what is the probability that 45% or more are under the control of non-government not-for-profit organizations? If you randomly selected 100 hospitals, what is the probability that less than 40% are under the control of non-government not-for-profit organizations?

DOES MONEY MAKE YOU HAPPY?

Common wisdom has always maintained that money can't buy you happiness. But science, as it turns out, says otherwise. According to the widely cited work of Richard Easterlin, money does make you happy but only up to a point. Wealth beyond a certain amount does not make us happier: once we've achieved a reasonable degree of financial security (internationally, an annual income of roughly $15,000 per year) our basic needs are met and our sense of well-being does not improve as income rises. Or so studies by Easterlin and his followers have suggested.

A recent paper by Daniel Sacks, Betsey Stevenson and Justin Wolfers suggests that Easterlin and his followers got it wrong. After poring over data from 140 countries, they concluded that rich people are happier than poor people, people in rich countries are happier than people in poor countries, and when countries get richer their people tend to get happier. Therefore, it appears that the happiness effect isn't relative; it is based on a person's absolute income.

These results stand in opposition to a number of previous studies on money and happiness that suggested that people were concerned only with the wealth of their next door neighbours. So, if you were keeping up with – or better yet, surpassing – the Joneses you were fine. By that logic, people in impoverished countries could be happy if they were just a little bit better off than those around them. That conclusion was very appealing to people from prosperous nations.

The data for these studies invariably come from large-scale surveys such as the World Values Survey or the Gallup World Poll, which include questions like 'All things considered, how satisfied are you with your life these days?' or 'Taking all things together, how would you say things are these days – would you say you're very happy, quite happy, not very happy, or not at all happy?'. These answers can then be transformed in a scale that allows for international comparisons and for the study of changing trends over time. An example of this is the findings in the World Values Survey, which has compiled data from over 350,000 people in 97 countries since 1981. This survey found that Denmark is home to the planet's most contented citizens with Zimbabwe being home to the most miserable.

DISCUSSION

1. Suppose you are asked to develop a sampling plan to determine the happiness of the public in your country. What sampling would you use? What is the target population? What frame would you use? Which of the four types of random sampling discussed in this chapter would you use? Could you use a combination of two or more of the types (two-stage sampling)? If so, how?

2. At least some of the studies mentioned here used some stratification in their sampling. What are some of the variables on which they could be stratified? If you were truly interested in ascertaining opinions from a variety of segments of the population with regard to opinions on 'happiness', what strata might make sense? Name at least five and justify why you would include them.

3. Suppose that in 1979 only 12% of the general adult public in a given country considered themselves to be 'happy'. Suppose further that you randomly selected 350 people from the general adult public of that country this year and 25% said that they considered themselves to be 'happy'. If only 12% of the general adult public still believes that they are 'happy', how likely is it that the 25% figure is a chance result in sampling 350 people? Hint: Use the techniques in this chapter to determine the probability of the 25% figure occurring by chance.

4. The World Values Survey includes the following question on happiness: 'Taking all things together, would you say you are: 4 – very happy, 3 – quite happy, 2 – not very happy, 1 – not at all happy'. Suppose that on the latest survey, the mean value for Brazil was 3.5 on the scale, with a 0.69 standard deviation. More recently, a new survey of 35 people was taken and a sample mean of 3.0 was recorded for Brazil. What is the probability of this sample mean or one greater occurring if the actual population mean is still just 3.5? Based on this probability, do you think that a sample mean of 3.0 is just a chance fluctuation on the 3.5 population

mean, or do you think that perhaps it indicates the population mean is now greater than 3.5? Support your conclusion.

Source: Adapted from 'Subjective Well-Being, Income, Economic Development and Growth', NBER Working Paper no. 16441 (October 2010); 'The Data Is In, More Money = More Happiness: Justin Wolfers' (January 2011), at http://blogs.cgdev.org/global_prosperity_wonkcast/2011/01/18/the-data-is-in-more-money-more-happiness-justin-wolfers/; 'Survey Says: People Are Happier' (August 2008), Bloomberg BusinessWeek at www.businessweek.com/globalbiz/content/aug2008/gb20080820_874593.htm.

USING THE COMPUTER

EXCEL

- Random numbers can be generated from Excel for several different distributions, including the binomial distribution, the Poisson distribution, the uniform distribution, and the normal distribution. To generate random numbers from a particular distribution, begin by selecting the **Data** tab on the Excel worksheet. From the **Analysis** panel at the right top of the **Data** tab worksheet, click on **Data Analysis**. If your Excel worksheet does not show the **Data Analysis** option, then you can load it as an add-in following directions given in Chapter 2. From the **Data Analysis** pulldown menu, select **Random Number Generation**.

- In the **Random Number Generation** dialog box, enter the number of columns of values you want to produce into **Number of Variables**.

- Next, enter the number of data points to be generated in each column into **Number of Random Numbers**.

- The third line of the dialog box, **Distribution**, contains the choices of distributions. Select from which one of the following distributions you want to generate random data: **discrete**, **uniform**, **normal**, **Bernoulli**, **binomial**, **Poisson**, and **patterned**.

- The options and required responses in the **Random Number Generation** dialog box will change with the chosen distribution.

MINITAB

- Random numbers can be generated from Minitab for many different distributions, including the binomial distribution, the Poisson distribution, the hypergeometric distribution, the uniform distribution, the normal distribution, and the exponential distribution. To generate random numbers from a particular distribution, select **Calc** on the menu bar. Select **Random Data** from the pulldown menu. From the long second pulldown menu, select the distribution from which you want to generate random numbers. A dialog box for the distribution selected will open, asking you to enter the number of rows of data that you want to generate. In this, it is asking you how many random numbers you want to generate. In the second space of the dialog box, **Store in column(s)**, enter the number of columns of random numbers you want to generate. Next, each individual distribution requires specific parameters. For example, the binomial distribution asks for **Number of trials:** (n) and **Event probability:** (p). The normal distribution asks for the value of the mean and the standard deviation. The Poisson distribution asks for the value of lambda in the box, **Mean**. The hypergeometric distribution requires the entry of three items, **Population size** (N), **Event count in population** (A), and the **Sample size** (n). The uniform distribution asks for the **Lower endpoint:** and the **Upper endpoint**. The exponential distribution requires a value for the **Scale:** and the **Threshold**.

Making Inferences About Population Parameters

THE ABILITY TO ESTIMATE POPULATION PARAMETERS or to test hypotheses about population parameters using sample statistics is one of the main applications of statistics in improving decision making in business. Whether estimating parameters or testing hypotheses about parameters, the inferential process consists of taking a random sample from a group or body (the population), analysing data from the sample, and reaching conclusions about the population using the sample data, as shown in Figure 1.1 of Chapter 1.

One widely used technique for estimating population measures (parameters) from a sample using statistics is the confidence interval. Confidence interval estimation is generally reserved for instances where a business researcher does not know what the population value is or does not have a very clear idea of it. For example, what is the mean euro amount spent by families per month at the cinema including concession expenditures, or what proportion of workers telecommute at least one day per week? Confidence intervals can be used to estimate these and many other useful and interesting population parameters, including means, proportions, and variances in the business world.

Sometimes, a business analyst already knows the value of a population parameter or has a good idea but would like to test to determine if the value has changed, if the value applies in other situations, or if the value is what other researchers say it is. In such cases, business researchers use hypothesis tests. In the hypothesis testing process, the known parameter is assumed to be true, data are gathered from random samples taken from the population, and the resulting data are analysed to determine if the parameter value is still true or has changed in some way. For example, does the average worker still work 40 hours per week? Are 65% of all workers unhappy with their job? Like with confidence intervals, hypothesis testing can be used to test hypotheses about means, proportions, variances, and other parameters.

Unit III of this textbook, from Chapter 8 to Chapter 11, contains a cadre of estimation and hypotheses testing techniques organized by usage and number of samples. Chapter 8 and Chapter 10 present confidence interval techniques for the estimation of parameters. Chapter 8 introduces the concept of a confidence interval and focuses on one-sample analyses, while Chapter 10 confidence intervals are for two-sample analyses. Chapter 9 introduces the concept of hypotheses testing and presents hypothesis tests for one sample. Chapter 10 contains hypothesis tests for two samples, while Chapter 11 presents hypothesis tests for three or more samples.

Because there is a plethora of confidence interval and hypothesis testing techniques presented in Unit III, it is easy to lose the big picture of when to use what technique. To assist you in sorting out these techniques, a taxonomy of techniques has been created and is presented in a tree diagram both here and at the start of the book for your convenience and consideration. Note that in determining which technique to use, there are several key questions that one should consider:

1. Are you estimating (using a confidence interval) or testing (using a hypothesis test)?
2. How many samples are you analysing?
3. Are you analysing means, proportions, or variances?
4. If you are analysing means, is (are) the standard deviation(s) or variance(s) known or not?
5. If you are analysing means from two samples, are the samples independent or related?
6. If you are analysing three or more samples, are you studying one or two independent variables, and is there a blocking variable?

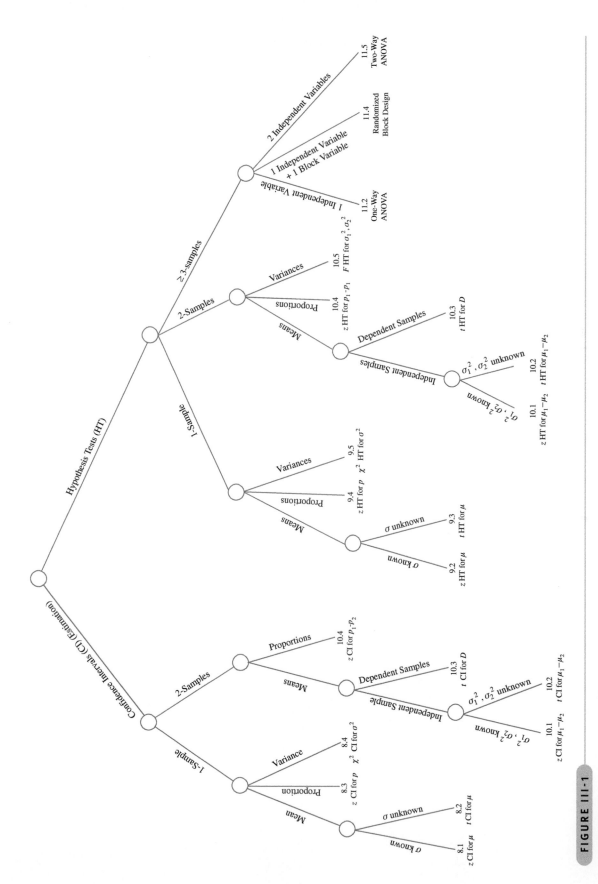

FIGURE III-1

Tree Diagram (CI) Taxonomy of Inferential Techniques

Statistical Inference: Estimation for Single Populations

LEARNING OBJECTIVES

The overall learning objective of Chapter 8 is to help you understand estimating parameters of single populations, thereby enabling you to:

1. Estimate the population mean with a known population standard deviation with the z statistic, correcting for a finite population if necessary
2. Estimate the population mean with an unknown population standard deviation using the t statistic and properties of the t distribution
3. Estimate a population proportion using the z statistic
4. Use the chi-square distribution to estimate the population variance given the sample variance
5. Determine the sample size needed in order to estimate the population mean and population proportion

Decision Dilemma

Compensation for Marketing Professionals

Who are marketing professionals and how much compensation do they receive? In an effort to answer these and others questions, a recent online questionnaire survey of 1,724 marketing professionals who were readers of *Marketing Week* magazine was taken. Demographic questions about gender, age, years of experience, job title, industry, company marketing budget, and others were asked along with compensation questions.

The results of the survey indicated that 61% of marketing professionals were female and 39% were male and that the majority (64%) were under 35 years of age. The mean salary of a marketing manager in the UK was £42,000. The range of marketing manager salaries was found to be quite significant. The top 10% of marketing managers were taking home more than twice the average (£84,600). Among brand managers, this spread was even greater with the top 10% earning an average of £83,300, while the overall mean was £40,500. This trend reached its height with communications managers, where the uppermost 10% average £87,000 against a mean of £40,300. The same effect was seen among marketing services managers (£85,000 at the top, £43,100 on average) and online marketing managers (£87,500 for the top 10%, £40,900 on average). Interestingly, at director level, the wage gap did not appear to be as extreme. The overall average for marketing directors was £78,500, while the top 10% of earners average £100,600. Similarly, among marketing services directors, the average salary was £74,000 and the top 10% got £102,900. For communications directors, the average was £74,900 and the upper band mean was £96,100. In other words, the differential at the top for directors was only about one-third more in income terms.

According to the survey, gender pay gap increased in the past decade for marketing professionals. In 2001, a female marketing manager earned £36,100 on average, compared with a male colleague's £40,600 (a gap of £4,500). For a marketing director, the pay differential between men and women was just £2,700. Pay discrimination has become worse since then. Female marketing managers now get nearly £10,000 less each year on average than their male peers. Among marketing directors, being male is worth £17,000 a year more than being female.

The event of the financial crisis seems to have changed the way marketing professions want to be compensated. Extras and benefits seem less important to marketers than what they perceive as a fair wage. Getting a high base salary with fewer benefits was scored 4.28 out of 5, the most favourable score for any aspect of the remuneration package except holiday entitlement, which scored 4.37. The survey also showed that job mobility of marketing professionals was high, with just under half of respondents (45%) expecting to change jobs during the coming year. Not surprisingly, the promise of a better remuneration was the drive for 27% of those who intended to change jobs but for a significant 26% of marketing professionals, a new challenge is the lure.

Managerial and Statistical Questions

1. Can the mean national salary for a marketing manager be estimated using sample data such as that reported in this study? If so, how much error is involved and how much confidence can we have in it?

2. The study reported that the majority of marketing professionals in the UK were under 35 years old. How can these sample figures be used to estimate a mean from the population? For example, if the survey established that the average age of a marketing professional was 32.2 years, is the population mean age for marketing professionals in the UK also 32.2 years, or is it different?

3. This Decision Dilemma reports that 45% of the responding marketing professionals intend to change jobs in the following year. Does this figure hold for all marketing professionals? Are 61% of all marketing professionals female as reported in this study? How can population proportions be estimated using sample data? How much error is involved? How confident can decision makers be in the results?

4. When survey data are broken down by regions, the sample size for each subgroup is naturally much lower. Does sample size affect the results of a study? Had the study reported that the mean salary for a marketing manager in the south-west region of the UK was £42,500 based on 25 respondents, would that information be less valid than the overall mean salary of £42,000 reported by 1,724 respondents? How do business decision makers discern between study results when sample sizes vary?

Source: Adapted from the *Marketing Week/Ball & Hoolahan* 2010 salary survey, available at www.marketingweek.co.uk/marketing-week/ball-and-hoolahan-salary-survey-2010/3008543.article.

Unit III of this text (Chapters 8 to 11) presents, discusses, and applies various statistical techniques for making inferential estimations and hypothesis tests to enhance decision making in business. Figure III-1 presents a tree diagram taxonomy of these techniques, organizing them by usage, number of samples, and type of statistic. Chapter 8 contains the portion of these techniques that can be used for estimating a mean, a proportion, or a variance for a population with a single sample. Figure 8.1 shows the leftmost branch of the tree diagram taxonomy in Figure III-1. This branch of the tree contains all statistical techniques for constructing confidence intervals from one-sample data presented in this text. Note that at the bottom of each tree branch in Figure 8.1, the title of the statistical technique along with its respective section number is given for ease of identification and use. In Chapter 8, techniques are presented that allow a business researcher to estimate a population mean, proportion, or variance by taking a sample from the population, analysing data from the sample, and projecting the resulting statistics back onto the population, thereby reaching conclusions about the population. Because it is often extremely difficult to obtain and analyse population data for a variety of reasons, mentioned in Chapter 7, the importance of the ability to estimate population parameters from sample statistics cannot be underestimated. Figure 8.2 depicts this process. If a business researcher is estimating a population mean and the population standard deviation is known, then

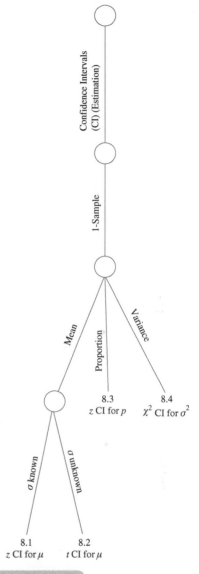

FIGURE 8.1

Chapter 8 Branch of the Tree Diagram Taxonomy of Inferential Techniques

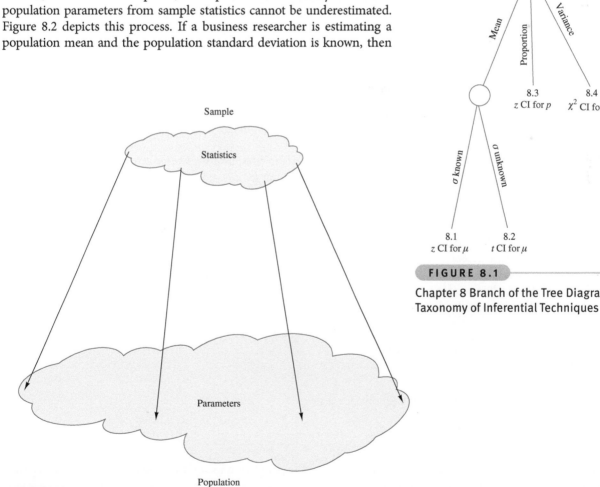

FIGURE 8.2

Using Sample Statistics to Estimate Population Parameters

he will use the z confidence interval for χ^2 contained in Section 8.1. If the population standard deviation is unknown and therefore the researcher is using the sample standard deviation, then the appropriate technique is the t confidence interval for μ contained in Section 8.2. If a business researcher is estimating a population proportion, then he will use the z confidence interval for p presented in Section 8.3. If the researcher desires to estimate a population variance with a single sample, then he will use the χ^2 confidence interval for σ^2 presented in Section 8.4. Section 8.5 contains techniques for determining how large a sample to take in order to ensure a given level of confidence within a targeted level of error.

8.1 ESTIMATING THE POPULATION MEAN USING THE z STATISTIC (σ KNOWN)

On many occasions estimating the population mean is useful in business research. For example, the manager of human resources in a company might want to estimate the average number of days of work an employee misses per year because of illness. If the firm has thousands of employees, direct calculation of a population mean such as this may be practically impossible. Instead, a random sample of employees can be taken, and the sample mean number of sick days can be used to estimate the population mean. Suppose another company developed a new process for prolonging the shelf life of a loaf of bread. The company wants to be able to date each loaf for freshness, but company officials do not know exactly how long the bread will stay fresh. By taking a random sample and determining the sample mean shelf life, they can estimate the average shelf life for the population of bread.

As the mobile phone industry matures, a mobile phone company is rethinking its pricing structure. Users appear to be spending more time on the phone and are shopping around for the best deals. To do better planning, the mobile company wants to ascertain the average number of minutes of time used per month by each of its residential users but does not have the resources available to examine all monthly bills and extract the information. The company decides to take a sample of customer bills and estimate the population mean from sample data. A researcher for the company takes a random sample of 85 bills for a recent month and from these bills computes a sample mean of 510 minutes. This sample mean, which is a statistic, is used to estimate the population mean, which is a parameter. If the company uses the sample mean of 510 minutes as an estimate for the population mean, then the sample mean is used as a *point estimate*.

A **point estimate** is *a statistic taken from a sample that is used to estimate a population parameter*. A point estimate is only as good as the representativeness of its sample. If other random samples are taken from the population, the point estimates derived from those samples are likely to vary. Because of variation in sample statistics, estimating a population parameter with an interval estimate is often preferable to using a point estimate. An **interval estimate** (confidence interval) is *a range of values within which the analyst can declare, with some confidence, the population parameter lies*. Confidence intervals can be two sided or one sided. This text presents only two-sided confidence intervals. How are confidence intervals constructed?

As a result of the central limit theorem, the z formula below for sample means can be used if the population standard deviation is known when sample sizes are large, regardless of the shape of the population distribution, or for smaller sizes if the population is normally distributed.

$$z = \frac{\bar{x} - \mu}{\frac{\sigma}{\sqrt{n}}}$$

Rearranging this formula algebraically to solve for μ gives

$$\mu = \bar{x} - z\frac{\sigma}{\sqrt{n}}$$

Because a sample mean can be greater than or less than the population mean, z can be positive or negative. Thus the preceding expression takes the following form.

$$\bar{x} \pm z\frac{\sigma}{\sqrt{n}}$$

Rewriting this expression yields the confidence interval formula for estimating μ with large sample sizes if the population standard deviation is known.

100$(1 - \alpha)$% CONFIDENCE INTERVAL TO ESTIMATE μ: σ KNOWN (8.1)

or

$$\bar{x} \pm z_{\alpha/2} \frac{\sigma}{\sqrt{n}}$$

$$\bar{x} - z_{\alpha/2} \frac{\sigma}{\sqrt{n}} \leq \mu \leq \bar{x} + z_{\alpha/2} \frac{\sigma}{\sqrt{n}}$$

where:

α = the area under the normal curve outside the confidence interval area
$\alpha/2$ = the area in one end (tail) of the distribution outside the confidence interval

Alpha (α) is the area under the normal curve in the tails of the distribution outside the area defined by the confidence interval. We will focus more on α in Chapter 9. Here we use α to locate the z value in constructing the confidence interval as shown in Figure 8.3. Because the standard normal table is based on areas between a z of 0 and $z_{\alpha/2}$, the table z value is found by locating the area of $0.5000 - \alpha/2$, which is the part of the normal curve between the middle of the curve and one of the tails. Another way to locate this z value is to change the confidence level from percentage to proportion, divide it in half, and go to the table with this value. The results are the same.

The confidence interval formula (8.1) yields a range (interval) within which we feel with some confidence that the population mean is located. It is not certain that the population mean is in the interval unless we have a 100% confidence interval that is infinitely wide. If we want to construct a 95% confidence interval, the level of confidence is 95%, or 0.95. If 100 such intervals are constructed by taking random samples from the population, it is likely that 95 of the intervals would include the population mean and five would not.

As an example, in the mobile phone company problem of estimating the population mean number of minutes called per residential user per month, from the sample of 85 bills it was determined that the sample mean is 510 minutes. Using this sample mean, a confidence interval can be calculated within which the researcher is relatively confident that the actual population mean is located. To make this calculation using formula 8.1, the value of the population standard deviation and the value of z (in addition to the sample mean, 510, and the sample size, 85) must be known. Suppose past history and similar studies indicate that the population standard deviation is 46 minutes.

The value of z is driven by the level of confidence. An interval with 100% confidence is so wide that it is meaningless. Some of the more common levels of confidence used by business researchers are 90%, 95%, 98%, and 99%. Why would a business researcher not just select the highest confidence and always use that level? The reason is that tradeoffs between sample size, interval width, and level of confidence must be considered. For example, as the level of confidence is increased, the interval gets wider, provided the sample size and standard deviation remain constant.

For the mobile phone problem, suppose the business researcher decided on a 95% confidence interval for the results. Figure 8.4 shows a normal distribution of sample means about the population mean. When using a 95% level of confidence, the researcher selects

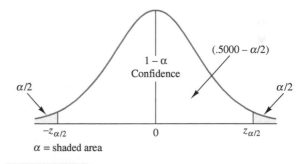

FIGURE 8.3

z Scores for Confidence Intervals in Relation to α

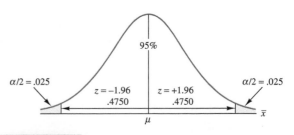

FIGURE 8.4

Distribution of Sample Means for 95% Confidence

an interval centred on μ within which 95% of all sample mean values will fall and then uses the width of that interval to create an interval around the *sample mean* within which he has some confidence the population mean will fall.

For 95% confidence, $\alpha = 0.05$ and $\alpha/2 = 0.025$. The value of $z_{\alpha/2}$ or $z_{0.025}$ is found by looking in the standard normal table under $0.5000 - 0.0250 = 0.4750$. This area in the table is associated with a z value of 1.96. Another way can be used to locate the table z value. Because the distribution is symmetric and the intervals are equal on each side of the population mean, 1/2(95%), or 0.4750, of the area is on each side of the mean. Table A.5 yields a z value of 1.96 for this portion of the normal curve. Thus the z value for a 95% confidence interval is always 1.96. In other words, of all the possible $\bar{x}$ values along the horizontal axis of the diagram, 95% of them should be within a z score of 1.96 from the population mean.

The business researcher can now complete the mobile phone problem. To determine a 95% confidence interval for $x = 510$, $\sigma = 46$, $n = 85$, and $z = 1.96$, the researcher estimates the average call length by including the value of z in formula 8.1.

$$510 - 1.96\frac{46}{\sqrt{85}} \leq \mu \leq 510 + 1.96\frac{46}{\sqrt{85}}$$

$$510 - 9.78 \leq \mu \leq 510 + 9.78$$

$$500.22 \leq \mu \leq 519.78$$

The confidence interval is constructed from the point estimate, which in this problem is 510 minutes, and the error of this estimate, which is ± 9.78 minutes. The resulting confidence interval is $500.22 \leq \mu \leq 519.78$. The mobile phone company researcher is 95%, confident that the average length of a call for the population is between 500.22 and 519.78 minutes.

What does being 95% confident that the population mean is in an interval actually indicate? It indicates that, if the company researcher were to randomly select 100 samples of 85 calls and use the results of each sample to construct a 95% confidence interval, approximately 95 of the 100 intervals would contain the population mean. It also indicates that 5% of the intervals would not contain the population mean. The company researcher is likely to take only a single sample and compute the confidence interval from that sample information. That interval either contains the population mean or it does not. Figure 8.5 depicts the meaning of a 95% confidence interval for the mean. Note that if 20 random samples are taken from the population, 19 of the 20 are likely to contain the population mean if a 95% confidence interval is used (19/20 = 95%). If a 90% confidence interval is constructed, only 18 of the 20 intervals are likely to contain the population mean.

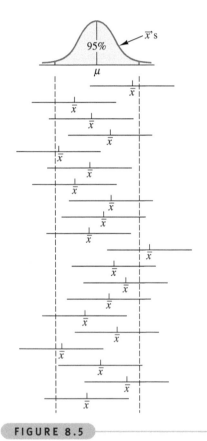

FIGURE 8.5

Twenty 95% Confidence Intervals of μ

DEMONSTRATION PROBLEM 8.1

A survey was taken of European companies that do business with firms in India. One of the questions on the survey was: approximately how many years has your company been trading with firms in India? A random sample of 44 responses to this question yielded a mean of 10.455 years. Suppose the population standard deviation for this question is 7.7 years. Using this information, construct a 90% confidence interval for the mean number of years that a company has been trading in India for the population of European companies trading with firms in India.

Solution

Here, $n = 44$, $\bar{x} = 10.455$, and $\sigma = 7.7$. To determine the value of $z_{\alpha/2}$, divide the 90% confidence in half, or take $0.5000 - \alpha/2 = 0.5000 - 0.0500$ where $\alpha = 10\%$. *Note:* The z distribution of $\bar{x}$ around μ contains

0.4500 of the area on each side of μ, or $1/2(90\%)$. Table A.5 yields a z value of 1.645 for the area of 0.4500 (interpolating between 0.4495 and 0.4505). The confidence interval is

$$\bar{x} - z\frac{\sigma}{\sqrt{n}} \leq \mu \leq \bar{x} + z\frac{\sigma}{\sqrt{n}}$$

$$10.455 - 1.645\frac{7.7}{\sqrt{44}} \leq \mu \leq 10.455 + 1.645\frac{7.7}{\sqrt{44}}$$

$$10.455 - 1.910 \leq \mu \leq 10.455 + 1.910$$

$$8.545 \leq \mu \leq 12.365$$

The analyst is 90% confident that if a census of all European companies trading with firms in India were taken at the time of this survey, the actual population mean number of years a company would have been trading with firms in India would be between 8.545 and 12.365. The point estimate is 10.455 years.

For convenience, Table 8.1 contains some of the more common levels of confidence and their associated z values.

Finite Correction Factor

Recall from Chapter 7 that if the sample is taken from a finite population, a finite correction factor may be used to increase the accuracy of the solution. In the case of interval estimation, the finite correction factor is used to reduce the width of the interval. As stated in Chapter 7, if the sample size is less than 5% of the population, the finite correction factor does not significantly alter the solution. If formula 8.1 is modified to include the finite correction factor, the result is formula 8.2.

TABLE 8.1

Values of z for Common Levels of Confidence

Confidence Level	z Value
90%	1.645
95%	1.96
98%	2.33
99%	2.575

CONFIDENCE INTERVAL TO ESTIMATE μ USING THE FINITE CORRECTION FACTOR (8.2)

$$\bar{x} - z_{\alpha/2}\frac{\sigma}{\sqrt{n}}\sqrt{\frac{N-n}{N-1}} \leq \mu \leq \bar{x} + z_{\alpha/2}\frac{\sigma}{\sqrt{n}}\sqrt{\frac{N-n}{N-1}}$$

Demonstration Problem 8.2 shows how the finite correction factor can be used.

DEMONSTRATION PROBLEM 8.2

A study is conducted in a company that employs 800 engineers. A random sample of 50 engineers reveals that the average sample age is 34.3 years. Historically, the population standard deviation of the age of the company's engineers is approximately eight years. Construct a 98% confidence interval to estimate the average age of all the engineers in this company.

Solution

This problem has a finite population. The sample size, 50, is greater than 5% of the population, so the finite correction factor may be helpful. In this case $N = 800$, $n = 50$, $x = 34.30$, and $\sigma = 8$. The z value for a 98% confidence interval is 2.33 (0.98 divided into two equal parts yields 0.4900; the z value is obtained from Table A.5 by using 0.4900). Substituting into formula 8.2 and solving for the confidence interval gives

$$34.30 - 2.33\frac{8}{\sqrt{50}}\sqrt{\frac{750}{799}} \leq \mu \leq 34.30 + 2.33\frac{8}{\sqrt{50}}\sqrt{\frac{750}{799}}$$

$$34.30 - 2.55 \leq \mu \leq 34.30 + 2.55$$
$$31.75 \leq \mu \leq 36.85$$

Without the finite correction factor, the result would have been

$$34.30 - 2.64 \leq \mu \leq 34.30 + 2.64$$
$$31.66 \leq \mu \leq 36.94$$

The finite correction factor takes into account the fact that the population is only 800 instead of being infinitely large. The sample, $n = 50$, is a greater proportion of the 800 than it would be of a larger population, and thus the width of the confidence interval is reduced.

Estimating the Population Mean Using the z Statistic when the Sample Size Is Small

In the formulas and problems presented so far in the section, sample size was large ($n \geq 30$). However, quite often in the business world, sample sizes are small. While the central limit theorem applies only when sample size is large, the distribution of sample means is approximately normal even for small sizes *if the population is normally distributed*. This is visually displayed in the bottom row of Figure 7.5 in Chapter 7. Thus, if it is known that the population from which the sample is being drawn is normally distributed and if σ is known, the z formulas presented in this section can still be used to estimate a population mean even if the sample size is small ($n < 30$).

As an example, suppose an Italian car rental firm wants to estimate the average number of kilometres travelled per day by each of its cars rented in Florence. A random sample of 20 cars rented in Florence reveals that the sample mean travel distance per day is 85.5 kilometres, with a population standard deviation of 19.3 kilometres. Compute a 99% confidence interval to estimate μ.

Here, $n = 20$, $\bar{x} = 85.5$, and $\sigma = 19.3$. For a 99% level of confidence, a z value of 2.575 is obtained. Assume that number of kilometres travelled per day is normally distributed in the population. The confidence interval is

$$\bar{x} - z_{\alpha/2}\frac{\sigma}{\sqrt{n}} \leq \mu \leq \bar{x} + z_{\alpha/2}\frac{\sigma}{\sqrt{n}}$$

$$85.5 - 2.575\frac{19.3}{\sqrt{20}} \leq \mu \leq 85.5 + 2.575\frac{19.3}{\sqrt{20}}$$

$$85.5 - 11.1 \leq \mu \leq 85.5 + 11.1$$
$$74.4 \leq \mu \leq 96.6$$

The point estimate indicates that the average number of kilometres travelled per day by a rental car in Florence is 85.5. With 99% confidence, we estimate that the population mean is somewhere between 74.4 and 96.6 kilometres per day.

Using the Computer to Construct z Confidence Intervals for the Mean

It is possible to construct a z confidence interval for the mean with either Excel or Minitab. Excel yields the $\pm$ error portion of the confidence interval that must be placed with the sample mean to construct the complete confidence interval. Minitab constructs the complete confidence interval. Figure 8.6 shows both the Excel output and the Minitab output for the mobile phone example.

Excel Output

The sample mean is:	510
The error of the interval is:	9.779
The confidence interval is:	510 ± 9.779
The confidence interval is:	$500.221 \leq \mu \leq 519.779$

Minitab Output

One-Sample Z

```
The assumed standard deviation = 46
  N    Mean  SE Mean      95% CI
 85  510.00     4.99  (500.22,  519.78)
```

FIGURE 8.6

Excel and Minitab Output for the Mobile Phone Example

8.1 PROBLEMS

8.1 Use the following information to construct the confidence intervals specified to estimate μ:
   a. 95% confidence for $\bar{x} = 25$, $\sigma = 3.5$, and $n = 60$
   b. 98% confidence for $\bar{x} = 119.6$, $\sigma = 23.89$, and $n = 75$
   c. 90% confidence for $\bar{x} = 3.419$, $\sigma = 0.974$, and $n = 32$
   d. 80% confidence for $\bar{x} = 56.7$, $\sigma = 12.1$, $N = 500$, and $n = 47$

8.2 For a random sample of 36 items and a sample mean of 211, compute a 95% confidence interval for μ if the population standard deviation is 23.

8.3 A random sample of 81 items is taken, producing a sample mean of 47. The population standard deviation is 5.89. Construct a 90% confidence interval to estimate the population mean.

8.4 A random sample of size 70 is taken from a population that has a variance of 49. The sample mean is 90.4 What is the point estimate of μ? Construct a 94% confidence interval for μ.

8.5 A random sample of size 39 is taken from a population of 200 members. The sample mean is 66 and the population standard deviation is 11. Construct a 96% confidence interval to estimate the population mean. What is the point estimate of the population mean?

8.6 A chocolates company fills a 200 gram box of Valentine's Day chocolates with individually wrapped chocolates. The number of chocolates per box varies because the box is sold by weight. The company wants to estimate the number of chocolates per box. Inspectors randomly sample 120 boxes of these chocolates and count the number of chocolates in each box. They find that the sample mean number of chocolates is 18.72. Assuming a population standard deviation of 0.8735, what is the point estimate of the number of chocolates per box? Construct a 99% confidence interval to estimate the mean number of chocolates per box for the population.

8.7 A small lawnmower company produced 1,500 lawnmowers in 1998. In an effort to determine how maintenance-free these units were, the company decided to conduct a multiyear study of the 1998 lawnmowers. A sample of 200 owners of these lawnmowers was drawn randomly from company records and contacted. The owners were given a toll-free telephone number and asked to call the company when the first major repair was required for the lawnmowers. Owners who no longer used the lawnmower to cut their grass were disqualified. After many years, 187 of the owners had reported. The other 13 disqualified themselves. The average number of years until the first major repair was 5.3 for the 187 owners reporting. It is believed that the population standard deviation was 1.28 years. If the company wants to advertise an average number of years of repair-free lawn mowing for this lawnmower, what is the point estimate? Construct a 95% confidence interval for the average number of years until the first major repair.

8.8   The average total value of purchases at a convenience store is less than that at a supermarket. Despite smaller item purchases, convenience stores can still be profitable because of the size of operation, volume of business, and the mark-up. A researcher is interested in estimating the average purchase amount for convenience stores in suburban Birmingham. To do so, she randomly sampled 24 purchases from several convenience stores in suburban Birmingham and tabulated the amounts to the nearest pound. Use the following data to construct a 90% confidence interval for the population average amount of purchases. Assume that the population standard deviation is 3.23 and the population is normally distributed.

£2	£11	£8	£7	£9	£3
5	4	2	1	10	8
14	7	6	3	7	2
4	1	3	6	8	4

8.9   A community health association is interested in estimating the average number of maternity days women stay in the local hospital. A random sample is taken of 36 women who had babies in the hospital during the past year. The following numbers of maternity days each woman was in the hospital are rounded to the nearest day.

3	3	4	3	2	5	3	1	4	3
4	2	3	5	3	2	4	3	2	4
1	6	3	4	3	3	5	2	3	2
3	5	4	3	5	4				

Use these data and a population standard deviation of 1.17 to construct a 98% confidence interval to estimate the average maternity stay in the hospital for all women who have babies in this hospital.

8.10   A meat-processing company in northern Europe produces and markets a package of eight small sausage sandwiches. The product is nationally distributed, and the company is interested in knowing the average retail price charged for this item in stores across the Eurozone. The company cannot justify a European census to generate this information. Based on the company information system's list of all retailers who carry the product, a researcher for the company contacts 36 of these retailers and ascertains the selling prices for the product. Use the following price data and a population standard deviation of 0.113 to determine a point estimate for the European retail price of the product. Construct a 90% confidence interval to estimate this price.

€2.23	€2.11	€2.12	€2.20	€2.17	€2.10
2.16	2.31	1.98	2.17	2.14	1.82
2.12	2.07	2.17	2.30	2.29	2.19
2.01	2.24	2.18	2.18	2.32	2.02
1.99	1.87	2.09	2.22	2.15	2.19
2.23	2.10	2.08	2.05	2.16	2.26

8.11   According to Eurostat, the average travel time per person per day in Norway is 68.2 minutes. Suppose a business researcher wants to estimate the average travel time in Norway using a 95% level of confidence. A random sample of 45 Norwegians is taken and the travel time per day is obtained from each. The data follow. Assuming a population standard deviation of 12.75, compute a 95% confidence interval on the data. What is the point estimate and what is the error of the interval? Explain what these results mean for Norwegians.

67	65	59	61	64	67	69	74	58	69	56	68
60	72	67	68	62	60	54	55	69	68	69	73
56	69	68	68	67	63	67	60	67	65	61	58
66	54	63	67	67	61	65	68	70			

8.12 Cereal prices have increased dramatically in recent years, triggered by a combination of production below the trend, high oil prices, and a strong growth in demand around the world. This acute price hike is hitting poor consumers in developing countries particularly badly as they have to spend an ever increasing share of their limited income on food. Suppose a random sample of prices of a kilogram of wheat is taken from across a developing nation in an effort to estimate the average wheat price for that country. Shown here is the Minitab output for such a sample. Examine the output. What is the point estimate? What is the value of the assumed population standard deviation? How large is the sample? What level of confidence is being used? What table value is associated with this level of confidence? What is the confidence interval? Often the portion of the confidence interval that is added and subtracted from the mean is referred to as the error of the estimate. How much is the error of the estimate in this problem?

```
One-Sample Z

The assumed standard deviation = 0.14
  N      Mean     SE Mean          95%  CI
  41    0.5765    0.0219     (0.5336,  0.6194)
```

8.2 ESTIMATING THE POPULATION MEAN USING THE t STATISTIC (σ UNKNOWN)

In Section 8.1, we learned how to estimate a population mean by using the sample mean when the population standard deviation is known. In most instances, if a business researcher desires to estimate a population mean, the population standard deviation will be unknown and thus techniques presented in Section 8.1 will not be applicable. When the population standard deviation is unknown, the sample standard deviation must be used in the estimation process. In this section, a statistical technique is presented to estimate a population mean using the sample mean when the population standard deviation is unknown.

Suppose a business researcher is interested in estimating the average flying time of a 767 jet from London to Moscow. Since the business researcher does not know the population mean or average time, it is likely that she also does not know the population standard deviation. By taking a random sample of flights, the researcher can compute a sample mean and a sample standard deviation from which the estimate can be constructed. Another business researcher is studying the impact of movie video advertisements on consumers using a random sample of people. The researcher wants to estimate the mean response for the population but has no idea what the population standard deviation is. He will have the sample mean and sample standard deviation available to perform this analysis.

The z formulas presented in Section 8.1 are inappropriate for use when the population standard deviation is unknown (and is replaced by the sample standard deviation). Instead, another mechanism to handle such cases was developed by a British statistician, William S. Gosset.

Gosset was born in 1876 in Canterbury, England. He studied chemistry and mathematics and in 1899 went to work for the Guinness Brewery in Dublin, Ireland. Gosset was involved in quality control at the brewery, studying variables such as raw materials and temperature. Because of the circumstances of his experiments, Gosset conducted many studies where the population standard deviation was unavailable. He discovered that using the standard z test with a sample standard deviation produced inexact and incorrect distributions. This finding led to his development of the distribution of the sample standard deviation and the t test.

Gosset was a student and close personal friend of Karl Pearson. When Gosset's first work on the t test was published, he used the pen name 'Student'. As a result, the t test is sometimes referred to as the Student's t test. Gosset's contribution was significant because it led to more exact statistical tests, which some scholars say marked the beginning of the modern era in mathematical statistics.*

---

*Adapted from Arthur L. Dudycha and Linda W. Dudycha, 'Behavioral Statistics: An Historical Perspective'. In Roger Kirk (ed.) *Statistical Issues: A Reader for the Behavioral Sciences*. Monterey, CA: Brooks/Cole, 1972.

The *t* Distribution

Gosset developed the **t distribution**, which is used instead of the *z* distribution for doing inferential statistics on the population mean when the population standard deviation is unknown and the population is normally distributed. The formula for the *t* statistic is

$$t = \frac{\bar{x} - \mu}{\dfrac{s}{\sqrt{n}}}$$

This formula is essentially the same as the *z* formula, but the distribution table values are different. The *t* distribution values are contained in Table A.6 in Appendix A and, for convenience, at the front of the book.

The *t* distribution actually is a series of distributions because every sample size has a different distribution, thereby creating the potential for many *t* tables. To make these *t* values more manageable, only select key values are presented; each line in the table contains values from a different *t* distribution. An assumption underlying the use of the *t* statistic is that the population is normally distributed. If the population distribution is not normal or is unknown, non-parametric techniques (presented in Chapter 17) should be used.

Robustness

Most statistical techniques have one or more underlying assumptions. If a statistical technique is relatively insensitive to minor violations in one or more of its underlying assumptions, the technique is said to be **robust** to that assumption. The *t* statistic for estimating a population mean is relatively robust to the assumption that the population is normally distributed.

Some statistical techniques are not robust, and a statistician should exercise extreme caution to be certain that the assumptions underlying a technique are being met before using it or interpreting statistical output resulting from its use. A business analyst should always beware of statistical assumptions and the robustness of techniques being used in an analysis.

Characteristics of the *t* Distribution

Figure 8.7 displays two *t* distributions superimposed on the standard normal distribution. Like the standard normal curve, *t* distributions are symmetric, unimodal, and a family of curves. The *t* distributions are flatter in the middle and have more area in their tails than the standard normal distribution.

An examination of *t* distribution values reveals that the *t* distribution approaches the standard normal curve as *n* becomes large. The *t* distribution is the appropriate distribution to use any time the population variance or standard deviation is unknown, regardless of sample size.

Reading the *t* Distribution Table

To find a value in the *t* distribution table requires knowing the degrees of freedom; each different value of degrees of freedom is associated with a different *t* distribution. The *t* distribution table used here is a compilation of many *t* distributions, with each line of the table having different degrees of freedom and containing *t* values for different *t* distributions. The degrees of freedom for the *t* statistic presented in this section are computed by $n - 1$. The term **degrees of freedom** refers to *the number of independent observations for a source of variation minus the number of independent parameters estimated in computing the variation.** In this case, one independent parameter, the population mean, μ, is being estimated by $\bar{x}$ in computing *s*. Thus, the degrees of freedom formula is *n* independent observations minus one independent

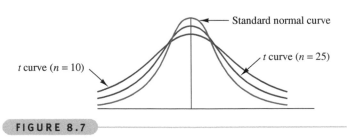

Standard normal curve

t curve (*n* = 25)

t curve (*n* = 10)

FIGURE 8.7

Comparison of Two *t* Distributions to the Standard Normal Curve

---

*Roger E. Kirk. *Experimental Design: Procedures for the Behavioral Sciences.* Belmont, California: Brooks/Cole, 1968.

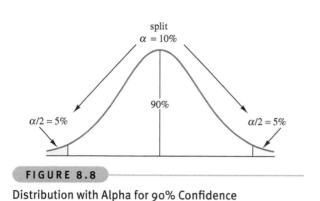

FIGURE 8.8

Distribution with Alpha for 90% Confidence

TABLE 8.2

t Distribution

Degrees of Freedom	$t_{.10}$	$t_{.05}$	$t_{.025}$	$t_{.01}$	$t_{.005}$	$t_{.001}$
.						
.						
.						
23						
㉔		1.711				
25						
.						
.						

parameter being estimated ($n - 1$). Because the degrees of freedom are computed differently for various *t* formulas, a degrees of freedom formula is given along with each *t* formula in the text.

In Table A.6, the degrees of freedom are located in the left column. The *t* distribution table in this text does not use the area between the statistic and the mean as does the *z* distribution (standard normal distribution). Instead, the *t* table uses the area in the tail of the distribution. The emphasis in the *t* table is on α, and each tail of the distribution contains $\alpha/2$ of the area under the curve when confidence intervals are constructed. For confidence intervals, the table *t* value is found in the column under the value of $\alpha/2$ and in the row of the degrees of freedom (df) value.

For example, if a 90% confidence interval is being computed, the total area in the two tails is 10%. Thus, α is 0.10 and $\alpha/2$ is 0.05, as indicated in Figure 8.8. The *t* distribution table shown in Table 8.2 contains only six values of $\alpha/2$ (0.10, 0.05, 0.025, 0.01, 0.005, 0.001). The *t* value is located at the intersection of the df value and the selected $\alpha/2$ value. So if the degrees of freedom for a given *t* statistic are 24 and the desired $\alpha/2$ value is 0.05, the *t* value is 1.711.

Confidence Intervals to Estimate the Population Mean Using the *t* Statistic

The *t* formula

$$t = \frac{\bar{x} - \mu}{\frac{s}{\sqrt{n}}}$$

can be manipulated algebraically to produce a formula for estimating the population mean when α is unknown and the population is normally distributed. The results are the formulas given next.

CONFIDENCE INTERVAL TO ESTIMATE μ: POPULATION STANDARD DEVIATION UNKNOWN AND THE POPULATION NORMALLY DISTRIBUTED (8.3)	$\bar{x} \pm t_{\alpha/2,n-1} \frac{s}{\sqrt{n}}$ $\bar{x} - t_{\alpha/2,n-1} \frac{s}{\sqrt{n}} \leq \mu \leq \bar{x} + t_{\alpha/2,n-1} \frac{s}{\sqrt{n}}$ $df = n - 1$

Formula 8.3 can be used in a manner similar to methods presented in Section 8.1 for constructing a confidence interval to estimate μ. For example, in the aerospace industry some companies allow their employees to accumulate extra working hours beyond their 40-hour week. These extra hours sometimes are referred to as *green* time, or *comp* time. Many managers work longer than the eight-hour workday preparing proposals, overseeing crucial tasks, and taking care of paperwork. Recognition of such overtime is important. Most managers are usually not paid extra for this work, but a record is kept of this time and occasionally the manager is allowed to use some of this comp time as extra leave or vacation time. Suppose a researcher wants to estimate the average amount of comp time accumulated per week for managers in the aerospace industry. He randomly samples 18 managers and measures the amount of extra time they work during a specific week and obtains the results shown (in hours).

| 6 | 21 | 17 | 20 | 7 | 0 | 8 | 16 | 29 |
| 3 | 8 | 12 | 11 | 9 | 21 | 25 | 15 | 16 |

He constructs a 90% confidence interval to estimate the average amount of extra time per week worked by a manager in the aerospace industry. He assumes that comp time is normally distributed in the population. The sample size is 18, so df = 17. A 90% level of confidence results in $\alpha/2 = 0.05$ area in each tail. The table t value is

$$t_{.05,17} = 1.740$$

The subscripts in the t value denote to other researchers the area in the right tail of the t distribution (for confidence intervals $\alpha/2$) and the number of degrees of freedom. The sample mean is 13.56 hours, and the sample standard deviation is 7.8 hours. The confidence interval is computed from this information as

$$\bar{x} \pm t_{\alpha/2,n-1} \frac{s}{\sqrt{n}}$$

$$13.56 \pm 1.740 \frac{7.8}{\sqrt{18}} = 13.56 \pm 3.20$$

$$10.36 \le \mu \le 16.76$$

The point estimate for this problem is 13.56 hours, with an error of ± 3.20 hours. The researcher is 90% confident that the average amount of comp time accumulated by a manager per week in this industry is between 10.36 and 16.76 hours.

From these figures, aerospace managers could attempt to build a reward system for such extra work or evaluate the regular 40-hour week to determine how to use the normal work hours more effectively and thus reduce comp time.

DEMONSTRATION PROBLEM 8.3

The owner of a large equipment rental company wants to make a rather quick estimate of the average number of days a mini-digger is rented out per person per time. The company has records of all rentals, but the amount of time required to conduct an audit of *all* accounts would be prohibitive. The owner decides to take a random sample of rental invoices. Fourteen different rentals of mini diggers are selected randomly from the files, yielding the following data. She uses these data to construct a 99% confidence interval to estimate the average number of days that a mini-digger is rented and assumes that the number of days per rental is normally distributed in the population.

| 3 | 1 | 3 | 2 | 5 | 1 | 2 | 1 | 4 | 2 | 1 | 3 | 1 | 1 |

As $n = 14$, the df $= 13$. The 99% level of confidence results in $\alpha/2 = 0.005$ area in each tail of the distribution. The table t value is

$$t_{.005,13} = 3.012$$

The sample mean is 2.14 and the sample standard deviation is 1.29. The confidence interval is

$$\bar{x} \pm t \frac{s}{\sqrt{n}}$$

$$2.14 \pm 3.012 \frac{1.29}{\sqrt{14}} = 2.14 \pm 1.04$$

$$1.10 \leq \mu \leq 3.18$$

The point estimate of the average length of time per rental is 2.14 days, with an error of ± 1.04. With a 99% level of confidence, the company's owner can estimate that the average length of time per rental is between 1.10 and 3.18 days. Combining this figure with variables such as frequency of rentals per year can help the owner estimate potential profit or loss per year for such a piece of equipment.

Using the Computer to Construct t Confidence Intervals for the Mean

Both Excel and Minitab can be used to construct confidence intervals for μ using the t distribution. Figure 8.9 displays Excel output and Minitab output for the aerospace comp time problem. The Excel output includes the mean, the standard error, the sample standard deviation, and the error of the confidence interval, referred to by Excel as the 'confidence level'. The standard error of the mean is computed by dividing the standard deviation (7.8006) by the square root of n (4.243). When using the Excel output, the confidence interval must be computed from the sample mean and the confidence level (error of the interval).

The Minitab output yields the confidence interval endpoints (10.36, 16.75). The 'SE Mean' is the standard error of the mean. The error of the confidence interval is computed by multiplying the standard error of the mean by the table value of t. Adding and subtracting this error from the mean yields the confidence interval endpoints produced by Minitab.

Excel Output

	Comp Time
Mean	13.56
Standard error	1.8386
Standard deviation	7.8006
Confidence level (90.0%)	3.20

Minitab Output

One-Sample T: Comp Time

Variable	N	Mean	StDev	SE Mean	90% CI
Comp Time	18	13.56	7.80	1.84	(10.36, 16.75)

FIGURE 8.9

Excel and Minitab Output for the Comp Time Example

Online Shopping Habits in Greater China

There are an estimated half a billion Internet users in China and among them over 100 million have shopped on the Internet, giving e-commerce a huge consumer base. With shopping websites sprouting, shopping habits are changing and that has directly stimulated usage and development of credit card products. MasterCard has recently produced an online shopping survey to better understand consumption trends and dynamics in the Greater China region. According to MasterCard's latest survey on online shopping habits, online shopping has become one of the most preferred shopping methods for consumers in Greater China. Consumers from China between the ages of 18 and 64 are the most frequent online shoppers in the region, with more than 80% having made purchases on the Internet, followed

by 66% in Taiwan and 49% in Hong Kong. Among all online shoppers in Greater China, female shoppers are the most enthusiastic, with those from China being aged between 25 and 34, and those from Taiwan aged between 35 and 49. Among the different online product categories, books, CDs, DVDs, household appliances and electronic products, fashion items and music downloads are the most popular. Special offers from restaurants, courier services, supermarkets, fashion retailers, and travel agent websites consistently appear as the most appealing to consumers. Since these statistics are based on a small sample of shoppers, it is virtually certain that the statistics given here are point estimates.

Source: MasterCard Greater China Customer Survey, available at www .mastercard.com/hk/personal/en/promotions/2011_collection/GCPC/ newsletter/4th/4.html. Accessed 2011.

8.2 PROBLEMS

8.13 Suppose the following data are selected randomly from a population of normally distributed values:

40	51	43	48	44	57	54
39	42	48	45	39	43	

Construct a 95% confidence interval to estimate the population mean.

8.14 Assuming x is normally distributed, use the following information to compute a 90% confidence interval to estimate μ:

313	320	319	340	325	310
321	329	317	311	307	318

8.15 If a random sample of 41 items produces $\bar{x} = 128.4$ and $s = 20.6$, what is the 98% confidence interval for μ? Assume x is normally distributed for the population. What is the point estimate?

8.16 A random sample of 15 items is taken, producing a sample mean of 2.364 with a sample variance of 0.81. Assume x is normally distributed and construct a 90% confidence interval for the population mean.

8.17 Use the following data to construct a 99% confidence interval for μ.

16.4	17.1	17.0	15.6	16.2
14.8	16.0	15.6	17.3	17.4
15.6	15.7	17.2	16.6	16.0
15.3	15.4	16.0	15.8	17.2
14.6	15.5	14.9	16.7	16.3

Assume x is normally distributed. What is the point estimate for μ?

8.18 According to Hotels.com's survey on hotel prices around the world, the average price per hotel room per night in Abu Dhabi was £116 in a recent year. Suppose another travel industry research company takes a random sample of 51 travellers and determines that the sample average cost of a hotel room per person per night in Abu Dhabi is £110, with a sample standard deviation of £28. Construct a 98% confidence interval for the population mean from these sample data. Assume that the data are normally distributed in the population. Now go back and examine the £116 figure published by Hotels.com. Does it fall into the confidence interval computed from the sample data? What does it tell you?

8.19 A valve manufacturer produces a butterfly valve composed of two semicircular plates on a common spindle that is used to permit flow in one direction only. The semicircular plates are supplied by a vendor with specifications that the plates be 2.37 millimetres thick and have a tensile strength of five pounds per millimetres. A random sample of 20 such plates is taken. Electronic calipers are used to measure the thickness of each plate; the measurements are given here. Assuming that the thicknesses of such plates are normally distributed, use the data to construct a 95% level of confidence for the population mean thickness of these plates. What is the point estimate? How much is the error of the interval?

2.4066	2.4579	2.6724	2.1228	2.3238
2.1328	2.0665	2.2738	2.2055	2.5267
2.5937	2.1994	2.5392	2.4359	2.2146
2.1933	2.4575	2.7956	2.3353	2.2699

8.20 Some fast-food chains offer a lower-priced combination meal in an effort to attract budget-conscious customers. One chain test-marketed a burger, fries, and a drink combination for £2.71. The weekly sales volume for these meals was impressive. Suppose the chain wants to estimate the average amount its customers spent on a meal at their restaurant while this combination offer was in effect. An analyst gathers data from 28 randomly selected customers. The following data represent the sample meal totals.

£3.21	5.40	3.50	4.39	5.60	8.65	5.02	4.20	1.25	7.64
3.28	5.57	3.26	3.80	5.46	9.87	4.67	5.86	3.73	4.08
5.47	4.49	5.19	5.82	7.62	4.83	8.42	9.10		

Use these data to construct a 90% confidence interval to estimate the population mean value. Assume the amounts spent are normally distributed.

8.21 The marketing director of a large department store wants to estimate the average number of customers who enter the store every five minutes. She randomly selects five-minute intervals and counts the number of arrivals at the store. She obtains the figures 58, 32, 41, 47, 56, 80, 45, 29, 32, and 78. The analyst assumes the number of arrivals is normally distributed. Using these data, the analyst computes a 95% confidence interval to estimate the mean value for all five-minute intervals. What interval values does she get?

8.22 Property Consultant DTZ publishes results of studies on global office occupancy costs around the world. According to DTZ's latest report, Hong Kong became the world's most expensive place to rent an office. Suppose as a part of one of these studies the following prices of office property per workstation (in dollars) are obtained for 14 businesses currently renting offices in Hong Kong, Use these data to construct a 98% confidence interval to estimate the average price of office property per workstation in Hong Kong. What is the point estimate? Assume the average price of office property per workstation for any locale are approximately normally distributed.

11407	11878	12770	13754	11752	13510	11355
12727	12506	11399	10342	12125	10630	14267

8.23 How much experience do supply-chain transportation managers have in their field? Suppose in an effort to estimate this, 41 supply-chain transportation managers are surveyed and asked how many years of managerial experience they have in transportation. Survey results (in years) are shown below. Use these data to construct a 99% confidence interval to estimate the mean number of years of experience in transportation. Assume that years of experience in transportation is normally distributed in the population.

5	8	10	21	20
25	14	6	19	3
1	9	11	2	3
13	2	4	9	4
5	4	21	7	6
3	28	17	32	2
25	8	13	17	27
7	3	15	4	16
6				

8.24 Cycle time in manufacturing can be viewed as the total time it takes to complete a product from the beginning of the production process. The concept of cycle time varies according to the industry and product or service being offered. Suppose a boat manufacturing company wants to estimate the mean cycle time it takes to produce a 16-foot skiff. A random sample of such skiffs is taken, and the cycle times (in hours) are recorded for each skiff in the sample. The data are analysed using Minitab and the results are shown below in hours. What is the point estimate for cycle time? How large was the sample size? What is the level of confidence and what is the confidence interval? What is the error of the confidence interval?

```
One-Sample T

  N      Mean     StDev     SE Mean        98% CI
 26     25.41      5.34        1.05     (22.81, 28.01)
```

8.3 ESTIMATING THE POPULATION PROPORTION

Business decision makers and researchers often need to be able to estimate a population proportion. For most businesses, estimating market share (their proportion of the market) is important because many company decisions evolve from market share information. Companies spend thousands estimating the proportion of produced goods that are defective. Market segmentation opportunities come from a knowledge of the proportion of various demographic characteristics among potential customers or clients.

Methods similar to those in Section 8.1 can be used to estimate the population proportion. The central limit theorem for sample proportions led to the following formula in Chapter 7:

$$z = \frac{\hat{p} - p}{\sqrt{\dfrac{p \cdot q}{n}}}$$

where $q = 1 - p$. Recall that this formula can be applied only when $n \cdot p$ and $n \cdot q$ are greater than 5.

Algebraically manipulating this formula to estimate p involves solving for p. However, p is in both the numerator and the denominator, which complicates the resulting formula. For this reason – for confidence interval purposes only and for large sample sizes – $\hat{p}$ is substituted for p in the denominator, yielding

$$z = \frac{\hat{p} - p}{\sqrt{\dfrac{\hat{p} \cdot \hat{q}}{n}}}$$

where $\hat{q} = 1 - \hat{p}$. Solving for p results in the confidence interval in formula (8.4).*

---

*Because we are not using the true standard deviation of $\hat{p}$, the correct divisor of the standard error of $\hat{p}$ is $n - 1$. However, for large sample sizes, the effect is negligible. Although technically the minimal sample size for the techniques presented in this section is $n \cdot p$ and $n \cdot q$ greater than 5, in actual practice sample sizes of several hundred are more commonly used. As an example, for $\hat{p}$ and $\hat{q}$ of 0.50 and $n = 300$, the standard error of $\hat{p}$ is 0.02887 using n and 0.02892 using $n - 1$, a difference of only 0.00005.

$$\hat{p} - z_{\alpha/2}\sqrt{\frac{\hat{p}\cdot\hat{q}}{n}} \leq p \leq \hat{p} + z_{\alpha/2}\sqrt{\frac{\hat{p}\cdot\hat{q}}{n}}$$

CONFIDENCE INTERVAL TO ESTIMATE p (8.4)

where:

$\hat{p}$ = sample proportion
$\hat{q} = 1 - \hat{p}$
p = population proportion
n = sample size

In this formula, $\hat{p}$ is the point estimate and $\pm z_{\alpha/2}\sqrt{\frac{\hat{p}\cdot\hat{q}}{n}}$ is the error of the estimation.

As an example, a study of 87 randomly selected companies with a telemarketing operation revealed that 39% of the sampled companies used telemarketing to assist them in order processing. Using this information, how could a researcher estimate the *population* proportion of telemarketing companies that use their telemarketing operation to assist them in order processing?

The sample proportion, $\hat{p} = 0.39$, is the *point estimate* of the population proportion, p. For $n = 87$ and $\hat{p} = 0.39$, a 95% confidence interval can be computed to determine the interval estimation of p. The z value for 95% confidence is 1.96. The value of $\hat{q} = 1 - \hat{p} = 1 - 0.39 = 0.61$. The confidence interval estimate is

$$0.39 - 1.96\sqrt{\frac{(0.39)(0.61)}{87}} \leq p \leq 0.39 + 1.96\sqrt{\frac{(0.39)(0.61)}{87}}$$

$$0.39 - 0.10 \leq p \leq 0.39 + 0.10$$

$$0.29 \leq p \leq 0.49$$

This interval suggests that the population proportion of telemarketing firms that use their operation to assist order processing is somewhere between 0.29 and 0.49, based on the point estimate of 0.39 with an error of ± 0.10. This result has a 95% level of confidence.

STATISTICS IN BUSINESS TODAY

Coffee Consumption in the United Kingdom

Although traditionally associated with tea drinking, the United Kingdom is, in fact, a coffee-drinking country. Although it is far from being the top country in terms of coffee consumption per capita (Finland leads with a massive per capita consumption of 12 kilograms per year), the United Kingdom's average coffee consumption figure of 2.8 kilograms per person per year still towers over the 2 kilograms of tea of annual per capita consumption.

A recent survey by Mintel has, however, made a worrying trend public. The over-55s represent the biggest coffee consumers but younger consumers do not seem to be developing a taste for the product. The age of 35 appears to be a major turning point for coffee

drinking in the UK, with coffee consumption rising considerably among 35- to 44-year-olds and steadily thereafter. Just 39% of 25- to 34-year-olds and 17% of 16- to 24-year-olds drink coffee, which they make at work, compared with as many as half (50%) of those aged 35 to 44. Meanwhile, 33% of those aged 16 to 24 and 57% of those aged 25 to 34 drink coffee at home; this compares to three-quarters (75%) of those aged 35 to 44 years old. The survey also showed that good old 'instant' remains the nation's favourite coffee, accounting for 80% of value sales. Roast or ground coffee account for 19% of sales and ready-to-drink coffee accounts for the remaining 1% of the market. Finally, looking at coffee sales as a whole, Fairtrade coffee accounted for around 5% of total coffee sales and is expected to continue growing at a very fast pace in the future. Fairtrade is becoming the norm in the coffee market as more and more of the major coffee and coffee shop brands switch their supply to Fairtrade sources.

Because much of the information presented here was gleaned from a survey, virtually all of the percentages and means are sample statistics and not population parameters. Thus, what are presented as coffee population statistics are actually point estimates.

Using the sample size and a level of confidence, confidence intervals can be constructed for the proportions. Confidence intervals for means can be constructed from these point estimates if the value of the standard deviation can be determined.

Source: Adapted from 'UK coffee sales full of beans' at www.mintel.com/press-centre/press-releases/498/uk-coffee-sales-full-of-beans. Other sources include the 'Countries by coffee consumption per capita', World Resource Institute and Wikipedia. Accessed 2011.

DEMONSTRATION PROBLEM 8.4

Coopers & Lybrand surveyed 210 chief executives of fast-growing small companies. Only 51% of these executives had a management succession plan in place. A spokesperson for Cooper & Lybrand said that many companies do not worry about management succession unless it is an immediate problem. However, the unexpected exit of a corporate leader can disrupt and unfocus a company for long enough to cause it to lose its momentum.

Use the data given to compute a 92% confidence interval to estimate the proportion of *all* fast-growing small companies that have a management succession plan.

Solution

The point estimate is the sample proportion given to be 0.51. It is estimated that 0.51, or 51%, of all fast-growing small companies have a management succession plan. Realizing that the point estimate might change with another sample selection, we calculate a confidence interval.

The value of n is 210; $\hat{p}$ is 0.51, and $\hat{q} = 1 - \hat{p} = 0.49$. Because the level of confidence is 92%, the value of $z_{0.04} = 1.75$. The confidence interval is computed as

$$0.51 - 1.75\sqrt{\frac{(0.51)(0.49)}{210}} \leq p \leq 0.51 + 1.75\sqrt{\frac{(0.51)(0.49)}{210}}$$
$$0.51 - 0.6 \leq p \leq 0.51 + 0.06$$
$$0.45 \leq p \leq 0.57$$

It is estimated with 92% confidence that the proportion of the population of fast-growing small companies that have a management succession plan is between 0.45 and 0.57.

DEMONSTRATION PROBLEM 8.5

A clothing company produces men's jeans. The jeans are made and sold with either a regular cut or a boot cut. In an effort to estimate the proportion of their men's jeans market in Dubai that prefers boot-cut jeans, the analyst takes a random sample of 212 jeans sales from the company's two Dubai retail outlets. Only 34 of the sales were for boot-cut jeans. Construct a 90% confidence interval to estimate the proportion of the population in Dubai who prefer boot-cut jeans.

Solution

The sample size is 212, and the number preferring boot-cut jeans is 34. The sample proportion is $\hat{p} = 34/212 = 0.16$. A point estimate for boot-cut jeans in the population is 0.16, or 16%. The z value for a 90% level of confidence is 1.645, and the value of $\hat{q} = 1 - \hat{p} = 1 - 0.16 = 0.84$. The confidence interval estimate is

$$0.16 - 1.645\sqrt{\frac{(0.16)(0.84)}{212}} \leq p \leq 0.16 + 1.645\sqrt{\frac{(0.16)(0.84)}{212}}$$

$$0.16 - 0.04 \leq p \leq 0.16 + 0.04$$

$$0.12 \leq p \leq 0.20$$

The analyst estimates that the population proportion of boot-cut jeans purchases is between 0.12 and 0.20. The level of confidence in this result is 90%.

Using the Computer to Construct Confidence Intervals of the Population Proportion

Minitab has the capability of producing confidence intervals for proportions. Figure 8.10 contains Minitab output for Demonstration Problem 8.5. The output contains the sample size (labelled as N), the number in the sample containing the characteristic of interest (X), the sample proportion, the level of confidence, and the endpoints of the confidence interval. Note that the endpoints of the confidence interval are essentially the same as those computed in Demonstration Problem 8.5.

```
Test and CI For One Proportion

Sample      X        N      Sample p             90%  CI
1          34       212     0.160377      (0.120328, 0.207718)
```

FIGURE 8.10

Minitab Output for Demonstration Problem 8.5

8.3 PROBLEMS

8.25 Use the information about each of the following samples to compute the confidence interval to estimate p:
  a. $n = 44$ and $\hat{p} = 0.51$; compute a 90% confidence interval.
  b. $n = 300$ and $\hat{p} = 0.82$; compute a 95% confidence interval.
  c. $n = 1{,}150$ and $\hat{p} = 0.48$; compute a 90% confidence interval.
  d. $n = 95$ and $\hat{p} = 0.32$; compute a 88% confidence interval.

8.26 Use the following sample information below to calculate the confidence interval to estimate the population proportion. Let x be the number of items in the sample having the characteristic of interest.
  a. $n = 116$ and $x = 57$, with 99% confidence
  b. $n = 800$ and $x = 479$, with 97% confidence
  c. $n = 240$ and $x = 106$, with 85% confidence
  d. $n = 60$ and $x = 21$, with 90% confidence

8.27 Suppose a random sample of 85 items has been taken from a population and 40 of the items contain the characteristic of interest. Use this information to calculate a 90% confidence interval to estimate the proportion of the population that has the characteristic of interest. Calculate a 95% confidence interval. Calculate a 99%

confidence interval. As the level of confidence changes and the other sample information stays constant, what happens to the confidence interval?

8.28 The Universal Music Group is the world's leading music company with wholly owned record operations or licensees in 77 countries. Suppose a researcher wants to determine what market share the company holds in Germany by randomly selecting 1,003 people who purchased a CD last month. In addition, suppose 25.5% of the purchases made by these people were for products manufactured and distributed by the Universal Music Group.

    a. Based on these data, construct a 99% confidence interval to estimate the proportion of the CD sales market in Germany that is held by the Universal Music Group.

    b. Suppose that the survey had been taken with 10,000 people. Recompute the confidence interval and compare your results with the first confidence interval. How did they differ? What might you conclude from this about sample size and confidence intervals?

8.29 According to the Stern Marketing Group, nine out of 10 professional women say that financial planning is more important today than it was five years ago. Where do these women go for help in financial planning? Forty-seven per cent use a financial adviser (broker, tax consultant, financial planner). Twenty-eight per cent use written sources such as magazines, books, and newspapers. Suppose these figures were obtained by taking a sample of 560 professional women who said that financial planning is more important today than it was five years ago.

    a. Construct a 95% confidence interval for the proportion of professional women who use a financial adviser. Use the percentage given in this problem as the point estimate.

    b. Construct a 90% confidence interval for the proportion of professional women who use written sources. Use the percentage given in this problem as the point estimate.

8.30 What proportion of pizza restaurants that are primarily for walk-in business have a salad bar? Suppose that, in an effort to determine this figure, a random sample of 1,250 of these restaurants across Europe based on an online business directory is contacted. If 997 of the restaurants sampled have a salad bar, what is the 98% confidence interval for the population proportion?

8.31 The highway department wants to estimate the proportion of vehicles on the M1 motorway between the hours of midnight and 5.00 A.M. that are 18-wheel tractor and trailer units. The estimate will be used to determine motorway repair and construction considerations and in motorway patrol planning. Suppose researchers for the motorway department counted vehicles at different locations on the M1 for several nights during this time period. Of the 3,481 vehicles counted, 927 were 18-wheelers.

    a. Determine the point estimate for the proportion of vehicles travelling on the M1 during this time period that are 18-wheelers.

    b. Construct a 99% confidence interval for the proportion of vehicles on the M1 during this time period that are 18-wheelers.

8.32 What proportion of commercial airline pilots are more than 40 years of age? Suppose a researcher has access to a list of all pilots who are members of the International Federation of Air Line Pilots' Associations. If this list is used as a frame for the study, she can randomly select a sample of pilots, contact them, and ascertain their ages. From 89 of these pilots so selected, she learns that 48 are more than 40 years of age. Construct an 85% confidence interval to estimate the population proportion of commercial airline pilots who are more than 40 years of age.

8.33 According to Runzheimer International, in a survey of relocation administrators 63% of all workers who rejected relocation offers did so for family considerations. Suppose this figure was obtained by using a random sample of the files of 672 workers who had rejected relocation offers. Use this information to construct a 95% confidence interval to estimate the population proportion of workers who reject relocation offers for family considerations.

8.34 Suppose a survey of 275 executives is taken in an effort to determine what qualities are most important for an effective CEO to possess. The survey participants are offered several qualities as options, one of which is 'communication'. Of the surveyed respondents, 121 select 'communicator' as the most important quality for an effective CEO. Use these data to construct a 98% confidence interval to estimate the population proportion of executives who believe that 'communicator' is the most important quality of an effective CEO.

8.4   ESTIMATING THE POPULATION VARIANCE

At times in statistical analysis, the researcher is more interested in the population variance than in the population mean or population proportion. For example, in the total quality movement, suppliers who want to earn world-class supplier status or even those who want to maintain customer contracts are often asked to show continual reduction of variation on supplied parts. Tests are conducted with samples in efforts to determine lot variation and to determine whether variability goals are being met.

Estimating the variance is important in many other instances in business. For example, variations between airplane altimeter readings need to be minimal. It is not enough just to know that, on average, a particular brand of altimeter produces the correct altitude. It is also important that the variation between instruments be small. Thus measuring the variation of altimeters is critical. Parts being used in engines must fit tightly on a consistent basis. A wide variability among parts can result in a part that is too large to fit into its slots or so small that it results in too much tolerance, which causes vibrations. How can variance be estimated?

You may recall from Chapter 3 that sample variance is computed by using the formula

$$s^2 = \frac{\Sigma(x - \bar{x})^2}{n - 1}$$

Because sample variances are typically used as estimators or estimations of the population variance, as they are here, a mathematical adjustment is made in the denominator by using $n - 1$ to make the sample variance an unbiased estimator of the population variance.

Suppose a researcher wants to estimate the population variance from the sample variance in a manner that is similar to the estimation of the population mean from a sample mean. The relationship of the sample variance to the population variance is captured by the **chi-square distribution** (χ^2). The ratio of the sample variance (s^2) multiplied by $n - 1$ to the population variance (σ^2) is approximately chi-square distributed, as shown in formula 8.5, if the population from which the values are drawn is normally distributed.

Caution: *Use of the chi-square statistic to estimate the population variance is extremely sensitive to violations of the assumption that the population is normally distributed. For that reason, some researchers do not include this technique among their statistical repertoire. Although the technique is still rather widely presented as a mechanism for constructing confidence intervals to estimate a population variance, you should proceed with extreme caution and apply the technique only in cases where the population is known to be normally distributed. We can say that this technique lacks robustness.*

Like the t distribution, the chi-square distribution varies by sample size and contains a degrees-of-freedom value. The number of degrees of freedom for the chi-square formula (8.5) is $n - 1$.

χ^2 FORMULA FOR SINGLE VARIANCE (8.5)	$\chi^2 = \dfrac{(n-1)s^2}{\sigma^2}$ $df = n - 1$

The chi-square distribution is not symmetrical, and its shape will vary according to the degrees of freedom. Figure 8.11 shows the shape of chi-square distributions for three different degrees of freedom.

Formula 8.5 can be rearranged algebraically to produce a formula that can be used to construct confidence intervals for population variances. This new formula is shown as formula 8.6.

CONFIDENCE INTERVAL TO ESTIMATE THE POPULATION VARIANCE (8.6)	$\dfrac{(n-1)s^2}{\chi^2_{\alpha/2}} \le \sigma^2 \le \dfrac{(n-1)s^2}{\chi^2_{1-\alpha/2}}$ $df = n - 1$

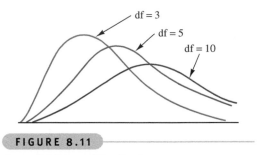

FIGURE 8.11

Three Chi-Square Distributions

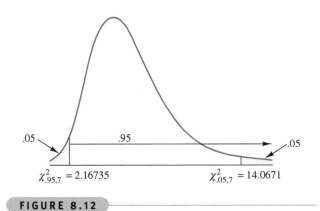

FIGURE 8.12

Two Table Values of Chi-Square

The value of alpha (α) is equal to 1 – (level of confidence expressed as a proportion). Thus, if we are constructing a 90% confidence interval, alpha is 10% of the area and is expressed in proportion form: $\alpha = 0.10$.

How can this formula be used to estimate the population variance from a sample variance? Suppose eight purportedly 7 cm aluminium cylinders in a sample are measured in diameter, resulting in the following values:

6.91 cm	6.93 cm	7.01 cm	7.02 cm
7.05 cm	7.00 cm	6.98 cm	7.01 cm

In estimating a population variance from these values, the sample variance must be computed. This value is $s^2 = 0.0022125$. If a point estimate is all that is required, the point estimate is the sample variance, 0.0022125. However, realizing that the point estimate will probably change from sample to sample, we want to construct an interval estimate. To do this, we must know the degrees of freedom and the table values of the chi-squares. Because $n = 8$, the degrees of freedom are df $= n - 1 = 7$. What are the chi-square values necessary to complete the information needed in formula 8.6? Assume the population of cylinder diameters is normally distributed.

Suppose we are constructing a 90% confidence interval. The value of α is $1 - 0.90 = 0.10$. It is the portion of the area under the chi-square curve that is outside the confidence interval. This outside area is needed because the chi-square table values given in Table A.8 are listed according to the area in the right tail of the distribution. In a 90% confidence interval, $\alpha/2$ or 0.05 of the area is in the right tail of the distribution and 0.05 is in the left tail of the distribution. The chi-square value for the 0.05 area on the right tail of the distribution can be obtained directly from the table by using the degrees of freedom, which in this case are 7. Thus the right-side chi-square, $\chi^2_{0.05,7}$, is 14.0671. Because Table A.8 lists chi-square values for areas in the right tail, the chi-square value for the left tail must be obtained by determining how much area lies to the right of the left tail. If 0.05 is to the left of the confidence interval, then $1 - 0.05 = 0.95$ of the area is to the right of the left tail. This calculation is consistent with the $1 - \alpha/2$ expression used in formula (8.6). Thus the chi-square for the left tail is $\chi^2_{0.95,7} = 2.16735$. Figure 8.12 shows the two table values of χ^2 on a chi-square distribution.

Incorporating these values into the formula, we can construct the 90% confidence interval to estimate the population variance of the 7-centimetre aluminium cylinders.

$$\frac{(n-1)s^2}{\chi^2_{\alpha/2}} \leq \sigma^2 \leq \frac{(n-1)s^2}{\chi^2_{1-\alpha/2}}$$

$$\frac{(7)(0.0022125)}{14.0671} \leq \sigma^2 \leq \frac{(7)(0.0022125)}{2.16735}$$

$$0.001101 \leq \sigma^2 \leq 0.007146$$

The confidence interval says that with 90% confidence, the population variance is somewhere between 0.001101 and 0.007146.

DEMONSTRATION PROBLEM 8.6

The US Bureau of Labor Statistics publishes data on the hourly compensation costs for production workers in manufacturing for various countries. The latest figures published for Greece show that the average hourly wage for a production worker in manufacturing is $19.23. Suppose the business council of Greece wants to know how consistent this figure is. They randomly select 25 production workers in manufacturing from across the country and determine that the standard deviation of hourly wages for such workers is $1.12. Use this information to develop a 95% confidence interval to estimate the population variance for the hourly wages of production workers in manufacturing in Greece. Assume that the hourly wages for production workers across the country in manufacturing are normally distributed.

Solution

By squaring the standard deviation, $s = 1.12$, we can obtain the sample variance, $s^2 = 1.2544$. This figure provides the point estimate of the population variance. Because the sample size, n, is 25, the degrees of freedom, $n - 1$, are 24. A 95% confidence means that alpha is $1 - 0.95 = 0.05$. This value is split to determine the area in each tail of the chi-square distribution: $\alpha/2 = 0.025$. The values of the chi-squares obtained from Table A.8 are

$$\chi^2_{0.025,24} = 39.3641 \text{ and } \chi^2_{0.975,24} = 12.40115$$

From this information, the confidence interval can be determined:

$$\frac{(n-1)s^2}{\chi^2_{\alpha/2}} \le \sigma^2 \le \frac{(n-1)s^2}{\chi^2_{1-\alpha/2}}$$

$$\frac{(24)(1.2544)}{39.3641} \le \sigma^2 \le \frac{(24)(1.2544)}{12.40115}$$

$$0.7648 \le \sigma^2 \le 2.4276$$

The business council can estimate with 95% confidence that the population variance of the hourly wages of production workers in manufacturing in Greece is between 0.7648 and 2.4276.

8.4 PROBLEMS

8.35 For each of the sample results below, construct the requested confidence interval. Assume the data come from normally distributed populations.
  a. $n = 12$, $\bar{x} = 28.4$, $s^2 = 44.9$; 99% confidence for σ^2
  b. $n = 7$, $\bar{x} = 4.37$, $s = 1.24$; 95% confidence for σ^2
  c. $n = 20$, $\bar{x} = 105$, $s = 32$; 90% confidence for σ^2
  d. $n = 17$, $s^2 = 18.56$; 80% confidence for σ^2

8.36 Use the sample data below to estimate the population variance. Produce a point estimate and a 98% confidence interval. Assume the data come from a normally distributed population.

| 27 | 40 | 32 | 41 | 45 | 29 | 33 | 39 |
| 30 | 28 | 36 | 32 | 42 | 40 | 38 | 46 |

8.37 The EU Labour Force Survey showed that UK employees spend an average of 42.1 hours per week in a working environment. Suppose this figure was obtained from a random sample of 20 workers and that the standard deviation of the sample was 4.3 hours. Assume hours worked per week are normally distributed in the population. Use this sample information to develop a 98% confidence interval for the population variance of the number of hours worked per week for a worker. What is the point estimate?

8.38 A manufacturing plant produces steel rods. During one production run of 20,000 such rods, the specifications called for rods that were 46 centimetres in length and 3.8 centimetres in width. Fifteen of these rods comprising a random sample were measured for length; the resulting measurements are shown here. Use these data to estimate the population variance of length for the rods. Assume rod length is normally distributed in the population. Construct a 99% confidence interval. Discuss the ramifications of the results.

44 cm	47 cm	43 cm	46 cm	46 cm
45 cm	43 cm	44 cm	47 cm	46 cm
48 cm	48 cm	43 cm	44 cm	45 cm

8.39 Suppose a random sample of 14 people 30–39 years of age produced the household incomes shown here. Use these data to determine a point estimate for the population variance of household incomes for people 30–39 years of age and construct a 95% confidence interval. Assume household income is normally distributed.

€37,500	44,800
33,500	36,900
42,300	32,400
28,000	41,200
46,600	38,500
40,200	32,000
35,500	36,800

8.5   ESTIMATING SAMPLE SIZE

In most business research that uses sample statistics to infer about the population, being able to *estimate the size of sample necessary to accomplish the purposes of the study* is important. The need for this **sample-size estimation** is the same for the large corporation investing tens of thousands of Euros in a massive study of consumer preference and for students undertaking a small case study and wanting to send questionnaires to local business people. In either case, such things as level of confidence, sampling error, and width of estimation interval are closely tied to sample size. If the large corporation is undertaking a market study, should it sample 40 people or 4,000 people? The question is an important one. In most cases, because of cost considerations, business researchers do not want to sample any more units or individuals than necessary.

Sample Size when Estimating μ

In research studies when μ is being estimated, the size of sample can be determined by using the z formula for sample means to solve for n. Consider,

$$z = \frac{\bar{x} - \mu}{\frac{\sigma}{\sqrt{n}}}$$

The difference between $\bar{x}$ and μ is the **error of estimation** resulting from the sampling process. Let $E = (\bar{x} - \mu) =$ the error of estimation. Substituting E into the preceding formula yields

$$z = \frac{E}{\frac{\sigma}{\sqrt{n}}}$$

Solving for n yields a formula that can be used to determine sample size.

SAMPLE SIZE WHEN ESTIMATING μ (8.7)
$$n = \frac{z_{\alpha/2}^2 \sigma^2}{E^2} = \left(\frac{z_{\alpha/2}\sigma}{E}\right)^2$$

Sometimes in estimating sample size the population variance is known or can be determined from past studies. Other times, the population variance is unknown and must be estimated to determine the sample size. In such cases, it is acceptable to use the following estimate to represent σ.

$$\sigma \approx \frac{1}{4}(range)$$

Using formula (8.7), the business researcher can estimate the sample size needed to achieve the goals of the study before gathering data. For example, suppose a researcher wants to estimate the average monthly expenditure on bread by a family in Brussels. She wants to be 90% confident of her results. How much error is she willing to tolerate in the results? Suppose she wants the estimate to be within €1.00 of the actual figure and the standard deviation of average monthly bread purchases is €4.00. What is the sample size estimation for this problem? The value of z for a 90% level of confidence is 1.645. Using formula (8.7) with $E = €1.00$, $\sigma = €4.00$, and $z = 1.645$ gives

$$n = \frac{z_{\alpha/2}^2 \sigma^2}{E^2} = \frac{(1.645)^2(4)^2}{1^2} = 43.30$$

That is, at least $n = 43.3$ must be sampled randomly to attain a 90% level of confidence and produce an error within €1.00 for a standard deviation of €4.00. Sampling 43.3 units is impossible, so this result should be rounded up to $n = 44$ units.

In this approach to estimating sample size, we view the error of the estimation as the amount of difference between the statistic (in this case, $\bar{x}$) and the parameter (in this case, μ). The error could be in either direction; that is, the statistic could be over or under the parameter. Thus, the error, E, is actually $\pm E$ as we view it. So when a problem states that the researcher wants to be within €1.00 of the actual monthly family expenditure for bread, it means that the researcher is willing to allow a tolerance within $\pm€1.00$ of the actual figure. Another name for this error is the **bounds** of the interval.

DEMONSTRATION PROBLEM 8.7

Suppose you want to estimate the average age of all Boeing 737-300 airplanes now in active European service. You want to be 95% confident, and you want your estimate to be within one year of the actual figure. The 737-300 was first placed in service about 24 years ago, but you believe that no active 737-300s in the European fleet are more than 20 years old. How large of a sample should you take?

Solution

Here, $E = 1$ year, the z value for 95% is 1.96, and σ is unknown, so it must be estimated by using $\sigma \approx (1/4) \cdot$ (range). As the range of ages is 0 to 20 years, $\sigma = (1/4)(20) = 5$. Use formula (8.7).

$$n = \frac{z^2 \sigma^2}{E^2} = \frac{(1.96)^2(5)^2}{1^2} = 96.04$$

Because you cannot sample 96.04 airplanes, the required sample size is 97. If you randomly sample 97 airplanes, you have an opportunity to estimate the average age of active 737-300s within one year and be 95% confident of the results.

Note: *Sample-size estimates for the population mean where σ is unknown using the t distribution are not shown here. Because a sample size must be known to determine the table value of t, which in turn is used to estimate the sample size, this procedure usually involves an iterative process.*

Determining Sample Size when Estimating p

Determining the sample size required to estimate the population proportion, p, also is possible. The process begins with the z formula for sample proportions:

$$z = \frac{\hat{p} - p}{\sqrt{\dfrac{p \cdot q}{n}}}$$

where $q = 1 - p$.

As various samples are taken from the population, $\hat{p}$ will rarely equal the population proportion, p, resulting in an error of estimation. The difference between $\hat{p}$ and p is the error of estimation, so $E = \hat{p} - p$.

$$z = \frac{E}{\sqrt{\dfrac{p \cdot q}{n}}}$$

Solving for n yields the formula for determining sample size.

SAMPLE SIZE WHEN ESTIMATING P (8.8)

$$n = \frac{z^2 pq}{E^2}$$

where:
p = population proportion
$q = 1 - p$
E = error of estimation
n = sample size

TABLE 8.3

$p \cdot q$ for Various Selected Values of p

p	$p \cdot q$
.9	.09
.8	.16
.7	.21
.6	.24
.5	.25
.4	.24
.3	.21
.2	.16
.1	.09

How can the value of n be determined prior to a study if the formula requires the value of p and the study is being done to estimate p? Although the actual value of p is not known prior to the study, similar studies might have generated a good approximation for p. If no previous value is available for use in estimating p, some possible p values, as shown in Table 8.3, might be considered.

Note that, as $p \cdot q$ is in the numerator of the sample size formula, $p = 0.5$ will result in the largest sample sizes. Often *if p is unknown, researchers use 0.5 as an estimate of p* in formula 8.8. This selection results in the largest sample size that could be determined from formula 8.8 for a given z value and a given error value.

DEMONSTRATION PROBLEM 8.8

Hewitt Associates conducted a survey to determine the extent to which employers are promoting health and fitness among their employees. One of the questions asked was: 'Does your company offer on-site exercise classes?' Suppose it was estimated before the study that no more than 40% of the companies would answer Yes. How large a sample would Hewitt Associates have to take in estimating the population proportion to ensure a 98% confidence in the results and to be within 0.03 of the true population proportion?

Solution

The value of E for this problem is 0.03. Because it is estimated that no more than 40% of the companies would say Yes, $p = 0.40$ can be used. A 98% confidence interval results in a z value of 2.33. Inserting these values into formula (8.8) yields

$$n = \frac{(2.33)^2(0.40)(0.60)}{(0.03)^2} = 1447.7$$

Hewitt Associates would have to sample 1,448 companies to be 98% confident in the results and maintain an error of 0.03.

8.5   PROBLEMS

8.40 Determine the sample size necessary to estimate μ for the following information:

    a. $\sigma = 36$ and $E = 5$ at 95% confidence.

    b. $\sigma = 4.13$ and $E = 1$ at 99% confidence.

    c. Values range from 80 to 500, error is to be within 10, and the confidence level is 90%.

    d. Values range from 50 to 108, error is to be within 3, and the confidence level is 88%.

8.41 Determine the sample size necessary to estimate p for the following information.

    a. $E = 0.02$, p is approximately 0.40, and the confidence level is 96%.

    b. E is to be within 0.04, p is unknown, and the confidence level is 95%.

    c. E is to be within 5%, p is approximately 55%, and the confidence level is 90%.

    d. E is to be no more than 0.01, p is unknown, and the confidence level is 99%.

8.42 A bank officer wants to determine the amount of the average total monthly deposits per customer at the bank. He believes an estimate of this average amount using a confidence interval is sufficient. How large a sample should he take to be within £200 of the actual average with 99% confidence? He assumes the standard deviation of total monthly deposits for all customers is about £1,000.

8.43 Suppose you have been following a particular airline stock for many years. You are interested in determining the average daily price of this stock in a 10-year period and you have access to the stock reports for these years. However, you do not want to average all the daily prices over 10 years because there are several thousand data points, so you decide to take a random sample of the daily prices and estimate the average. You want to be 90% confident of your results, you want the estimate to be within €2.00 of the true average, and you believe the standard deviation of the price of this stock is about €12.50 over this period of time. How large a sample should you take?

8.44 A group of investors wants to develop a chain of fast-food restaurants. In determining potential costs for each facility, they must consider, among other expenses, the average monthly electric bill. They decide to sample some fast-food restaurants currently operating to estimate the monthly cost of electricity. They want to be 90% confident of their results and want the error of the interval estimate to be no more than $100. They estimate that such bills range from $600 to $2,500. How large a sample should they take?

8.45 Suppose a production facility purchases a particular component part in large lots from a supplier. The production manager wants to estimate the proportion of defective parts received from this supplier. She believes the proportion defective is no more than 0.20 and wants to be within 0.02 of the true proportion of defective parts with a 90% level of confidence. How large a sample should she take?

8.46 What proportion of secretaries of *FT Global* 500 companies has a personal computer at his or her workstation? You want to answer this question by conducting a random survey. How large a sample should you take if you want to be 95% confident of the results and you want the error of the confidence interval to be no more than 0.05? Assume no one has any idea of what the proportion actually is.

8.47 What proportion of shoppers at a large appliance store actually makes a big-ticket purchase? To estimate this proportion within 10% and be 95% confident of the results, how large a sample should you take? Assume you have no idea what proportion of all shoppers actually make a large-ticket purchase.

Decision Dilemma SOLVED

Compensation for Marketing Professionals

Published national management salary and demographic information, such as mean salary, mean age, and mean years experience, is most likely based on random samples of data. Such is the case in the Decision Dilemma where the salary, age, and experience parameters are actually point estimates based on a survey of 1,724 marketing managers. For example, the study states that the average salary for a marketing manager is £42,000. This is a point estimate of the population mean salary based on the sample mean of 1,724 marketing professionals. Suppose it is known that the population standard deviation for marketing manager salaries is £16,000. From this information, a 95% confidence interval using the *z* statistic can be constructed as follows:

$$£42,000 \pm 1.96 \frac{£16,000}{\sqrt{1,724}} = £42,000 \pm £755.28$$

$$£41,244.72 \leq \mu \leq £42,755,28$$

This confidence interval, constructed to estimate the mean annual salary for marketing managers across the United Kingdom, shows that the estimated mean salary is £42,000 with an error of £755.28. The reason that this error is quite small is that the sample size is very large. In fact, the sample size is so large that if the £16,000 had actually been a sample standard deviation, the table *t* value would have been 1.961342 (compared to the *z* = 1.96), resulting in a confidence interval error of

estimate of £755.80. Note that because of the large sample size, this *t* value is not found in Table A.6 in Appendix A, and it was obtained using Excel's TINV function within the Paste function. Of course, in using the *t* statistic, one would have to assume that salaries are normally distributed in the population. Because the sample size is so large and the corresponding *t* table value is so close to the *z* value, the error using the *t* statistic is only £0.52 more than that produced using the *z* statistic. Confidence intervals for the population mean age can be computed similarly.

Had the study reported that the mean annual salary of a marketing manager in the south-west of the UK to be £42,400 based on a sample of only 25 respondents, the confidence interval would be much wider. Suppose annual salaries of marketing managers in the south-west are normally distributed and that the sample standard deviation for such managers is also £16,000. A 95% confidence interval for estimating the population mean annual salary for South-west marketing managers can be computed using the *t* statistic as follows:

$$£42,400 \pm 2.064 \frac{£16,000}{\sqrt{25}} = £42,000 \pm £6,604.80$$

$$£35,395.20 \leq \mu \leq £48,604.80$$

Note that the point estimate mean annual salary for south-west marketing managers is £42,400 as reported in the study. However, the error of estimation in the interval is £6,604.80, indicating that the actual population mean annual salary could be as low as £35,395.20 or as high as £48,604.80. Observe that the error of this interval, £6,604.80, is nearly 10 times as big as the error in the confidence interval used to estimate the UK figure. This is due to the fact that the sample size used in the UK estimate is about 70 times as large. Since sample size is under the radical sign in confidence interval computation, taking the square root of this (70) indicates that the error in the south-west estimate is almost nine times as large as it is in the UK estimate (with a slight adjustment for the fact that a t value is used in the south-west estimate).

The study reported that 45% of the respondents expected to change jobs in the coming year. Using methods presented in Section 8.3, a 99% confidence interval can be computed assuming that the sample size is 1,724, the table z value for a 99% confidence interval is 2.575, and is 0.45. The resulting confidence interval is:

$$0.45 \pm 2.575 \sqrt{\frac{(0.45)(0.55)}{1,724}} = 0.45 \pm 0.031$$

$$0.419 \leq p \leq 0.481$$

While the point estimate is 0.45 or 45%, the error of the estimate is 0.031 or 3.1%, and therefore we are 99% confident that the actual population proportion of marketing managers who expected to change jobs in the coming year is between 0.419 and 0.481.

ETHICAL CONSIDERATIONS

Using sample statistics to estimate population parameters poses a couple of ethical concerns. Many survey reports and advertisers use point estimates as the values of the population parameter. Often, no error value is stated, as would have been the case if a confidence interval had been computed. These point estimates are subject to change if another sample is taken. It is probably unethical to state as a conclusion that a point estimate is the population parameter without some sort of disclaimer or explanation about what a point estimate is.

The misapplication of t formulas when data are not normally distributed in the population is also of concern. Although some studies have shown that the t formula analyses are robust, a researcher should be careful not to violate the assumptions underlying the use of the t formulas. An even greater potential for misuse lies in using the chi-square for the estimation of a population variance because this technique is highly sensitive to violations of the assumption that the data are normally distributed.

SUMMARY

Techniques for estimating population parameters from sample statistics are important tools for business research. These tools include techniques for estimating population means, techniques for estimating the population proportion and the population variance, and methodology for determining how large a sample to take.

At times in business research a product is new or untested or information about the population is unknown. In such cases, gathering data from a sample and making estimates about the population is useful and can be done with a point estimate or an interval estimate. A point estimate is the use of a statistic from the sample as an estimate for a parameter of the population. Because point estimates vary with each sample, it is usually best to construct an interval estimate. An interval estimate is a range of values computed from the sample within which the researcher believes with some confidence that the population parameter lies. Certain levels of confidence seem to be used more than others: 90%, 95%, 98%, and 99%.

If the population standard deviation is known, the z statistic is used to estimate the population mean. If the population standard deviation is unknown, the t distribution should be used instead of the z distribution. It is assumed when using the t distribution that the population from which the samples are drawn is normally distributed. However, the technique for estimating a population mean by using the t test is robust, which means it is relatively insensitive to minor violations to the assumption. The population variance can be estimated by using sample variance and the chi-square distribution. The chi-square technique for estimating the population variance is not robust; it is sensitive to violations of the assumption that the population is normally distributed. Therefore, extreme caution must be exercised in using this technique.

The formulas in Chapter 7 resulting from the central limit theorem can be manipulated to produce formulas for estimating sample size for large samples. Determining the sample size necessary to estimate a population mean, if the population standard deviation is unavailable, can be based on one-fourth the range as an approximation of the population standard deviation. Determining sample size when estimating a population proportion requires the value of the population proportion. If the population proportion is unknown, the population proportion from a similar study can be used. If none is available, using a value of 0.50 will result in the largest sample size estimation for the problem if other variables are held constant. Sample size determination is used mostly to provide a ballpark figure to give researchers some guidance. Larger sample sizes usually result in greater costs.

KEY TERMS

bounds
chi-square distribution
degrees of freedom (df)
error of estimation
interval estimate

point estimate
Robust
sample-size estimation
t distribution
t value

GO ONLINE TO DISCOVER THE EXTRA FEATURES FOR THIS CHAPTER

The Student Study Guide containing solutions to the odd-numbered questions, additional Quizzes and Concept Review Activities, Excel and Minitab databases, additional data files in Excel and Minitab, and more worked examples.
www.wiley.com/college/cortinhas

FORMULAS

$100(1 - \alpha)\%$ confidence interval to estimate μ: population standard deviation known

$$\bar{x} - z_{\alpha/2}\frac{\sigma}{\sqrt{n}} \leq \mu \leq \bar{x} + z_{\alpha/2}\frac{\sigma}{\sqrt{n}}$$

Confidence interval to estimate μ using the finite correction factor

$$\bar{x} - z_{\alpha/2}\frac{\sigma}{\sqrt{n}}\sqrt{\frac{N-n}{N-1}} \leq \mu \leq \bar{x} + z_{\alpha/2}\frac{\sigma}{\sqrt{n}}\sqrt{\frac{N-n}{N-1}}$$

Confidence interval to estimate μ: population standard deviation unknown

$$\bar{x} - t_{\alpha/2,n-1}\frac{s}{\sqrt{n}} \leq \mu \leq \bar{x} + t_{\alpha/2,n-1}\frac{s}{\sqrt{n}}$$

Confidence interval to estimate p

$$\hat{p} - z_{\alpha/2}\sqrt{\frac{\hat{p}\cdot\hat{q}}{n}} \leq p \leq \hat{p} + z_{\alpha/2}\sqrt{\frac{\hat{p}\cdot\hat{q}}{n}}$$

χ^2 formula for single variance

$$\chi^2 = \frac{(n-1)s^2}{\sigma^2}$$

$$\text{df} = n - 1$$

Confidence interval to estimate the population variance

$$\frac{(n-1)s^2}{\chi^2_{\alpha/2}} \le \sigma^2 \le \frac{(n-1)s^2}{\chi^2_{1-\alpha/2}}$$

$$\text{df} = n - 1$$

Sample size when estimating μ

$$n = \frac{z^2_{\alpha/2}\sigma^2}{E^2} = \left(\frac{z_{\alpha/2}\sigma}{E}\right)^2$$

Sample size when estimating p

$$n = \frac{z^2 pq}{E^2}$$

SUPPLEMENTARY PROBLEMS

CALCULATING THE STATISTICS

8.48 Use the data below to construct 80%, 94%, and 98% confidence intervals to estimate μ. Assume that σ is 7.75. State the point estimate.

44	37	49	30	56	48	53	42	51
38	39	45	47	52	59	50	46	34
39	46	27	35	52	51	46	45	58
51	37	45	52	51	54	39	48	

8.49 Construct 90%, 95%, and 99% confidence intervals to estimate μ from the data below. State the point estimate. Assume the data come from a normally distributed population.

12.3 11.6 11.9 12.8 12.5 11.4 12.0 11.7 11.8 12.3

8.50 Use the following information to compute the confidence interval for the population proportion:
a. $n = 715$ and $x = 329$, with 95% confidence
b. $n = 284$ and $\hat{p} = 0.71$, with 90% confidence
c. $n = 1,250$ and $\hat{p} = 0.48$, with 95% confidence
d. $n = 457$ and $x = 270$, with 98% confidence

8.51 Use the data below to construct 90% and 95% confidence intervals to estimate the population variance. Assume the data come from a normally distributed population.

212 229 217 216 223 219 208 214 232 219

8.52 Determine the sample size necessary under the following conditions:
a. To estimate μ with $\sigma = 44$, $E = 3$, and 95% confidence.
b. To estimate μ with a range of values from 20 to 88 with $E = 2$, and 90% confidence.

c. To estimate p with p unknown, $E = 0.04$, and 98% confidence.
d. To estimate p with $E = 0.03$, 95% confidence, and p thought to be approximately 0.70.

TESTING YOUR UNDERSTANDING

8.53 In planning both market opportunity and production levels, being able to estimate the size of a market can be important. Suppose a nappy manufacturer wants to know how many nappies a one-month-old baby uses during a 24-hour period. To determine this usage, the manufacturer's analyst randomly selects 17 parents of one-month-olds and asks them to keep track of nappy usage for 24 hours. The results are shown. Construct a 99% confidence interval to estimate the average daily nappy usage of a one-month-old baby. Assume nappy usage is normally distributed.

12	8	11	9	13	14	10
10	9	13	11	8	11	15
10	7	12				

8.54 Suppose you want to estimate the proportion of cars that are sport utility vehicles (SUVs) being driven in Shanghai, China, at rush hour by standing at the end of the Shanghai-Nanjin expressway and counting SUVs. You believe the figure is no higher than 0.40. If you want the error of the confidence interval to be no greater than 0.03, how many cars should you randomly sample? Use a 90% level of confidence.

8.55 Use the data in Problem 8.53 to construct a 99% confidence interval to estimate the population variance for the number of nappies used during a 24-hour period for one-month-olds. How could information about the population variance be used by a manufacturer or marketer in planning?

8.56 What is the average length of a company's health and safety policy statement? Suppose health and safety policy statements are sampled from 45 medium-sized companies. The average number of pages in the sample policy statements is 213, and the population standard deviation is 48. Use this information to construct a 98% confidence interval to estimate the mean number of pages for the population of medium-sized company health and safety policy statement.

8.57 A random sample of small-business managers was given a leadership-style questionnaire. The results were scaled so that each manager received a score for initiative. Suppose the following data are a random sample of these scores:

37	42	40	39	38	31	40
37	35	45	30	33	35	44
36	37	39	33	39	40	41
33	35	36	41	33	37	38
40	42	44	35	36	33	38
32	30	37	42			

Assuming σ is 3.891, use these data to construct a 90% confidence interval to estimate the average score on initiative for all small-business managers.

8.58 A European beauty salon chain wants to estimate the number of times per year a woman has her hair done at a beauty salon if she uses one at least once a year. The chain's researcher estimates that, of those women who use a beauty salon at least once a year, the standard deviation of number of times of usage is approximately 6. The beauty salon chain wants the estimate to be within one time of the actual mean value. How large a sample should the researcher take to obtain a 98% confidence level?

8.59 Is the environment a major issue with South-East Asians? To answer that question, a researcher conducts a survey of 1,255 randomly selected South-East Asians. Suppose 714 of the sampled people replied that the environment is a major issue with them. Construct a 95% confidence interval to estimate the proportion of South-East Asians who feel that the environment is a major issue with them. What is the point estimate of this proportion?

8.60 According to a survey by Topaz Enterprises, a travel auditing company, the average error by travel agents is $128. Suppose this figure was obtained from a random sample of 41 travel agents and the sample standard deviation is $21. What is the point estimate of the national average error for all travel

agents? Compute a 98% confidence interval for the national average error based on these sample results. Assume the travel agent errors are normally distributed in the population. How wide is the interval? Interpret the interval.

8.61 A Europe-wide survey on telemarketing was undertaken. One of the questions asked was: 'How long has your organization had a telemarketing operation?' Suppose the data below represent some of the answers received to this question. Suppose further that only 300 telemarketing firms comprised the population when this survey was taken. Use the data below to compute a 98% confidence interval to estimate the average number of years a telemarketing organization has had a telemarketing operation. The population standard deviation is 3.06.

5	5	6	3	6	7	5
5	6	8	4	9	6	4
10	5	10	11	5	14	7
5	9	6	7	3	4	3
7	5	9	3	6	8	16
12	11	5	4	3	6	5
8	3	5	9	7	13	4
6	5	8	3	5	8	7
11	5	14	4			

8.62 An entrepreneur wants to open an appliance service repair shop. She would like to know the average home repair bill, including the charge for the service call for appliance repair in the area. She wants the estimate to be within £20 of the actual figure. She believes the range of such bills is between £30 and £600. How large a sample should the entrepreneur take if she wants to be 95% confident of the results?

8.63 A survey of insurance offices was taken, resulting in a random sample of 245 companies. Of these 245 companies, 189 responded that they were going to purchase new software for their offices in the next year. Construct a 90% confidence interval to estimate the population proportion of insurance offices that intend to purchase new software during the next year.

8.64 A survey of companies included a question that asked whether the company had at least one bilingual telephone operator. The sample results of 90 companies follow (Y denotes that the company does have at least one bilingual operator; N denotes that it does not).

N	N	N	N	Y	N	Y	N	N
Y	N	N	N	Y	Y	N	N	N
N	N	Y	N	Y	N	Y	N	Y
Y	Y	N	Y	N	N	N	Y	N
N	Y	N	N	N	N	N	N	N
Y	N	Y	Y	N	N	Y	N	Y
N	N	Y	Y	N	N	N	N	N
Y	N	N	N	N	Y	N	N	N
Y	Y	Y	N	N	Y	N	N	N
N	N	N	Y	Y	N	N	Y	N

Use this information to estimate with 95% confidence the proportion of the population that does have at least one bilingual operator.

8.65 A cinema has had a poor accounting system. The manager has no idea how many large containers of popcorn are sold per movie showing. She knows that the amounts vary by day of the week and hour of the day. However, she wants to estimate the overall average per movie showing. To do so, she randomly selects 12 movie performances and counts the number of large containers of popcorn sold between 30 minutes before the movie showing and 15 minutes after the movie showing. The sample average was 43.7 containers, with a variance of 228. Construct a 95% confidence interval to estimate the mean number of large containers of popcorn sold during a movie showing. Assume the number of large containers of popcorn sold per movie is normally distributed in the population. Use this information to construct a 98% confidence interval to estimate the population variance.

8.66 According to a survey by Horizons, the average price of a three-course adult meal in the United Kingdom is £18.94. Suppose this figure was based on a sample of 27 different establishments and the standard deviation was £1.45. Construct a 95% confidence interval for the population mean cost for all fast-food meals in the United Kingdom. Assume the costs of a fast-food meal in the UK are normally distributed. Using the interval as a guide, is it likely that the population mean is really £18.00? Why or why not?

8.67 A survey of 77 commercial airline flights of under two hours resulted in a sample average late time for a flight of 2.48 minutes. The population standard deviation was 12 minutes. Construct a 95% confidence interval for the average time that a commercial flight of under two hours is late. What is the point estimate? What does the interval tell about whether the average flight is late?

8.68 A regional survey of 560 companies asked the director of operations how satisfied he or she was with the software support received from the computer staff of the company. Suppose 33% of the 560 directors said they were satisfied. Construct a 99% confidence interval for the proportion of the population of directors who would have said they were satisfied with the software support if a census had been taken.

8.69 A research firm has been asked to determine the proportion of all restaurants in Prague that serve alcoholic beverages. The firm wants to be 98% confident of its results but has no idea of what the actual proportion is. The firm would like to report an error of no more than 0.05. How large a sample should it take?

8.70 A marketing magazine firm attempts to win subscribers with a mail campaign that involves a contest using magazine stickers. Often when people subscribe to magazines in this manner they sign up for multiple magazine subscriptions. Suppose the marketing firm wants to estimate the average number of subscriptions per customer of those who purchase at least one subscription. To do so, the marketing firm's researcher randomly selects 65 returned contest entries. Twenty-seven contain subscription requests. Of the 27, the average number of subscriptions is 2.10, with a standard deviation of 0.86. The researcher uses this information to compute a 98% confidence interval to estimate μ and assumes that x is normally distributed. What does the researcher find?

8.71 The average price of a pint of lager now stands at £3.08, according to the London *Evening Standard*. The price of beer varies considerably across the UK, with the cheapest part of the country being the West Midlands and the most expensive being Surrey. Suppose a national survey of 23 pubs was taken and the price per pint of lager was ascertained. If the following data represent these prices, what is a 90% confidence interval for the population variance of these prices? Assume prices are normally distributed in the population.

2.98	3.02	3.05	2.99	3.1
2.97	2.95	3.08	3	2.94
2.85	2.99	3.06	3.03	2.99
3.02	2.88	3.01	3.04	3.13
3.02	2.99	2.99		

8.72 The price of a head of iceberg lettuce varies greatly with the season and the geographic location of a store. During February a researcher contacts a random sample of 39 supermarkets across Scandinavia and asks the produce manager of each to state the current price charged for a head of iceberg lettuce.

Using the researcher's results that follow, construct a 99% confidence interval to estimate the mean price of a head of iceberg lettuce in February in Scandinavia. Assume that σ is 0.205.

1.59	1.25	1.65	1.40	0.89
1.19	1.50	1.49	1.30	1.39
1.29	1.60	0.99	1.29	1.19
1.20	1.50	1.49	1.29	1.35
1.10	0.89	1.10	1.39	1.39
1.50	1.50	1.55	1.20	1.15
0.99	1.00	1.30	1.25	1.10
1.00	1.55	1.29	1.39	

INTERPRETING THE OUTPUT

8.73 A soft drink company produces cola in a 330-millilitre can. Even though their machines are set to fill the cans with 330-millilitre, variation due to calibration, operator error, and other things sometimes precludes the cans having the correct fill. To monitor the can fills, a quality team randomly selects some filled 330-millilitre cola cans and measures their fills in the lab. A confidence interval for the population mean is constructed from the data. Shown here is the Minitab output from this effort. Discuss the output.

One-Sample Z

The assumed standard deviation = 3.33

N	Mean	SE Mean	99% CI
58	330.414	0.437	(329.288, 331.540)

8.74 A company has developed a new light bulb that seems to burn longer than most residential bulbs.

To determine how long these bulbs burn, the company randomly selects a sample of these bulbs and burns them in the laboratory. The Excel output shown here is a portion of the analysis from this effort. Discuss the output.

Bulb Burn

Mean	2198.217
Standard deviation	152.9907
Count	84
Confidence level (90.0%)	27.76691

8.75 Suppose a researcher wants to estimate the average age of a person who is a first-time home buyer. A random sample of first-time home buyers is taken and their ages are ascertained. The Minitab output shown here is an analysis of that data. Study the output and explain its implication.

One Sample T

N	Mean	StDev	SE Mean	99% CI
21	27.63	6.54	1.43	(23.57, 31.69)

8.76 What proportion of all Japanese workers drive their cars to work? Suppose a poll of Japanese workers is taken in an effort to answer that question, and the Minitab output shown here is an analysis of the data from the poll. Explain the meaning of the output in light of the question.

Test and CI One Proportion

Sample	X	N	Sample p	95% CI
1	506	781	0.647887	(0.613240, 0.681413)

ANALYSING THE DATABASES

1. Construct a 95% confidence interval for the population mean number of production workers using the Manufacturing database as a sample. What is the point estimate? How much is the error of the estimation? Comment on the results.

2. Construct a 90% confidence interval to estimate the average census for hospitals using the Hospital database. State the point estimate and the error of the estimation. Change the level of confidence to 99%. What happened to the interval? Did the point estimate change?

3. The Financial database contains financial data on 100 companies. Use this database as a sample and estimate the earnings per share for all corporations from these data. Select several levels of confidence and compare the results.

4. Using the tally or frequency feature of the computer software, determine the sample proportion of the Hospital database under the variable 'service' that are 'general medical' (category 1). From this statistic, construct a 95% confidence interval to estimate the population proportion of hospitals that are 'general medical'. What is the point estimate? How much error is there in the interval?

ARTNET AG

artnet AG is a German-based online portal for researching, buying, and selling fine art and design. artnet Galleries has a large network, with over 2,200 galleries in over 250 cities worldwide and more than 166,000 artworks by over 39,000 artists from around the globe. artnet Galleries serves dealers and art buyers alike by providing a survey of the market and its pricing trends, as well as the means to communicate instantly, inexpensively, and globally. This service is free of charge for collectors while galleries pay a monthly fee. Other key services include *artnet Magazine*, the insider's guide to the art market, with daily news, reviews, and features by renowned writers in the art community and the Price Database. The artnet Price Database is a colour-illustrated online archive of fine art auction results worldwide. Representing auction results from over 500 international auction houses since 1985, the Price Database covers more than four million auction results by over 188,000 artists, ranging from Old Masters to Contemporary Art.

artnet.com AG was incorporated under the laws of Germany in 1998. In 1999, management took the company public on the Frankfurt Stock Exchange. Its principal holding is its wholly owned subsidiary, artnet Worldwide Corp., a New York corporation that was founded in 1989. In 2006, the group launched two new products, the Market Performance Reports and 20th Century Design database. A year later, artnet Worldwide Corp. established artnet UK Ltd, a sales agent for artnet Worldwide Corp., in the UK. In 2008, the group launched artnet Online Auctions, which offers modern and contemporary fine art, prints, and photographs by renowned artists. In France in the same year, artnet Worldwide Corp. established artnet France SARL, a sales agent for artnet Worldwide Corp. The group launched Price Database Fine Art and Design and the Price Database Decorative Art in 2009.

artnet's philosophy has been to make every effort to create transparency in the art market and create an overview of the artworks on offer. More recently it has also dedicated itself to simplifying, expediting, and reducing the cost of transactions. Back when the company was being established, artnet positioned itself in the space between art, the market, and the Internet. The company has been collecting detailed data on the art market for several years and processes this in such a way that it can easily be retrieved by users. This long-term view of the market has allowed the company to establish itself successfully in a market led by major players such as Southeby's, Christie's International, and eBay.

DISCUSSION

1. artnet AG has grown and flourished because of its good customer relationships, delivering a range of innovative products and listening to the customer's needs. Suppose company management wants to formally measure customer satisfaction on its Online Auctions at least once a year and develops a brief survey that includes the four questions below. Suppose 115 customers participated in this survey with the results shown. Use techniques presented in this chapter to analyse the data to estimate population responses to these questions.

Question	Yes	No
1. In general, were you satisfied with the payment and shipping policy?	63	52
2. Were the contact people at artnet AG helpful and courteous?	86	29
3. Was the commission paid to artnet justified?	101	14
4. Would you recommend artnet AG to other customers?	105	10

2. Now suppose artnet AG officers want to ascertain employee satisfaction with the company. They randomly sample nine employees and ask them to complete a satisfaction survey under the supervision of an independent testing organization. As part of this survey, employees are asked to respond to questions by providing a score from 0 to 50 along a continuous scale where 0 denotes no satisfaction and 50 denotes the upmost satisfaction. Assume that the data are normally distributed in the population.

The questions and the results of the survey are shown in the following table. Analyse the results by using techniques from this chapter.

Question	Mean	Standard Deviation		Question	Mean	Standard Deviation
1. Are you treated fairly as an employee?	37.9	8.6		4. Is your physical work environment acceptable?	33.4	8.1
2. Has the company given you the training you need to do the job adequately?	27.4	12.7		5. Is the compensation for your work adequate and fair?	39.5	2.1
3. Does management seriously consider your input in making decisions about production?	41.8	6.3				

(continued)

Source: Adapted from arnet AG, available at www.artnet.com/about/aboutindex.asp (Company Background and Company Overview) and http://en.wikipedia.org/wiki/artnet. Accessed 2012.

USING THE COMPUTER

EXCEL

- Excel has some capability to construct confidence intervals to estimate a population mean using the z statistic when σ is known and using the t statistic when σ is unknown.

- To construct confidence intervals of a single population mean using the z statistic (σ is known), begin with the **Insert Function** (f_x). To access the **Insert Function**, go to the **Formulas** tab on an Excel worksheet (top centre tab). The **Insert Function** is on the far left of the menu bar. In the **Insert Function** dialog box at the top, there is a pulldown menu where it says **Or select a category**. From the pulldown menu associated with this command, select **Statistical**. Select **CONFIDENCE** from the **Insert Function's Statistical** menu. In the **CONFIDENCE** dialog box, place the value of alpha (a number between 0 and 1), which equals 1 – level of confidence. (*Note:* level of confidence is given as a proportion and not as a per-cent.) For example, if the level of confidence is 95%, enter 0.05 as alpha. Insert the value of the population standard deviation in **Standard_dev**. Insert the size of the sample in **Size**. The output is the ± error of the confidence interval.

- To construct confidence intervals of a single population mean using the t statistic (σ is unknown), begin by selecting the **Data** tab on the Excel worksheet. From the **Analysis** panel at the right top of the **Data** tab worksheet, click on **Data Analysis**. If your Excel worksheet does not show the **Data Analysis** option, then you can load it as an add-in following directions given in Chapter 2. From the **Data Analysis** pulldown menu, select **Descriptive Statistics**. In the **Descriptive Statistics** dialog box, enter the location of the observations from the single sample in **Input Range**. Check **Labels** if you have a label for your data. Check **Summary Statistics**. Check **Confidence Level for Mean:** (required to get confidence interval output). If you want to change the level of confidence from the default value of 95%, enter it (in per cent, between 0 and 100) in the box with the % sign beside it. The output is a single number that is the ± error portion of the confidence interval and is shown at the bottom of the **Descriptive Statistics** output as **Confidence Level**.

MINITAB

- Minitab has the capability for constructing confidence intervals about a population mean either when σ is known or it is unknown and for constructing confidence intervals about a population proportion.

- To begin constructing confidence intervals of a single population mean using z statistic (σ known), select **Stat** on the menu bar. Select **Basic Statistics** from the pulldown menu. From the second pulldown menu, select **1-sample Z**. Check **Samples in columns:** if you have raw data and enter the location of the column containing the observations. Check **Summarized data** if you wish to use the summarized statistics of the sample mean and the sample size rather than raw data. Enter the size of the sample in the box beside **Sample size**. Enter the sample mean in the box beside **Mean**. Enter the value of the population standard deviation in the box beside **Standard deviation**. Click on **Options** if you want to enter a level of confidence. To insert a level of confidence, place the confidence level as a percentage (between 0 and 100) in the box beside **Confidence level**. *Note:* to construct a two-sided confidence interval (only type of confidence intervals presented in this text), the selection in the box beside **Alternative** must be **not equal**.

- To begin constructing confidence intervals of a single population mean using t statistic (σ unknown), select **Stat** on the menu bar. Select **Basic Statistics** from the pulldown menu. From the second pulldown menu, select **1-sample t**. Check **Samples in columns:** if you have raw data and enter the location of the column containing the observations. Check **Summarized data** if you wish to use the summarized statistics of the sample size, sample mean, and the sample standard deviation rather than raw data. Enter the size of the sample in the box beside **Sample size:**. Enter the sample mean in the box beside **Mean:**. Enter the sample standard deviation in the box beside **Standard deviation**. Click on **Options** if you want to enter a level of confidence. To insert a level of confidence, place the confidence level as a percentage (between 0

and 100) in the box beside **Confidence level**. *Note:* to construct a two-sided confidence interval (only type of confidence intervals presented in this text), the selection in the box beside **Alternative** must be **not equal**.

- To begin constructing confidence intervals of a single population proportion using a z statistic, select **Stat** on the menu bar. Select **Basic Statistics** from the pulldown menu. From the second pulldown menu, select **1 Proportion**. Check **Samples in columns** if you have raw data and enter the location of the column containing the observations. Note that the data in the column must contain one of only two values (e.g., 1 or 2). Check **Summarized data** if you wish to use the summarized statistics of the number of trials and number of events rather than raw data. Enter the size of the sample in the box beside **Number of trials**. Enter the number of observed events (having the characteristic that you are testing) in the box beside **Number of events**. Click on **Options** if you want to enter a level of confidence. To insert a level of confidence, place the confidence level as a percentage (between 0 and 100) in the box beside **Confidence level**. *Note:* to construct a two-sided confidence interval (only type of confidence intervals presented in this text), the selection in the box beside **Alternative** must be **not equal**.

Statistical Inference: Hypothesis Testing for Single Populations

LEARNING OBJECTIVES

The main objective of Chapter 9 is to help you to learn how to test hypotheses on single populations, thereby enabling you to:

1. Develop both one- and two-tailed null and alternative hypotheses that can be tested in a business setting by examining the rejection and non-rejection regions in light of Type I and Type II errors

2. Reach a statistical conclusion in hypothesis testing problems about a population mean with a known population standard deviation using the z statistic

3. Reach a statistical conclusion in hypothesis testing problems about a population mean with an unknown population standard deviation using the t statistic

4. Reach a statistical conclusion in hypothesis-testing problems about a population proportion using the z statistic

5. Reach a statistical conclusion in hypothesis-testing problems about a population variance using the chi-square statistic

6. Solve for possible Type II errors when failing to reject the null hypothesis

Decision Dilemma

Word-of-Mouth Business: Recommendations and *Influentials*

Word-of-mouth information about products and services is exchanged on a daily basis by millions of consumers. It is important for businesses to understand the impact of such 'business referrals' because a happy customer can potentially steer dozens of new customers to a business and unhappy customers can direct customers away from a business or product. Word-of-mouth advertising is one of the most credible forms of advertising because the person making the recommendation puts their reputation on the line and that person has nothing to gain but the appreciation of those who are listening. A newer mode of word-of-mouth advertising is the Internet forum or message board, which is an online discussion site. In this format, group members can post messages citing opinions and offering information about a variety of topics. Many consumers now go to such message boards to obtain and share information about various products, services, and companies. According to an online study of over 25,000 consumers across more than 50 countries around the world by Nielsen, about 90% of all respondents find peer recommendation the most trustworthy form of advertisement. In addition, 70% say that they completely or somewhat trust consumer opinions posted online. Furthermore, another survey by BrightLocal.com revealed that 69% of UK consumers trust online reviews as much as personal recommendations. The research also showed that consumers seek the most advice when looking for restaurants/cafés and hotels/pubs/bars.

Businesses are increasingly aware of the importance of consumer recommendations and referrals. Linkdex recently conducted a survey in the UK and the US to find out which tools were most important to small and medium-sized enterprises (SMEs) when it came to marketing the products and services they sell. Perhaps to validate the recent increase in online networking platforms, which work on the basis of peer-to-peer recommendation, the most important marketing tool for SMEs was established as being word-of-mouth. In fact a massive 81% of the companies polled said that referrals and recommendations were the most important marketing method. The second most important promotional method for SMEs was 'Google or another search engine', followed by direct sales teams and PR.

Some advice givers are referred to as *influentials*. *Influentials* are 'trend-setting opinion leaders whose

activism and expertise make them the natural source for word-of-mouth referrals'. A report issued by Roper Starch Worldwide and co-sponsored by *The Atlantic Monthly* stated that *influentials* tend to be among the first to try new products. They are looking for new restaurants and vacation spots to try, are activists on the job and in their community, and are self-indulgent. Businesses would do well to seek out such *influentials* and win them over to the company's products, thereby tapping into the word-of-mouth pipeline. On average, an *influential* recommends restaurants to 5.0 people a year. The following chart shows the average number of recommendations made by *influentials* per year on other items. These data were compiled and released by Roper Starch Worldwide.

Product or Service	Average Number of Recommendations
Office equipment	5.8
Vacation destination	5.1
TV show	4.9
Retail store	4.7
Clothing	4.5
Consumer electronics	4.5
Car	4.1
Stocks, mutual funds, CDs, etc.	3.4

Managerial and Statistical Questions

1. Each of the figures enumerated in this Decision Dilemma were derived by studies conducted on samples and published as fact. If we want to challenge these figures by conducting surveys of our own, then how would we go about testing these results? Are these studies dated now? Do they apply to all market segments (geographically, economically, etc.)? How could we test to determine whether these results apply to our market segment today?

2. The Roper Starch Worldwide study listed the mean number of recommendations made by *influentials* per year for different products or services. If these figures become accepted by industry users, how can we conduct our own tests to determine whether they are actually true? If we randomly sampled some *influentials*

(*continued*)

and our mean figures did not match these figures, then could we automatically conclude that their figures are not true? How much difference would we have to obtain to reject their claims? Is there a possibility that we could make an error in conducting such research?

3. The studies by Nielsen and BrightLocal.com produced a variety of proportions about word-of-mouth advertising and advice seeking. Are these figures necessarily true? Since these figures are based on sample information and are probably point estimates, could there be error in the estimations? How can we test to determine whether these figures that become accepted in the media as population param-

eters are actually true? Could there be differences in various population subgroups?

4. Suppose you have theories regarding word-of-mouth advertising, business referrals, or *influentials*. How would you test the theories to determine whether they are true?

Source: Adapted from 'Nielsen Global Online Consumer Survey', July 2009, available at http://blog.nielsen.com/nielsenwire/wp-content/uploads/2009/07/trustinadvertising0709.pdf; BrightLocal.com 'UK Results: Local Consumer Review Survey', available at www.brightlocal.com/blog/2010/12/06/uk-results-local-consumer-review-survey/; Chip Walker, 'Word of Mouth', *American Demographics*, July 1995, pp. 38–45; and 'Word-of-Mouth Advertising', *Entrepreneur* magazine, 2009, available at: www.entrepreneur.com/encyclopedia/term/82660.html.

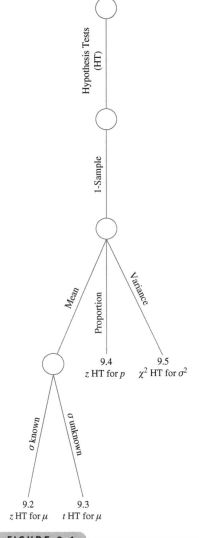

Chapter 9 Branch of the Tree Diagram Taxonomy of Inferential Techniques

A foremost statistical mechanism for decision making is the hypothesis test. The concept of hypothesis testing lies at the heart of inferential statistics, and the use of statistics to 'prove' or 'disprove' claims hinges on it. With **hypothesis testing**, business researchers are able *to structure problems in such a way that they can use statistical evidence to test various theories about business phenomena*. Business applications of statistical hypothesis testing run the gamut from determining whether a production line process is out of control to providing conclusive evidence that a new management leadership approach is significantly more effective than an old one.

Figure III-1 in the Unit III Introduction displays a tree diagram taxonomy of inferential techniques, organizing them by usage, number of samples, and type of statistic. While Chapter 8 contains the portion of these techniques that can be used for estimating a mean, a proportion, or a variance for a population with a single sample, Chapter 9 contains techniques used for testing hypotheses about a population mean, a population proportion, and a population variance using a single sample. The entire right side of the tree diagram taxonomy displays various hypothesis-testing techniques. The leftmost branch of this right side contains Chapter 9 techniques (for single samples), and this branch is displayed in Figure 9.1. Note that at the bottom of each tree branch in Figure 9.1, the title of the statistical technique along with its respective section number is given for ease of identification and use. If a business researcher is testing a population mean and the population standard deviation is known, then she will use the z test for μ contained in Section 9.2. If the population standard deviation is unknown and therefore the researcher is using the sample standard deviation, then the appropriate technique is the t test for m contained in Section 9.3. If a business researcher is testing a population proportion, then she will use the z test for p presented in Section 9.4. If the researcher desires to test a population variance from a single sample, then she will use the χ^2 test for σ^2 presented in Section 9.5. Section 9.6 contains techniques for solving for Type II errors.

9.1 INTRODUCTION TO HYPOTHESIS TESTING

In the field of business, decision makers are continually attempting to find answers to questions such as the following:

- What container shape is most economical and reliable for shipping a product?
- Which management approach best motivates employees in the retail industry?
- How can the company's retirement investment financial portfolio be diversified for optimum performance?
- What is the best way to link client databases for fast retrieval of useful information?
- Which indicator best predicts the general state of the economy in the next six months?
- What is the most effective means of advertising in a business-to-business setting?

Business researchers are often called upon to provide insights and information to decision makers to assist them in answering such questions. In searching for answers to questions and in attempting to find explanations for business phenomena, business researchers often develop 'hypotheses' that can be studied and explored. Paraphrasing the definition published in the 10th edition of *Merriam Webster's Collegiate Dictionary*, **hypotheses** are *tentative explanations of a principle operating in nature*. In this text, we will explore various types of hypotheses, how to test them, and how to interpret the results of such tests so that useful information can be brought to bear on the business decision-making process.

Types of Hypotheses

Three types of hypotheses that will be explored here:

1. *Research* hypotheses
2. *Statistical* hypotheses
3. *Substantive* hypotheses

Although much of the focus will be on testing statistical hypotheses, it is also important for business decision makers to have an understanding of both research and substantive hypotheses.

Research Hypotheses

Research hypotheses are most nearly like hypotheses defined earlier. A **research hypothesis** is *a statement of what the researcher believes will be the outcome of an experiment or a study*. Before studies are undertaken, business researchers often have some idea or theory based on experience or previous work as to how the study will turn out. These ideas, theories, or notions established before an experiment or study is conducted are research hypotheses. Some examples of research hypotheses in business might include:

- Older workers are more loyal to a company.
- Companies with more than €1 billion in assets spend a higher percentage of their annual budget on advertising than do companies with less than €1 billion in assets.
- The implementation of a Six Sigma quality approach in manufacturing will result in greater productivity.
- The price of scrap metal is a good indicator of the industrial production index six months later.
- Airline company stock prices are positively correlated with the volume of OPEC oil production.

Virtually all inquisitive, thinking business people have similar research hypotheses concerning relationships, approaches, and techniques in business. Such hypotheses can lead decision makers to new and better ways to accomplish business goals. However, to formally test research hypotheses, it is generally best to state them as statistical hypotheses.

Statistical Hypotheses

In order to scientifically test research hypotheses, a more formal hypothesis structure needs to be set up using **statistical hypotheses**. Suppose business researchers want to 'prove' the research hypothesis that older workers are more loyal to a company. A 'loyalty' survey instrument is either developed or obtained. If this instrument is administered to both older and younger workers, how much higher do older workers have to score on the 'loyalty' instrument (assuming higher scores indicate more loyalty) than younger workers to prove the research hypothesis? What is the 'proof threshold'?

Instead of attempting to prove or disprove research hypotheses directly in this manner, business researchers convert their research hypotheses to statistical hypotheses and then test the statistical hypotheses using standard procedures.

All statistical hypotheses consist of two parts, a null hypothesis and an alternative hypothesis. These two parts are constructed to contain all possible outcomes of the experiment or study. Generally, the **null hypothesis** *states that the 'null' condition exists; that is, there is nothing new happening, the old theory is still true, the old standard is correct, and the system is in control.* The **alternative hypothesis**, on the other hand, *states that the new theory is true, there are new standards, the system is out of control, and/or something is happening.* As an example, suppose flour packaged by a manufacturer is sold by weight; and a particular size of package is supposed to average 1,000 grams. Suppose the manufacturer wants to test to determine whether their packaging process is out of control as determined by the weight of the flour packages. The null hypothesis for this experiment is that the average weight of the flour packages is 1,000 grams (no problem). The alternative hypothesis is that the average is not 1,000 grams (process is out of control).

It is common symbolism to represent the null hypothesis as H_0 and the alternative hypothesis as H_a. The null and alternative hypotheses for the flour example can be restated using these symbols and μ for the population mean as:

$$H_0: \mu = 1,000 \text{ g}$$
$$H_a: \mu \neq 1,000 \text{ g}$$

As another example, suppose a company has held an 18% share of the market. However, because of an increased marketing effort, company officials believe the company's market share is now greater than 18%, and the officials would like to prove it. The null hypothesis is that the market share is still 18% or perhaps it has even dropped below 18%. Converting the 18% to a proportion and using p to represent the population proportion, results in the following null hypothesis:

$$H_0: p \leq 0.18$$

The alternative hypothesis is that the population proportion is now greater than 0.18:

$$H_a: p > 0.18$$

Note that the 'new idea' or 'new theory' that company officials want to 'prove' is stated in the alternative hypothesis. The null hypothesis states that the old market share of 18% is still true.

Generally speaking, new hypotheses that business researchers want to 'prove' are stated in the alternative hypothesis. Because many business researchers only undertake an experiment to determine whether their new hypothesis is correct, they are hoping that the alternative hypothesis will be 'proven' true. However, if a manufacturer is testing to determine whether his process is out of control as shown in the flour-packaging example, then he is most likely hoping that the alternative hypothesis is not 'proven' true thereby demonstrating that the process is still in control.

Note in the market share example that the null hypothesis also contains the 'less than' case ($<$) because between the two hypotheses (null and alternative), all possible outcomes must be included ($<$, $>$, and $=$) One could say that the null and alternative hypotheses are mutually exclusive (no overlap) and collectively exhaustive (all cases included). Thus, whenever a decision is made about which hypothesis is true, logically either one is true or the other but not both. Even though the company officials are not interested in 'proving' that their market share is less than 18%, logically it should be included as a possibility. On the other hand, many researchers and statisticians leave out the 'less than' portion of the null hypothesis on the market share problem because company officials are only interested in 'proving' that the market share has increased and the inclusion of the 'less than' sign in the null hypothesis is confusing. This approach can be justified in the way that statistical hypotheses are tested. If the equal part of the null hypothesis is rejected because the market share is seemingly greater, then certainly the 'less than' portion of the null hypothesis is also rejected because it is further away from 'greater than' than is 'equal'. Using this logic, the null hypothesis for the market share problem can be written as

$$H_0: p = 0.18$$

rather than

$$H_0: p \leq 0.18$$

Thus, in this form, the statistical hypotheses for the market share problem can be written as

$$H_0: p = 0.18$$
$$H_a: p > 0.18$$

Even though the 'less than' sign, $<$, is not included in the null hypothesis, it is implied that it is there. We will adopt such an approach in this book; and thus, all *null* hypotheses presented in this book will be written with an equal sign only ($=$) rather than with a directional sign ($\leq$) or ($\geq$).

Statistical hypotheses are written so that they will produce either a one-tailed or a two-tailed test. The hypotheses shown already for the flour package manufacturing problem are two-tailed:

$$H_0: \mu = 1{,}000 \text{ g}$$
$$H_a: \mu \neq 1{,}000 \text{ g}$$

Two-tailed tests always use $=$ and $\neq$ in the statistical hypotheses and are directionless in that the alternative hypothesis allows for either the greater than ($>$) or less than ($<$) possibility. In this particular example, if the process is 'out of control', plant officials might not know whether machines are overfilling or under filling packages and are interested in testing for either possibility.

The hypotheses shown for the market share problem are one-tailed:

$$H_0: p = 0.18$$
$$H_a: p > 0.18$$

One-tailed tests are always directional, and the alternative hypothesis uses either the greater than ($>$) or the less than ($<$) sign. A one-tailed test should only be used when the researcher knows for certain that the outcome of an experiment is going to occur only in one direction or the researcher is only interested in one direction of the experiment as in the case of the market share problem. In one-tailed problems, the researcher is trying to 'prove' that something is older, younger, higher, lower, more, less, greater, and so on. These words are considered 'directional' words in that they indicate the direction of the focus of the research. Without these words, the alternative hypothesis of a one-tailed test cannot be established.

In business research, the conservative approach is to conduct a two-tailed test because sometimes study results can be obtained that are in opposition to the direction that researchers thought would occur. For example, in the market share problem, it might turn out that the company had actually lost market share; and even though company officials were not interested in 'proving' such a case, they may need to know that it is true. It is recommended that, if in doubt, business researchers should use a two-tailed test.

STATISTICS IN BUSINESS TODAY

Rising Cost of Healthcare

With nearly $2.5 trillion spent annually on healthcare, rising healthcare costs in the United States have become a concern to most Americans. It is estimated that per capita total expenditure on healthcare in the US now averages $7,410 per person. This is a dramatic increase from 1970 when the average was $356. Around the world, this value is second only to the health expenditure in Norway, where the per capita expenditure is $7,662. In most other countries, the cost of healthcare is much lower, with the average expenditures in Japan running at about $3,321 and in the United Kingdom at about $3,285. The average American family healthcare premium in 2008 was $12,680, of which covered workers paid an average of $3,354. A study by Harvard University researchers found that the average out-of-pocket medical debt for those who filed for bankruptcy was $12,000. This study noted that 68% of those who filed for bankruptcy had health insurance and that 50% of all bankruptcy filings were partly the result of medical expenses. A survey of Iowa consumers found that in order to cope with rising health insurance costs, 86% said they had cut back on how much they could save, and 44% said that they had cut back on food and heating expenses. Hypothesis tests of these means and proportions could be done in particular parts of the United States or in ensuing years to determine if these figures still hold true.

Sources: National Coalition on Health Care website: www.nchc.org/facts/cost.shtml; the Kaiser Family Foundation website: http://facts.kff.org/?CFID=38303438&CFTOKEN=93461690; Healthcare Purchasing News Fast Stats, available at www.highbeam.com/doc/1G1-168547543.html; World Health Organization Stats, available at www.who.int/gho/health_financing/per_capita_expenditure/en/index.html.

Substantive Hypotheses

In testing a statistical hypothesis, a business researcher reaches a conclusion based on the data obtained in the study. If the null hypothesis is rejected and therefore the alternative hypothesis is accepted, it is common to say that a statistically significant result has been obtained. For example, in the market share problem, if the null hypothesis is rejected, the result is that the market share is 'significantly greater' than 18%. The word *significant* to statisticians and business researchers merely means that the result of the experiment is unlikely due to chance and a decision has been made to reject the null hypothesis. However, in everyday business life, the word *significant* is more likely to connote 'important' or 'a large amount'. One problem that can arise in testing statistical hypotheses is that particular characteristics of the data can result in a statistically significant outcome that is not a significant business outcome.

As an example, consider the market share study. Suppose a large sample of potential customers is taken, and a sample market share of 18.2% is obtained. Suppose further that a statistical analysis of these data results in statistical significance. We would conclude statistically that the market share is significantly higher than 18%. This finding actually means that it is unlikely that the difference between the sample proportion and the population proportion of 0.18 is due just to chance. However, to the business decision maker, a market share of 18.2% might not be significantly higher than 18%. Because of the way the word *significant* is used to denote rejection of the null hypothesis rather than an important business difference, business decision makers need to exercise caution in interpreting the outcomes of statistical tests.

In addition to understanding a statistically significant result, business decision makers need to determine what, to them, is a *substantive* result. A substantive result is *when the outcome of a statistical study produces results that are important to the decision maker.* The importance to the researcher will vary from study to study. As an example, in a recent year, one healthcare administrator was excited because patient satisfaction had significantly increased (statistically) from one year to the next. However, an examination of the data revealed that on a five-point scale, their satisfaction ratings had gone up from 3.61 to only 3.63. Is going from a 3.61 rating to a 3.63 rating in one year really a substantive increase? On the other hand, increasing the average purchase at a large, high-volume store from £55.45 to £55.50 might be substantive as well as significant if volume is large enough to drive profits higher. Both business researchers and decision makers should be aware that statistically significant results are not always substantive results.

Using the HTAB System to Test Hypotheses

In conducting business research, the process of testing hypotheses involves four major tasks:

- Task 1: Establishing the hypotheses.
- Task 2: Conducting the test.
- Task 3: Taking statistical action.
- Task 4: Determining the business implications.

This process, depicted in Figure 9.2, is referred to here as the HTAB system (**H**ypothesize, **T**est, **A**ction, **B**usiness implications).

Task 1, establishing the hypotheses, encompasses all activities that lead up to the establishment of the statistical hypotheses being tested. These activities might include investigating a business opportunity or problem, developing theories about possible solutions, and establishing research hypotheses. Task 2, conducting the test, involves the selection of the proper statistical test, setting the value of alpha, establishing a decision rule, gathering sample data, and computing the statistical analysis. Task 3, taking statistical action, is making a statistical decision about whether or not to reject the null hypothesis based on the outcome of the statistical test. Task 4, determining the business implications, is deciding what the statistical action means in business terms – that is, interpreting the statistical outcome in terms of business decision making.

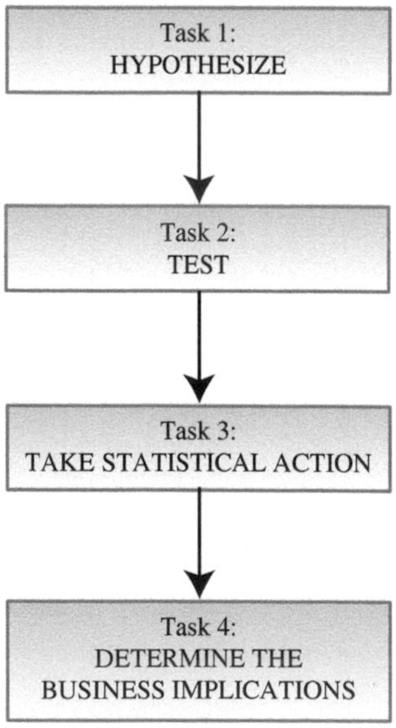

Task 1:
HYPOTHESIZE

Task 2:
TEST

Task 3:
TAKE STATISTICAL ACTION

Task 4:
DETERMINE THE
BUSINESS IMPLICATIONS

FIGURE 9.2

HTAB System of Testing Hypotheses

Typically, statisticians and researchers present the hypothesis-testing process in terms of an eight-step approach:

- Step 1-Establish a null and alternative hypothesis.
- Step 2-Determine the appropriate statistical test.
- Step 3-Set the value of alpha, the Type I error rate.
- Step 4-Establish the decision rule.
- Step 5-Gather sample data.
- Step 6-Analyse the data.
- Step 7-Reach a statistical conclusion.
- Step 8-Make a business decision.

These eight steps fit nicely into the four HTAB tasks as a part of the HTAB paradigm. Figure 9.3 presents the HTAB paradigm incorporating, the eight steps into the four HTAB tasks.

Task 1 of the HTAB system, hypothesizing, includes step 1, which is establishing a null and alternative hypothesis. In establishing the null and alternative hypotheses, it is important that the business researcher clearly identifies what is being tested and whether the hypotheses are one tailed or two tailed. In the hypothesis-testing process, it is *always assumed that the null hypothesis is true* at the beginning of the study. In other words, it is assumed that the process is in control (no problem), that the market share has not increased, that older workers are not more loyal to a company than younger workers, and so on. This process is analogous to the trial system in which the accused is presumed innocent at the beginning of the trial.

Task 2 of the HTAB system, testing, includes steps 2 to 6. Step 2 is to select the most appropriate statistical test to use for the analysis. In selecting such a test, the business researcher needs to consider the type, number, and level of data being used in the study along with the statistic used in the analysis (mean, proportion, variance, etc.). In addition, business researchers should consider the assumptions underlying certain statistical tests and determine whether they can be met in the study before using such tests.

At step 3, the value of alpha is set. Alpha is the probability of committing a Type I error and will be discussed later. Common values of alpha include 0.05, 0.01, 0.10, and 0.001.

A decision rule should be established before the study is undertaken (step 4). Using alpha and the test statistic, critical values can be determined. These **critical values** are *used at the decision step to determine whether the null hypothesis is rejected* or not. If the *p*-value method (discussed later) is used, the value of alpha is used as a critical probability value. The process begins by assuming that the null hypothesis is true. Data are gathered and statistics computed. If the evidence is away from the null hypothesis, the business researcher begins to doubt that the null hypothesis is really true. If the evidence is far enough away from the null hypothesis that the critical value is surpassed, the business researcher will reject the null hypothesis and declare that a statistically significant result has been attained. Here again, it is analogous to the court of law system. Initially, a defendant is assumed to be innocent. The prosecution presents evidence against the defendant (analogous to data gathered and analysed in a study). At some point, if enough evidence is presented against the defendant such that the jury no longer believes the defendant is innocent, a critical level of evidence has been reached and the jury finds

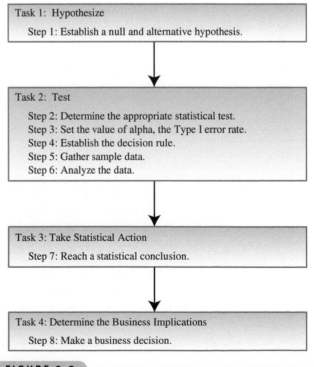

FIGURE 9.3

HTAB Paradigm Incorporating the Eight Steps

the defendant guilty. The first four steps in testing hypotheses should *always* be completed *before* the study is undertaken. It is not sound research to gather data first and then try to determine what to do with the data.

Step 5 is to gather sample data. This step might include the construction and implementation of a survey, conducting focus groups, randomly sampling items from an assembly line, or even sampling from secondary data sources (e.g., financial databases). In gathering data, the business researcher is cautioned to recall the proper techniques of random sampling (presented in Chapter 7). Care should be taken in establishing a frame, determining the sampling technique, and constructing the measurement device. A strong effort should be made to avoid all non-sampling errors. After the data are sampled, the test statistic can be calculated (step 6).

Task 3 of the HTAB system, take statistical action, includes step 7. Using the previously established decision rule (in step 4) and the value of the test statistic, the business researcher can draw a statistical conclusion. In *all* hypothesis tests, the business researcher needs to conclude whether the null hypothesis is rejected or is not rejected (step 7).

Task 4 of the HTAB system, determining the business implications, incorporates step 8. After a statistical decision is made, the business researcher or decision maker decides what business implications the study results contain (step 8). For example, if the hypothesis-testing procedure results in a conclusion that train passengers are significantly older today than they were in the past, the manager may decide to cater to these older customers or to draw up a strategy to make travelling by train more appealing to younger people. It is at this step that the business decision maker must decide whether a statistically significant result is really a substantive result.

Rejection and Non-rejection Regions

Using the critical values established at step 4 of the hypothesis-testing process, the possible statistical outcomes of a study can be divided into two groups:

1. Those that cause the rejection of the null hypothesis.
2. Those that do not cause the rejection of the null hypothesis.

Conceptually and graphically, statistical outcomes that result in the rejection of the null hypothesis lie in what is termed the **rejection region**. Statistical outcomes that fail to result in the rejection of the null hypothesis lie in what is termed the **non-rejection region**.

As an example, consider the flour-packaging manufacturing example. The null hypothesis is that the average fill for the population of packages is 1,000 grams. Suppose a sample of 100 such packages is randomly selected, and a sample mean of 1,001 grams is obtained. Because this mean is not 1,001 grams, should the business researcher decide to reject the null hypothesis? In the hypothesis test process we are using sample statistics (in this case, the sample mean of 1,001 grams) to make decisions about population parameters (in this case, the population mean of 1,000 grams). It makes sense that in taking random samples from a population with a mean of 1,000 grams not all sample means will equal 1,000 grams. In fact, the central limit theorem (see Chapter 7) states that for large sample sizes, sample means are normally distributed around the population mean. Thus, even when the population mean is 1,000 grams, a sample mean might still be 1,001, 995, or even 1,010. However, suppose a sample mean of 1,100 grams is obtained for 100 packages. This sample mean may be so far from what is reasonable to expect for a population with a mean of 1,000 grams that the decision is made to reject the null

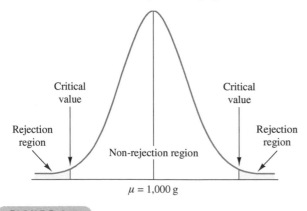

$\mu = 1,000$ g

hypothesis. This prompts the question: when is the sample mean so far away from the population mean that the null hypothesis is rejected? The critical values established at step 4 of the hypothesis-testing process are used to divide the means that lead to the rejection of the null hypothesis from those that do not. Figure 9.4 displays a normal distribution of sample means around a population mean of 1,000 grams. Note the critical values in each end (tail) of the distribution. In each direction beyond the critical values lie the rejection regions. Any sample mean that falls in that region will lead the business researcher to reject the null hypothesis. Sample means that fall between the two critical values are close enough to the population mean that the business researcher will decide not to reject the null hypothesis. These means are in the non-rejection region.

FIGURE 9.4

Rejection and Non-rejection Regions

Type I and Type II Errors

Because the hypothesis-testing process uses sample statistics calculated from random data to reach conclusions about population parameters, it is possible to make an incorrect decision about the null hypothesis. In particular, two types of errors can be made in testing hypotheses: Type I error and Type II error.

A **Type I error** is committed by *rejecting a true null hypothesis*. With a Type I error, the null hypothesis is true, but the business researcher decides that it is not. As an example, suppose the flour-packaging process actually is 'in control' and is averaging 1,000 grams of flour per package. Suppose also that a business researcher randomly selects 100 packages, weighs the contents of each, and computes a sample mean. It is possible, by chance, to randomly select 100 of the more extreme packages (mostly heavy weighted or mostly light weighted) resulting in a mean that falls in the rejection region. The decision is to reject the null hypothesis even though the population mean is actually 1,000 grams. In this case, the business researcher has committed a Type I error.

The notion of a Type I error can be used outside the realm of statistical hypothesis testing in the business world. For example, if a manager fires an employee because some evidence indicates that she is stealing from the company and if she really is not stealing from the company, then the manager has committed a Type I error. As another example, suppose a worker on the assembly line of a large manufacturer hears an unusual sound and decides to shut the line down (reject the null hypothesis). If the sound turns out not to be related to the assembly line and no problems are occurring with the assembly line, then the worker has committed a Type I error. An analogous courtroom example of a Type I error is when an innocent person is sent to jail.

In Figure 9.4, the rejection regions represent the possibility of committing a Type I error. Means that fall beyond the critical values will be considered so extreme that the business researcher chooses to reject the null hypothesis. However, if the null hypothesis is true, any mean that falls in a rejection region will result in a decision that produces a Type I error. The *probability of committing a Type I error* is called **alpha** (α) or **level of significance**. Alpha equals the area under the curve that is in the rejection region beyond the critical value(s). The value of alpha is always set before the experiment or study is undertaken. As mentioned previously, common values of alpha are 0.05, 0.01, 0.10, and 0.001.

A **Type II error** is committed when a business researcher *fails to reject a false null hypothesis*. In this case, the null hypothesis is false, but a decision is made to not reject it. Suppose in the case of the flour problem that the packaging process is actually producing a population mean of 1,100 grams even though the null hypothesis is 1,000 grams. A sample of 100 packages yields a sample mean of 1,010 grams, which falls in the non-rejection region. The business decision maker decides not to reject the null hypothesis. A Type II error has been committed. The packaging procedure is out of control and the hypothesis-testing process does not identify it.

Suppose in the business world an employee is stealing from the company. A manager sees some evidence that the stealing is occurring but lacks enough evidence to conclude that the employee is stealing from the company. The manager decides not to fire the employee based on theft. The manager has committed a Type II error. Consider the manufacturing line with the noise. Suppose the worker decides not enough noise is heard to shut the line down, but in actuality, one of the cords on the line is unravelling, creating a dangerous situation. The worker is committing a Type II error. Beginning in the 1980s, many manufacturers started protecting more against Type II errors. They found that in many cases, it was more costly to produce a bad product (e.g., scrap/rework costs and loss of market share due to poor quality) than it was to make it right the first time. They encouraged workers to 'shut down' the line if the quality of work was seemingly not what it should be (risking a Type I error) rather than allowing poor-quality product to be shipped. In a court of law, a Type II error is committed when a guilty person is declared innocent.

The probability of committing a Type II error is **beta** (β). Unlike alpha, beta is not usually stated at the beginning of the hypothesis-testing procedure. Actually, because beta occurs only when the null hypothesis is not true, the computation of beta varies with the many possible alternative parameters that might occur. For example, in the flour-packaging problem, if the population mean is not 1,000 grams, then what is it? It could be 1,010, 980, or 1,075 grams. A value of beta is associated with each of these alternative means.

How are alpha and beta related? First of all, because alpha can only be committed when the null hypothesis is rejected and beta can only be committed when the null hypothesis is not rejected, a business researcher cannot commit both a Type I error and a Type II error at the same time on the same hypothesis test. Generally, alpha and beta are inversely related. If alpha is reduced, then beta is increased, and vice versa. In terms of the manufacturing assembly line, if management makes it harder for workers to shut down the assembly line (reduce Type I error), then there is a

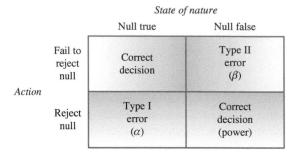

FIGURE 9.5

Alpha, Beta, and Power

greater chance that bad product will be made or that a serious problem with the line will arise (increase Type II error). Legally, if the courts make it harder to send innocent people to jail, then they have made it easier to let guilty people go free. One way to reduce both errors is to increase the sample size. If a larger sample is taken, it is more likely that the sample is representative of the population, which translates into a better chance that a business researcher will make the correct choice. Figure 9.5 shows the relationship between the two types of error. The 'state of nature' is how things actually are and the 'action' is the decision that the business researcher actually makes. Note that each action alternative contains only one of the errors along with the possibility that a correct decision has been made. **Power**, which is equal to $1 - \beta$, is *the probability of a statistical test rejecting the null hypothesis when the null hypothesis is false.* Figure 9.5 shows the relationship between α, β, and power.

9.2 TESTING HYPOTHESES ABOUT A POPULATION MEAN USING THE z STATISTIC (σ KNOWN)

One of the most basic hypothesis tests is a test about a population mean. A business researcher might be interested in testing to determine whether an established or accepted mean value for an industry is still true or in testing a hypothesized mean value for a new theory or product. As an example, a computer products company sets up a telephone service to assist customers by providing technical support. The average wait time during weekday hours is 37 minutes. However, a recent hiring effort added technical consultants to the system, and management believes that the average wait time decreased, and they want to prove it. Other business scenarios resulting in hypothesis tests of a single mean might include the following:

- A financial investment firm wants to test to determine whether the average hourly change in the Euro Stoxx 50 Average over a 10-year period is +0.25.
- A manufacturing company wants to test to determine whether the average thickness of a plastic bottle is 2.4 millimetres.
- A retail store wants to test to determine whether the average age of its customers is less than 40 years.

Formula 9.1 can be used to test hypotheses about a single population mean when σ is known if the sample size is large ($n \geq 30$) for any population and for small samples ($n < 30$) if x is known to be normally distributed in the population.

z TEST FOR A SINGLE MEAN (9.1)

$$z = \frac{\bar{x} - \mu}{\dfrac{\sigma}{\sqrt{n}}}$$

A labour market survey of gross earnings by major occupational groups across the United Kingdom found that the average basic gross salary of marketing and sales managers was £48,984 in 2002.* Because this survey is now more than 10 years old, an economics researcher wants to test this figure by taking a random sample of 112 marketing and sales managers in the United Kingdom to determine whether the basic gross income figure has changed. The researcher could use the eight steps of hypothesis testing to do so. Assume the population standard deviation of gross incomes for marketing and sales managers is £14,530.

---

*Adapted from the New Earnings Survey (NES) 2002, Office for National Statistics, available at www.statistics.gov.uk/statbase/Product.asp?vlnk=5752.

HYPOTHESIZE

At step 1, the hypotheses must be established. Because the researcher is testing to determine whether the figure has changed, the alternative hypothesis is that the mean net income is not £48,984. The null hypothesis is that the mean still equals £48,984. These hypotheses follow:

$$H_0: \mu = £48,984$$
$$H_a: \mu \neq £48,984$$

TEST

Step 2 is to determine the appropriate statistical test and sampling distribution. Because the population standard deviation is known (£14,530) and the researcher is using the sample mean as the statistic, the z test in formula (9.1) is the appropriate test statistic.

$$z = \frac{\bar{x} - \mu}{\frac{\sigma}{\sqrt{n}}}$$

Step 3 is to specify the Type I error rate, or alpha, which is 0.05 in this problem. Step 4 is to state the decision rule. Because the test is two tailed and alpha is 0.05, there is $\alpha/2$ or 0.025 area in each of the tails of the distribution. Thus, the rejection region is in the two ends of the distribution with 2.5% of the area in each. There is a 0.4750 area between the mean and each of the critical values that separate the tails of the distribution (the rejection region) from the non-rejection region. By using this 0.4750 area and Table A.5 in Appendix A, the critical z value can be obtained.

$$z_{\alpha/2} = \pm 1.96$$

Figure 9.6 displays the problem with the rejection regions and the critical values of z. The decision rule is that if the data gathered produce a z value greater than 1.96 or less than −1.96, the test statistic is in one of the rejection regions and the decision is to reject the null hypothesis. If the observed z value calculated from the data is between −1.96 and +1.96, the decision is to not reject the null hypothesis because the observed z value is in the non-rejection region.

Step 5 is to gather the data. Suppose the 112 Marketing and Sales Managers who respond produce a sample mean of £52,695. At step 6, the value of the test statistic is calculated by using $\bar{x} = £52,695$, $n = 112$, $\sigma = £14,530$ and a hypothesized $\mu = £48,984$:

$$z = \frac{52,695 - 48,984}{\frac{14,530}{\sqrt{112}}} = 2.70$$

ACTION

Because this test statistic, $z = 2.70$, is greater than the critical value of z in the upper tail of the distribution, $z = +1.96$, the statistical conclusion reached at step 7 of the hypothesis-testing process is to reject the null hypothesis. *The calculated test statistic* is often referred to as the **observed value.** Thus, the observed value of z for this problem is 2.70 and the critical value of z for this problem is 1.96.

BUSINESS IMPLICATIONS

Step 8 is to make a managerial decision. What does this result mean? Statistically, the researcher has enough evidence to reject the figure of £48,984 as the

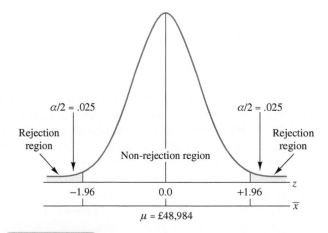

FIGURE 9.6

Marketing and Sales Manager Gross Income Example

true national average gross income for marketing and sales managers. Although the researcher conducted a two-tailed test, the evidence gathered indicates that the national average may have increased. The sample mean of £52,695 is £3,711 higher than the national mean being tested. The researcher can conclude that the national average is more than before, but because the £52,695 is only a sample mean, it offers no guarantee that the national average for all marketing and sales managers is £3,711 more. If a confidence interval were constructed with the sample data, £52,695 would be the point estimate. Other samples might produce different sample means. Managerially, this statistical finding may mean that marketing and sales managers will be more expensive to hire either as full-time employees or as consultants. It may mean that consulting services have gone up in price. For new marketing and sales managers, it may mean the potential for greater earning power.

Testing the Mean with a Finite Population

If the hypothesis test for the population mean is being conducted with a known finite population, the population information can be incorporated into the hypothesis-testing formula. Doing so can increase the potential for rejecting the null hypothesis. However, remember from Chapter 7 that if the sample size is less than 5% of the population, the finite correction factor does not significantly alter the solution. Formula 9.1 can be amended to include the population information.

FORMULA TO TEST HYPOTHESES ABOUT μ WITH A FINITE POPULATION (9.2)	$z = \dfrac{\bar{x} - \mu}{\dfrac{\sigma}{\sqrt{n}}\sqrt{\dfrac{N-n}{N-1}}}$

In the marketing and sales manager gross income example, suppose only 600 marketing and sales managers practise in the United Kingdom. A sample of 112 marketing and sales managers taken from a population of only 600 marketing and sales managers is 18.67% of the population and therefore is much more likely to be representative of the population than a sample of 112 marketing and sales managers taken from a population of 20,000 marketing and sales managers (0.56% of the population). The finite correction factor takes this difference into consideration and allows for an increase in the observed value of z. The observed z value would change to

$$z = \frac{\bar{x} - \mu}{\dfrac{\sigma}{\sqrt{n}}\sqrt{\dfrac{N-n}{N-1}}} = \frac{52,695 - 48,984}{\dfrac{14,530}{\sqrt{112}}\sqrt{\dfrac{600-112}{600-1}}} = \frac{3711}{1239.2} = 2.99$$

Use of the finite correction factor increased the observed z value from 2.70 to 2.99. The decision to reject the null hypothesis does not change with this new information. However, on occasion, the finite correction factor can make the difference between rejecting and failing to reject the null hypothesis.

Using the *p*-Value to Test Hypotheses

Another way to reach a statistical conclusion in hypothesis-testing problems is by using the **p-value**, sometimes referred to as **observed significance level.** The p-value is growing in importance with the increasing use of statistical computer packages to test hypotheses. No preset value of α is given in the p-value method. Instead, the probability of getting a test statistic at least as extreme as the observed test statistic (computed from the data) is computed under the assumption that the null hypothesis is true. Virtually every statistical computer program yields this probability (p-value). *The p-value defines the smallest value of alpha for which the null hypothesis can be rejected.* For example, if the p-value of a test is 0.038, the null hypothesis cannot be rejected at $\alpha = 0.01$ because 0.038 is the smallest value of alpha for which the null hypothesis can be rejected. However, the null hypothesis can be rejected for $\alpha = 0.05$.

Suppose a researcher is conducting a one-tailed test with a rejection region in the upper tail and obtains an observed test statistic of $z = 2.04$ from the sample data. Using the standard normal table, Table A.5, we find that the probability of randomly obtaining a z value this great or greater by chance is $0.5000 - 0.4793 = 0.0207$. The p-value is 0.0207. Using this information, the researcher would reject the null hypothesis for $\alpha = 0.05$ or 0.10 or any value more than 0.0207. The researcher would not reject the null hypothesis for any alpha value less than or equal to 0.0207 (in particular, $\alpha = 0.01, 0.001$, etc.).

For a two-tailed test, recall that we split alpha to determine the critical value of the test statistic. With the *p*-value, the probability of getting a test statistic at least as extreme as the observed value is computed. This *p*-value is then compared to $\alpha/2$ for two-tailed tests to determine statistical significance.

Note: The business researcher should be cautioned that some statistical computer packages are programmed to double the observed probability and report that value as the *p*-value when the user signifies that a two-tailed test is being requested. The researcher then compares this *p*-value to alpha values to decide whether to reject the null hypothesis. The researcher must be sure she understands what the computer software package does to the *p*-value for a two-tailed test before she reaches a statistical conclusion.

As an example of using *p*-values with a two-tailed test, consider the marketing and sales managers gross income problem. The observed test statistic for this problem is $z = 2.70$. Using Table A.5, we know that the probability of obtaining a test statistic at least this extreme if the null hypothesis is true is $0.5000 - 0.4965 = 0.0035$. Observe that in the Minitab output in Figure 9.7 the *p*-value is 0.007. Minitab doubles the *p*-value on a two-tailed test so that the researcher can compare the *p*-value to reach a statistical conclusion. On the other hand, when Excel yields a *p*-value in its output, it always gives the one-tailed value, which in this case is 0.003 (see output in Figure 9.7). To reach a statistical conclusion from an Excel-produced *p*-value when doing a two-tailed test, the researcher must compare the *p*-value to $\alpha/2$.

Minitab Output
One-Sample Z

```
Test of mu = 48984 vs not = 48984
The assumed standard deviation = 14530
```

N	Mean	SE Mean	95% CI	Z	P
112	52695	1373	(50004, 55386)	2.70	0.007

Excel Output

Sample mean	52695
Standard Deviation	14530
Count (*n*)	112
Hypothesized value of μ	48984
p-value	0.003

FIGURE 9.7

Minitab and Excel Output with *p*-Values

Range of *p*-Values	Rejection Range
p-value > .10	Cannot reject the null hypothesis for commonly accepted values of alpha
.05 < *p*-value ≤ .10	Reject the null hypothesis for α = .10
.01 < *p*-value ≤ .05	Reject the null hypothesis for α = .05
.001 < *p*-value ≤ .01	Reject the null hypothesis for α = .01
.0001 < *p*-value ≤ .001	Reject the null hypothesis for α = .001

FIGURE 9.8

Rejecting the Null Hypothesis Using *p*-Values

Figure 9.8 summarizes the decisions that can be made using various *p*-values for a one-tailed test. To use this for a two-tailed test, compare the *p*-value in one tail to $\alpha/2$ in a similar manner. Because $\alpha = 0.10$ is usually the largest value of alpha used by most researchers, if a *p*-value is not less than 0.10, then the decision is to fail to reject the null hypothesis. In other words, the null hypothesis cannot be rejected if *p*-values are 0.579 or 0.106 or 0.283, etc. If a *p*-value is less than 0.10 but not less than 0.05, then the null hypothesis can be rejected for $\alpha = 0.10$. If it is less than 0.05, but not less than 0.01, then the null hypothesis can be rejected for $\alpha = 0.05$ and so on.

Using the Critical Value Method to Test Hypotheses

Another method of testing hypotheses is the critical value method. In the marketing and sales managers income example, the null hypothesis was rejected because the computed value of *z* was in the rejection zone. What mean income would it take to cause the observed *z* value to be in the rejection zone? The **critical value method** *determines the critical mean value required for z to be in the rejection region and uses it to test the hypotheses.*

This method also uses Formula 9.1. However, instead of an observed *z*, a critical $\bar{x}$ value, $\bar{x}_c$, is determined. The critical table value of z_c is inserted into the formula, along with μ and σ. Thus,

$$z_c = \frac{\bar{x}_c - \mu}{\frac{\sigma}{\sqrt{n}}}$$

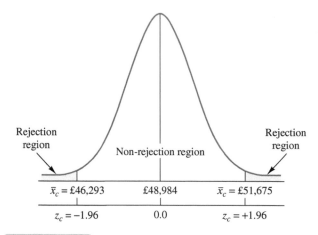

Rejection region

Non-rejection region

Rejection region

$\bar{x}_c = £46,293$ £48,984 $\bar{x}_c = £51,675$

$z_c = -1.96$ 0.0 $z_c = +1.96$

FIGURE 9.9

Rejection and Non-rejection Regions for Critical Value Method

Substituting values from the marketing and sales managers income example gives

$$\pm 1.96 = \frac{\bar{x}_c - 48,984}{\frac{14,530}{\sqrt{112}}}$$

or

$$\bar{x}_c = 48,984 \pm 1.96\frac{14,530}{\sqrt{112}} = 48,984 \pm 2,691$$

lower $\bar{x}_c = 46,293$ and upper $\bar{x}_c = 51,675$

Figure 9.9 depicts graphically the rejection and non-rejection regions in terms of means instead of z scores.

With the critical value method, most of the computational work is done ahead of time. In this problem, before the sample means are computed, the analyst knows that a sample mean value of greater than £51,675 or less than £46,293 must be attained to reject the hypothesized population mean. Because the sample mean for this problem was £52,695, which is greater than £51,675, the analyst rejects the null hypothesis. This method is particularly attractive in industrial settings where standards can be set ahead of time and then quality-control technicians can gather data and compare actual measurements of products to specifications.

DEMONSTRATION PROBLEM 9.1

In an attempt to determine why customer service is important to managers, researchers surveyed managing directors of manufacturing plants in Scotland.* One of the reasons proposed was that customer service is a means of retaining customers. On a scale from 1 to 5, with 1 being low and 5 being high, the survey respondents rated this reason more highly than any of the others, with a mean response of 4.30. Suppose Asian researchers believe Asian manufacturing managers would not rate this reason as highly and conduct a hypothesis test to prove their theory. Alpha is set at 0.05. Data are gathered and the results below are obtained. Use these data and the eight steps of hypothesis testing to determine whether Asian managers rate this reason significantly lower than the 4.30 mean ascertained in Scotland. Assume from previous studies that the population standard deviation is 0.574.

3	4	5	5	4	5	5	4	4	4	4
4	4	4	4	5	4	4	4	3	4	4
4	3	5	4	4	5	4	4	4	5	

Solution

HYPOTHESIZE

STEP 1: Establish hypotheses. Because the Asian researchers are interested only in 'proving' that the mean figure is lower in Asia, the test is one tailed. The alternative hypothesis is that the population mean is lower than 4.30. The null hypothesis states the equality case.

$$H_0: \mu = 4.30$$
$$H_a: \mu < 4.30$$

*William G. Donaldson, 'Manufacturers Need to Show Greater Commitment to Customer Service', *Industrial Marketing Management*, 24 (October 1995), pp. 421–430. The 1-to-5 scale has been reversed here for clarity of presentation.

TEST

STEP 2: Determine the appropriate statistical test. The test statistic is

$$z = \frac{\bar{x} - \mu}{\frac{\sigma}{\sqrt{n}}}$$

STEP 3: Specify the Type I error rate.

$$\alpha = 0.05$$

STEP 4: State the decision rule. Because this test is a one-tailed test, the critical z value is found by looking up $0.5000 - 0.0500 = 0.4500$ as the area in Table A.5. The critical value of the test statistic is $z_{0.05} = -1.645$. An observed test statistic must be less than -1.645 to reject the null hypothesis. The rejection region and critical value can be depicted as in the following diagram:

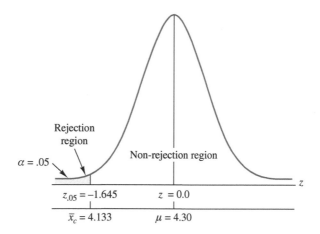

STEP 5: Gather the sample data. The data are shown.
STEP 6: Calculate the value of the test statistic.

$$\bar{x} = 4.156 \quad \sigma = 0.574$$

$$z = \frac{4.156 - 4.30}{\frac{0.574}{\sqrt{32}}} = -1.42$$

ACTION

STEP 7: State the statistical conclusion. Because the observed test statistic is not less than the critical value and is not in the rejection region, the statistical conclusion is that the null hypothesis cannot be rejected. The same result is obtained using the p-value method. The observed test statistic is $z = -1.42$. From Table A.5, the probability of getting a z value at least this extreme when the null hypothesis is true is $0.5000 - 0.4222 = 0.0778$. Hence, the null hypothesis cannot be rejected at $\alpha = 0.05$ because the smallest value of alpha for which the null hypothesis can be rejected is 0.0778. Had $\alpha = 0.10$, the decision would have been to reject the null hypothesis.

BUSINESS IMPLICATIONS

STEP 8: Make a managerial decision. The test does not result in enough evidence to conclude that Asian managers think it is less important to use customer service as a means of retaining customers than do Scottish managers. Customer service is an important tool for retaining customers in both regions according to managers.

Using the critical value method: for what sample mean (or more extreme) value would the null hypothesis be rejected? This critical sample mean can be determined by using the critical z value associated with alpha, $z_{0.05} = -1.645$.

$$z_c = \frac{\bar{x}_c - \mu}{\frac{\sigma}{\sqrt{n}}}$$

$$-1.645 = \frac{\bar{x}_c - 4.30}{\frac{0.574}{\sqrt{32}}}$$

$$\bar{x}_c = 4.133$$

The decision rule is that a sample mean less than 4.133 would be necessary to reject the null hypothesis. Because the mean obtained from the sample data is 4.156, the researchers fail to reject the null hypothesis. The preceding diagram includes a scale with the critical sample mean and the rejection region for the critical value method.

Minitab Output

One-Sample Z

```
Test of mu = 4.3 vs < 4.3
The assumed standard deviation = 0.574


N      Mean   SE Mean        95% CI          Z       P
32     4.156    0.101   (3.957, 4.355) -1.42   0.078
```

Excel Output

The *p*-value for the ratings problem is **0.078339**

FIGURE 9.10

Minitab and Excel Output for Demonstration Problem 9.1

Using the Computer to Test Hypotheses About a Population Mean Using the z Statistic

Both Minitab and Excel can be used to test hypotheses about a single population mean using the z statistic. Figure 9.10 contains output from both Minitab and Excel for Demonstration Problem 9.1. For z tests, Minitab requires knowledge of the population standard deviation. Note that the standard Minitab output includes a statement of the one-tailed hypothesis, the observed z value, and the *p*-value. Because this test is a one-tailed test, the *p*-value was not doubled.

The Excel output contains only the right-tailed *p*-value of the z statistic. With a negative observed z for Demonstration Problem 9.1, the *p*-value was calculated by taking 1 − (Excel's answer).

9.2 PROBLEMS

9.1 **a.** Use the data given to test the following hypotheses (using the z statistic):

$$H_0: \mu = 25; \ H_a: \mu \neq 25$$
$$\bar{x} = 28.1; \ n = 57; \ \sigma = 8.46; \ \alpha = 0.01$$

b. Use the *p*-value to reach a statistical conclusion.

c. Using the critical value method, what are the critical sample mean values?

9.2 Use the data given to test the hypotheses below. Assume the data are normally distributed in the population.

$$H_0: \mu = 7.48; \ H_a: \mu < 7.48$$
$$\bar{x} = 6.91; \ n = 24; \ \sigma = 1.21; \ \alpha = 0.01$$

9.3 **a.** Use the data given to test the following hypotheses:

$$H_0: \mu = 1,200; \ H_a: \mu > 1,200$$
$$\bar{x} = 1,215; \ n = 113; \ \sigma = 100; \ \alpha = 0.10$$

b. Use the p-value to obtain the results.

c. Solve for the critical value required to reject the mean.

9.4 The World Health Organization (WHO) releases figures on urban air pollution in selected cities around the world. For the city of Tianjin in China, the WHO claims that the average number of micrograms of suspended particles per cubic metre of air was 306 in a recent year. Suppose Tianjin officials have been working with businesses, commuters, and industries to reduce this figure. These city officials hire an environmental company to take random measures of air soot over a period of several weeks. The resulting data follow. Assume that the population standard deviation is 9.184. Use these data to determine whether the urban air soot in Tianjin is significantly lower than it was when the World Health Organization conducted its measurements. Let $\alpha = 0.01$. If the null hypothesis is rejected, discuss the substantive hypothesis.

320	316	269	265	263	282	292
279	263	276	291	318	257	279
250	265	259	284	269	278	263
271	268	293	304	277	300	250
309	309	290	259			

9.5 According to a report by the Bill & Melinda Gates Foundation on Postsecondary Success, the 2005 average annual earnings for someone with a graduate degree was $75,000, which was three times higher than for someone not completing high school. The finding that 'the more education you have the higher your earnings' is, according to the report, valid across all racial and ethnic groups and for both men and women. Suppose a researcher wants to test to determine whether this figure is still accurate today. The researcher randomly selects 54 people that hold a graduate degree from across the United States and obtains an earnings statement from each. The resulting sample average is $77,234. Assuming a population standard deviation of $9,390, and a 5% level of significance, determine whether the mean annual earnings of people with a graduate degree has changed.

9.6 According to a January 2011 study by Visa, respondents from six key e-commerce markets (China, India, Indonesia, Malaysia, Taiwan, and Thailand) reported spending an average of $2,086 online on travel, share trading, and electrical appliances over the previous 12 months. Suppose a researcher believes that the average spending on travel, share trading, and electrical appliances is now lower, and he sets up a study in an attempt to prove his theory. He randomly samples 18 online shoppers from the six key markets and finds out that the average annual spending for this sample is $1,897, with a population standard deviation of $781. Use $\alpha = 0.01$ to test the researcher's theory. Assume online shopping spending in these markets are normally distributed.

9.7 A manufacturing company produces valves in various sizes and shapes. One particular valve plate is supposed to have a tensile strength of 5 pounds per millimetre (lbs/mm). The company tests a random sample of 42 such valve plates from a lot of 650 valve plates. The sample mean is a tensile strength of 5.0611 lbs/mm, and the population standard deviation is 0.2803 lbs/mm. Use $\alpha = 0.10$ and test to determine whether the lot of valve plates has an average tensile strength of 5 lbs/mm.

9.8 According to a report released by CIBC entitled 'Women Entrepreneurs: Leading the Charge', the average age of Canadian businesswomen in 2008 was 41. In the report, there was some indication that researchers believed that this mean age will increase. Suppose now, a couple of years later, business researchers in Canada want to test to determine if, indeed, the mean age of a Canadian businesswoman has increased. The researchers randomly sample 97 Canadian businesswomen and ascertain that the sample mean age is 43.4. From past experience, it is known that the population standard deviation is 8.95. Test to determine if the mean age of a Canadian businesswoman has increased using a 1% level of significance. What is the p-value for this test? What is the decision? If the null hypothesis is rejected, is the result substantive?

9.9 According to a recent survey by *China Briefing*, an average worker in China costs more than the average worker in any other emerging Asian economy, except Malaysia and Thailand, when considered in terms of combined salary and welfare payments. The survey, conducted in-house, took samples of minimum wage levels from each of China's provinces and 40 cities, and based its figures on the mean. The total labour costs in China (annual salary plus mandatory welfare payments) was $2,250 international dollars per year. Suppose an economics researcher wants to test whether this figure is correct and decided to randomly contact 24 representative Chinese workers and asked them about their total annual earnings. Suppose the resulting data are given below and that the population standard deviation total earnings is $349. Use a 10% level of significance to test her hypothesis. Assume that such earnings are normally distributed in the population. What is the observed value? What is the p-value? What is the decision? If the null hypothesis is rejected, is the result substantive?

| 2,227 | 2,102 | 2,346 | 2,175 | 2,373 | 2,392 | 2,197 | 2,317 | 2,292 | 2,365 | 2,121 | 2,387 |
| 2,195 | 2,280 | 2,123 | 2,278 | 2,148 | 2,266 | 2,383 | 2,237 | 2,204 | 2,370 | 2,366 | 2,267 |

9.10 A 2007 Capgemini World Retail Banking Survey of 25 nations and 180 banks estimated that the average person paid €77 for banking services per year. Suppose some researchers believe that the average fee paid for banking services has increased since the Financial Crisis starting 2007–2008 and want to test to determine whether it is so. They randomly select a representative sample of banking customers and determine how much they paid in banking fees in the previous year. The output is given here. Assume $\alpha = 0.05$. How many people were sampled? What were the sample mean? Was this a one- or two-tailed test? What was the result of the study? What decision could be stated about the null hypothesis from these results?

```
One-Sample Z: C1

Test of mu = 77 vs. > 77
The assumed standard deviation = 27.68

                                         95% Lower
Variable    N    Mean    StDev   SE Mean    Bound      Z       P
C1          40   91.83   18.63    4.38      84.63    3.39   0.000
```

9.3 TESTING HYPOTHESES ABOUT A POPULATION MEAN USING THE t STATISTIC (σ UNKNOWN)

Very often when a business researcher is gathering data to test hypotheses about a single population mean, the value of the population standard deviation is unknown and the researcher must use the sample standard deviation as an estimate of it. In such cases, the z test cannot be used.

Chapter 8 presented the t distribution, which can be used to analyse hypotheses about a single population mean when s is unknown if the population is normally distributed for the measurement being studied. In this section, we will examine the t test for a single population mean. In general, this t test is applicable whenever the researcher is drawing a single random sample to test the value of a population mean (μ), the population standard deviation is unknown, and the population is normally distributed for the measurement of interest. Recall from Chapter 8 that the assumption that the data be normally distributed in the population is rather robust.

The formula for testing such hypotheses follows.

t TEST FOR μ (9.3)

$$t = \frac{\bar{x} - \mu}{\frac{s}{\sqrt{n}}}$$

$$df = n - 1$$

The Jiangsu Black Leopard Machinery Manufacturing Co., Ltd builds a range of carbonated beverage PET filling machines. One of its more popular models has a maximum capacity of filling 3,000 bottles per hour. For a carbonated beverage filling machine to be operating properly, each PET bottle should be filled with exactly 500 millilitres. The amount of liquid in each bottle can, however, vary slightly for a number of reasons, including machinery malfunction or maladjustment. The distribution of liquid filled by the machine is normal. However, the factory quality supervisor is worried that a specific machine is out of adjustment and is filling bottles that do not average 500 millilitres. To test this concern, he randomly selects 20 of the bottles filled the day before by the machine and measures the liquid in them. Table 9.1 shows the liquid contents in millilitres obtained, along with the computed sample mean and sample standard deviation.

The test is to determine whether the machine is out of control, and the factory quality supervisor has not specified whether he believes the machine is producing bottles that have too much or too little liquid in them. Thus a two-tailed test is appropriate. The following hypotheses are tested.

$$H_0: \mu = 500 \text{ ml}$$
$$H_a: \mu \neq 500 \text{ ml}$$

TABLE 9.1

Liquid in a Sample of 20 Bottles (ml)

506	492	495	492	499
507	495	490	492	492
505	498	510	495	492
506	499	492	510	504

$\bar{x} = 498.55$, $s = 6.825$, $n = 20$

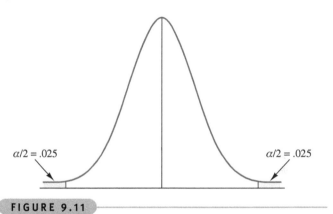

$\alpha/2 = .025$ $\alpha/2 = .025$

FIGURE 9.11

Rejection Regions for the Bottle-Filling Example

An α of 0.05 is used. Figure 9.11 shows the rejection regions.

Because $n = 20$, the degrees of freedom for this test are 19 $(20 - 1)$. The t distribution table is a one-tailed table but the test for this problem is two tailed, so alpha must be split, which yields $\alpha/2 = 0.025$, the value in each tail. (To obtain the table t value when conducting a two-tailed test, always split alpha and use $\alpha/2$.) The table t value for this example is 2.093. Table values such as this one are often written in the following form:

$$t_{0.025,19} = 2.093$$

Figure 9.12 depicts the t distribution for this example, along with the critical values, the observed t value, and the rejection regions. In this case, the decision rule is to reject the null hypothesis if the observed value of t is less than -2.093 or greater than $+2.093$ (in the tails of the distribution). Computation of the test statistic yields

$$t = \frac{\bar{x} - \mu}{\frac{s}{\sqrt{n}}} = \frac{498.55 - 500}{\frac{6.825}{\sqrt{20}}} = -0.95 \text{ (observed } t \text{ values)}$$

Because the observed t value is -0.95, the null hypothesis is not rejected. Not enough evidence is found in this sample to reject the hypothesis that the population mean is 500 millilitres.

Figure 9.13 shows Minitab and Excel output for this example. Note that the Minitab output includes the observed t value (-0.95) and the p-value (0.354). Since this test is a two-tailed test, Minitab has doubled the one-tailed p-value for $t = -0.95$. Thus the p-value of 0.354 can be compared directly to $\alpha = 0.05$ to reach the conclusion to fail to reject the null hypothesis.

The Excel output contains the observed t value (-0.95) plus the p-value and the critical table t value for both a one-tailed and a two-tailed test. Excel also gives the table value of $t = 2.09$ for a two-tailed test, which allows one to verify that the statistical conclusion is to fail to reject the null hypothesis because the observed t value is only -0.95, which is less than -2.09.

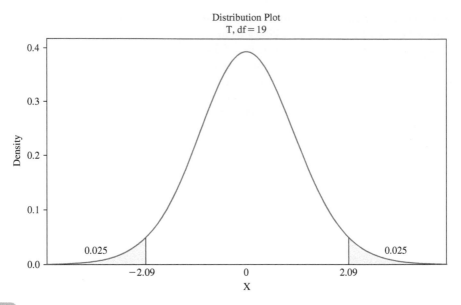

Graph of Observed and Critical *t* Values for the Bottle-Filling Example

Minitab Output

One-Sample T: Liquid

```
Test of mu = 500 vs not = 500

Variable    N    Mean    StDev   SE Mean      95% CI            T       P
C1         20   498.55   6.82      1.53   (495.36, 501.74)  -0.95   0.354
```

Excel Output

t-Test: Two-Sample Assuming Unequal Variances

	liquid	mean
Mean	498.55	500
Variance	46.57631579	0
Observations	20	20
Hypothesized Mean Difference	0	
Df	19	
t Stat	−0.95016807	
$P(T<=t)$ one-tail	0.17697887	
t Critical one-tail	1.729132792	
$P(T<=t)$ two-tail	0.35395774	
t Critical two-tail	2.09302405	

Minitab and Excel Output for the Bottle-Filling Example

DEMONSTRATION PROBLEM 9.2

Figures released by the Eurostat Farm Structure Survey show that the average size of farms in the UK is much larger than in the rest of the EU and has been increasing in recent years. In 2000, the mean size of a farm was 67.7 hectares, up from 67.3 hectares in 1993. Between those years, the number of farms decreased from 244,000 to 233,000 but the amount of tillable land remained relatively constant, so now farms are bigger. This trend might be explained, in part, by the inability of small farms to compete with the prices and costs of large-scale operations and to produce a level of income necessary to support the farmers' desired standard of living. Suppose an agribusiness researcher believes the average size of farms has now increased from the 2000 mean figure of 67.7 hectares. To test this notion, she randomly sampled 23 farms across the United Kingdom and ascertained the size of each farm from county records. The data she gathered are shown below. Use a 5% level of significance to test her hypothesis. Assume that number of hectares per farm is normally distributed in the population.

63.6	69.9	67.7	72.1	79.0	68.1	64.9	66.1	66.6
79.6	71.7	64.1	62.6	71.4	66.6	68.1	79.6	61.9
77.9	73.0	84.3	80.1	80.0				

Solution

HYPOTHESIZE

STEP 1: The researcher's hypothesis is that the average size of a UK farm is more than 67.7 hectares. Because this theory is unproven, it is the alternative hypothesis. The null hypothesis is that the mean is still 67.7 hectares.

$$H_0: \mu = 67.7$$
$$H_a: \mu > 67.7$$

TEST

STEP 2: The statistical test to be used is

$$t = \frac{\bar{x} - \mu}{\dfrac{s}{\sqrt{n}}}$$

STEP 3: The value of alpha is 0.05.

STEP 4: With 23 data points, $df = n - 1 = 23 - 1 = 22$. This test is one tailed, and the critical table t value is

$$t_{0.05,22} = 1.717$$

The decision rule is to reject the null hypothesis if the observed test statistic is greater than 1.717.

STEP 5: The gathered data are shown.

STEP 6: The sample mean is 71.3 and the sample standard deviation is 6.705. The observed t value is

$$t = \frac{\bar{x} - \mu}{\dfrac{s}{\sqrt{n}}} = \frac{71.3 - 67.7}{\dfrac{6.705}{\sqrt{23}}} = 2.54$$

ACTION

STEP 7: The observed t value of 2.54 is greater than the table t value of 1.717, so the business researcher rejects the null hypothesis. She accepts the alternative hypothesis and concludes that the average size of a UK farm is now more than 67.7 hectares. The following graph represents this analysis pictorially:

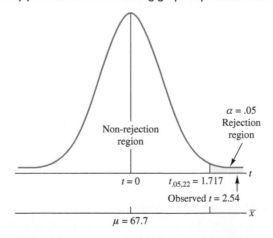

BUSINESS IMPLICATIONS

STEP 8: Agribusiness researchers can speculate about what it means to have larger farms. If the average size of a farm has increased from 67.7 hectares to 71.3 hectares, it may represent a substantive increase.

It could mean that small farms are not financially viable. It might mean that corporations are buying out small farms and that large company farms are on the increase. Such a trend might spark legislative movements to protect the small farm. Larger farm sizes might also affect commodity trading.

Using the Computer to Test Hypotheses About a Population Mean Using the *t* Test

Minitab has the capability of computing a one-sample *t* test for means. Figure 9.14 contains Minitab output for Demonstration Problem 9.2. The output contains the hypotheses being tested, the sample statistics, the observed *t* value (2.54), and the *p*-value (0.009). Because the *p*-value is less than $\alpha = 0.05$, the decision is to reject the null hypothesis.

Excel does not have a one-sample *t* test function. However, by using the two-sample *t* test for means with unequal variances, the results for a one-sample test can be obtained. This is accomplished by inputting the sample data for the first sample and the value of the parameter being tested (in this case, $\mu = 67.7$) for the second sample. The output includes the observed *t* value (2.54) and both the table *t* values and *p*-values for one- and two-tailed tests. Because Demonstration Problem 9.2 was a one-tailed test, the *p*-value of which is the same value obtained using Minitab, is used.

Minitab Output

One-Sample T: Hectares

Test of mu = 67.7 vs >67.7

Variable	N	Mean	StDev	SE Mean	95% Lower Bound	T	P
Hectares	23	71.26	6.71	1.40	68.86	2.54	0.009

Excel Output

t-Test: Two-Sample Assuming Unequal Variances

	Hectares
Mean	71.25652
Variance	44.96257
Observations	23
df	22
t Stat	2.543688
P(T<=t) one-tail	0.009254
t Critical one-tail	1.717144
P(T<=t) two-tail	0.018508
t Critical two-tail	2.073873

FIGURE 9.14

Minitab and Excel Output for Demonstration Problem 9.2

9.3 PROBLEMS

9.11 A random sample of size 20 is taken, resulting in a sample mean of 16.45 and a sample standard deviation of 3.59. Assume x is normally distributed and use this information and $\alpha = 0.05$ to test the following hypotheses:

$$H_0: \mu = 16 \quad H_a: \mu \neq 16$$

9.12 A random sample of 51 items is taken, with $\bar{x} = 58.42$ and $s^2 = 25.68$. Use these data to test the hypotheses below, assuming you want to take only a 1% risk of committing a Type I error and that x is normally distributed.

$$H_0: \mu = 60 \quad H_a: \mu < 60$$

9.13 The following data were gathered from a random sample of 11 items:

1,200	1,175	1,080	1,275	1,201	1,387
1,090	1,280	1,400	1,287	1,225	

Use these data and a 5% level of significance to test the hypotheses below, assuming that the data come from a normally distributed population.

$$H_0: \mu = 1160 \quad H_a: \mu > 1160$$

9.14 The following data (in pounds), which were selected randomly from a normally distributed population of values, represent measurements of a machine part that is supposed to weigh, on average, 8.3 pounds:

8.1	8.4	8.3	8.2	8.5	8.6	8.4	8.3	8.4	8.2
8.8	8.2	8.2	8.3	8.1	8.3	8.4	8.5	8.5	8.7

Use these data and $\alpha = 0.01$ to test the hypothesis that the parts average 8.3 pounds.

9.15 A hole-punch machine is set to punch a hole 1.84 centimetres in diameter in a strip of sheet metal in a manufacturing process. The strip of metal is then creased and sent on to the next phase of production, where a metal rod is slipped through the hole. It is important that the hole be punched to the specified diameter of 1.84 centimetres. To test punching accuracy, technicians have randomly sampled 12 punched holes and measured the diameters. The data (in centimetres) is shown below. Use an alpha of 0.10 to determine whether the holes are being punched an average of 1.84 centimetres. Assume the punched holes are normally distributed in the population.

1.81	1.89	1.86	1.83
1.85	1.82	1.87	1.85
1.84	1.86	1.88	1.85

9.16 Suppose a study reports that the average price for a litre of self-serve regular unleaded petrol in Europe is €1.37. You believe that the figure is higher in Ireland. You decide to test this claim for this part of Europe by randomly calling petrol stations. Your random survey of 25 stations produces the following prices:

1.36	1.37	1.32	1.33	1.40
1.33	1.35	1.33	1.33	1.35
1.32	1.28	1.36	1.29	1.40
1.31	1.35	1.31	1.27	1.40
1.34	1.31	1.31	1.28	1.29

Assume petrol prices for a region are normally distributed. Do the data you obtained provide enough evidence to reject the claim? Use a 1% level of significance.

9.17 Suppose that in past years the average price per square metre for warehouses in Italy has been €32.28. A national real estate investor wants to determine whether that figure has changed now. The investor hires a researcher who randomly samples 49 warehouses that are for sale across Italy and finds that the mean price per square metre is €31.67, with a standard deviation of €1.29. Assume that prices of a warehouse's square metres are normally distributed in population. If the researcher uses a 5% level of significance, what statistical conclusion can be reached? What are the hypotheses?

9.18 Major cities around the world compete with each other in an effort to attract new businesses. Some of the criteria that businesses use to judge cities as potential locations for their headquarters might include the labour pool; the environment, including work, governmental, and living; the tax structure, the availability of skilled/educated labour, housing, education, medical care; and others. Suppose in a study done several years ago, the city of Valencia, Spain, received a mean rating of 3.51 (on a scale of 1 to 5 and assuming an interval level of data) on housing, but that since that time, considerable residential building has occurred in the Valencia area such that city leaders feel the mean might now be higher. They hire a team of researchers to conduct a survey of businesses around the world to determine how businesses now rate the city on housing (and other variables). Sixty-one businesses take part in the new survey, with a result that Valencia receives a mean response of 3.72 on housing with a sample standard deviation of 0.65. Assuming that such responses are normally distributed, use a 1% level of significance and these data to test to determine if the mean housing rating for the city of Valencia by businesses has significantly increased.

9.19 Based on population figures and other general information on the German population, suppose it has been estimated that, on average, a family of four in Germany spends about €1,135 annually on dental expenditures. Suppose further that a regional dental association wants to test to determine if this figure is accurate for their area of the country. To test this, 22 families of four are randomly selected from the population in that area of the country and a log is kept of the family's dental expenditures for one year. The resulting data are given below. Assuming that dental expenditures are normally distributed in the population, use the data and an alpha of 0.05 to test the dental association's hypothesis.

1,008	812	1,117	1,323	1,308	1,415
831	1,021	1,287	851	930	730
699	872	913	944	954	987
1,695	995	1,003	994		

9.20 According to data released by the World Bank, the mean PM10 (particulate matter) concentration for the city of Kabul, Afghanistan, in 2008 was 37. Suppose that because of efforts to improve air quality in Kabul, increases in modernization, and efforts to establish environmentally-friendly businesses, city leaders believe rates of particulate matter in Kabul have decreased. To test this notion, they randomly sample 12 readings over a one-year period with the resulting readings shown below. Do these data present enough evidence to determine that PM10 readings are significantly less now in Kabul? Assume that particulate readings are normally distributed and that $\alpha = 0.01$.

26	37	28	46	50	40
24	33	24	31	46	38

9.21 According to Eurostat, the daily round trip average commuting time in Europe for people who commute to a city with a metropolitan population of more than five million is 64 minutes. Suppose a researcher lives in a city with a population of 5.4 million and wants to test this claim in her city. Assume that commuter times are normally distributed in the population. She takes a random sample of commuters and gathers data. The data are analysed using both Minitab and Excel, and the output is shown here. What are the results of the study? What are the hypotheses?

Minitab Output

One-Sample T

```
Test of mu = 19 vs not = 19

  N    Mean    StDev   SE Mean       95% CI         T       P
 24   66.083   3.550    0.725   (64.584, 67.582)   2.88   0.009
```

Excel Output

Mean	66.08333
Variance	12.60145
Observations	24
df	23
t Stat	2.8751
$P(T <= t)$ one-tail	0.0042
t Critical one-tail	1.7128
$P(T <= t)$ two-tail	0.0085
t Critical two-tail	2.0686

9.4   TESTING HYPOTHESES ABOUT A PROPORTION

Data analysis used in business decision making often contains proportions to describe such aspects as market share, consumer types, quality defects, on-time delivery rate, profitable stocks, and others. Business surveys often produce information expressed in proportion form, such as 0.45 of all businesses offer flexible hours to employees or 0.88 of all businesses have websites. Business researchers conduct hypothesis tests about such proportions to determine whether they have changed in some way. As an example, suppose a company held a 26%, or 0.26, share of the market for several years. Due to a massive marketing effort and improved product quality, company officials believe that the market share increased, and they want to prove it. Other examples of hypothesis testing about a single population proportion might include:

- A market researcher wants to test to determine whether the proportion of new car purchasers who are female has increased.
- A financial researcher wants to test to determine whether the proportion of companies that were profitable last year in the average investment officer's portfolio is 0.60.
- A quality manager for a large manufacturing firm wants to test to determine whether the proportion of defective items in a batch is less than 0.04.

Formula 9.4 for inferential analysis of a proportion was introduced in Section 7.3 of Chapter 7. Based on the central limit theorem, this formula makes possible the testing of hypotheses about the population proportion in a manner similar to that of the formula used to test sample means. Recall that $\hat{p}$ denotes a sample proportion and p denotes the population proportion. To validly use this test, the sample size must be large enough such that $n \cdot p \geq 5$ and $n \cdot q \geq 5$.

z TEST OF A POPULATION
PROPORTION (9.4)

$$z = \frac{\hat{p} - p}{\sqrt{\dfrac{p \cdot q}{n}}}$$

where:

$\hat{p}$ = sample proportion
p = population proportion
$q = 1 - p$

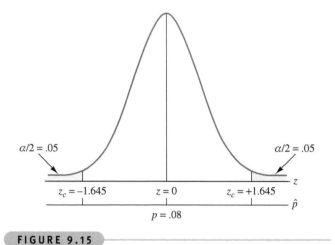

FIGURE 9.15

Distribution with Rejection Regions for Flawed-Product Example

A manufacturer believes exactly 8% of its products contain at least one minor flaw. Suppose a company researcher wants to test this belief. The null and alternative hypotheses are

$$H_0: p = 0.08$$
$$H_a: p \neq 0.08$$

This test is two-tailed because the hypothesis being tested is whether the proportion of products with at least one minor flaw is 0.08. Alpha is selected to be 0.10. Figure 9.15 shows the distribution, with the rejection regions and $z_{0.05}$. Because α is divided for a two-tailed test, the table value for an area of $(1/2)(0.10) = 0.05$ is $z_{0.05} = \pm 1.645$.

For the business researcher to reject the null hypothesis, the observed z value must be greater than 1.645 or less than -1.645. The business researcher randomly selects a sample of 200 products, inspects each item for flaws, and determines that 33 items have at least one minor flaw. Calculating the sample proportion gives

$$\hat{p} = \frac{33}{200} = 0.165$$

The observed z value is calculated as:

$$z = \frac{\hat{p} - p}{\sqrt{\dfrac{p \cdot q}{n}}} = \frac{0.165 - 0.080}{\sqrt{\dfrac{(0.08)(0.92)}{200}}} = \frac{0.085}{0.019} = 4.43$$

Note that the denominator of the z formula contains the population proportion. Although the business researcher does not actually know the population proportion, he is testing a population proportion value. Hence he uses the hypothesized population value in the denominator of the formula as well as in the numerator. This method contrasts with the confidence interval formula, where the sample proportion is used in the denominator.

The observed value of z is in the rejection region (observed $z = 4.43 >$ table $z_{0.05} = +1.645$), so the business researcher rejects the null hypothesis. He concludes that the proportion of items with at least one minor flaw in the population from which the sample of 200 was drawn is not 0.08. With an alpha of 0.10, the risk of committing a Type I error in this example is 0.10.

The observed value of $z = 4.43$ is outside the range of most values in virtually all z tables. Thus if the researcher were using the p-value to arrive at a decision about the null hypothesis, the probability would be 0.0000, and he would reject the null hypothesis.

The Minitab output shown in Figure 9.16 displays a p-value of 0.000 for this problem, underscoring the decision to reject the null hypothesis.

Test and CI for One Proportion

Test of p = 0.08 vs p not = 0.08

Sample	X	N	Sample p	90% CI	Exact P-Value
1	33	200	0.165000	(0.123279, 0.214351)	0.000

FIGURE 9.16

Minitab Output for the Flawed-Product Example

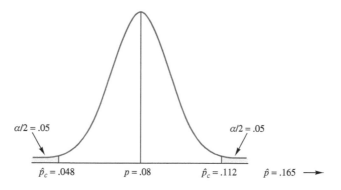

Distribution Using Critical Value Method for the Flawed-Product Example

Suppose the researcher wanted to use the critical value method. He would enter the table value of $z_{.05} = 1.645$ in the z formula for single sample proportions, along with the hypothesized population proportion and n, and solve for the critical value of rejection denoted as $\hat{p}_c$. The result is

$$z_{\alpha/2} = \frac{\hat{p}_c - p}{\sqrt{\dfrac{p \cdot q}{n}}}$$

$$\pm 1.645 = \frac{\hat{p}_c - 0.08}{\sqrt{\dfrac{(0.08)(0.92)}{200}}}$$

$$\hat{p}_c = 0.08 \pm 1.645 \sqrt{\frac{(0.08)(0.92)}{200}} = 0.08 \pm 0.032$$

$$= 0.048 \text{ and } 0.112$$

Using the critical value method, if the sample proportion is less than 0.048 or greater than 0.112, the decision will be to reject the null hypothesis. Since the sample proportion, $\hat{p}$, is 0.165, which is greater than 0.112, the decision here is to reject the null hypothesis. The proportion of products with at least one flaw is not 0.08. Figure 9.17 shows these critical values, the observed value, and the rejection regions.

STATISTICS IN BUSINESS TODAY

Testing Hypotheses about Commuting

How do the British commute to work? An Office for National Statistics survey taken a few years ago indicated that 76% of UK commuters drive to work, 11% walk, 6% use a bus, and the remaining 6% used other means of transportation (bicycle, train, among others). Using hypothesis-testing methodology presented in this chapter, researchers can test whether these proportions still hold true today as well as how these figures vary by region. For example, in London the proportion of commuters using public transportation is much higher (50%) than for the rest of the UK (9%). In rural parts of the country where public transportation is unavailable, the proportion of commuters using public transportation would be zero.

What is the average travel time of a commute to work in the United Kingdom? According to the Office for National Statistics, travel time varies according to the region of the country. For example, people working in London, in particular central London, tend to travel longer to get to work, with more than half, 56%, needing to commute for more than 30 minutes to get to work every day. In contrast, of those working in the rest of the UK, only 20% need to travel as long to reach their workplace. It is possible to test any of these means using hypothesis testing techniques presented in this chapter to either validate the figures or to determine whether the figures are no longer true.

DEMONSTRATION PROBLEM 9.3

A survey of the breakfast habits in Europe by the European Breakfast Cereal Association shows that the primary breakfast food for 34% of Europeans is breakfast cereals. A cereal producer in France, where cereals are plentiful, believes the figure is higher for France. To test this idea, she contacts a random sample of 550 French residents and asks which primary breakfast food they consumed for breakfast that day. Suppose 212 replied that breakfast cereal was the primary food. Using a level of significance of 0.05, test the idea that the cereal figure is higher for France.

Solution

HYPOTHESIZE

STEP 1: The cereal producer's theory is that the proportion of French residents who eat cereal for breakfast is higher than the European proportion, which is the alternative hypothesis. The null hypothesis is that the proportion in France does not differ from the European average. The hypotheses for this problem are

$$H_0: p = 0.34$$
$$H_a: p > 0.34$$

TEST

STEP 2: The test statistic is

$$z = \frac{\hat{p} - p}{\sqrt{\dfrac{p \cdot q}{n}}}$$

STEP 3: The Type I error rate is 0.05.

STEP 4: This test is a one-tailed test, and the table value is $z_{0.05} = +1.645$. The sample results must yield an observed z value greater than 1.645 for the cereal producer to reject the null hypothesis. The following diagram shows $z_{0.05}$ and the rejection region for this problem.

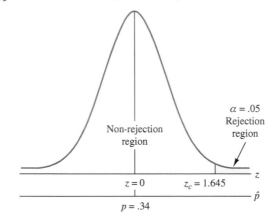

STEP 5: $n = 550$ and $x = 212$

STEP 6:

$$\hat{p} = \frac{212}{550} = 0.385$$

$$z = \frac{\hat{p} - p}{\sqrt{\dfrac{p \cdot q}{n}}} = \frac{0.385 - 0.34}{\sqrt{\dfrac{(0.34)(0.66)}{550}}} = \frac{0.045}{0.0202} = 2.23$$

ACTION

STEP 7: Because $z = 2.23$ is beyond $z_{0.05} = 1.645$ in the rejection region, the cereal producer rejects the null hypothesis. The probability of obtaining a $z \geq 2.23$ by chance is 0.013. Because this probability is less than $\alpha = 0.05$, the null hypothesis is also rejected with the p-value. On the basis of the random sample, the producer is ready to conclude that the proportion of French residents who eat cereal as the primary food for breakfast is higher than the European proportion.

BUSINESS IMPLICATIONS

STEP 8: If the proportion of residents who eat cereal for breakfast is higher in France than in other parts of Europe, cereal producers might have a market opportunity in France that is not available in other parts of the continent. Perhaps French residents are being loyal to home-country products, in which case marketers of other French products might be successful in appealing to residents to support their products. The fact that more cereal is sold in France might mean that if French cereal producers appealed to markets outside France in the same way they do inside the country, they might increase their market share of the breakfast food market in other countries. Is a proportion of almost 0.385 really a substantive increase over 0.34? Certainly in a market of any size at all, an increase of almost 5% of the market share could be worth millions of euros and in such a case, would be substantive.

A critical proportion can be solved for by

$$z_{0.05} = \frac{\hat{p}_c - p}{\sqrt{\dfrac{p \cdot q}{n}}}$$

$$1.645 = \frac{\hat{p}_c - 0.34}{\sqrt{\dfrac{(0.34)(0.66)}{550}}}$$

$$\hat{p}_c = 0.34 + 1.645\sqrt{\frac{(0.34)(0.66)}{550}} = 0.34 + 0.0202 = 0.3602$$

With the critical value method, a sample proportion greater than 0.3602 must be obtained to reject the null hypothesis. The sample proportion for this problem is 0.385, so the null hypothesis is also rejected with the critical value method.

Using the Computer to Test Hypotheses About a Population Proportion

Minitab has the capability of testing hypotheses about a population proportion. Figure 9.18 shows the Minitab output for Demonstration Problem 9.3. Notice that the output includes a restatement of the hypotheses, the sample proportion, and the p-value. From this information, a decision regarding the null hypothesis can be made by comparing the p-value (0.014) to a (0.050). Because the p-value is less than a, the decision is to reject the null hypothesis.

Test and CI for One Proportion

Test of p = 0.34 vs p > 0.34

Sample	X	N	Sample p	95% Lower Bound	Exact P-Value
1	212	550	0.385455	0.350928	0.014

FIGURE 9.18

Minitab Output for Demonstration Problem 9.3

9.4 PROBLEMS

9.22 Suppose you are testing H_0: $p = 0.45$ versus H_a: $p > 0.45$. A random sample of 310 people produces a value of $\hat{p} = 0.465$. Use $\alpha = 0.05$ to test this hypothesis.

9.23 Suppose you are testing H_0: $p = 0.63$ versus H_a: $p < 0.63$. For a random sample of 100 people, $x = 55$, where x denotes the number in the sample that have the characteristic of interest. Use a 0.01 level of significance to test this hypothesis.

9.24 Suppose you are testing H_0: $p = 0.29$ versus H_a: $p \neq 0.29$. A random sample of 740 items shows that 207 have this characteristic. With a 0.05 probability of committing a Type I error, test the hypothesis. For the p-value method, what is the probability of the observed z value for this problem? If you had used the critical value method, what would the two critical values be? How do the sample results compare with the critical values?

9.25 The European Commission recently conducted a survey on passengers' satisfaction with rail services and discovered that in the case of Bulgarians, 10.1% were very satisfied with the quality of facilities and services, 31.9% were rather satisfied, 29% were rather dissatisfied, and 22.3% were very dissatisfied. Suppose the largest train company in Bulgaria decides to invest considerable time and money to improve the quality of its facilities and services. After using the new policies for a year, company managers want to determine whether the new policy has changed perceptions of the service they offer. They contact 408 of the company's rail consumers who travelled with the company in the past year and ask them how satisfied they are with the rail services. Fifty-two respond that they are very satisfied. Use a 1% level of significance to test the hypothesis.

9.26 A study by Hewitt Associates showed that 79% of companies offer employees flexible scheduling. Suppose a researcher believes that in accounting firms this figure is lower. The researcher randomly selects 415 accounting firms and through interviews determines that 303 of these firms have flexible scheduling. With a 1% level of significance, does the test show enough evidence to conclude that a significantly lower proportion of accounting firms offer employees flexible scheduling?

9.27 A survey was undertaken by Bruskin/Goldring Research for Quicken to determine how people plan to meet their financial goals in the next year. Respondents were allowed to select more than one way to meet their goals. Thirty-one per cent said that they were using a financial planner to help them meet their goals. Twenty-four per cent were using family/friends to help them meet their financial goals followed by broker/accountant (19%), computer software (17%), and books (14%). Suppose another researcher takes a similar survey of 600 people to test these results. If 200 people respond that they are going to use a financial planner to help them meet their goals, is this proportion enough evidence to reject the 31% figure generated in the Bruskin/Goldring survey using $\alpha = 0.10$? If 158 respond that they are going to use family/friends to help them meet their financial goals, is this result enough evidence to declare that the proportion is significantly higher than Bruskin/Goldring's figure of 0.24 if $\alpha = 0.05$?

9.28 A 2011 survey by Mercer India Monitor asked 91 firms across various industries in India about their recruitment and training budgets for the following year. The survey found that 94% of respondents were planning to increase recruitment levels in the following year and only 6% were looking at maintaining or reducing the existing headcount level. Suppose a researcher thinks that Indian companies have changed their hiring plans in light of the recent worsening of international economic conditions. To test this hypothesis, a new study is conducted by contacting 376 Indian companies from a variety of industries. Two hundred and ninety two companies respond that they plan to increase recruitment in the following year. Does the test show enough evidence to declare that the proportion of companies that plan to increase recruitment is now significantly different? Let $\alpha = 0.01$.

9.29 A large manufacturing company investigated the service it received from suppliers and discovered that, in the past, 32% of all materials shipments were received late. However, the company recently installed a just-in-time system in which suppliers are linked more closely to the manufacturing process. A random sample of 118 deliveries since the just-in-time system was installed reveals that 22 deliveries were late. Use this sample information to test whether the proportion of late deliveries was reduced significantly. Let $\alpha = 0.05$.

9.30 Where do CFOs get their money news? According to Robert Half International, 47% get their money news from newspapers, 15% get it from communication/colleagues, 12% get it from television, 11% from the Internet, 9% from magazines, 5% from radio, and 1% don't know. Suppose a researcher wants to test these results. She randomly samples 67 CFOs and finds that 40 of them get their money news from newspapers. Does the test show enough evidence to reject the findings of Robert Half International? Use $\alpha = 0.05$.

9.5   TESTING HYPOTHESES ABOUT A VARIANCE

At times a researcher needs to test hypotheses about a population variance. For example, in the area of statistical quality control, manufacturers try to produce equipment and parts that are consistent in measurement. Suppose a company produces industrial wire that is specified to be a particular thickness. Because of the production process, the thickness of the wire will vary slightly from one end to the other and from lot to lot and batch to batch. Even if the average thickness of the wire as measured from lot to lot is on specification, the variance of the measurements might be too great to be acceptable. In other words, on the average the wire is the correct thickness, but some portions of the wire might be too thin and others unacceptably thick. By conducting hypothesis tests for the variance of the thickness measurements, the quality control people can monitor for variations in the process that are too great.

The procedure for testing hypotheses about a population variance is similar to the techniques presented in Chapter 8 for estimating a population variance from the sample variance. Formula 9.5 used to conduct these tests assumes a normally distributed population.

FORMULA FOR TESTING HYPOTHESES ABOUT POPULATION VARIANCE (9.5)	$$\chi^2 = \frac{(n-1)s^2}{\sigma^2}$$ $$df = n - 1$$

Note: *As was mentioned in Chapter 8, the chi-square test of a population variance is extremely sensitive to violations of the assumption that the population is normally distributed.*

As an example, a manufacturing firm has been working diligently to implement a just-in-time inventory system for its production line. The final product requires the installation of a pneumatic tube at a particular station on the assembly line. With the just-in-time inventory system, the company's goal is to minimize the number of pneumatic tubes that are piled up at the station waiting to be installed. Ideally, the tubes would arrive just as the operator needs them. However, because of the supplier and the variables involved in getting the tubes to the line, most of the time there will be some build-up of tube inventory. The company expects that, on average, about 20 pneumatic tubes will be at the station. However, the production superintendent does not want the variance of this inventory to be greater than 4. On a given day, the number of pneumatic tubes piled up at the workstation is determined eight different times and the following number of tubes are recorded:

23   17   20   29   21   18   19   24

Using these sample data, we can test to determine whether the variance is greater than 4. The hypothesis test is one tailed. Assume the number of tubes is normally distributed. The null hypothesis is that the variance is acceptable with no problems – the variance is equal to (or less than) 4. The alternative hypothesis is that the variance is greater than 4.

$$H_0: \sigma^2 = 4$$
$$H_a: \sigma^2 > 4$$

Suppose alpha is 0.05. Because the sample size is 8, the degrees of freedom for the critical table chi-square value are $8 - 1 = 7$. Using Table A.8 in Appendix A, we find the critical chi-square value.

$$\chi^2_{0.05,7} = 14.0671$$

Because the alternative hypothesis is greater than 4, the rejection region is in the upper tail of the chi-square distribution. The sample variance is calculated from the sample data to be

$$s^2 = 15.125$$

The observed chi-square value is calculated as

$$\chi^2 = \frac{(8-1)(15.125)}{4} = 26.47$$

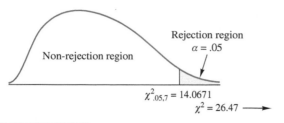

FIGURE 9.19

Hypothesis Test Distribution for Pneumatic Tube Example

Because this observed chi-square value, $\chi^2 = 26.47$ is greater than the critical chi-square table value, $\chi^2_{0.05,7} = 14.0671$ the decision is to reject the null hypothesis. On the basis of this sample of eight data measurements, the population variance of inventory at this workstation is greater than 4. Company production personnel and managers might want to investigate further to determine whether they can find a cause for this unacceptable variance. Figure 9.19 shows a chi-square distribution with the critical value, the rejection region, the non-rejection region, the value of α, and the observed chi-square value.

Using Excel, the p-value of the observed chi-square, 26.47, is determined to be 0.00042. Because this value is less than $\alpha = 0.05$, the conclusion is to reject the null hypothesis using the p-value. In fact, using this p-value, the null hypothesis could be rejected for

$$\alpha = 0.001$$

This null hypothesis can also be tested by the critical value method. Instead of solving for an observed value of chi-square, the critical chi-square value for alpha is inserted into formula 9.5 along with the hypothesized value of σ^2 and the degrees of freedom $(n - 1)$.

Solving for s^2 yields a critical sample variance value, s_c^2.

$$\chi_c^2 = \frac{(n-1)s_c^2}{\sigma^2}$$

$$s_c^2 = \frac{\chi_c^2 \cdot \sigma^2}{(n-1)} = \frac{(14.0671)(4)}{7} = 8.038$$

The critical value of the sample variance is $s_c^2 = 8.038$. Because the observed sample variance actually was 15.125, which is larger than the critical variance, the null hypothesis is rejected.

DEMONSTRATION PROBLEM 9.4

A small business has 37 employees. Because of the uncertain demand for its product, the company usually pays overtime on any given week. The company assumed that about 50 total hours of overtime per week is required and that the variance on this figure is about 25. Company officials want to know whether the variance of overtime hours has changed. Given here is a sample of 16 weeks of overtime data (in hours per week). Assume hours of overtime are normally distributed. Use these data to test the null hypothesis that the variance of overtime data is 25. Let $\alpha = 0.05$.

57	56	52	44
46	53	44	44
48	51	55	48
63	53	51	50

Solution

HYPOTHESIZE

STEP 1: This is a two-tailed test. The null and alternative hypotheses are

$$H_0: \sigma^2 = 25$$
$$H_a: \sigma^2 \neq 25$$

TEST

STEP 2: The test statistic is

$$\chi^2 = \frac{(n-1)s^2}{\sigma^2}$$

STEP 3: Because this test is two tailed, $\alpha = 0.10$ must be split: $\alpha/2 = 0.05$.

STEP 4: The degrees of freedom are $16 - 1 = 15$. The two critical chi-square values are

$$\chi^2_{(1-0.05),15} = \chi^2_{0.95,15} = 7.26093$$

$$\chi^2_{0.05,15} = 24.9958$$

The decision rule is to reject the null hypothesis if the observed value of the test statistic is less than 7.26093 or greater than 24.9958.

STEP 5: The data are as listed previously.

STEP 6: The sample variance is

$$s^2 = 28.0625$$

The observed chi-square value is calculated as

$$\chi^2 = \frac{(n-1)s^2}{\sigma^2} = \frac{(15)(28.0625)}{25} = 16.84$$

ACTION

STEP 7: This observed chi-square value is in the non-rejection region because $\chi^2_{0.95,15} = 7.26094 < \chi^2_{observed} = 16.84 < \chi^2_{0.05,15} = 24.9958$. The company fails to reject the null hypothesis. The population variance of overtime hours per week is 25.

BUSINESS IMPLICATIONS

STEP 8: This result indicates to the company managers that the variance of weekly overtime hours is about what they expected.

9.5 PROBLEMS

9.31 Test each of the following hypotheses by using the given information. Assume the populations are normally distributed.

a. $H_0: \sigma^2 = 20$
$H_a: \sigma^2 > 20$
$\alpha = 0.05, n = 15, s^2 = 32$

b. $H_0: \sigma^2 = 8.5$
$H_a: \sigma^2 \neq 8.5$
$\alpha = 0.10, n = 22, s^2 = 17$

c. $H_0: \sigma^2 = 45$
$H_a: \sigma^2 < 45$
$\alpha = 0.01, n = 8, s^2 = 4.12$

d. $H_0: \sigma^2 = 5$
$H_a: \sigma^2 \neq 5$
$\alpha = 0.05, n = 11, s^2 = 1.2$

9.32 Previous experience shows the variance of a given process to be 14. Researchers are testing to determine whether this value has changed. They gather the following dozen measurements of the process. Use these data and $\alpha = 0.05$ to test the null hypothesis about the variance. Assume the measurements are normally distributed.

52	44	51	58	48	49
38	49	50	42	55	51

9.33   A manufacturing company produces bearings. One line of bearings is specified to be 1.64 centimetres in diameter. A major customer requires that the variance of the bearings be no more than 0.001 cm^2. The producer is required to test the bearings before they are shipped, and so the diameters of 16 bearings are measured with a precise instrument, resulting in the following values. Assume bearing diameters are normally distributed. Use the data and $\alpha = 0.01$ to test the data to determine whether the population of these bearings is to be rejected because of too high a variance.

1.69	1.62	1.63	1.70
1.66	1.63	1.65	1.71
1.64	1.69	1.57	1.64
1.59	1.66	1.63	1.65

9.34   A given branch of a major bank averages about €100,000 in deposits per week. However, because of the way pay periods fall, seasonality, and erratic fluctuations in the local economy, deposits are subject to a wide variability. In the past, the variance for weekly deposits has been about €199,996,164. In terms that make more sense to managers, the standard deviation of weekly deposits has been €14,142. Shown here are data from a random sample of 13 weekly deposits for a recent period. Assume weekly deposits are normally distributed. Use these data and $\alpha = 0.01$ to test to determine whether the variance for weekly deposits has changed.

€93,000	€135,000	€112,000
68,000	46,000	104,000
128,000	143,000	131,000
104,000	96,000	71,000
87,000		

9.35   A company produces industrial wiring. One batch of wiring is specified to be 2.16 centimetres thick. A company inspects the wiring in seven locations and determines that, on the average, the wiring is about 2.16 centimetres thick. However, the measurements vary. It is unacceptable for the variance of the wiring to be more than 0.04 cm^2. The standard deviation of the seven measurements on this batch of wiring is 0.34 centimetres. Use $\alpha = 0.01$ to determine whether the variance on the sample wiring is too great to meet specifications. Assume wiring thickness is normally distributed.

9.6   SOLVING FOR TYPE II ERRORS

If a researcher reaches the statistical conclusion to fail to reject the null hypothesis, he makes either a correct decision or a Type II error. If the null hypothesis is true, the researcher makes a correct decision. If the null hypothesis is false, then the result is a Type II error.

In business, failure to reject the null hypothesis may mean staying with the status quo, not implementing a new process, or not making adjustments. If a new process, product, theory, or adjustment is not significantly better than what is currently accepted practice, the decision maker makes a correct decision. However, if the new process, product, theory, or adjustment would significantly improve sales, the business climate, costs, or morale, the decision maker makes an error in judgement (Type II). In business, Type II errors can translate to lost opportunities, poor product quality (as a result of failure to discern a problem in the process), or failure to react to the marketplace. Sometimes the ability to react to changes, new developments, or new opportunities is what keeps a business moving and growing. The Type II error plays an important role in business statistical decision making.

Determining the probability of committing a Type II error is more complex than finding the probability of committing a Type I error. The probability of committing a Type I error either is given in a problem or is stated by the researcher before proceeding with the study. A Type II error, β, varies with possible values of the alternative parameter. For example, suppose a researcher is conducting a statistical test on the following hypotheses:

$$H_0: \mu = 12 \text{ grams}$$
$$H_a: \mu < 12 \text{ grams}$$

A Type II error can be committed only when the researcher fails to reject the null hypothesis and the null hypothesis is false. In these hypotheses, if the null hypothesis, $\mu = 12$ grams, is false, what is the true value for the population mean? Is the mean really 11.99 or 11.90 or 11.5 or 10 grams? For each of these possible values of the population mean, the researcher can compute the probability of committing a Type II error. Often, when the null hypothesis is false, the value of the alternative mean is unknown, so the researcher will compute the probability of committing Type II errors for several possible values. How can the probability of committing a Type II error be computed for a specific alternative value of the mean?

Suppose that, in testing the preceding hypotheses, a sample of 60 bottles of cinnamon powder yields a sample mean of 11.985 grams. Assume that the population standard deviation is 0.10 grams. From $\alpha = 0.05$ and a one-tailed test, the table $z_{0.05}$ value is -1.645. The observed z value from sample data is

$$z = \frac{11.985 - 12.00}{\frac{0.10}{\sqrt{60}}} = -1.16$$

From this observed value of z, the researcher determines not to reject the null hypothesis. By not rejecting the null hypothesis, the researcher either makes a correct decision or commits a Type II error. What is the probability of committing a Type II error in this problem if the population mean actually is 11.99?

The first step in determining the probability of a Type II error is to calculate a critical value for the sample mean, $\bar{x}_c$. In testing the null hypothesis by the critical value method, this value is used as the cut-off for the non-rejection region. For any sample mean obtained that is less than $\bar{x}_c$ (or greater for an upper-tail rejection region), the null hypothesis is rejected. Any sample mean greater than $\bar{x}_c$ (or less for an upper-tail rejection region) causes the researcher to fail to reject the null hypothesis. Solving for the critical value of the mean gives

$$z_c = \frac{\bar{x}_c - \mu}{\frac{\sigma}{\sqrt{n}}}$$

$$-1.645 = \frac{\bar{x}_c - 12}{\frac{0.10}{\sqrt{60}}}$$

$$\bar{x}_c = 11.979$$

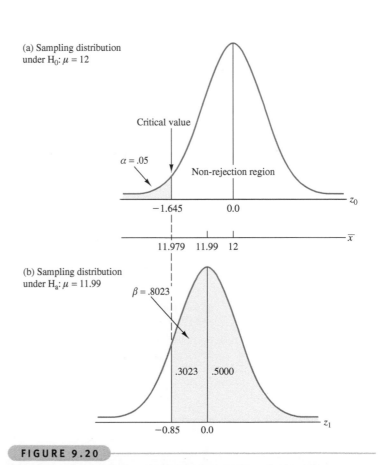

(a) Sampling distribution under H_0: $\mu = 12$

(b) Sampling distribution under H_a: $\mu = 11.99$

FIGURE 9.20

Type II Error for the Cinnamon Bottles Example with Alternative Mean = 11.99 grams rejected

Figure 9.20(a) shows the distribution of values when the null hypothesis is true. It contains a critical value of the mean, $\bar{x}_c = 11.979$ grams, below which the null hypothesis will be. Figure 9.20(b) shows the distribution when the alternative mean, $\mu_1 = 11.99$ grams, is true. How often will the business researcher fail to reject the top distribution as true when, in reality, the bottom distribution is true? If the null hypothesis is false, the researcher will fail to reject the null hypotheses whenever $\bar{x}$ is in the non-rejection region, $\bar{x}_c \geq 11.979$ grams. If μ actually equals 11.99 grams, what is the probability of failing to reject $\mu = 12$ grams when 11.979 grams is the critical value? The business researcher calculates this probability by extending the critical value ($\bar{x}_c = 11.979$ grams) from distribution (a) to distribution (b) and solving for the area to the right of $\bar{x}_c = 11.979$.

$$z_1 = \frac{\bar{x}_c - \mu_1}{\frac{\sigma}{\sqrt{n}}} = \frac{11.979 - 11.99}{\frac{0.10}{\sqrt{60}}} = -0.85$$

This value of z yields an area of 0.3023. The probability of committing a Type II error is all the area to the right of $\bar{x}_c = 11.979$ in distribution (b), or $0.3023 + 0.5000 = 0.8023$. Hence there is an 80.23% chance of committing a Type II error if the alternative mean is 11.99 grams.

DEMONSTRATION PROBLEM 9.5

Recompute the probability of committing a Type II error for the cinnamon bottle example if the alternative mean is 11.96 grams.

Solution

Everything in distribution (a) of Figure 9.20 stays the same. The null hypothesized mean is still 12 grams, the critical value is still 11.979 grams, and $n = 60$. However, distribution (b) of Figure 9.20 changes with $\mu_1 = 11.96$ grams, as the diagram below shows.
    The z formula used to solve for the area of distribution (b), $\mu_1 = 11.96$, to the right of 11.979 is

$$z_1 = \frac{\bar{x}_c - \mu_1}{\dfrac{\sigma}{\sqrt{n}}} = \frac{11.979 - 11.96}{\dfrac{0.10}{\sqrt{60}}} = 1.47$$

From Table A.5, only 0.0708 of the area is to the right of the critical value. Thus the probability of committing a Type II error is only 0.0708, as illustrated in the following diagram:

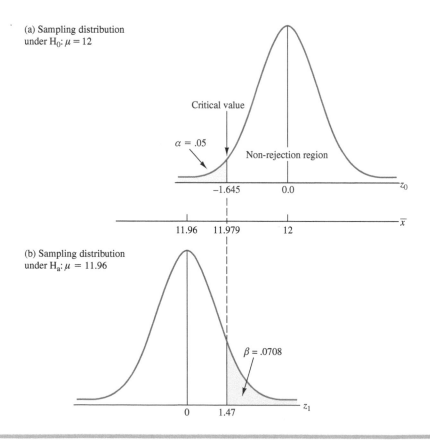

DEMONSTRATION PROBLEM 9.6

Suppose you are conducting a two-tailed hypothesis test of proportions. The null hypothesis is that the population proportion is 0.40. The alternative hypothesis is that the population proportion is not 0.40. A random sample of 250 produces a sample proportion of 0.44. With alpha of 0.05, the table z value for $\alpha/2$ is ± 1.96. The observed z from the sample information is

$$z = \frac{\hat{p} - p}{\sqrt{\dfrac{p \cdot q}{n}}} = \frac{0.44 - 0.40}{0.031} = 1.29$$

Thus the null hypothesis is not rejected. Either a correct decision is made or a Type II error is committed. Suppose the alternative population proportion really is 0.36. What is the probability of committing a Type II error?

Solution

Solve for the critical value of the proportion:

$$z_c = \frac{\hat{p}_c - p}{\sqrt{\dfrac{p \cdot q}{n}}}$$

$$\pm 1.96 = \frac{\hat{p}_c - 0.40}{\sqrt{\dfrac{(0.40)(0.60)}{250}}}$$

$$\hat{p}_c = 0.40 \pm 0.06$$

The critical values are 0.34 on the lower end and 0.46 on the upper end. The alternative population proportion is 0.36. The following diagram illustrates these results and the remainder of the solution to this problem:

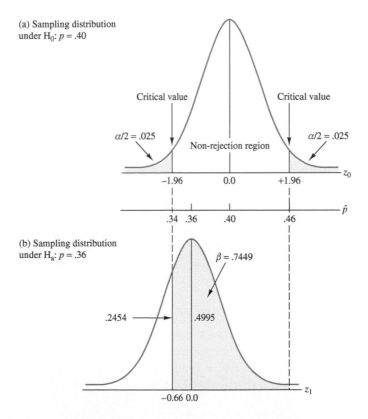

(a) Sampling distribution under H_0: $p = .40$

Critical value   Critical value

$\alpha/2 = .025$ Non-rejection region $\alpha/2 = .025$

-1.96 0.0 $+1.96$ z_0

.34 .36   .40   .46 $\hat{p}$

(b) Sampling distribution under H_a: $p = .36$

$\beta = .7449$

.2454   .4995

$-0.66\ 0.0$ z_1

Solving for the area between $\hat{p}_c = 0.34$ and $p_1 = 0.36$ yields

$$z_1 = \frac{0.34 - 0.36}{\sqrt{\dfrac{(0.36)(0.64)}{250}}} = -0.66$$

The area associated with $z_1 = -0.66$ is 0.2454.

The area between 0.36 and 0.46 of the sampling distribution under H_a: $p = 0.36$ (graph (b)) can be solved for by using the following z value:

$$z = \frac{0.46 - 0.36}{\sqrt{\dfrac{(0.36)(0.64)}{250}}} = 3.29$$

The area from Table A.5 associated with $z = 3.29$ is 0.4995. Combining this value with the 0.2454 obtained from the left side of the distribution in graph (b) yields the total probability of committing a Type II error:

$$0.2454 + 0.4995 = 0.7449$$

With two-tailed tests, both tails of the distribution contain rejection regions. The area between the two tails is the non-rejection region and the region where Type II errors can occur. If the alternative hypothesis is true, the area of the sampling distribution under H_a between the locations where the critical values from H_0 are located is β. In theory, both tails of the sampling distribution under H_a would be non-β area. However, in this problem, the right critical value is so far away from the alternative proportion ($p_1 = 0.36$) ($p_1 = 0.36$) that the area between the right critical value and the alternative proportion is near 0.5000 (0.4995) and virtually no area falls in the upper right tail of the distribution (0.0005).

Some Observations About Type II Errors

Type II errors are committed only when the researcher fails to reject the null hypothesis but the alternative hypothesis is true. If the alternative mean or proportion is close to the hypothesized value, the probability of committing a Type II error is high. If the alternative value is relatively far away from the hypothesized value, as in the problem with $\mu = 12$ grams and $\mu_a = 11.96$ grams, the probability of committing a Type II error is small. The implication is that when a value is being tested as a null hypothesis against a true alternative value that is relatively far away, the sample statistic obtained is likely to show clearly which hypothesis is true. For example, suppose a researcher is testing to determine whether a company really is filling 2-litre bottles of cola with an average of 2 litres. If the company decides to underfill the bottles by filling them with only 1 litre, a sample of 50 bottles is likely to average a quantity near the 1-litre fill rather than near the 2-litre fill. Committing a Type II error is highly unlikely. Even a customer probably could see by looking at the bottles on the shelf that they are underfilled. However, if the company fills 2-litre bottles with 1.99 litres, the bottles are close in fill volume to those filled with 2.00 litres. In this case, the probability of committing a Type II error is much greater. A customer probably could not catch the underfill just by looking.

In general, if the alternative value is relatively far from the hypothesized value, the probability of committing a Type II error is smaller than it is when the alternative value is close to the hypothesized value. The probability of committing a Type II error decreases as alternative values of the hypothesized parameter move farther away from the hypothesized value. This situation is shown graphically in operating characteristic curves and power curves.

Operating Characteristic and Power Curves

Because the probability of committing a Type II error changes for each different value of the alternative parameter, it is best in managerial decision making to examine a series of possible alternative values. For example, Table 9.2 shows the probabilities of committing a Type II error (β) for several different possible alternative means for the cinnamon bottle example discussed in Demonstration Problem 9.5, in which the null hypothesis was H_0: $\mu = 12$ grams and $\alpha = 0.05$.

As previously mentioned, power is the probability of rejecting the null hypothesis when it is false and represents the correct decision of selecting the alternative hypothesis when it is true. Power is equal to $1 - \beta$. Note that Table 9.2 also contains the power values for the alternative means and that the β and power probabilities sum to 1 in each case.

These values can be displayed graphically as shown in Figures 9.21 and 9.22. Figure 9.21 is a Minitab-generated **operating characteristic (OC) curve** *constructed by plotting the β values against the various values of the alternative hypothesis.* Notice that when the alternative means are near the value of the null hypothesis, $\mu = 12$, the probability

TABLE 9.2

β Values and Power Values for the Cinnamon Bottles Example

Alternative Mean	Probability of Committing A Type II Error, β	Power
$\mu_a = 11.999$	.94	.06
$\mu_a = 11.995$	.89	.11
$\mu_a = 11.99$	.80	.20
$\mu_a = 11.98$	.53	.47
$\mu_a = 11.97$	.24	.76
$\mu_a = 11.96$	.07	.93
$\mu_a = 11.95$	.01	.99

of committing a Type II error is high because it is difficult to discriminate between a distribution with a mean of 12 and a distribution with a mean of 11.999. However, as the values of the alternative means move away from the hypothesized value, $\mu = 12$, the values of β drop. This visual representation underscores the notion that it is easier to discriminate between a distribution with $\mu = 12$ and a distribution with $\mu = 11.95$ than between distributions with $\mu = 12$ and $\mu = 11.999$.

Figure 9.22 is an Excel **power curve** constructed by *plotting the power values $(1 - \beta)$ against the various values of the alternative hypotheses.* Note that the power increases as the alternative mean moves away from the value of μ in the null hypotheses. This relationship makes sense. As the alternative mean moves farther and farther away from the null hypothesized mean, a correct decision to reject the null hypothesis becomes more likely.

Effect of Increasing Sample Size on the Rejection Limits

The size of the sample affects the location of the rejection limits. Consider the soft drink example in which we were testing the following hypotheses.

$$H_0: \mu = 12 \text{ grams}$$
$$H_a: \mu < 12 \text{ grams}$$

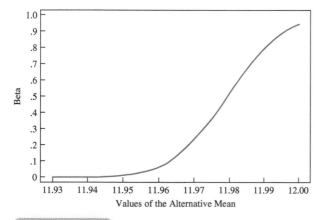

FIGURE 9.21

Minitab Operating-Characteristic Curve for the Cinnamon Bottles Example

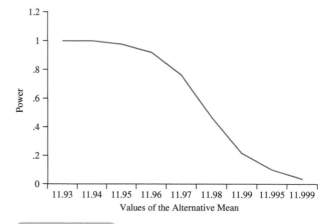

FIGURE 9.22

Excel Power Curve for the Cinnamon Bottles Example

Sample size was 60 ($n = 60$) and the standard deviation was 0.10 ($\sigma = 0.10$). With $\alpha = 0.05$, the critical value of the test statistic was $z_{0.05} = -1.645$. From this information, a critical raw score value was computed:

$$z_c = \frac{\bar{x}_c - \mu}{\dfrac{\sigma}{\sqrt{n}}}$$

$$-1.645 = \frac{\bar{x}_c - 12}{\dfrac{0.10}{\sqrt{60}}}$$

$$\bar{x}_c = 11.979$$

Any sample mean obtained in the hypothesis-testing process that is less than 11.979 will result in a decision to reject the null hypothesis.

Suppose the sample size is increased to 100. The critical raw score value is

$$-1.645 = \frac{\bar{x}_c - 12}{\dfrac{0.10}{\sqrt{100}}}$$

$$\bar{x}_c = 11.984$$

Notice that the critical raw score value is nearer to the hypothesized value ($\mu = 12$) for the larger sample size than it was for a sample size of 60. Because n is in the denominator of the standard error of the mean ($\sigma/\sqrt{n}$), an increase in n results in a decrease in the standard error of the mean, which when multiplied by the critical value of the test statistic ($z_{\alpha/2}$) results in a critical raw score that is closer to the hypothesized value. For $n = 500$ the critical raw score value for this problem is 11.993.

Increased sample size not only affects the distance of the critical raw score value from the hypothesized value of the distribution, but also can result in reducing β for a given value of α. Examine Figure 9.20. Note that the critical raw score value is 11.979 with alpha equal to 0.05 for $n = 60$. The value of β for an alternative mean of 11.99 is 0.8023. Suppose the sample size is 100. The critical raw score value (already solved) is 11.984. The value of β is now 0.7257. The computation is

$$z = \frac{11.984 - 11.99}{\dfrac{0.10}{\sqrt{100}}} = -0.60$$

The area under the standard normal curve for $z = -0.60$ is 0.2257. Adding $0.2257 + 0.5000$ (from the right half of the H_a sampling distribution) results in a β of 0.7257. Figure 9.23 shows the sampling distributions with α and β for this problem. In addition, by increasing sample size a business researcher could reduce alpha without necessarily increasing beta. It is possible to reduce the probabilities of committing Type I and Type II errors simultaneously by increasing sample size.

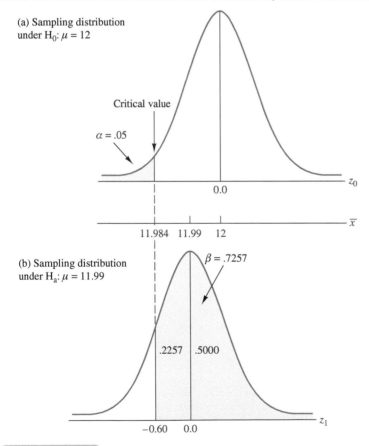

FIGURE 9.23

Type II Error for Cinnamon Bottles Example with n Increased to 100

9.6   PROBLEMS

9.36 Suppose a null hypothesis is that the population mean is greater than or equal to 100. Suppose further that a random sample of 48 items is taken and the population standard deviation is 14. For each of the following a values, compute the probability of committing a Type II error if the population mean actually is 99:

a. $\alpha = 0.10$
b. $\alpha = 0.05$
c. $\alpha = 0.01$
d. Based on the answers to parts (a), (b), and (c), what happens to the value of β as α gets smaller?

9.37 For Problem 9.36, use $\alpha = 0.05$ and solve for the probability of committing a Type II error for the following possible true alternative means:

a. $\mu_a = 98.5$
b. $\mu_a = 98$
c. $\mu_a = 97$
d. $\mu_a = 96$
e. What happens to the probability of committing a Type II error as the alternative value of the mean gets farther from the null hypothesized value of 100?

9.38 Suppose a hypothesis states that the mean is exactly 50. If a random sample of 35 items is taken to test this hypothesis, what is the value of β if the population standard deviation is 7 and the alternative mean is 53? Use $\alpha = 0.01$.

9.39 An alternative hypothesis is that $p < 0.65$. To test this hypothesis, a random sample of size 360 is taken. What is the probability of committing a Type II error if $\alpha = 0.05$ and the alternative proportion is as follows?

a. $p_a = 0.60$
b. $p_a = 0.55$
c. $p_a = 0.50$

9.40 The London Stock Exchange recently reported that the average age of a female shareholder is 44 years. A broker in Manchester wants to know whether this figure is accurate for the female shareholders in Manchester. The broker secures a master list of shareholders in Manchester and takes a random sample of 58 women. Suppose the average age for shareholders in the sample is 45.1 years, with a population standard deviation of 8.7 years. Test to determine whether the broker's sample data differ significantly enough from the 44-years figure released by the London Stock Exchange to declare that Manchester female shareholders are different in age from female shareholders in general. Use $\alpha = 0.05$. If no significant difference is noted, what is the broker's probability of committing a Type II error if the average age of a female Manchester shareholder is actually 45 years? 46 years? 47 years? 48 years? Construct an OC curve for these data. Construct a power curve for these data.

9.41 A *Reader's Digest* European Trusted Brands Survey is regularly taken to determine which brands consumers in European countries most trust. The most trusted brands in the 16 European Countries included in the survey were Nivea and Nokia. In the category of most trusted Mobile Phone Handset, the brand that received most votes by Russian consumers was Nokia with 54% of preferences. Suppose due to the entrance of new competitors in the market, a researcher feels that this figure is now too high. He takes a poll of 463 Russians, and 245 say that their most trusted mobile phone handset is Nokia. Does the survey show enough evidence to declare that the proportion of Russians saying that the their most trusted brand of mobile phone handset is Nokia is significantly lower than stated in the *Reader's Digest* poll? Let alpha equal 0.10. If the researcher fails to reject the null hypothesis and if the figure is actually 52% now, what is the probability of committing a Type II error? What is the probability of committing a Type II error if the figure is really 49%? 40%?

Decision Dilemma SOLVED

Word-of-Mouth Business: Recommendations and *Influentials*

In the Decision Dilemma, many data facts are reported from various surveys about consumers seeking advice from others before purchasing items or services. Most of the statistics are stated as though they are facts about the population. For example, one study reports that 70% of all consumers say that they completely or somewhat trust consumer opinions posted online Suppose a business researcher believes that this figure is not true, has changed over time, is not true for a particular region of the country, or is different for a particular type of website. Using hypothesis techniques presented in Section 9.4 of this chapter, this figure (70%) can be tested as a population proportion. Because the figures presented in the Decision Dilemma have been published and widely disseminated, the researcher who wants to test them would likely place these figures in the null hypothesis (e.g., H_0: $p = 0.70$), gather a random sample from whichever population is to be studied, and conduct a hypothesis test.

It was reported by Roper Starch Worldwide that *influentials* make recommendations about office equipment an average of 5.8 times per year. These and any of the other means reported in this study could be tested. The researcher would need to scientifically identify *influentials* in the population and randomly select a sample. A research mechanism could be set up whereby the number of referrals by each influential could be recorded

for a year and averaged, thereby producing a sample mean and a sample standard deviation. Using a selected value of alpha, the sample mean could be statistically tested against the population mean (in this case, H_0: $\mu = 5.8$). The probability of falsely rejecting a true null would be alpha. If the null was actually false (μ 5.8), the probability (β) of failing to reject the false null hypothesis would depend upon what the true number of mean referrals per year was for *influentials* on office equipment.

If a researcher has theories on *influentials* and these research theories can be stated as statistical hypotheses, the theory should be formulated as an alternate hypothesis; and the null hypothesis should be that the theory is not true. Samples are randomly selected. If the statistic of choice is a mean, then a z test or t test for a population mean should be used in the analysis dependent on whether or not the population standard deviation is known or unknown. In many studies, the sample standard deviation is used in the analysis instead of the unknown population standard deviation. In these cases, a t test should be used when the assumption that the population data are normally distributed can be made. If the statistic of interest is a proportion, then the z test for a population proportion is the technique to apply. Techniques presented in Chapter 8, Section 8.5, can be used to assist the researcher in determining how large the sample size should be in each case.

ETHICAL CONSIDERATIONS

The process of hypothesis testing encompasses several areas that could potentially lead to unethical activity, beginning with the null and alternative hypotheses. In the hypothesis-testing approach, the preliminary assumption is that the null hypothesis is true. If a researcher has a new theory or idea that he or she is attempting to prove, it is somewhat unethical to express that theory or idea as the null hypothesis. In doing so, the researcher is assuming that what he or she is trying to prove is true and the burden of proof is on

the data to reject this idea or theory. The researcher must take great care not to assume that what he or she is attempting to prove is true.

Hypothesis testing through random sampling opens up many possible unethical situations that can occur in sampling, such as identifying a frame that is favourable to the outcome the researcher is seeking or using non-random sampling techniques to test hypotheses. In addition, the researcher should be careful to use the proper test statistic for tests of a population mean, particularly when s is unknown. If *t* tests are used, or in testing a population variance, the researcher should be careful to apply the techniques only when it can be shown with some confidence that the population is normally distributed. The chi-square test of a population variance has been shown to be extremely sensitive to the assumption that the population is normally distributed. Unethical usage of this technique occurs when the statistician does not carefully check the population distribution shape for compliance with this assumption. Failure to do so can easily result in the reporting of spurious conclusions.

It can be unethical from a business decision-making point of view to knowingly use the notion of statistical significance to claim business significance when the results are not substantive. Therefore, it is unethical to intentionally attempt to mislead the business user by inappropriately using the word *significance*.

SUMMARY

Three types of hypotheses were presented in this chapter: research hypotheses, statistical hypotheses, and substantive hypotheses. Research hypotheses are statements of what the researcher believes will be the outcome of an experiment or study. In order to test hypotheses, business researchers formulate their research hypotheses into statistical hypotheses. All statistical hypotheses consist of two parts: a null hypothesis and an alternative hypothesis. The null and alternative hypotheses are structured so that either one or the other is true but not both. In testing hypotheses, the researcher assumes that the null hypothesis is true. By examining the sampled data, the researcher either rejects or does not reject the null hypothesis. If the sample data are significantly in opposition to the null hypothesis, the researcher rejects the null hypothesis and accepts the alternative hypothesis by default.

Hypothesis tests can be one tailed or two tailed. Two-tailed tests always utilize = and $\neq$ in the null and alternative hypotheses. These tests are non-directional in that significant deviations from the hypothesized value that are either greater than or less than the value are in rejection regions. The one-tailed test is directional, and the alternative hypothesis contains < or > signs. In these tests, only one end or tail of the distribution contains a rejection region. In a one-tailed test, the researcher is interested only in deviations from the hypothesized value that are either greater than or less than the value but not both.

Not all statistically significant outcomes of studies are important business outcomes. A substantive result is when the outcome of a statistical study produces results that are important to the decision maker.

When a business researcher reaches a decision about the null hypothesis, the researcher either makes a correct decision or an error. If the null hypothesis is true, the researcher can make a Type I error by rejecting the null hypothesis. The probability of making a Type I error is alpha (α). Alpha is usually set by the researcher when establishing the hypotheses. Another expression sometimes used for the value of a is level of significance.

If the null hypothesis is false and the researcher fails to reject it, a Type II error is committed. Beta (β) is the probability of committing a Type II error. Type II errors must be computed from the hypothesized value of the parameter, a, and a specific alternative value of the parameter being examined. As many possible Type II errors in a problem exist as there are possible alternative statistical values.

If a null hypothesis is true and the researcher fails to reject it, no error is committed, and the researcher makes a correct decision. Similarly, if a null hypothesis is false and it is rejected, no error is committed. Power $(1 - \beta)$ is the probability of a statistical test rejecting the null hypothesis when the null hypothesis is false.

An operating characteristic (OC) curve is a graphical depiction of values of β that can occur as various values of the alternative hypothesis are explored. This graph can be studied to determine what happens to β as one moves away from the value of the null hypothesis. A power curve is used in conjunction with an operating characteristic curve. The power curve is a graphical depiction of the values of power as various values of the alternative hypothesis are examined. The researcher can view the increase in power as values of the alternative hypothesis diverge from the value of the null hypothesis.

Included in this chapter were hypothesis tests for a single mean when σ is known and when σ unknown, a test of a single population proportion, and a test for a population variance. Three different analytic approaches were presented: (1) standard method, (2) the *p*-value, and (3) critical value method.

KEY TERMS

alpha (α)
alternative hypothesis
beta (β)
critical value
critical value method
hypothesis
hypothesis testing
level of significance
Non-rejection region
null hypothesis
observed significance
   level
observed value

one-tailed test
operating characteristic (OC) curve
p-value
power
power curve
rejection region
research hypothesis
statistical hypothesis
substantive result
two-tailed test
Type I error
Type II error

GO ONLINE TO DISCOVER THE EXTRA FEATURES FOR THIS CHAPTER

The Student Study Guide containing solutions to the odd-numbered questions, additional Quizzes and Concept Review Activities, Excel and Minitab databases, additional data files in Excel and Minitab, and more worked examples.
www.wiley.com/college/cortinhas

FORMULAS

z test for a single mean (9.1)

$$z = \frac{\bar{x} - \mu}{\frac{\sigma}{\sqrt{n}}}$$

Formula to test hypotheses about μ with a finite population (9.2)

$$z = \frac{\bar{x} - \mu}{\frac{\sigma}{\sqrt{n}}\sqrt{\frac{N-n}{N-1}}}$$

t test for a single mean (9.3)

$$t = \frac{\bar{x} - \mu}{\frac{s}{\sqrt{n}}}$$

$$\text{df} = n - 1$$

z test of a population proportion (9.4)

$$z = \frac{\hat{p} - p}{\sqrt{\frac{p \cdot q}{n}}}$$

Formula for testing hypotheses about a population variance (9.5)

$$\chi^2 = \frac{(n-1)s^2}{\sigma^2}$$

$$\text{df} = n - 1$$

SUPPLEMENTARY PROBLEMS

CALCULATING THE STATISTICS

9.42 Use the information given and the HTAB system to test the hypotheses. Let $\alpha = 0.01$.

H_a: $\mu = 36$ H_a: $\mu \neq 36$ $n = 63$ $\bar{x} = 38.4$
$\sigma = 5.93$

9.43 Use the information given and the HTAB system to test the hypotheses. Let alpha be 0.05. Assume the population is normally distributed.

H_0: $\mu = 7.82$ H_a: $\mu < 7.82$ $n = 17$ $\bar{x} = 17.1$
$s = 1.7$

9.44 For each of the problems below, test the hypotheses. Incorporate the HTAB system with its eight-step process.
a. H_0: $p = 0.28$ H_a: $p > 0.28$ $n = 783$ $x = 230$
$\alpha = 0.10$
b. H_0: $p = 0.61$ H_a: $p \neq 0.61$ $n = 401$ $\hat{p} = 0.56$
$\alpha = 0.05$

9.45 Test the following hypotheses by using the information given and the HTAB system. Let alpha be 0.01. Assume the population is normally distributed.

H_0: $\sigma^2 = 15.4$ H_a: $\sigma^2 > 15.4$ $n = 18$ $s^2 = 29.6$

9.46 Solve for the value of beta in each of the following problems:

a. H_0: $\mu = 130$ H_a: $\mu < 130$ $n = 75$ $\sigma = 12$
$\alpha = 0.01$

The alternative mean is actually 135.

b. H_0: $p = 44$ H_a: $p < 44$ $n = 1,095$ $\alpha = 0.05$

The alternative proportion is actually 0.42.

TESTING YOUR UNDERSTANDING

9.47 According to one online survey taken a few years ago, 27% of companies said they did not pay either wages or expenses to students and recent graduates hired by them on work experience and internship programmes. Suppose that business researchers want to test to determine if this figure is still accurate today by taking a new survey of 80 companies who hire students and graduates in work experience or internship programmes. Suppose further that of these 80 companies, 25% say they do not pay anything to students and recent graduates hired by them on work experience and internship programmes. Is this result enough evidence to state that a significantly different proportion of companies do not pay either wages or expenses to students and recent graduates on work experience or internship programmes? Let $\alpha = 0.01$.

9.48 According to the Food and Agriculture Organization, the average person consumes 17 kilograms of fish per year. Is this figure accurate for residents of landlocked European countries? Suppose 64 landlocked European. residents are identified by a random procedure and their average fish consumption per year is 16.2 kilograms. Assume a population variance of 5.31 kilograms of fish per year. Use a 5% level of significance to determine whether the Food and Agriculture Organization figure for global consumption is also true for landlocked European residents on the basis of the sample data.

9.49 Brokers generally agree that bonds are a better investment during times of low interest rates than during times of high interest rates. A survey of executives during a time of low interest rates showed that 57% of them had some retirement funds invested in bonds. Assume this percentage is constant for bond market investment by executives with retirement funds. Suppose interest rates have risen lately and the proportion of executives with retirement investment money in the bond market may have dropped. To test this idea, a researcher randomly samples 210 executives who have retirement funds. Of these, 93 now have retirement funds invested in bonds. For $\alpha = 0.10$, does the test show enough evidence to declare that the proportion of executives with retirement fund investments in the bond market is significantly lower than 0.57?

9.50 Motorway engineers in Spain are painting white stripes on an *Autopista*. The stripes are supposed to be approximately 3 metres long. However, because of the machine, the operator, and the motion of the vehicle carrying the equipment, considerable variation occurs among the stripe lengths. Engineers claim that the variance of stripes is not more than 4 centimetres. Use the sample lengths given here from 12 measured stripes to test the variance claim. Assume stripe length is normally distributed. Let $\alpha = 0.05$.

Stripe Lengths in Metres			
3.14	2.87	2.99	3.08
2.80	3.17	3.26	3.02
2.83	2.99	3.20	3.17

9.51 A computer manufacturer estimates that its line of minicomputers has, on average, 8.4 days of downtime per year. To test this claim, a researcher contacts seven companies that own one of these computers and is allowed to access company computer records. It is determined that, for the sample, the average number of downtime days is 5.6, with a sample standard deviation of 1.3 days. Assuming that the number of downtime days is normally distributed, test to determine whether these minicomputers actually average 8.4 days of downtime in the entire population. Let $\alpha = 0.01$.

9.52 Life insurance experts have been claiming that the average worker in the city of Frankfurt has no more than €25,000 of personal life insurance. An insurance researcher believes that this is not true and sets out to prove that the average worker in Frankfurt has more than €25,000 of personal life insurance. To test this claim, she randomly samples 100 workers in Frankfurt and interviews them about their personal life insurance coverage. She discovers that the average amount of personal life insurance coverage for this sample group is €26,650. The population standard deviation is €12,000.

   a. Determine whether the test shows enough evidence to reject the null hypothesis posed by the salesperson. Assume the probability of committing a Type I error is 0.05.
   b. If the actual average for this population is €30,000, what is the probability of committing a Type II error?

9.53 A financial analyst watched a particular stock for several months. The price of this stock remained fairly stable during this time. In fact, the financial analyst claims that the variance of the price of this stock did not exceed £4 for the entire period. Recently, the market heated up, and the price of this stock appears more volatile. To determine whether it is more volatile, a sample of closing prices of this stock for eight days is taken randomly. The sample mean price is £36.25, with a sample standard deviation of £7.80. Using a level of significance of 0.10, test to determine whether the financial analyst's previous variance figure is now too low. Assume stock prices are normally distributed.

9.54 A recent study claimed that one in 10 moviegoers experienced headaches while watching 3D movies. Suppose a new type of 3D glasses is developed by a company that claims it will have a significant impact in reducing the number of people experiencing headaches or discomfort while watching 3D movies. A

cinema decides to conduct a study to see whether the proportion of people experiencing headaches while watching 3D movies is reduced by adopting the new glasses. If 125 moviegoers are randomly selected and if 10 experience a headache while watching a 3D movie, does this result provide enough evidence to declare that a significantly lower proportion of movie goers are affected by headaches while watching 3D movies? Let $\alpha = 0.05$. If the proportion really is 0.09, what is the probability of committing a Type II error?

9.55 Suppose the number of beds filled per day in a medium-sized hospital is normally distributed. A hospital administrator tells the board of directors that, on average, at least 185 beds are filled on any given day. One of the board members believes that average is less than 185 and she sets out to test to determine if she is correct. She secures a random sample of 16 days of data (shown below). Use $\alpha = 0.05$ and the sample data to test the board member's theory. Assume the number of filled beds per day is normally distributed in the population.

Number of Beds Occupied per Day			
173	149	166	180
189	170	152	194
177	169	188	160
199	175	172	187

9.56 According to Gartner Inc., the largest share of the worldwide PC market is held by Hewlett-Packard with 18.4%. Suppose that a market researcher believes that Hewlett Packard holds a higher share of the market in the northern region of Europe. To verify this theory, he randomly selects 428 people who purchased a personal computer in the last month in the northern region of Europe. Eighty-four of these purchases were Hewlett-Packard computers. Using a 1% level of significance, test the market researcher's theory. What is the probability of making a Type I error? If the market share is really 0.21 in the north-west region of Europe, what is the probability of making a Type II error?

9.57 A national publication reported that a university student living away from home spends, on average, no more than €15 per month on coffee. You believe this figure is too low and want to disprove the claim. To conduct the test, you randomly select 17 university students and ask them to keep track of the amount of money they spend during a given month on coffee. The sample produces an average expenditure on coffee of €19.34, with a population standard deviation of €4.52. Use these sample data to conduct the hypothesis test. Assume you are willing to take a 10% risk of making a Type I error and that spending on coffee per month is normally distributed in the population.

9.58 A local company installs natural-gas grills. As part of the installation, a ditch is dug to lay a small natural-gas line from the grill to the main line. On average, the depth of these lines seems to run about 1 metre. The company claims that the depth does not vary by more than 40 square centimetres (the variance). To test this claim, a researcher randomly took 22 depth measurements at different locations. The sample average depth was 1.05 metres with a standard deviation of 6 centimetres. Is this enough evidence to reject the company's claim about the variance? Assume line depths are normally distributed. Let alpha = 0.05.

9.59 A study of pollutants showed that certain industrial emissions should not exceed 2.5 parts per million. You believe a particular company may be exceeding this average. To test this supposition, you randomly take a sample of nine air tests. The sample average is 3.4 parts per million, with a sample standard deviation of 0.6. Does this result provide enough evidence for you to conclude that the company is exceeding the safe limit? Use alpha = 0.01. Assume emissions are normally distributed.

9.60 According to property consultant Cushman & Wakefield, the average cost per square metre for office rental space in Tokyo is €1,441, making it the most expensive city in the world to rent offices. A large real estate company wants to confirm this figure. The firm conducts a telephone survey of 95 offices in the central business district of Tokyo and asks the office managers how much they pay in rent per square metre. Suppose the sample average is €1,383 per square metre. The population standard deviation is €322.5.
   a. Conduct a hypothesis test using $\alpha = 0.05$ to determine whether the cost per square metre reported by Cushman & Wakefield should be rejected.
   b. If the decision in part (a) is to fail to reject and if the actual average cost per square metre is €1,350, what is the probability of committing a Type II error?

9.61 The site uSwitch.com reports that, on average, a washing machine uses 65 litres of water per wash. Suppose the following data are the numbers of litres of water used in a wash by 12 randomly selected households and the data come from a normal distribution of data. Use these data and a 5% level of significance to test to determine whether the population variance for such water usage is 125 litres.

76	65	84	58	86	61
76	64	66	65	50	56

9.62 Downtime in manufacturing is costly and can result in late deliveries, backlogs, failure to meet orders, and even loss of market share. Suppose a manufacturing plant has been averaging 23 minutes of downtime per day for the past several years, but during the past year, there has been a significant effort by both management and production workers to reduce downtime. In an effort to determine if downtime has been significantly reduced, company productivity researchers have randomly sampled 31 days over the past several months from company records and have recorded the daily downtimes shown below in minutes. Use these data and an alpha of 0.01 to test to determine if downtime has been significantly reduced. Assume that daily downtimes are normally distributed in the population.

19	22	17	19	32	24	16	18	27	17
24	19	23	27	28	19	17	18	26	22
19	15	18	25	23	19	26	21	16	21
24									

INTERPRETING THE OUTPUT

9.63 According to the US Census Bureau, the average American generates 4.4 pounds of garbage per day. Suppose we believe that because of recycling and a greater emphasis on the environment, the figure is now lower. To test this notion, we take a random sample of Americans and have them keep a log of their garbage for a day. We record and analyse the results by using a statistical computer package. The output is shown below. Describe the sample. What statistical decisions can be made on the basis of this analysis? Let alpha be 0.05. Assume that pounds of garbage per day are normally distributed in the population. Discuss any substantive results.

One-Sample Z

```
Test of mu = 4.4 vs < 4.4
The assumed standard deviation = 0.866
                        95%
                      Upper
  N   Mean  SE Mean  Bound      Z     P
 22  3.969   0.185  4.273  -2.33  0.010
```

9.64 One survey conducted by Direct Line Car Insurance determined that the Aston Martin DB9 is the UK's favourite car of the past 25 years. Suppose a researcher conducts his own survey in an effort to determine whether this figure is correct. He uses an alpha of 0.05. Below is the Minitab output with the results of the survey. Discuss the findings, including the hypotheses, one- or two-tailed tests, sample statistics, and the conclusion. Explain from the data why you reached the conclusion you did. Are these results substantive?

Test and CI for One Proportion

Test of p = 0.1 vs p not = 0.1

Sample	X	N	Sample p	95% CI	Exact P-Value
1	52	384	0.135417	(0.102816, 0.173762)	0.033

9.65 In a recent year, published statistics by energy.eu claimed that the average retail price of a litre of diesel in Spain was €1.287. Suppose a survey of Spanish petrol retailers is conducted this year to determine whether the price per litre of diesel has increased. The Excel output of the results of the survey are shown here. Analyse the output and explain what it means in this study. An alpha of 0.05 was used in this analysis. Assume that diesel prices are normally distributed in the population. Comment on any substantive results.

Mean	1.324
Variance	0.0019
Observations	37
df	36
t Stat	5.064
$P(T<=t)$ one-tail	0.0000062
t Critical one-tail	1.688
$P(T<=t)$ two-tail	0.000012
t Critical two-tail	2.028

9.66 According to a study by the Centre for Retail Research, Dutch households spent an average of €442.65 last year on Christmas shopping. Suppose a large Dutch retail company wants to test that figure in Amsterdam, theorizing that the average might be higher in the capital. The research firm hired to conduct the study arrives at the results shown here. Analyse the data and explain the results. Comment on any substantive findings.

One-Sample T: C1

Test of mu = 442.65 vs. > 442.65

Variable	N	Mean	StDev	SE Mean	95% Lower Bound	T	P
C1	433	473.83	43.02	2.07	470.42	15.08	0.000

ANALYSING THE DATABASES

1. Suppose the average number of employees per industry group in the Manufacturing database is believed to be less than 150 (in 000s). Test this belief as the alternative hypothesis by using the 140 SIC Code industries given in the database as the sample. Let alpha be 0.10. Assume that the numbers of employees per industry group are normally distributed in the population. What did you decide and why?

2. Examine the Hospital database. Suppose you want to 'prove' that the average hospital in the United States averages more than 700 births per year. Use the Hospital database as your sample and test this hypothesis. Let alpha be 0.01. On average, do hospitals in the United States employ fewer than 900 personnel? Use the Hospital database as your sample and an alpha of 0.10 to test this figure as the alternative hypothesis. Assume that the number of births and number of employees in the hospitals are normally distributed in the population.

3. Consider the Financial database. Are the average earnings per share for companies in the stock market less than $2.50? Use the sample of companies represented by this database to test that hypothesis. Let alpha be 0.05. Test to determine whether the average return on equity for all companies is equal to 21.

Use this database as the sample and let alpha be 0.10. Assume that the earnings per share and return on equity are normally distributed in the population.

4. Suppose a researcher wants to test to determine if the average unemployment rate in the Netherlands is more than 6.0 per cent. Use the unemployment data from the International Labour database and a 1% level of significance to test this hypothesis. Assume that the unemployment rate is normally distributed in the population.

CASE

EUROPEAN BANKS FACE INCREASINGLY COMPLEX CHANNEL CHALLENGES

Europe's retail banks are offering customers an increasingly sophisticated choice of remote (mostly electronic) interactions and traditional face-to-face connections through branches and other physical meeting places in response to increased consumer demand for more and varied channels for financial services. A large joint study by McKinsey and the European Financial Management and Marketing Association (EFMA) investigated what channels for financial services consumers most want. The research included interviews with 3,000 consumers, an online survey of more than 150 banks, and in-depth discussions with leading European executives. Consumers say they want both electronic and physical contact points, but the extent to which they embrace electronic banking, and consequently the pace of change, varies significantly from country to country and generation to generation.

The report reveals a complex, fast-moving European retail-banking sector in the throes of deep-rooted change, with some players already moving fast down the multichannel route and others just starting. It confirms that customers increasingly use face-to-face channels for sales and advice, and remote ones (the Internet and ATMs) for most transactions. But at a country and product level, and during different phases of a product purchase, the picture is more complex.

Spanish customers, for example, already seem quite comfortable using direct channels (for instance, the Internet) to buy even complex products such as mortgages. In the Netherlands, 50% of consumers told us that they had not visited a bank branch in the previous year, compared with just 10 to 20% in France, Germany, Italy, and the United Kingdom. Even more striking is the fact that older Dutch consumers (aged 55 and above) use Internet banking more than 20- to 35-year-olds in other European countries: 67% versus 58%, respectively.

Overall, the study found that consumers in Western Europe use an average of 2.6 channels to interact with their banks. The Dutch and French lead with an average of 3.0 while Italians score the lowest with an average of 2.0. Mass-market and affluent customers use comparable numbers of channels, but younger people (aged 20 to 35) use more of them than older ones (aged 55 and above) – an average of 2.9 and 2.3, respectively.

This research identified three important implications for the future of European banks. The first is that the density of branches is likely to fall considerably in overbanked areas of Europe (mostly in Southern Europe). The second is that banks will need to develop smaller, more customer-focused branches. Finally, banks will have to make sure that the integration of all the numerous channels for interaction with customers is seamlessly integrated.

DISCUSSION

In the research process for the study on consumer channels for financial services in Europe, many different numerical questions were raised regarding consumers preferences for contacting their financial services provider by country and by age group. In each of these areas, statistics – in particular, hypothesis testing – plays a central role. Using the case information and the concepts of statistical hypothesis testing, discuss the following:

1. Many proportions were generated in the market research that were conducted for this project, including the proportion of consumers that use direct channels, proportion of consumers that did not visit a bank in the previous year, proportion of older consumers that use Internet banking, and

so on. Use techniques presented in this chapter to analyse each of the following and discuss how the results might affect marketing decision makers regarding Europe's banks channel strategy:

a. The case information stated that 50% of all Dutch consumers have not visited a bank in the previous year. How might we test that figure? Suppose 850 Dutch consumers are randomly selected using the Dutch Virtual Census information. Suppose 475 state that they have not visited a bank in the previous year. Test the 63% percentage using an alpha of 0.05.

b. Suppose that in the past, 64% of all European financial services consumers who used Internet banking to buy mortgages were men. Perhaps due to changing cultural values, we believe that more European women are now using Internet banking to purchase mortgages. We randomly sample 689 European consumers that bought a mortgage via Internet banking and 386 are women. Does this result provide enough evidence to conclude that men now represent a lower proportion of consumers who used Internet banking to buy mortgages?

c. What proportion of Spanish consumers regularly use Internet banking as their main channel to interact with their financial services provider? Suppose one source says that in the past the proportion has been about 0.73. We want to test to determine whether this figure is true. A random sample of 438 Spaniards is selected, and the Minitab results of testing this hypothesis are shown here. Discuss and explain this output and the implications of this study using alpha = 0.05.

Test and CI for One Proportion

Test of P = .73 vs. p not =.73

Sample	X	N	Sample p	95% CI	Exact P-Value
1	286	438	0.652968	(0.606329, 0.697525)	0.000

2. The statistical mean can be used to measure various aspects of the Hispanic culture and the Hispanic market, including size of purchase, frequency of purchase, age of consumer, size of store, and so on. Use techniques presented in this chapter to analyse

each of the following and discuss how the results might affect marketing decisions:

a. What is the average age of applicants for a personal loan through Internet banking in Italy? Suppose initial tests indicate that the mean age is 31. Is this figure really correct? To test whether it is, a researcher randomly contacts 24 Italians who recently applied for a personal loan through Internet banking with results shown in the Excel output below. Discuss the output in terms of a hypothesis test to determine whether the mean age is actually 31. Let alpha be 0.01. Assume that ages of purchasers are normally distributed in the population.

Mean	28.81
Variance	50.2651
Observations	24
df	23
t Stat	−1.52
$P(T<=t)$ one-tail	0.0716
t Critical one-tail	2.50
$P(T \neq t)$ two-tail	0.1431
t Critical two-tail	2.81

b. What is the average expenditure of a German customer on travel insurance per year using face-to-face channels? Suppose it is hypothesized that the figure is €45 per year. A researcher who knows the German market believes that this figure is too high and wants to prove her case. She randomly selects 18 German bank customers, has them keep a log of travel insurance for one year, and obtains the figures below. Analyse the data using techniques from this chapter and an alpha of 0.05. Assume that expenditures per customer are normally distributed in the population.

€55	37	59	57	27	28
16	46	34	62	9	34
4	25	38	58	3	50

Source: Adapted from 'How Europe's retail banks handle channel strategy', *McKinsey Quarterly*, available at www.mckinseyquarterly.com/Europe/How_Europes_retail_banks_handle_channel_strategy_2836; 'Face-to-face: A €15–20 billion multichannel opportunity', available at www.mckinsey.com/clientservice/Financial_Services/Knowledge_Highlights/Recent_Reports/~/media/Reports/Financial_Services/Retail_face_to_face.ashx.

USING THE COMPUTER

EXCEL

- Excel has limited capability for conducting hypothesis testing with single samples. By piecing together various Excel commands, it is possible to compute a z test of a single population mean and a t test of a single population mean.

- To conduct a z test of a single population mean, begin with the **Insert Function** (f_x). To access **the Insert Function**, go to the **Formulas** tab on an Excel worksheet (top centre tab). The Insert Function is on the far left of the menu bar. In the **Insert Function** dialog box at the top, there is a pulldown menu where it says **Or select a category**. From the pulldown menu associated with this command, select **Statistical**. Select **ZTEST** from the **Insert Function**'s Statistical menu. In the **ZTEST** dialog box, place the location of the observed values in **Array**. Place the hypothesized value of the mean in **X**. Record the value of the population standard deviation in **Sigma**. The output is the right-tailed p-value for the test statistic. If the z value is negative, subtract 1−Excel output to obtain the p-value for the left tail.

- To perform a t test of a single mean in Excel, one needs to 'fool' Excel by using a two-sample t test. To do this, enter the location of the single sample observations as one of the two requested samples and enter the location of the hypothesized mean repeated as many times as there are observations as the other sample.

- Begin this t test by selecting the **Data** tab on the Excel worksheet. From the **Analysis** panel at the right top of the **Data** tab worksheet, click on **Data Analysis**. If your Excel worksheet does not show the **Data Analysis** option, then you can load it as an add-in following directions given in Chapter 2. From the **Data Analysis** pulldown menu, select **t-Test: Two-Sample Assuming Unequal Variances** from the dialog box. Enter the location of the observations from the single sample of data in **Variable 1 Range:** Enter the location of the repeated hypothesized mean values in **Variable 2 Range:**. Enter the value of zero in **Hypothesized Mean Difference**. Check **Labels** if you have labels. Select **Alpha**. The output includes the observed t value, p-values for both one- and two-tailed tests, and critical t values for both one- and two-tailed tests.

MINITAB

- Minitab has the capability for testing hypotheses about a population mean either when σ is known or it is unknown and for testing hypotheses about a population proportion. The commands, pulldown menus, and dialog boxes are the same as those used to construct confidence intervals shown in Chapter 8.

- To begin a z test of a single population mean, select **Stat** on the menu bar. Select **Basic Statistics** from the pulldown menu. From the second pulldown menu, select **1-sample Z.** Check **Samples in columns** if you have raw data and enter the location of the column containing the observations. Check **Summarized data** if you wish to use the summarized statistics of the sample mean and the sample size rather than raw data. Enter the size of the sample in the box beside **Sample size**. Enter the sample mean in the box beside **Mean**. Enter the value of the population standard deviation in the box beside **Standard deviation**. After checking the box **Perform hypothesis test**, enter the value of the hypothesized test mean in the box beside **Hypothesized mean:** Click on Options if you want to enter a value for alpha and/or enter the direction of the alternative hypothesis (greater than, less than, or not equal). *Note:* **Options** does not allow you to directly enter a value of alpha but rather allows you to insert a level of confidence, which is $1 - \alpha$ for a two-tailed test. Both the observed z and its associated p-value for the hypothesis test and the confidence interval are given in the output.

- To begin a t test of a single population mean, select Stat on the menu bar. Select Basic Statistics from the pulldown menu. From the second pulldown menu, select **1−sample t**. Check **Samples in columns** if you have raw data and enter the location of the column containing the observations. Check **Summarized data** if you wish to use the summarized statistics of the sample size, sample mean, and the sample standard deviation rather than raw data. Enter the size of the sample in the box beside **Sample size**. Enter the sample mean in the box beside **Mean**. Enter the sample standard deviation in the box beside **Standard deviation**. After checking the box **Perform hypothesis test**, enter the value of the hypothesized test mean in the box beside

Hypothesized mean: Click on **Options** if you want to enter a value for alpha and/or enter the direction of the alternative hypothesis (greater than, less than, or not equal). *Note:* **Options** does not allow you to directly enter a value of alpha but rather allows you to insert a level of confidence, which is $1 - \alpha$ for a two-tailed test. Both the observed t and its associated p-value for the hypothesis test and the confidence interval are given in the output.

- To begin a z test of a single population proportion, select Stat on the menu bar. Select Basic Statistics from the pulldown menu. From the second pulldown menu, select 1 Proportion. Check Samples in columns if you have raw data and enter the location of the column containing the observations. Note that the data in the column must contain one of only two values (e.g., 1 or 2). Check **Summarized data** if you wish to use the summarized statistics of the number of trials and number of events rather than raw data. Enter the size of the sample in the box beside **Number of trials**. Enter the number of observed events (having the characteristic that you are testing) in the box beside **Number of events**. Click on **Options** if you want to enter a value for alpha, and/or enter the direction of the alternative hypothesis (greater than, less than, or not equal). *Note:* **Options** does not allow you to directly enter a value of alpha but rather allows you to insert a level of confidence, which is $1 - \alpha$ for a two-tailed test. After checking the box **Perform hypothesis test**, enter the value of the hypothesized test proportion in the box beside **Hypothesized proportion:** Both p-value associated with the observed z for the hypothesis test and the confidence interval are given in the output.

CHAPTER 10

Statistical Inferences About Two Populations

Learning Objectives

The general focus of Chapter 10 is on testing hypotheses and constructing confidence intervals about parameters from two populations, thereby enabling you to:

1. Test hypotheses and develop confidence intervals about the difference in two means with known population variances using the z statistic

2. Test hypotheses and develop confidence intervals about the difference in two means of independent samples with unknown population variances using the t test

3. Test hypotheses and develop confidence intervals about the difference in two dependent populations

4. Test hypotheses and develop confidence intervals about the difference in two population proportions

5. Test hypotheses about the difference in two population variances using the F distribution

Online Shopping

The use of online shopping has grown exponentially in the past decade. The Pew Internet and American Life Project surveyed 2,400 American adults and reported that about 50% of adult Americans have purchased an item on the Internet at one time or another. A Nielsen survey of over 26,000 Internet users across the globe reported that 875 million consumers around the world have shopped online. In addition, they determined that more than 85% of the world's online population has used the Internet to make a purchase, increasing the market for online shopping by 40% in the past two years. The highest percentage of Internet users who shop online is 99%, found in South Korea. This figure is followed by 97% in the United Kingdom, Germany, and Japan. The United States is eighth at 94%. A Gallup household survey of 1,043 adults taken in a recent year broke down online shopping by household income and type of store. The study reported that while only 16% of households with incomes less than $35,000 made a purchase at an online retailer or on the Internet, 48% of households with more than $100,000 did so, followed by 37% of households in the $75,000 to $99,999 level, 30% in the $50,000 to $74,999 level, and 25% of the $35,000 to $49,999 level. The average amount spent in the past 30 days at an online retailer was $130. Broken down by types of stores, survey results included $166 at electronics stores, $123 at speciality apparel stores, $121 at department stores, $94 at general merchandise stores, $80 at office supply stores, and $63 at discount retailers. The European Interactive Advertising Association (EIAA) conducted a study of over 7,000 people across Europe with regard to online shopping. They discovered that European online shoppers spent an average of €750 and purchased 10 items online over a six-month period. By country, the average number of items purchased over a six-month period were: 18 in the United Kingdom; 11 in Denmark; 10 in Germany; 9 in Sweden; 8 in France; 7 in Norway, the Netherlands, and Italy; 6 in Belgium; and 5 in Spain. In terms of the average amount spent shopping online over a six-month period, shoppers in Norway spent €1,406, followed by the United Kingdom, Denmark, Sweden, Belgium, the Netherlands, Germany, France, Italy, and Spain with €1,201, €1,159, €1,013, €790, €681, €521, €509, €454, and €452, respectively.

1. One study reported that the average amount spent by online American shoppers in the past 30 days is $123 at speciality stores and $121 at department stores. These figures are relatively close to each other and were derived from sample information. Suppose a researcher wants to test to determine if there is actually any significant difference in the average amount spent by online American shoppers in the past 30 days at specialty stores vs. department stores. How does she go about conducting such a test?

2. The EIAA study reported that the average number of items purchased over a six-month period for online shoppers was 11 in Denmark and 10 in Germany. These figures were derived from a survey of 7,000 people. Is the average number of items purchased over a six-month period for Denmark significantly higher than the average number for Germany? How would one go about determining this?

3. According to the Nielsen survey, 97% of Internet users in Japan shop online. This compares to only 94% in the United States. However, these figures were obtained through a sample of Internet users. If a researcher wants to conduct a similar survey to test to determine if the proportion of Japanese Internet users who shop online is significantly higher than the proportion of American Internet users who shop online, how would he go about doing so?

Sources: The Pew Internet & American Life Project, available at www.pewinternet.org/pdfs/PIP_Online%20Shopping.pdf; the Gallup poll with results is located at www.gallup.com/poll/20527/There-Digital-Divide-Online-Shopping.aspx; the Nielsen report is available at www.nielsenmedia.com/nc/portal/site/Public/menuitem.55dc65b4a7d5adff3f65936147a062a0/?vgnextoid=0bfef273110c7110VgnVCM100000ac0a260aRCRD; the EIAA press release is available at www.eiaa.net/news/eiaa-articles-details.asp?id=121&lang=9.

To this point, all discussion of confidence intervals and hypothesis tests has centred on single population parameters. That is, a single sample is randomly drawn from a population, and using data from that sample, a population mean, proportion, or variance is estimated or tested. Chapter 8 presents statistical techniques for constructing confidence intervals to estimate a population mean, a population proportion, or a population variance. Chapter 9 presents statistical techniques for testing hypotheses about a population mean, a population proportion, or a population variance. Often, it is of equal interest to make inferences about two populations. A retail analyst might want to compare per person annual expenditures on shoes in the year 2009 with those in the year 2006 to determine whether a change has occurred over time. A market researcher might want to estimate or test to determine the proportion of market share of one company in two different regions.

In this chapter, we will consider several different techniques for analysing data that come from two samples. One technique is used with proportions, one is used with variances, and the others are used with means. The techniques for analysing means are separated into those using the z statistic and those using the t statistic. In four of the five techniques presented in this chapter, the two samples are assumed to be **independent samples**. The samples are independent because *the items or people sampled in each group are in no way related to those in the other group.* Any similarity between items or people in the two samples is coincidental and due to chance. One of the techniques presented in the chapter is for analysing data from dependent, or related, samples in which items or persons in one sample are matched in some way with items or persons in the other sample. For four of the five techniques, we will examine both hypothesis tests and confidence intervals.

Figure III-1 in the Introduction to Unit III displays a tree diagram taxonomy of inferential techniques, organizing them by usage, number of samples, and level of data. Chapter 10 contains techniques for constructing confidence intervals and testing hypotheses about the differences in two population means and two population proportions and, in addition, testing hypotheses about two population variances. The entire left side of the tree diagram taxonomy displays various confidence interval estimation techniques. The rightmost branch of this side contains Chapter 10 techniques and is displayed in Figure 10.1. The entire right side of the tree diagram taxonomy displays various hypothesis-testing techniques. The central branch of this contains Chapter 10 techniques (2 samples) for testing hypotheses, and this branch is displayed in Figure 10.2. Note that at the bottom of each tree branch in Figures 10.1 and 10.2, the title of the statistical technique along with its respective section number is given for ease of identification and use. If a business researcher is constructing confidence intervals or testing hypotheses about the difference in two population means and the population standard deviations or variances are known, then he will use the z test for $\mu_1 - \mu_2$ contained in Section 10.1. If the population standard deviations or variances are unknown, then the appropriate technique is the t test for $\mu_1 - \mu_2$ contained in Section 10.2. If a business researcher is constructing confidence intervals or testing hypotheses about the difference in two related populations, then he will use the t test presented in Section 10.3. If a business researcher is constructing a confidence interval or testing a hypothesis about the difference in two population proportions, then he will use the z test for $p_1 - p_2$ presented in Section 10.4. If the researcher desires to test a hypothesis about two population variances, then he will use the F test presented in Section 10.5.

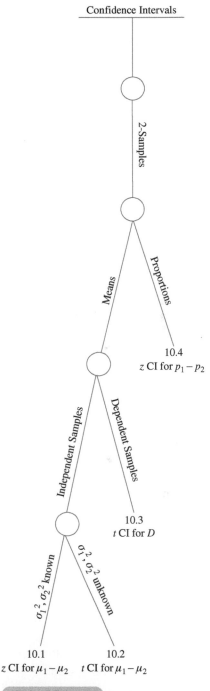

Confidence Intervals

2-Samples

Means

Proportions

10.4
z CI for $p_1 - p_2$

Independent Samples

Dependent Samples

10.3
t CI for D

σ_1^2, σ_2^2 known

σ_1^2, σ_2^2 unknown

10.1
z CI for $\mu_1 - \mu_2$

10.2
t CI for $\mu_1 - \mu_2$

FIGURE 10.1

Branch of the Tree Diagram Taxonomy of Inferential Techniques: Confidence Intervals

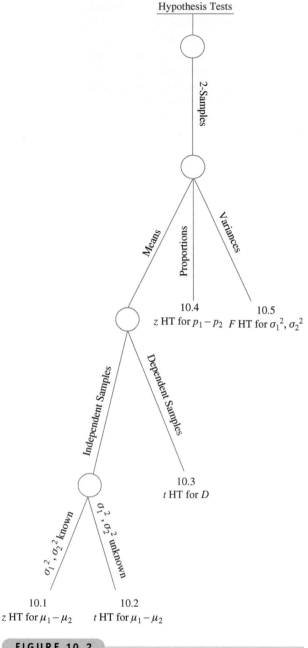

Hypothesis Tests

2-Samples

Means

Proportions

Variances

10.4
z HT for $p_1 - p_2$

10.5
F HT for σ_1^2, σ_2^2

Independent Samples

Dependent Samples

10.3
t HT for D

σ_1^2, σ_2^2 known

σ_1^2, σ_2^2 unknown

10.1
z HT for $\mu_1 - \mu_2$

10.2
t HT for $\mu_1 - \mu_2$

FIGURE 10.2

Branch of the Tree Diagram Taxonomy of Inferential
Techniques: Hypothesis Tests

10.1 HYPOTHESIS TESTING AND CONFIDENCE INTERVALS ABOUT THE DIFFERENCE IN TWO MEANS USING THE *Z* STATISTIC (POPULATION VARIANCES KNOWN)

In some research designs, the sampling plan calls for selecting two independent samples, calculating the sample means and using the difference in the two sample means to estimate or test the difference in the two population means. The object might be to determine whether the two samples come from the same population or, if they come from different populations, to determine the amount of difference in the populations. This type of analysis can be used to determine, for example, whether the effectiveness of two brands of toothpaste differs or whether two brands of tyres wear differently. Business research might be conducted to study the difference in the productivity of men and women on an assembly line under certain conditions. An engineer might want to determine differences in the strength of aluminium produced under two different temperatures. Does the average cost of a two-bedroom, one-storey house differ between Manchester and Leeds? If so, how much is the difference? These and many other interesting questions can be researched by comparing the difference in two sample means.

How does a researcher analyse the difference in two samples by using sample means? The central limit theorem states that the difference in two sample means, $\bar{x}_1 - \bar{x}_2$, is normally distributed for large sample sizes (both n_1 and $n_2 \geqslant 30$) regardless of the shape of the populations. It can also be shown that

$$\mu_{\bar{x}_1 - \bar{x}_2} = \mu_1 - \mu_2$$

$$\sigma_{\bar{x}_1 - \bar{x}_2} = \sqrt{\frac{\sigma_1^2}{n_1} + \frac{\sigma_2^2}{n_2}}$$

These expressions lead to a z formula for the difference in two sample means.

z FORMULA FOR THE DIFFERENCE IN TWO SAMPLE MEANS (INDEPENDENT SAMPLES AND POPULATION VARIANCES KNOWN) (10.1)	$$z = \frac{(\bar{x}_1 - \bar{x}_2) - (\mu_1 - \mu_2)}{\sqrt{\frac{\sigma_1^2}{n_1} + \frac{\sigma_2^2}{n_2}}}$$ where: μ_1 = the mean of population 1 μ_2 = the mean of population 2 n_1 = size of sample 1 n_2 = size of sample 2

This formula is the basis for statistical inferences about the difference in two means using two random independent samples.

Note: *If the populations are normally distributed on the measurement being studied and if the population variances are known, formula 10.1 can be used for small sample sizes.*

Hypothesis Testing

In many instances, a business researcher wants to test the differences in the mean values of two populations. As an example, a consumer organization might want to test two brands of light bulbs to determine whether one burns longer than the other. A company wanting to relocate might want to determine whether a significant difference separates the average price of a home in Barcelona, from house prices in Madrid. Formula 10.1 can be used to test the difference between two population means.

As a specific example, suppose we want to conduct a hypothesis test to determine whether the average annual wage for an advertising manager is different from the average annual wage of an auditing manager. Because we are testing to determine whether the means are different, it might seem logical that the null and alternative hypotheses would be

$$H_0: \mu_1 = \mu_2$$

$$H_a: \mu_1 \neq \mu_2$$

where advertising managers are population 1 and auditing managers are population 2. However, statisticians generally construct these hypotheses as

$$H_0: \mu_1 - \mu_2 = \delta$$

$$H_a: \mu_1 - \mu_2 \neq \delta$$

TABLE 10.1

Wages for Advertising Managers and Auditing Managers (€000)

Advertising Managers		Auditing Managers	
74.256	64.276	69.962	67.160
96.234	74.194	55.052	37.386
89.807	65.360	57.828	59.505
93.261	73.904	63.362	72.790
103.030	54.270	37.194	71.351
74.195	59.045	99.198	58.653
75.932	68.508	61.254	63.508
80.742	71.115	73.065	43.649
39.672	67.574	48.036	63.369
45.652	59.621	60.053	59.676
93.083	62.483	66.359	54.449
63.384	69.319	61.261	46.394
57.791	35.394	77.136	71.804
65.145	86.741	66.035	72.401
96.767	57.351	54.335	56.470
77.242		42.494	67.814
67.056		83.849	71.492
$n_1 = 32$		$n_2 = 32$	
$\bar{x}_1 = 70.700$		$\bar{x}_2 = 62.187$	
$\sigma_1 = 16.253$		$\sigma_2 = 12.900$	
$\sigma_1^2 = 264.160$		$\sigma_2^2 = 166.410$	

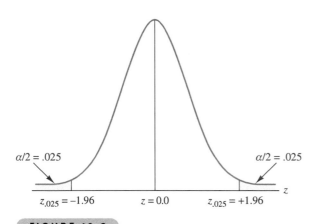

$\alpha/2 = .025$ $\alpha/2 = .025$

$z_{.025} = -1.96$ $z = 0.0$ $z_{.025} = +1.96$

FIGURE 10.3

Critical Values and Rejection Regions for the Wage Example

This format allows the business analyst not only to test if the population means are equal but also affords her the opportunity to hypothesize about a particular difference in the means (δ). Generally speaking, most business analysts are only interested in testing whether the difference in the means is different. Thus, δ is set equal to zero, resulting in the following hypotheses, which we will use for this problem and most others.

$$H_0: \mu_1 - \mu_2 = 0$$

$$H_a: \mu_1 - \mu_2 \neq 0$$

Note, however, that a business researcher could be interested in testing to determine if there is, for example, a difference of means equal to, say, 10, in which case, $\delta = 10$.

A random sample of 32 advertising managers from across Europe is taken. The advertising managers are contacted by telephone and asked what their annual salary is. A similar random sample is taken of 34 auditing managers. The resulting salary data are listed in Table 10.1, along with the sample means, the population standard deviations, and the population variances.

In this problem, the business analyst is testing whether there is a difference in the average wage of an advertising manager and an auditing manager; therefore the test is two tailed. If the business analyst had hypothesized that one was paid more than the other, the test would have been one tailed.

Suppose $\alpha = 0.05$. Because this test is two tailed, each of the two rejection regions has an area of 0.025, leaving 0.475 of the area in the distribution between each critical value and the mean of the distribution. The associated critical table value for this area is $z_{0.025} = \pm 1.96$. Figure 10.3 shows the critical table z value along with the rejection regions.

Formula 10.1 and the data in Table 10.1 yield a z value to complete the hypothesis test

$$z = \frac{(70.700 - 62.187) - (0)}{\sqrt{\dfrac{264.160}{32} + \dfrac{166.410}{34}}} = 2.35$$

The observed value of 2.35 is greater than the critical value obtained from the z table, 1.96. The business researcher rejects the null hypothesis and can say that there is a significant difference between the average annual wage of an advertising manager and the average annual wage of an auditing manager. The business researcher then examines the sample means (70.700 for advertising managers and 62.187 for auditing managers) and uses common sense to conclude that advertising managers earn more, on average, than do auditing managers. Figure 10.4 shows the relationship between the observed z and $z_{\alpha/2}$.

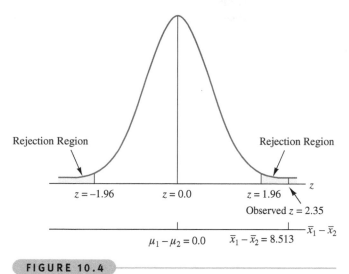

FIGURE 10.4

Location of Observed z Value for the Wage Example

This conclusion could have been reached by using the *p*-value. Looking up the probability of $z \geqslant 2.35$ in the *z* distribution table in Table A.5 (Appendix A) yields an area of $0.5000 - 0.4906 = 0.0094$. This *p*-value (0.0094) is less than $\alpha/2 = 0.025$. The decision is to reject the null hypothesis.

DEMONSTRATION PROBLEM 10.1

A sample of 87 professional working women showed that the average amount paid annually into a private pension fund per person was €3,352. The population standard deviation is €1,100. A sample of 76 professional working men showed that the average amount paid annually into a private pension fund per person was €5,727, with a population standard deviation of €1,700. A women's activist group wants to 'prove' that women do not pay as much per year as men into private pension funds. If they use $\alpha = 0.001$ and these sample data, will they be able to reject a null hypothesis that women annually pay the same as or more than men into private pension funds? Use the eight-step hypothesis-testing process.

Solution

HYPOTHESIZE

STEP 1: This test is one tailed. Because the women's activist group wants to prove that women pay less than men into private pension funds annually, the alternative hypothesis should be $\mu_w - \mu_m < 0$, and the null hypothesis is that women pay the same as or more than men, $\mu_w - \mu_m = 0$.

TEST

STEP 2: The test statistic is

$$z = \frac{(\bar{x}_1 - \bar{x}_2) - (\mu_1 - \mu_2)}{\sqrt{\dfrac{\sigma_1^2}{n_1} + \dfrac{\sigma_2^2}{n_2}}}$$

STEP 3: Alpha has been specified as 0.001.

step 4: By using this value of alpha, a critical $z_{0.001} = -3.08$ can be determined. The decision rule is to reject the null hypothesis if the observed value of the test statistic, z, is less than −3.08.

STEP 5: The sample data follow.

Women	Men
$\bar{x}_1 = €3,352$	$\bar{x}_1 = €5,7247$
$\sigma_1 = €1,100$	$\sigma_2 = €1,700$
$n_1 = 87$	$n_2 = 76$

STEP 6: Solving for z gives

$$z = \frac{(3352 - 5727) - (0)}{\sqrt{\frac{1100^2}{87} + \frac{1700^2}{76}}} = \frac{-2375}{227.9} = -10.42$$

ACTION

STEP 7: The observed z value of −10.42 is deep in the rejection region, well past the table value of $z_c = -3.08$. Even with the small $\alpha = 0.001$, the null hypothesis is rejected.

BUSINESS IMPLICATIONS

STEP 8: The evidence is substantial that women, on average, pay less than men into private pension funds annually. The following diagram displays these results:

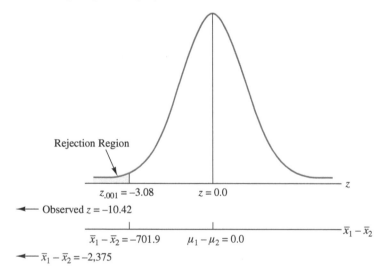

The probability of obtaining an observed z value of −10.42 by chance is virtually zero, because the value is beyond the limits of the z table. By the p-value, the null hypothesis is rejected because the probability is 0.0000, or less than $\alpha = 0.001$.

If this problem were worked by the critical value method, what critical value of the difference in the two means would have to be surpassed to reject the null hypothesis for a table z value of −3.08? The answer is

$$(\bar{x}_1 - \bar{x}_2)_c = (\mu_1 - \mu_2) - z\sqrt{\frac{\sigma_1^2}{n_1} + \frac{\sigma_2^2}{n_2}}$$

$$= 0 - 3.08(227.9) = -701.9$$

The difference in sample means would need to be at least 701.9 to reject the null hypothesis. The actual sample difference in this problem was −2,375 (3,352−5,727), which is considerably larger than the critical value of difference. Thus, with the critical value method also, the null hypothesis is rejected.

Confidence Intervals

Sometimes being able to estimate the difference in the means of two populations is valuable. By how much do two populations differ in size or weight or age? By how much do two products differ in effectiveness? Do two different manufacturing or training methods produce different mean results? The answers to these questions are often difficult to obtain through census techniques. The alternative is to take a random sample from each of the two populations and study the difference in the sample means.

Algebraically, Formula 10.1 can be manipulated to produce a formula for constructing confidence intervals for the difference in two population means.

CONFIDENCE INTERVAL TO ESTIMATE $\mu_1 - \mu_2$ (10.2)	$(\bar{x}_1 - \bar{x}_2) - z\sqrt{\frac{\sigma_1^2}{n_1} + \frac{\sigma_2^2}{n_2}} \leq \mu_1 - \mu_2 \leq (\bar{x}_1 - \bar{x}_2) + z\sqrt{\frac{\sigma_1^2}{n_1} + \frac{\sigma_2^2}{n_2}}$

Suppose a study is conducted to estimate the difference between middle-income shoppers and low-income shoppers in terms of the average amount saved on grocery bills per week by using coupons. Random samples of 60 middle-income shoppers and 80 low-income shoppers are taken, and their purchases are monitored for one week. The average amounts saved with coupons, as well as sample sizes and population standard deviations are in the table below.

Middle-Income Shoppers	Low-Income Shoppers
$n_1 = 60$	$n_1 = 80$
$\bar{x}_1 = £5.84$	$\bar{x}_2 = £2.67$
$\sigma_1 = £1.41$	$\sigma_2 = £0.54$

This information can be used to construct a 98% confidence interval to estimate the difference between the mean amount saved with coupons by middle-income shoppers and the mean amount saved with coupons by low-income shoppers.

The z_c value associated with a 98% level of confidence is 2.33. This value, the data shown, and formula (10.2) can be used to determine the confidence interval.

$$(5.84 - 2.67) - 2.33\sqrt{\frac{1.41^2}{60} + \frac{0.54^2}{80}} \leq \mu_1 - \mu_2 \leq (5.84 - 2.67) + 2.33\sqrt{\frac{1.41^2}{60} + \frac{0.54^2}{80}}$$

$$3.17 - 0.45 \leq \mu_1 - \mu_2 \leq 3.17 + 0.45$$

$$2.72 \leq \mu_1 - \mu_2 \leq 3.62$$

There is a 98% level of confidence that the actual difference in the population mean coupon savings per week between middle-income and low-income shoppers is between £2.72 and £3.62. That is, the difference could be as little as £2.72 or as great as £3.62. The point estimate for the difference in mean savings is £3.17. Note that a zero difference in the population means of these two groups is unlikely, because zero is not in the 98% range.

DEMONSTRATION PROBLEM 10.2

A consumer test group wants to determine the difference in petrol consumption of cars using regular unleaded petrol and cars using premium unleaded petrol. Researchers for the group divided a fleet of 100 cars of the same make in half and tested each car on one tank of petrol. Fifty of the cars were filled with regular unleaded petrol and 50 were filled with premium unleaded petrol. The sample average for the regular petrol group was 21.45 miles per gallon (mpg), and the sample average for the premium petrol group was 24.6 mpg. Assume that the population standard deviation of the regular unleaded petrol population is 3.46 mpg, and that the population standard deviation of the premium unleaded petrol population is 2.99 mpg. Construct a 95% confidence interval to estimate the difference in the mean petrol mileage between the cars using regular petrol and the cars using premium petrol.

Solution

The z value for a 95% confidence interval is 1.96. The other sample information follows.

Regular	Premium
$n_r = 50$	$n_p = 50$
$\bar{X}_r = 21.45$	$\bar{X}_p = 24.6$
$\sigma_r = 3.46$	$\sigma_p = 2.99$

Based on this information, the confidence interval is

$$(21.45 - 24.6) - 1.96\sqrt{\frac{3.46^2}{50} + \frac{2.99^2}{50}} \le \mu_1 - \mu_2 \le (21.45 - 24.6) + 1.96\sqrt{\frac{3.46^2}{50} + \frac{2.99^2}{50}}$$

$$-3.15 - 1.27 \le \mu_1 - \mu_2 \le -3.15 + 1.27$$

$$-4.42 \le \mu_1 - \mu_2 \le -1.88$$

We are 95% confident that the actual difference in mean petrol mileage between the two types of petrol is between -1.88 mpg and -4.42 mpg. The point estimate is -3.15 mpg.

Designating one group as group 1 and another as group 2 is an arbitrary decision. If the two groups in Demonstration Problem 10.2 were reversed, the confidence interval would be the same, but the signs would be reversed and the inequalities would be switched. Thus the researcher must interpret the confidence interval in light of the sample information. For the confidence interval in Demonstration Problem 10.2, the population difference in mean mileage between regular and premium could be as much as -4.42 mpg. This result means that the premium petrol could average 4.42 mpg more than regular petrol. The other side of the interval shows that, on the basis of the sample information, the difference in favour of premium petrol could be as little as 1.88 mpg.

If the confidence interval were being used to test the hypothesis that there is a difference in the average number of miles per gallon between regular and premium petrol, the interval would tell us to reject the null hypothesis because the interval does *not* contain zero. When both ends of a confidence interval have the same sign, zero is not in the interval. In Demonstration Problem 10.2, the interval signs are both negative. We are 95% confident that the true difference in population means is negative. Hence, we are 95% confident that there is a non-zero difference in means. For such a test, $\alpha = 1 - 0.95 = 0.05$. If the signs of the confidence interval for the difference of the sample means are different, the interval includes zero, and finding no significant difference in population means is possible.

Using the Computer to Test Hypotheses About the Difference in Two Population Means Using the z Test

Excel has the capability of testing hypotheses about two population means using a z test, but Minitab does not. Figure 10.5 shows Excel output for the advertising manager and auditing manager wage problem. For z tests, Excel requires knowledge of the population variances. The standard output includes the sample means and population variances, the sample sizes,

z-Test: Two Sample for Means

	Ad Mgr	Aud Mgr
Mean	70.700	62.187
Known Variance	264.160	166.411
Observations	32	34
Hypothesized Mean Difference	0	
z	2.35	
P (Z<=z) one-tail	0.0094	
z Critical one-tail	1.64	
P (Z<=z) two-tail	0.0189	
z Critical two-tail	1.96	

FIGURE 10.5

Excel Output for the Advertising Manager and Auditing Manager Wage Problem

the hypothesized mean difference (which here, as in most cases, is zero), the observed z value, and the p-values and critical table z values for both a one-tailed and a two-tailed test. Note that the p-value for this two-tailed test is 0.0189, which is less than $\alpha = 0.05$ and thus indicates that the decision should be to reject the null hypothesis.

10.1 PROBLEMS

10.1 **a.** Test the following hypotheses of the difference in population means by using the following data ($\alpha = 0.10$) and the eight-step process:

$$H_0: \mu_1 - \mu_2 = 0 \qquad H_a: \mu_1 - \mu_2 < 0$$

Sample 1	Sample 2
$\bar{x}_1 = 51.3$	$\bar{x}_2 = 53.2$
$\sigma_1^2 = 52$	$\sigma_2^2 = 60$
$n_1 = 31$	$n_2 = 32$

b. Use the critical value method to find the critical difference in the mean values required to reject the null hypothesis.

c. What is the p-value for this problem?

10.2 Use the following sample information to construct a 90% confidence interval for the difference in the two population means:

Sample 1	Sample 2
$n_1 = 32$	$n_2 = 31$
$\bar{x}_1 = 70.4$	$\bar{x}_2 = 68.7$
$\sigma_1 = 5.76$	$\sigma_2 = 6.1$

10.3 Examine the data below. Assume the variances for the two populations are 22.74 and 26.65 respectively.

a. Use the data to test the following hypotheses ($\alpha = 0.02$):

$$H_0: \mu_1 - \mu_2 = 0 \qquad H_a: \mu_1 - \mu_2 \neq 0$$

Sample 1						Sample 2					
90	88	80	88	83	94	78	85	82	81	75	76
88	87	91	81	83	88	90	80	76	83	88	77
81	84	84	87	87	93	77	75	79	86	90	75
88	90	91	88	84	83	82	83	88	80	80	74
89	95	97	95	93	97	80	90	74	89	84	79

b. Construct a 98% confidence interval to estimate the difference in population means using these data. How does your result validate the decision you reached in part (a)?

10.4 The Trade Show Bureau conducted a survey to determine why people go to trade shows. The respondents were asked to rate a series of reasons on a scale from 1 to 5, with 1 representing little importance and 5 representing great importance. One of the reasons suggested was general curiosity. The following responses for 50 people from the computers/electronics industry and 50 people from the food/beverage industry were recorded for general curiosity. Use these data and $\alpha = 0.01$ to determine whether there is a significant difference between people in these two industries on this question. Assume the variance for the computer/electronics population is 1.0188 and the variance for the food/beverage population is 0.9180.

Computers/Electronics					Food/Beverage				
1	2	1	3	2	3	3	2	4	3
0	3	3	2	1	4	5	2	4	3
3	3	1	2	2	3	2	3	2	3
3	2	2	2	2	4	3	3	3	3
1	2	3	2	1	2	4	2	3	3
1	1	3	3	2	2	4	4	4	4
2	1	4	1	4	3	5	3	3	2
2	3	0	1	0	2	0	2	2	5
3	3	2	2	3	4	3	3	2	3
2	1	0	2	3	4	3	3	3	2

10.5 Suppose you own a plumbing repair business and employ 15 plumbers. You are interested in estimating the difference in the average number of calls completed per day between two of the plumbers. A random sample of 40 days of plumber A's work results in a sample average of 5.3 calls, with a population variance of 1.99. A random sample of 37 days of plumber B's work results in a sample mean of 6.5 calls, with a population variance of 2.36. Use this information and a 95% level of confidence to estimate the difference in population mean daily efforts between plumber A and plumber B. Interpret the results. Is it possible that, for these populations of days, the average number of calls completed between plumber A and plumber B do not differ?

10.6 Eurostat's labour market statistics show that the average cost of labour to a company per employee per hour is €35.09 in Belgium and €33.91 in Denmark. Suppose these figures were obtained from surveying 14 Belgian workers and 15 Danish workers and that their respective population standard deviations are €3.38 and €3.51. Assume that hourly labour costs per employee are normally distributed in the population.

a. Calculate a 98% confidence interval to estimate the difference in the mean hourly company labour costs for these two groups. What is the value of the point estimate?

b. Test to determine whether there is a significant difference in the hourly costs employers pay per employee between Belgian and Danish workers. Use a 2% level of significance.

10.7 A company's auditor believes the per diem cost in Vienna, Austria, rose significantly between 1999 and 2009. To test this belief, the auditor samples 51 business trips from the company's records for 1999; the sample average was €190 per day, with a population standard deviation of €18.50. The auditor selects a second random sample of 47 business trips from the company's records for 2009; the sample average was €198 per day, with a population standard deviation of €15.60. If he uses a risk of committing a Type I error of 0.01, does the auditor find that the per diem average expense in Vienna has gone up significantly?

10.8 Suppose a market analyst wants to determine the difference in the average hotel price of a cappuccino in Hannover, Germany, and Toulouse, France. To do so, he takes a telephone survey of 21 randomly selected consumers in Hanover who have purchased a cappuccino in one of the city's hotels and asks how much they paid for it. The analyst undertakes a similar survey in Toulouse with 18 respondents. Assume the population variance for Hannover is 0.03, the population variance for Toulouse is 0.015, and that the price of cappuccinos is normally distributed. Using the resulting sample information that follows,

a. Compute a 99% confidence interval to estimate the difference in the mean price of a cappuccino between the two cities.

b. Using a 1% level of significance, test to determine if there is a significant difference in the price of a cappuccino between the two cities.

Hannover			Toulouse		
€2.55	€2.36	€2.43	€2.25	€2.40	€2.39
2.67	2.54	2.43	2.30	2.33	2.40
2.50	2.54	2.38	2.49	2.29	2.23
2.61	2.80	2.49	2.41	2.48	2.29
2.43	2.61	2.57	2.39	2.59	2.53
2.36	2.56	2.71	2.26	2.38	2.45
2.50	2.64	2.27			

10.9 Employee suggestions can provide useful and insightful ideas for management. Some companies solicit and receive employee suggestions more than others, and company culture influences the use of employee suggestions. Suppose a study is conducted to determine whether there is a significant difference in the mean number of suggestions a month per employee between the Canon Corporation and the Pioneer Electronic Corporation. The study shows that the average number of suggestions per month is 5.8 at Canon and 5.0 at Pioneer. Suppose these figures were obtained from random samples of 36 and 45 employees, respectively. If the population standard deviations of suggestions per employee are 1.7 and 1.4 for Canon and Pioneer, respectively, is there a significant difference in the population means? Use $\alpha = 0.05$.

10.10 Two processes in a manufacturing line are performed manually: operation A and operation B. A random sample of 50 different assemblies using operation A shows that the sample average time per assembly is 8.05 minutes, with a population standard deviation of 1.36 minutes. A random sample of 38 different assemblies using operation B shows that the sample average time per assembly is 7.26 minutes, with a population standard deviation of 1.06 minutes. For $\alpha = 0.10$, is there enough evidence in these samples to declare that operation A takes significantly longer to perform than operation B?

10.2   HYPOTHESIS TESTING AND CONFIDENCE INTERVALS ABOUT THE DIFFERENCE IN TWO MEANS: INDEPENDENT SAMPLES AND POPULATION VARIANCES UNKNOWN

The techniques presented in Section 10.1 are for use whenever the population variances are known. On many occasions, statisticians test hypotheses or construct confidence intervals about the difference in two population means and the population variances are not known. If the population variances are not known, the z methodology is not appropriate. This section presents methodology for handling the situation when the population variances are unknown.

Hypothesis Testing

The hypothesis test presented in this section is a test that compares the means of two samples to determine whether there is a difference in the two population means from which the samples come. This technique is used whenever the population variances are unknown (and hence the sample variances must be used) and the samples are independent (not related in any way). *An assumption underlying this technique is that the measurement or characteristic being studied is normally distributed for both populations.* In Section 10.1, the difference in large sample means was analysed by Formula 10.1:

$$z = \frac{(\bar{x}_1 - \bar{x}_2) - (\mu_1 - \mu_2)}{\sqrt{\dfrac{\sigma_1^2}{n_1} + \dfrac{\sigma_2^2}{n_2}}}$$

If $\sigma_1^2 = \sigma_2^2$, Formula 10.1 algebraically reduces to

$$z = \frac{(\bar{x}_1 - \bar{x}_2) - (\mu_1 - \mu_2)}{\sigma \sqrt{\dfrac{1}{n_1} + \dfrac{1}{n_2}}}$$

If σ is unknown, it can be estimated by *pooling* the two sample variances and computing a pooled sample standard deviation.

$$\sigma \approx s_p = \sqrt{\frac{s_1^2(n_1 - 1) + s_2^2(n_2 - 1)}{n_1 + n_2 - 2}}$$

s_p^2 is the weighted average of the two sample variances, s_1^2 and s_1^2. Substituting this expression for σ and changing z to t produces a formula to test the difference in means.

| t FORMULA TO TEST THE DIFFERENCE IN MEANS ASSUMING σ_1^2, σ_2^2 ARE EQUAL (10.3) | $$t = \frac{(\bar{x}_1 - \bar{x}_2) - (\mu_1 - \mu_2)}{\sqrt{\dfrac{s_1^2(n_1 - 1) + s_2^2(n_2 - 1)}{n_1 + n_2 - 2}}\sqrt{\dfrac{1}{n_1} + \dfrac{1}{n_2}}}$$ $$\mathrm{df} = n_1 + n_2 - 2$$ |

Formula 10.3 is constructed by assuming that the two population variances, σ_1^2 and σ_2^2 are equal. Thus, when using formula 10.3 to test hypotheses about the difference in two means for small independent samples when the population variances are unknown, we must assume that the two samples come from populations in which the variances are essentially equal.

At the Adela Manufacturing Company, an application of this test arises. New employees are expected to attend a three-day induction seminar to learn about the company. At the end of the seminar, they are tested to measure their knowledge about the company. The traditional training method has been a lecture and a question-and-answer session. Management decided to experiment with a different training procedure, which processes new employees in two days by using DVDs and having no question-and-answer session. If this procedure works, it could save the company thousands of euros over a period of several years. However, there is some concern about the effectiveness of the two-day method, and company managers would like to know whether there is any difference in the effectiveness of the two training methods.

To test the difference in the two methods, the managers randomly select one group of 15 newly hired employees to take the three-day seminar (method A) and a second group of 12 new employees for the two-day DVD method (method B). Table 10.2 shows the test scores of the two groups. Using $\alpha = 0.05$, the managers want to determine whether there is a significant difference in the mean scores of the two groups. They assume that the scores for this test are normally distributed and that the population variances are approximately equal.

HYPOTHESIZE

STEP 1: The hypotheses for this test follow.

$$H_0: \mu_1 - \mu_2 = 0$$
$$H_a: \mu_1 - \mu_2 \neq 0$$

TEST

STEP 2. The statistical test to be used is formula 10.3.

STEP 3. The value of alpha is 0.05.

STEP 4. Because the hypotheses are = and $\neq$, this test is two tailed. The degrees of freedom are 25 ($15 + 12 - 2 = 25$) and alpha is 0.05. The t table requires an alpha value for one tail only, and, because it is a two-tailed test, alpha is split from 0.05 to 0.025 to obtain the table t value: $t_{0.25,25} = \pm 2.060$.

The null hypothesis will be rejected if the observed t value is less than -2.060 or greater than $+2.060$.

STEP 5: The sample data are given in Table 10.2. From these data, we can calculate the sample statistics. The sample means and variances follow.

TABLE 10.2
Test Scores for New Employees After Training

Training Method A					Training Method B			
56	50	52	44	52	59	54	55	65
47	47	53	45	48	52	57	64	53
42	51	42	43	44	53	56	53	57

Method A	Method B
$\bar{x}_1 = 47.73$	$\bar{x}_1 = 56.5$
$s_1^2 = 19.495$	$s_2^2 = 18.273$
$n_1 = 15$	$n_2 = 12$

NOTE: If the equal variances assumption can not be met the following formula should be used:

t FORMULA TO TEST THE DIFFERENCE IN MEANS	$t = \dfrac{(\bar{x}_1 - \bar{x}_2) - (\mu_1 - \mu_2)}{\sqrt{\dfrac{s_1^2}{n_1} + \dfrac{s_2^2}{n_2}}}$	$df = \dfrac{\left[\dfrac{s_1^2}{n_1} + \dfrac{s_2^2}{n_2}\right]^2}{\dfrac{\left(\dfrac{s_1^2}{n_1}\right)^2}{n_1 - 1} + \dfrac{\left(\dfrac{s_2^2}{n_2}\right)^2}{n_2 - 1}}$

Because this formula requires a more complex degrees-of-freedom component, it may be unattractive to some users. Many statistical computer software packages offer the user a choice of the 'pooled' formula or the 'unspooled' formula. The 'pooled' formula in the computer packages is Formula 10.3, in which equal population variances are assumed. This is the 'unspooled' formula used when population variances cannot be assumed to equal. Excel refers to this as a t-Test: Two-Sample Assuming Unequal Variances. Again, in each of these formulas, the populations from which the two samples are drawn are assumed to be normally distributed for the phenomenon being measured.

STEP 6: The observed value of t is

$$t = \frac{(47.73 - 56.50) - (0)}{\sqrt{\dfrac{(19.495)(14) + (18.273)(11)}{(15 + 12 - 2)}}\sqrt{\dfrac{1}{15} + \dfrac{1}{12}}} = -5.20$$

ACTION

STEP 7: Because the observed value, $t = -5.20$, is less than the lower critical table value, $t = -2.06$, the observed value of t is in the rejection region. The null hypothesis is rejected. There is a significant difference in the mean scores of the two tests.

BUSINESS IMPLICATIONS

STEP 8: Figure 10.6 shows the critical areas and the observed t value. Note that the computed t value is –5.20, which is enough to cause the managers of the Adela Manufacturing Company to reject the null hypothesis. Their conclusion is that there is a significant difference in the effectiveness of the training methods. Upon examining the sample means, they realize that method B (the two-day DVD method) actually produced an average score that was more than eight points higher than that for the group trained with method A. Given that training method B scores are significantly higher and the fact that the seminar is a day shorter than method A (thereby saving both time and money), it makes business sense to adopt method B as the standard training method.

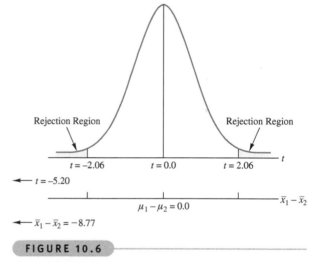

FIGURE 10.6

t Values for the Training Methods Example

In a test of this sort, which group is group 1 and which is group 2 is an arbitrary decision. If the two samples had been designated in reverse, the observed t value would have been $t = +5.20$ (same magnitude but different sign), and the decision would have been the same.

Using the Computer to Test Hypotheses and Construct Confidence Intervals About the Difference in Two Population Means Using the t Test

Both Excel and Minitab have the capability of analysing t tests for the difference in two means. The two computer packages yield similar output. Figure 10.7 contains Excel and Minitab output for the Adela Manufacturing Company training methods example. Notice that both outputs contain the same sample means, the degrees of freedom (df = 25), the observed t value –5.20, and the p-value (0.000022 on Excel as two-tailed p and 0.000 on Minitab). This p-value can be compared directly with $\alpha = 0.05$ for decision-making purposes (reject the null hypothesis).

Each package offers other information. Excel displays the sample variances whereas Minitab displays sample standard deviations. Excel displays the pooled variance whereas Minitab displays the pooled standard deviation. Excel

Excel Output

t-Test: Two-Sample Assuming Equal Variances

	Method A	Method B
Mean	47.73	56.50
Variance	19.495	18.273
Observations	15	12
Pooled Variance	18.957	
Hypothesized Mean Difference	0	
df	25	
t Stat	−5.20	
P(*T*<=*t*) one-tail	0.000011	
t Critical one-tail	1.71	
P(*T*<=*t*) two-tail	0.000022	
t Critical two-tail	2.06	

Minitab Output

Two-Sample T-Test and CI: Method A, Method B

```
Two-sample T for Method A vs Method B

              N         Mean       StDev      SE Mean
Method A      15        47.73      4.42       1.1
Method B      12        56.50      4.27       1.2

Difference = mu (Method A) - mu (Method B)
Estimate for difference: -8.77
95% CI for difference: (-12.24, -5.29)
T-Test of difference = 0 (vs not =): T-Value = -5.20
P-Value = 0.000 DF = 25
Both use Pooled StDev = 4.3540
```

FIGURE 10.7

Excel and Minitab Output for the Training Methods Example

prints out *p*-values for both a one-tailed test and a two-tailed test, and the user must select the appropriate value for his or her test. Excel also prints out the critical *t* values for both one- and two-tailed tests. Notice that the critical *t* value for a two-tailed test (2.06) is the same as the critical *t* value obtained by using the *t* table (±2.060). Minitab yields the standard errors of the mean for each sample and the 95% confidence interval. Minitab uses the same command for hypothesis testing and confidence interval estimation for the two-sample case. For this reason, Minitab output for this type of problem always contains both the hypothesis-testing and confidence interval results.

DEMONSTRATION PROBLEM 10.3

Is there a difference in the way Chinese cultural values affect the purchasing strategies of industrial buyers in Taiwan and mainland China? A study by researchers at the National Chiao-Tung University in Taiwan attempted to determine whether there is a significant difference in the purchasing strategies of industrial buyers between Taiwan and mainland China based on the cultural dimension labelled 'integration'. Integration is being in harmony with one's self, family, and associates. For the study, 46 Taiwanese buyers and 26 mainland Chinese buyers were contacted and interviewed. Buyers were asked to respond to 35 items using a nine-point scale with possible answers ranging from no importance (1) to extreme importance (9). The resulting statistics for the two groups are shown in step 5. Using $\alpha = 0.01$, test to determine whether there is a significant difference between buyers in Taiwan and buyers in mainland China on integration. Assume that integration scores are normally distributed in the population.

HYPOTHESIZE

STEP 1: If a two-tailed test is undertaken, the hypotheses and the table t value are as follows.

$$H_0: \mu_1 - \mu_2 = 0$$

$$H_a: \mu_1 - \mu_2 \neq 0$$

TEST

STEP 2. The appropriate statistical test is formula 10.3.

STEP 3: The value of alpha is 0.01.

STEP 4: The sample sizes are 46 and 26. Thus, there are 70 degrees of freedom. With this figure and $\alpha/2 = 0.005$, critical table t values can be determined.

$$t_{0.005,70} = 2.648$$

STEP 5: The sample data follow.

	Integration
Taiwanese Buyers	**Mainland Chinese Buyers**
$n_1 = 46$	$n_2 = 26$
$\bar{X}_1 = 5.42$	$\bar{X}_2 = 5.04$
$s_1^2 = (0.58)^2 = 0.3346$	$s_2^2 = (0.49)^2 = 0.2401$
$df = n_2 + n_1 - 2 = 46 + 260 - 2 = 70$	

STEP 6. The observed t value is

$$t = \frac{(5.42 - 5.04) - (0)}{\sqrt{\dfrac{(0.3364)(45) + (0.2401)(25)}{46 + 26 - 2}}\sqrt{\dfrac{1}{46} + \dfrac{1}{26}}} = 2.82$$

ACTION

STEP 7: Because the observed value of $t = 2.82$ is greater than the critical table value of $t = 2.648$ the decision is to reject the null hypothesis.

BUSINESS IMPLICATIONS

STEP 8. The Taiwan industrial buyers scored significantly higher than the mainland China industrial buyers on integration. Managers should keep in mind in dealing with Taiwanese buyers that they may be more likely to place worth on personal virtue and social hierarchy than do the mainland Chinese buyers.

The following graph shows the critical t values, the rejection regions, the observed t value, and the difference in the raw means:

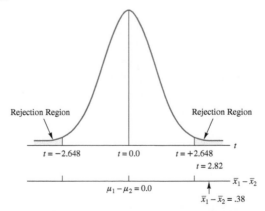

Confidence Intervals

Confidence interval formulas can be derived to estimate the difference in the population means for independent samples when the population variances are unknown. The focus in this section is only on confidence intervals when approximately equal population variances and normally distributed populations can be assumed.

CONFIDENCE INTERVAL TO ESTIMATE $\mu_1 - \mu_2$ ASSUMING THE POPULATION VARIANCES ARE UNKNOWN AND EQUAL (10.4)	$(\bar{x}_1 - \bar{x}_2) - t\sqrt{\dfrac{s_1^2(n_1 - 1) + s_2^2(n_2 - 1)}{n_1 + n_2 - 2}}\sqrt{\dfrac{1}{n_1} + \dfrac{1}{n_2}} \le \mu_1 - \mu_2 \le$ $(\bar{x}_1 - \bar{x}_2) + t\sqrt{\dfrac{s_1^2(n_1 - 1) + s_2^2(n_2 - 1)}{n_1 + n_2 - 2}}\sqrt{\dfrac{1}{n_1} + \dfrac{1}{n_2}}$ $\mathrm{df} = n_1 + n_2 - 2$

TABLE 10.3

Conscientiousness Data on Phone Survey Respondents and Average Citizens

Phone Survey Respondents	Average Citizens
35.38	35.03
37.06	33.90
37.74	34.56
36.97	36.24
37.84	34.59
37.50	34.95
40.75	33.30
35.31	34.73
35.30	34.79
	37.83
$n_1 = 9$	$n_2 = 10$
$\bar{x}_1 = 37.09$	$\bar{x}_2 = 34.99$
$s_1 = 1.727$	$s_2 = 1.253$
$\mathrm{df} = 9 + 10 - 2 = 17$	

One group of researchers set out to determine whether there is a difference between 'average citizens' and those who are 'phone survey respondents'.* Their study was based on a well-known personality survey that attempted to assess the personality profile of both average citizens and phone survey respondents. Suppose they sampled nine phone survey respondents and 10 average citizens in this survey and obtained the results on one personality factor, conscientiousness, which are displayed in Table 10.3. Assume that conscientiousness scores are normally distributed in the population.

The table t value for a 99% level of confidence and 17 degrees of freedom is $t_{0.005,17} = 2.898$. The confidence interval is

$$(37.09 - 34.99) \pm 2.898\sqrt{\frac{(1.727)^2(8) + (1.253)^2(9)}{9 + 10 - 2}}\sqrt{\frac{1}{9} + \frac{1}{10}}$$

$$2.10 \pm 1.99$$

$$0.11 \le \mu_1 - \mu_2 \le 4.09$$

The researchers are 99% confident that the true difference in population mean personality scores for conscientiousness between phone survey respondents and average citizens is between 0.11 and 4.09. Zero is not in this interval, so they can conclude that there is a significant difference in the average scores of the two groups. Higher scores indicate more conscientiousness. Therefore, it is possible to conclude from Table 10.3 and this confidence interval that phone survey respondents are significantly more conscientious than average citizens. These results indicate that researchers should be careful in using phone survey results to reach conclusions about average citizens.

Figure 10.8 contains Minitab output for this problem. Note that the Minitab output includes both the confidence interval (0.112 to 4.093) and the observed t value (3.06) for hypothesis testing. Because the p-value is 0.007, which is less than 0.01, the Minitab hypothesis-testing information validates the conclusion reached that there is a significant difference in the scores of the two groups.

---

* Note that the results on this portion of the actual study are about the same as those shown here except that in the actual study the sample sizes were in the 500–600 range.

Two-Sample T-Test and CI: Phone Survey Respondent, Average Citizen

Two-Sample T for Phone Survey Respondent vs Average Citizen

	N	Mean	StDev	SE Mean
Phone Survey Respondent	9	37.09	1.73	0.58
Average Citizen	10	34.99	1.25	0.40

Difference = mu (Phone Survey Respondent) - mu (Average Citizen)
Estimate for difference: 2.102
99% CI for difference: (0.112, 4.093)
T-Test of difference = 0 (vs not =): T-Value = 3.06
P-Value = 0.007 DF = 17
Both use Pooled StDev = 1.4949

FIGURE 10.8

Minitab Output for the Phone Survey Respondent and Average Citizen Example

DEMONSTRATION PROBLEM 10.4

A coffee manufacturer is interested in estimating the difference in the average daily coffee consumption of regular-coffee drinkers and decaffeinated-coffee drinkers. Its researcher randomly selects 13 regular-coffee drinkers and asks how many cups of coffee per day they drink. He randomly locates 15 decaffeinated-coffee drinkers and asks how many cups of coffee per day they drink. The average for the regular-coffee drinkers is 4.35 cups, with a standard deviation of 1.20 cups. The average for the decaffeinated-coffee drinkers is 6.84 cups, with a standard deviation of 1.42 cups. The researcher assumes, for each population, that the daily consumption is normally distributed, and he constructs a 95% confidence interval to estimate the difference in the averages of the two populations.

Solution

The table t value for this problem is $t_{0.025,26} = 2.056$. The confidence interval estimate is

$$(4.35-6.84)\pm2.056\sqrt{\frac{(1.20)^2(12)+(1.42)^2(14)}{13+15-2}}\sqrt{\frac{1}{13}+\frac{1}{15}}$$

$$-2.49\pm1.03$$

$$-3.52\leq\mu_1-\mu_2\leq-1.46$$

The researcher is 95% confident that the difference in population average daily consumption of cups of coffee between regular- and decaffeinated-coffee drinkers is between 1.46 cups and 3.52 cups. The point estimate for the difference in population means is 2.49 cups, with an error of 1.03 cups.

Ethical Differences Between Men and Women

Is there a difference between men and women in making ethical managerial decisions? One study attempted to answer this question by studying 164 managers of a large financial conglomerate. A questionnaire was constructed using vignettes (brief, focused cases) to depict four ethical questions under two different scenarios. Vignettes dealt with (1) sale of an unsafe product, (2) bribery, (3) product misrepresentation, and (4) industrial espionage. These vignettes were to be considered under the scenarios of (a) enhancing the firm's profit position and (b) the individual's own economic gain. The questionnaire was structured to produce a score for each respondent on each ethical question without regard to scenario and a score for each respondent on each ethical question with regard to each scenario. The null hypothesis that there is no significant difference in the mean ethical scores of men and women was tested for each question using a t test for independent samples.

The results were mixed. In considering the responses to the four vignettes without regard to either of the two scenarios, there was a significant difference between men and women on the sale of an unsafe product at ($\alpha = 0.01$). On this question, women scored significantly higher (more ethical) than men, indicating that women are less likely to sell an unsafe product. On the questions of product misrepresentation and industrial espionage, women scored significantly higher on both ($\alpha = 0.10$). There was no significant difference between men and women on the question of bribery.

The results were somewhat different when the two scenarios were considered (firm profit position and personal economic gain). On the question of selling an unsafe product, women were significantly more ethical ($\alpha = 0.01$) when considered in light of enhancing the firm's profit position and significantly more ethical ($\alpha = 0.05$) when considering one's personal economic gain. On the question of bribery, there was no significant difference in the ethics scores of men and women in light of enhancing the firm's profit position but women were significantly more ethical ($\alpha = 0.10$) when considering one's personal economic gain. On the question of product misrepresentation, there was no significant difference between men and women when considering one's personal economic gain but women were significantly more ethical than men in light of enhancing the firm's profit position ($\alpha = 0.10$). On the question of industrial espionage, women were significantly more ethical than men ($\alpha = 0.10$) in light of enhancing the firm's profit position, and women were also significantly more ethical than men ($\alpha = 0.01$) when considering one's personal gain.

This study used two-sample hypothesis testing in an effort to determine whether there is a difference between men and women on ethical management issues. The results here can assist decision makers in assigning managers to various tasks that involve any of these four ethical questions. Interesting questions can be studied about why women might be more ethical than men in some managerial situations and what might be done to foster stronger ethics among men.

Source: Adapted from James J. Hoffman, 'Are Women Really More Ethical than Men? Maybe It Depends on the Situation', *Journal of Managerial Issues*, vol. X, no. 1 (Spring 1998), pp. 60–73.

10.2 PROBLEMS

10.11 Use the data given and the eight-step process to test the following hypotheses:

$$H_0: \mu_1 - \mu_2 = 0 \qquad H_a: \mu_1 - \mu_2 < 0$$

Sample 1	Sample 2
$n_1 = 8$	$n_2 = 11$
$\bar{x}_1 = 24.56$	$\bar{x}_2 = 26.42$
$s_1^2 = 12.4$	$s_2^2 = 12.4$

Use a 1% level of significance, and assume that x is normally distributed.

10.12 a. Use the data below and $\alpha = 0.10$ to test the stated hypotheses. Assume x is normally distributed in the populations and the variances of the populations are approximately equal.

$$H_0: \mu_1 - \mu_2 = 0 \qquad H_a: \mu_1 - \mu_2 \neq 0$$

Sample 1	Sample 2
$n_1 = 20$	$n_2 = 20$
$\bar{x}_1 = 118$	$\bar{x}_1 = 113$
$s_1 = 23.9$	$s_2 = 21.6$

b. Use these data to construct a 90% confidence interval to estimate $\mu_1 - \mu_2$.

10.13 Suppose that for years the mean of population 1 has been accepted to be the same as the mean of population 2, but that now population 1 is believed to have a greater mean than population 2. Letting $\alpha = 0.05$ and assuming the populations have equal variances and x is approximately normally distributed, use the following data to test this belief.

Sample 1		Sample 2	
43.6	45.7	40.1	36.4
44.0	49.1	42.2	42.3
45.2	45.6	43.1	38.8
40.8	46.5	37.5	43.3
48.3	45.0	41.0	40.2

10.14 a. Suppose you want to determine whether the average values for populations 1 and 2 are different, and you randomly gather the following data:

Sample 1						Sample 2					
2	10	7	8	2	5	10	12	8	7	9	11
9	1	8	0	2	8	9	8	9	10	11	10
11	2	4	5	3	9	11	10	7	8	10	10

Test your conjecture, using a probability of committing a Type I error of 0.01. Assume the population variances are the same and x is normally distributed in the populations.

b. Use these data to construct a 98% confidence interval for the difference in the two population means.

10.15 Suppose a real estate broker is interested in comparing the asking prices of flats in the British cities of Exeter and Cardiff. The broker conducts a small telephone survey in the two cities, asking the prices of flats. A random sample of 21 listings in Exeter resulted in a sample average price of £116,900, with a standard deviation of £2,300. A random sample of 26 listings in Cardiff resulted in a sample average price of £114,000, with a standard deviation of £1,750. The broker assumes that the prices of flats are normally distributed and that the variance in prices in the two cities is about the same.

a. What would he obtain for a 90% confidence interval for the difference in mean prices of mid-range homes between Exeter and Cardiff?

b. Test whether there is any difference in the mean prices of mid-range homes of the two cities for $\alpha = 0.10$.

10.16 According to the Office for National Statistics' *Statistical Bulletin*, the average weekly wage of a professional is £704.1 a week and the average weekly wage of a manager and senior official is £721.7. Assume that such wages are normally distributed in the population and that the population variances are equal. Suppose these figures were actually obtained from the data below.

a. Use these data and $\alpha = 0.10$ to test to determine if there is a significant difference in the mean weekly wage of a professional and the mean weekly wage of manager and senior official.

b. Using these same data, construct a 90% confidence interval to estimate the difference in the population mean weekly wages of professionals and managers and senior officials.

Professionals	Managers and Senior Officials
729	715
730	704
706	712
735	670
700	702
734	706
716	701
707	709
727	705
705	733
737	691
734	701

10.17 Based on an indication that mean daily car rental rates may be higher for Helsinki than for Luxembourg, a survey of eight car rental companies in Helsinki is taken and the sample mean car rental rate is €47, with a standard deviation of €3. Further, suppose a survey of nine car rental companies in Luxembourg results in a sample mean of €44 and a standard deviation of €3. Use $\alpha = 0.05$ to test to determine whether the average daily car rental rates in Helsinki are significantly higher than those in Luxembourg. Assume car rental rates are normally distributed and the population variances are equal.

10.18 What is the difference in average daily hotel room rates between Faro, Portugal, and Malaga, Spain? Suppose we want to estimate this difference by taking hotel rate samples from each city and using a 98% confidence level. The data for such a study follow. Use these data to produce a point estimate for the mean difference in the hotel rates for the two cities. Assume the population variances are approximately equal and hotel rates in any given city are normally distributed.

Faro	Malaga
$n_M = 22$	$n_{NO} = 20$
$\bar{x}_M = €112$	$\bar{x}_{NO} = €122$
$s_M = €11$	$s_{NO} = €12$

10.19 A study was made to establish how much European people are planning to spend on Valentine's Day gifts. The study showed that 55% of people in Europe plan to buy a Valentine's Day present for a loved one. The survey also found that men are the big spenders with an average of €70 each, compared with women who plan to spend around €60 each. Suppose a market researcher wants to determine whether there is any difference between men and women in the average amount they spend on Valentine's Day gifts. She uses the data below, randomly gathered from 11 men and women, and an alpha of 0.01 to test this difference. She assumes the spending on Valentine's Day gifts is normally distributed and the population variances are equal. What does the researcher find?

Men	Women
€75	€51
70	72
75	71
77	69
54	80
54	66
52	77
75	78
71	75
58	56
73	52

Use the data from the previous table to construct a 95% confidence interval to estimate the difference in men and women's average spending on Valentine's Day gifts.

10.20   A study was made to compare the costs of supporting a family of four for a year in different foreign cities. The standard against which living in foreign cities was compared was a lifestyle of living at home on an annual income of €75,000. A comparable living standard in Hannover and Rome was attained for about €64,000. Suppose an executive wants to determine whether there is any difference in the average annual cost of supporting her family of four in the manner to which they are accustomed between Hannover and Rome. She uses the data below, randomly gathered from 11 families in each city, and an alpha of 0.01 to test this difference. She assumes the annual cost is normally distributed and the population variances are equal. What does the executive find?

Hannover, Germany	Rome
€69,000	€65,000
64,500	64,000
67,500	66,000
64,500	64,900
66,700	62,000
68,000	60,500
65,000	62,500
69,000	63,000
71,000	64,500
68,500	63,500
67,500	62,400

Use the data from the previous table to construct a 95% confidence interval to estimate the difference in average annual costs between the two cities.

10.3   STATISTICAL INFERENCES FOR TWO RELATED POPULATIONS

In the preceding section, hypotheses were tested and confidence intervals constructed about the difference in two population means when the samples are independent. In this section, a method is presented to analyse **dependent samples** or related samples. Some researchers refer to this test as the **matched-pairs test**. Others call it the *t test for related measures* or the *correlated t test*.

What are some types of situations in which the two samples being studied are related or dependent? Let's begin with the before-and-after study. Sometimes as an experimental control mechanism, the same person or object is measured both before and after a treatment. Certainly, the after measurement is *not* independent of the before measurement because the measurements are taken on the same person or object in both cases. Table 10.4 gives data from a hypothetical study in which people were asked to rate a company before and after one week of viewing a 15-minute DVD of the company twice a day. The 'before' scores are one sample and the 'after' scores are a second sample, but each pair of scores is related because the two measurements apply to the same person. The before scores and the after scores are not likely to vary from each other as much as scores gathered from independent samples because individuals bring their biases about businesses and the company to the study. These individual biases affect both the before scores and the after scores in the same way because each pair of scores is measured on the same person.

Other examples of related measures samples include studies in which twins, siblings, or spouses are matched and placed in two different groups. For example, a fashion designer might be interested in comparing men's and women's perceptions of women's clothing. If the men and women selected for the study are spouses or siblings, a built-in relatedness to the measurements of the two groups in the study is likely. Their scores are more apt to be alike or related than those of randomly chosen independent groups of men and women because of similar backgrounds or tastes.

TABLE 10.4

Rating of a Company
(on a Scale from 0 to 50)

Individual	Before	After
1	32	39
2	11	15
3	21	35
4	17	13
5	30	41
6	38	39
7	14	22

Hypothesis Testing

To ensure the use of the proper hypothesis-testing techniques, the researcher must determine whether the two samples being studied are dependent or independent. The approach to analysing two *related* samples is different from the techniques used to analyse independent samples. Use of the techniques in Section 10.2 to analyse related group data can result in a loss of power and an increase in Type II errors.

The matched-pairs test for related samples requires that the two samples be the same size and that the individual related scores be matched. Formula 10.5 is used to test hypotheses about dependent populations.

t FORMULA TO TEST THE DIFFERENCE IN TWO DEPENDENT POPULATIONS (10.5)

$$t = \frac{\bar{d} - D}{\frac{s_d}{\sqrt{n}}}$$

$$df = n - 1$$

where:
n = number of pairs
d = sample difference in pairs
D = mean population difference
s_d = standard deviation of sample difference
$\bar{d}$ = mean sample difference

This t test for dependent measures uses the sample difference, d, between individual matched sample values as the basic measurement of analysis instead of individual sample values. Analysis of the d values effectively converts the problem from a two-sample problem to a single sample of differences, which is an adaptation of the single-sample means formula. This test utilizes the sample mean of differences, and the standard deviation of differences, s_d, which can be computed by using formulas 10.6 and 10.7.

FORMULAS FOR $\bar{d}$ AND s_d (10.6 AND 10.7)

$$\bar{d} = \frac{\sum d}{n}$$

$$s_d = \sqrt{\frac{\sum (d - \bar{d})^2}{n - 2}} = \sqrt{\frac{\sum d^2 - \frac{\left(\sum d\right)^2}{n}}{n - 2}}$$

An assumption for this test is that the differences of the two populations are normally distributed.

Analysing data by this method involves calculating a t value with formula 10.5 and comparing it with a critical t value obtained from the table. The critical t value is obtained from the t distribution table in the usual way, with the exception that, in the degrees of freedom $(n - 1)$, n is the number of matched pairs of scores.

Suppose a stock market investor is interested in determining whether there is a significant difference in the P/E (price to earnings) ratio for companies from one year to the next. In an effort to study this question, the investor randomly samples nine companies from the *Handbook of Common Stocks* and records the P/E ratios for each of these companies at the end of year 1 and at the end of year 2. The data are shown in Table 10.5.

These data are related data because each P/E value for year 1 has a corresponding year 2 measurement on the same company. Because no prior information indicates whether P/E ratios have gone up or down, the hypothesis tested is two tailed. Assume $\alpha = 0.01$. Assume that differences in P/E ratios are normally distributed in the population.

TABLE 10.5

P/E Ratios for Nine Randomly Selected Companies

Company	Year 1 P/E Ratio	Year 2 P/E Ratio
1	8.9	12.7
2	38.1	45.4
3	43.0	10.0
4	34.0	27.2
5	34.5	22.8
6	15.2	24.1
7	20.3	32.3
8	19.9	40.1
9	61.9	106.5

HYPOTHESIZE

STEP 1:

$$H_0: D = 0$$

$$H_a: D \neq 0$$

TEST

STEP 2: The appropriate statistical test is

$$t = \frac{\bar{d} - D}{\dfrac{s_d}{\sqrt{n}}}$$

STEP 3: $\alpha = 0.01$.

TABLE 10.6			
Analysis of P/E Ratio Data			
Company	Year 1 P/E	Year 2 P/E	d
1	8.9	12.7	−3.8
2	38.1	45.4	−7.3
3	43.0	10.0	33.0
4	34.0	27.2	6.8
5	34.5	22.8	11.7
6	15.2	24.1	−8.9
7	20.3	32.3	−12.0
8	19.9	40.1	−20.2
9	61.9	106.5	−44.6

STEP 4: Because $\alpha = 0.01$ and this test is two tailed, $\alpha/2 = 0.005$ is used to obtain the table t value. With nine pairs of data, $n = 9$, df $= n - 1 = 8$. The table t value is $t_{0.055,8} = \pm 3.3555$. If the observed test statistic is greater than 3.355 or less than $t_{0.005,8} = \pm 3.355$, the null hypothesis will be rejected.

STEP 5: The sample data are given in Table 10.5.

STEP 6: Table 10.6 shows the calculations to obtain the observed value of the test statistic, which is $t = -0.70$.

ACTION

STEP 7: Because the observed t value is greater than the critical table t value in the lower tail ($t = -0.70 > t = -3.355$) value it is in the non-rejection region.

BUSINESS **I**MPLICATIONS

STEP 8: There is not enough evidence from the data to declare a significant difference in the average P/E ratio between year 1 and year 2. Figure 10.9 depicts the rejection regions, the critical values of t, and the observed value of t for this example.

Using the Computer to Make Statistical Inferences About Two Related Populations

Both Minitab and Excel can be used to make statistical inferences about two related populations. Figure 10.10 shows Minitab and Excel output for the P/E Ratio problem. The Minitab output contains summary data for each sample and the difference of the two samples along with a confidence interval of the difference, a restating of the tested hypotheses,

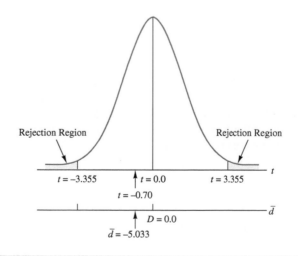

FIGURE 10.9

Graphical Depiction of P/E Ratio Analysis

Minitab Output

Paired T-Test and CI: Year 1, Year 2

Paired T for Year 1 - Year 2

	N	Mean	StDev	SE Mean
Year 1	9	30.64	16.37	5.46
Year 2	9	35.68	28.94	9.65
Difference	9	-5.03	21.60	7.20

99% CI for mean difference: (-29.19, 19.12)
T-Test of mean difference = 0 (vs not = 0):
T-Value = -0.70 P-Value = 0.504

Excel Output

t-Test: Paired Two Sample for Means

	Year 1	Year 2
Mean	30.64	35.68
Variance	268.135	837.544
Observations	9	9
Pearson Correlation	0.674	
Hypothesized Mean Difference	0	
df	8	
t Stat	-0.70	
P (T<=t) one-tail	0.252	
t Critical one-tail	2.90	
P (T<=t) two-tail	0.504	
t Critical two-tail	3.36	

FIGURE 10.10

Minitab and Excel Output for the P/E Ratio Example

the observed t value, and the p-value. Because the p-value (0.504) is greater than the value of alpha (0.01), the decision is to fail to reject the null hypothesis.

The Excel output contains the hypothesized mean difference, the observed t value (−0.70), and the critical t values and their associated p-values for both a one-tailed and a two-tailed test. The p-value for a two-tailed test is the same as that produced by Minitab, indicating that the decision is to fail to reject the null hypothesis.

DEMONSTRATION PROBLEM 10.5

Let us revisit the hypothetical study discussed earlier in the section in which consumers are asked to rate a company both before and after viewing a video on the company twice a day for a week. The data from Table 10.4 are displayed again here. Use an alpha of 0.05 to test to determine whether there is a significant increase in the ratings of the company after the one-week video treatment. Assume that differences in ratings are normally distributed in the population.

Individual	Before	After
1	32	39
2	11	15
3	21	35
4	17	13
5	30	41
6	38	39
7	14	22

Solution

Because the same individuals are being used in a before-and-after study, it is a related measures study. The desired effect is to increase ratings, which means the hypothesis test is one tailed.

HYPOTHESIZE

STEP 1:

$$H_0: D = 0$$

$$H_a: D < 0$$

Because the researchers want to 'prove' that the ratings increase from Before to After and because the difference is computed by subtracting After ratings from the Before ratings, the desired alternative hypothesis is $D < 0$.

TEST

STEP 2: The appropriate test statistic is formula (10.5).

STEP 3: The Type I error rate is 0.05.

STEP 4: The degrees of freedom are $n - 1 = 7 - 1 = 6$. For $\alpha = 0.05$ the table t value is, $t_{0.05,6} = -1.943$. The decision rule is to reject the null hypothesis if the observed value is less than −1.943.

STEP 5: The sample data and some calculations follow.

Individual	Before	After	d
1	32	39	−7
2	11	15	−4
3	21	35	−14
4	17	13	+4
5	30	41	−11
6	38	39	−1
7	14	22	−8
$\bar{d} = -5.87$		$s_d = 6.0945$	

STEP 6: The observed t value is:

$$t = \frac{-5.857 - 0}{\frac{6.0945}{\sqrt{7}}} = -2.54$$

Computer analysis of this problem reveals that the p-value is 0.022.

ACTION

STEP 7: Because the observed value of −2.54 is less than the critical, table value of −1.943 and the p-value (0.022) is less than alpha (0.05), the decision is to reject the null hypothesis.

BUSINESS IMPLICATIONS

STEP 8. There is enough evidence to conclude that, on average, the ratings have increased significantly. This result might be used by managers to support a decision to continue using the videos or to expand the use of such videos in an effort to increase public support for their company.

The graph, right, depicts the observed value, the rejection region, and the critical t value for the problem.

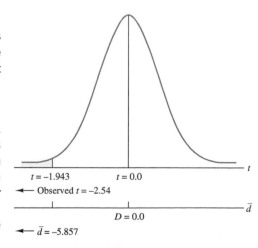

Confidence Intervals

Sometimes a researcher is interested in estimating the mean difference in two populations for related samples. A confidence interval for D, the mean population difference of two related samples, can be constructed by algebraically rearranging formula (10.5), which was used to test hypotheses about D. Again the assumption is that the differences are normally distributed in the population.

CONFIDENCE INTERVAL FORMULA TO ESTIMATE THE DIFFERENCE IN RELATED POPULATIONS, D (10.8)	$\bar{d} - t \dfrac{s_d}{\sqrt{n}} \leq D \leq \bar{d} + t \dfrac{s_d}{\sqrt{n}}$ $df = n - 1$

The following housing industry example demonstrates the application of formula 10.8. The sale of new houses apparently fluctuates seasonally. Superimposed on the seasonality are economic and business cycles that also influence the sale of new houses. In certain parts of the country, new-house sales increase in the spring and early summer and drop off in the autumn. Suppose a national real estate association wants to estimate the average difference in the number of new-house sales per company in Nottingham between 2011 and 2012. To do so, the association randomly selects 18 real estate firms in the Nottingham area and obtains their new-house sales figures for May 2011 and May 2012. The numbers of sales per company are shown in Table 10.7. Using these data, the association's analyst estimates the average difference in the number of sales per real estate company in Nottingham for May 2011 and May 2012 and constructs a 99% confidence interval. The analyst assumes that differences in sales are normally distributed in the population.

The number of pairs, n, is 18, and the degrees of freedom are 17. For a 99% level of confidence and these degrees of freedom, the table t value is $t_{0.005,17} = 2.898$. The values for $\bar{d}$, and s_d are shown in Table 10.8.

TABLE 10.7

Number of New House Sales in Nottingham

Real Estate Company	May 2011	May 2012
1	8	11
2	19	30
3	5	6
4	9	13
5	3	5
6	0	4
7	13	15
8	11	17
9	9	12
10	5	12
11	8	6
12	2	5
13	11	10
14	14	22
15	7	8
16	12	15
17	6	12
18	10	10

TABLE 10.8

Differences in Number of New House Sales

Real Estate Company	May 2011	May 2012	d
1	8	11	−3
2	19	30	−11
3	5	6	−1
4	9	13	−4
5	3	5	−2
6	0	4	−4
7	13	15	−2
8	11	17	−6
9	9	12	−3
10	5	12	−7
11	8	6	+2
12	2	5	−3
13	11	10	+1
14	14	22	−8
15	7	8	−1
16	12	15	−3
17	6	12	−6
18	10	10	0
$\bar{d} = -3.389$	and	$s_d = 3.274$	

```
Paired T-Test and CI:

Paired T for - 2011
                      N           Mean         StDev        SE Mean
2010                 18           8.44          4.64         1.09
2011                 18          11.83          6.54         1.54
Difference           18          -3.389         3.274        0.772

99% CI for mean difference: (-5.626, -1.152)
T-Test of mean difference = 0 (vs not = 0):
T-Value = -4.39 P-Value = 0.000
```

FIGURE 10.11

Minitab Output for the New House Sales Example

The point estimate of the difference is $\bar{d} = -3.389$. The 99% confidence interval is

$$\bar{d} - t\frac{s_d}{\sqrt{n}} \le D \le \bar{d} + t\frac{s_d}{\sqrt{n}}$$

$$-3.389 - 2.898\frac{3.274}{\sqrt{18}} \le D \le -3.389 + 2.898\frac{3.274}{\sqrt{18}}$$

$$-3.389 - 2.236 \le D \le -3.389 + 2.236$$

$$-5.625 \le D \le -1.153$$

The analyst estimates with a 99% level of confidence that the average difference in new-house sales for a real estate company in Nottingham between 2011 and 2012 in May is somewhere between –5.625 and –1.153 houses. Because 2012 sales were subtracted from 2011 sales, the minus signs indicate more sales in 2012 than in 2011. Note that both ends of the confidence interval contain negatives. This result means that the analyst can be 99% confident that zero difference is not the average difference. If the analyst were using this confidence interval to test the hypothesis that there is no significant mean difference in average new-house sales per company in Nottingham between May 2011 and May 2012, the null hypothesis would be rejected for $\alpha = 0.01$. The point estimate for this example is –3.389 houses, with an error of 2.236 houses. Figure 10.11 is the Minitab computer output for the confidence interval.

10.3 PROBLEMS

10.21 Use the data given and a 1% level of significance to test the hypotheses below. Assume the differences are normally distributed in the population.

$H_0: D = 0 \qquad H_a: D > 0$

Pair	Sample 1	Sample 2
1	38	22
2	27	28
3	30	21
4	41	38
5	36	38
6	38	26
7	33	19
8	35	31
9	44	35

10.22   Use the data given to test the hypotheses below $\alpha = 0.05$. Assume the differences are normally distributed in the population.

$$H_0: D = 0 \qquad H_a: D \neq 0$$

Individual	Before	After
1	107	102
2	99	98
3	110	100
4	113	108
5	96	89
6	98	101
7	100	99
8	102	102
9	107	105
10	109	110
11	104	102
12	99	96
13	101	100

10.23   Construct a 98% confidence interval to estimate D from the following sample information. Assume the differences are normally distributed in the population.

$$\bar{d} = 40.56,\ s_d = 26.58,\ n = 22$$

10.24   Construct a 90% confidence interval to estimate D from the sample information below. Assume the differences are normally distributed in the population.

Client	Before	After
1	32	40
2	28	25
3	35	36
4	32	32
5	26	29
6	25	31
7	37	39
8	16	30
9	35	31

10.25   Because of uncertainty in real estate markets, many homeowners are considering remodelling and constructing additions rather than selling. Probably the most expensive room in the house to remodel is the kitchen, with an average cost of about €23,400. In terms of resale value, is remodelling the kitchen worth the cost? The following cost and resale figures are published for 11 cities. Use these data to construct a 99% confidence interval for the difference between cost and added resale value of kitchen remodelling. Assume the differences are normally distributed in the population.

City	Cost	Resale
A	20,427	25,163
B	27,255	24,625
C	22,115	12,600
D	23,256	24,588
E	21,887	19,267
F	24,255	20,150
G	19,852	22,500
H	23,624	16,667
I	25,885	26,875
J	28,999	35,333
K	20,836	16,292

10.26 The marketing director brought to the attention of sales managers that most of the company's sales representatives contacted clients and maintained client relationships in a disorganized, haphazard way. The sales managers brought the reps in for a three-day seminar and training session on how to use an organizer to schedule visits and recall pertinent information about each client more effectively. Sales reps were taught how to schedule visits most efficiently to maximize their efforts. Sales managers were given data on the number of site visits by sales reps on a randomly selected day both before and after the seminar. Use the data below to test whether significantly more site visits were made after the seminar ($\alpha = 0.05$). Assume the differences in the number of site visits are normally distributed.

Rep	Before	After
1	2	4
2	4	5
3	1	3
4	3	3
5	4	3
6	2	5
7	2	6
8	3	4
9	1	5

10.27 Eleven employees were put under the care of the company nurse because of high cholesterol readings. The nurse lectured them on the dangers of this condition and put them on a new diet. Shown are the cholesterol readings of the 11 employees both before the new diet and one month after use of the diet began. Construct a 98% confidence interval to estimate the population mean difference of cholesterol readings for people who are involved in this programme. Assume differences in cholesterol readings are normally distributed in the population.

Employee	Before	After
1	255	197
2	230	225
3	290	215
4	242	215
5	300	240
6	250	235
7	215	190
8	230	240
9	225	200
10	219	203
11	236	223

10.28 Lawrence and Glover published the results of a study in the *Journal of Managerial Issues* in which they examined the effects of accounting firm mergers on auditing delay. Auditing delay is the time between a company's fiscal year-end and the date of the auditor's report. The hypothesis is that with the efficiencies gained through mergers the length of the audit delay would decrease. Suppose to test their hypothesis, they examined the audit delays on 27 clients of six firms from both before and after the merger of the companies (a span of five years). Suppose further that the mean difference in audit delay for these clients from before merger to after merger was a decrease in 3.71 days and the standard deviation of difference was five days. Use these data and $\alpha = 0.01$ to test whether the audit delays after the merger were significantly lower than before the merger. Assume that the differences in auditing delay are normally distributed in the population.

10.29 A nationally known supermarket decided to promote its own brand of soft drinks on TV for two weeks. Before the ad campaign, the company randomly selected 21 of its stores across the United States to be part of a study to measure the campaign's effectiveness. During a specified half-hour period on a certain Monday morning, all the stores in the sample counted the number of cans of its own brand of soft drink sold. After the campaign, a similar count was made. The average difference was an increase of 75 cans, with a standard deviation of difference of 30 cans. Using this information, construct a 90% confidence interval to estimate the population average difference in soft drink sales for this company's brand before and after the ad campaign. Assume the differences in soft drink sales for the company's brand are normally distributed in the population.

10.30 Is there a significant difference in the petrol consumption of a car for regular unleaded and premium unleaded? To test this question, a researcher randomly selected 15 drivers for a study. They were to drive their cars for one month on regular unleaded and for one month on premium unleaded petrol. The participants drove their own cars for this experiment. The average sample difference was 2.85 miles per gallon in favour of the premium unleaded, and the sample standard deviation of difference was 1.9 miles per gallon. For $\alpha = 0.01$, does the test show enough evidence for the researcher to conclude that there is a significant difference in mileage between regular unleaded and premium unleaded petrol? Assume the differences in petrol consumption figures are normally distributed in the population.

10.4 STATISTICAL INFERENCES ABOUT TWO POPULATION PROPORTIONS, $P_1 - P_2$

Sometimes a researcher wishes to make inferences about the difference in two population proportions. This type of analysis has many applications in business, such as comparing the market share of a product for two different markets, studying the difference in the proportion of female customers in two different geographic regions, or comparing the proportion of defective products from one period to another. In making inferences about the difference in two population proportions, the statistic normally used is the difference in the sample proportions $\hat{p}_1 - \hat{p}_2$. This statistic is computed by taking random samples and determining $\hat{p}$ for each sample for a given characteristic, then calculating the difference in these sample proportions.

The central limit theorem states that for large samples (each of $n_1 \cdot \hat{p}_1$, $n_1 \cdot \hat{q}_1$, $n_2 \cdot \hat{p}_2$, and $n_2 \cdot \hat{q}_2 > 5$, where $\hat{q} = 1 - \hat{p}$), the difference in sample proportions is normally distributed with a mean difference of

$$\mu_{\hat{p}_1 - \hat{p}_2} = p_1 - p_2$$

and a standard deviation of the difference of sample proportions of

$$\sigma_{\hat{p}_1 - \hat{p}_2} = \sqrt{\frac{p_1 \cdot q_1}{n_1} + \frac{p_2 \cdot q_2}{n_2}}$$

From this information, a z formula for the difference in sample proportions can be developed.

$$z = \frac{(\hat{p}_1 - \hat{p}_2) - (p_1 - p_2)}{\sqrt{\frac{p_1 \cdot q_1}{n_1} + \frac{p_2 \cdot q_2}{n_2}}}$$

z FORMULA FOR THE DIFFERENCE IN TWO POPULATION PROPORTIONS (10.9)

where:
$\hat{p}_1$ = proportion from sample 1
$\hat{p}_2$ = proportion from sample 2
n_1 = size of sample 1
n_2 = size of sample 2
p_1 = proportion from population 1
p_2 = proportion from population 2
$q_1 = 1 - p_1$
$q_2 = 1 - p_2$

Hypothesis Testing

Formula 10.9 is the formula that can be used to determine the probability of getting a particular difference in two sample proportions when given the values of the population proportions. In testing hypotheses about the difference in two population proportions, particular values of the population proportions are not usually known or assumed. Rather, the hypotheses are about the difference in the two population proportions $(p_1 - p_2)$. Note that formula 10.9 requires knowledge of the values of p_1 and p_2. Hence, a modified version of formula 10.9 is used when testing hypotheses about $p_1 - p_2$. This formula utilizes a pooled value obtained from the sample proportions to replace the population proportions in the denominator of formula (10.9).

The denominator of formula 10.9 is the standard deviation of the difference in two sample proportions and uses the population proportions in its calculations. However, the population proportions are unknown, so an estimate of the standard deviation of the difference in two sample proportions is made by using sample proportions as point estimates of the population proportions. The sample proportions are combined by using a weighted average to produce $\hat{p}$, which, in conjunction with $\hat{p}$ and the sample sizes, produces a point estimate of the standard deviation of the difference in sample proportions. The result is formula 10.10, which we shall use to test hypotheses about the difference in two population proportions.

z FORMULA TO TEST THE DIFFERENCE IN POPULATION PROPORTIONS (10.10)

$$z = \frac{(\hat{p}_1 - \hat{p}_2) - (p_1 - p_2)}{\sqrt{(\bar{p}.\bar{q})\left(\dfrac{1}{n_1} + \dfrac{1}{n_2}\right)}}$$

where:

$$\bar{p} = \frac{x_1 + x_2}{n_1 + n_2} = \frac{n_1\hat{p}_1 + n_2\hat{p}_2}{n_1 + n_2} \text{ and } \bar{q} = 1 - \bar{p}$$

Testing the difference in two population proportions is useful whenever the researcher is interested in comparing the proportion of one population that has a certain characteristic with the proportion of a second population that has the same characteristic. For example, a researcher might be interested in determining whether the proportion of people driving new cars (less than one year old) in Oslo is different from the proportion in Copenhagen. A study could be conducted with a random sample of Oslo drivers and a random sample of Copenhagen drivers to test this idea. The results could be used to compare the new-car potential of the two markets and the propensity of drivers in these areas to buy new cars.

As another example, do consumers and CEOs have different perceptions of ethics in business? A group of researchers attempted to determine whether there was a difference in the proportion of consumers and the proportion of CEOs who believe that fear of getting caught or losing one's job is a strong influence of ethical behaviour. In their study, they found that 57% of consumers said that fear of getting caught or losing one's job was a strong influence on ethical behaviour, but only 50% of CEOs felt the same way.

Suppose these data were determined from a sample of 755 consumers and 616 CEOs. Does this result provide enough evidence to declare that a significantly higher proportion of consumers than of CEOs believe fear of getting caught or losing one's job is a strong influence on ethical behaviour?

HYPOTHESIZE

STEP 1: Suppose sample 1 is the consumer sample and sample 2 is the CEO sample. Because we are trying to prove that a higher proportion of consumers than of CEOs believe fear of getting caught or losing one's job is a strong influence on ethical behaviour, the alternative hypothesis should be $p_1 - p_2 > 0$. The following hypotheses are being tested:

$$H_0: p_1 - p_2 = 0$$

$$H_a: p_1 - p_2 > 0$$

where:

p_1 is the proportion of consumers who select the factor
p_2 is the proportion of CEOs who select the factor

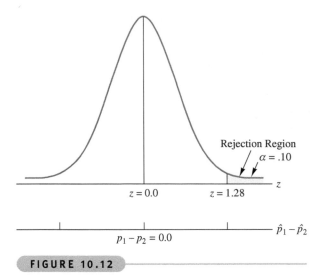

FIGURE 10.12

Rejection Region for the Ethics Example

TEST

STEP 2:   The appropriate statistical test is formula 10.10.

STEP 3:   Let $\alpha = 0.10$.

STEP 4:   Because this test is a one-tailed test, the critical table z value is $z_c = 1.28$. If an observed value of z of more than 1.28 is obtained, the null hypothesis will be rejected. Figure 10.12 shows the rejection region and the critical value for this problem.

STEP 5:   The sample information follows.

Consumers	CEOs
$n_1 = 755$	$n_2 = 616$
$\hat{p}_1 = .57$	$\hat{p}_2 = .50$

STEP 6:

$$\bar{p} = \frac{n_1\hat{p}_1 + n_2\hat{p}_2}{n_1 + n_2} = \frac{(755)(0.57) + (616)(0.50)}{755 + 616} = 0.539$$

If the statistics had been given as raw data instead of sample proportions, we would have used the following formula:

$$\bar{p} = \frac{x_1 + x_2}{n_1 + n_2}$$

The observed z value is

$$z = \frac{(.57 - .50) - (0)}{\sqrt{(0.539)(0.461)\left(\dfrac{1}{755} + \dfrac{1}{616}\right)}} = 2.59$$

ACTION

STEP 7: Because $z = 2.59$ is greater than the critical table z value of 1.28 and is in the rejection region, the null hypothesis is rejected.

BUSINESS IMPLICATIONS

STEP 8: A significantly higher proportion of consumers than of CEOs believe fear of getting caught or losing one's job is a strong influence on ethical behaviour. CEOs might want to take another look at ways to influence ethical behaviour. If employees are more like consumers than CEOs, CEOs might be able to use fear of getting caught or losing one's job as a means of ensuring ethical behaviour on the job. By transferring the idea of ethical behaviour to the consumer, retailers might use fear of being caught and prosecuted to retard shoplifting in the retail trade.

DEMONSTRATION PROBLEM 10.6

A study of female entrepreneurs was conducted to determine their definition of success. The women were offered optional choices such as happiness/self-fulfilment, sales/profit, and achievement/challenge. The women were divided into groups according to the gross sales of their businesses. A significantly higher proportion of female entrepreneurs in the €100,000 to €500,000 category than in the less than €100,000 category seemed to rate sales/profit as a definition of success.

Suppose you decide to test this result by taking a survey of your own and identify female entrepreneurs by gross sales. You interview 100 female entrepreneurs with gross sales of less than €100,000, and 24 of them define sales/profit as success. You then interview 95 female entrepreneurs with gross sales of €100,000 to €500,000, and 39 cite sales/profit as a definition of success. Use this information to test to determine whether there is a significant difference in the proportions of the two groups that define success as sales/profit. Use $\alpha = 0.01$.

Solution

HYPOTHESIZE

STEP 1: You are testing to determine whether there is a difference between two groups of entrepreneurs, so a two-tailed test is required. The hypotheses follow.

$$H_0: p_1 - p_2 = 0$$

$$H_a: p_1 - p_2 \neq 0$$

STEP 2: The appropriate statistical test is formula 10.10.
STEP 3: Alpha has been specified as 0.01.
STEP 4: With $\alpha = 0.01$, you obtain a critical z value from Table A.5 in Appendix A for $\alpha/2 = 0.005$, $z_{0.005} = \pm 2.575$. If the observed z value is more than 2.575 or less than −2.575, the null hypothesis is rejected.
STEP 5: The sample information follows.

Less than €100,000	€100,000 to €500,000
$n_1 = 100$	$n_2 = 95$
$x_1 = 24$	$x_2 = 39$
$\hat{p}_1 = \dfrac{24}{100} = 0.24$	$\hat{p}_2 = \dfrac{39}{95} = 0.41$

where:

$$\bar{p} = \frac{x_1 + x_2}{n_1 + n_2} = \frac{24 + 39}{100 + 95} = \frac{63}{195} = 0.323$$

$x =$ the number of entrepreneurs who define sales/profits as success
STEP 6: The observed z value is

$$z = \frac{(\hat{p}_1 - \hat{p}_2) - (p_1 - p_2)}{\sqrt{(\bar{p} \cdot \bar{q})\left(\dfrac{1}{n_1} + \dfrac{1}{n_2}\right)}} = \frac{(0.24 - 0.41) - (0)}{\sqrt{(0.323)(0.677)\left(\dfrac{1}{100} + \dfrac{1}{95}\right)}} = \frac{-0.17}{0.067} = -2.54$$

ACTION

STEP 7: Although this observed value is near the rejection region, it is in the non-rejection region. The null hypothesis is not rejected. The test did not show enough evidence here to reject the null hypothesis and declare that the responses to the question by the two groups are different statistically. Note that alpha was small and that a two-tailed test was conducted. If a one-tailed test had been used, z_c would have been $z_{0.01} = -2.33$, and the null hypothesis would have been rejected. If alpha had been 0.05, z_c would have been $z_{0.025} = \pm 1.96$, and the null hypothesis would have been rejected. This result underscores the crucial importance of selecting alpha and determining whether to use a one-tailed or two-tailed test in hypothesis testing.

The following diagram shows the critical values, the rejection regions, and the observed value for this problem:

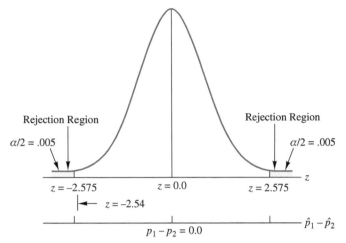

BUSINESS IMPLICATIONS

STEP 8: We cannot statistically conclude that a greater proportion of female entrepreneurs in the higher gross sales category define success as sales/profit. One of the payoffs of such a determination is to find out what motivates the people with whom we do business. If sales/profits motivate people, offers or promises of greater sales and profits can be a means of attracting their services, their interest, or their business. If sales/profits do not motivate people, such offers would not generate the kind of response wanted and we would need to look for other ways to motivate them.

Confidence Intervals

Sometimes in business research the investigator wants to estimate the difference in two population proportions. For example, what is the difference, if any, in the population proportions of workers in the United Kingdom who favour union membership and workers in Germany who favour union membership? In studying two different suppliers of the same part, a large manufacturing company might want to estimate the difference between suppliers in the proportion of parts that meet specifications. These and other situations requiring the estimation of the difference in two population proportions can be solved by using confidence intervals.

The formula for constructing confidence intervals to estimate the difference in two population proportions is a modified version of formula 10.9. Formula 10.9 for two proportions requires knowledge of each of the population proportions. Because we are attempting to estimate the difference in these two proportions, we obviously do not know their value. To overcome this lack of knowledge in constructing a confidence interval formula, we substitute the sample proportions in place of the population proportions and use these sample proportions in the estimate, as follows:

$$z = \frac{(\hat{p}_1 - \hat{p}_2) - (p_1 - p_2)}{\sqrt{\dfrac{\hat{p}_1 \cdot \hat{q}_1}{n_1} + \dfrac{\hat{p}_2 \cdot \hat{q}_2}{n_2}}}$$

Solving this equation for $p_1 - p_2$ produces the formula for constructing confidence intervals for $p_1 - p_2$.

CONFIDENCE INTERVAL TO ESTIMATE $p_1 - p_2$ (10.11)

$$(\hat{p}_1 - \hat{p}_2) - z\sqrt{\frac{\hat{p}_1 \cdot \hat{q}_1}{n_1} + \frac{\hat{p}_2 \cdot \hat{q}_2}{n_2}} \le p_1 - p_2 \le (\hat{p}_1 - \hat{p}_2) + z\sqrt{\frac{\hat{p}_1 \cdot \hat{q}_1}{n_1} + \frac{\hat{p}_2 \cdot \hat{q}_2}{n_2}}$$

To see how this formula is used, suppose that in an attempt to target its clientele, managers of a supermarket chain want to determine the difference between the proportion of morning shoppers who are men and the proportion of afternoon shoppers who are men. Over a period of two weeks, the chain's researchers conduct a systematic random sample survey of 400 morning shoppers, which reveals that 352 are women and 48 are men. During this same period, a systematic random sample of 480 afternoon shoppers reveals that 293 are women and 187 are men. Construct a 98% confidence interval to estimate the difference in the population proportions of men.

The sample information is shown here.

Morning Shoppers	Afternoon Shoppers
$n_1 = 400$	$n_2 = 480$
$x_1 = 48$ men	$x_2 = 187$ men
$\hat{p}_1 = .12$	$\hat{p}_2 = .39$
$\hat{q}_1 = .88$	$\hat{q}_2 = .61$

For a 98% level of confidence, $z = 2.33$. Using formula 10.11 yields

$$(0.12 - 0.39) - 2.33\sqrt{\frac{(0.12)(0.88)}{400} + \frac{(0.39)(0.61)}{480}} \leq p_1 - p_2$$

$$\leq (0.12 - 0.39) + 2.33\sqrt{\frac{(0.12)(0.88)}{400} + \frac{(0.39)(0.61)}{480}}$$

$$-0.27 - 0.64 \geq p_1 - p_2 \geq -0.27 + 0.64$$

$$-0.334 \geq p_1 - p_2 \geq -0.206$$

There is a 98% level of confidence that the difference in population proportions is between −0.334 and −0.206. Because the afternoon shopper proportion was subtracted from the morning shoppers, the negative signs in the interval indicate a higher proportion of men in the afternoon shoppers than in the morning shoppers. Thus the confidence level is 98% that the difference in proportions is at least 0.206 and may be as much as 0.334.

Using the Computer to Analyse the Difference in Two Proportions

Minitab has the capability of testing hypotheses or constructing confidence intervals about the difference in two proportions. Figure 10.13 shows Minitab output for the shopping example. Notice that the output contains a summary of sample information along with the difference in sample proportions, the confidence interval, the computer z value for a hypothesis test, and the p-value. The confidence interval shown here is the same as the one we just computed except for rounding differences.

```
            TEST AND CI FOR TWO PROPORTIONS

Sample        X              N          Sample p
1            48             400         0.120000
2           187             480         0.389583
Difference = p(1) - p(2)
Estimate for difference: -0.269583
98% CI for difference: (-0.333692, -0.205474)
Test for difference = 0 (vs not = 0):
Z = -9.78   P-Value = 0.000
Fisher's exact test: P-Value = 0.000
```

FIGURE 10.13

Minitab Output for the Shopping Example

10.4 PROBLEMS

10.31 Using the given sample information, test the following hypotheses:

a. $H_0: p_1 - p_2 = 0$; $H_a: p_1 - p_2 \neq 0$

Sample 1	Sample 2	
$n_1 = 368$	$n_2 = 405$	
$x_1 = 175$	$x_2 = 182$	Let $\alpha = .05$.

Note that x is the number in the sample having the characteristic of interest.

b. $H_0: p_1 - p_2 = 0$; $H_a: p_1 - p_2 > 0$

Sample 1	Sample 2	
$n_1 = 649$	$n_2 = 558$	
$\hat{p}_1 = .38$	$\hat{p}_2 = .25$	Let $\alpha = .10$.

10.32 In each of the following cases, calculate a confidence interval to estimate $p_1 - p_2$.

a. $n_1 = 85$, $n_2 = 90$, $\hat{p}_1 = 0.75$; $\hat{p}_2 = 0.67$; level of confidence $= 90\%$

b. $n_1 = 1{,}100$, $n_2 = 1{,}300$, $\hat{p}_1 = 0.19$; $\hat{p}_2 = 0.17$; level of confidence $= 95\%$

c. $n_1 = 430$, $n_2 = 399$, $x_1 = 275$, $x_2 = 275$; level of confidence $= 85\%$

d. $n_1 = 1{,}500$, $n_2 = 1{,}500$, $x_1 = 1{,}050$, $x_2 = 1{,}100$; level of confidence $= 80\%$

10.33 According to a study conducted for Gateway Computers, 59% of men and 70% of women say that weight is an extremely/very important factor in purchasing a laptop computer. Suppose this survey was conducted using 374 men and 481 women. Do these data show enough evidence to declare that a significantly higher proportion of women than men believe that weight is an extremely/very important factor in purchasing a laptop computer? Use a 5% level of significance.

10.34 Does age make a difference in the amount of savings a worker feels is needed to be secure at retirement? A study by CommSciences for Transamerica Asset Management found that 0.24 of workers in the 25–33 age category feel that $250,000 to $500,000 is enough to be secure at retirement. However, 0.35 of the workers in the 34–52 age category feel that this amount is enough. Suppose 210 workers in the 25–33 age category and 176 workers in the 34–52 age category were involved in this study. Use these data to construct a 90% confidence interval to estimate the difference in population proportions on this question.

10.35 Companies that recently developed new products were asked to rate which activities are most difficult to accomplish with new products. Options included such activities as assessing market potential, market testing, finalizing the design, developing a business plan, and the like. A researcher wants to conduct a similar study to compare the results between two industries: the computer hardware industry and the banking industry. He takes a random sample of 56 computer firms and 89 banks. The researcher asks whether market testing is the most difficult activity to accomplish in developing a new product. Some 48% of the sampled computer companies and 56% of the sampled banks respond that it is the most difficult activity. Use a level of significance of 0.20 to test whether there is a significant difference in the responses to the question from these two industries.

10.36 A large production facility uses two machines to produce a key part for its main product. Inspectors have expressed concern about the quality of the finished product. Quality control investigation has revealed that the key part made by the two machines is defective at times. The inspectors randomly sampled 35 units of the key part from each machine. Of those produced by machine A, five were defective. Seven of the 35 sampled parts from machine B were defective. The production manager is interested in estimating the difference in proportions of the populations of parts that are defective between machine A and machine B. From the sample information, compute a 98% confidence interval for this difference.

10.37 According to a CCH Unscheduled Absence survey, 9% of small businesses use telecommuting of workers in an effort to reduce unscheduled absenteeism. This proportion compares to 6% for all businesses. Is there really a significant difference between small businesses and all businesses on this issue? Use these data and an alpha of 0.10 to test this question. Assume that there were 780 small businesses and 915 other businesses in this survey.

10.38   Many Americans spend time worrying about paying their bills. A survey by Fleishman-Hilliard Research for MassMutual discovered that 60% of Americans with kids say that paying bills is a major concern. This proportion compares to 52% of Americans without kids. Suppose 850 Americans with kids and 910 without kids were contacted for this study. Use these data to construct a 95% confidence interval to estimate the difference in population proportions between Americans with kids and Americans without kids on this issue.

10.5   TESTING HYPOTHESES ABOUT TWO POPULATION VARIANCES

Sometimes we are interested in studying the variance of a population rather than a mean or proportion. Section 9.5 discussed how to test hypotheses about a single population variance, but on some occasions business researchers are interested in testing hypotheses about the difference in two population variances. In this section, we examine how to conduct such tests. When would a business researcher be interested in the variances from two populations?

In quality control, analysts often examine both a measure of central tendency (mean or proportion) and a measure of variability. Suppose a manufacturing plant made two batches of an item, produced items on two different machines, or produced items on two different shifts. It might be of interest to management to compare the variances from two batches or two machines to determine whether there is more variability in one than another.

Variance is sometimes used as a measure of the risk of a stock in the stock market. The greater the variance, the greater the risk. By using techniques discussed here, a financial researcher could determine whether the variances (or risk) of two stocks are the same.

In testing hypotheses about two population variances, the sample variances are used. It makes sense that if two samples come from the same population (or populations with equal variances), the ratio of the sample variances, s_1^2/s_2^2, should be about 1. However, because of sampling error, sample variances even from the same population (or from two populations with equal variances) will vary. This *ratio of two sample variances* formulates what is called an *F* value.

$$F = \frac{s_1^2}{s_2^2}$$

These ratios, if computed repeatedly for pairs of sample variances taken from a population, are distributed as an *F* distribution. The *F* distribution will vary by the sizes of the samples, which are converted to degrees of freedom.

With the *F* distribution, there are degrees of freedom associated with the numerator (of the ratio) and the denominator. An assumption underlying the *F* distribution is that the populations from which the samples are drawn are normally distributed for *x*. *The F test of two population variances is extremely sensitive to violations of the assumption that the populations are normally distributed.* The statistician should carefully investigate the shape of the distributions of the populations from which the samples are drawn to be certain the populations are normally distributed. The formula used to test hypotheses comparing two population variances follows.

F TEST FOR TWO POPULATION
VARIANCES (10.12)

$$F = \frac{s_1^2}{s_2^2}$$

$$df_{numerator} = v_1 = n_1 - 1$$

$$df_{denominator} = v_2 = n_2 - 1$$

Table A.7 in Appendix A contains *F* distribution table values for $\alpha = 0.10, 0.05, 0.025, 0.01$, and 0.005. Figure 10.14 shows an *F* distribution for $v_1 = 6$ and $v_2 = 30$. Notice that the distribution is non-symmetric, which can be a problem

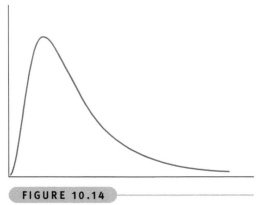

FIGURE 10.14

An F Distribution for $v_1 = 6$, $v_2 = 30$

when we are conducting a two-tailed test and want to determine the critical value for the lower tail. Table A.7 contains only F values for the upper tail. However, the F distribution is not symmetric nor does it have a mean of zero as do the z and t distributions; therefore, we cannot merely place a minus sign on the upper-tail critical value and obtain the lower-tail critical value (in addition, the F ratio is always positive – it is the ratio of two variances). This dilemma can be solved by using formula 10.13, which essentially states that the critical F value for the lower tail $(1 - \alpha)$ can be solved for by taking the inverse of the F value for the upper tail (α). The degrees of freedom numerator for the upper-tail critical value is the degrees of freedom denominator for the lower-tail critical value, and the degrees of freedom denominator for the upper-tail critical value is the degrees of freedom numerator for the lower-tail critical value.

FORMULA FOR DETERMINING THE CRITICAL VALUE FOR THE LOWER-TAIL F (10.13)

$$F_{1-\alpha,v_2,v_1} = \frac{1}{F_{\alpha,v_1,v_2}}$$

A hypothesis test can be conducted using two sample variances and Formula 10.12. The following example illustrates this process:

Suppose a machine produces metal sheets that are specified to be 22 millimetres thick. Because of the machine, the operator, the raw material, the manufacturing environment, and other factors, there is variability in the thickness. Two machines produce these sheets. Operators are concerned about the consistency of the two machines. To test consistency, they randomly sample 10 sheets produced by machine 1 and 12 sheets produced by machine 2. The thickness measurements of sheets from each machine are given in the table on the following page. Assume sheet thickness is normally distributed in the population. How can we test to determine whether the variance from each sample comes from the same population variance (population variances are equal) or from different population variances (population variances are not equal)?

HYPOTHESIZE

STEP 1: Determine the null and alternative hypotheses. In this case, we are conducting a two-tailed test (variances are the same or not), and the following hypotheses are used:

$$H_0 : \sigma_1^2 = \sigma_2^2$$
$$H_a : \sigma_1^2 \neq \sigma_2^2$$

TEST

STEP 2: The appropriate statistical test is

$$F = \frac{s_1^2}{s_2^2}$$

STEP 3: Let $\alpha = 0.05$.

STEP 4: Because we are conducting a two-tailed test, $\alpha/2 = 0.025$. Because $n_1 = 10$ and $n_2 = 12$, the degrees of freedom numerator for the upper-tail critical value is $v_1 = n_1 - 1 = 10 - 1 = 9$ and the degrees of freedom denominator for the upper-tail critical value is $v_2 = n_2 - 1 = 12 - 1 = 11$. The critical F value for the upper tail obtained from Table A.7 is

$$F_{.025,9,11} = 3.59$$

Table 10.9 is a copy of the F distribution for a one-tailed $\alpha = 0.025$ (which yields equivalent values for two-tailed $\alpha = 0.05$ where the upper tail contains 0.025 of the area). Locate $F_{0.025,9,11} = 3.59$ in the table. The lower-tail critical value can be calculated from the upper-tail value by using Formula 10.13.

TABLE 10.9

A Portion of the *F* Distribution Table

Percentage Points of the *F* Distribution

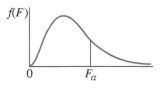

$\alpha = 0.025$

ν_2	ν_1	*Numerator Degrees of Freedom*							
	1	2	3	4	5	6	7	8	9
1	647.8	799.5	864.2	899.6	921.8	937.1	948.2	956.7	963.3
2	38.51	39.00	39.17	39.25	39.30	39.33	39.36	39.37	39.39
3	17.44	16.04	15.44	15.10	14.88	14.73	14.62	14.54	14.47
4	12.22	10.65	9.98	9.60	9.36	9.20	9.07	8.98	8.90
5	10.01	8.43	7.76	7.39	7.15	6.98	6.85	6.76	6.68
6	8.81	7.26	6.60	6.23	5.99	5.82	5.70	5.60	5.52
7	8.07	6.54	5.89	5.52	5.29	5.12	4.99	4.90	4.82
8	7.57	6.06	5.42	5.05	4.82	4.65	4.53	4.43	4.36
9	7.21	5.71	5.08	4.72	4.48	4.32	4.20	4.10	4.03
10	6.94	5.46	4.83	4.47	4.24	4.07	3.95	3.85	3.78
11	6.72	5.26	4.63	4.28	4.04	3.88	3.76	3.66	3.59
12	6.55	5.10	4.47	4.12	3.89	3.73	3.61	3.51	3.44
13	6.41	4.97	4.35	4.00	3.77	3.60	3.48	3.39	3.31
14	6.30	4.86	4.24	3.89	3.66	3.50	3.38	3.29	3.21
15	6.20	4.77	4.15	3.80	3.58	3.41	3.29	3.20	3.12
16	6.12	4.69	4.08	3.73	3.50	3.34	3.22	3.12	3.05
17	6.04	4.62	4.01	3.66	3.44	3.28	3.16	3.06	2.98
18	5.98	4.56	3.95	3.61	3.38	3.22	3.10	3.01	2.93
19	5.92	4.51	3.90	3.56	3.33	3.17	3.05	2.96	2.88
20	5.87	4.46	3.86	3.51	3.29	3.13	3.01	2.91	2.84
21	5.83	4.42	3.82	3.48	3.25	3.09	2.97	2.87	2.80
22	5.79	4.38	3.78	3.44	3.22	3.05	2.93	2.84	2.76
23	5.75	4.35	3.75	3.41	3.18	3.02	2.90	2.81	2.73
24	5.72	4.32	3.72	3.38	3.15	2.99	2.87	2.78	2.70
25	5.69	4.29	3.69	3.35	3.13	2.97	2.85	2.75	2.68
26	5.66	4.27	3.67	3.33	3.10	2.94	2.82	2.73	2.65
27	5.63	4.24	3.65	3.31	3.08	2.92	2.80	2.71	2.63
28	5.61	4.22	3.63	3.29	3.06	2.90	2.78	2.69	2.61
29	5.59	4.20	3.61	3.27	3.04	2.88	2.76	2.67	2.59
30	5.57	4.18	3.59	3.25	3.03	2.87	2.75	2.65	2.57
40	5.42	4.05	3.46	3.13	2.90	2.74	2.62	2.53	2.45
60	5.29	3.93	3.34	3.01	2.79	2.63	2.51	2.41	2.33
120	5.15	3.80	3.23	2.89	2.67	2.52	2.39	2.30	2.22
∞	5.02	3.69	3.12	2.79	2.57	2.41	2.29	2.19	2.11

Freedom
Denominator Degrees of

$F_{0.025,9,11}$

$$F_{.975,11,9} = \frac{1}{F_{.025,9,11}} = \frac{1}{3.95} = 0.28$$

The decision rule is to reject the null hypothesis if the observed F value is greater than 3.59 or less than 0.28.

STEP 5: Next we compute the sample variances. The data are shown here.

Machine 1		Machine 2	
22.3	21.9	22.0	21.7
21.8	22.4	22.1	21.9
22.3	22.5	21.8	22.0
21.6	22.2	21.9	22.1
21.8	21.6	22.2	21.9
		22.0	22.1
$s_1^2 = 0.11378$		$s_2^2 = 0.02023$	
$n_1 = 10$		$n_2 = 12$	

STEP 6:

$$F = \frac{s_1^2}{s_2^2} = \frac{0.11378}{0.02023} = 5.62$$

The ratio of sample variances is 5.62.

ACTION

STEP 7: The observed F value is 5.62, which is greater than the upper-tail critical value of 3.59. As Figure 10.15 shows, this F value is in the rejection region. Thus, the decision is to reject the null hypotheses. The population variances are not equal.

BUSINESS IMPLICATIONS

STEP 8: An examination of the sample variances reveals that the variance from machine 1 measurements is greater than that from machine 2 measurements. The operators and process managers might want to examine machine 1 further; an adjustment may be needed or some other reason may be causing the seemingly greater variations on that machine.

Using the Computer to Test Hypotheses About Two Population Variances

Both Excel and Minitab have the capability of directly testing hypotheses about two population variances. Figure 10.16 shows Minitab and Excel output for the sheet metal example. The Minitab output contains the observed F value and its associated p-value. The Excel output contains the two sample means, the two sample variances, the observed

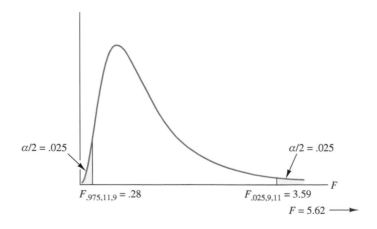

FIGURE 10.15

Minitab Graph of F Values and Rejection Region for the Sheet Metal Example

Minitab Output

```
Test for Equal Variances: Machine 1, Machine 2
```

95% Bonferroni confidence intervals for standard deviations

	N	Lower	StDev	Upper
Machine 1	10	0.220642	0.337310	0.679225
Machine 2	12	0.096173	0.142223	0.262900

```
F-Test (Normal Distribution)
Test statistic = 5.62, p-value = 0.009
```

EXCEL Output

F- Test Two-Sample for Variances

	Machine 1	Machine 2
Mean	22.04	21.975
Variance	0.11378	0.02023
Observations	10	12
df	9	11
F	5.62	
P (F<=f) one-tail	0.0047	
F Critical one-tail	3.59	

FIGURE 10.16

Minitab and Excel Output for the Sheet Metal Example

F value, the p-value for a one-tailed test, and the critical F value for a one-tailed test. Because the sheet metal example is a two-tailed test, compare the p-value of 0.0047 to $\alpha/2 = 0.025$. Because this value (0.0047) is less than $\alpha/2 = 0.025$, the decision is to reject the null hypothesis.

DEMONSTRATION PROBLEM 10.7

According to a study, a family of four in France with €50,000 annual income spends more than €18,100 a year on basic goods and services. In contrast, a family of four in Spain with the same annual income spends only €15,460 on the same items. Suppose we want to test to determine whether the variance of money spent per year on the basics by families across Europe is greater than the variance of money spent on the basics by families in France – that is, whether the amounts spent by families of four in France are more homogeneous than the amounts spent by such families across the continent. Suppose a random sample of eight French families produces the figures in the table below, which are given along with those reported from a random sample of seven families across Europe. Complete a hypothesis-testing procedure to determine whether the variance of values taken from across Europe can be shown to be greater than the variance of values obtained from families in France. Let $\alpha = 0.01$. Assume the amount spent on the basics is normally distributed in the population.

Amount Spent on Basics by Family of Four with €50,000 Annual Income

Across Europe	France
€15,417	€19,167
16,042	18,250
13,667	18,750
17,292	17,667
14,667	17,500
18,167	19,000
12,292	19,250
	17,750

Solution

HYPOTHESIZE

STEP 1: This is a one-tailed test with the following hypotheses:

$$H_o: \sigma_1^2 = \sigma_2^2$$
$$H_a: \sigma_1^2 > \sigma_2^2$$

Note that what we are trying to prove – that the variance for the European population is greater than the variance for families in France – is in the alternative hypothesis.

TEST

STEP 2: The appropriate statistical test is

$$F = \frac{s_1^2}{s_2^2}$$

STEP 3: The Type I error rate is 0.01.

STEP 4: This test is a one-tailed test, so we will use the F distribution table in Table A.7 (Appendix A) with $\alpha = 0.01$. The degrees of freedom for $n_1 = 7$ and $n_2 = 8$ are $v_1 = 6$ and $v_2 = 7$. The critical F value for the upper tail of the distribution is

$$F_{0.01,6,7} = 7.19$$

The decision rule is to reject the null hypothesis if the observed value of F is greater than 7.19.

STEP 5: The following sample variances are computed from the data.

$$s_1^2 = 4{,}139{,}881$$
$$n_1 = 7$$
$$s_2^2 = 511.904.8$$
$$n_2 = 8$$

STEP 6: The observed F value can be determined by

$$F = \frac{s_1^2}{s_2^2} = \frac{4{,}139{,}881}{511{,}904.8} = 8.09$$

ACTION

STEP 7: Because the observed value of $F = 8.09$ is greater than the table critical F value of 7.19, the decision is to reject the null hypothesis.

BUSINESS IMPLICATIONS

STEP 8: The variance for families in Europe is greater than the variance for families in France. Families in France are more homogeneous in the amount spent on basics than families across Europe. Marketing managers need to understand this homogeneity as they attempt to find niches in the French population. France may not contain as many subgroups as can be found across Europe. The task of locating market niches may be easier in France than in the rest of the region because fewer possibilities are likely. The following Minitab graph shows the rejection region as well as the critical and calculated values of F:

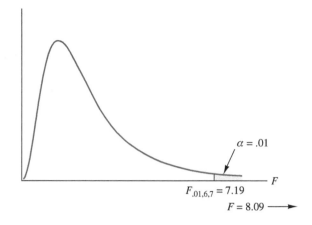

$\alpha = .01$

$F_{.01,6,7} = 7.19$

$F = 8.09 \longrightarrow$

Note: *Some authors recommend the use of this F test to determine whether the data being analysed by a t test for two population means are meeting the assumption of equal population variances. However, some statistical researchers suggest that for equal sample sizes, the t test is insensitive to the equal variance assumption, and therefore the F test is not needed in that situation. For unequal sample sizes, the F test of variances is 'not generally capable of detecting assumption violations that lead to poor performance' with the t test.* This text does not present the application of the F test to determine whether variance assumptions for the t test have been met.*

10.5   PROBLEMS

10.39   Test the hypotheses below by using the given sample information and $\alpha = 0.01$. Assume the populations are normally distributed.

$$H_0: \sigma_1^2 = \sigma_2^2 \quad H_a: \sigma_1^2 < \sigma_2^2$$

$$n_1 = 10, \, n_2 = 12, \, s_1^2 = 562, \, s_2^2 = 1013$$

10.40   Test the hypotheses below by using the given sample information and $\alpha = 0.01$. Assume the populations are normally distributed.

$$H_0: \sigma_1^2 = \sigma_2^2 \quad H_a: \sigma_1^2 \neq \sigma_2^2$$

$$n_1 = 5, \, n_2 = 19, \, s_1 = 4.68, \, s_2 = 2.78$$

10.41   Suppose the data shown here are the results of a survey to investigate petrol prices. Ten service stations were selected randomly in each of two cities and the figures represent the prices of a litre of unleaded regular petrol on a given day. Use the *F* test to determine whether there is a significant difference in the variances of the prices of unleaded regular petrol between these two cities. Let $\alpha = 0.01$. Assume petrol prices are normally distributed.

City 1			City 2		
1.429	1.383	1.408	1.388	1.375	1.433
1.417	1.413	1.408	1.425	1.442	1.404
1.413	1.408	1.367	1.413	1.413	1.408
	1.392			1.400	

---

* Carol A. Markowski and Edward P. Markowski, 'Conditions for the Effectiveness of a Preliminary Test of Variance', *The American Statistician*, vol. 44 (November 1990), pp. 322–326.

10.42 How long does it take to sell a house on the market? One survey by the property portal Home.co.uk reported that in the United Kingdom, houses for sale are on the market an average of 109 days. Of course, the length of time varies by market. Suppose random samples of 13 houses in Canterbury and 11 houses in Coventry that are for sale are traced. The data shown here represent the number of days each house was on the market before being sold. Use the given data and a 1% level of significance to determine whether the population variances for the number of days until sale are different in Canterbury than in Coventry. Assume the numbers of days for which houses are on the market are normally distributed.

Canterbury		Coventry	
132	126	118	56
138	94	85	69
131	161	113	67
127	133	81	54
99	119	94	137
126	88	93	
134			

10.43 Data published by the World Resources Institute show that the average annual per capita energy consumption in watts was 4,473.2 for Slovenia compared with an average of 4,592.8 for Slovakia. Suppose a random sample of 12 Slovenian households showed that the standard deviation of these consumption figures was 1,716, whereas a random sample of 15 Slovenian households resulted in a standard deviation of 1,389. Do these samples provide enough evidence to conclude that the variance of per capita energy consumption for Slovenians is greater than the variance of per capita energy consumption for Slovakians? Let alpha be 0.05. Assume the per capita energy consumption amounts are normally distributed.

10.44 According to the Office for National Statistics, the average age of a public sector worker is 44 years old. Another source puts the average age of a worker in the private sector at 37.3 years. Is there any difference in the variation of ages of public sector and private sector workers? Suppose a random sample of 15 public sector workers is taken and the variance of their ages is 91.5. Suppose also a random sample of 15 private sector workers is taken and the variance of their ages is 67.3. Use these data and alpha = 0.01 to answer the question. Assume ages are normally distributed.

Decision Dilemma SOLVED

Online Shopping

Various techniques in Chapter 10 can be used to analyse the online shopping data presented in the Decision Dilemma if the data are actually only sample statistics. The Pew Internet and American Life Project surveyed 2,400 American adults. However, suppose only 245 of these shopped at speciality stores, spending an average of $123, and 212 shopped at department stores, spending an average of $121. Suppose further that the population standard deviation of such spending at speciality stores is $35 and the population standard deviation of such spending at department stores is $32. Using the z test presented in Section 10.1, we can test to determine if there is a significant difference in the average amounts spent by online shoppers in the past 30 days between these two types of stores. Let alpha equal 0.05. This is a two-tailed test in which the null hypothesis is that there is no difference between the mean amounts for the two types of stores, and the alternative hypothesis is that there is a difference.

The critical z value for this test is ± 1.96. Applying formula 10.1 to the data, an observed z value of 0.64 is obtained. Because this observed value is less than the critical value of ± 1.96, the decision is to fail to reject the null hypothesis. This is confirmed by the p-value, which is 0.2611. If the reported averages of \$123 and \$121 were sample averages obtained from samples of 245 and 212, respectively, with these population standard deviations, there is likely no real difference in the population mean amounts spent at speciality stores and department stores.

The EIAA study reported that the mean number of items purchased over a six-month period for online shoppers was 11 in Denmark and 10 in Germany. However, these were sample means obtained from a survey of 7,000 people. Is the mean of 11 obtained for Denmark really significantly higher than the mean of 10 obtained from Germany? Suppose that 843 Danish people participated in this study, as did 1,064 Germans. In addition, suppose that the sample standard deviation of number of items was 2.7 for Denmark and 3.1 for Germany. If we want to test to determine if the mean number of Denmark is significantly higher than the mean number for Germany, the null hypothesis is that there is no difference and the alternative hypothesis is that the mean for Denmark is higher than the mean for Germany (a one-tailed test). Let alpha be 0.01. Since the population variances are unknown and the samples are independent, we can use formula 10.3 in Section 10.2 of this chapter to perform the analysis. The degrees of freedom for this analysis are 1,905 (843 + 1,064 − 2). The critical table t value is approximately 2.326. The observed t value computed using formula 10.3 is $t = 7.40$, which is significant because it is bigger than the critical table value. Thus, even though the difference in the mean number of items between Denmark and Germany purchased over a six-month period for online shoppers is only 1, given the large sample sizes and the population standard deviations,

there is sufficient evidence to conclude that a significantly greater number of items are purchased by people in Denmark than by people in Germany. Whether a difference of 1 is substantive is a matter of interpretation by online businesses and business researchers.

The Nielsen survey of 26,000 Internet users from around the world revealed that 97% of Japan's Internet users shop online compared to 94% in the United States. Since these two values are taken from a sample of users, they are actually sample proportions. Suppose a business researcher wants to use these figures to test to determine if the proportion of Japanese Internet users who shop online is significantly higher than the proportion of American Internet users who shop online. The null hypothesis is that there is no difference in the population proportions. The alternate hypothesis is that the population proportion for Japan is higher than the population proportion for the United States. Suppose the 0.97 sample proportion came from 561 Japanese participants in the Nielsen survey and that the 0.94 sample proportion came from 685 American participants in the survey. Using a 5% level of significance, we can test these hypotheses using Formula 10.9 from Section 10.4. Since this is a one-tailed z test with an alpha of 0.05, the critical table z value is 1.645. Using formula 10.9 to analyse the data, we obtain an observed z value of 2.50. Because this observed value is greater than the critical table value, the decision is to reject the null hypothesis. This conclusion is confirmed by examining the p-value, which is 0.0062 and is less than 0.05. Thus, our analysis indicates that a significantly higher proportion of Japanese Internet users shop online than do Americans. An important business question might be, is 0.03 (the difference between 0.97 and 0.94) really a substantive or important difference? A 3% difference can sometimes be a lot in the business world. On the other hand, 94% of American Internet users shopping online is still an extremely high proportion.

ETHICAL CONSIDERATIONS

The statistical techniques presented in this chapter share some of the pitfalls of confidence interval methodology and hypothesis-testing techniques mentioned in preceding chapters. Included among these pitfalls are assumption violations. Remember, if small sample sizes are used in analysing means, the z tests are valid only when the population is normally distributed and the population variances are known. If the population variances are unknown, a t test can be used if the population is normally distributed and if the population variances can be assumed to be equal. The z tests

and confidence intervals for two population proportions also have a minimum sample-size requirement that should be met. In addition, it is assumed that both populations are normally distributed when the F test is used for two population variances.

Use of the t test for two independent populations is not unethical when the populations are related, but it is likely to result in a loss of power. As with any hypothesis-testing procedure, in determining the null and alternative hypotheses, make certain you are not assuming true what you are trying to prove.

SUMMARY

Business research often requires the analysis of two populations. Three types of parameters can be compared: means, proportions, and variances. Except for the F test for population variances, all techniques presented contain both confidence intervals and hypothesis tests. In each case, the two populations are studied through the use of sample data randomly drawn from each population.

The population means are analysed by comparing two sample means. When sample sizes are large ($n \geqslant 30$) and population variances are known, a z test is used. When sample sizes are small, the population variances are known, and the populations are normally distributed, the z test is used to analyse the population means. If the population variances are unknown, and the populations are normally distributed, the t test of means for independent samples is used. For populations that are related on some measure, such as twins or before-and-after, a t test for dependent measures (matched pairs) is used. The difference in two population proportions can be tested or estimated using a z test.

The population variances are analysed by an F test when the assumption that the populations are normally distributed is met. The F value is a ratio of the two variances. The F distribution is a distribution of possible ratios of two sample variances taken from one population or from two populations containing the same variance.

KEY TERMS

dependent samples
F distribution
F value

independent samples
matched-pairs test
related measures

GO ONLINE TO DISCOVER THE EXTRA FEATURES FOR THIS CHAPTER

The Student Study Guide containing solutions to the odd-numbered questions, additional Quizzes and Concept Review Activities, Excel and Minitab databases, additional data files in Excel and Minitab, and more worked examples.
www.wiley.com/college/cortinhas

FORMULAS

z test for the difference in two independent sample means

$$z = \frac{(\bar{x}_1 - \bar{x}_2) - (\mu_1 - \mu_2)}{\sqrt{\dfrac{\sigma_1^2}{n_1} = \dfrac{\sigma_2^2}{n_2}}}$$

Confidence interval for estimating the difference in two independent population means using z

$$(\bar{x}_1 - \bar{x}_2) - z\sqrt{\frac{\sigma_1^2}{n_1} = \frac{\sigma_2^2}{n_2}} \leq \mu_1 - \mu_2 \leq (\bar{x}_1 - \bar{x}_2)$$

$$+ z\sqrt{\frac{\sigma_1^2}{n_1} = \frac{\sigma_2^2}{n_2}}$$

t test for two independent sample means, and population variances unknown but assumed to be equal (assume also that the two populations are normally distributed)

$$t = \frac{(\bar{x}_1 - \bar{x}_2) - (\mu_1 - \mu_2)}{\sqrt{\dfrac{s_1^2(n_1 - 1) + s_2^2(n_2 - 1)}{n_1 + n_2 - 2}} \sqrt{\dfrac{1}{n_1} + \dfrac{1}{n_2}}}$$

$$\mathrm{df} = n_1 + n_2 - 2$$

Confidence interval for estimating the difference in two independent means, and population variances unknown but assumed to be equal (assume also that two populations are normally distributed)

$$(\bar{x}_1 - \bar{x}_2) \pm t \sqrt{\frac{s_1^2(n_1 - 1) + s_2^2(n_2 - 1)}{n_1 + n_2 - 2}} \sqrt{\frac{1}{n_1} + \frac{1}{n_2}}$$

$$\mathrm{df} = n_1 + n_2 - 2$$

t test for the difference in two related samples (the differences are normally distributed in the population)

$$t = \frac{\bar{d} - D}{\frac{s_d}{\sqrt{n}}}$$

$$\mathrm{df} = n - 1$$

Formulas for $\bar{d}$ and s_2

$$\bar{d} = \frac{\sum d}{n}$$

$$s_d = \sqrt{\frac{\sum (d - \bar{d})^2}{n - 1}} = \sqrt{\frac{\sum d^2 - \frac{(\sum d)^2}{n}}{n - 1}}$$

Confidence interval formula for estimating the difference in related samples (the differences are normally distributed in the population)

$$\bar{d} - t\frac{s_d}{\sqrt{n}} \le D \le \bar{d} + t\frac{s_d}{\sqrt{n}}$$

$$\mathrm{df} = n - 1$$

z formula for testing the difference in population proportions

$$z = \frac{(\hat{p}_1 - \hat{p}_2) - (p_1 - p_2)}{\sqrt{(\bar{p}\cdot\bar{q})\left(\frac{1}{n_1} + \frac{1}{n_2}\right)}}$$

where:

$$\bar{p} = \frac{x_1 + x_2}{n_1 + n_2} = \frac{n_1\hat{p}_1 + n_2\hat{p}_2}{n_1 + n_2} \text{ and } \bar{q} = 1 - \bar{p}$$

Confidence interval to estimate $p_1 - p_2$

$$(\hat{p}_1 - \hat{p}_2) - z\sqrt{\frac{\hat{p}_1\cdot\hat{q}_1}{n_1} + \frac{\hat{p}_2\cdot\hat{q}_2}{n_2}} \le p_1 - p_2$$

$$\le (\hat{p}_1 - \hat{p}_2) + z\sqrt{\frac{\hat{p}_1\cdot\hat{q}_1}{n_1} + \frac{\hat{p}_2\cdot\hat{q}_2}{n_2}}$$

F test for two population variances (assume the two populations are normally distributed)

$$F = \frac{s_1^2}{s_2^2}$$

$$\mathrm{df}_{\mathrm{numerator}} = v_1 = n_1 - 1$$
$$\mathrm{df}_{\mathrm{denominator}} = v_2 = n_2 - 1$$

Formula for determining the critical value for the lower-tail *F*

$$F_{1-\alpha,v_2,v_1} = \frac{1}{F_{\alpha,v_1,v_2}}$$

SUPPLEMENTARY PROBLEMS

CALCULATING THE STATISTICS

10.45 Test the hypotheses below with the data given. Let

$\sigma = 0.10$

$H_0: \mu_1 - \mu_2 = 0 \qquad H_a: \mu_1 - \mu_2 \ne 0$

Sample 1	Sample 2
$\bar{x}_1 = 138.4$	$\bar{x}_2 = 142.5$
$\sigma_1 = 6.71$	$\sigma_2 = 8.92$
$n_1 = 48$	$n_2 = 39$

10.46 Use the following data to construct a 98% confidence interval to estimate the difference between μ_1 and μ_2:

Sample 1	Sample 2
$\bar{x}_1 = 34.9$	$\bar{x}_2 = 27.6$
$\sigma_1^2 = 2.97$	$\sigma_2^2 = 3.50$
$n_1 = 34$	$n_2 = 31$

10.47 The data below come from independent samples drawn from normally distributed populations. Use these data to test the hypotheses below. Let the Type I error rate be 0.05.

$$H_0: \mu_1 - \mu_2 = 0$$
$$H_a: \mu_1 - \mu_2 > 0$$

Sample 1	Sample 2
$\bar{x}_1 = 2.06$	$\bar{x}_2 = 1.93$
$s_1^2 = .176$	$s_2^2 = .143$
$n_1 = 12$	$n_2 = 15$

10.48 Construct a 95% confidence interval to estimate $\mu_1 - \mu_2$ by using the data below. Assume the populations are normally distributed.

Sample 1	Sample 2
$\bar{x}_1 = 74.6$	$\bar{x}_2 = 70.9$
$s_1^2 = 10.5$	$s_2^2 = 11.4$
$n_1 = 18$	$n_2 = 19$

10.49 The following data have been gathered from two related samples. The differences are assumed to be normally distributed in the population. Use these data and alpha of 0.01 to test the following hypotheses.

$$H_0 : D = 0$$

$$H_a : D < 0$$

$$n = 21, \bar{d} = -1.16, s_d = 1.01$$

10.50 Use the data below to construct a 99% confidence interval to estimate D. Assume the differences are normally distributed in the population.

Respondent	Before	After
1	47	63
2	33	35
3	38	36
4	50	56
5	39	44
6	27	29
7	35	32
8	46	54
9	41	47

10.51 Test the following hypotheses by using the given data and alpha equal to 0.05:

$$H_0: p_1 - p_2 = 0$$

$$H_a: p_1 - p_2 \neq 0$$

Sample 1	Sample 2
$\bar{n}_1 = 783$	$\bar{n}_2 = 896$
$x_1 = 345$	$x_2 = 421$

10.52 Use the following data to construct a 99% confidence interval to estimate $p_1 - p_2$:

Sample 1	Sample 2
$n_1 = 409$	$n_2 = 378$
$\hat{p}_1 = .71$	$\hat{p}_2 = .67$

10.53 Test the following hypotheses by using the given data. Let $\alpha = 0.05$

$$H_0: \sigma^2_1 - \sigma^2_2 = 0$$

$$H_a: \sigma^2_1 - \sigma^2_2 \neq 0$$

$$n_1 = 8,\ n_2 = 10,\ s_1^2 = 46,\ s_2^2 = 37$$

TESTING YOUR UNDERSTANDING

10.54 Suppose a large insurance company wants to estimate the difference between the average amount of term life insurance purchased per family and the average amount of whole life insurance purchased per family. To obtain an estimate, one of the company's actuaries randomly selects 27 families who have term life insurance only and 29 families who have whole life policies only. Each sample is taken from families in which the leading provider is younger than 45 years of age. Use the data obtained to construct a 95% confidence interval to estimate the difference in means for these two groups. Assume the amount of insurance is normally distributed.

Term	Whole Life
$\bar{x}_T = €75,000$	$\bar{x}_W = €45,000$
$\bar{x}_T = €22,000$	$\bar{s}_W = €15,500$
$n_T = 27$	$n_W = 29$

10.55 A study is conducted to estimate the average difference in bus usage for a large city during the morning and afternoon rush hours. The transit authority's researcher randomly selects nine buses because of the variety of routes they represent. On a given day the number of riders on each bus is counted at 7.45 a.m. and at 4.45 p.m., with the following results:

Bus	Morning	Afternoon
1	43	41
2	51	49
3	37	44
4	24	32
5	47	46
6	44	42
7	50	47
8	55	51
9	46	49

Use the data to compute a 90% confidence interval to estimate the population average difference. Assume the passenger numbers are normally distributed.

10.56 There are several methods used by people to organize their lives in terms of keeping track of appointments, meetings, and deadlines. Some of these include using a desk calendar, using informal notes of scrap paper, keeping them 'in your head', using a day planner, and keeping a formal 'to do' list. Suppose a business researcher wants to test the hypothesis that a greater proportion of marketing managers keep track of such obligations 'in their head' than do accountants. To test this, a business researcher samples 400 marketing managers and 450 accountants. Of those sampled, 220 marketing managers keep track 'in their head' while 216 of the accountants do so. Using a 1% level of significance, what does the business researcher find?

10.57 A study was conducted to compare the salaries of accounting clerks and data entry operators. One of the hypotheses to be tested is that the variability of salaries among accounting clerks is the same as the variability of salaries of data entry operators. To test this hypothesis, a random sample of 16 accounting clerks was taken, resulting in a sample mean salary of €26,400 and a sample standard deviation of €1,200. A random sample of 14 data entry operators was taken as well, resulting in a sample mean of €25,800 and a sample standard deviation of €1,050. Use these data and alpha = 0.05 to test to determine whether the population variance of salaries is the same for accounting clerks as it is for data entry operators. Assume that salaries of data entry operators and accounting clerks are normally distributed in the population.

10.58 A study was conducted to develop a scale to measure stress in the workplace. Respondents were asked to rate 26 distinct work events. Each event was to be compared with the stress of the first week on the job, which was awarded an arbitrary score of 500. Sixty professional men and 41 professional women participated in the study. One of the stress events was 'lack of support from the boss'. The men's sample average rating of this event was 631 and the women's sample average rating was 848. Suppose the population standard deviations for men and for women both were about 100. Construct a 95% confidence interval to estimate the difference in the population mean scores on this event for men and women.

10.59 A national grocery store chain wants to test the difference in the average weight of turkeys sold in Westminster and the average weight of turkeys sold in Portsmouth. According to the chain's researcher, a random sample of 20 turkeys sold at the chain's stores in Westminster yielded a sample mean of 7.95 kilograms, with a standard deviation of 1.45 kilograms. Her random sample of 24 turkeys sold at the chain's stores in Portsmouth yielded a sample mean of 6.75 kilograms, with a standard deviation of 1.22 kilograms. Use a 1% level of significance to determine whether there is a difference in the mean weight of turkeys sold in these two cities. Assume the population variances are approximately the same and that the weights of turkeys sold in the stores are normally distributed.

10.60 A tree nursery has been experimenting with fertilizer to increase the growth of seedlings. A sample of 35 two-year-old pine trees is grown for three more years with a cake of fertilizer buried in the soil near the trees' roots. A second sample of 35 two-year-old pine trees is grown for three more years under identical conditions (soil, temperature, water) as the first group, but not fertilized. Tree growth is measured over the three-year period with the following results.

Trees with Fertilizer	Trees Without Fertilizer
$n_1 = 35$	$n_2 = 35$
$\bar{x}_1 = 38.4$ inches	$\bar{x}_1 = 23.1$ inches
$\sigma_1 = 9.8$ inches	$\sigma_2 = 7.4$ inches

Do the data support the theory that the population of trees with the fertilizer grew significantly larger during the period in which they were fertilized than the non-fertilized trees? Use $\alpha = 0.01$.

10.61 One of the most important aspects of a store's image is the perceived quality of its merchandise. Other factors include merchandise pricing, assortment of products, convenience of location, and service. Suppose image perceptions of shoppers of speciality stores and shoppers of discount stores are being compared. A random sample of shoppers is taken at each type of store, and the shoppers are asked whether the quality of merchandise is a determining factor in their perception of the store's image. Some 75% of the 350 shoppers at the speciality stores say 'Yes', but only 52% of the 500 shoppers at the discount store say 'Yes'. Use these data to test to determine if there is a significant difference between the proportion of shoppers at speciality stores and the proportion of shoppers at discount stores who say that quality of merchandise is a determining factor in their perception of a store's image. Let alpha equal 0.10.

10.62 Is there more variation in the output of one shift in a manufacturing plant than in another shift? In an effort to study this question, plant managers gathered productivity reports from the 8 a.m. to 4 p.m. shift for eight days. The reports indicated that the following numbers of units were produced on each day for this shift:

5,528   4,779   5,112   5,380
4,918   4,763   5,055   5,106

Productivity information was also gathered from seven days for the 4 p.m. to midnight shift, resulting in the following data:

4,325   4,016   4,872   4,559
3,982   4,754   4,116

Use these data and $\alpha = 0.01$ to test so as to determine whether the variances of productivity for the two shifts are the same. Assume productivity is normally distributed in the population.

10.63 What is the average difference between the price of name-brand soup and the price of store-brand soup? To obtain an estimate, an analyst randomly samples eight stores. Each store sells its own brand and a national name brand. The prices of a can of name-brand tomato soup and a can of the store-brand tomato soup follow.

Store	Name Brand	Store Brand
1	£0.54	£0.49
2	0.55	0.50
3	0.59	0.52
4	0.53	0.51
5	0.54	0.50
6	0.61	0.56
7	0.51	0.47
8	0.53	0.49

Construct a 90% confidence interval to estimate the average difference. Assume that the differences in prices of tomato soup are normally distributed in the population.

10.64 As the prices of heating oil and natural gas increase, consumers become more careful about heating their homes. Researchers want to know how warm homeowners keep their houses in January and how the results from Groningen, Netherlands, and Bremen, Germany, compare. The researchers

randomly call 23 Groningen households between 7 p.m. and 9 p.m. on January 15 and ask the respondent how warm the house is according to the thermostat. The researchers then call 19 households in Bremen the same night and ask the same question. The results follow.

Groningen					Bremen			
21.7	21.7	18.3	20.0	22.8	23.9	23.3	21.7	
21.1	16.1	19.4	20.6	23.3	22.8	23.3	21.1	
23.9	20.0	21.7	22.8	22.2	21.7	20.6	22.2	
23.3	20.0	19.4	20.6	23.3	22.8	21.1	22.2	
20.6	22.2	19.4	22.2	20.6	21.1	19.4		
21.1	22.8	22.2						

For $\alpha = 0.01$, is the average temperature of a house in Bremen significantly higher than that of a house in Groningen on the evening of January 15? Assume the population variances are equal and the house temperatures are normally distributed in each population.

10.65 In manufacturing, does worker productivity drop on Friday? In an effort to determine whether it does, a company's personnel analyst randomly selects from a manufacturing plant five workers who make the same part. He measures their output on Wednesday and again on Friday and obtains the following results.

Worker	Wednesday Output	Friday Output
1	71	53
2	56	47
3	75	52
4	68	55
5	74	58

The analyst uses $\alpha = 0.05$ and assumes the difference in productivity is normally distributed. Do the samples provide enough evidence to show that productivity drops on Friday?

10.66 A manufacturer uses two machines to drill holes in pieces of sheet metal used in engine construction. The workers who attach the sheet metal to the engine become inspectors in that they reject sheets so poorly drilled that they cannot be attached. The production manager is interested in knowing whether one machine produces more defective drillings than the other machine. As an experiment, employees mark the sheets so that the manager can determine which machine was used to drill the holes. A random sample of 191 sheets of metal drilled by machine 1 is taken, and 38 of the sheets are defective. A random sample of 202 sheets of metal drilled by machine 2 is taken, and 21 of the sheets are defective. Use $\alpha = 0.05$ to determine whether there is a significant difference in the proportion of sheets drilled with defective holes between machine 1 and machine 2.

10.67 Is there a difference in the proportion of construction workers who are under 35 years of age and the proportion of telephone repair people who are under 35 years of age? Suppose a study is conducted in Dundee, Scotland, using random samples of 338 construction workers and 281 telephone repair people. The sample of construction workers includes 297 people below 35 years of age and the sample of telephone repair people includes 192 people under that age. Use these data to construct a 90% confidence interval to estimate the difference in proportions of people under 35 years of age among construction workers and telephone repair people.

10.68 Executives often spend so many hours in meetings that they have relatively little time to manage their individual areas of operation. What is the difference in mean time spent in meetings by executives of the aerospace industry and executives of the automobile industry? Suppose random samples of 33 aerospace executives and 35 automobile executives are monitored for a week to determine how much time they spend in meetings. The results follow.

Aerospace	Automobile
$n_1 = 33$	$n_2 = 35$
$\bar{x}_1 = 12.4$ hours	$\bar{x}_2 = 4.6$ hours
$\sigma_1 = 2.9$ hours	$\sigma_2 = 1.8$ hours

Use the data to estimate the difference in the mean time per week executives in these two industries spend in meetings. Use a 99% level of confidence.

10.69 Various types of retail outlets sell toys during the holiday season. Among them are speciality toy stores, large discount toy stores, and other retailers that carry toys as only one part of their stock of goods. Is there any difference in the dollar amount of a customer purchase between a large discount toy store and a speciality toy store if they carry relatively comparable types of toys? Suppose in December a random sample of 60 sales slips is selected from a large discount toy outlet and a random sample of 40 sales slips is selected from a speciality toy store. The data gathered from these samples follow.

Large Discount Toy Store	Speciality Toy Store
$\bar{x}_D = \$47.20$	$\bar{x}_2 = \$27.40$
$\sigma_D = \$12.45$	$\sigma_s = \$9.82$

Use $\alpha = 0.01$ and the data to determine whether there is a significant difference in the average size of purchases at these stores.

10.70 One of the new thrusts of quality control management is to examine the process by which a product is produced. This approach also applies to paperwork. In industries where large long-term projects are undertaken, days and even weeks may elapse as a change order makes its way through a maze of approvals before receiving final approval. This process can result in long delays and stretch schedules to the breaking point. Suppose a quality control consulting group claims that it can significantly reduce the number of days required for such paperwork to receive approval. In an attempt to 'prove' its case, the group selects five jobs for which it revises the paperwork system. The data below show the number of days required for a change order to be approved before the group intervened and the number of days required for a change order to be approved after the group instituted a new paperwork system.

Before	After
12	8
7	3
10	8
16	9
8	5

Use $\alpha = 0.01$ to determine whether there was a significant drop in the number of days required to process paperwork to approve change orders. Assume that the differences in days are normally distributed.

10.71 For the two large newspapers in your city, you are interested in knowing whether there is a significant difference in the average number of pages in each dedicated solely to advertising. You randomly select 10 editions of newspaper A and 6 editions of newspaper B (excluding weekend editions). The data are shown on the following page. Use $\alpha = 0.01$ to test whether there is a significant difference in averages. Assume the number of pages of advertising per edition is normally distributed and the population variances are approximately equal.

A		B	
17	17	8	14
21	15	11	10
11	19	9	6
19	22		
26	16		

INTERPRETING THE OUTPUT

10.72 A study by Colliers International presented the highest and the lowest global rental rates per year per square foot of office space. Among the cities with the lowest rates were Stamford, USA, Shanghai, China, and Buenos Aires, Argentina, with rates of $37.02, $37.86, and $38.59, respectively. At the high end were Hong Kong, London (West End), and Tokyo, with rates over $100. Suppose a researcher conducted her own survey of businesses renting office space to determine whether one city is significantly more expensive than another. The data are tallied and analysed by using Minitab. The results are shown below. Discuss the output. Assume that rental rates are normally distributed in the population. What cities were studied? How large were the samples? What were the sample statistics? What was the value of alpha? What were the hypotheses, and what was the conclusion?

```
Two-Sample T-Test and CI
─────────────────────────────────────────────
Sample        N    Mean   StDev   SE Mean
Hong Kong    19   130.4   12.9     3.0
London       23   128.4   13.9     2.9

Difference=mu (Hong Kong)-mu (Mumbai)
Estimate for difference: 2.00
98% CI for difference: (-8.11, 12.11)
T-Test of difference=0 (vs not =):
T-Value=0.48 P-Value=0.634 DF=40
Both use Pooled StDev=13.4592
─────────────────────────────────────────────
```

10.73 Why do employees 'blow the whistle' on other employees for unethical or illegal behaviour? One study conducted by the AICPA reported the likelihood that employees would blow the whistle on another employee for such things as unsafe working conditions, unsafe products, and poorly managed operations. On a scale from 1 to 7, with 1 denoting highly improbable and 7 denoting highly probable, unnecessary purchases received a 5.72 in the study. Suppose this study was administered at a company and then all employees were subjected to a one-month series of seminars on reasons to blow the whistle on fellow employees. One month later the study was administered again to the same employees at the company in an effort to determine whether the treatment had any effect. The following Excel output shows the results of the study. What were the sample sizes? What might the hypotheses have been? If $\alpha = 0.05$, what conclusions could be made? Which of the statistical tests presented in this chapter is likely to have been used? Assume that differences in scores are normally distributed.

t-Test: Paired Two Sample for Means

	Variable 1	Variable 2
Mean	3.991	5.072
Variance	1.898	0.785
Observations	14	14
Pearson Correlation	-.04585	
Hypothesized Mean Difference	0	
df	13	
t Stat	-2.47	
P(T<=t) one-tail	0.0102	
t Critical one-tail	1.77	
P(T<=t) two-tail	0.00204	
t Critical two-tail	2.16	

10.74 A large manufacturing company produces computer printers that are distributed and sold all over Europe. Due to lack of industry information, the company has a difficult time ascertaining its market share in different parts of the continent. They hire a market research firm to estimate their market share in a northern country and a southern country. They would also like to know whether there is a difference in their market shares in these two countries; if so, they want to estimate how much. The market research firm randomly selects printer customers from different locales across both countries and determines what brand of computer printer they purchased. The following Minitab output shows the results from this study. Discuss the results including sample sizes, estimation of the difference in proportions, and any significant differences determined. What were the hypotheses tested?

```
Test and CI for Two Proportions
────────────────────────────────────────────────────────
Sample              X       N        Sample P
Northern Country   147     473       0.310782
Southern Country   104     385       0.270130

Difference=p (Northern Country)-p (Southern
Country)
Estimate for difference: 0.0406524
99% CI for difference: (-0.0393623, 0.120667)
Test for difference=0 (vs not=0):
Z=1.31 P-Value= 0.191
────────────────────────────────────────────────────────
```

10.75 A manufacturing company produces plastic pipes that are specified to be 10 inches long and ⅛ inch thick with an opening of ¾ inch. These pipes are moulded on two different machines. To maintain consistency, the company periodically randomly selects pipes for testing. In one specific test, pipes were randomly sampled from each machine and the lengths were measured. A statistical test was computed using Excel in an effort to determine whether the variance for machine 1 was significantly greater than the variance for machine 2. The results are shown opposite. Discuss the outcome of this test along with some of the other information given in the output.

F-Test Two-Sample for Variances

	Variable 1	Variable 2
Mean	10.02876	9.97050
Variance	0.02920	0.01965
Observations	26	28
dF	25	27
F	1.49	
P(F<=f) one-tail	0.15766	
F Critical one-tail	1.92	

ANALYSING THE DATABASES

1. Test to determine whether there is a significant difference between mean Value Added by the Manufacturer and the mean Cost of Materials in manufacturing. Use the Manufacturing database as the sample data and let alpha be .01.

2. Use the Manufacturing database to test to determine whether there is a significantly greater variance among the values of End-of-Year Inventories than among Cost of Materials. Let $\alpha = 0.05$.

3. Is there a difference between the average Number of Admissions at a general medical hospital and a psychiatric hospital? Use the Hospital database to test this hypothesis with $\alpha = 0.10$. The variable Service in the Hospital database differentiates general medical hospitals (coded 1) and psychiatric hospitals (coded 2). Now test to determine whether there is a difference between these two types of hospitals on the variables Beds and Total Expenses.

4. Use the Financial database to test whether there is a significant difference in the proportion of companies whose earnings per share are more than €2.00 and the proportion of companies whose dividends per share are between €1.00 and €2.00. Let $\alpha = 0.05$.

5. Using the appropriate technique selected from this chapter and the International Stock Market database, test to determine if there is a significant difference between the monthly average of NASDAQ stock market index and the monthly average of the Hang Seng stock market index. Let $\alpha = 0.01$.

CASE

MANUTAN INTERNATIONAL MAIL ORDER EUROPEAN SUCCESS

Manutan International S.A. is a specialist mail-order business providing industrial and office equipment and supplies to the business-to-business, collective, and public sectors. Based in France, Manutan has built up a pan-European network of 24 subsidiaries in 20 countries, from Portugal to Russia and from Sweden to Italy. Manutan also has a strong online presence with 18 e-commerce websites and a workforce of around 1,500 staff. Manutan's more than 200 catalogues feature over 200,000 items and serve more than 600,000 customers throughout Europe.

While mail-order houses had been in operation in France for many years, they have tended to focus primarily on the consumer market. In 1966, however, André Guichard, joined by son Jean-Pierre, set up a company dedicated to providing mail-order services to the business sector. The company's major innovation was in its choice of goods, that of industrial equipment, especially materials handling equipment.

Manutan quickly recognized the potential of entering other European markets. The company's first choice was the United Kingdom, where, as in France in the 1960s, the market for mail-order materials handling, lifting, and storage equipment was more or less non-existent. Manutan's success in the United Kingdom led it to quickly move into a large number of new foreign markets after 1974.

Manutan focused on its mail-order operations until the early 2000s. In 2001, however, the company expanded and established a presence on the Internet, launching its first e-commerce-enabled website. That site later provided the platform for the rollout of 18 websites targeting each of the company's markets. By 2005, the company had posted more than €1 million in sales through its e-commerce sites.

During the recent global financial crisis, Manutan made a strategic decision not to cut its marketing and commercial investments. On the contrary, Manutan maintained its promotional expenditure, and refocused part of its staff on direct selling, thereby increasing the multi-channel contact with their customers. This strategic insight seems to have paid off: the company recorded revenues of €563 million during the 2010 financial year, an increase of 15.2% over the previous year.

DISCUSSION

1. Manutan's list of several hundred thousand business-to-business customers continues to grow. Managers would like to know whether the average euro amount of sales per transaction per customer has changed from last year to this year. Suppose company accountants sampled 20 customers randomly from last year's records and determined that the mean sales per customer was €2,300, with a standard deviation of €500. They sampled 25 customers randomly from this year's files and determined that the mean sales per customer for this sample was €2,450, with a standard deviation of €540. Analyse these data and summarize your findings for managers. Explain how this information can be used by decision makers. Assume that sales per customer are normally distributed.

2. One common approach to measuring the quality of a company's services is through the use of customer satisfaction surveys. Suppose in a random sample, Manutan's customers are asked whether the last office equipment transition with Manutan was outstanding (Yes or No). Assume Manutan supplies these equipments from two different locations and that the office equipment customers can be divided according to where their equipment was supplied from. Suppose a random sample of 45 customers who bought office equipment from warehouse 1 results in 18 saying the service was of excellent quality and a random sample of 51 customers who bought office equipment that came from warehouse 2 results in 12 saying the service was of excellent quality. Use a confidence interval to express the estimated difference in population proportions of excellent ratings between the two groups of customers. Does it seem to matter which warehouse the office equipment comes from in terms of the quality rating received from customers? What would you report from these data?

3. Suppose the customer satisfaction survey included a question on the overall quality of Manutan measured on a scale from 0 to 10 where higher numbers indicate greater quality. Company managers monitor the figures from year to year to help determine whether Manutan is improving customers' perceptions of its quality. Suppose random samples of the responses from 2011 customers and 2012 customers are taken and analysed on this question, and the following Minitab analysis of the data results. Help managers interpret this analysis so that comparisons can be made between 2011 and 2012. Discuss the samples, the statistics, and the conclusions.

```
Two-Sample T-Test and CI: 2011, 2012

Two-sample T for 2011 vs 2012

          N       Mean      StDev    SE Mean
2011      75      6.466     0.352    0.041
2012      93      6.604     0.398    0.041

Difference = mu (2011) - mu (2012)
Estimate for difference: -0.1376
95% CI for difference: (-0.2535, -0.0217)
T-Test of difference = 0 (vs not =):
T-Value = -2.34   P-Value = 0.020   DF = 166
Both use Pooled StDev = 0.3782
```

4. Suppose Manutan supplies pulleys that are specified to be 50 millimetres in diameter. A large batch of pulleys is sold in week 1 and another is sold in week 5. Quality control people want to determine whether there is a difference in the variance of the

diameters of the two batches. Assume that a sample of six pulleys from the week 1 batch results in the following diameter measurements (in mm): 51, 50, 48, 50, 49, 51. Assume that a sample of seven pulleys from the week 5 batch results in the following diameter measurements (in mm): 50, 48, 48, 51, 52, 50, 52. Conduct a test to determine whether the variance in diameters differs between these two populations. Why would the quality control people be interested in such a test? What results from this test would you report to them? What about the means of these two batches? Analyse these data in terms of the means and report on the results. Assume that pulley diameters are normally distributed in the population.

Source: Adapted from The Manutan Group website at www.manutaninternational.com/main/core.php?pag_id=161; and 'Manutan International SA', available at www.fundinguniverse.com/company-histories/Manutan-International-SA-Company-History.html.

USING THE COMPUTER

EXCEL

- Excel has the capability of performing any of the statistical techniques presented in this chapter with the exception of testing the difference in two population proportions.

- Each of the tests presented here in Excel is accessed through the **Data Analysis** feature.

- To conduct a z test for the difference in two means, begin by selecting the **Data** tab on the Excel worksheet. From the **Analysis** panel at the right top of the **Data** tab worksheet, click on **Data Analysis**. If your Excel worksheet does not show the **Data Analysis** option, then you can load it as an add-in following directions given in Chapter 2. From the **Data Analysis** pulldown menu, select **z-Test: Two Sample for Means** from the dialog box. Enter the location of the observations from the first group in **Variable 1 Range**. Enter the location of the observations from the second group in **Variable 2 Range**. Enter the hypothesized value for the mean difference in **Hypothesized Mean Difference**. Enter the known variance of population 1 in **Variable 1 Variance (known)**. Enter the known variance of population 2 in **Variable 2 Variance (known)**. Check **Labels** if you have labels. Select **Alpha**.

- To conduct a t test for the difference in two means, begin by selecting the **Data** tab on the Excel worksheet. From the **Analysis** panel at the right top of the **Data** tab worksheet, click on **Data Analysis**. If your Excel worksheet does not show the **Data Analysis** option, then you can load it as an add-in following directions given in Chapter 2. From the **Data Analysis** pulldown menu, select **t-Test: Two-Sample Assuming Equal Variances** from the dialog box if you are assuming that the population variances are equal. Select **t-Test: Two-Sample Assuming Unequal Variances** from the dialog box if you are assuming that the population variances are not equal. Input is the same for either test. Enter the location of the observations from the first group in **Variable 1 Range**. Enter the location of the observations from the second group in **Variable 2 Range**. Enter the hypothesized value for the mean difference in **Hypothesized Mean Difference**. Check **Labels** if you have labels. Select **Alpha**.

- To conduct a t test for related populations, begin by selecting the **Data** tab on the Excel worksheet. From the **Analysis** panel at the right top of the **Data** tab worksheet, click on **Data Analysis**. If your Excel worksheet does not show the **Data Analysis** option, then you can load it as an add-in following directions given in Chapter 2. From the **Data Analysis** pulldown menu, select **t-Test: Paired Two-Sample for Means** from the dialog box. Enter the location of the observations from the first group in **Variable 1 Range**. Enter the location of the observations from the second group in **Variable 2 Range**. Enter the hypothesized value for the mean difference in **Hypothesized Mean Difference**. Check **Labels** if you have labels. Select **Alpha**.

- To conduct an F test for two population variances, begin by selecting the **Data** tab on the Excel worksheet. From the **Analysis** panel at the right top of the **Data** tab worksheet, click on **Data Analysis**. If your Excel worksheet does not show the **Data Analysis** option, then you can load it as an add-in following directions given in Chapter 2. From the **Data Analysis** pulldown menu, select **F-Test: Two-Sample for Variances** from the dialog box. Enter the location of the observations from the first group in **Variable 1 Range**. Enter the location of the observations from the second group in **Variable 2 Range**. Check **Labels** if you have labels. Select **Alpha**.

MINITAB

- With the exception of the two-sample z test and confidence interval, Minitab has the capability to perform any of the statistical techniques presented in this chapter.

- To begin a t test for the difference in two means or confidence intervals about the difference of two means from independent populations and population variances unknown, select **Stat** on the menu bar. Select **Basic Statistics** from the pulldown menu. From the second pulldown menu, select **2-Sample t**. Check **Samples in one column** if the data are 'stacked' in one column: (1) Place the location of the column with the stacked observations in the box labelled **Samples**. (2) Place the location of the column containing the group identifiers in the box labelled **Subscripts**. Check **Samples in different columns** if the data are located in two separate columns (unstacked): (1) Place the location of one column of observations in **First**, and (2) Place the location of the other column of observations in **Second**. Check **Summarized data** if you want to enter statistics on the two samples rather than the raw data: (1) In the row beside **First**, enter the values of the sample size, the sample mean, and the sample standard deviation for the first sample in **Sample size, Mean**, and **Standard deviation**, respectively, and (2) In the row beside **Second**, enter the values of the sample size, the sample mean, and the sample standard deviation for the second sample in **Sample size, Mean**, and **Standard deviation**, respectively. Check the box beside **Assume equal variances** if you want to use the equal variances model. Leave it blank if you do not want to assume equal variances. Click on **Options** if you want to enter a confidence level, the mean difference being tested, and/or the direction of the alternative hypothesis (greater than, less than, or not equal). Both the observed t for the

hypothesis test and the confidence interval are given in the output.

- To begin t tests or confidence intervals about the difference in two related populations, select **Stat** on the menu bar. Select **Basic Statistics** from the pulldown menu. From the second pulldown menu, select **Paired t**. Check **Samples in columns** if the data are in columns: (1) Place the location of one sample of observations in the box beside **First sample**, and (2) Place the location of the other sample of observations in the box beside **Second sample**. Check **Summarized data (differences)** if you want to enter sample statistics on the pairs of data rather than the raw data. Enter the values of the sample size, the sample mean, and the sample standard deviation for the pairs in the boxes beside **Sample size, Mean**, and **Standard deviation**, respectively. Click on **Options** if you want to enter a confidence level, the test mean difference, and/or the direction of the alternative hypothesis (greater than, less than, or not equal). Both the observed t for the hypothesis test and the confidence interval are given in the output.

- To begin a z test or confidence interval about the difference in two population proportions, select **Stat** on the menu bar. Select **Basic Statistics** from the pulldown menu. From the second pulldown menu, select **2 Proportions**. Check **Samples in one column** if the dichotomous data are in one column and the group identifiers are in another column: (1) Place the location of the raw data in the box beside **Samples**, and (2) Place the location of the group identifiers in the box beside **Subscripts**. Check **Samples in different columns** if the dichotomous data are located in two separate columns (unstacked): (1) Place the location of one column of observations in **First**, and (2) Place the location of the other column of observations in **Second**. Check **Summarized data (differences)** if you want to enter the values of x and n for each group rather than the raw data: (1) Enter the sample sizes for each sample under **Trials** in the boxes beside **First** and **Second** denoting the two samples, and (2) Enter the number of items that possessed the desired characteristics (x) for each sample under **Events** in the boxes beside **First** and **Second** denoting the two samples. Click on **Options** if you want to enter a confidence level, the test difference in population proportions, and/or the direction of the alternative hypothesis (greater than, less than, or not equal). You can check the box beside **Use pooled estimate of p for test** if you want to use a pooled estimate of p in conducting the z test for population proportions. Both the observed z for the hypothesis test and the confidence interval are given in the output.

- To begin an F test about the difference in two population variances, select **Stat** on the menu bar. Select **Basic Statistics** from the pulldown menu. From the second pulldown menu, select **2 Variances**. Check **Samples in one column** if the data are in one column and the group identifiers are in another column: (1) Place the location of the raw data in the box beside **Samples**, and (2) Place the location of the group identifiers in the box beside **Subscripts**. Check **Samples in different columns** if the data are

located in two separate columns (unstacked): (1) Place the location of one column of observations in **First**, and (2) Place the location of the other column of observations in **Second**. Check **Summarized data** if you want to enter the values of the sample size and the sample variance for each group rather than the raw data: (1) Enter the sample sizes for each sample under **Sample size:** in the boxes beside **First** and **Second** designating the two samples. (2) Enter the values of the respective sample for each sample under **Variance** in the boxes beside **First** and **Second** designating the two samples. Click on **Options** if you want to enter a confidence level or your own title. Both the observed F for the hypothesis test and the confidence interval are given in the output.

CHAPTER 11

Analysis of Variance and Design of Experiments

LEARNING OBJECTIVES

The focus of this chapter is the design of experiments and the analysis of variance, thereby enabling you to:

1. Describe an experimental design and its elements, including independent variables – both treatment and classification – and dependent variables
2. Test a completely randomized design using a one-way analysis of variance
3. Use multiple comparison techniques, including Tukey's honestly significant difference test and the Tukey–Kramer procedure, to test the difference in two treatment means when there is overall significant difference between treatments
4. Test a randomized block design that includes a blocking variable to control for confounding variables
5. Test a factorial design using a two-way analysis of variance, noting the advantages and applications of such a design and accounting for possible interaction between two treatment variables

Decision Dilemma

Job and Career Satisfaction of Foreign Self-Initiated Expatriates

Because of worker shortages in some industries, in a global business environment, firms around the world sometimes must compete with each other for workers. This is especially true in industries and job designations where skilled workers are required. In order to fill such needs, companies sometimes turn to self-initiated expatriates. Self-initiated expatriates are defined as workers who are hired as individuals on a contractual basis to work in a foreign country – in contrast to individuals who are given overseas transfers by a parent organization; that is, they are 'guest workers' as compared to 'organizational expatriates'. Some examples could be computer experts from India, China, and Japan being hired by Silicon Valley companies; British engineers working with Russian companies to extract oil and gas; or financial experts from Germany who are hired by Singapore companies to help manage the stock market. How satisfied are self-initiated expatriates with their jobs and their careers?

In an attempt to answer that question, suppose a study was conducted by randomly sampling self-initiated expatriates in five industries: information technology (IT), finance, education, healthcare, and consulting. Each is asked to rate his or her present job satisfaction on a seven-point Likert scale, with 7 being very satisfied and 1 being very dissatisfied. Suppose the data shown below are a portion of the study.

IT	Finance	Education	Healthcare	Consulting
5	3	2	3	6
6	4	3	2	7
5	4	3	4	5
7	5	2	3	6
	4	2	5	
		3		

Suppose in addition, self-initiated expatriates are asked to report their overall satisfaction with their career on the same seven-point scale. The ratings are broken down by the respondent's experience in the host country and age and the resultant data are shown below.

		Time in Host Country			
		<1 year	1–2 years	3–4 years	≥5 years
		3	4	3	6
	30–39	2	5	4	4
		3	3	5	5
		4	3	4	4
Age	40–49	3	4	4	6
		2	3	5	5
		4	4	5	6
	Over 50	3	4	4	5
		4	5	5	6

Managerial and Statistical Questions

1. Is there a difference in the job satisfaction ratings of self-initiated expatriates by industry? If we were to use the t test for the difference between two independent population means presented in Chapter 10 to analyse these data, we would need to do 10 different t tests since there are five different industries. Is there a better, more parsimonious way to analyse this data? Can the analysis be done simultaneously using one technique?

2. The second table in the Decision Dilemma displays career satisfaction data broken down two different ways, age and time in country. How does a researcher analyse such data when there are two different types of groups or classifications? What if one variable, such as age, acts on another variable, such as time in the country, such that there is an interaction? That is, time in the country might matter more in one category than in another. Can this effect be measured and, if so, how?

Source: Concepts adapted from Chay Hoon Lee, 'A Study of Underemployment Among Self-Initiated Expatriates,' *Journal of World Business*, vol. 40, no. 2 (May 2005), pp. 172–187.

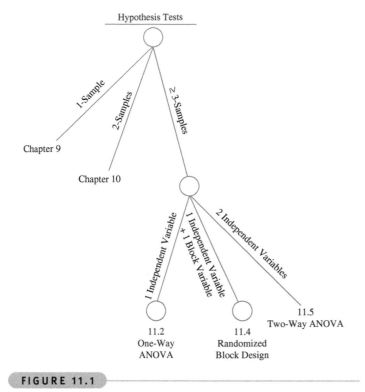

Hypothesis Tests

1-Sample

Chapter 9

2-Samples

Chapter 10

≥ 3-Samples

1 Independent Variable

1 Independent Variable + 1 Block Variable

2 Independent Variables

11.2
One-Way
ANOVA

11.4
Randomized
Block Design

11.5
Two-Way ANOVA

FIGURE 11.1

Branch of the Tree Diagram Taxonomy of Inference Techniques

Sometimes business research entails more complicated hypothesis-testing scenarios than those presented to this point in the text. Instead of comparing the wear of tyre tread for two brands of tyres to determine whether there is a significant difference between the brands, as we could have done by using Chapter 10 techniques, a tyre researcher may choose to compare three, four, or even more brands of tyres at the same time. In addition, the researcher may want to include different levels of quality of tyres in the experiment, such as low-quality, medium-quality, and high-quality tyres. Tests may be conducted under varying conditions of temperature, precipitation, or road surface. Such experiments involve selecting and analysing more than two samples of data.

Figure III-1 in the Introduction to Unit III displays the tree diagram taxonomy of inferential techniques, organizing the techniques by usage and number of samples. The entire right side of the tree diagram taxonomy contains various hypothesis-testing techniques. Techniques for testing hypotheses using a *single* sample are presented in Chapter 9; and techniques for testing hypotheses about the differences in two populations using *two* samples are presented in Chapter 10. The far right branch of the tree diagram taxonomy contains techniques for analysing *three or more* samples. This branch, shown in Figure 11.1, represents the techniques presented in Chapter 11.

11.1 INTRODUCTION TO DESIGN OF EXPERIMENTS

An **experimental design** is *a plan and a structure to test hypotheses in which the researcher either controls or manipulates one or more variables*. It contains independent and dependent variables. In an experimental design, an **independent variable** may be either a treatment variable or a classification variable. A **treatment variable** is *a variable the experimenter controls or modifies in the experiment*. A **classification variable** is *some characteristic of the experimental subject that was present prior to the experiment and is not a result of the experimenter's manipulations or control*. Independent variables are sometimes also referred to as **factors**. Spar executives might sanction an in-house study to compare daily sales volumes for a given size store in four different demographic settings: (1) inner-city stores (large city), (2) suburban stores (large city), (3) stores in a medium-sized city, and (4) stores in a small town. Managers might also decide to compare sales on the five different weekdays (Monday to Friday). In this study, the independent variables are store demographics and day of the week.

A finance researcher might conduct a study to determine whether there is a significant difference in application fees for home loans in five geographic regions of Europe and might include three different types of lending organizations. In this study, the independent variables are geographic region and types of lending organizations. Or suppose a manufacturing organization produces a valve that is specified to have an opening of 6.37 centimetres. Quality controllers within the company might decide to test to determine how the openings for produced valves vary among four different machines on three different shifts. This experiment includes the independent variables of type of machine and work shift.

Whether an independent variable can be manipulated by the researcher depends on the concept being studied. Independent variables such as work shift, gender of employee, geographic region, type of machine, and quality of tyre are classification variables with conditions that existed prior to the study. The business researcher cannot change the characteristic of the variable, so he or she studies the phenomenon being explored under several conditions of the various aspects of the variable. As an example, the valve experiment is conducted under the conditions of all three work shifts.

However, some independent variables can be manipulated by the researcher. For example, in the well-known Hawthorne studies of the Western Electric Company in the 1920s in the United States, the amount of light in production areas was varied to determine the effect of light on productivity. In theory, this independent variable could be manipulated by the researcher to allow any level of lighting. Other examples of independent variables that can be manipulated include the amount of bonuses offered to workers, level of humidity, and temperature. These are examples of treatment variables.

TABLE 11.1

Valve Opening Measurements (in cm) for 24 Valves Produced on an Assembly Line

6.26	6.19	6.33	6.26	6.50
6.19	6.44	6.22	6.54	6.23
6.29	6.40	6.23	6.29	6.58
6.27	6.38	6.58	6.31	6.34
6.21	6.19	6.36	6.56	

$\bar{x} = 6.34$ Total Sum of Squares Deviation $= \text{SST} = \Sigma(x_i - \bar{x})^2 = .3915$

Each independent variable has two or more levels, or classifications. **Levels**, or **classifications**, of independent variables are *the subcategories of the independent variable used by the researcher in the experimental design*. For example, the different demographic settings listed for the Spar study are four levels, or classifications, of the independent variable store demographics: (1) inner-city store, (2) suburban store, (3) store in a medium-sized city, and (4) store in small town. In the valve experiment, four levels or classifications of machines within the independent variable machine type are used: machine 1, machine 2, machine 3, and machine 4.

The other type of variable in an experimental design is a **dependent variable**. A dependent variable is *the response to the different levels of the independent variables*. It is the measurement taken under the conditions of the experimental design that reflect the effects of the independent variable(s). In the Spar study, the dependent variable is the euro amount of daily total sales. For the study on loan application fees, the fee charged for a loan application is probably the dependent variable. In the valve experiment, the dependent variable is the size of the opening of the valve.

Experimental designs in this chapter are analysed statistically by a group of techniques referred to as **analysis of variance**, or **ANOVA**. The analysis of variance concept begins with the notion that individual items being studied, such as employees, machine-produced products, regional offices, hospitals, and so on, are not all the same. Note the measurements for the openings of 24 valves randomly selected from an assembly line that are given in Table 11.1. The mean opening is 6.34 centimetres. Only one of the 24 valve openings is actually the mean. Why do the valve openings vary? The total sum of squares of deviation of these valve openings around the mean is 0.3915 cm². Why is this value not zero? Using various types of experimental designs, we can explore some possible reasons for this variance with analysis of variance techniques. As we explore each of the experimental designs and their associated analysis, note that the statistical technique is attempting to 'break down' the total variance among the objects being studied into possible causes. In the case of the valve openings, this variance of measurements might be due to such variables as machine, operator, shift, supplier, and production conditions, among others.

Many different types of experimental designs are available to researchers. In this chapter, we will present and discuss three specific types of experimental designs: completely randomized design, randomized block design, and factorial experiments.

11.1 PROBLEMS

11.1 Some New York Stock Exchange analysts believe that 24-hour trading on the stock exchange is the wave of the future. As an initial test of this idea, the New York Stock Exchange opened two after-hour 'crossing sections' in the early 1990s and studied the results of these extra-hour sessions for one year.

   a. State an independent variable that could have been used for this study.

   b. List at least two levels, or classifications, for this variable.

   c. Give a dependent variable for this study.

11.2 Ryanair is able to keep fares low in part because of relatively low maintenance costs on its airplanes. One of the main reasons for the low maintenance costs is that Ryanair flies only one type of aircraft, the Boeing 737. However, Ryanair flies three different versions of the 737. Suppose the company decides to conduct a study to determine whether there is a significant difference in the average annual maintenance costs for the three types of 737s used.

a. State an independent variable for such a study.

b. What are some of the levels or classifications that might be studied under this variable?

c. Give a dependent variable for this study.

11.3 A large multinational banking company wants to determine whether there is a significant difference in the average euro amounts purchased by users of different types of credit cards. Among the credit cards being studied are MasterCard, Visa, and American Express.

a. If an experimental design were set up for such a study, what are some possible independent variables?

b. List at least three levels, or classifications, for each independent variable.

c. What are some possible dependent variables for this experiment?

11.4 Is there a difference in the family demographics of people who stay at hotels? Suppose a study is conducted in which three categories of hotels are used: economy hotels, modestly priced chain hotels, and exclusive hotels. One of the dependent variables studied might be the number of children in the family of the person staying in the hotel. Name three other dependent variables that might be used in this study.

11.2 THE COMPLETELY RANDOMIZED DESIGN (ONE-WAY ANOVA)

One of the simplest experimental designs is the completely randomized design. In the **completely randomized design**, *subjects are assigned randomly to treatments*. The completely randomized design contains only one independent variable, with two or more treatment levels, or classifications. If only two treatment levels, or classifications, of the independent variable are present, the design is the same one used to test the difference in means of two independent populations presented in Chapter 10, which used the *t* test to analyse the data.

In this section, we will focus on completely randomized designs with three or more classification levels. Analysis of variance, or ANOVA, will be used to analyse the data that result from the treatments.

A completely randomized design could be structured for a tyre-quality study in which tyre quality is the independent variable and the treatment levels are low, medium, and high quality. The dependent variable might be the number of miles driven before the tread fails state inspection. A study of daily sales volumes for Spar stores could be undertaken by using a completely randomized design with demographic setting as the independent variable. The treatment levels, or classifications, would be inner-city stores, suburban stores, stores in medium-sized cities, and stores in small towns. The dependent variable would be sales dollars.

As an example of a completely randomized design, suppose a researcher decides to analyse the effects of the machine operator on the valve opening measurements of valves produced in a manufacturing plant, like those shown in Table 11.1. The independent variable in this design is machine operator. Suppose further that four different operators operate the machines. These four machine operators are the levels of treatment, or classification, of the independent variable. The dependent variable is the opening measurement of the valve. Figure 11.2 shows the

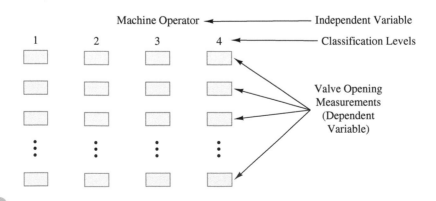

FIGURE 11.2

Completely Randomized Design

structure of this completely randomized design. Is there a significant difference in the mean valve openings of 24 valves produced by the four operators? Table 11.2 contains the valve opening measurements for valves produced under each operator.

One-Way Analysis of Variance

In the machine operator example, is it possible to analyse the four samples by using a t test for the difference in two sample means? These four samples would require $_4C_2 = 6$ individual t tests to accomplish the analysis of two groups at a time. Recall that if $\alpha = 0.05$ for a particular test, there is a 5% chance of rejecting a null hypothesis that is true (i.e., committing a Type I error). If enough tests are done, eventually one or more null hypothesis will be falsely rejected by chance. Hence, $\alpha = 0.05$ is valid only for one t test. In this problem, with six t tests, the error rate compounds, so when the analyst is finished with the problem there is a much greater than 0.05 chance of committing a Type I error. Fortunately, a technique has been developed that analyses all of the sample means at one time and thus precludes the build-up of error rate: analysis of variance (ANOVA). A completely randomized design is analysed by a **one-way analysis of variance**.

In general, if k samples are being analysed, the following hypotheses are being tested in a one-way ANOVA:

$$H_0: \mu_1 = \mu_2 = \mu_3 = \ldots = \mu_k$$
$$H_a: \text{At least one of the means is different from the others}$$

The null hypothesis states that the population means for all treatment levels are equal. Because of the way the alternative hypothesis is stated, if even one of the population means is different from the others, the null hypothesis is rejected.

Testing these hypotheses by using one-way ANOVA is accomplished by partitioning the total variance of the data into the following two variances.

1. The variance resulting from the treatment (columns)
2. The error variance, or that portion of the total variance unexplained by the treatment

As part of this process, the total sum of squares of deviation of values around the mean can be divided into two additive and independent parts.

$$\text{SST} \quad = \quad \text{SSC} \quad + \quad \text{SSE}$$
$$\sum_{i=1}^{n_j}\sum_{j=1}^{C}(x_{ij} - \bar{x})^2 = \sum_{j=1}^{C}n_j(\bar{x}_j - \bar{x})^2 + \sum_{i=1}^{n_j}\sum_{j=1}^{C}(x_{ij} - \bar{x}_j)^2$$

where:
  SST = total sum of squares
  SSC = sum of squares column (treatment)
  SSE = sum of squares error
    i = particular member of a treatment level
    j = a treatment level
    C = number of treatment levels
    n_j = number of observations in a given treatment level
    $\bar{x}$ = grand mean
    $\bar{x}_j$ = mean of a treatment group or level
    x_{ij} = individual value

This relationship is shown in Figure 11.3. Observe that the total sum of squares of variation is partitioned into the sum of squares of treatment (columns) and the sum of squares of error.

TABLE 11.2

Valve Openings by Operator

1	2	3	4
6.33	6.26	6.44	6.29
6.26	6.36	6.38	6.23
6.31	6.23	6.58	6.19
6.29	6.27	6.54	6.21
6.40	6.19	6.56	
	6.50	6.34	
	6.19	6.58	
	6.22		

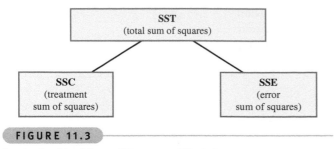

FIGURE 11.3

Portioning Total Sum of Squares of Variation

The formulas used to accomplish one-way analysis of variance are developed from this relationship. The double summation sign indicates that the values are summed within a treatment level and across treatment levels. Basically, ANOVA compares the relative sizes of the *treatment* variation and the *error* variation (within-group variation). The error variation is unaccounted-for variation and can be viewed at this point as variation due to individual differences within treatment groups. If a significant difference in treatments is present, the treatment variation should be large relative to the error variation.

Figure 11.4 displays the data from the machine operator example in terms of treatment level. Note the variation of values (x) *within* each treatment level. Now examine the variation between levels 1 through 4 (the difference in the machine operators). In particular, note that values for treatment level 3 seem to be located differently from those of levels 2 and 4. This difference also is underscored by the mean values for each treatment level:

$$\bar{x}_1 = 6.3180 \quad \bar{x}_2 = 6.2775 \quad \bar{x}_3 = 6.488571 \quad \bar{x}_4 = 6.23$$

Analysis of variance is used to determine statistically whether the variance between the treatment level means is greater than the variances within levels (error variance). Several important assumptions underlie analysis of variance:

1. Observations are drawn from normally distributed populations.
2. Observations represent random samples from the populations.
3. Variances of the populations are equal.

These assumptions are similar to those for using the t test for independent samples in Chapter 10. It is assumed that the populations are normally distributed and that the population variances are equal. These techniques should be used only with random samples.

An ANOVA is computed with the three sums of squares: total, treatment (columns), and error. Shown here are the formulas to compute a one-way analysis of variance. The term SS represents sum of squares, and the term MS represents mean square. SSC is the sum of squares columns, which yields the sum of squares between treatments. It measures the variation between columns or between treatments since the independent variable treatment levels are presented as columns. SSE is the sum of squares of error, which yields the variation within treatments (or columns). Some say that it is a measure of the individual differences unaccounted for by the treatments. SST is the total sum of

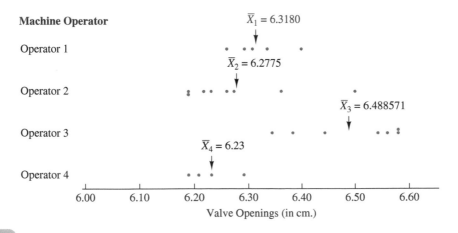

FIGURE 11.4

Location of Mean Value Opening by Operator

squares and is a measure of all variation in the dependent variable. As shown previously, SST contains both SSC and SSE and can be partitioned into SSC and SSE. MSC, MSE, and MST are the mean squares of column, error, and total, respectively. Mean square is an average and is computed by dividing the sum of squares by the degrees of freedom. Finally, the F value is determined by dividing the treatment variance (MSC) by the error variance (MSE). As discussed in Chapter 10, the F is a ratio of two variances. In the ANOVA situation, the **F value** is *a ratio of the treatment variance to the error variance.*

FORMULAS FOR COMPUTING A ONE-WAY ANOVA

$$SSC = \sum_{j=1}^{C} n_j (\bar{x}_j - \bar{x})^2$$

$$SSE = \sum_{i=1}^{n_j} \sum_{j=1}^{C} (x_{ij} - \bar{x}_j)^2$$

$$SST = \sum_{i=1}^{n_j} \sum_{j=1}^{C} (x_{ij} - \bar{x})^2$$

$$df_C = C - 1$$
$$df_E = N - C$$
$$df_T = N - 1$$

$$MSC = \frac{SSC}{df_C}$$

$$MSE = \frac{SSE}{df_E}$$

$$F = \frac{MSC}{MSE}$$

where:

i = a particular member of a treatment level
j = a treatment level
C = number of treatment levels
n_j = number of observations in a given treatment level
$\bar{x}$ = grand mean
$\bar{x}_j$ = column mean
x_{ij} = individual value

Performing these calculations for the machine operator example yields the following.

Machine Operator			
1	2	3	4
6.33	6.26	6.44	6.29
6.26	6.36	6.38	6.23
6.31	6.23	6.58	6.19
6.29	6.27	6.54	6.21
6.40	6.19	6.56	
	6.50	6.34	
	6.19	6.58	
	6.22		

$$T_j: \quad T_1 = 31.59 \qquad T_2 = 50.22 \qquad T_3 = 45.42 \qquad T_4 = 24.92 \qquad T = 152.15$$
$$n_j: \quad n_1 = 5 \qquad n_2 = 8 \qquad n_3 = 7 \qquad n_4 = 4 \qquad N = 24$$
$$\bar{x}_j: \quad \bar{x}_1 = 6.318 \qquad \bar{x}_2 = 6.2775 \qquad \bar{x}_3 = 6.488571 \qquad \bar{x}_4 = 6.230 \qquad \bar{x} = 6.339583$$

$$
\begin{aligned}
\text{SSC} = \sum_{j=1}^{C} n_j (\bar{x}_j - \bar{x})^2 &= [5(6.318 - 6.339583)^2 + 8(6.2775 - 6.339583)^2 \\
&\quad + 7(6.488571 - 6.339583)^2 + 4(6.230 - 6.339583)^2] \\
&= 0.00233 + 0.03083 + 0.15538 + 0.04803 \\
&= 0.23658
\end{aligned}
$$

$$
\begin{aligned}
\text{SSE} = \sum_{i=1}^{n_j} \sum_{j=1}^{C} (x_{ij} - \bar{x}_j)^2 &= [(6.33 - 6.318)^2 + (6.26 - 6.318)^2 + (6.31 - 6.318)^2 \\
&\quad + (6.29 - 6.318)^2 + (6.40 - 6.318)^2 + (6.26 - 6.2775)^2 \\
&\quad + (6.36 - 6.2775)^2 + \cdots + (6.19 - 6.230)^2 + (6.21 - 6.230)^2] \\
&= 0.15492
\end{aligned}
$$

$$
\begin{aligned}
\text{SST} = \sum_{i=1}^{n_j} \sum_{j=1}^{C} (x_{ij} - \bar{x})^2 &= [(6.33 - 6.339583)^2 + (6.26 - 6.339583)^2 \\
&\quad + (6.31 - 6.339583)^2 + \cdots + (6.19 - 6.339583)^2 \\
&\quad + (6.21 - 6.339583)^2] \\
&= 0.39150
\end{aligned}
$$

$$
\begin{aligned}
\text{df}_C &= C - 1 = 4 - 1 = 3 \\
\text{df}_E &= N - C = 24 - 4 = 20 \\
\text{df}_T &= N - 1 = 24 - 1 = 23
\end{aligned}
$$

$$\text{MSC} = \frac{\text{SSC}}{\text{df}_C} = \frac{0.23658}{3} = 0.078860$$

$$\text{MSE} = \frac{\text{SSE}}{\text{df}_E} = \frac{0.15492}{20} = 0.007746$$

$$F = \frac{0.078860}{0.00746} = 10.18$$

From these computations, an analysis of variance chart can be constructed, as shown in Table 11.3. The observed *F* value is 10.18. It is compared to a critical value from the *F* table to determine whether there is a significant difference in treatment or classification.

Reading the *F* Distribution Table

The **F distribution** table is Table A.7 in Appendix A. Associated with every *F* value in the table are two unique df values: degrees of freedom in the numerator (df_C) and degrees of freedom in the denominator (df_E). To look up a value in the *F* distribution table, the researcher must know both degrees of freedom. Because each *F* distribution is determined by a unique pair of degrees of freedom, many *F* distributions are possible. Space constraints limit Table A.7 in Appendix A to *F* values for only $\alpha = 0.005, 0.01, 0.025, 0.05$, and 0.10. However, statistical computer software packages for computing ANOVAs usually give a probability for the *F* value, which allows a hypothesis-testing decision for any alpha based on the *p*-value method.

In the one-way ANOVA, the df_C values are the treatment (column) degrees of freedom, $C - 1$. The df_E values

TABLE 11.3

Analysis of Variance for the Machine Operator Example

Source of Variance	df	SS	MS	F
Between	3	0.23658	0.078860	10.18
Error	20	0.15492	0.007746	
Total	23	0.39150		

An Abbreviated F Table for $\alpha = 0.05$

					Numerator Degrees of Freedom				
		1	2	3	4	5	6	7	8	9
Denominator Degrees of Freedom	. . . 19	4.38	3.52	3.13	2.90	2.74	2.63	2.54	2.48	2.42
	20	4.35	3.49	3.10	2.87	2.71	2.60	2.51	2.45	2.39
	21	4.32	3.47	3.07	2.84	2.68	2.57	2.49	2.42	2.37

are the error degrees of freedom, $N - C$. Table 11.4 contains an abbreviated F distribution table for $\alpha = 0.05$. For the machine operator example, $df_C = 3$ and $df_E = 20$, $F_{0.05,3,20}$ from Table 11.4 is 3.10. This value is the critical value of the F test. Analysis of variance tests are always one-tailed tests with the rejection region in the upper tail. The decision rule is to reject the null hypothesis if the observed F value is greater than the critical F value ($F_{0.05,3,20} = 3.10$). For the machine operator problem, the observed F value of 10.18 is larger than the table F value of 3.10. The null hypothesis is rejected. Not all means are equal, so there is a significant difference in the mean valve openings by machine operator. Figure 11.5 is a Minitab graph of an F distribution showing the critical F value for this example and the rejection region. Note that the F distribution begins at zero and contains no negative values because the F value is the ratio of two variances, and variances are always positive.

Using the Computer for One-Way ANOVA

Many researchers use the computer to analyse data with a one-way ANOVA. Figure 11.6 shows the Minitab and Excel output of the ANOVA computed for the machine operator example. The output includes the analysis of variance table presented in Table 11.3. Both Minitab and Excel ANOVA tables display the observed F value, mean squares, sum of squares, degrees of freedom, and a value of p. The value of p is the probability of an F value of 10.18 occurring by chance in an ANOVA with this structure (same degrees of freedom) even if there is no difference between means of the treatment levels. Using the p-value method of testing hypotheses presented in Chapter 9, we can easily see that because this p-value is only 0.000279, the null hypothesis would be rejected using $\alpha = 0.05$.

Most computer output yields the value of p, so there is no need to look up a table value of F against which to compare the observed F value. The Excel output also includes the critical table F value for this problem, $F_{0.05,3,20} = 3.10$.

The second part of the Minitab output in Figure 11.6 contains the size of samples and sample means for each of the treatment levels. Displayed graphically are the 95% confidence levels for the population means of each treatment level group. These levels are computed by using a pooled standard deviation from all the treatment level groups. The researcher can visually observe the confidence intervals and make a subjective determination about the relative difference in the population means. More rigorous statistical techniques for testing the differences in pairs of groups are given in Section 11.3.

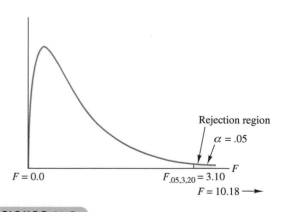

Minitab Graph of F Values for the Machine Operator Example

Minitab Output

One-way ANOVA: Operator 1, Operator 2, Operator 3, Operator 4

```
Source    DF       SS        MS       F       P
Factor     3    0.23658   0.07886   10.18   0.000
Error     20    0.15492   0.00775
Total     23    0.39150


S = 0.08801 R-Sq = 60.43% R-Sq(adj) = 54.49%
```

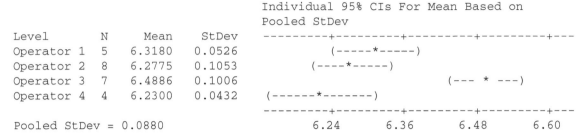

```
                                     Individual 95% CIs For Mean Based on
                                     Pooled StDev
Level        N    Mean     StDev    ---------+---------+---------+---------+---
Operator 1   5   6.3180   0.0526            (-----*-----)
Operator 2   8   6.2775   0.1053         (----*-----)
Operator 3   7   6.4886   0.1006                               (--- * ---)
Operator 4   4   6.2300   0.0432    (------*-------)
                                     ---------+---------+---------+---------+---
Pooled StDev = 0.0880                     6.24      6.36      6.48      6.60
```

Excel Output

Anova: Single Factor

SUMMARY

Groups	Count	Sum	Average	Variance
Operator 1	5	31.59	6.318000	0.002770
Operator 2	8	50.22	6.277500	0.011079
Operator 3	7	45.42	6.488571	0.010114
Operator 4	4	24.92	6.230000	0.001867

ANOVA

Source of Variation	SS	df	MS	F	P-value	F crit
Between Groups	0.236580	3	0.078860	10.18	0.000279	3.10
Within Groups	0.154916	20	0.007746			
Total	0.391496	23				

FIGURE 11.6

Minitab and Excel Analysis of the Machine Operator Problem

Comparison of F and t Values

Analysis of variance can be used to test hypotheses about the difference in two means. Analysis of data from two samples by both a t test and an ANOVA shows that the observed F value equals the observed t value squared.

$$F = t^2 \quad \text{for } df_c = 1$$

The t test of independent samples actually is a special case of one-way ANOVA when there are only two treatment levels ($df_c = 1$). The t test is computationally simpler than ANOVA for two groups. However, some statistical computer software packages do not contain a t test. In these cases, the researcher can perform a one-way ANOVA and then either take the square root of the F value to obtain the value of t or use the generated probability with the p-value method to reach conclusions.

DEMONSTRATION PROBLEM 11.1

A company has three manufacturing plants, and company officials want to determine whether there is a difference in the average age of workers at the three locations. The following data are the ages of five randomly selected workers at each plant. Perform a one-way ANOVA to determine whether there is a significant difference in the mean ages of the workers at the three plants. Use $\alpha = 0.01$ and note that the sample sizes are equal.

Solution

HYPOTHESIZE

STEP 1: The hypotheses follow.

$$H_0: \mu_1 = \mu_2 = \mu_3$$
$$H_a: \text{At least one of the means is different from the others.}$$

TEST

STEP 2: The appropriate test statistic is the F test calculated from ANOVA.

STEP 3: The value of α is 0.01.

STEP 4: The degrees of freedom for this problem are $3 - 1 = 2$ for the numerator and $15 - 3 = 12$ for the denominator. The critical F value is $F_{0.01, 2, 12} = 6.93$.

Because ANOVAs are always one tailed with the rejection region in the upper tail, the decision rule is to reject the null hypothesis if the observed value of F is greater than 6.93.

STEP 5:

Plant (Employee Ages)

1	2	3
29	32	25
27	33	24
30	31	24
27	34	25
28	30	26

STEP 6:

$$T_j: \quad T_1 = 141 \qquad T_2 = 160 \qquad T_3 = 124 \qquad T = 425$$
$$n_j: \quad n_1 = 5 \qquad n_2 = 5 \qquad n_3 = 5 \qquad N = 15$$
$$\bar{X}_j: \quad \bar{X}_1 = 28.2 \qquad \bar{X}_2 = 32.0 \qquad \bar{X}_3 = 24.8 \qquad \bar{X} = 28.33$$

$$SSC = 5(28.2 - 28.33)^2 + 5(32.0 - 28.33)^2 + 5(24.8 - 28.33)^2 = 129.73$$

$$SSE = (29 - 28.2)^2 + (27 - 28.2)^2 + \cdots + (25 - 24.8)^2 + (26 - 24.8)^2 = 19.60$$

$$SST = (29 - 28.33)^2 + (27 - 28.33)^2 + \cdots + (25 - 28.33)^2 + (26 - 28.33)^2 = 149.33$$

$$df_C = 3 - 1 = 2$$
$$df_E = 15 - 3 = 12$$
$$df_T = 15 - 1 = 14$$

Source of Variance	SS	df	MS	F
Between	129.73	2	64.87	39.80
Error	19.60	12	1.63	
Total	149.33	14		

ACTION

STEP 7: The decision is to reject the null hypothesis because the observed F value of 39.80 is greater than the critical table F value of 6.93.

BUSINESS IMPLICATIONS

STEP 8: There is a significant difference in the mean ages of workers at the three plants. This difference can have hiring implications. Company leaders should understand that because motivation, discipline, and experience may differ with age, the differences in ages may call for different managerial approaches in each plant.

The chart below displays the dispersion of the ages of workers from the three samples, along with the mean age for each plant sample. Note the difference in group means. The significant F value says that the difference between the mean ages is relatively greater than the differences between ages within each group.

```
                                    x
Plant 1                           x   x   x   x                    x̄₁ = 28.2

Plant 2                                      x__x__x__x__x          x̄₂ = 32.0

Plant 3        x   x   x                                            x̄₃ = 24.8
               x   x

          23  24  25  26  27  28  29  30  31  32  33  34
                              Age
```

Following are the Minitab and Excel output for this problem.

Minitab Output

```
One-way ANOVA: Plant 1, Plant 2, Plant 3

Source    DF      SS       MS       F        P
Factor     2    129.73   64.87    39.71    0.000
Error     12     19.60    1.63
Total     14    149.33

S = 1.278  R-Sq = 86.88%  R-Sq(adj) = 84.69%
                                 Individual 95% CIs For Mean
                                 Based on Pooled StDev
Level     N    Mean    StDev   --+---------+---------+---------+----
Plant 1   5   28.200   1.304                   (---*---)
Plant 2   5   32.000   1.581                            (---*---)
Plant 3   5   24.800   0.837    (---*---)
                               --+---------+---------+---------+----
Pooled StDev = 1.278           25.0      27.5      30.0      32.5
```

Excel Output

Anova: Single Factor

SUMMARY

Groups	Count	Sum	Average	Variance
Plant 1	5	141	28.2	1.7
Plant 2	5	160	32	2.5
Plant 3	5	124	24.8	0.7

ANOVA

Source of Variation	SS	df	MS	F	P-value	F crit
Between Groups	129.73333	2	64.8667	39.71	0.0000051	3.89
Within Groups	19.6	12	1.6333			
Total	149.33333	14				

> ## 11.2 PROBLEMS

11.5 Compute a one-way ANOVA on the following data:

1	2	3
2	5	3
1	3	4
3	6	5
3	4	5
2	5	3
1		5

Determine the observed F value. Compare the observed F value with the critical table F value and decide whether to reject the null hypothesis. Use $\alpha = 0.05$.

11.6 Compute a one-way ANOVA on the following data:

1	2	3	4	5
14	10	11	16	14
13	9	12	17	12
10	12	13	14	13
	9	12	16	13
	10		17	12
				14

Determine the observed F value. Compare the observed F value with the critical table F value and decide whether to reject the null hypothesis. Use $\alpha = 0.01$.

11.7 Develop a one-way ANOVA on the following data:

1	2	3	4
113	120	132	122
121	127	130	118
117	125	129	125
110	129	135	125

Determine the observed F value. Compare it to the critical F value and decide whether to reject the null hypothesis. Use a 1% level of significance.

11.8 Compute a one-way ANOVA on the following data:

1	2
27	22
31	27
31	25
29	23
30	26
27	27
28	23

Determine the observed F value. Compare it to the critical table F value and decide whether to reject the null hypothesis. Perform a t test for independent measures on the data. Compare the t and F values. Are the results different? Use $\alpha = 0.05$.

11.9 Suppose you are using a completely randomized design to study some phenomenon. There are five treatment levels and a total of 55 people in the study. Each treatment level has the same sample size. Complete the following ANOVA:

Source of Variance	SS	df	MS	F
Treatment	583.39			
Error	972.18			
Total	1,555.57			

11.10 Suppose you are using a completely randomized design to study some phenomenon. There are three treatment levels and a total of 17 people in the study. Complete the ANOVA table below. Use $\alpha = 0.05$ to find the table F value and use the data to test the null hypothesis.

Source of Variance	SS	df	MS	F
Treatment	29.64			
Error	68.42			
Total				

11.11 A milk company has four machines that fill litre jugs with milk. The quality control manager is interested in determining whether the average fill for these machines is the same. The data below represent random samples of fill measures (in millilitres) for 18 jugs of milk filled by the different machines. Use $\alpha = 0.01$ to test the hypotheses. Discuss the business implications of your findings.

Machine 1	Machine 2	Machine 3	Machine 4
1,012.5	997.5	992.5	1,000.0
1,002.5	1,005.0	995.0	1,005.0
1,005.0	1,002.5	992.5	997.5
1,010.0	997.5	987.5	1,002.5
	1,000.0		
	1,000.0		

11.12 It seems logical that the starting salaries of new accounting graduates would differ according to geographic regions of the United Kingdom. A random selection of accounting firms is taken from three geographic regions, and each is asked to state the starting salary for a new accounting graduate who is going to work in auditing. The data obtained are below. Use a one-way ANOVA to analyse these data. Note that the data can be restated to make the computations more reasonable (example: £42,500 = 4.25). Use a 1% level of significance. Discuss the business implications of your findings.

North	South	West
40,500	51,000	45,500
41,500	49,500	43,500
40,000	49,000	45,000
41,000	48,000	46,500
41,500	49,500	46,000

11.13 A management consulting company presents a three-day seminar on project management to various clients. The seminar is basically the same each time it is given. However, sometimes it is presented to high-level managers, sometimes to mid-level managers, and sometimes to low-level managers. The seminar facilitators believe evaluations of the seminar may vary with the audience. Suppose the data below are some randomly selected evaluation scores from different levels of managers who attended the seminar. The ratings are on a scale from 1 to 10, with 10 being the highest. Use a one-way ANOVA to determine whether there is a significant difference in the evaluations according to manager level. Assume $\alpha = 0.05$. Discuss the business implications of your findings.

High Level	Mid-level	Low Level
7	8	5
7	9	6
8	8	5
7	10	7
9	9	4
	10	8
	8	

11.14 Family transportation costs are usually higher than most people believe because those costs include car payments, insurance, fuel costs, repairs, parking, and public transportation. Twenty randomly selected families in four major European cities are asked to use their records to estimate a monthly figure for transportation cost. Use the data obtained and ANOVA to test whether there is a significant difference in monthly transportation costs for families living in these cities. Assume that $\alpha = 0.05$. Discuss the business implications of your findings.

Amsterdam	Madrid	Frankfurt	Dublin
650	250	850	540
480	525	700	450
550	300	950	675
600	175	780	550
675	500	600	600

11.15 Shown here is the Minitab output for a one-way ANOVA. Analyse the results. Include the number of treatment levels, the sample sizes, the F value, the overall statistical significance of the test, and the values of the means.

One-Way Analysis of Variance

```
Analysis of Variance
```

Source	df	SS	MS	F	P
Factor	3	1,701	567	2.95	0.040
Error	61	11,728	192		
Total	64	13,429			

Individual 95% CIs For Mean Based on Pooled StDev

Level	N	Mean	StDev
C1	18	226.73	13.59
C2	15	238.79	9.41
C3	21	232.58	12.16
C4	11	239.82	20.96

Pooled StDev = 13.87

11.16   Business is very good for a chemical company. In fact, it is so good that workers are averaging more than 40 hours per week at each of the chemical company's five plants. However, management is not certain whether there is a difference between the five plants in the average number of hours worked per week per worker. Random samples of data are taken at each of the five plants. The data are analysed using Excel. The results follow below. Explain the design of the study and determine whether there is an overall significant difference between the means at $\alpha = 0.05$. Why or why not? What are the values of the means? What are the business implications of this study to the chemical company?

Anova: Single Factor

SUMMARY

Groups	Count	Sum	Average	Variance
Plant 1	11	636.5577	57.87	63.5949
Plant 2	12	601.7648	50.15	62.4813
Plant 3	8	491.7352	61.47	47.4772
Plant 4	5	246.0172	49.20	65.6072
Plant 5	7	398.6368	56.95	140.3540

ANOVA

Source of Variation	SS	df	MS	F	P-value	F crit
Between Groups	900.0863	4	225.0216	3.10	0.026595	2.62
Within Groups	2760.136	38	72.63516			
Total	3660.223	42				

11.3   MULTIPLE COMPARISON TESTS

Analysis of variance techniques are particularly useful in testing hypotheses about the differences of means in multiple groups because ANOVA utilizes only one single overall test. The advantage of this approach is that the probability of committing a Type I error, α, is controlled. As noted in Section 11.2, if four groups are tested two at a time, it takes six t tests ($_4C_2$) to analyse hypotheses between all possible pairs. In general, if k groups are tested two at a time, $_kC_2 = k(k-1)/2$ paired comparisons are possible.

Suppose alpha for an experiment is 0.05. If two different pairs of comparisons are made in the experiment using alpha of 0.05 in each, there is a 0.95 probability of not making a Type I error in each comparison. This approach results in a 0.9025 probability of not making a Type I error in either comparison (0.95×0.95), and a 0.0975 probability of committing a Type I error in at least one comparison ($1 - 0.9025$). Thus, the probability of committing a Type I error for this experiment is not 0.05 but 0.0975. In an experiment where the means of four groups are being tested two at a time, six different tests are conducted. If each is analysed using $\alpha = 0.05$, the probability that no Type I error will be committed in any of the six tests is $0.95 \times 0.95 \times 0.95 \times 0.95 \times 0.95 \times 0.95 = 0.735$ and the probability of committing at least one Type I error in the six tests is $1 - 0.735 = 0.265$. If an ANOVA is computed on all groups simultaneously using $\alpha = 0.05$, the value of alpha is maintained in the experiment.

Sometimes the researcher is satisfied with conducting an overall test of differences in groups such as the one ANOVA provides. However, when it is determined that there is an overall difference in population means, it is often desirable to go back to the groups and determine from the data which pairs of means are significantly different. Such pairwise analyses can lead to the build-up of the Type I experimental error rate, as mentioned. Fortunately, several techniques, referred to as **multiple comparisons**, have been developed to handle this problem.

Multiple comparisons are to be used only when an overall significant difference between groups has been obtained by using the F value of the analysis of variance. Some of these techniques protect more for Type I errors and others protect more for Type II errors. Some multiple comparison techniques require equal sample sizes. There seems to be some difference of opinion in the literature about which techniques are most appropriate. Here we will consider only a posteriori or post hoc pairwise comparisons.

A posteriori or **post hoc** pairwise comparisons are made *after the experiment when the researcher decides to test for any significant differences in the samples based on a significant overall F value.* In contrast, **a priori** comparisons are made when the researcher *determines before the experiment which comparisons are to be made.* The error rates for these two types of comparisons are different, as are the recommended techniques. In this text, we only consider pairwise (two-at-a-time) multiple comparisons. Other types of comparisons are possible but belong in a more advanced presentation. The two multiple comparison tests discussed here are Tukey's HSD test for designs with equal sample sizes and the Tukey–Kramer procedure for situations in which sample sizes are unequal. Minitab yields computer output for each of these tests.

Tukey's Honestly Significant Difference (HSD) Test: The Case of Equal Sample Sizes

Tukey's honestly significant difference test, sometimes known as Tukey's T method, is a popular test for pairwise a posteriori multiple comparisons. This test, developed by John W. Tukey and presented in 1953, is somewhat limited by the fact that it requires equal sample sizes.

Tukey's HSD test takes into consideration the number of treatment levels, the value of mean square error, and the sample size. Using these values and a table value, q, the HSD determines the critical difference necessary between the means of any two treatment levels for the means to be significantly different. Once the HSD is computed, the researcher can examine the absolute value of any or all differences between pairs of means from treatment levels to determine whether there is a significant difference. The formula to compute a Tukey's HSD test follows.

$$HSD = q_{\alpha,C,N-C}\sqrt{\frac{MSE}{n}}$$

TUKEY'S HSD TEST where:

$$MSE = \text{mean square error}$$
$$n = \text{sample size}$$
$$q_{\alpha,C,N-C} = \text{critical value of the Studentized range distribution from Table A.10 in Appendix A}$$

In Demonstration Problem 11.1, an ANOVA test was used to determine that there was an overall significant difference in the mean ages of workers at the three different plants, as evidenced by the F value of 39.8. The sample data for this problem follow.

	PLANT		
	1	2	3
	29	32	25
	27	33	24
	30	31	24
	27	34	25
	28	30	26
Group Means	28.2	32.0	24.8
n_j	5	5	5

Because the sample sizes are equal in this problem, Tukey's HSD test can be used to compute multiple comparison tests between groups 1 and 2, 2 and 3, and 1 and 3. To compute the HSD, the values of MSE, n, and q must be determined. From the solution presented in Demonstration Problem 11.1, the value of MSE is 1.63. The sample size, n_j, is 5. The value of q is obtained from Table A.10 by using

$$\text{Number of Populations} = \text{Number of Treatment Means} = C$$

along with $df_E = N - C$.

In this problem, the values used to look up q are

$$C = 3$$
$$df_B = N - C = 12$$

TABLE 11.5

q Values for $\alpha = 0.01$

Degrees of Freedom	Number of Populations			
	2	3	4	5
1	90	135	164	186
2	14	19	22.3	24.7
3	8.26	10.6	12.2	13.3
4	6.51	8.12	9.17	9.96
11	4.39	5.14	5.62	5.97
12	4.32	5.04	5.50	5.84

Table A.10 has a *q* table for $\alpha = 0.05$ and one for $\alpha = 0.01$. In this problem, $\alpha = 0.01$. Shown in Table 11.5 is a portion of Table A.10 for $\alpha = 0.01$.

For this problem, $q_{0.01,3,12} = 5.04$. HSD is computed as

$$\text{HSD} = q\sqrt{\frac{\text{MSE}}{n}} = 5.04\sqrt{\frac{1.63}{5}} = 2.88$$

Using this value of HSD, the business researcher can examine the differences between the means from any two groups of plants. Any of the pairs of means that differ by more than 2.88 are significantly different at $\alpha = 0.01$. Here are the differences for all three possible pairwise comparisons:

$$|\bar{x}_1 - \bar{x}_2| = |28.2 - 32.0| = 3.8$$
$$|\bar{x}_1 - \bar{x}_3| = |28.2 - 24.8| = 3.4$$
$$|\bar{x}_2 - \bar{x}_3| = |32.0 - 24.8| = 7.2$$

All three comparisons are greater than the value of HSD, which is 2.88. Thus, the mean ages between any and all pairs of plants are significantly different.

Using the Computer to Do Multiple Comparisons

Table 11.6 shows the Minitab output for computing a Tukey's HSD test. The computer output contains the confidence intervals for the differences in pairwise means for pairs of treatment levels. If the confidence interval includes zero, there is no significant difference in the pair of means. (If the interval contains zero, there is a possibility of no difference in the means.) Note in Table 11.6 that all three pairs of confidence intervals contain the same sign throughout the interval. For example, the confidence interval for estimating the difference in means from 1 and 2 is $0.914 \le \mu_1 - \mu_2 \le 6.686$. This interval does not contain zero, so we are confident that there is more than a zero difference in the two means. The same holds true for levels 1 and 3 and levels 2 and 3.

TABLE 11.6

Minitab Output for Tukey's HSD

```
Tukey 99% Simultaneous Confidence Intervals
All Pairwise Comparisons

Individual confidence level = 99.62%

Plant 1 subtracted from:

          Lower    Center    Upper    -------+---------+--------+---------+--
Plant 2   0.914    3.800     6.686                         (----*----)
Plant 3  -6.286   -3.400    -0.514            (----*----)
                                     -------+---------+--------+---------+--
                                        -6.0      0.0      6.0      12.0

Plant 2 subtracted from:

          Lower    Center    Upper    -------+---------+--------+---------+--
Plant 3 -10.086   -7.200    -4.314   (----*----)
                                     -------+---------+--------+---------+--
                                        -6.0      0.0      6.0      12.0
```

DEMONSTRATION PROBLEM 11.2

A metal-manufacturing firm wants to test the tensile strength of a given metal under varying conditions of temperature. Suppose that in the design phase, the metal is processed under five different temperature conditions and that random samples of size five are taken under each temperature condition. The data follow.

Tensile Strength of Metal Produced Under Five Different Temperature Settings

1	2	3	4	5
2.46	2.38	2.51	2.49	2.56
2.41	2.34	2.48	2.47	2.57
2.43	2.31	2.46	2.48	2.53
2.47	2.40	2.49	2.46	2.55
2.46	2.32	2.50	2.44	2.55

A one-way ANOVA is performed on these data by using Minitab, with the resulting analysis shown here.

```
One-way ANOVA: Tensile Strength versus Temp. Setting

Source           DF        SS         MS        F      P
Temp. Setting     4   0.108024   0.027006   43.70   0.000
Error            20   0.012360   0.000618
Total            24   0.120384

S = 0.02486   R-Sq = 89.73%   R-Sq(adj) = 87.68%
```

Note from the ANOVA table that the F value of 43.70 is statistically significant at $\alpha = 0.01$. There is an overall difference in the population means of metal produced under the five temperature settings. Use the data to compute a Tukey's HSD to determine which of the five groups are significantly different from the others.

Solution

From the ANOVA table, the value of MSE is 0.000618. The sample size, n_j, is 5. The number of treatment means, C, is 5 and the df_E are 20. With these values and $\alpha = 0.01$, the value of q can be obtained from Table A.10 in Appendix A.

$$q_{0.01,5,20} = 5.29$$

HSD can be computed as

$$HSD = q\sqrt{\frac{MSE}{n}} = 5.29\sqrt{\frac{0.000618}{5}} = 0.0588$$

The treatment group means for this problem follow.

Group 1 = 2.446
Group 2 = 2.350
Group 3 = 2.488
Group 4 = 2.468
Group 5 = 2.552

Computing all pairwise differences between these means (in absolute values) produces the following data.

	Group				
	1	2	3	4	5
1	–	.096	.042	.022	.106
2	.096	–	.138	.118	.202
3	.042	.138	–	.020	.064
4	.022	.118	.020	–	.084
5	.106	.202	.064	.084	–

Comparing these differences to the value of $HSD = 0.0588$, we can determine that the differences between groups 1 and 2 (0.096), 1 and 5 (0.106), 2 and 3 (0.138), 2 and 4 (0.118), 2 and 5 (0.202), 3 and 5 (0.064), and 4 and 5 (0.084) are significant at $\alpha = 0.01$.

Not only is there an overall significant difference in the treatment levels as shown by the ANOVA results, but there is a significant difference in the tensile strength of metal between seven pairs of levels. By studying the magnitudes of the individual treatment levels' means, the steel-manufacturing firm can determine which temperatures result in the greatest tensile strength. The Minitab output for this Tukey's HSD is shown below. Note that the computer analysis shows significant differences between pairs 1 and 2, 1 and 5, 2 and 3, 2 and 4, 2 and 5, 3 and 5, and 4 and 5 because these confidence intervals do not contain zero. These results are consistent with the manual calculations.

```
Tukey 99% Simultaneous Confidence Intervals
All Pairwise Comparisons among Levels of Temp. Setting

Individual confidence level = 99.87%

Temp. Setting = 1 subtracted from:
Temp.
Setting    Lower    Center    Upper      ------+---------+-------+---------+---
2       -0.15481  -0.09600  -0.03719         (--*--)
3       -0.01681   0.04200   0.10081                  (--*--)
4       -0.03681   0.02200   0.08081                (--*--)
5        0.04719   0.10600   0.16481                     (--*--)
                                           ------+---------+-------+---------+---
                                             -0.15     0.00    0.15       0.30


Temp. Setting = 2 subtracted from:
Temp.
Setting    Lower    Center    Upper      ------+---------+-------+---------+---
3        0.07919   0.13800   0.19681                       (--*--)
4        0.05919   0.11800   0.17681                     (--*--)
5        0.14319   0.20200   0.26081                          (--*--)
                                           ------+---------+-------+---------+---
                                             -0.15     0.00    0.15       0.30


Temp. Setting = 3 subtracted from:
Temp.
Setting    Lower    Center    Upper      ------+---------+-------+---------+---
4       -0.07881  -0.02000   0.03881              (--*--)
5        0.00519   0.06400   0.12281                  (--*--)
                                           ------+---------+-------+---------+---
                                             -0.15     0.00    0.15       0.30


Temp. Setting = 4 subtracted from:
Temp.
Setting    Lower    Center    Upper      ------+---------+-------+---------+---
5        0.02519   0.08400   0.14281                    (--*--)
                                           ------+---------+-------+---------+---
                                             -0.15     0.00    0.15       0.30
```

Tukey–Kramer Procedure: The Case of Unequal Sample Sizes

Tukey's HSD was modified by C. Y. Kramer in the mid-1950s to handle situations in which the sample sizes are unequal. The modified version of HSD is sometimes referred to as the **Tukey–Kramer procedure**. The formula for computing the significant differences with this procedure is similar to that for the equal sample sizes, with the exception that the mean square error is divided in half and weighted by the sum of the inverses of the sample sizes under the root sign.

TUKEY–KRAMER FORMULA

$$q_{\alpha,C,N-C}\sqrt{\frac{MSE}{2}\left(\frac{1}{n_r}+\frac{1}{n_s}\right)}$$

where:

MSE = mean square error
n_r = sample size for rth sample
n_s = sample size for sth sample
$q_{\alpha,C,N-C}$ = critical value of the Studentized range distribution from Table A.10

As an example of the application of the Tukey–Kramer procedure, consider the machine operator example in Section 11.2. A one-way ANOVA was used to test for any difference in the mean valve openings produced by four different machine operators. An overall F of 10.18 was computed, which was significant at $\alpha = 0.05$. Because the ANOVA hypothesis test is significant and the null hypothesis is rejected, this problem is a candidate for multiple comparisons. Because the sample sizes are not equal, Tukey's HSD cannot be used to determine which pairs are significantly different. However, the Tukey–Kramer procedure can be applied. Shown in Table 11.7 are the means and sample sizes for the valve openings for valves produced by the four different operators.

The mean square error for this problem, MSE, is shown in Table 11.3 as 0.007746. The four operators in the problem represent the four levels of the independent variable, machine operator. Thus, $C = 4$, $N = 24$, and $N - C = 20$. The value of alpha in the problem is 0.05. With this information, the value of q is obtained from Table A.10 as

$$q_{0.05,4,20} = 3.96$$

The distance necessary for the difference in the means of two samples to be statistically significant must be computed by using the Tukey–Kramer procedure for each pair because the sample sizes differ. In this problem with $C = 4$, there are $C(C - 1)2$ or six possible pairwise comparisons. The computations follow.

For operators 1 and 2,

$$3.96\sqrt{\frac{0.007746}{2}\left(\frac{1}{5}+\frac{1}{8}\right)}=0.1405$$

The difference between the means of operator 1 and operator 2 is

$$6.3180 - 6.2775 = 0.0405.$$

Because this result is less than the critical difference of 0.1405, there is no significant difference between the average valve openings of valves produced by machine operators 1 and 2.

Table 11.8 reports the critical differences for each of the six pairwise comparisons as computed by using the Tukey–Kramer procedure, along with the absolute value of the actual distances between the means. Any actual distance between means that is greater than the critical distance is significant. As shown in the table, the means of three pairs of samples, operators 1 and 3, operators 2 and 3, and operators 3 and 4 are significantly different.

TABLE 11.7

Means and Sample Sizes for the Valves Produced by Four Operators

Operator	Sample Size	Mean
1	5	6.3180
2	8	6.2775
3	7	6.4886
4	4	6.2300

TABLE 11.8

Results of Pairwise Comparisons for the Machine Operators Example Using the Tukey–Kramer Procedure

Pair	Critical Difference	Actual Difference
1 and 2	.1405	.0405
1 and 3	.1443	.1706*
1 and 4	.1653	.0880
2 and 3	.1275	.2111*
2 and 4	.1509	.0475
3 and 4	.1545	.2586*

*Significant at $\alpha = 0.05$.

TABLE 11.9

Minitab Multiple Comparisons in the Machine Operator Example Using the Tukey–Kramer Procedure

```
Tukey 95% Simultaneous Confidence Intervals
All Pairwise Comparisons

Individual confidence level = 98.89%

Operator 1 subtracted from:
                  Lower            Center           Upper
Operator 2      -0.18099         -0.04050         -0.09999
Operator 3       0.02627          0.17057          0.31487
Operator 4      -0.25332          0.08800          0.07732

                  ---------+---------+---------+---------+---
Operator 2                      (----*----)
Operator 3                        (----*----)
Operator 4                  (-----*-----)
                  ---------+---------+---------+---------+---
                       -0.25      0.00      0.25      0.50

Operator 2 subtracted from:
                  Lower            Center           Upper
Operator 3       0.08353          0.21107          0.33862
Operator 4      -0.19841         -0.04750          0.10341
                  ---------+---------+---------+---------+---
Operator 3                          (----*----)
Operator 4                  (----*----)
                  ---------+---------+---------+---------+---
                       -0.25      0.00      0.25      0.50

Operator 3 subtracted from:
                  Lower            Center           Upper
Operator 4      -0.41304         -0.25857         -0.10411
                  ---------+---------+---------+---------+---
Operator 4      (--------*--------)
                  ---------+---------+---------+---------+---
                       -0.25      0.00      0.25      0.50
```

Table 11.9 shows the Minitab output for this problem. Minitab uses the Tukey–Kramer procedure for unequal values of n. As before with the HSD test, Minitab produces a confidence interval for the differences in means for pairs of treatment levels. If the confidence interval includes zero, there is no significant difference in the pairs of means. If the signs over the interval are the same (zero is not in the interval), there is a significant difference in the means. Note that the signs over the intervals for pairs (1, 3), (2, 3), and (3, 4) are the same, indicating a significant difference in the means of those two pairs. This conclusion agrees with the results determined through the calculations reported in Table 11.8.

11.3 PROBLEMS

11.17 Suppose an ANOVA has been performed on a completely randomized design containing six treatment levels. The mean for group 3 is 15.85, and the sample size for group 3 is eight. The mean for group 6 is 17.21, and the sample size for group 6 is seven. MSE is 0.3352. The total number of observations is 46. Compute the significant difference for the means of these two groups by using the Tukey–Kramer procedure. Let $\alpha = 0.05$.

11.18   A completely randomized design has been analysed by using a one-way ANOVA. There are four treatment groups in the design, and each sample size is six. MSE is equal to 2.389. Using $\alpha = 0.05$, compute Tukey's HSD for this ANOVA.

11.19   Using the results of problem 11.5, compute a critical value by using the Tukey–Kramer procedure for groups 1 and 2. Use $\alpha = 0.05$. Determine whether there is a significant difference between these two groups.

11.20   Use the Tukey–Kramer procedure to determine whether there is a significant difference between the means of groups 2 and 5 in problem 11.6. Let $\alpha = 0.01$.

11.21   Using the results from problem 11.7, compute a Tukey's HSD to determine whether there are any significant differences between group means. Let $\alpha = 0.01$.

11.22   Using problem 11.8, compute Tukey's HSD and determine whether there is a significant difference in means by using this methodology. Let $\alpha = 0.05$.

11.23   Use the Tukey–Kramer procedure to do multiple comparisons for problem 11.11. Let $\alpha = 0.01$. State which pairs of machines, if any, produce significantly different mean fills.

11.24   Use Tukey's HSD test to compute multiple comparisons for the data in problem 11.12. Let $\alpha = 0.01$. State which regions, if any, are significantly different from other regions in mean starting salary figures.

11.25   Using $\alpha = 0.05$, compute critical values using the Tukey–Kramer procedure for the pairwise groups in problem 11.13. Determine which pairs of groups are significantly different, if any.

11.26   Do multiple comparisons on the data in problem 11.14 using Tukey's HSD test and $\alpha = 0.05$. State which pairs of cities, if any, have significantly different mean costs.

11.27   Problem 11.16 analysed the number of weekly hours worked per person at five different plants. An F value of 3.10 was obtained with a probability of 0.0266. Because the probability is less than 0.05, the null hypothesis is rejected at $\alpha = 0.05$. There is an overall difference in the mean weekly hours worked in each plant. Which pairs of plants have significant differences in the means, if any? To answer this question, a Minitab computer analysis was done. The data follow. Study the output in light of problem 11.16 and discuss the results.

```
Tukey 95% Simultaneous Confidence Intervals
All Pairwise Comparisons

Individual confidence level = 99.32%

Plant 1 subtracted from:

          Lower    Center    Upper   -----+--------+------+-------+--
Plant 2  -17.910   -7.722    2.466        (----*----)
Plant 3   -7.743    3.598   14.939             (----*----)
Plant 4  -21.830   -8.665    4.499     (------*------)
Plant 5  -12.721   -0.921   10.880          (-----*-----)
                                      -----+--------+------+-------+--
Plant 2 subtracted from:              -15       0      15      30

          Lower    Center    Upper   -----+--------+------+-------+--
Plant 3    0.180   11.320   22.460              (------*------)
Plant 4  -13.935   -0.944   12.048     (------*------)
Plant 5   -4.807    6.801   18.409         (-----*-----)
                                      -----+--------+------+-------+--
Plant 3 subtracted from:              -15       0      15      30

          Lower    Center    Upper   -----+--------+------+-------+--
Plant 4  -26.178  -12.263    1.651  (-------*-------)
Plant 5  -17.151   -4.519    8.113     (------*------)
                                      -----+--------+------+-------+--
Plant 4 subtracted from:              -15       0      15      30
                                      -----+--------+------+-------+--
          Lower    Center    Upper             (------*------)
Plant 5   -6.547    7.745   22.036   -----+--------+------+-------+--
                                      -15       0      15      30
```

Does National Ideology Affect a Firm's Definition of Success?

One researcher, G. C. Lodge, proposed that companies pursue different performance goals based on the ideology of their home country. L. Thurow went further by suggesting that such national ideologies drive US firms to be short-term profit maximizers, Japanese firms to be growth maximizers, and European firms to be a mix of the two.

Three other researchers, J. Katz, S. Werner, and L. Brouthers, decided to test these suggestions by studying 114 international banks from the United States, the European Union (EU), and Japan listed in the Global 1,000. Specifically, there were 34 banks from the United States, 45 banks from the European Union, and 35 banks from Japan in the study. Financial and market data were gathered and averaged on each bank over a five-year period to limit the effect of single-year variations. All statistics were converted by Morgan Stanley Capital International to US dollar denominations on the same day of each year to ensure consistency of measurement.

The banks were compared on general measures of success such as profitability, capitalization, growth, size, risk, and earnings distribution by specifically examining 11 measures. Eleven one-way analyses of variance designs were computed, one for each dependent variable. These included return on equity, return on assets, yield, capitalization, assets, market value, growth, Tobin's Q, price-to-earnings ratio, payout ratio, and risk. The independent variable in each ANOVA was country, with three levels: US, EU, and Japan.

In all 11 ANOVAs, there was a significant difference between banks in the three countries ($\alpha = 0.01$)

supporting the theme of different financial success goals for different national cultures. Because of the overall significant difference attained in the ANOVAs, each analysis of variance was followed by a Duncan's multiple range test (multiple comparison) to determine which, if any, of the pairs were significantly different. These comparisons revealed that US and EU banks maintained significantly higher levels than Japanese banks on return on equity, return on assets, and yield. This result underscores the notion that US and EU banks have more of a short-term profit orientation than do Japanese banks. There was a significant difference in banks from each of the three countries on amount of capitalization. US banks had the highest level of capitalization followed by EU banks and then Japanese banks. This result may reflect the cultural attitude about how much capital is needed to ensure a sound economy, with US banks maintaining higher levels of capital.

The study found that Japanese banks had significantly higher levels on growth, Tobin's Q, and price-to-earnings ratio than did the other two national entities. This result confirms the hypothesis that Japanese firms are more interested in growth. In addition, Japanese banks had a significantly higher asset size and market value of equity than did US banks. The researchers had hypothesized that EU banks would have a greater portfolio risk than that of US or Japanese banks. They found that EU banks did have significantly higher risk and paid out significantly higher dividends than did either Japanese or US banks.

Source: Adapted from Jeffrey P. Katz, Steve Werner, and Lance Brouthers, 'Does Winning Mean the Same Thing Around the World?' National Ideology and the Performance of Global Competitors', *Journal of Business Research*, vol. 44, no. 2 (February 1999), pp. 117–126.

11.4   THE RANDOMIZED BLOCK DESIGN

A second research design is the **randomized block design**. The randomized block design is similar to the completely randomized design in that it focuses on one independent variable (treatment variable) of interest. However, the randomized block design also includes a second variable, referred to as a blocking variable, that can be used to control for confounding or concomitant variables.

Confounding variables, or **concomitant variables**, are *variables that are not being controlled by the researcher in the experiment but can have an effect on the outcome of the treatment being studied*. For example, Demonstration

Problem 11.2 showed how a completely randomized design could be used to analyse the effects of temperature on the tensile strengths of metal. However, other variables not being controlled by the researcher in this experiment may affect the tensile strength of metal, such as humidity, raw materials, machine, and shift. One way to control for these variables is to include them in the experimental design. The randomized block design has the capability of adding one of these variables into the analysis as a blocking variable. A **blocking variable** is *a variable that the researcher wants to control but is not the treatment variable of interest.*

One of the first people to use the randomized block design was Sir Ronald A. Fisher. He applied the design to the field of agriculture, where he was interested in studying the growth patterns of varieties of seeds for a given type of plant. The seed variety was his independent variable. However, he realized that as he experimented on different plots of ground, the 'block' of ground might make some difference to the experiment. Fisher designated several different plots of ground as blocks, which he controlled as a second variable. Each of the seed varieties was planted on each of the blocks. The main thrust of his study was to compare the seed varieties (independent variable). He merely wanted to control for the difference in plots of ground (blocking variable).

In Demonstration Problem 11.2, examples of blocking variables might be machine number (if several machines are used to make the metal), worker, shift, or day of the week. The researcher probably already knows that different workers or different machines will produce at least slightly different metal tensile strengths because of individual differences. However, designating the variable (machine or worker) as the blocking variable and computing a randomized block design affords the potential for a more powerful analysis. In other experiments, some other possible variables that might be used as blocking variables include sex of subject, age of subject, intelligence of subject, economic level of subject, brand, supplier, or vehicle.

A special case of the randomized block design is the repeated measures design. The **repeated measures design** is a randomized block design in which each block level is an individual item or person, and that person or item is measured across all treatments. Thus, where a block level in a randomized block design is night shift and items produced under different treatment levels on the night shift are measured, in a repeated measures design, a block level might be an individual machine or person; items produced by that person or machine are then randomly chosen across all treatments. Thus, a repeated measure of the person or machine is made across all treatments. This repeated measures design is an extension of the t test for dependent samples presented in Section 10.3.

The sum of squares in a completely randomized design is

$$SST = SSC + SSE$$

In a randomized block design, the sum of squares is

$$SST = SSC + SSR + SSE$$

where:

    SST = sum of squares total
    SSC = sum of squares columns (treatment)
    SSR = sum of squares rows (blocking)
    SSE = sum of squares error

SST and SSC are the same for a given analysis whether a completely randomized design or a randomized block design is used. For this reason, the SSR (blocking effects) comes out of the SSE; that is, some of the error variation in the completely randomized design is accounted for in the blocking effects of the randomized block design, as shown in Figure 11.7. By reducing the error term, it is possible that the value of F for treatment will increase (the denominator of the F value is decreased). However, if there is not sufficient difference between levels of the blocking variable, the use of a randomized block design can lead to a less powerful result than would a completely randomized design computed on the same problem. Thus, the researcher should seek out blocking variables that he or she believes are significant contributors to variation among measurements of the dependent variable. Figure 11.8 shows the layout of a randomized block design.

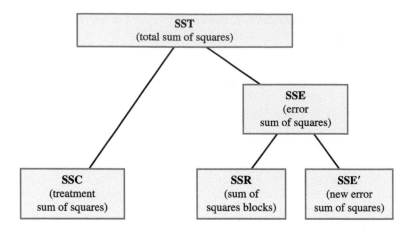

FIGURE 11.7

Partitioning the Total Sum of Squares in a Randomized Block Design

In each of the intersections of independent variable and blocking variable in Figure 11.8, one measurement is taken. In the randomized block design, one measurement is given for each treatment level under each blocking level.

The null and alternate hypotheses for the treatment effects in the randomized block design are

$$H_0: \mu_1 = \mu_2 = \mu_3 = \ldots = \mu_C$$
H_a: At least one of the treatment means is different from the others

For the blocking effects, they are

$$H_0: \mu_1 = \mu_2 = \mu_3 = \ldots = \mu_R$$
H_a: At least one of the blocking means is different from the others

Essentially, we are testing the null hypothesis that the population means of the treatment groups are equal. If the null hypothesis is rejected, at least one of the population means does not equal the others.

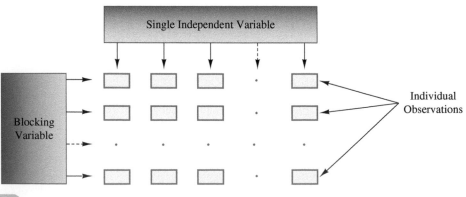

FIGURE 11.8

A Randomized Block Design

The formulas for computing a randomized block design follow.

$$SSC = n \sum_{j=1}^{C} (\bar{x}_j - \bar{x})^2$$

$$SSR = C \sum_{i-1}^{n} (\bar{x}_i - \bar{x})^2$$

$$SSE = \sum_{i=1}^{n} \sum_{j=1}^{C} (x_{ij} - \bar{x}_j - \bar{x}_i + \bar{x})^2$$

$$SST = \sum_{i=1}^{n} \sum_{j=1}^{C} (x_{ij} - \bar{x})^2$$

FORMULAS FOR COMPUTING A RANDOMIZED BLOCK DESIGN

where:

i = block group (row)
j = treatment level (column)
C = number of treatment levels (columns)
n = number of observations in each treatment level (number of blocks or rows)
x_{ij} = individual observation
$\bar{x}_j$ = treatment (column) mean
$\bar{x}_i$ = treatment (row) mean
$\bar{x}$ = grand mean
N = total number of observations

$$df_C = C - 1$$

$$df_R = n - 1$$

$$df_E = (C - 1)(n - 1) = N - n - C + 1$$

$$MSC = \frac{SSC}{C - 1}$$

$$MSR = \frac{SSR}{N - 1}$$

$$MSE = \frac{SSE}{N - n - C + 1}$$

$$F_{treatments} = \frac{MSC}{MSE}$$

$$F_{blocks} = \frac{MSR}{MSE}$$

The observed F value for treatments computed using the randomized block design formula is tested by comparing it to a table F value, which is ascertained from Table A.7 in Appendix A by using α, df_C (treatment), and df_E (error). If the observed F value is greater than the table value, the null hypothesis is rejected for that alpha value. Such a result would indicate that not all population treatment means are equal. At this point, the business researcher has the option of computing multiple comparisons if the null hypothesis has been rejected.

Some researchers also compute an F value for blocks even though the main emphasis in the experiment is on the treatments. The observed F value for blocks is compared to a critical table F value determined from Table A.7 by using α, df_R (blocks), and df_E (error). If the F value for blocks is greater than the critical F value, the null hypothesis that all block population means are equal is rejected. This result tells the business researcher that including the blocking in the design was probably worthwhile and that a significant amount of variance was drawn off from the error term, thus increasing the power of the treatment test. In this text, we have omitted F_{blocks} from the normal presentation and problem solving. We leave the use of this F value to the discretion of the reader.

As an example of the application of the randomized block design, consider a tyre company that developed a new tyre. The company conducted tread-wear tests on the tyre to determine whether there is a significant difference in tread wear if the average speed with which the vehicle is driven varies. The company set up an experiment in which the independent variable was speed of vehicle. There were three treatment levels: slow speed (car is driven 20 miles per hour), medium speed (car is driven 40 miles per hour), and high speed (car is driven 60 miles per hour). Company researchers realized that several possible variables could confound the study. One of these variables was supplier. The company uses five suppliers to provide a major component of the rubber from which the tyres are made. To control for this variable experimentally, the researchers used supplier as a blocking variable. Fifteen tyres were randomly selected for the study, three from each supplier. Each of the three was assigned to be tested under a different speed condition. The data are given here, along with treatment and block totals. These figures represent tyre wear in units of 10,000 miles.

	Speed			Block
Supplier	Slow	Medium	Fast	Means $\bar{x}_i$
1	3.7	4.5	3.1	3.77
2	3.4	3.9	2.8	3.37
3	3.5	4.1	3.0	3.53
4	3.2	3.5	2.6	3.10
5	3.9	4.8	3.4	4.03
Treatment Means $\bar{x}$	3.54	4.16	2.98	$\bar{u} = 3.56$

To analyse this randomized block design using $\alpha = 0.01$, the computations are as follows.

$$C = 3$$
$$n = 5$$
$$N = 15$$

$$SSC = n\sum_{j=1}^{C}(\bar{x}_j - \bar{x})^2$$
$$= 5[(3.54 - 3.56)^2 + (4.16 - 3.56)^2 + (2.98 - 3.56)^2]$$
$$= 3.484$$

$$SSR = C\sum_{i=1}^{n}(\bar{x}_i - \bar{x})^2$$
$$= 3[(3.77 - 3.56)^2 + (3.37 - 3.56)^2 + (3.53 - 3.56)^2 + (3.10 - 3.56)^2$$
$$+ (4.03 - 3.56)^2]$$
$$= 1.549$$

$$SSE = \sum_{i=1}^{n}\sum_{j=1}^{C}(x_{ij} - \bar{x}_j - \bar{x}_i + \bar{x})^2$$
$$= (3.7 - 3.54 - 3.77 + 3.56)^2 + (3.4 - 3.54 - 3.37 + 3.56)^2$$
$$+ \cdots + (2.6 - 2.98 - 3.10 + 3.56)^2 + (3.4 - 2.98 - 4.03 + 3.56)^2$$
$$= 0.143$$

$$SST = \sum_{i=1}^{n}\sum_{j=1}^{C}(x_{ij} - \bar{x})^2$$
$$= (3.7 - 3.56)^2 + (3.4 - 3.56)^2 + \cdots + (2.6 - 3.56)^2 + (3.4 - 3.56)^2$$
$$= 5.176$$

$$MSC = \frac{SSC}{C-1} = \frac{3.484}{2} = 1.742$$

$$MSR = \frac{SSR}{n-1} = \frac{1.549}{4} = 0.38725$$

$$MSE = \frac{SSE}{N-n-C+1} = \frac{0.143}{8} = 0.017875$$

$$F = \frac{MSC}{MSE} = \frac{1.742}{0.017875} = 97.45$$

Source of Variation	SS	df	MS	F
Treatment	3.484	2	1.742	97.45
Block	1.549	4	.38725	
Error	.143	8	.017875	
Total	5.176	14		

For alpha of 0.01, the critical F value is

$$F_{0.01,2,8} = 8.65$$

Because the observed value of F for treatment (97.45) is greater than this critical F value, the null hypothesis is rejected. At least one of the population means of the treatment levels is not the same as the others; that is, there is a significant difference in tread wear for cars driven at different speeds. If this problem had been set up as a completely randomized design, the SSR would have been a part of the SSE. The degrees of freedom for the blocking effects would have been combined with degrees of freedom of error. Thus, the value of SSE would have been $1.549 + 0.143 = 1.692$, and df_E would have been $4 + 8 = 12$. These would then have been used to recompute $MSE = 1.692/12 = 0.141$. The value of F for treatments would have been

$$F = \frac{MSC}{MSE} = \frac{1.742}{0.141} = 12.35$$

Thus, the F value for treatment with the blocking was 97.45 and *without* the blocking was 12.35. By using the random block design, a much larger observed F value was obtained.

Using the Computer to Analyse Randomized Block Designs

Both Minitab and Excel have the capability of analysing a randomized block design. The computer output from each of these software packages for the tyre tread wear example is displayed in Table 11.10. The randomized block design analysis is done on Minitab by using the same process as the two-way ANOVA, which will be discussed in Section 11.5.

The Minitab output includes F values and their associated p-values for both the treatment and the blocking effects. As with most standard ANOVA tables, the sum of squares, mean squares, and degrees of freedom for each source of variation are included.

Excel treats a randomized block design like a two-way ANOVA (Section 11.5) that has only one observation per cell. The Excel output includes sums, averages, and variances for each row and column. The Excel ANOVA table displays the observed F values for the treatment (columns) and the blocks (rows). An important inclusion in the Excel output is the p-value for each F, along with the critical (table) F values.

TABLE 11.10

Minitab and Excel Output for the Tread Wear Example

Minitab Output

```
Two-way ANOVA: Mileage versus Supplier, Speed

Source     DF       SS        MS       F        P
Supplier   4    1.54933   0.38733   21.72    0.000
Speed      2    3.48400   1.74200   97.68    0.000
Error      8    0.14267   0.01783
Total     14    5.17600

S = 0.1335  R-Sq = 97.24%  R-Sq(adj) = 95.18%
```

Excel Output

Anova: Two-Factor Without Replication

SUMMARY	Count	Sum	Average	Variance
Supplier 1	3	11.3	3.767	0.4933
Supplier 2	3	10.1	3.367	0.3033
Supplier 3	3	10.6	3.533	0.3033
Supplier 4	3	9.3	3.100	0.2100
Supplier 5	3	12.1	4.033	0.5033
Slow	5	17.7	3.54	0.073
Medium	5	20.8	4.16	0.258
Fast	5	14.9	2.98	0.092

ANOVA

Source of Variation	SS	df	MS	F	P-value	F crit
Rows	1.549333	4	0.38733333	21.72	0.0002357	7.01
Columns	3.484000	2	1.742000	97.68	0.0000024	8.65
Error	0.142667	8	0.01783333			
Total	5.176000	14				

DEMONSTRATION PROBLEM 11.3

Suppose a European business association studied the cost of a pair of men's black cotton socks in the European Union during the summer of 2012. From experience, association directors believed there was a significant difference in the average cost of a pair of men's black cotton socks among urban areas in different parts of Europe. To test this belief, they placed random calls to identical department stores in five different European cities. In addition, the researchers realized that the brand of socks might make a difference. They were mostly interested in the differences between cities, so they made city their treatment variable. To control for the fact that pricing varies with brand, the researchers included brand as a blocking variable and selected six different brands to participate. The researchers randomly telephoned one department store for each brand in each city, resulting in 30 measurements (five cities and six brands). Each till operator was asked to report the current cost of a pair of men's black cotton socks at that department store. The data are shown here. Test these data by using a randomized block design analysis to determine whether there is a significant difference in the average cost of a pair of men's black cotton socks by city. Let $\alpha = 0.01$.

			Geographic Region			
Brand	**Vienna**	**Brussels**	**Luxembourg**	**Lisbon**	**Tallinn**	$\bar{x}_i$
A	3.47	3.40	3.38	3.32	3.50	3.414
B	3.43	3.41	3.42	3.35	3.44	3.410
C	3.44	3.41	3.43	3.36	3.45	3.418
D	3.46	3.45	3.40	3.30	3.45	3.412
E	3.46	3.40	3.39	3.39	3.48	3.424
F	3.44	3.43	3.42	3.39	3.49	3.434
$\bar{x}_i$	3.450	3.4167	3.4067	3.3517	3.4683	$\bar{x} = 3.4187$

Solution

HYPOTHESIZE

STEP 1: The hypotheses follow.
For treatments,

$$H_0: \mu_1 = \mu_2 = \mu_3 = \mu_4 = \mu_5$$
$$H_a: \text{At least one of the treatment means is different from the others}$$

For blocks,

$$H_0: \mu_1 = \mu_2 = \mu_3 = \mu_4 = \mu_5 = \mu_6$$
$$H_a: \text{At least one of the blocking means is different from the others}$$

TEST

STEP 2: The appropriate statistical test is the F test in the ANOVA for randomized block designs.

STEP 3: Let $\alpha = 0.01$.

STEP 4: There are four degrees of freedom for the treatment ($C - 1 = 5 - 1 = 4$), five degrees of freedom for the blocks ($n - 1 = 6 - 1 = 5$), and 20 degrees of freedom for error [$(C - 1)(n - 1) = (4)(5) = 20$]. Using these, $\alpha = 0.01$, and Table A.7 in Appendix A, we find the critical F values.

$$F_{0.01,4,20} = 4.43 \text{ for treatments}$$
$$F_{0.01,5,20} = 4.10 \text{ for blocks}$$

The decision rule is to reject the null hypothesis for treatments if the observed F value for treatments is greater than 4.43 and to reject the null hypothesis for blocking effects if the observed F value for blocks is greater than 4.10.

STEP 5: The sample data including row and column means and the grand mean are given in the preceding table.

STEP 6:

$$SSC = n \sum_{j=1}^{c} (\bar{x}_j - \bar{x})^2$$
$$= 6[(3.450 - 3.4187)^2 + (3.4167 - 3.4187)^2 + (3.4067 - 3.4187)^2$$
$$+ (3.3517 - 3.4187)^2 + (3.4683 - 34187)^2]$$
$$= 0.04846$$

$$SSR = C \sum_{i=1}^{n} (\bar{x}_i - \bar{x})^2$$
$$= 5[(3.414 - 3.4187)^2 + (3.410 - 3.4187)^2 + (3.418 - 3.4187)^2$$
$$+ (3.412 - 3.4187)^2 + (3.424 - 3.4187)^2 + (3.434 - 3.4187)^2]$$
$$= 0.00203$$

$$SSE = \sum_{i=1}^{n} \sum_{j=1}^{c} (x_{ij} - \bar{x}_j - \bar{x}_i - \bar{x})^2$$
$$= (3.47 - 3.450 - 3.414 + 3.4187)^2 + (3.43 - 3.450 - 3.410 + 3.4187)^2 + \cdots$$
$$+ (3.48 - 3.4683 - 3.424 + 3.4187)^2 + (3.49 - 3.4683 - 3.434 + 3.4187)^2 = 0.01281$$

$$SST = \sum_{i=1}^{n} \sum_{j=1}^{c} (x_{ij} - \bar{x})^2$$
$$= (3.47 - 3.4187)^2 + (3.43 - 3.4187)^2 + \cdots + (3.48 - 3.4187)^2 + (3.49 - 3.4187)^2$$
$$= 0.06330$$

$$MSC = \frac{SSC}{C-1} = \frac{0.04846}{4} = 0.01213$$

$$MSR = \frac{SSR}{n-1} = \frac{0.00203}{5} = 0.00041$$

$$MSE = \frac{SSE}{(C-1)(n-1)} = \frac{0.01281}{20} = 0.00064$$

$$F = \frac{MSC}{MSE} = \frac{0.01213}{0.00064} = 18.95$$

Source of Variance	SS	df	MS	F
Treatment	.04846	4	.01213	18.95
Block	.00203	5	.00041	
Error	.01281	20	.00064	
Total	.06330	29		

ACTION

STEP 7:   Because $F_{treat} = 18.95 > F_{0.01,4,20} = 4.43$, the null hypothesis is rejected for the treatment effects. There is a significant difference in the average price of a pair of men's black cotton socks in various cities.

A glance at the MSR reveals that there appears to be relatively little blocking variance. The result of determining an F value for the blocking effects is

$$F = \frac{MSR}{MSE} = \frac{0.00041}{0.00064} = 0.64$$

The value of F for blocks is not significant at $\alpha = 0.01$ ($F_{0.01,5,20} = 4.10$). This result indicates that the blocking portion of the experimental design did not contribute significantly to the analysis. If the blocking effects (SSR) are added back into SSE and the df_R are included with df_E, the MSE becomes 0.00059 instead of 0.00064. Using the value 0.00059 in the denominator for the treatment F increases the observed treatment F value to 20.56. Thus, including non-significant blocking effects in the original analysis caused a loss of power.

Shown here are the Minitab and Excel ANOVA table outputs for this problem.

Minitab Output

```
Two-way ANOVA: Men's Black Cotton Socks Prices versus Brand, City

Source    DF          SS          MS        F       P
Brand      5   0.0020267   0.0004053     0.63   0.677
City       4   0.0485133   0.0121283    18.94   0.000
Error     20   0.0128067   0.0006403
Total     29   0.0633467
```

Excel Output

Anova: Two-Factor Without Replication

ANOVA

Source of Variation	SS	df	MS	F	P-value	F crit
Rows	0.002027	5	0.000405	0.63	0.6768877	4.10
Columns	0.048513	4	0.012128	18.94	0.0000014	4.43
Error	0.012807	20	0.000640			
Total	0.063347	29				

BUSINESS IMPLICATIONS

STEP 8: The fact that there is a significant difference in the price of a pair of men's black cotton socks in different parts of Europe can be useful information to decision makers. Price differences in garments and other goods can sometimes be indications of cost-of-living differences or distribution problems, which can affect a company's relocation decision or cost-of-living increases given to employees who transfer to the higher-priced locations. Knowing that the price of a pair of men's black cotton socks varies around Europe can generate interest among market researchers who might want to study why the differences are there and what drives them. This information can sometimes result in a better understanding of the marketplace.

11.4 PROBLEMS

11.28 Use ANOVA to analyse the data from the randomized block design given here. Let $\alpha = 0.05$. State the null and alternative hypotheses and determine whether the null hypothesis is rejected.

		Treatment Level			
		1	2	3	4
	1	23	26	24	24
	2	31	35	32	33
Block	3	27	29	26	27
	4	21	28	27	22
	5	18	25	27	20

11.29 The data below were gathered from a randomized block design. Use $\alpha = 0.01$ to test for a significant difference in the treatment levels. Establish the hypotheses and reach a conclusion about the null hypothesis.

		Treatment Level		
		1	2	3
	1	1.28	1.29	1.29
	2	1.40	1.36	1.35
Block	3	1.15	1.13	1.19
	4	1.22	1.18	1.24

11.30 A randomized block design has a treatment variable with six levels and a blocking variable with 10 blocks. Using this information and $\alpha = 0.05$, complete the following table and reach a conclusion about the null hypothesis:

Source of Variance	SS	df	MS	F
Treatment	2,477.53			
Blocks	3,180.48			
Error	11,661.38			
Total				

11.31 A randomized block design has a treatment variable with four levels and a blocking variable with seven blocks. Using this information and $\alpha = 0.01$, complete the following table and reach a conclusion about the null hypothesis:

Source of Variance	SS	df	MS	F
Treatment	199.48			
Blocks	265.24			
Error	306.59			
Total				

11.32 Safety in hotels is a growing concern among travellers. Suppose a survey was conducted by the Hotels, Restaurants & Cafés in Europe Association to determine European travellers' perception of safety in various hotel chains. The association chose four different European chains from the economy lodging sector and randomly selected 10 people who had stayed overnight in a hotel in each of the four chains in the past two years. Each selected traveller was asked to rate each hotel chain on a scale from 0 to 100 to indicate how safe he or she felt at that hotel. A score of 0 indicates completely unsafe and a score of 100 indicates perfectly safe. The scores follow. Test this randomized block design to determine whether there is a significant difference in the safety ratings of the four hotels. Use $\alpha = 0.05$.

Traveller	Hotel 1	Hotel 2	Hotel 3	Hotel 4
1	40	30	55	45
2	65	50	80	70
3	60	55	60	60
4	20	40	55	50
5	50	35	65	60
6	30	30	50	50
7	55	30	60	55
8	70	70	70	70
9	65	60	80	75
10	45	25	45	50

11.33 In recent years, the debate over the British economy has been constant. The electorate seems somewhat divided as to whether the economy is in a recovery or not. Suppose a survey was undertaken to ascertain whether the perception of economic recovery differs according to political affiliation. People were selected for the survey from the Labour Party, the Conservative Party, and the Liberal Democratic Party. A 25-point scale was developed in which respondents gave a score of 25 if they felt the economy was definitely in complete recovery, a 0 if the economy was definitely not in a recovery, and some value in between for more uncertain responses. To control for differences in socioeconomic class, a blocking variable was maintained using five different socioeconomic categories. The data are given here in the form of a randomized block design. Use $\alpha = 0.01$ to determine whether there is a significant difference in mean responses according to political affiliation.

Socioeconomic Class	Political Affiliation		
	Labour	Conservative	Liberal Democrat
Upper	11	5	8
Upper middle	15	9	8
Middle	19	14	15
Lower middle	16	12	10
Lower	9	8	7

11.34   As part of a manufacturing process, a plastic container is supposed to be filled with 46 centilitres of saltwater solution. The plant has three machines that fill the containers. Managers are concerned that the machines might not be filling the containers with the same amount of saltwater solution, so they set up a randomized block design to test this concern. A pool of five machine operators operates each of the three machines at different times. Company technicians randomly select five containers filled by each machine (one container for each of the five operators). The measurements are gathered and analysed. The Minitab output from this analysis follows. What was the structure of the design? How many blocks were there? How many treatment classifications? Is there a statistical difference in the treatment means? Are the blocking effects significant? Discuss the implications of the output.

```
Two-way ANOVA: Measurement versus Machine, Operator
Source      DF       SS       MS       F       P
Machine      2    78.30    39.15    6.72    .019
Operator     4     5.09     1.27    0.22    .807
Error        8    46.66     5.83
Total       14   130.06
```

11.35   The comptroller of a company is interested in determining whether the average length of long-distance calls by managers varies according to type of telephone. A randomized block design experiment is set up in which a long-distance call by each of five managers is sampled for four different types of telephone: mobile, Voice over Internet Protocol (VoIP), regular, and cordless. The treatment is type of telephone and the blocks are the managers. The results of analysis by Excel are shown here. Discuss the results and any implications they might have for the company.

Anova: Two-Factor Without Replication

Source of Variation	SS	df	MS	F	P-value	F crit
Managers	11.3346	4	2.8336	12.74	0.00028	3.26
Phone Type	10.6043	3	3.5348	15.89	0.00018	3.49
Error	2.6696	12	0.2225			
Total	24.6085	19				

11.5   A FACTORIAL DESIGN (TWO-WAY ANOVA)

Some experiments are designed so that *two or more treatments* (independent variables) *are explored simultaneously.* Such experimental designs are referred to as **factorial designs**. In factorial designs, *every level of each treatment is studied under the conditions of every level of all other treatments.* Factorial designs can be arranged such that three, four, or n treatments or independent variables are studied simultaneously in the same experiment. As an example, consider the valve opening data in Table 11.1. The mean valve opening for the 24 measurements is 6.34 centimetres. However, every valve but one in the sample measures something other than the mean. Why? Company management realizes that valves at this firm are made on different machines, by different operators, on different shifts, on different days, with raw materials from different suppliers. Business researchers who are interested in finding the sources of variation might decide to set up a factorial design that incorporates all five of these independent variables in one study. In this text, we explore the factorial designs with two treatments only.

Advantages of the Factorial Design

If two independent variables are analysed by using a completely randomized design, the effects of each variable are explored separately (one per design). Thus, it takes two completely randomized designs to analyse the effects of the two independent variables. By using a factorial design, the business researcher can analyse both variables at the same time in one design, saving the time and effort of doing two different analyses and minimizing the experiment-wise error rate.

Some business researchers use the factorial design as a way to control confounding or concomitant variables in a study. By building variables into the design, the researcher attempts to control for the effects of multiple variables *in* the experiment. With the completely randomized design, the variables are studied in isolation. With the factorial design, there is potential for increased power over the completely randomized design because the additional effects of the second variable are removed from the error sum of squares.

The researcher can explore the possibility of interaction between the two treatment variables in a two-factor factorial design if multiple measurements are taken under every combination of levels of the two treatments. Interaction will be discussed later.

Factorial designs with two treatments are similar to randomized block designs. However, whereas randomized block designs focus on one treatment variable and control for a blocking effect, a two-treatment factorial design focuses on the effects of both variables. Because the randomized block design contains only one measure for each (treatment-block) combination, interaction cannot be analysed in randomized block designs.

Factorial Designs with Two Treatments

The structure of a two-treatment factorial design is featured in Figure 11.9. Note that there are two independent variables (two treatments) and that there is an intersection of each level of each treatment. These intersections are referred to as *cells*. One treatment is arbitrarily designated as *row* treatment (forming the rows of the design) and the other treatment is designated as *column* treatment (forming the columns of the design). Although it is possible to analyse factorial designs with unequal numbers of items in the cells, the analysis of unequal cell designs is beyond the scope of this text. All factorial designs discussed here have cells of equal size.

Treatments (independent variables) of factorial designs must have at least two levels each. The simplest factorial design is a 2×2 factorial design, where each treatment has two levels. If such a factorial design were represented diagrammatically in the manner of Figure 11.9, it would include two rows and two columns, forming four cells.

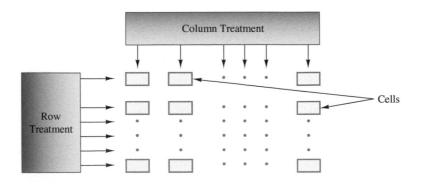

FIGURE 11.9

Two-Way Factorial Design

In this section, we study only factorial designs with $n > 1$ measurements for each combination of treatment levels (cells). This approach allows us to attempt to measure the interaction of the treatment variables. As with the completely randomized design and the randomized block design, a factorial design contains only one dependent variable.

Applications

Many applications of the factorial design are possible in business research. For example, the natural gas industry can design an experiment to study usage rates and how they are affected by temperature and precipitation. Theorizing that the outside temperature and type of precipitation make a difference in natural gas usage, industry researchers can gather usage measurements for a given community over a variety of temperature and precipitation conditions. At the same time, they can make an effort to determine whether certain types of precipitation, combined with certain temperature levels, affect usage rates differently than other combinations of temperature and precipitation (interaction effects).

Stock market analysts can select a company from an industry such as the construction industry and observe the behaviour of its stock under different conditions. A factorial design can be set up by using volume of the stock market and prime interest rate as two independent variables. For volume of the market, business researchers can select some days when the volume is up from the day before, some days when the volume is down from the day before, and some other days when the volume is essentially the same as on the preceding day. These groups of days would constitute three levels of the independent variable, market volume. Business researchers can do the same thing with prime rate. Levels can be selected such that the prime rate is (1) up, (2) down, and (3) essentially the same. For the dependent variable, the researchers would measure how much the company's stock rises or falls on those randomly selected days (stock change). Using the factorial design, the business researcher can determine whether stock changes are different under various levels of market volume, whether stock changes are different under various levels of the prime interest rate, and whether stock changes react differently under various combinations of volume and prime rate (interaction effects).

Statistically Testing the Factorial Design

Analysis of variance is used to analyse data gathered from factorial designs. For factorial designs with two factors (independent variables), a **two-way analysis of variance (two-way ANOVA)** is used to test hypotheses statistically. The following hypotheses are tested by a two-way ANOVA.

Row effects:	H_0: Row means all are equal.
	H_a: At least one row mean is different from the others.
Column effects:	H_0: Column means are all equal.
	H_a: At least one column mean is different from the others.
Interaction effects:	H_0: The interaction effects are zero.
	H_a: An interaction effect is present.

Formulas for computing a two-way ANOVA are given below. These formulas are computed in a manner similar to computations for the completely randomized design and the randomized block design. F values are determined for three effects:

1. Row effects
2. Column effects
3. Interaction effects

The row effects and the column effects are sometimes referred to as the main effects. Although F values are determined for these main effects, an F value is also computed for interaction effects. Using these observed F values, the researcher can make a decision about the null hypotheses for each effect.

Each of these observed F values is compared to a table F value. The table F value is determined by α, df_{num}, and df_{denom}. The degrees of freedom for the numerator (df_{num}) are determined by the effect being studied. If the observed F value is for columns, the degrees of freedom for the numerator are $C - 1$. If the observed F value is for rows, the degrees of freedom for the numerator are $R - 1$. If the observed F value is for interaction, the degrees of freedom for the numerator are $(R - 1)(C - 1)$. The number of degrees of freedom for the denominator of the table value for each of the three effects is the same, the error degrees of freedom, $RC(n - 1)$. The table F values (critical F) for a two-way ANOVA follow.

TABLE *F* VALUES FOR A TWO-WAY ANOVA	Row effects:	$F_{\alpha, R-1, RC(n-1)}$
	Column effects:	$F_{\alpha, C-1, RC(n-1)}$
	Interaction effects:	$F_{\alpha, (R-1)(C-1), RC(n-1)}$

FORMULAS FOR COMPUTING A TWO-WAY ANOVA

$$SSR = nC \sum_{i=1}^{R} (\bar{x}_i - \bar{x})^2$$

$$SSC = nR \sum_{j=1}^{C} (\bar{x}_j - \bar{x})^2$$

$$SSI = n \sum_{i=1}^{R} \sum_{j=5}^{C} (\bar{x}_{ij} - \bar{x}_i - \bar{x}_j - \bar{x})^2$$

$$SSE = \sum_{i=1}^{R} \sum_{j=1}^{C} \sum_{k=1}^{n} (x_{ijk} - \bar{x}_{ij})^2$$

$$SST = \sum_{i=1}^{R} \sum_{j=1}^{C} \sum_{k=1}^{n} (x_{ijk} - \bar{x})^2$$

$$df_R = R - 1$$
$$df_C = C - 1$$
$$df_I = (R - 1)(C - 1)$$
$$df_E = RC(n - 1)$$
$$df_T = N - 1$$

$$MSR = \frac{SSR}{R - 1}$$

$$MSC = \frac{SSC}{C - 1}$$

$$MSI = \frac{SSI}{(R - 1)(C - 1)}$$

$$MSE = \frac{SSE}{RC(n - 1)}$$

$$F_R = \frac{MSR}{MSE}$$

$$F_C = \frac{MSC}{MSE}$$

$$F_I = \frac{MSI}{MSE}$$

where:
n = number of observations per cell
C = number of column treatments
R = number of row treatments
i = row treatment level

j = column treatment level
k = cell member
x_{ijk} = individual observation
$\bar{x}_{ij}$ = cell mean
$\bar{x}_i$ = row mean
$\bar{x}_j$ = column mean
$\bar{\bar{x}}$ = grand mean

Interaction

As noted before, along with testing the effects of the two treatments in a factorial design, it is possible to test for the interaction effects of the two treatments whenever multiple measures are taken in each cell of the design. **Interaction** occurs *when the effects of one treatment vary according to the levels of treatment of the other effect.* For example, in a study examining the impact of temperature and humidity on a manufacturing process, it is possible that temperature and humidity will interact in such a way that the effect of temperature on the process varies with the humidity. Low temperatures might not be a significant manufacturing factor when humidity is low but might be a factor when humidity is high. Similarly, high temperatures might be a factor with low humidity but not with high humidity.

As another example, suppose a business researcher is studying the amount of red meat consumed by families per month and is examining economic class and religion as two independent variables. Class and religion might interact in such a way that with certain religions, economic class does not matter in the consumption of red meat, but with other religions, class does make a difference.

In terms of the factorial design, interaction occurs when the pattern of cell means in one row (going across columns) varies from the pattern of cell means in other rows. This variation indicates that the differences in column effects depend on which row is being examined. Hence, an interaction of the rows and columns occurs. The same thing can happen when the pattern of cell means within a column is different from the pattern of cell means in other columns.

Interaction can be depicted graphically by plotting the cell means within each row (and can also be done by plotting the cell means within each column). The means within each row (or column) are then connected by a line. If the broken lines for the rows (or columns) are parallel, no interaction is indicated.

Figure 11.10 is a graph of the means for each cell in each row in a 2 × 3 (2 rows, 3 columns) factorial design with interaction. Note that the lines connecting the means in each row cross each other. In Figure 11.11 the lines converge, indicating the likely presence of some interaction. Figure 11.12 depicts a 2 × 3 factorial design with no interaction.

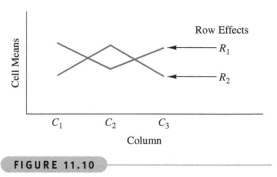

FIGURE 11.10

A 2 × 3 Factorial Design with Interaction

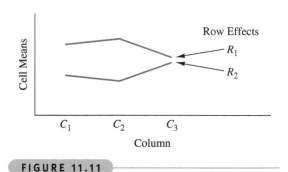

FIGURE 11.11

A 2 × 3 Factorial Design with Some Interaction

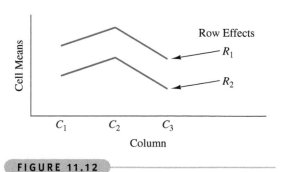

FIGURE 11.12

A 2 × 3 Factorial Design with No Interaction

When the interaction effects are significant, the main effects (row and column) are confounded and should not be analysed in the usual manner. In this case, it is not possible to state unequivocally that the row effects or the column effects are significantly different because the difference in means of one main effect varies according to the level of the other main effect (interaction is present). Some specific procedures are recommended for examining main effects when significant interaction is present. However, these techniques are beyond the scope of material presented here. Hence, in this text, whenever interaction effects are present (F_{inter} is significant), the researcher should *not* attempt to interpret the main effects (F_{row} and F_{col}).

As an example of a factorial design, consider the fact that at the end of a financially successful fiscal year, CEOs often must decide whether to award a dividend to stockholders or to make a company investment. One factor in this decision would seem to be whether attractive investment opportunities are available.* To determine whether this factor is important, business researchers randomly select 24 CEOs and ask them to rate how important 'availability of profitable investment opportunities' is in deciding whether to pay dividends or invest. The CEOs are requested to respond to this item on a scale from 0 to 4, where 0 = no importance, 1 = slight importance, 2 = moderate importance, 3 = great importance, and 4 = maximum importance. The 0–4 response is the dependent variable in the experimental design.

The business researchers are concerned that where the company's stock is traded (NYSE Euronext Stock Exchange, NASDAQ OMX Stock Exchange, and over-the-counter) might make a difference in the CEOs' response to the question. In addition, the business researchers believe that how stockholders are informed of dividends (annual reports versus presentations) might affect the outcome of the experiment. Thus, a two-way ANOVA is set up with 'where the company's stock is traded' and 'how stockholders are informed of dividends' as the two independent variables. The variable 'how stockholders are informed of dividends' has two treatment levels, or classifications.

1. Annual/quarterly reports
2. Presentations to analysts

The variable 'where company stock is traded' has three treatment levels, or classifications.

1. NYSE Euronext
2. NASDAQ OMX
3. Over-the-counter

This factorial design is a 2×3 design (2 rows, 3 columns) with four measurements (ratings) per cell, as shown in the following table.

Where Company Stock Is Traded

		NYSE Euronext	NASDAQ OMX	Over the Counter	$\bar{X}_i =$
	Annual/Quarterly Reports	2 1 2 1	2 3 3 2	4 3 4 3	2.5
		$\bar{X}_{11} = 1.5$	$\bar{X}_{12} = 2.5$	$\bar{X}_{13} = 3.5$	
How Stockholders Are Informed of Dividends	Presentations to Analysts	2 3 1 2	3 3 2 4	4 4 3 4	2.9167
		$\bar{X}_{21} = 2.0$	$\bar{X}_{22} = 3.0$	$\bar{X}_{23} = 3.75$	
	$\bar{X}_j =$	1.75	2.75	3.625	

$$\bar{X} = 2.7083$$

*Adapted from H. 'Kent Baker,' 'Why Companies Pay No Dividends,' *Akron Business and Economic Review*, vol. 20 (Summer 1989), pp. 48–61.

These data are analysed by using a two-way analysis of variance and $\alpha = 0.05$.

$$SSC = nC\sum_{i=1}^{R}(\bar{x}_i - \bar{x})^2$$
$$= 4(3)[(2.5 - 2.7083)^2 + (2.9167 - 2.7083)^2] = 1.0418$$

$$SSC = nR\sum_{j=1}^{C}(\bar{x}_j - \bar{x})^2$$
$$= 4(2)[(1.75 - 2.7083)^2 + (2.75 - 2.7083)^2 + (3.625 - 2.7083)^2] = 14.0833$$

$$SSI = n\sum_{i=1}^{R}\sum_{j=1}^{C}(\bar{x}_{ij} - \bar{x}_i - \bar{x}_j + \bar{x})^2$$
$$= 4[(1.5 - 2.5 - 1.75 + 2.7083)^2 + (2.5 - 2.5 - 2.75 + 2.7083)^2$$
$$+ (3.5 - 2.5 - 3.625 + 2.7083)^2 + (2.0 - 2.9167 - 1.75 + 2.7083)^2$$
$$+ (3.0 - 2.9167 - 2.75 + 2.7083)^2 + (3.75 - 2.9167 - 3.625 + 2.7083)^2] = 0.0833$$

$$SSE = \sum_{i=1}^{R}\sum_{j=1}^{C}\sum_{k=1}^{n}(\bar{x}_{ijk} - \bar{x})^2$$
$$= (2 - 1.5)^2 + (1 - 1.5)^2 + \cdots + (3 - 3.75)^2 + (4 - 3.75)^2 = 7.7500$$

$$SST = \sum_{i=1}^{R}\sum_{j=1}^{C}\sum_{k=1}^{n}(x_{ijk} - \bar{x})^2$$
$$= (2 - 2.7083)^2 + (1 - 2.7083)^2 + \cdots + (3 - 2.7083)^2 + (4 - 2.7083)^2 = 22.9583$$

$$MSR = \frac{SSR}{R-1} = \frac{1.0418}{1} = 1.0418$$

$$MSC = \frac{SSC}{C-1} = \frac{14.0833}{2} = 7.0417$$

$$MSI = \frac{SSI}{(R-1)(C-1)} = \frac{0.0833}{2} = 0.0417$$

$$MSE = \frac{SSE}{RC(n-1)} = \frac{7.7500}{18} = 0.4306$$

$$F_R = \frac{MSR}{MSE} = \frac{1.0418}{0.4306} = 2.42$$

$$F_C = \frac{MSC}{MSE} = \frac{7.0417}{0.4306} = 16.35$$

$$F_I = \frac{MSI}{MSE} = \frac{0.0417}{0.4306} = 0.10$$

Source of Variation	SS	df	MS	F
Row	1.0418	1	1.0418	2.42
Column	14.0833	2	7.0417	16.35*
Interaction	.0833	2	.0417	0.10
Error	7.7500	18	.4306	
Total	22.9583	23		

*Denotes significance at $\alpha = 0.05$.

The critical F value for the interaction effects at $\alpha = 0.05$ is

$$F_{0.05,2,18} = 3.55$$

The observed F value for interaction effects is 0.10. Because this value is less than the critical table value (3.55), no significant interaction effects are evident. Because no significant interaction effects are present, it is possible to examine the main effects.

The critical F value of the row effects at $\alpha = 0.05$ is $F_{0.05,1,18} = 4.41$. The observed F value of 2.42 is less than the table value. Hence, no significant row effects are present.

The critical F value of the column effects at $\alpha = 0.05$ is $F_{0.05,2,18} = 3.55$. This value is coincidently the same as the critical table value for interaction because in this problem the degrees of freedom are the same for interaction and column effects. The observed F value for columns (16.35) is greater than this critical value. Hence, a significant difference in row effects is evident at $\alpha = 0.05$.

A significant difference is noted in the CEOs' mean ratings of the item 'availability of profitable investment opportunities' according to where the company's stock is traded. A cursory examination of the means for the three levels of the column effects (where stock is traded) reveals that the lowest mean was from CEOs whose company traded stock on the NYSE Euronext. The highest mean rating was from CEOs whose company traded stock over-the-counter. Using multiple comparison techniques, the business researchers can statistically test for differences in the means of these three groups.

Because the sample sizes within each column are equal, Tukey's HSD test can be used to compute multiple comparisons. The value of MSE is 0.431 for this problem. In testing the column means with Tukey's HSD test, the value of n is the number of items in a column, which is eight. The number of treatments is $C = 3$ for columns and $N - C = 24 - 3 = 21$.

With these two values and $\alpha = 0.05$, a value for q can be determined from Table A.10 in Appendix A:

$$q_{0.05,3,21} = 3.58$$

From these values, the honestly significant difference can be computed:

$$HSD = q\sqrt{\frac{MSE}{n}} = 3.58\sqrt{\frac{0.431}{8}} = 0.831$$

The mean ratings for the three columns are

$$\bar{x}_1 = 1.75, \bar{x}_2 = 2.75, \bar{x}_3 = 3.625$$

The absolute value of differences between means are as follows:

$$|\bar{x}_1 - \bar{x}_2| = |1.75 - 2.75| = 1.00$$
$$|\bar{x}_1 - \bar{x}_3| = |1.75 - 3.625| = 1.875$$
$$|\bar{x}_2 - \bar{x}_3| = |2.75 - 3.625| = 0.875$$

All three differences are greater than 0.831 and are therefore significantly different at $\alpha = 0.05$ by the HSD test. Where a company's stock is traded makes a difference in the way a CEO responds to the question.

Using a Computer to Do a Two-Way ANOVA

A two-way ANOVA can be computed by using either Minitab or Excel. Figure 11.13 displays the Minitab and Excel output for the CEO example. The Minitab output contains an ANOVA table with each of the three F values and their associated p-values. In addition, there are individual 95% confidence intervals for means of both row and column effects. These intervals give the researcher a visual idea of differences between means. A more formal test of multiple comparisons of the column means is done with Minitab by using Tukey's HSD test. This output is displayed in Figure 11.14. Observe that in all three comparisons the signs on each end of the particular confidence interval are the same (and thus zero is not included); hence there is a significant difference in the means in each of the three pairs.

The Excel output for two-way ANOVA with replications on the CEO dividend example is included in Figure 11.13. The Excel output contains cell, column, and row means along with observed F values for rows (sample), columns, and interaction. The Excel output also contains p-values and critical F values for each of these F's. Note that the output here is virtually identical to the findings obtained by the manual calculations.

Minitab Output

```
Two-way ANOVA: Rating versus How Reported, Where Traded

Source              DF        SS         MS         F        P
How Reported         1     1.0417    1.04167      2.42    0.137
Where Traded         2    14.0833    7.04167     16.35    0.000
Interaction          2     0.0833    0.04167      0.10    0.908
Error               18     7.7500    0.43056
Total               23    22.9583
S = 0.6562    R-Sq = 66.24%    R-Sq(adj) = 56.87%
```

```
                           Individual 95% CIs For Mean Based on Pooled StDev
How Reported Mean    + ——— + ——— + ——— + ———
1          2.50000  (————————*————————)
2          2.91667            (————————*—————————)
                     + ——— + ——— + ——— + ———
                     2.10    2.45    2.80    3.15
```

```
                           Individual 95% CIs For Mean Based on Pooled StDev
Where Traded Mean    — + ——— + ——— + ——— + ———
1           1.750   (———*———)
2           2.750          (———*———)
3           3.625                      (———*———)
                     — + ——— + ——— + ——— + ———
                      1.60    2.40    3.20    4.00
```

Excel Output

ANOVA: Two-Factor With Replication

SUMMARY	NYSE Euronext	NASDAQ OMX	OTC	Total
A.Q. Reports				
Count	4	4	4	12
Sum	6	10	14	30
Average	1.5	2.5	3.5	2.5
Variance	0.33333	0.33333	0.33333	1
Pres. to Analysts				
Count	4	4	4	12
Sum	8	12	15	35
Average	2	3	3.75	2.91667
Variance	0.66667	0.66667	0.25	0.99242
Total				
Count	8	8	8	
Sum	14	22	29	
Average	1.75	2.75	3.625	
Variance	0.5	0.5	0.26786	

ANOVA

Source of variation	SS	df	MS	F	P-value	F crit
Sample	1.04167	1	1.04167	2.42	0.137251	4.41
Columns	14.08333	2	7.04167	16.35	0.000089	3.55
Interaction	0.08333	2	0.04167	0.10	0.90823	3.55
Within	7.75	18	0.43056			
Total	22.95833	23				

FIGURE 11.13

Minitab and Excel Output for the CEO Dividend Problem

```
Tukey 95% Simultaneous Confidence Intervals
All Pairwise Comparisons among levels of where Traded

Individual confidence level = 98.00%
Where Traded = 1 subtracted from:

Where
traded      Lower    Center    Upper     ----+------+-------+-------+----
2          0.1818    1.0000   1.8182                   (----*-----)
3          1.0568    1.8750   2.6932                      (-----*-----)
                                          ----+------+-------+-------+----
                                            -1.2     0.0     1.2     2.4


Where Traded = 2 subtracted from:

Where
traded      Lower    Center    Upper     ----+------+-------+-------+----
3          0.0568    0.8750   1.6932                   (-----*-----)
                                          ----+------+-------+-------+----
                                            -1.2     0.0     1.2     2.4
```

FIGURE 11.14

Tukey's Pairwise Comparisons for Column Means

DEMONSTRATION PROBLEM 11.4

Some theorists believe that training warehouse workers can reduce absenteeism. Suppose an experimental design is structured to test this belief. Warehouses in which training sessions have been held for workers are selected for the study. The four types of warehouses are (1) general merchandise, (2) commodity, (3) bulk storage, and (4) cold storage. The training sessions are differentiated by length. Researchers identify three levels of training sessions according to the length of sessions: (1) 1–20 days, (2) 21–50 days, and (3) more than 50 days. Three warehouse workers are selected randomly for each particular combination of type of warehouse and session length. The workers are monitored for the next year to determine how many days they are absent. The resulting data are in the following 4 × 3 design (4 rows, 3 columns) structure. Using this information, calculate a two-way ANOVA to determine whether there are any significant differences in effects. Use $\alpha = 0.05$.

Solution

HYPOTHESIZE

STEP 1:   The following hypotheses are being tested:
For row effects:

$$H_0: \mu_1 = \mu_2 = \mu_3 = \mu_4$$
$$H_a: \text{At least one of the row means is different from the others}$$

For column effects:

$$H_0: \mu_1 = \mu_2 = \mu_3$$
$$H_a: \text{At least one of the column means is different from the others}$$

For interaction effects:

$$H_0: \text{The interaction effects are zero.}$$
$$H_a: \text{There is an interaction effect.}$$

TEST

STEP 2:   The two-way ANOVA with the F test is the appropriate statistical test.
STEP 3: $\alpha = 0.05$.

STEP 4:

$$df_{rows} = 4 - 1 = 3$$
$$df_{columns} = 3 - 1 = 2$$
$$df_{interaction} = (3)(2) = 6$$
$$df_{error} = (4)(3)(2) = 24$$

For row effects, $F_{0.05,3,24} = 3.01$; for column effects, $F_{0.05,2,24} = 3.40$; and for interaction effects, $F_{0.05,6,24} = 2.51$. For each of these effects, if any observed F value is greater than its associated critical F value, the respective null hypothesis will be rejected.

STEP 5:

		Length of Training Session (Days)			
		1–20	21–50	More than 50	X_r
	General Merchandise	3 4.5 4	2 2.5 2	2.5 1 1.5	2.5556
Types of Warehouses	Commodity	5 4.5 4	1 3 2.5	0 1.5 2	2.6111
	Bulk Storage	2.5 3 3.5	1 3 1.5	3.5 3.5 4	2.8333
	Cold Storage	2 2 3	5 4.5 2.5	4 4.5 5	3.6111
	X_c	3.4167	2.5417	2.75	

$$X = 2.9028$$

STEP 6: The Minitab and Excel (ANOVA table only) output for this problem follows.

Minitab Output

```
Two-way ANOVA: Absences versus Type of Ware, Length
Source          DF        SS          MS        F        P
Type of ware     3     6.4097     2.13657     3.46     0.032
Length           2     5.0139     2.50694     4.06     0.030
Interaction      6    33.1528     5.52546     8.94     0.000
Error           24    14.8333     0.61806
Total           35    59.4097
```

Excel Output

ANOVA

Source of Variation	SS	df	MS	F	P-value	F crit
Types of Warehouses	6.40972	3	2.136574	3.46	0.032205	3.01
Length of Training Session	5.01389	2	2.506944	4.06	0.030372	3.40
Interaction	33.15278	6	5.525463	8.94	0.000035	2.51
Within	14.83333	24	0.618056			
Total	59.40972	35				

ACTION

STEP 7: Looking at the source of variation table, we must first examine the interaction effects. The observed F value for interaction is 8.94 for both Excel and Minitab. The observed F value for interaction is greater than the critical F value. The interaction effects are statistically significant at $\alpha = 0.05$. The p-value for interaction shown in Excel is 0.000035. The interaction effects are significant at $\alpha = 0.0001$. The business researcher should not bother to examine the main effects because the significant interaction confounds the main effects.

BUSINESS IMPLICATIONS

STEP 8: The significant interaction effects indicate that certain warehouse types in combination with certain lengths of training session result in different absenteeism rates than do other combinations of levels for these two variables. Using the cell means shown here, we can depict the interactions graphically.

Minitab produces the following graph of the interaction.

		Length of Training Session (Days)		
		1–20	21–50	More than 50
Types of Warehouses	General Merchandise	3.8	2.2	1.7
	Commodity	4.5	2.2	1.2
	Bulk Storage	3.0	1.8	3.7
	Cold Storage	2.3	4.0	4.5

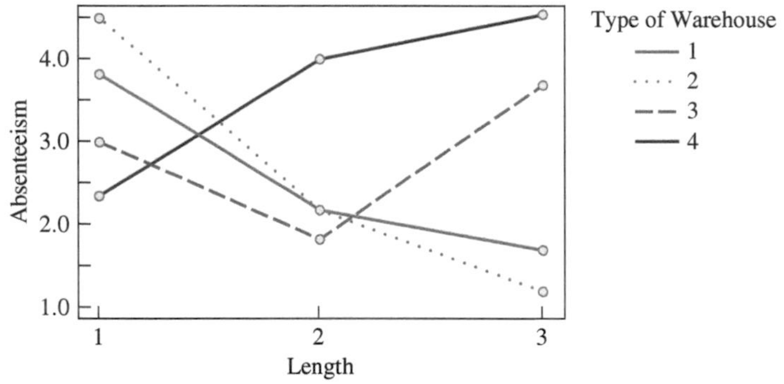

Note the intersecting and crossing lines, which indicate interaction. Under the short-length training sessions, 1, cold-storage workers had the lowest rate of absenteeism and workers at commodity warehouses had the highest. However, for medium-length sessions, 2, cold-storage workers had the highest rate of absenteeism and bulk-storage had the lowest. For the longest training sessions, 3, commodity warehouse workers had the lowest rate of absenteeism, even though these workers had the highest rate of absenteeism for short-length sessions. Thus, the rate of absenteeism for workers at a particular type of warehouse depended on length of session. There was an interaction between type of warehouse and length of session. This graph could be constructed with the row levels along the bottom axis instead of column levels.

Source: Adapted from Paul R. Murphy and Richard E. Poist, 'Managing the Human Side of Public Warehousing: An Overview of Modern Practices', *Transportation Journal*, vol. 31 (Spring 1992) pp. 54–63.

11.5   PROBLEMS

11.36 Describe the factorial design below. How many independent and dependent variables are there? How many levels are there for each treatment? If the data were known, could interaction be determined from this design? Compute all degrees of freedom. Each data value is represented by an x.

	Variable 1			
	x_{111}	x_{121}	x_{131}	x_{141}
	x_{112}	x_{122}	x_{132}	x_{142}
	x_{113}	x_{123}	x_{133}	x_{143}
Variable 2				
	x_{211}	x_{221}	x_{231}	x_{241}
	x_{212}	x_{222}	x_{232}	x_{242}
	x_{213}	x_{223}	x_{233}	x_{243}

11.37 Describe the factorial design below. How many independent and dependent variables are there? How many levels are there for each treatment? If the data were known, could interaction be determined from this design? Compute all degrees of freedom. Each data value is represented by an x.

	Variable 1		
	x_{111}	x_{121}	x_{131}
	x_{112}	x_{122}	x_{132}
	x_{211}	x_{221}	x_{231}
	x_{212}	x_{222}	x_{232}
Variable 2			
	x_{311}	x_{321}	x_{331}
	x_{312}	x_{322}	x_{332}
	x_{411}	x_{421}	x_{431}
	x_{412}	x_{422}	x_{432}

11.38 Complete the two-way ANOVA table below. Determine the critical table F values and reach conclusions about the hypotheses for effects. Let $\alpha = 0.05$.

Source of Variance	SS	df	MS	F
Row	126.98	3		
Column	37.49	4		
Interaction	380.82			
Error	733.65	60		
Total				

11.39 Complete the following two-way ANOVA table. Determine the critical table F values and reach conclusions about the hypotheses for effects. Let $\alpha = 0.05$.

Source of Variance	SS	df	MS	F
Row	1.047	1		
Column	3.844	3		
Interaction	0.773			
Error				
Total	12.632	23		

11.40 The data gathered from a two-way factorial design follow. Use the two-way ANOVA to analyse these data. Let $\alpha = 0.01$.

		Treatment 1		
		A	B	C
Treatment 2	A	23	21	20
		25	21	22
	B	27	24	26
		28	27	27

11.41 Suppose the data below have been gathered from a study with a two-way factorial design. Use $\alpha = 0.05$ and a two-way ANOVA to analyse the data. State your conclusions.

		Treatment 2			
		A	B	C	D
Treatment 1	A	1.2 1.3 1.3 1.5	2.2 2.1 2.0 2.3	1.7 1.8 1.7 1.6	2.4 2.3 2.5 2.4
	B	1.9 1.6 1.7 2.0	2.7 2.5 2.8 2.8	1.9 2.2 1.9 2.0	2.8 2.6 2.4 2.8

11.42 Children are generally believed to have considerable influence over their parents in the purchase of certain items, particularly food and beverage items. To study this notion further, a study is conducted in which parents are asked to report how many food and beverage items purchased by the family per week are purchased mainly because of the influence of their children. Because the age of the child may have an effect on the study, parents are asked to focus on one particular child in the family for the week, and to report the age of the child. Four age categories are selected for the children: 4–5 years, 6–7 years, 8–9 years, and 10–12 years. Also, because the number of children in the family might make a difference, three different sizes of families are chosen for the study: families with one child, families with two children, and families with three or more children. Suppose the following data represent the reported number of child-influenced buying incidents per week. Use the data to compute a two-way ANOVA. Let $\alpha = 0.05$.

		Number of Children in Family		
		1	2	3 or more
Age of Child (years)	4–5	2	1	1
		4	2	1
	6–7	5	3	2
		4	1	1
	8–9	8	4	2
		6	5	3
	10–12	7	3	4
		8	5	3

11.43 A shoe retailer conducted a study to determine whether there is a difference in the number of pairs of shoes sold per day by stores according to the number of competitors within a 1-mile radius and the location of the store. The company researchers selected three types of stores for consideration in the study: stand-alone sub-urban stores, mall stores, and downtown stores. These stores vary in the numbers of competing stores within a 1-mile radius, which have been reduced to four categories: 0 competitors, 1 competitor, 2 competitors, and 3 or more competitors. Suppose the data below represent the number of pairs of shoes sold per day for each of these types of stores with the given number of competitors. Use $\alpha = 0.05$ and a two-way ANOVA to analyse the data.

| | | Number of Competitors | | | |
		0	1	2	3 or more
	Stand-Alone	41	38	59	47
		30	31	48	40
		45	39	51	39
Store	Mall	25	29	44	43
Location		31	35	48	42
		22	30	50	53
	Downtown	18	22	29	24
		29	17	28	27
		33	25	26	32

11.44 Study the analysis of variance table below that was produced by using Minitab. Describe the design (number of treatments, sample sizes, etc.). Are there any significant effects? Discuss the output.

```
Two-way ANOVA:DV versus RowEffect, ColEffect
Source         DF        SS        MS        F        P
RowEffect       2     92.31    46.156    13.23    0.000
ColEffect       4    998.80   249.700    71.57    0.000
Interaction     8    442.13    55.267    15.84    0.000
Error          30    104.67     3.489
Total          44   1637.91
```

11.45 Consider the valve opening data displayed in Table 11.1. Suppose the data represent valves produced on four different machines on three different shifts and that the quality controllers want to know whether there is any difference in the mean measurements of valve openings by shift or by machine. The data are given here, organized by machine and shift. In addition, Excel has been used to analyse the data with a two-way ANOVA. What are the hypotheses for this problem? Study the output in terms of significant differences. Discuss the results obtained. What conclusions might the quality controllers reach from this analysis?

| | | Valve Openings (cm) | | |
		Shift 1	Shift 2	Shift 3
	1	6.56	6.38	6.29
		6.40	6.19	6.23
	2	6.54	6.26	6.19
Machine		6.34	6.23	6.33
	3	6.58	6.22	6.26
		6.44	6.27	6.31
	4	6.36	6.29	6.21
		6.50	6.19	6.58

ANOVA: Two-Factor with Replication

ANOVA

Source of Variation	SS	df	MS	F	P-value	F crit
Sample	0.00538	3	0.00179	0.14	0.9368	3.49
Columns	0.19731	2	0.09865	7.47	0.0078	3.89
Interaction	0.03036	6	0.00506	0.38	0.8760	3.00
Within	0.15845	12	0.01320			
Total	0.39150	23				

11.46 Finish the computations in the Minitab ANOVA table shown below and determine the critical table F values. Interpret the analysis. Examine the associated Minitab graph and interpret the results. Discuss this problem, including the structure of the design, the sample sizes, and decisions about the hypotheses.

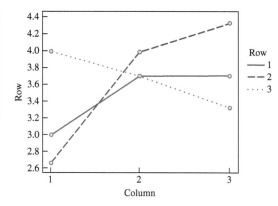

```
Two-way ANOVA: depvar versus row, column
Source          DF        SS         MS
Row              2     0.296      0.148
Column           2     1.852      0.926
Interaction      4     4.370      1.093
Error           18    14.000      0.778
Total           26    20.519
```

Decision Dilemma SOLVED

Job and Career Satisfaction of Foreign and Self-Initiated Expatriates

Is there a difference in the job satisfaction ratings of self-initiated expatriates by industry? The data presented in the Decision Dilemma to study this question represent responses on a seven-point Likert scale by 24 self-initiated expatriates from five different industries. The Likert scale score is the dependent variable. There is only one independent variable, industry, with five classification levels: IT, finance, education, healthcare, and consulting. If a series of t tests for the difference of two means from independent populations were used to analyse these data, there would be $_5C_2$ or 10 different t tests on this one problem.

Using α for each test, the probability of at least one of the 10 tests being significant by chance when the null hypothesis is true is $1 - (0.95)^{10} = 0.4013$. That is, performing 10 t tests on this problem could result in an overall probability of committing a Type I error equal to 0.4013, not 0.05. In order to control the overall error, a one-way ANOVA is used on this completely randomized design to analyse these data by producing a single value of F and holding the probability of committing a Type I error at 0.05. Both Excel and Minitab have the capability of analysing these data, and Minitab output for this problem is shown below.

```
One-way ANOVA: IT, Finance, Education, Healthcare, Consulting
Source    DF        SS        MS        F        P
Factor     4     43.175    10.794    15.25    0.000
Error     19     13.450     0.708
Total     23     56.625
S = 0.8414    R-Sq = 76.25%    R-Sq(adj) = 71.25%
                                   Individual 95% CIs For Mean Based on
                                   Pooled StDev
Level         N      Mean     StDev  -----+-------+------+-------+-
IT            4    5.7500    0.9574                    (----*----)
Finance       5    4.0000    0.7071             (----*----)
Education     6    2.5000    0.5477  (----*----)
Healthcare    5    3.4000    1.1402        (----*----)
Consulting    4    6.0000    0.8165                       (-----*-----)
                                     -----+-------+------+-------+-
                                        3.0      4.5    6.0     7.5
```

With an *F* value of 15.25 and a *p*-value of 0.000, the results of the one-way ANOVA show that there is an overall significant difference in job satisfaction between the five industries. Examining the Minitab confidence intervals shown graphically suggests that there might be a significant difference between some pairs of industries. Because there was an overall significant difference in the industries, it is appropriate to use Tukey's HSD test to determine which of the pairs of industries are significantly different. Tukey's test controls for the overall error so that the problem mentioned previously arising from computing ten *t* tests is avoided. The Minitab output for Tukey's test is shown below.

```
Tukey 95% Simultaneous Confidence Intervals
All Pairwise Comparisons
Individual confidence level = 99.27%
IT subtracted from:
             Lower     Center     Upper   -----+-------+------+-------+-
Finance    -3.4462    -1.7500   -0.0538       (----*----)
Education  -4.8821    -3.2500   -1.6179   (----*----)
Healthcare -4.0462    -2.3500   -0.6538     (------*-----)
Consulting -1.5379     0.2500    2.0379           (------*-----)
                                          -----+-------+------+-------+-
                                            -3.0     0.0    3.0     6.0

Finance subtracted from:
             Lower     Center     Upper   -----+-------+------+-------+-
Education  -3.0311    -1.5000    0.0311       (----*----)
Healthcare -2.1991    -0.6000    0.9991       (----*----)
Consulting  0.3038     2.0000    3.6962            (----*----)
                                          -----+-------+------+-------+-
                                            -3.0     0.0    3.0     6.0

Education subtracted from:
             Lower     Center     Upper   -----+-------+------+-------+-
Healthcare -0.6311     0.9000    2.4311            (----*----)
Consulting  1.8679     3.5000    5.1321                 (----*----)
                                          -----+-------+------+-------+-
                                            -3.0     0.0    3.0     6.0

Healthcare subtracted from:
             Lower     Center     Upper   -----+-------+------+-------+-
Consulting  0.9038     2.6000    4.2962                 (----*----)
                                          -----+-------+------+-------+-
                                            -3.0     0.0    3.0     6.0
```

(continued)

Any confidence interval in which the sign of the value for the lower end of the interval is the same as the sign of the value for the upper end indicates that zero is not in the interval and that there is a significant difference between the pair in that case. Examining the Minitab output reveals that IT and Finance, IT and Education, IT and Healthcare, Finance and Consulting, Education and Consulting, and Healthcare and Consulting all are significantly different pairs of industries.

In analysing career satisfaction, self-initiated expatriates were sampled from three age categories and four categories of time in the host country. This experimental design is a two-way factorial design with age and time in the host country as independent variables and individual scores on the seven-point Likert scale being the dependent variable. There are three classification levels under the independent variable Age: 30–39, 40–49, and over 50, and there are four classifications under the independent variable Time in Host Country: <1 year, 1 to 2 years, 3 to 4 years, and 5 or more years. Because there is more than one score per cell, interaction can be analysed. A two-way ANOVA with replication is run in Excel to analyse the data and the result is shown below.

Excel Output:

ANOVA

Source of Variation	SS	df	MS	F	P-value	F crit
Age	3.5556	2	1.7778	2.91	0.073920	3.40
Time in Host Country	20.9722	3	6.9907	11.44	0.000075	3.01
Interaction	1.1111	6	0.1852	0.30	0.929183	2.51
Within	14.6667	24	0.6111			
Total	40.3056	35				

An examination of this output reveals no significant interaction effects (p-value of 0.929183). Since there are no significant interaction effects, it is appropriate to examine the main effects. Because of a p-value of 0.000075, there is a significant difference in Time in Host Country using an alpha of 0.0001. Multiple comparison analysis could be done to determine which, if any, pairs of Time in Host Country are significantly different. The p-value for Age is 0.073920 indicating that there is a significant difference between Age classifications at alpha 0.10 but not at alpha of 0.05.

ETHICAL CONSIDERATIONS

In theory, any phenomenon that affects the dependent variable in an experiment should be either entered into the experimental design or controlled in the experiment. Researchers will sometimes report statistical findings from an experiment and fail to mention the possible concomitant variables that were neither controlled by the experimental setting nor controlled by the experimental design. The findings from such studies are highly questionable and often lead to spurious conclusions. Scientifically, the researcher needs to conduct the experiment in an environment such that as many concomitant variables are controlled as possible. To the extent that they are not controlled, the researcher has an ethical responsibility to report that fact in the findings.

Other ethical considerations enter into conducting research with experimental designs. Selection of treatment levels should be done with fairness or even randomness in cases where the treatment has several possibilities for levels. A researcher can build in skewed views of the treatment effects by erroneously selecting treatment levels to be studied. Some researchers believe that reporting significant main effects from a factorial design when there are confounding interaction effects is unethical or at least misleading.

Another ethical consideration is the levelling of sample sizes. Some designs, such as the two-way factorial design or completely randomized design with Tukey's HSD, require equal sample sizes. Sometimes unequal sample sizes arise either through the selection process or through attrition. A number of techniques for approaching this problem are not presented in this book. It remains highly unethical to make up data values or to eliminate values arbitrarily to produce equal sample sizes.

SUMMARY

Sound business research requires that the researcher plan and establish a design for the experiment before a study is undertaken. The design of the experiment should encompass the treatment variables to be studied, manipulated, and controlled. These variables are often referred to as the independent variables. It is possible to study several independent variables and several levels, or classifications, of each of those variables in one design. In addition, the researcher selects one measurement to be taken from sample items under the conditions of the experiment. This measurement is referred to as the dependent variable because if the treatment effect is significant, the measurement of the dependent variable will 'depend' on the independent variable(s) selected. This chapter explored three types of experimental designs: completely randomized design, randomized block design, and the factorial experimental designs.

The completely randomized design is the simplest of the experimental designs presented in this chapter. It has only one independent, or treatment, variable. With the completely randomized design, subjects are assigned randomly to treatments. If the treatment variable has only two levels, the design becomes identical to the one used to test the difference in means of independent populations presented in Chapter 10. The data from a completely randomized design are analysed by a one-way analysis of variance (ANOVA). A one-way ANOVA produces an F value that can be compared to table F values in Table A.7 in Appendix A to determine whether the ANOVA F value is statistically significant. If it is, the null hypothesis that all population means are equal is rejected and at least one of the means is different from the others. Analysis of variance does not tell the researcher which means, if any, are significantly different from others. Although the researcher can visually examine means to determine which ones are greater and lesser, statistical techniques called multiple comparisons must be used to determine statistically whether pairs of means are significantly different.

Two types of multiple comparison techniques are presented and used in this chapter: Tukey's HSD test and the Tukey–Kramer procedure. Tukey's HSD test requires that equal sample sizes be used. It utilizes the mean square of error from the ANOVA, the sample size, and a q value that is obtained from Table A.10 to solve for the least difference between a pair of means that would be significant (HSD). The absolute value of the difference in sample means is compared to the HSD value to determine statistical significance. The Tukey–Kramer procedure is used in the case of unequal sample sizes.

A second experimental design is the randomized block design. This design contains a treatment variable (independent variable) and a blocking variable. The independent variable is the main variable of interest in this design. The blocking variable is a variable the researcher is interested in controlling rather than studying. A special case of randomized block design is the repeated measures design, in which the blocking variable represents subjects or items for which repeated measures are taken across the full range of treatment levels.

In randomized block designs, the variation of the blocking variable is removed from the error variance. This approach can potentially make the test of treatment effects more powerful. If the blocking variable contains no significant differences, the blocking can make the treatment effects test less powerful. Usually an F is computed only for the treatment effects in a randomized block design. Sometimes an F value is computed for blocking effects to determine whether the blocking was useful in the experiment.

A third experimental design is the factorial design. A factorial design enables the researcher to test the effects of two or more independent variables simultaneously. In complete factorial designs, every treatment level of each independent variable is studied under the conditions of every other treatment level for all independent variables. This chapter focused only on factorial designs with two independent variables. Each independent variable can have two or more treatment levels. These two-way factorial designs are analysed by two-way analysis of variance (ANOVA). This analysis produces an F value for each of the two treatment effects and for interaction. Interaction is present when the results of one treatment vary significantly according to the levels of the other treatment. At least two measurements per cell must be present in order to compute interaction. If the F value for interaction is statistically significant, the main effects of the experiment are confounded and should not be examined in the usual manner.

KEY TERMS

a posteriori
a priori
analysis of variance
  (ANOVA)
blocking variable
classification variable
classifications
completely randomized
  design
concomitant variables
confounding variables
dependent variable
experimental design
F distribution
F value

factorial design
factors
independent variable
interaction
levels
multiple comparisons
one-way analysis of variance
post hoc
randomized block design
repeated measures design
treatment variable
Tukey–Kramer procedure
Tukey's HSD test
two-way analysis of variance

GO ONLINE TO DISCOVER THE EXTRA FEATURES FOR THIS CHAPTER

The Student Study Guide containing solutions to the odd-numbered questions, additional Quizzes and Concept Review Activities, Excel and Minitab databases, additional data files in Excel and Minitab, and more worked examples.
www.wiley.com/college/cortinhas

FORMULAS

Formulas for computing a one-way ANOVA

$$SSC = \sum_{j=1}^{C} n_j(\bar{x}_j - \bar{x})^2$$

$$SSE = \sum_{i=1}^{n_i} \sum_{j=1}^{C} (x_{ij} - \bar{x}_j)^2$$

$$SST = \sum_{i=1}^{n_i} \sum_{j=1}^{C} (x_{ij} - \bar{x})^2$$

$$df_C = C - 1$$
$$df_E = N - C$$
$$df_T = N - 1$$

$$MSC = \frac{SSC}{df_C}$$

$$MSE = \frac{SSE}{df_E}$$

$$F = \frac{MSC}{MSE}$$

Tukey's HSD test

$$HSD = q_{\alpha,C,N-C} \sqrt{\frac{MSE}{n}}$$

Tukey–Kramer formula

$$q_{\alpha,C,N-C} \sqrt{\frac{MSE}{2} \left(\frac{1}{n_r} + \frac{1}{n_s} \right)}$$

Formulas for computing a randomized block design

$$SSC = n \sum_{j=1}^{C} n_j(\bar{x}_j - \bar{x})^2$$

$$SSR = C \sum_{i=1}^{n} (\bar{x}_i - \bar{x})^2$$

$$SSE = \sum_{i=1}^{n} \sum_{j=1}^{C} (x_{ij} - \bar{x}_j - \bar{x}_i + \bar{x})^2$$

$$df_C = C - 1$$
$$df_R = n - 1$$
$$df_E = (C - 1)(n - 1) = N - n - C + 1$$

$$MSC = \frac{SSC}{C - 1}$$

$$MSR = \frac{SSR}{n - 1}$$

$$MSE = \frac{SSE}{N - n - C + 1}$$

$$F_{treatments} = \frac{MSC}{MSE}$$

$$F_{blocks} = \frac{MSR}{MSE}$$

Formulas for computing a two-way ANOVA

$$SSR = nC \sum_{i=1}^{R} (\bar{x}_i - \bar{x})^2$$

$$SSC = nR \sum_{j=1}^{C} (\bar{x}_j - \bar{x})^2$$

$$SSI = n \sum_{i=1}^{R} \sum_{j=1}^{C} (\bar{x}_{ij} - \bar{x}_i - \bar{x}_j + \bar{x})^2$$

$$SSE = \sum_{i=1}^{R} \sum_{j=1}^{C} \sum_{k=1}^{n} (x_{ijk} - \bar{x}_{ij})^2$$

$$SST = \sum_{i=1}^{R} \sum_{j=1}^{C} \sum_{k=1}^{n} (x_{ijk} - \bar{x})^2$$

$$df_R = R - 1$$
$$df_C = C - 1$$
$$df_I = (R - 1)(C - 1)$$
$$df_E = RC(n - 1)$$
$$df_T = N - 1$$

$$MSR = \frac{SSR}{R - 1}$$

$$MSC = \frac{SSC}{C - 1}$$

$$MSI = \frac{SSI}{(R - 1)(C - 1)}$$

$$MSE = \frac{SSE}{RC(n - 1)}$$

$$F_R = \frac{MSR}{MSE}$$

$$F_C = \frac{MSC}{MSE}$$

$$F_I = \frac{MSI}{MSE}$$

SUPPLEMENTARY PROBLEMS

CALCULATING THE STATISTICS

11.47 Compute a one-way ANOVA on the data below. Use $\alpha = 0.05$. If there is a significant difference in treatment levels, use Tukey's HSD to compute multiple comparisons. Let $\alpha = 0.05$ for the multiple comparisons.

	Treatment		
1	*2*	*3*	*4*
10	9	12	10
12	7	13	10
15	9	14	13
11	6	14	12

11.48 Complete the following ANOVA table:

Source of Variance	SS	df	MS	F
Treatment				
Error	249.61	19		
Total	317.80	25		

11.49 You are asked to analyse a completely randomized design that has six treatment levels and a total of 42 measurements. Complete the following table, which contains some information from the study:

Source of Variance	SS	df	MS	F
Treatment	210			
Error	655			
Total				

11.50 Compute a one-way ANOVA of the data below. Let $\alpha = 0.01$. Use the Tukey–Kramer procedure to conduct multiple comparisons for the means.

Treatment		
1	*2*	*3*
7	11	8
12	17	6
9	16	10
11	13	9
8	10	11
9	15	7
11	14	10
10	18	
7		
8		

11.51 Examine the structure of the experimental design below. Determine which of the three designs presented in the chapter would be most likely to characterize this structure. Discuss the variables and the levels of variables. Determine the degrees of freedom.

Methodology			
Person	*Method 1*	*Method 2*	*Method 3*
1	x_{11}	x_{12}	x_{13}
2	x_{21}	x_{22}	x_{23}
3	x_{31}	x_{32}	x_{33}
4	x_{41}	x_{42}	x_{43}
5	x_{51}	x_{52}	x_{53}
6	x_{61}	x_{62}	x_{63}

11.52 Complete the ANOVA table below and determine whether there is any significance in treatment effects. Let $\alpha = 0.05$.

Source of Variance	SS	df	MS	F
Treatment	20,994	3		
Blocking		9		
Error	33,891			
Total	71,338			

11.53 Analyse the data below, gathered from a randomized block design using $\alpha = 0.05$. If there is a significant difference in the treatment effects, use Tukey's HSD test to do multiple comparisons.

		Treatment			
		A	*B*	*C*	*D*
	1	17	10	9	21
	2	13	9	8	16
Blocking	3	20	17	18	22
Variable	4	11	6	5	10
	5	16	13	14	22
	6	23	19	20	28

11.54 A two-way ANOVA has been computed on a factorial design. Treatment 1 has five levels and treatment 2 has two levels. Each cell contains four measures. Complete the ANOVA table below. Use $\alpha = 0.05$ to test to determine significance of the effects. Comment on your findings.

Source of Variance	SS	df	MS	F
Treatment 1	29.13			
Treatment 2	12.67			
Interaction	73.49			
Error	110.38			
Total				

11.55 Compute a two-way ANOVA on the data below ($\alpha = 0.01$).

		Treatment 1		
		A	*B*	*C*
		5	2	2
	A	3	4	3
		6	4	5
		11	9	13
	B	8	10	12
Treatment 2		12	8	10
		6	7	4
	C	4	6	6
		5	7	8
		9	8	8
	D	11	12	9
		9	9	11

TESTING YOUR UNDERSTANDING

11.56 A company conducted a consumer research project to ascertain customer service ratings from its customers. The customers were asked to rate the company on a scale from 1 to 7 on various quality characteristics. One question was the promptness of company response to a repair problem. The data below represent customer responses to this question. The customers were divided by geographic region and by age. Use analysis of variance to analyse the responses. Let $\alpha = 0.05$. Compute multiple comparisons where they are appropriate. Graph the cell means and observe any interaction.

		Geographic Region			
		South	West	East	North
		3	2	3	2
	21–35	2	4	3	3
		3	3	2	2
		5	4	5	6
Age	36–50	5	4	6	4
		4	6	5	5
		3	2	3	3
	Over 50	1	2	2	2
		2	3	3	1

11.57 A major auto manufacturer wants to know whether there is any difference in the average mileage of four different brands of tyres (A, B, C, and D) because the manufacturer is trying to select the best supplier in terms of tyre durability. The manufacturer selects comparable levels of tyres from each company and tests some on comparable cars. The mileage results follow.

A	B	C	D
31,000	24,000	30,500	24,500
25,000	25,500	28,000	27,000
28,500	27,000	32,500	26,000
29,000	26,500	28,000	21,000
32,000	25,000	31,000	25,500
27,500	28,000		26,000
	27,500		

Use $\alpha = 0.05$ to test whether there is a significant difference in the mean mileage of these four brands. Assume tyre mileage is normally distributed.

11.58 Agricultural researchers are studying three different ways of planting organic potatoes to determine whether significantly different levels of production yield will result. The researchers have access to a large organic potato farm on which to conduct their tests. They identify six blocks of land. In each block of land, organic potatoes are planted in each of the

three different ways. At the end of the growing season, the organic potatoes are harvested and the average number of thousand of kilograms per acre is determined for organic potatoes planted under each method in each block. Using the following data and $\alpha = 0.01$ test to determine whether there is a significant difference in yields among the planting methods.

Block	Method 1	Method 2	Method 3
1	1,310	1,080	850
2	1,275	1,100	1,020
3	1,280	1,050	780
4	1,225	1,020	870
5	1,190	990	805
6	1,300	1,030	910

11.59 A number of construction sector jobs that seem to pay approximately the same wages per hour. Some of these are bricklaying, iron working, and crane operation. Suppose a labour researcher takes a random sample of workers from each of these types of construction jobs and from across the country and asks what are their hourly wages. If this survey yields the data below, is there a significant difference in mean hourly wages for these three jobs? If there is a significant difference, use the Tukey–Kramer procedure to determine which pairs, if any, are also significantly different. Let $\alpha = 0.05$.

	Job Type	
Bricklaying	Iron Working	Crane Operation
19.25	26.45	16.20
17.80	21.10	23.30
20.50	16.40	22.90
24.33	22.86	19.50
19.81	25.55	27.00
22.29	18.50	22.95
21.20		25.52
		21.20

11.60 Why are mergers attractive to CEOs? One of the reasons might be a potential increase in market share that can come with the pooling of company markets. Suppose a random survey of CEOs is taken, and they are asked to respond on a scale from 1 to 5 (5 representing strongly agree) whether increase in market share is a good reason for considering a merger of their company with another. Suppose also that the data are as given here and that CEOs have been categorized by size of company and years they have been with their company. Use a two-way ANOVA to determine whether there are any significant differences in the responses to this question. Let $\alpha = 0.05$.

	Company Size (€ million per year in sales)			
	0-5	6-20	21-100	>100
	2	2	3	3
	3	1	4	4
0-2	2	2	4	4
	2	3	5	3
Years with the	2	2	3	3
	1	3	2	3
Company 3-5	2	2	4	3
	3	3	4	4
	2	2	3	2
	1	3	2	3
Over 5	1	1	3	2
	2	2	3	3

11.61 Are some unskilled office jobs viewed as having more status than others? Suppose a study is conducted in which eight unskilled, unemployed people are interviewed. The people are asked to rate each of five positions on a scale from 1 to 10 to indicate the status of the position, with 10 denoting most status and 1 denoting least status. The resulting data are given below. Use $\alpha = 0.05$ to analyse the repeated measures randomized block design data.

	Job				
	Mail Clerk	Typist	Receptionist	Secretary	Telephone Operator
1	4	5	3	7	6
2	2	4	4	5	4
3	3	3	2	6	7
Respondent 4	4	4	4	5	4
5	3	5	1	3	5
6	3	4	2	7	7
7	2	2	2	4	4
8	3	4	3	6	6

INTERPRETING THE OUTPUT

11.62 Analyse the Minitab output below. Describe the design of the experiment. Using $\alpha = 0.05$ determine whether there are any significant effects; if so, explain why. Discuss any other ramifications of the output.

```
One-way ANOVA: Dependent Variable versus
Factor Analysis of Variance
Source   DF      SS       MS      F      p
Factor    3    876.6    292.2   3.01   0.045
Error    32   3107.5     97.1
Total    35   3984.1

                   Individual 95% CIs for
                   Mean Based on Pooled
                   StDev
Level  N   Mean    StDev  -+----+----+----+--
C1     8  307.73   5.98   (----*----)
C2     7  313.20   9.71          (----*----)
C3    11  308.60   9.78     (----*----)
C4    10  319.74  12.18               (---*---)
                          -+----+----+----+--
Pooled StDev = 9.85   301.0 308.0 315.0 322.0
```

11.63 Following is Excel output for an ANOVA problem. Describe the experimental design. The given value of alpha was 0.05. Discuss the output in terms of significant findings.

ANOVA: Two-Factor Without Replication

ANOVA

Source of Variation	SS	df	MS	F	P-value	F crit
Rows	48.278	5	9.656	3.16	0.057	3.33
Columns	10.111	2	5.056	1.65	0.23	4.10
Error	30.556	10	3.056			
Total	88.944	17				

11.64 Study the following Minitab output and graph. Discuss the meaning of the output.

```
Two-Way ANOVA: Dependent Variable
Versus
Row Effects, Column Effects
Source        DF    SS     MS     F      p
Row Eff        4   4.70   1.17  0.98  0.461
Col. Eff       1   3.20   3.20  2.67  0.134
Interaction    4  22.30   5.57  4.65  0.022

Error         10  12.00   1.20
Total         19  42.20
```

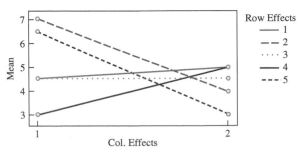

11.65 Interpret the following Excel output. Discuss the structure of the experimental design and any significant effects. Alpha is 0.05.

ANOVA: Two-Factor With Replication

ANOVA

Source of Variation	SS	df	MS	F	P-value	F crit
Sample	2913.889	3	971.296	4.30	0.0146	3.01
Columns	240.389	2	120.194	0.53	0.5940	3.40
Interaction	1342.944	6	223.824	0.99	0.4533	2.51
Within	5419.333	24	225.806			
Total	9916.556	35				

11.66 Study the Minitab output below. Determine whether there are any significant effects and discuss the results. What kind of design was used and what was the size of it?

```
Two-Way Analysis of Variance

Source        df      SS       MS

Blocking       4     41.44    10.36
Treatment      4    143.93    35.98
Error         16    117.82     7.36
Total         24    303.19
```

11.67 Discuss the following Minitab output.

```
One-Way Analysis of Variance
Source         df        SS       MS        F      P
Treatment    3138.0     46.0     3.51    0.034
Error         20      262.2     13.1
Total         23      400.3
Individual 95% CIs For Mean Based on
Pooled StDev
Level N Mean   StDev   ----+----+----+----+
1     6 53.778 5.470   (----*----)
2     6 54.665 1.840       (----*----)
3     6 59.911 3.845             (----*----)
4     6 57.293 2.088         (----*----)
                       ----+----+----+----+
Pooled StDev = 3.621   52.5  56.0 59.5 63.0

Tukey's pairwise comparisons
Family error rate = 0.0500
Individual error rate = 0.0111
Critical value = 3.96
Intervals for (column level mean) - (row
level mean)
              1         2         3
2    -6.741
      4.967
3   -11.987   -11.100
     -0.279     0.608
4    -9.369    -8.482    -3.236
      2.339     3.225     8.472
```

ANALYSING THE DATABASES

1. The International Labour database contains data on the unemployment rates in per cent from 10 countries presented yearly over a 38-year period. Compute a one-way ANOVA to determine whether there is any significant difference in the unemployment rates of the six European countries in the database. Did you find any significant differences by country? If so, which pairs are significantly different?

2. Do various financial indicators differ significantly according to type of company? Use a one-way ANOVA and the financial database to answer this question. Let Type of Company be the independent variable with seven levels (Apparel, Chemical, Electric Power, Grocery, Healthcare Products, Insurance, and Petroleum). Compute three one-way ANOVAs, one for each of the following dependent variables: Earnings Per Share, Dividends Per Share, and Average P/E Ratio. On each ANOVA, if there is a significant overall difference between Type of Company, compute multiple comparisons to determine which pairs of types of companies, if any, are significantly different.

3. In the Manufacturing database, the Value of Industrial Shipments has been recoded into four classifications (1–4) according to magnitude of value. Let this value be the independent variable with four levels of classifications. Compute a one-way ANOVA to determine whether there is any significant difference

in classification of the Value of Industrial Shipments on the Number of Production Workers (dependent variable). Perform the same analysis using End-of-Year Inventories as the dependent variable. Now change the independent variable to Industry Group, of which there are 20, and perform first a one-way ANOVA using Number of Production Workers as the dependent variable and then a one-way ANOVA using End-of-Year Inventory as the dependent variable.

4. The Hospital database contains data on hospitals from seven different geographic regions. Let this variable be the independent variable. Determine whether there is a significant difference in Admissions for these geographic regions using a one-way ANOVA. Perform the same analysis using Births as the dependent variable. Control is a variable with four levels of classification denoting the type of control the hospital is under (such as federal government or for-profit). Use this variable as the independent variable and test to determine whether there is a significant difference in the Admissions of a hospital by Control. Perform the same test using Births as the dependent variable.

CASE

THE TOMOE VALVE COMPANY

Since its formation in Japan over 50 years ago, TOMOE has become a global leader in the production of high-quality and technically advanced ranges of butterfly valves, actuators, and control systems. TOMOE products can be found operating in a wide range of industries worldwide including oil and gas, petrochemical, marine, water treatment, vehicle manufacture, and power generation, forming a trusted and reliable part of the process in each and every application. The European headquarters of TOMOE was established in Newport, South Wales, in 1986 to develop the business in the UK, Europe, Africa, the Middle East, and Canada. In 1997, TOMOE Tritec Ltd was formed to design and manufacture high-performance triple offset rotary process valves.

In April 2006, TOMOE Tritec Ltd was merged with TOMOE Valve Ltd. The newly combined company is called TOMOE Valve Limited, located at the European Headquarters in Newport, and handles operations in the UK, Europe, Russia, Africa, USA and Canada, and South America. From this site, TOMOE implements a full programme of design, manufacture, and valve assembly, as well as maintaining comprehensive stocks for all TOMOE process valves.

The company maintains large stocks of butterfly valves ready for fast delivery and a support team of engineers ready to provide technical help and practical advice. The UK production facility also incorporates a Valve Actuation Centre offering a 24-hour turnaround of stock item valves and actuators that can be customized to exact specifications.

The success of TOMOE's strategy for product development and global expansion has since led to the opening of new manufacturing operations in China (2001), Indonesia (2004), and Japan (2006).

DISCUSSION

1. TOMOE claims that its range of butterfly valves offers unparalleled levels of leak tightness, valve longevity, and wear resistance. Suppose TOMOE engineers have developed a new butterfly valve that is even more resistant than its existing product line. They want to set up an experimental design to test the strength of the valve (in pounds per square inch or psi) but they want to conduct the tests under three different temperature conditions: 70°, 110°, and 150°. In addition, suppose TOMOE uses two different suppliers (company A and company B) of the synthetic materials that are used to manufacture the valves. Some valves are made primarily of raw materials supplied by company A and some are made primarily of raw materials from company B. Thus, the engineers have set up a 23 factorial design with temperature and supplier as the independent variables and pressure (measured in psi) as the dependent

variable. Data are gathered and are shown here. Analyse the data and discuss the business implications of the findings. If you were conducting the study, what would you report to the engineers?

	Temperature		
	70°	*110°*	*150°*
Supplier A	163	157	146
	159	162	137
	161	155	140
Supplier B	158	159	150
	154	157	142
	164	160	155

2. Pipeline operators estimate that it costs between US$75 and US$500 to replace each seal, thus making the longer-lasting TOMOE valves more attractive. TOMOE does business with pipeline companies around the world. Suppose in an attempt to develop marketing materials, TOMOE marketers are interested in determining whether there is a significant difference in the cost of replacing pipeline seals in different countries. Four countries – Canada, Colombia, Taiwan, and the United States – are chosen for the study. Pipeline operators from equivalent operations are selected from companies in each country. The operators keep a cost log of seal replacements. A random sample of the data follows. Use these data to help TOMOE determine whether there is a difference in the cost of seal replacements in the various countries. Explain your answer and say how TOMOE might use the information in its marketing materials.

Canada	Colombia	Taiwan	United States
215	355	170	230
205	280	190	190
245	300	235	225
270	330	195	220
290	360	205	215
260	340	180	245
225	300	190	230

3. In the late 1980s, TOMOE developed a 24-hour turnaround of stock item valves system. Using this and other quality improvement approaches, the company was able to greatly reduce lead-times for all of its valves. Suppose that TOMOE wants to test to determine whether lead-times differ significantly according to the type of valve it is manufacturing. As a control of the experiment, they are including in the study, as a blocking variable, the day of the week the valve was ordered. One lead-time was selected per valve per day of the week. The data are given here in weeks. Analyse the data and discuss your findings.

	Type of Valve					
	Control	*Tritec*	*300 Series*	*700 Series*	*800 Series*	*HLV*
Monday	1.6	2.2	1.3	1.8	2.5	0.8
Tuesday	1.8	2.0	1.4	1.5	2.4	1.0
Wednesday	1.0	1.8	1.0	1.6	2.0	0.8
Thursday	1.8	2.2	1.4	1.6	1.8	0.6
Friday	2.0	2.4	1.5	1.8	2.2	1.2

Source: Adapted from TOMOE Europe website at www.tomoeeurope .co.uk/index.html, July 2011, and 'Valve Automation Centre promises 24 hour turnaround', News and Case Studies, available at www .tomoeeurope.co.uk/TomoeNewsCasestudies_new.html, 21.10.2009.

USING THE COMPUTER

EXCEL

- Excel has the capability of performing a completely randomized design (one-way ANOVA), a randomized block design, and a two-way factorial design (two-way ANOVA).

- Each of the tests presented here in Excel is accessed through the **Data Analysis** feature.

- To conduct a one-way ANOVA, begin by selecting the **Data** tab on the Excel worksheet. From the **Analysis** panel at the right top of the **Data** tab worksheet, click on **Data Analysis**. If your Excel worksheet does not show the **Data Analysis** option, then you can load it as an add-in following directions given in Chapter 2. From the **Data Analysis** pulldown menu, select **Anova: Single Factor**. Click and drag over the data and enter in **Input Range**. Check **Labels in the First Row** if you included labels in the data. Insert the value of alpha in **Alpha**.

- To conduct a randomized block design, load the treatment observations into columns. Data may be loaded either with or without labels. Select the **Data** tab on the Excel worksheet. From the **Analysis** panel at the right top of the **Data** tab worksheet, click on **Data Analysis**. If your Excel worksheet does not show the **Data Analysis** option, then you can load it as an add-in following directions given in Chapter 2. From the **Data Analysis** pulldown menu, select **Anova: Two-Factor Without Replication**. Click and drag over the data under **Input Range**. Check **Labels in the First Row** if you have included labels in the data. Insert the value of alpha in **Alpha**.

- To conduct a two-way ANOVA, load the treatment observations into columns. Excel is quite particular about how the data are entered for a two-way ANOVA. Data must be loaded in rows and columns as with most two-way designs. However, two-way ANOVA in Excel requires labels for both rows and columns; and if labels are not supplied, Excel will incorrectly use some of the data for labels. Since cells will have multiple values, there need only be a label for each new row (cell). Select the **Data** tab on the Excel worksheet. From the **Analysis** panel at the right top of the **Data** tab worksheet, click on **Data Analysis**. If your Excel worksheet does not show the **Data Analysis** option, then you can load it as an add-in following directions given in Chapter 2. From the **Data Analysis** pulldown menu, select **Anova: Two-Factor With Replication**. Click and drag over the data under **Input Range**. Enter the number of values per cell in **Rows per sample**. Insert the value of alpha in **Alpha**.

MINITAB

- Minitab also has the capability to perform a completely randomized design (one-way ANOVA), a randomized block design, and a two-way factorial design along with multiple comparisons.

- There are two ways to compute a one-way ANOVA in Minitab, stacking all observations in one column with group identifiers in another column, or entering the observations unstacked in separate columns.

- To begin a one-way ANOVA with **Unstacked Data**, select **Stat** from the menu bar. Select **ANOVA** from the pulldown menu. Select **One-Way** (**Unstacked**). In the slot, **Responses (in separate columns)**, list

the columns containing the data. For multiple comparisons, select **Comparisons** and make your selection from the dialog box that appears. The multiple comparison options are Tukey's, Fisher's, Dunnett's, or Hsu's MCB tests. In the multiple comparison dialog box, you can insert the family error rate in the box on the right as a whole number.

- To begin a one-way ANOVA with **Stacked Data**, select **Stat** from the menu bar. Select **ANOVA** from the pulldown menu. Select **One-Way**. In the slot **Response**, list the column containing the observations. In the slot **Factor**, list the column containing the group identifiers. For multiple comparisons, select **Comparisons** and make your selection from the dialog box that appears. The multiple comparison options are Tukey's, Fisher's, Dunnett's, or Hsu's MCB tests. In the multiple comparison dialog box, you can insert the family error rate in the box on the right as a whole number.

- There are two ways to compute a randomized block design or a two-way ANOVA in Minitab, using the **Two-Way** procedure or using the **Balanced ANOVA** procedure. Both the randomized block design and the two-way ANOVA are analysed in the same way and are presented together here.

- To begin using the **Two-Way** procedure, select **Stat** from the menu bar. Select **ANOVA** from the pulldown menu. Select **Two-Way**. The observations should be 'stacked', that is, listed in one column. Place the location of these observations in the **Response** box. The group identifiers for the row effects should be listed in another column. Place the location of the row identifiers in **Row factor**. The group identifiers for the column effects should be listed in another column. Place the location of the column identifiers in **Column factor**.

- To begin using the **Balanced ANOVA** procedure, select **Stat** from the menu bar. Select **ANOVA** from the pulldown menu. Select **Balanced ANOVA** from the pulldown menu. Place the location of the observations in the **Responses** box. Place the columns containing the group identifiers to be analysed in the **Model** box. To compute a two-way ANOVA with interaction, place the location of the column containing the row effects identifiers and the location of the column containing column effects identifiers in the **Model** box. To test for interaction also place a third term in the **Model** box that is the product of the row and column effects. For example, if the row effects are X and the column effects are Y, place a X*Y in the **Model** box along with X and Y.

Regression Analysis and Forecasting

IN THE FIRST THREE UNITS OF THE TEXT, you were introduced to basic statistics, distributions, and how to make inferences through confidence interval estimation and hypothesis testing. In Unit IV, we explore relationships between variables through regression analysis and learn how to develop models that can be used to predict one variable by another variable or even multiple variables. We will examine a cadre of statistical techniques that can be used to forecast values from time-series data and how to measure how well the forecast is.

CHAPTER 12

Simple Regression Analysis and Correlation

LEARNING OBJECTIVES

The overall objective of this chapter is to give you an understanding of bivariate linear regression analysis, thereby enabling you to:

1. Calculate the Pearson product-moment correlation coefficient to determine if there is a correlation between two variables

2. Explain what regression analysis is and the concepts of independent and dependent variable

3. Calculate the slope and y-intercept of the least squares equation of a regression line and from those, determine the equation of the regression line

4. Calculate the residuals of a regression line and from those determine the fit of the model, locate outliers, and test the assumptions of the regression model

5. Calculate the standard error of the estimate using the sum of squares of error, and use the standard error of the estimate to determine the fit of the model

6. Calculate the coefficient of determination to measure the fit for regression models, and relate it to the coefficient of correlation

7. Use the t and F tests to test hypotheses for both the slope of the regression model and the overall regression model

8. Calculate confidence intervals to estimate the conditional mean of the dependent variable and prediction intervals to estimate a single value of the dependent variable

9. Determine the equation of the trend line to forecast outcomes for time periods in the future, using alternate coding for time periods if necessary

10. Use a computer to develop a regression analysis, and interpret the output that is associated with it

Decision Dilemma

Predicting International Hourly Wages by the Price of a Big Mac

The McDonald's Corporation is the leading global food-service retailer with more than 31,000 local restaurants serving nearly 60 million people in more than 119 countries each day. This global presence, in addition to its consistency in food offerings and restaurant operations, makes McDonald's a unique and attractive setting for economists to make salary and price comparisons around the world. Because the Big Mac hamburger is a standardized hamburger produced and sold in virtually every McDonald's around the world, the *Economist*, a weekly newspaper focusing on international politics and business news and opinion, as early as 1986 was compiling information about Big Mac prices as an indicator of exchange rates. Building on this idea, researchers Ashenfelter and Jurajda proposed comparing wage rates across countries and the price of a Big Mac hamburger. Shown below are Big Mac prices and net hourly wage figures (in US dollars) for 27 countries. Note that net hourly wages are based on a weighted average of 12 professions.

Country	Big Mac Price (US$)	Wage (US$)
Argentina	1.42	1.70
Australia	1.86	7.80
Brazil	1.48	2.05
Britain	3.14	12.30
Canada	2.21	9.35
Chile	1.96	2.80
China	1.20	2.40
Czech Republic	1.96	2.40
Denmark	4.09	14.40
Euro area	2.98	9.59
Hungary	2.19	3.00
Indonesia	1.84	1.50
Japan	2.18	13.60
Malaysia	1.33	3.10
Mexico	2.18	2.00
New Zealand	2.22	6.80

(*continued*)

Country	Big Mac Price (US$)	Wage (US$)
Philippines	2.24	1.20
Poland	1.62	2.20
Russia	1.32	2.60
Singapore	1.85	5.40
South Africa	1.85	3.90
South Korea	2.70	5.90
Sweden	3.60	10.90
Switzerland	4.60	17.80
Thailand	1.38	1.70
Turkey	2.34	3.20
United States	2.71	14.30

Managerial and Statistical Questions

1. Is there a relationship between the price of a Big Mac and the net hourly wages of workers around the world? If so, how strong is the relationship?
2. Is it possible to develop a model to predict or determine the net hourly wage of a worker around the world by the price of a Big Mac hamburger in that country? If so, how good is the model?
3. If a model can be constructed to determine the net hourly wage of a worker around the world by the price of a Big Mac hamburger, what would be the predicted net hourly wage of a worker in a country if the price of a Big Mac hamburger was $3.00?

Sources: McDonald's website, at www.mcdonalds.com/corp/about. html; Michael R. Pakko and Patricia S. Pollard, 'Burgernomics: A Big Mac Guide to Purchasing Power Parity', research publication by the St Louis Federal Reserve Bank, at http://research.stlouisfed.org/publications/review/03/11/pakko.pdf; Orley Ashenfelter and Stepán Jurajda, 'Cross-Country Comparisons of Wage Rates: The Big Mac Index', unpublished manuscript, Princeton University and CERGEEI/Charles University, October 2001; *The Economist*, at www.economist.com/index.html.

In business, the key to decision making often lies in the understanding of the relationships between two or more variables. For example, a company in the distribution business may determine that there is a relationship between the price of crude oil and their own transportation costs. Financial experts, in studying the behaviour of the bond market, might find it useful to know if the interest rates on bonds are related to the prime interest rate set by the Central Bank. A marketing executive might want to know how strong the relationship is between advertising euros and sales euros for a product or a company.

In this chapter, we will study the concept of correlation and how it can be used to estimate the relationship between two variables. We will also explore simple regression analysis through which mathematical models can be developed to predict one variable by another. We will examine tools for testing the strength and predictability of regression models, and we will learn how to use regression analysis to develop a forecasting trend line.

12.1 CORRELATION

Correlation is *a measure of the degree of relatedness of variables.* It can help a business researcher determine, for example, whether the stocks of two airlines rise and fall in any related manner. For a sample of pairs of data, correlation analysis can yield a numerical value that represents the degree of relatedness of the two stock prices over time. In the transportation industry, is a correlation evident between the price of transportation and the weight of the object being shipped? If so, how strong are the correlations? In economics, how strong is the correlation between the producer price index and the unemployment rate? In retail sales, are sales related to population density, number of competitors, size of the store, amount of advertising, or other variables?

Several measures of correlation are available, the selection of which depends mostly on the level of data being analysed. Ideally, researchers would like to solve for ρ, the population coefficient of correlation. However, because researchers virtually always deal with sample data, this section introduces a widely used sample **coefficient of correlation**, r. This measure is applicable only if both variables being analysed have at least an interval level of data. Chapter 17 presents a correlation measure that can be used when the data are ordinal.

The statistic r is known as the **Pearson product-moment correlation coefficient**, named after Karl Pearson (1857–1936), an English statistician who developed several coefficients of correlation along with other significant statistical concepts. The term r is a *measure of the linear correlation of two variables.* It is a number that ranges from -1 to 0 to $+1$, representing the strength of the relationship between the variables. An r value of $+1$ denotes a perfect positive relationship between two sets of numbers. An r value of -1 denotes a perfect negative correlation, which indicates an inverse relationship between two variables: as one variable gets larger, the other gets smaller. An r value of 0 means no linear relationship is present between the two variables.

PEARSON PRODUCT-MOMENT CORRELATION COEFFICIENT (12.1)	$$r = \frac{\Sigma(x-\bar{x})(y-\bar{y})}{\sqrt{\Sigma(x-\bar{x})^2\,\Sigma(y-\bar{y})^2}} = \frac{\Sigma xy - \dfrac{(\Sigma x \Sigma y)}{n}}{\sqrt{\left[\Sigma x^2 - \dfrac{(\Sigma x)^2}{n}\right]\left[\Sigma y^2 - \dfrac{(\Sigma y)^2}{n}\right]}}$$

Figure 12.1 depicts five different degrees of correlation: (a) represents strong negative correlation, (b) represents moderate negative correlation, (c) represents moderate positive correlation, (d) represents strong positive correlation, and (e) contains no correlation.

What is the measure of correlation between the Euro OverNight Index Average (Eonia) interest rate and the commodities futures index? With data such as those shown in Table 12.1, which represent the values for Eonia interest rates and commodities futures indexes for a sample of 12 days, a correlation coefficient, r, can be computed.

(a) Strong Negative Correlation ($r = -.933$)

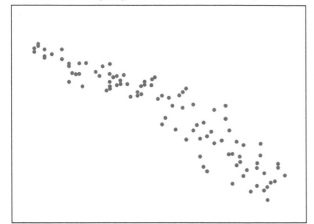

(b) Moderate Negative Correlation ($r = -.674$)

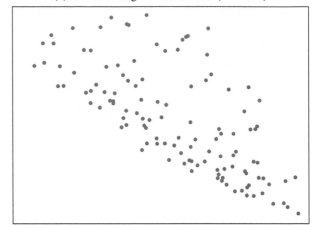

(c) Moderate Positive Correlation ($r = .518$)

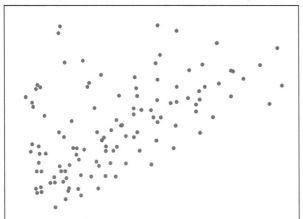

(d) Strong Positive Correlation ($r = .909$)

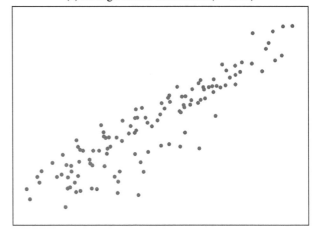

(e) Virtually No Correlation ($r = -.004$)

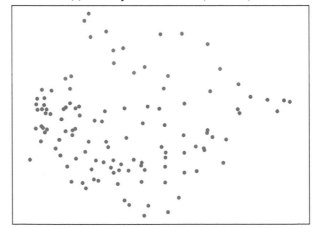

FIGURE 12.1

Five Correlations

Examination of the formula for computing a Pearson product-moment correlation coefficient (12.1) reveals that the following values must be obtained to compute r: Σx, Σx^2, Σy, Σy^2, Σxy and n. In correlation analysis, it does not matter which variable is designated x and which is designated y. For this example, the correlation coefficient is computed as shown in Table 12.2. The r value obtained ($r = 0.815$) represents a relatively strong positive relationship between interest rates and commodities futures index over this 12-day period.

Figure 12.2 shows both Excel and Minitab output for this problem.

TABLE 12.1

Data for the Economics Example

Day	Interest Rate	Futures Index
1	7.43	221
2	7.48	222
3	8.00	226
4	7.75	225
5	7.60	224
6	7.63	223
7	7.68	223
8	7.67	226
9	7.59	226
10	8.07	235
11	8.03	233
12	8.00	241

Excel Output

	Interest Rate	Futures Index
Interest Rate	1	
Futures Index	0.815	1

Minitab Output

Correlations: Interest Rate, Futures Index
Pearson correlation of Interest Rate and Futures
Index = 0.815
p-Value = 0.001

FIGURE 12.2

Excel and Minitab Output for the Economics Example

TABLE 12.2

Computation of r for the Economics Example

Day	Interest x	Futures Index y	x^2	y^2	xy
1	7.43	221	55.205	48,841	1,642.03
2	7.48	222	55.950	49,284	1,660.56
3	8.00	226	64.000	51,076	1,808.00
4	7.75	225	60.063	50,625	1,743.75
5	7.60	224	57.760	50,176	1,702.40
6	7.63	223	58.217	49,729	1,701.49
7	7.68	223	58.982	49,729	1,712.64
8	7.67	226	58.829	51,076	1,733.42
9	7.59	226	57.608	51,076	1,715.34
10	8.07	235	65.125	55,225	1,896.45
11	8.03	233	64.481	54,289	1,870.99
12	8.00	241	64.000	58,081	1,928.00
	$\Sigma x = 92.93$	$\Sigma y = 2,725$	$\Sigma x^2 = 720.220$	$\Sigma y^2 = 619,207$	$\Sigma xy = 21,115.07$

$$r = \frac{(21{,}115.07) - \frac{(92.93)(2725)}{12}}{\sqrt{\left[(720.22) - \frac{(92.93)^2}{12}\right]\left[(619{,}207) - \frac{(2725)^2}{12}\right]}} = 0.815$$

12.1   PROBLEMS

12.1   Determine the value of the coefficient of correlation, r, for the following data:

X	4	6	7	11	14	17	21
Y	18	12	13	8	7	7	4

12.2   Determine the value of r for the following data:

X	158	296	87	110	436
Y	349	510	301	322	550

12.3   In an effort to determine whether any correlation exists between the price of stocks of airlines, an analyst sampled six days of activity of the stock market. Using the prices of Ryanair stock and Air Berlin stock below, compute the coefficient of correlation. Stock prices have been rounded off to the nearest tenth for ease of computation.

Ryanair	Air Berlin
27.6	3.01
26.3	3.48
20.6	3.94
22.6	3.69
22.4	3.42
22.7	3.18

12.4   The following data are the euro to US dollar exchange rate (EUR/USD) and the euro to British pound exchange rate (EUR/GBP) for nine consecutive days.

Date	EUR/USD	EUR/GBP
13/07/2011	1.4073	0.882299
14/07/2011	1.4202	0.88095
15/07/2011	1.4146	0.8775
18/07/2011	1.4045	0.87315
19/07/2011	1.416	0.878899
20/07/2011	1.4207	0.88065
21/07/2011	1.4222	0.878701
22/07/2011	1.4391	0.882699
25/07/2011	1.438	0.882501

Use the data to compute a correlation coefficient, r, to determine the correlation between the EUR/USD and the EUR/GBP exchange rates.

12.5   The Health and Safety Executive released the following data on the incidence rates for fatal injuries in the United Kingdom for several industries in three separate periods:

Industry	1986/87	1996/97	2009/10
Agriculture	27	20	17
Extractive industries	30	9	6
Manufacturing industries	109	53	23
Construction	99	66	30
Service industries	80	59	35
Other/Unclassified	10	0	0

Compute r for each pair of years and determine which years are most highly correlated.

12.2   INTRODUCTION TO SIMPLE REGRESSION ANALYSIS

Regression analysis is *the process of constructing a mathematical model or function that can be used to predict or determine one variable by another variable or other variables.* The most elementary regression model is called **simple regression** (or bivariate regression) involving two variables in which one variable is predicted by another variable. In simple regression, *the variable to be predicted* is called the **dependent variable** and is designated as *y*. The *predictor* is called the **independent variable**, or *explanatory variable*, and is designated as *x*. In simple regression analysis, only a straight-line relationship between two variables is examined. Non-linear relationships and regression models with more than one independent variable can be explored by using multiple regression models, which are presented in Chapters 13 and 14.

Can the cost of flying a commercial airliner be predicted using regression analysis? If so, what variables are related to such cost? A few of the many variables that can potentially contribute are type of plane, distance, number of passengers, amount of luggage/freight, weather conditions, direction of destination, and perhaps even pilot skill. Suppose a study is conducted using only Airbus A320s travelling 500 miles on comparable routes during the same season of the year. Can the number of passengers predict the cost of flying such routes? It seems logical that more passengers result in more weight and more baggage, which could, in turn, result in increased fuel consumption and other costs. Suppose the data displayed in Table 12.3 are the costs and associated number of passengers for twelve 500-mile commercial airline flights using Airbus A320s during the same season of the year. We will use these data to develop a regression model to predict cost by number of passengers.

Usually, the first step in simple regression analysis is to construct a **scatter plot** (or scatter diagram), discussed in Chapter 2. Graphing the data in this way yields preliminary information about the shape and spread of the data. Figure 12.3 is an Excel scatter plot of the data in Table 12.3. Figure 12.4 is a close-up view of the scatter plot produced by Minitab. Try to imagine a line passing through the points. Is a linear fit possible? Would a curve fit the data better? The scatter plot gives some idea of how well a regression line fits the data. Later in the chapter, we present statistical techniques that can be used to determine more precisely how well a regression line fits the data.

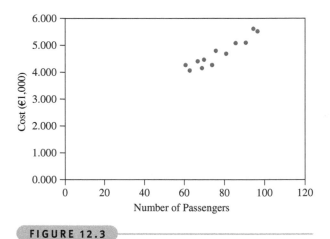

FIGURE 12.3

Excel Scatter Plot of Airline Cost Data

TABLE 12.3

Airline Cost Data

Number of Passengers	Cost (€000s)
61	4.280
63	4.080
67	4.420
69	4.170
70	4.480
74	4.300
76	4.820
81	4.700
86	5.110
91	5.130
95	5.640
97	5.560

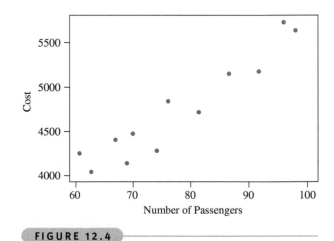

FIGURE 12.4

Close-Up Minitab Scatter Plot of Airline Cost Data

12.3   DETERMINING THE EQUATION OF THE REGRESSION LINE

The first step in determining the equation of the regression line that passes through the sample data is to establish the equation's form. Several different types of equations of lines are discussed in algebra, finite math, or analytic geometry courses. Recall that among these equations of a line are the two-point form, the point-slope form, and the slope-intercept form. In regression analysis, researchers use the slope-intercept equation of a line. In maths courses, the slope-intercept form of the equation of a line often takes the form

$$y = mx + b$$

where:

   $m =$ slope of the line
   $b = y$ intercept of the line

In statistics, the slope-intercept form of the equation of the regression line through the population points is

$$\hat{y} = \beta_0 + \beta_1 x$$

where:

   $\hat{y} =$ the predicted value of y
   $\beta_0 =$ the population y intercept
   $\beta_1 =$ the population slope

For any specific dependent variable value, y_i,

$$y_i = \beta_0 + \beta_1 x_i + \varepsilon_i$$

where:

   $x_i =$ the value of the independent variable for the ith value
   $y_i =$ the value of the dependent variable for the ith value
   $\beta_0 =$ the population y intercept
   $\beta_1 =$ the population slope
   $\varepsilon_i =$ the error of prediction for the ith value

Unless the points being fitted by the regression equation are in perfect alignment, the regression line will miss at least some of the points. In the preceding equation, ε_i represents the error of the regression line in fitting these points. If a point is on the regression line, $\varepsilon_i = 0$.

These mathematical models can be either deterministic models or probabilistic models. **Deterministic models** are *mathematical models that produce an 'exact' output for a given input.* For example, suppose the equation of a regression line is

$$y = 1.68 + 2.40x$$

For a value of $x = 5$, the exact predicted value of y is

$$y = 1.68 + 2.40(5) = 13.68$$

We recognize, however, that most of the time the values of y will not equal exactly the values yielded by the equation. Random error will occur in the prediction of the y values for values of x because it is likely that the variable x does not explain all the variability of the variable y. For example, suppose we are trying to predict the volume of sales (y) for a company through regression analysis by using the annual expenditure amount of advertising (x) as the predictor. Although sales are often related to advertising, other factors related to sales are not accounted for by amount of advertising. Hence, a regression model to predict sales volume by amount of advertising probably involves some error. For this reason, in regression, we present the general model as a probabilistic model. A **probabilistic model** is *one that includes an error term that allows for the y values to vary for any given value of x.*

A deterministic regression model is

$$y = \beta_0 + \beta_1 x$$

The probabilistic regression model is

$$y = \beta_0 + \beta_1 x + \varepsilon$$

$\beta_0 + \beta_1 x$ is the deterministic portion of the probabilistic model, $\beta_0 + \beta_1 x + \varepsilon$. In a deterministic model, all points are assumed to be on the line and in all cases ε is zero.

Virtually all regression analyses of business data involve sample data, not population data. As a result, β_0 and β_1 are unattainable and must be estimated by using the sample statistics, b_0 and b_1. Hence the equation of the regression line contains the sample y intercept, b_0, and the sample slope, b_1.

$$\hat{y} = b_0 + b_1 x$$

EQUATION OF THE SIMPLE REGRESSION LINE

where:

b_0 = the sample intercept
b_1 = the sample slope

To determine the equation of the regression line for a sample of data, the researcher must determine the values for b_0 and b_1. This process is sometimes referred to as least squares analysis. **Least squares analysis** is *a process whereby a regression model is developed by producing the minimum sum of the squared error values.* On the basis of this premise and calculus, a particular set of equations has been developed to produce components of the regression model. The derivation of these formulas is beyond the scope of the discussion here.

Examine the regression line fit through the points in Figure 12.5. Observe that the line does not actually pass through any of the points. The vertical distance from each point to the line is the error of the prediction. In theory, an infinite number of lines could be constructed to pass through these points in some manner. The least squares regression line is the regression line that results in the smallest sum of errors squared.

Formula 12.2 is an equation for computing the value of the sample slope. Several versions of the equation are given to afford latitude in doing the computations.

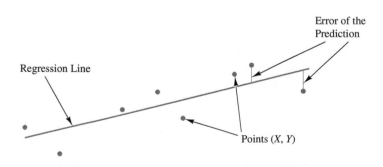

FIGURE 12.5

Minitab Plot of a Regression Line

SLOPE OF THE REGRESSION LINE (12.2)	$$b_1 = \frac{\Sigma(x-\bar{x})(y-\bar{y})}{\Sigma(x-\bar{x})^2} = \frac{\Sigma xy - n\bar{x}\,\bar{y}}{\Sigma x^2 - n\bar{x}^2} = \frac{\Sigma xy - \dfrac{(\Sigma x)(\Sigma y)}{n}}{\Sigma x^2 - \dfrac{(\Sigma x)^2}{n}}$$

The expression in the numerator of the slope formula 12.2 appears frequently in this chapter and is denoted as SS_{xy}.

$$SS_{xy} = \Sigma(x-\bar{x})(y-\bar{y}) = \Sigma xy - \frac{(\Sigma x)(\Sigma y)}{n}$$

The expression in the denominator of the slope formula 12.2 also appears frequently in this chapter and is denoted as SS_{xx}.

$$SS_{xx} = \Sigma(x-\bar{x})^2 = \Sigma x^2 - \frac{(\Sigma x)^2}{n}$$

With these abbreviations, the equation for the slope can be expressed as in Formula 12.3.

ALTERNATIVE FORMULA FOR SLOPE (12.3)	$$b_1 = \frac{SS_{xy}}{SS_{xx}}$$

Formula 12.4 is used to compute the sample y intercept. The slope must be computed before the y intercept.

y INTERCEPT OF THE REGRESSION LINE (12.4)	$$b_0 = \bar{y} - b_1\bar{x} = \frac{\Sigma y}{n} - b_1\frac{(\Sigma x)}{n}$$

Formulas 12.2, 12.3, and 12.4 show that the following data are needed from sample information to compute the slope and intercept: Σx, Σy, Σx^2, and, Σxy, unless sample means are used. Table 12.4 contains the results of solving for the slope and intercept and determining the equation of the regression line for the data in Table 12.3.

The least squares equation of the regression line for this problem is

$$\hat{y} = 1.57 + 0.0407x$$

The slope of this regression line is 0.0407. Because the x values were recoded for the ease of computation and are actually in €1,000 denominations, the slope is actually €40.70. One interpretation of the slope in this problem is that for every unit increase in x (every person added to the flight of the airplane), there is a €40.70 increase in the cost of the flight. The y-intercept is the point where the line crosses the y-axis (where x is zero). Sometimes in regression analysis, the y-intercept is meaningless in terms of the variables studied. However, in this problem, one interpretation of the y-intercept, which is 1.570 or €1,570, is that even if there were no passengers on the commercial flight, it would still cost €1,570. In other words, there are costs associated with a flight that carries no passengers.

Superimposing the line representing the least squares equation for this problem on the scatter plot indicates how well the regression line fits the data points, as shown in the Excel graph in Figure 12.6. The next several sections explore mathematical ways of testing how well the regression line fits the points.

TABLE 12.4

Solving for the Slope and the y Intercept of the Regression Line for the Airline Cost Example

Number of Passengers	Cost (€000s)		
x	y	x^2	xy
61	4.280	3,721	261.080
63	4.080	3,969	257.040
67	4.420	4,489	296.140
69	4.170	4,761	287.730
70	4.480	4,900	313.600
74	4.300	5,476	318.200
76	4.820	5,776	366.320
81	4.700	6,561	380.700
86	5.110	7,396	439.460
91	5.130	8,281	466.830
95	5.640	9,025	535.800
97	5.560	9,409	539.320
$\Sigma x = 930$	$\Sigma y = 56.690$	$\Sigma x^2 = 73,764$	$\Sigma xy = 4462.220$

$$SS_{xy} = \Sigma xy - \frac{(\Sigma x)(\Sigma y)}{n} = 4462.22 - \frac{(930)(56.69)}{12} = 68.745$$

$$SS_{xx} = \Sigma x^2 - \frac{(\Sigma x)^2}{n} = 73,764 - \frac{(930)^2}{12} = 1689$$

$$b_1 = \frac{SS_{xy}}{SS_{xx}} = \frac{68.745}{1689} = 0.0407$$

$$b_0 = \frac{\Sigma y}{n} - b_1\frac{\Sigma x}{n} = \frac{56.19}{12} - (0.0407)\frac{930}{12} = 1.57$$

$$\hat{y} = 1.57 + 0.0407x$$

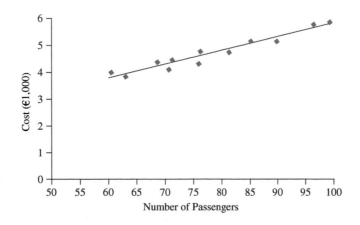

FIGURE 12.6

Excel Graph of the Regression Line for the Airline Cost Example

DEMONSTRATION PROBLEM 12.1

A specialist in hospital administration stated that the number of full-time equivalent employees (FTEs) in a hospital can be estimated by counting the number of beds in the hospital (a common measure of hospital size). A healthcare business researcher decided to develop a regression model in an attempt to predict the number of FTEs of a hospital by the number of beds. She surveyed 12 hospitals and obtained the data below. The data are presented in sequence, according to the number of beds.

Number of Beds	FTEs	Number of Beds	FTEs
23	69	50	138
29	95	54	178
29	102	64	156
35	118	66	184
42	126	76	176
46	125	78	225

Solution

The following Minitab graph is a scatter plot of these data. Note the linear appearance of the data.

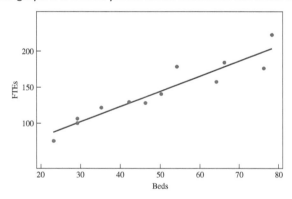

Next, the researcher determined the values of Σx, Σy, Σx^2, and Σxy.

Hospital	Number of Beds	FTEs		
	x	y	x^2	xy
1	23	69	529	1,587
2	29	95	841	2,755
3	29	102	841	2,958
4	35	118	1,225	4,130
5	42	126	1,764	5,292
6	46	125	2,116	5,750
7	50	138	2,500	6,900
8	54	178	2,916	9,612
9	64	156	4,096	9,984
10	66	184	4,356	12,144
11	76	176	5,776	13,376
12	78	225	6,084	17,550
	$\Sigma x = 592$	$\Sigma y = 1,692$	$\Sigma x^2 = 33,044$	$\Sigma xy = 92,038$

Using these values, the researcher solved for the sample slope (b_1) and the sample y-intercept (b_0).

$$SS_{xy} = \sum xy - \frac{(\sum x)(\sum y)}{n} = 92{,}038 - \frac{(592)(1682)}{12} = 8566$$

$$SS_{xx} = \sum x^2 - \frac{(\sum x)^2}{n} = 33{,}044 - \frac{(592)^2}{12} = 3838.667$$

$$b_1 = \frac{SS_{xy}}{SS_{xx}} = \frac{8566}{3838.667} = 2.232$$

$$b_0 = \frac{\sum y}{n} - b_1 \frac{\sum x}{12} = \frac{1692}{12} - (2.232)\frac{592}{12} = 30.888$$

The least squares equation of the regression line is

$$\hat{y} = 30.888 + 2.232x$$

The slope of the line, $b_1 = 2.232$, means that for every unit increase of x (every bed), y (number of FTEs) is predicted to increase by 2.232. Even though the y-intercept helps the researcher sketch the graph of the line by being one of the points on the line (0, 30.888), it has limited usefulness in terms of this solution because $x = 0$ denotes a hospital with no beds. On the other hand, it could be interpreted that a hospital has to have at least 31 FTEs to open its doors even with no patients – a sort of 'fixed cost' of personnel.

12.3 PROBLEMS

12.6 Sketch a scatter plot from the following data, and determine the equation of the regression line:

x	12	21	28	8	20
y	17	15	22	19	24

12.7 Sketch a scatter plot from the following data, and determine the equation of the regression line:

x	140	119	103	91	65	29	24
y	25	29	46	70	88	112	128

12.8 A corporation owns several companies. The strategic planner for the corporation believes euros spent on advertising can to some extent be a predictor of total euro sales. As an aid in long-term planning, she gathers the following sales and advertising information from several of the companies for 2009 (€ millions).

Advertising	Sales
12.5	148
3.7	55
21.6	338
60.0	994
37.6	541
6.1	89
16.8	126
41.2	379

Develop the equation of the simple regression line to predict sales from advertising expenditures using these data.

12.9 Investment analysts generally believe the interest rate on bonds is inversely related to the prime interest rate for loans; that is, bonds perform well when lending rates are down and perform poorly when interest rates are up. Can the bond rate be predicted by the prime interest rate? Use the following data to construct a least squares regression line to predict bond rates by the prime interest rate.

Bond Rate	Prime Interest Rate
5%	16%
12	6
9	8
15	4
7	7

12.10 The Swedish research institute Berg Insight estimated that mobile marketing and advertising sales would grow from 1.3 billion euros to 10 billion euros between 2009 and 2015. But is mobile advertising effective in increasing sales? Suppose the data below are pairs of the amount spent on mobile advertisements and the amount of sales for a nine-year period for an online retailer. Use these data to develop the equation of the regression model to predict the amount of sales by the amount spent on mobile advertisements. Discuss the meaning of the slope.

Sales (€ millions)	Mobile Advertisements (€000s)
34.3	58.1
35.0	55.4
38.5	57.0
40.1	58.5
35.5	57.4
37.9	56.7
41.3	58.9
38.7	57.8
37.9	58.0

12.11 The following data provided by the House of Commons Library show total milk production and the number of registered dairy production holdings in the United Kingdom from 1995 to 2009. Can total milk production in the UK be predicted by the number of registered dairy production holdings? Use these data to develop the equation of a regression line to predict the average milk production by the number of dairy production holdings. Discuss the slope and y-intercept of the model.

Year	Total milk production (millions of litres)	Number of registered dairy production holdings
1995	14,259	35,741
1996	14,256	34,570
1997	14,426	33,528
1998	14,220	31,753
1999	14,587	30,221
2000	14,078	28,422
2001	14,291	26,556
2002	14,447	24,930
2003	14,583	22,992
2004	14,134	21,616
2005	14,059	20,313
2006	13,909	19,011
2007	13,626	17,915
2008	13,326	17,060
2009	13,215	16,404

12.12 Can the annual revenues of Google be predicted by the annual number of searches that go through Google .com? Shown below are the annual revenues of Google for 10 years according to Google investor relations and the number of searches conducted through Google.com per year for the same 10 years as estimated using a number of Internet sources. Use these data to develop a regression model to predict annual revenues by number of searches per year. Construct a scatter plot and draw the regression line through the points.

	Annual Returns (US$ million)	Number of searches (billion)
2001	86	35
2002	440	54
2003	1,466	65
2004	3,189	73
2005	6,139	100
2006	10,605	250
2007	16,593	400
2008	21,795	700
2009	23,651	1,044
2010	29,321	1,072

12.4 RESIDUAL ANALYSIS

How does a business researcher test a regression line to determine whether the line is a good fit of the data other than by observing the fitted line plot (regression line fit through a scatter plot of the data)? One particularly popular approach is to use the *historical data* (x and y values used to construct the regression model) to test the model. With this approach, the values of the independent variable (x values) are inserted into the regression model and a predicted value ($\hat{y}$) is obtained for each x value. These predicted values ($\hat{y}$) are then compared to the actual y values to determine how much error the equation of the regression line produced. *Each difference between the actual y values and the predicted y values is the error of the regression line at a given point, $y - \hat{y}$,* and is referred to as the **residual**. It is the sum of squares of these residuals that is minimized to find the least squares line.

Table 12.5 shows values and the residuals for each pair of data for the airline cost regression model developed in Section 12.3. The predicted values are calculated by inserting an x value into the equation of the regression line and solving for $\hat{y}$. For example, when $x = 61$, $\hat{y} = 1.57 + 0.0407(61) = 4.053$, as displayed in column 3 of the table. Each of these predicted y values is subtracted from the actual y value to determine the error, or residual. For example, the first y value listed in the table is 4.280 and the first predicted value is 4.053, resulting in a residual of $4.280 - 4.053 = 0.227$. The residuals for this problem are given in column 4 of the table.

Note that the sum of the residuals is approximately zero. Except for rounding error, the sum of the residuals is *always zero.* The reason is that a residual is geometrically the vertical distance from the regression line to a data

TABLE 12.5

Predicted Values and Residuals for the Airline Cost Example

Number of Passengers x	Cost (€000s) y	Predicted Value $\hat{y}$	Residual $y - \hat{y}$
61	4.280	4.053	.227
63	4.080	4.134	−.054
67	4.420	4.297	.123
69	4.170	4.378	−.208
70	4.480	4.419	.061
74	4.300	4.582	−.282
76	4.820	4.663	.157
81	4.700	4.867	−.167
86	5.110	5.070	.040
91	5.130	5.274	−.144
95	5.640	5.436	.204
97	5.560	5.518	.042
		$\Sigma(y-\hat{y}) =$	−.001

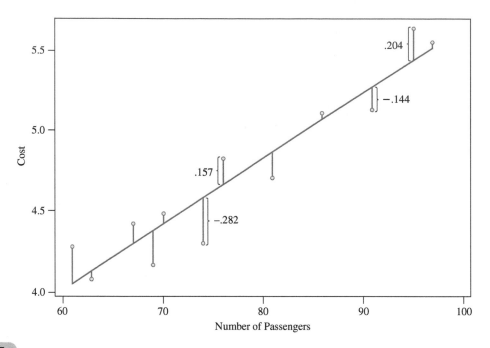

Close-Up Minitab Scatter Plot with Residuals for the Airline Cost Example

point. The equations used to solve for the slope and intercept place the line geometrically in the middle of all points. Therefore, vertical distances from the line to the points will cancel each other and sum to zero. Figure 12.7 is a Minitab-produced scatter plot of the data and the residuals for the airline cost example.

An examination of the residuals may give the researcher an idea of how well the regression line fits the historical data points. The largest residual for the airline cost example is −0.282, and the smallest is .040. Because the objective of the regression analysis was to predict the cost of flight in thousands of euros, the regression line produces an error of €282 when there are 74 passengers and an error of only €40 when there are 86 passengers. This result presents the *best* and *worst* cases for the residuals. The researcher must examine other residuals to determine how well the regression model fits other data points.

Sometimes residuals are used to locate outliers. **Outliers** are *data points that lie apart from the rest of the points*. Outliers can produce residuals with large magnitudes and are usually easy to identify on scatter plots. Outliers can be the result of misrecorded or miscoded data, or they may simply be data points that do not conform to the general trend. The equation of the regression line is influenced by every data point used in its calculation in a manner similar to the arithmetic mean. Therefore, outliers sometimes can unduly influence the regression line by 'pulling' the line toward the outliers. The origin of outliers must be investigated to determine whether they should be retained or whether the regression equation should be recomputed without them.

Residuals are usually plotted against the x-axis, which reveals a view of the residuals as x increases. Figure 12.8 shows the residuals plotted by Excel against the x-axis for the airline cost example.

Using Residuals to Test the Assumptions of the Regression Model

One of the major uses of residual analysis is to test some of the assumptions underlying regression. The following are the assumptions of simple regression analysis.

1. The model is linear.
2. The error terms have constant variances.
3. The error terms are independent.
4. The error terms are normally distributed.

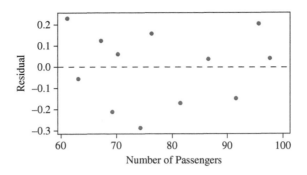

Excel Graph of Residuals for the Airline Cost Example

A particular method for studying the behaviour of residuals is the residual plot. The **residual plot** is *a type of graph in which the residuals for a particular regression model are plotted along with their associated value of x as an ordered pair* $(x, y - \hat{y})$. Information about how well the regression assumptions are met by the particular regression model can be gleaned by examining the plots. Residual plots are more meaningful with larger sample sizes. For small sample sizes, residual plot analyses can be problematic and subject to over-interpretation. Hence, because the airline cost example is constructed from only 12 pairs of data, one should be cautious in reaching conclusions from Figure 12.8. The residual plots in Figures 12.9, 12.10, and 12.11, however, represent large numbers of data points and therefore are more likely to depict overall trends accurately.

If a residual plot such as the one in Figure 12.9 appears, the assumption that the model is linear does not hold. Note that the residuals are negative for low and high values of x and are positive for middle values of x. The graph of these residuals is parabolic, not linear. The residual plot does not have to be shaped in this manner for a non-linear relationship to exist. Any significant deviation from an approximately linear residual plot may mean that a non-linear relationship exists between the two variables.

The assumption of *constant error variance* sometimes is called **homoscedasticity**. If *the error variances are not constant* (called **heteroscedasticity**), the residual plots might look like one of the two plots in Figure 12.10. Note in Figure 12.10(a) that the error variance is greater for small values of x and smaller for large values of x. The situation is reversed in Figure 12.10(b).

If the error terms are not independent, the residual plots could look like one of the graphs in Figure 12.11. According to these graphs, instead of each error term being independent of the one next to it, the value of the residual is a function of the residual value next to it. For example, a large positive residual is next to a large positive residual and a small negative residual is next to a small negative residual.

The graph of the residuals from a regression analysis that meets the assumptions – a *healthy residual graph* – might look like the graph in Figure 12.12. The plot is relatively linear; the variances of the errors are about equal for each value of x, and the error terms do not appear to be related to adjacent terms.

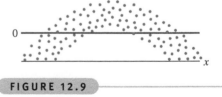

Non-linear Residual Plot

(a)   (b)

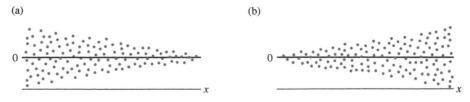

Non-constant Error Variance

(a) (b)

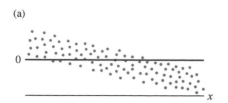
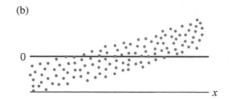

FIGURE 12.11

Graphs of Non-independent Error Terms

FIGURE 12.12

Healthy Residual Graph

Using the Computer for Residual Analysis

Some computer programs contain mechanisms for analysing residuals for violations of the regression assumptions. Minitab has the capability of providing graphical analysis of residuals. Figure 12.13 displays Minitab's residual graphic analyses for a regression model developed to predict the production of carrots in the United States per month by the total production of sweet corn. The data were gathered over a time period of 168 consecutive months.

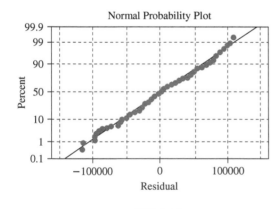
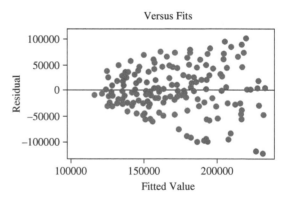
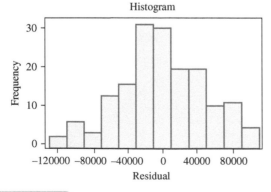

FIGURE 12.13

Minitab Residual Analyses

These Minitab residual model diagnostics consist of three different plots. The graph on the upper right is a plot of the residuals versus the fits. Note that this residual plot 'flares-out' as x gets larger. This pattern is an indication of heteroscedasticity, which is a violation of the assumption of constant variance for error terms. The graph in the upper left is a normal probability plot of the residuals. A straight line indicates that the residuals are normally distributed. Observe that this normal plot is relatively close to being a straight line, indicating that the residuals are nearly normal in shape. This normal distribution is confirmed by the graph on the lower left, which is a histogram of the residuals. The histogram groups residuals in classes so the researcher can observe where groups of the residuals lie without having to rely on the residual plot and to validate the notion that the residuals are approximately normally distributed. In this problem, the pattern is indicative of at least a mound-shaped distribution of residuals.

DEMONSTRATION PROBLEM 12.2

Compute the residuals for Demonstration Problem 12.1 in which a regression model was developed to predict the number of full-time equivalent employees (FTEs) by the number of beds in a hospital. Analyse the residuals by using Minitab graphic diagnostics.

Solution

The data and computed residuals are shown in the following table.

Hospital	Number of Beds x	FTEs y	Predicted value $\hat{y}$	Residuals $y - \hat{y}$
1	23	69	82.22	−13.22
2	29	95	95.62	−0.62
3	29	102	95.62	6.38
4	35	118	109.01	8.99
5	42	126	124.63	1.37
6	46	125	133.56	−8.56
7	50	138	142.49	−4.49
8	54	178	151.42	26.58
9	64	156	173.74	−17.74
10	66	184	178.20	5.80
11	76	176	200.52	−24.52
12	78	225	204.98	20.02
				$\Sigma(y - \hat{y}) = -.01$

Note that the regression model fits these particular data well for hospitals 2 and 5, as indicated by residuals of −0.62 and 1.37 FTEs, respectively. For hospitals 1, 8, 9, 11, and 12, the residuals are relatively large, indicating that the regression model does not fit the data for these hospitals well. The residuals versus the fitted values graph indicates that the residuals seem to increase as x increases, indicating a potential problem with heteroscedasticity. The normal plot of residuals indicates that the residuals are nearly normally distributed. The histogram of residuals shows that the residuals pile up in the middle, but are somewhat skewed towards the larger positive values.

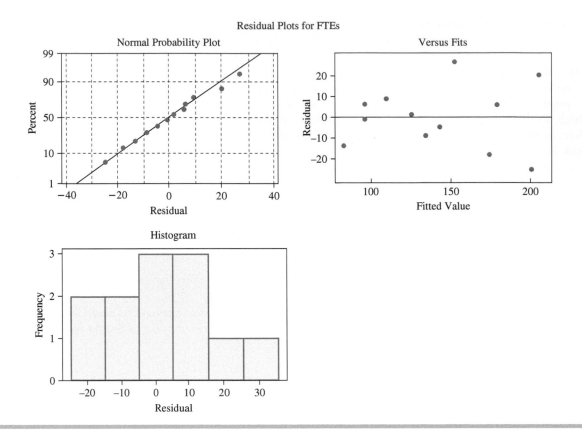

Residual Plots for FTEs

12.4 PROBLEMS

12.13 Determine the equation of the regression line for the following data, and compute the residuals:

x	15	8	19	12	5
y	47	36	56	44	21

12.14 Solve for the predicted values of y and the residuals for the data in Problem 12.6. The data are provided here again:

x	12	21	28	8	20
y	17	15	22	19	24

12.15 Solve for the predicted values of y and the residuals for the data in Problem 12.7. The data are provided here again:

x	140	119	103	91	65	29	24
y	25	29	46	70	88	112	128

12.16 Solve for the predicted values of y and the residuals for the data in Problem 12.8. The data are provided here again:

Advertising	12.5	3.7	21.6	60.0	37.6	6.1	16.8	41.2
Sales	148	55	338	994	541	89	126	379

12.17 Solve for the predicted values of *y* and the residuals for the data in Problem 12.9. The data are provided here again:

Bond Rate	5%	12%	9%	15%	7%
Prime Interest Rate	16%	6%	8%	4%	7%

12.18 In problem 12.10, you were asked to develop the equation of a regression model to predict the sales by the amount spent on mobile advertisements. Using this regression model and the data given in problem 12.10 (and provided here again), solve for the predicted values of *y* and the residuals. Comment on the size of the residuals.

Sales (€ millions)	Mobile Advertisements (€000s)
34.3	58.1
35.0	55.4
38.5	57.0
40.1	58.5
35.5	57.4
37.9	56.7
41.3	58.9
38.7	57.8
37.9	58.0

12.19 The equation of a regression line is

$$\hat{y} = 50.506 - 1.646x$$

and the data are as follows.

x	5	7	11	12	19	25
y	47	38	32	24	22	10

Solve for the residuals and graph a residual plot. Do these data seem to violate any of the assumptions of regression?

12.20 Bordeaux is an important wine-producing region of France. Bordeaux wine is made by more than 8,500 producers or *châteaux* spread over more than 1,000 square kilometres. Some people might argue that because of transportation costs, the cost of wine increases with the distance of markets from the city of Bordeaux. Suppose the wine prices in eight towns are as follows.

Cost of Wine (€ per bottle)	Distance from the city of Bordeaux (kilometres)
18.48	73
16.17	25
17.15	79
17.64	57
15.33	15
17.85	51
16.8	64
16.59	17

Use the prices along with the distance of each town from Bordeaux to develop a regression line to predict the price of a bottle of wine by the number of kilometres the town is from Bordeaux. Use the data and the regression equation to compute residuals for this model. Sketch a graph of the residuals in the order of the *x* values. Comment on the shape of the residual graph.

12.21 Graph the following residuals, and indicate which of the assumptions underlying regression appear to be in jeopardy on the basis of the graph:

x	$y - \hat{y}$
213	−11
216	−5
227	−2
229	−1
237	+6
247	+10
263	+12

12.22 Graph the following residuals, and indicate which of the assumptions underlying regression appear to be in jeopardy on the basis of the graph:

x	$y - \hat{y}$
10	+6
11	+3
12	−1
13	−11
14	−3
15	+2
16	+5
17	+8

12.23 Study the Minitab residuals versus fits graphic below for a simple regression analysis. Comment on the residual evidence of lack of compliance with the regression assumptions.

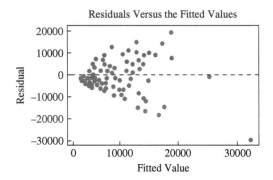

12.5 STANDARD ERROR OF THE ESTIMATE

Residuals represent errors of estimation for individual points. With large samples of data, residual computations become laborious. Even with computers, a researcher sometimes has difficulty working through pages of residuals in an effort to understand the error of the regression model. An alternative way of examining the error of the model is the standard error of the estimate, which provides a single measurement of the regression error.

Because the sum of the residuals is zero, attempting to determine the total amount of error by summing the residuals is fruitless. This zero-sum characteristic of residuals can be avoided by squaring the residuals and then summing them.

TABLE 12.6

Determining SSE for the Airline Cost Example

Number of Passengers x	Cost (€000s) y	Residual $y - \hat{y}$	$(y - \hat{y})^2$
61	4.280	.227	.05153
63	4.080	−.054	.00292
67	4.420	.123	.01513
69	4.170	−.208	.04326
70	4.480	.061	.00372
74	4.300	−.282	.07952
76	4.820	.157	.02465
81	4.700	−.167	.02789
86	5.110	.040	.00160
91	5.130	−.144	.02074
95	5.640	.204	.04162
97	5.560	.042	.00176
		$\Sigma(y - \hat{y}) = 0.001$	$\Sigma(y - \hat{y})^2 = 0.31434$

Sum of squares of error = SSE = .31434

Table 12.6 contains the airline cost data from Table 12.3, along with the residuals and the residuals squared. The *total of the residuals squared* column is called the **sum of squares of error** (SSE).

SUM OF SQUARES OF ERROR $\qquad$ $SSE = \Sigma(y - \hat{y})^2$

In theory, infinitely many lines can be fit to a sample of points. However, formulas 12.2 and 12.4 produce a line of best fit for which the SSE is the smallest for any line that can be fit to the sample data. This result is guaranteed, because formulas 12.2 and 12.4 are derived from calculus to minimize SSE. For this reason, the regression process used in this chapter is called *least squares* regression.

A computational version of the equation for computing SSE is less meaningful in terms of interpretation than $\Sigma(y - \hat{y})^2$ but it is usually easier to compute. The computational formula for SSE follows.

COMPUTATIONAL FORMULA FOR SSE $\qquad$ $SSE = \Sigma y^2 - b_0\Sigma y - b_1\Sigma xy$

For the airline cost example,

$$\Sigma y^2 = \Sigma[(4.280)^2 + (4.080)^2 + (4.420)^2 + (4.170)^2 + (4.480)^2 + (4.300)^2 + (4.820)^2 \\ + (4.700)^2 + (5.110)^2 + (5.130)^2 + (5.640)^2 + (5.560)^2] = 270.9251$$

$$b_0 = 1.5697928$$

$$b_1 = 0.0407016*$$

$$\Sigma y = 56.69$$

*Note: In previous sections, the values of the slope and intercept were rounded off for ease of computation and interpretation. They are shown here with more precision in an effort to reduce rounding error.

$$\Sigma xy = 4462.22$$

$$SSE = \Sigma y^2 - b_0 \Sigma y - b_1 \Sigma xy$$

$$= 270.9251 - (1.5697928)(56.69) - (0.0407016)(4462.22) = 0.31405$$

The slight discrepancy between this value and the value computed in Table 12.6 is due to rounding error.

The sum of squares error is in part a function of the number of pairs of data being used to compute the sum, which lessens the value of SSE as a measurement of error. A more useful measurement of error is the standard error of the estimate. The **standard error of the estimate**, denoted s_e, is *a standard deviation of the error of the regression model* and has a more practical use than SSE. The standard error of the estimate follows.

STANDARD ERROR OF THE ESTIMATE	$s_e = \sqrt{\dfrac{SSE}{n-2}}$

The standard error of the estimate for the airline cost example is

$$s_e = \sqrt{\frac{SSE}{n-2}} = \sqrt{\frac{0.31434}{10}} = 0.1773$$

How is the standard error of the estimate used? As previously mentioned, the standard error of the estimate is a standard deviation of error. Recall from Chapter 3 that if data are approximately normally distributed, the empirical rule states that about 68% of all values are within $\mu \pm 1\sigma$ and that about 95% of all values are within $\mu \pm 2\sigma$. One of the assumptions for regression states that for a given x the error terms are normally distributed. Because the error terms are normally distributed, s_e is the standard deviation of error, and the average error is zero, approximately 68% of the error values (residuals) should be within $0 \pm 1s_e$, and 95% of the error values (residuals) should be within $0 \pm 2s_e$. By having knowledge of the variables being studied and by examining the value of s_e, the researcher can often make a judgement about the fit of the regression model to the data by using s_e. How can the s_e value for the airline cost example be interpreted?

The regression model in that example is used to predict airline cost by number of passengers. Note that the range of the airline cost data in Table 12.3 is from 4.08 to 5.64 (€4,080 to €5,640). The regression model for the data yields an s_e of 0.1773. An interpretation of s_e is that the standard deviation of error for the airline cost example is €177.30. If the error terms were normally distributed about the given values of x, approximately 68% of the error terms would be within $\pm$€177.30 and 95% would be within $\pm 2(\text{€}177.30) = \pm\text{€}354.60$. Examination of the residuals reveals that 100% of the residuals are within $2s_e$. The standard error of the estimate provides a single measure of error, which, if the researcher has enough background in the area being analysed, can be used to understand the magnitude of errors in the model. In addition, some researchers use the standard error of the estimate to identify outliers. They do so by looking for data that are outside $\pm 2s_e$ or $\pm 3s_e$.

DEMONSTRATION PROBLEM 12.3

Compute the sum of squares of error and the standard error of the estimate for Demonstration Problem 12.1, in which a regression model was developed to predict the number of FTEs at a hospital by the number of beds.

Solution

Hospital	Number of Beds x	FTEs y	Residuals $y - \hat{y}$	$(y - \hat{y})^2$
1	23	69	−13.22	174.77
2	29	95	−.62	−0.38
3	29	102	6.38	40.70
4	35	118	8.99	80.82
5	42	126	1.37	1.88
6	46	125	−8.56	73.27
7	50	138	−4.49	20.16
8	54	178	26.58	706.50
9	64	156	−17.74	314.71
10	66	184	5.80	33.64
11	76	176	−24.52	601.23
12	78	225	20.02	400.80
	$\Sigma x = 592$	$\Sigma y = 1692$	$\Sigma(y - \hat{y}) = -.01$	$\Sigma(y - \hat{y})^2 = 2448.86$

$$SSE = 2{,}448.86$$

$$s_e = \sqrt{\frac{SSE}{n-2}} = \sqrt{\frac{2448.86}{10}} = 15.65$$

The standard error of the estimate is 15.65 FTEs. An examination of the residuals for this problem reveals that 8 of 12 (67%) are within $\pm 1s_e$ and 100% are within $\pm 2s_e$. Is this size of error acceptable? Hospital administrators probably can best answer that question.

12.5   PROBLEMS

12.24 Determine the sum of squares of error (SSE) and the standard error of the estimate (s_e) for Problem 12.6. Determine how many of the residuals computed in Problem 12.14 (for Problem 12.6) are within one standard error of the estimate. If the error terms are normally distributed, approximately how many of these residuals should be within $\pm 1s_e$?

12.25 Determine the SSE and the s_e for Problem 12.7. Use the residuals computed in Problem 12.15 (for Problem 12.7) and determine how many of them are within $\pm 1s_e$ and $\pm 2s_e$. How do these numbers compare with what the empirical rule says should occur if the error terms are normally distributed?

12.26 Determine the SSE and the s_e for Problem 12.8. Think about the variables being analysed by regression in this problem and comment on the value of s_e.

12.27 Determine the SSE and s_e for Problem 12.9. Examine the variables being analysed by regression in this problem and comment on the value of s_e.

12.28 In problem 12.10, you were asked to develop the equation of a regression model to predict the amount of sales by the amount spent on mobile advertisements. For this regression model, solve for the standard error of the estimate and comment on it.

12.29 Use the data from problem 12.19 and determine the s_e.

12.30   Determine the SSE and the s_e for Problem 12.20. Comment on the size of s_e for this regression model, which is used to predict the cost of Bordeaux wine.

12.31   Determine the equation of the regression line to predict the return on investment (ROI) of a company from the ratio of revenue per employee in a recent year as reported by Reuters. Compute the standard error of the estimate for this model. Does the revenue per employee appear to be a good predictor of a company's return on investment? Why or why not?

Company	ROI (%)	Revenue per Employee (£)
3i Group PLC	3.45	1,224,033
Aggreko PLC	23.66	331,152
AMEC PLC	16.85	134,688
Anglo American PLC	15.82	171,208
Autofagasta PLC	22.02	609,679
ARM Holdings PLC	10.14	237,247
Associated British Foods PLC	8.20	109,147

12.6   COEFFICIENT OF DETERMINATION

A widely used measure of fit for regression models is the **coefficient of determination**, or r^2. The coefficient of determination is *the proportion of variability of the dependent variable (y) accounted for or explained by the independent variable (x).*

The coefficient of determination ranges from 0 to 1. An r^2 of zero means that the predictor accounts for none of the variability of the dependent variable and that there is no regression prediction of y by x. An r^2 of 1 means perfect prediction of y by x and that 100% of the variability of y is accounted for by x. Of course, most r^2 values are between the extremes. The researcher must interpret whether a particular r^2 is high or low, depending on the use of the model and the context within which the model was developed.

In exploratory research where the variables are less understood, low values of r^2 are likely to be more acceptable than they are in areas of research where the parameters are more developed and understood. One Russian space programme researcher who uses vehicular weight to predict mission cost searches for the regression models to have an r^2 of 0.90 or higher. However, a business researcher who is trying to develop a model to predict the motivation level of employees might be pleased to get an r^2 near 0.50 in the initial research.

The dependent variable, y, being predicted in a regression model has a variation that is measured by the sum of squares of y (SS_{yy}):

$$SS_{yy} = \Sigma(y - \bar{y})^2 = \Sigma y^2 - \frac{(\Sigma y)^2}{n}$$

and is the sum of the squared deviations of the y values from the mean value of y. This variation can be broken into two additive variations: the *explained variation,* measured by the sum of squares of regression (SSR), and the *unexplained variation,* measured by the sum of squares of error (SSE). This relationship can be expressed in equation form as

$$SS_{yy} = SSR + SSE$$

If each term in the equation is divided by SS_{yy}, the resulting equation is

$$1 = \frac{SSR}{SS_{yy}} + \frac{SSE}{SS_{yy}}$$

The term r^2 is the proportion of the y variability that is explained by the regression model and represented here as

$$r^2 = \frac{SSR}{SS_{yy}}$$

Substituting this equation into the preceding relationship gives

$$1 = r^2 + \frac{SSE}{SS_{yy}}$$

Solving for r^2 yields formula 12.5.

COEFFICIENT OF DETERMINATION (12.5)

$$r^2 = 1 - \frac{SSE}{SS_{yy}} = 1 - \frac{SSE}{\sum y^2 - \frac{(\sum y)^2}{n}}$$

Note: $0 \le r^2 \le 1$

The value of r^2 for the airline cost example is solved as follows:

$$SSE = 0.31434$$

$$SS_{yy} = \sum y^2 - \frac{(\sum y)^2}{n} = 270.9251 - \frac{(56.69)^2}{12} = 3.11209$$

$$r^2 = 1 - \frac{SSE}{SS_{yy}} = 1 - \frac{0.31434}{3.11209} = 0.899$$

That is, 89.9% of the variability of the cost of flying an Airbus A320 airplane on a commercial flight is explained by variations in the number of passengers. This result also means that 11.1% of the variance in airline flight cost, y, is unaccounted for by x or unexplained by the regression model.

The coefficient of determination can be solved for directly by using

$$r^2 = \frac{SSR}{SS_{yy}}$$

It can be shown through algebra that

$$SSR = b_1^2 SS_{xx}$$

From this equation, a computational formula for r^2 can be developed.

COMPUTATIONAL FORMULA FOR r^2

$$r^2 = \frac{b_1^2 SS_{xx}}{SS_{yy}}$$

For the airline cost example, $b_1 = 0.0407016$, $SS_{xx} = 1{,}689$, and $SS_{yy} = 3.11209$. Using the computational formula for r^2 yields

$$r^2 = \frac{(0.0407016)^2(1689)}{3.11209} = 0.899$$

DEMONSTRATION PROBLEM 12.4

Compute the coefficient of determination (r^2) for Demonstration Problem 12.1, in which a regression model was developed to predict the number of FTEs of a hospital by the number of beds.

Solution

$$SSE = 2448.86$$

$$SS_{yy} = 260{,}136 - \frac{(1692)^2}{12} = 21{,}564$$

$$r^2 = 1 - \frac{SSE}{SS_{yy}} = 1 - \frac{2448.86}{21{,}564} = 0.886$$

This regression model accounts for 88.6% of the variance in FTEs, leaving only 11.4% unexplained variance.

Using $SS_{xx} = 3{,}838.667$ and $b_1 = 2.232$ from Demonstration Problem 12.1, we can solve for r^2 with the computational formula:

$$r^2 = \frac{b_1^2 SS_{xx}}{SS_{yy}} = \frac{(2.232)^2(3838.667)}{21{,}564} = 0.886$$

Relationship Between r and r^2

Is r, the coefficient of correlation (introduced in Section 12.1), related to r^2, the coefficient of determination in linear regression? The answer is yes: r^2 equals $(r)^2$. The coefficient of determination is the square of the coefficient of correlation. In Demonstration Problem 12.1, a regression model was developed to predict FTEs by number of hospital beds. The r^2 value for the model was 0.886. Taking the square root of this value yields $r = 0.941$, which is the correlation between the sample number of beds and FTEs. A word of caution here: because r^2 is always positive, solving for r by taking $\sqrt{r^2}$ gives the correct magnitude of r but may give the wrong sign. The researcher must examine the sign of the slope of the regression line to determine whether a positive or negative relationship exists between the variables and then assign the appropriate sign to the correlation value.

12.6 PROBLEMS

12.32 Compute r^2 for Problem 12.24 (Problem 12.6). Discuss the value of r^2 obtained.

12.33 Compute r^2 for Problem 12.25 (Problem 12.7). Discuss the value of r^2 obtained.

12.34 Compute r^2 for Problem 12.26 (Problem 12.8). Discuss the value of r^2 obtained.

12.35 Compute r^2 for Problem 12.27 (Problem 12.9). Discuss the value of r^2 obtained.

12.36 In problem 12.10, you were asked to develop the equation of a regression model to predict the amount of sales by the amount spent on mobile advertisements. For this regression model, solve for the coefficient of determination and comment on it.

12.37 The Office for National Statistics produces a monthly report on car production in the United Kingdom. Some researchers may feel that the level of car production is a function of the level of confidence consumers have in the economy based on existing employment opportunities and their own income prospects. Shown here is the total car production index for a 20-month period and the consumer confidence index for the same 20 months published by the building society Nationwide. Determine the equation of the regression line to predict the total car production in the UK from the consumer confidence index. Compute the standard error of the estimate for this model. Compute the value of r^2. Does the consumer confidence index appear to be a good predictor of the total car production index? Why or why not?

Nationwide Consumer Confidence Index	Total Car Production Final Index
100	91.8
101	88.6
96	91
95	93
92	93
89	89.5
92	89.4
95	87.1
86	80.8
88	81.9
94	86.5
90	83
86	84.2
86	85.3
87	82.9
88	87
89	90.5
94	88.5
93	91.2
98	94.8

12.7 TESTS FOR THE SLOPE OF THE REGRESSION MODEL AND TESTING THE OVERALL MODEL

Testing the Slope

A hypothesis test can be conducted on the sample slope of the regression model to determine whether the population slope is significantly different from zero. This test is another way to determine how well a regression model fits the data. Suppose a researcher decides that it is not worth the effort to develop a linear regression model to predict y from x. An alternative approach might be to average the y values and use $\bar{y}$ as the predictor of y for all values of x. For the airline cost example, instead of using number of passengers as the predictor, the researcher would use the average value of airline cost, $\bar{y}$, as the predictor. In this case the average value of y is

$$\bar{y} = \frac{56.69}{12} = 4.7242, \text{ or } €4,724.20$$

Using this result as a model to predict y, if the number of passengers is 61, 70, or 95 – or any other number – the predicted value of y is still 4.7242. Essentially, this approach fits the line of $\bar{y} = 4.7242$, through the data, which is a horizontal line with a slope of zero. Would a regression analysis offer anything more than the $\bar{y}$ model? Using this non-regression model (the $\bar{y}$ model) as a worst case, the researcher can analyse the regression line to determine whether it adds a more significant amount of predictability of y than does the $\bar{y}$ model. Because the slope of the $\bar{y}$ line is zero, one way to determine whether the regression line adds significant predictability is to test the population slope of the regression line to find out whether the slope is different from zero. As the slope of the regression line diverges from zero, the regression model is adding predictability that the $\bar{y}$ line is not generating. For this reason, testing the slope of the regression line to determine whether the slope is different from zero is important. If the slope is not different from zero, the regression line is doing nothing more than the line in predicting y.

How does the researcher go about testing the slope of the regression line? Why not just examine the observed sample slope? For example, the slope of the regression line for the airline cost data is 0.0407. This value is obviously not zero. The problem is that this slope is obtained from a sample of 12 data points; and if another sample was taken, it is likely that a different slope would be obtained. For this reason, the population slope is statistically tested using the sample slope. The question is: if all the pairs of data points for the population were available, would the slope of that regression line be different from zero? Here the sample slope, b_1, is used as evidence to test whether the population slope is different from zero. The hypotheses for this test follow.

$$H_0: \beta_1 = 0$$
$$H_a: \beta_1 = 0$$

Note that this test is two tailed. The null hypothesis can be rejected if the slope is either negative or positive. A negative slope indicates an inverse relationship between x and y. That is, larger values of x are related to smaller values of y, and vice versa. Both negative and positive slopes can be different from zero. To determine whether there is a significant *positive* relationship between two variables, the hypotheses would be one tailed, or

$$H_0: \beta_1 = 0$$
$$H_a: \beta_1 > 0$$

To test for a significant *negative* relationship between two variables, the hypotheses also would be one tailed, or

$$H_0: \beta_1 = 0$$
$$H_a: \beta_1 < 0$$

In each case, testing the null hypothesis involves a t test of the slope.

t TEST OF SLOPE

$$t = \frac{b_1 - \beta_1}{s_b}$$

where:

$$s_b = \frac{s_e}{\sqrt{SS_{xx}}}$$

$$s_e = \sqrt{\frac{SSE}{n-2}}$$

$$SS_{xx} = \Sigma x^2 - \frac{(\Sigma x)^2}{n}$$

$\beta_1 = $ the hypothesized slope
df $= n - 2$

The test of the slope of the regression line for the airline cost regression model for $\alpha = 0.05$ follows. The regression line derived for the data is

$$\hat{y} = 1.57 + 0.0407x$$

The sample slope is $0.0407 = b_1$. The value of s_e is 0.1773, $\Sigma x = 930$, $\Sigma x^2 = 73{,}764$, and $n = 12$. The hypotheses are

$$H_0: \beta_1 = 0$$

$$H_a: \beta_1 \text{ different } 0$$

The $df = n - 2 = 12 - 2 = 10$. As this test is two tailed, $\alpha/2 = 0.025$. The table t value is $t_{0.025,10} = \pm 2.228$. The observed t value for this sample slope is

$$t = \frac{0.0407 - 0}{0.173 \Big/ \sqrt{73{,}764 - \dfrac{(930)^2}{12}}} = 9.43$$

As shown in Figure 12.14, the t value calculated from the sample slope falls in the rejection region and the p-value is 0.00000014. The null hypothesis that the population slope is zero is rejected. This linear regression model is adding significantly more predictive information to the $\hat{y}$ model (no regression).

It is desirable to reject the null hypothesis in testing the slope of the regression model. In rejecting the null hypothesis of a zero population slope, we are stating that the regression model is adding something to the explanation of the variation of the dependent variable that the average value of y model does not. Failure to reject the null hypothesis in

---

STATISTICS IN BUSINESS TODAY

Predicting the Price of a Car

What variables are good predictors of the base price of a new car? We can consider several variables as potentially good predictors of the base price: power (bhp), engine size, weight, and luggage capacity. The following data on 6 new diesel cars were obtained from a car comparison website. The car models are Audi A6 2.0 TDi, BMW 520d SE, Mercedes B180 CDi SE, VW Passat S Tdi 1.6, Citroen C5 Tourer Hdi 110 VTR Estate, and Volvo C70 D3. The base prices of these six models ranged from £19,655 to £32,495. Suppose a business researcher wanted to develop a regression model to predict the base price of these cars. What variable would be the strongest predictor and how strong would the prediction be?

Using a correlation matrix constructed from the data for the five variables, it was determined that power (bhp) was most correlated with base price and had the greatest potential as a predictor. Engine size had the second highest correlation with base price, followed by gross weight. Luggage capacity was negatively related to base price, indicating that the more expensive cars tended to have smaller boots.

A regression model was developed using power (bhp) as a predictor of base price. The Minitab output

from the data follows. Excel output contains similar items.

Regression Analysis: Base Price versus Weight

```
The regression equation is
Base Price = 5201 + 148 Power(bhp)
Predictor      Coef   SE Coef      T       P
Constant       5201      3497    1.49   0.211
Power(bhp)   147.94     24.74    5.98   0.004
S = 1941.31  R-Sq = 89.9%  R-Sq(adj) = 87.4%
```

Note that the r^2 for this model is almost 90% and that the t statistic is significant at $\alpha = 0.01$. In the regression equation, the slope indicates that for every unit of break horse power increase there is a £147.95 increase in the price. The y-intercept indicates that if the car had no power at all, it would still cost £5,201! The standard error of the estimate is £1,941.31.

Regression models were developed for each of the other possible predictor variables. Engine size was the next best predictor variable producing an r^2 of 53.7%. Gross weight produced an r^2 of 39.1%, and luggage capacity produced an r^2 of 18%.

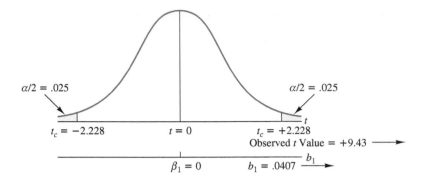

FIGURE 12.14

t Test of Slope from Airline Cost Example

this test causes the researcher to conclude that the regression model has no predictability of the dependent variable, and the model, therefore, has little or no use.

DEMONSTRATION PROBLEM 12.5

Test the slope of the regression model developed in Demonstration Problem 12.1 to predict the number of FTEs in a hospital from the number of beds to determine whether there is a significant positive slope. Use $\alpha = 0.01$.

Solution

The hypotheses for this problem are

$$H_0: \beta_1 = 0$$

$$H_a: \beta_1 > 0$$

The level of significance is 0.01. With 12 pairs of data, df = 10. The critical table *t* value is $t_{0.01,10} = 2.764$. The regression line equation for this problem is

$$\hat{y} = 30.888 + 2.232x$$

The sample slope, b_1, is 2.232, and $s_e = 15.65$, $\Sigma x = 592$, $\Sigma x^2 = 33,044$, and $n = 12$. The observed *t* value for the sample slope is

$$t = \frac{2.232 - 0}{15.65 / \sqrt{33,044 - \frac{(592)^2}{12}}} = 8.84$$

The observed *t* value (8.84) is in the rejection region because it is greater than the critical table *t* value of 2.764 and the *p*-value is 0.0000024. The null hypothesis is rejected. The population slope for this regression line is significantly different from zero in the positive direction. This regression model is adding significant predictability over the $\bar{y}$ model.

Testing the Overall Model

It is common in regression analysis to compute an *F* test to determine the overall significance of the model. Most computer software packages include the *F* test and its associated ANOVA table as standard regression output. In multiple regression (Chapters 13 and 14), this test determines whether at least one of the regression coefficients (from multiple predictors) is different from zero. Simple regression provides only one predictor and only one regression coefficient to test. Because the regression coefficient is the slope of the regression line, the F test for overall significance is testing the same thing as the t test in simple regression. The hypotheses being tested in simple regression by the F test for overall significance are

$$H_0: \beta_1 = 0$$

$$H_a: \beta_1 \text{ different } 0$$

In the case of simple regression analysis, $F = t^2$. Thus, for the airline cost example, the F value is

$$F = t^2 = (9.43)^2 = 88.92$$

The F value is computed directly by

$$F = \frac{SS_{reg}\big/df_{reg}}{SS_{err}\big/df_{err}} = \frac{MS_{reg}}{MS_{err}}$$

where:

$df_{reg} = k$
$df_{err} = n - k - 1$
$k = $ the number of independent variables

The values of the sum of squares (SS), degrees of freedom (df), and mean squares (MS) are obtained from the analysis of variance table, which is produced with other regression statistics as standard output from statistical software packages. Shown here is the analysis of variance table produced by Minitab for the airline cost example.

```
Analysis of Variance
Source            DF         SS         MS         F          p
Regression        1          2.7980     2.7980     89.09      0.000
Residual Error    10         0.3141     0.0314
Total             11         3.1121
```

The F value for the airline cost example is calculated from the analysis of variance table information as

$$F = \frac{2.7980\big/1}{0.3141\big/10} = \frac{2.7980}{0.03141} = 89.09$$

The difference between this value (89.09) and the value obtained by squaring the t statistic (88.92) is due to rounding error. The probability of obtaining an F value this large or larger by chance if there is no regression prediction in this model is 0.000 according to the ANOVA output (the p-value). This output value means it is highly unlikely that the population slope is zero and that there is no prediction due to regression from this model given the sample statistics obtained. Hence, it is highly likely that this regression model adds significant predictability of the dependent variable.

Note from the ANOVA table that the degrees of freedom due to regression are equal to 1. Simple regression models have only one independent variable; therefore, $k = 1$. The degrees of freedom error in simple regression analysis is always $n - k - 1 = n - 1 - 1 = n - 2$. With the degrees of freedom due to regression (1) as the numerator degrees of freedom and the degrees of freedom due to error ($n - 2$) as the denominator degrees of freedom, Table A.7 in Appendix A can be used to obtain the critical F value ($F_{\alpha,1,n-2}$) to help make the hypothesis testing decision about the overall regression model if the p-value of F is not given in the computer output. This critical F value is always found in the right tail of the distribution. In simple regression, the relationship between the critical t value to test the slope and the critical F value of overall significance is

$$t^2_{\alpha/2,n-2} = F_{\alpha,1,n-2}$$

For the airline cost example with a two-tailed test and $\alpha = 0.05$, the critical value of $t_{0.025,10}$ is ± 2.228 and the critical value of $F_{0.205,10}$ is 4.96.

$$t^2_{0.225,10} = (\pm 2.228)^2 = 4.96 = F_{0.05,1,10}$$

12.7 PROBLEMS

12.38   Test the slope of the regression line determined in Problem 12.6. Use $\alpha = 0.05$.

12.39   Test the slope of the regression line determined in Problem 12.7. Use $\alpha = 0.01$.

12.40   Test the slope of the regression line determined in Problem 12.8. Use $\alpha = 0.10$.

12.41   Test the slope of the regression line determined in Problem 12.9. Use a 5% level of significance.

12.42   Test the slope of the regression line developed in Problem 12.10. Use a 5% level of significance.

12.43   Study the following analysis of variance table, which was generated from a simple regression analysis. Discuss the F test of the overall model. Determine the value of t and test the slope of the regression line.

```
Analysis of Variance
Source        DF        SS         MS        F        p
Regression     1      116.65     116.65     8.26     0.021
Error          8      112.95      14.12
Total          9      229.60
```

12.8 ESTIMATION

One of the main uses of regression analysis is as a prediction tool. If the regression function is a good model, the researcher can use the regression equation to determine values of the dependent variable from various values of the independent variable. For example, financial brokers would like to have a model with which they could predict the selling price of a stock on a certain day by a variable such as unemployment rate or producer price index. Marketing managers would like to have a site location model with which they could predict the sales volume of a new location by variables such as population density or number of competitors. The airline cost example presents a regression model that has the potential to predict the cost of flying an airplane by the number of passengers.

In simple regression analysis, a point estimate prediction of y can be made by substituting the associated value of x into the regression equation and solving for y. From the airline cost example, if the number of passengers is $x = 73$, the predicted cost of the airline flight can be computed by substituting the x value into the regression equation determined in Section 12.3:

$$\hat{y} = 1.57 + 0.0407x = 1.57 + 0.0407(73) = 4.5411$$

The point estimate of the predicted cost is 4.5411 or €4,541.10.

Confidence Intervals to Estimate the Conditional Mean of y: $\mu_{y|x}$

Although a point estimate is often of interest to the researcher, the regression line is determined by a sample set of points; and if a different sample is taken, a different line will result, yielding a different point estimate. Hence computing a *confidence interval* for the estimation is often useful. Because for any value of x (independent variable) there can be many values of y (dependent variable), one type of **confidence interval** is *an estimate of the average value of y for a given x*. This average value of y is denoted $E(y_x)$ – the expected value of y and can be computed using formula (12.6).

$$\hat{y} \pm t_{\alpha/2,n-2}s_e\sqrt{\frac{1}{n}+\frac{(x_0-\bar{x})^2}{SS_{xx}}}$$

CONFIDENCE INTERVAL TO ESTIMATE $E(y_x)$
FOR A GIVEN VALUE OF x (12.6)

where:
x_0 = a particular value of x

$$SS_{xx} = \Sigma x^2 - \frac{(\Sigma x)^2}{n}$$

The application of this formula can be illustrated with construction of a 95% confidence interval to estimate the average value of y (airline cost) for the airline cost example when x (number of passengers) is 73. For a 95% confidence interval, $\alpha = 0.05$ and $\alpha/2 = 0.025$. The df $= n - 2 = 12 - 2 = 10$. The table t value is $t_{0.025,10} = 2.228$. Other needed values for this problem, which were solved for previously, are

$$s_e = 0.1773 \quad \Sigma x = 930 \quad \bar{x} = 77.5 \quad \Sigma x^2 = 73{,}764$$

For $x_0 = 73$, the value of $\hat{y}$ is 4.5411. The computed confidence interval for the average value of y, $E(y_{73})$, is

$$4.5411 \pm (2.228)(0.1773)\sqrt{\frac{1}{12}+\frac{(73-77.5)^2}{73{,}764-\frac{(930)^2}{12}}} = 4.5411 \pm 0.1220$$

$$4.4191 \leq E(y_{73}) \leq 4.6631$$

That is, with 95% confidence the average value of y for $x = 73$ is between 4.4191 and 4.6631.

Table 12.7 shows confidence intervals computed for the airline cost example for several values of x to estimate the average value of y. Note that as x values get farther from the mean x value (77.5), the confidence intervals get wider; as the x values get closer to the mean, the confidence intervals narrow. The reason is that the numerator of the second term under the radical sign approaches zero as the value of x nears the mean and increases as x departs from the mean.

TABLE 12.7

Confidence Intervals to Estimate the Average Value of y for Some x Values in the Airline Cost Example

x	Confidence Interval	
62	$4.0934 \pm .1876$	3.9058 to 4.2810
68	$4.3376 \pm .1461$	4.1915 to 4.4837
73	$4.5411 \pm .1220$	4.4191 to 4.6631
85	$5.0295 \pm .1349$	4.8946 to 5.1644
90	$5.2230 \pm .1656$	5.0574 to 5.3886

Prediction Intervals to Estimate a Single Value of y

A second type of interval in regression estimation is a **prediction interval** to *estimate a single value of y for a given value of x.*

$$\hat{y} \pm t_{\alpha/2,n-2}s_e\sqrt{1+\frac{1}{n}+\frac{(x_0-\bar{x})^2}{SS_{xx}}}$$

PREDICTION INTERVAL TO ESTIMATE y
FOR A GIVEN VALUE OF x (12.7)

where:
x_0 = a particular value of x

$$SS_{xx} = \Sigma x^2 - \frac{(\Sigma x)^2}{n}$$

Formula 12.7 is virtually the same as formula 12.6, except for the additional value of 1 under the radical. This additional value widens the prediction interval to estimate a single value of y from the confidence interval to estimate the average value of y. This result seems logical because the average value of y is towards the middle of a group of y values. Thus the confidence interval to estimate the average need not be as wide as the prediction interval produced by Formula 12.7, which takes into account all the y values for a given x.

A 95% prediction interval can be computed to estimate the single value of y for $x = 73$ from the airline cost example by using formula 12.7. The same values used to construct the confidence interval to estimate the average value of y are used here.

$$t_{0.025,10} = 2.228, \ s_e = 0.1773, \ \Sigma x = 930, \ \bar{x} = 77.5, \ \Sigma x^2 = 73{,}764$$

For $x_0 = 73$, the value of $\hat{y} = 4.5411$. The computed prediction interval for the single value of y is

$$4.5411 \pm (2.228)(0.1773) \sqrt{1 + \frac{1}{12} + \frac{(73 - 77.5)^2}{73{,}764 - \frac{(930)^2}{12}}} = 4.5411 \pm 0.4134$$

$$4.1277 \leq y \leq 4.9543$$

Prediction intervals can be obtained by using the computer. Shown in Figure 12.15 is the computer output for the airline cost example. The output displays the predicted value for $x = 73$ ($\hat{y} = 4.5411$), a 95% confidence interval for the average value of y for $x = 73$, and a 95% prediction interval for a single value of y for $x = 73$. Note that the resulting values are virtually the same as those calculated in this section.

Figure 12.16 displays Minitab confidence intervals for various values of x for the average y value and the prediction intervals for a single y value for the airline example. Note that the intervals flare out towards the ends, as the values of x depart from the average x value. Note also that the intervals for a single y value are always wider than the intervals for the average y value for any given value of x.

Fit	StDev Fit	95.0% CI	95.0 PI
4.5410	0.0547	(4.4191, 4.6629)	(4.1278, 4.9543)

FIGURE 12.15

Minitab Output for Prediction Intervals

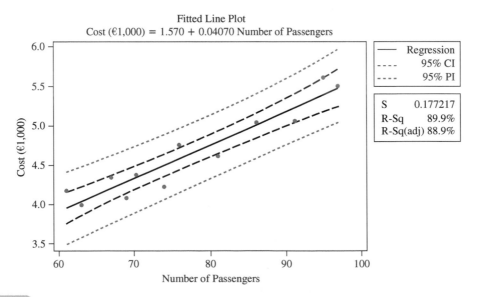

FIGURE 12.16

Minitab Intervals for Estimation

An examination of the prediction interval formula to estimate y for a given value of x explains why the intervals flare out.

$$\hat{y} \pm t_{\alpha/2,n-2} s_e \sqrt{1 + \frac{1}{n} + \frac{(x_0 - \bar{x})^2}{SS_{xx}}}$$

As we enter different values of x_0 from the regression analysis into the equation, the only thing that changes in the equation is $(x - \bar{x})^2$. This expression increases as individual values of x_0 get farther from the mean, resulting in an increase in the width of the interval. The interval is narrower for values of x_0 nearer $\bar{x}$ and wider for values of x_0 further from $\bar{x}$. A comparison of formulas 12.6 and 12.7 reveals them to be identical except that Formula 12.7 – to compute a prediction interval to estimate y for a given value of x – contains a 1 under the radical sign. This distinction ensures that Formula 12.7 will yield wider intervals than 12.6 for otherwise identical data.

Caution: *A regression line is determined from a sample of points. The line, the r2, the se, and the confidence intervals change for different sets of sample points. That is, the linear relationship developed for a set of points does not necessarily hold for values of x outside the domain of those used to establish the model. In the airline cost example, the domain of x values (number of passengers) varied from 61 to 97. The regression model developed from these points may not be valid for flights of say 40, 50, or 100 because the regression model was not constructed with x values of those magnitudes. However, decision makers sometimes extrapolate regression results to values of x beyond the domain of those used to develop the formulas (often in time-series sales forecasting). Understanding the limitations of this type of use of regression analysis is essential.*

DEMONSTRATION PROBLEM 12.6

Construct a 95% confidence interval to estimate the average value of y (FTEs) for Demonstration Problem 12.1 when $x = 40$ beds. Then construct a 95% prediction interval to estimate the single value of y for $x = 40$ beds.

Solution

For a 95% confidence interval, $\alpha = 0.05$, $n = 12$, and df $= 10$. The table t value is $t_{0.025,10} = 2.228$; $s_e = 15.65$, $\Sigma x = 592$, $\bar{x} = 49.33$, and $\Sigma x^2 = 33{,}044$. For $x_0 = 40$, $\hat{y} = 120.17$. The computed confidence interval for the average value of y is

$$120.17 \pm (2.228)(15.65) \sqrt{\frac{1}{12} + \frac{(40 - 49.33)^2}{33{,}044 - \frac{(592)^2}{12}}} = 120.17 \pm 11.35$$

$$108.82 \leq E(y_{40}) \leq 131.52$$

With 95% confidence, the statement can be made that the average number of FTEs for a hospital with 40 beds is between 108.82 and 131.52.

The computed prediction interval for the single value of y is

$$120.17 \pm (2.228)(15.65) \sqrt{1 + \frac{1}{12} + \frac{(40 - 49.33)^2}{33{,}044 - \frac{(592)^2}{12}}} = 120.17 \pm 36.67$$

$$83.5 \leq Y \leq 156.84$$

With 95% confidence, the statement can be made that a single number of FTEs for a hospital with 40 beds is between 83.5 and 156.84. Obviously this interval is much wider than the 95% confidence interval for the average value of y for $x = 40$.

The following Minitab graph depicts the 95% interval bands for both the average y value and the single y values for all 12 x values in this problem. Note once again the flaring out of the bands near the extreme values of x.

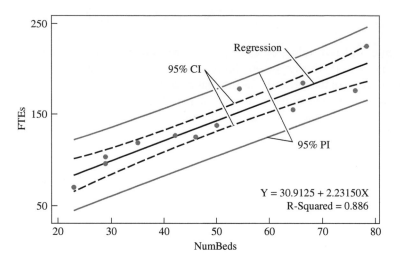

12.8 PROBLEMS

12.44 Construct a 95% confidence interval for the average value of y for Problem 12.6. Use $x = 25$.

12.45 Construct a 90% prediction interval for a single value of y for Problem 12.7; use $x = 100$. Construct a 90% prediction interval for a single value of y for Problem 12.7; use $x = 130$. Compare the results. Which prediction interval is greater? Why?

12.46 Construct a 98% confidence interval for the average value of y for Problem 12.8; use $x = 20$. Construct a 98% prediction interval for a single value of y for Problem 12.8; use $x = 20$. Which is wider? Why?

12.47 Construct a 99% confidence interval for the average bond rate in Problem 12.9 for a prime interest rate of 10%. Discuss the meaning of this confidence interval.

12.9 USING REGRESSION TO DEVELOP A FORECASTING TREND LINE

TABLE 12.8

Ten-Year Sales Data for DWD Chemicals

Year	Sales (£ millions)
2003	7.84
2004	12.26
2005	13.11
2006	15.78
2007	21.29
2008	25.68
2009	23.80
2010	26.43
2011	29.16
2012	33.06

Business researchers often use historical data with measures taken over time in an effort to forecast what might happen in the future. A particular type of data that often lends itself well to this analysis is **time-series data** defined as *data gathered on a particular characteristic over a period of time at regular intervals*. Some examples of time-series data are 10 years of weekly Dow Jones Industrial Averages, 12 months of daily oil production, or monthly consumption of coffee over a two-year period. To be useful to forecasters, time-series measurements need to be made in regular time intervals and arranged according to time of occurrence. As an example, consider the time-series sales data over a 10-year time period for the DWD Chemicals Company shown in Table 12.8. Note that the measurements (sales) are taken over time and that the sales figures are given on a yearly basis. Time-series data can also be reported daily, weekly, monthly, quarterly, semi-annually, or for other defined time periods.

It is generally believed that time-series data contain any one or combination of four elements: trend, cyclicality, seasonality, and irregularity. While each of these four elements will be discussed in greater deal in Chapter 15, Time-Series Forecasting and Index Numbers, here we examine **trend** and define it as *the long-term general direction of data*. Observing the scatter plot of the DWD Chemical Company's sales data shown in Figure 12.17, it is apparent that there is positive trend in the data. That is, there appears to be a long-term upward general direction of sales over time. How can trend be expressed in mathematical terms? In the field of forecasting, it is common to attempt to fit a trend line through time-series data by determining the equation of the trend line and then using the equation of the trend line to predict future data points. How does one go about developing such a line?

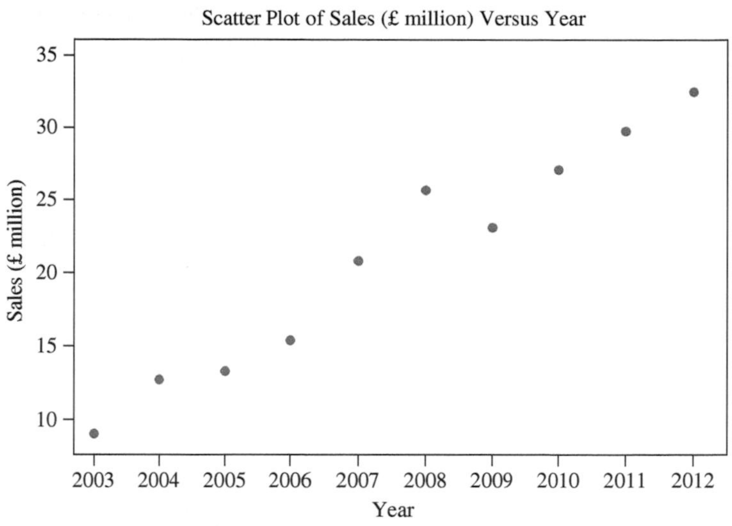

Scatter Plot of Sales (£ million) Versus Year

Determining the Equation of the Trend Line

FIGURE 12.17

Minitab Scatter Plot of DWD Sales Data

Developing the equation of a linear trend line in forecasting is actually a special case of simple regression where the y or dependent variable is the variable of interest that a business analyst wants to forecast and for which a set of measurements has been taken over a period of time. For example, with the DWD Chemicals Company data, if company forecasters want to predict sales for the year 2015 using these data, sales would be the dependent variable in the simple regression analysis. In linear trend analysis, the time period is used as the x, the independent or predictor variable, in the analysis to determine the equation of the trend line. In the case of the DWD Chemicals Company, the x variable represents the years 2003–2012.

Using sales as the y variable and time (year) as the x variable, the equation of the trend line can be calculated in the usual way as shown in Table 12.9 and is determined to be: $\hat{y} = -5,320.56 + 2.6687x$. The slope, 2.6687, means that for

TABLE 12.9

Determining the Equation of the Trend Line for the DWD Chemicals Company Sales Data

Year x	Sales y	x^2	xy
2003	7.84	4,000,000	15,680.00
2004	12.26	4,004,001	24,532.26
2005	13.11	4,008,004	26,246.22
2006	15.78	4,012,009	31,607.34
2007	21.29	4,016,016	42,665.16
2008	25.68	4,020,025	51,488.40
2009	23.80	4,024,036	47,742.80
2010	26.43	4,028,049	53,045.01
2011	29.16	4,032,064	58,553.28
2012	33.06	4,036,081	66,417.54
$\Sigma x = 20,045$	$\Sigma y = 208.41$	$\Sigma x^2 = 40,180,285$	$\Sigma xy = 417,978.01$

$$b_1 = \frac{\Sigma xy - \frac{(\Sigma x)(\Sigma y)}{n}}{\Sigma x^2 - \frac{(\Sigma x)^2}{n}} = \frac{(417,978.01) - \frac{(20,045)(208.41)}{10}}{40,180,285 - \frac{(20,045)^2}{10}} = \frac{220.17}{82.5} = 2.6687$$

$$b_0 = \frac{\Sigma y}{n} - b_1 \frac{\Sigma x}{n} = \frac{208.41}{10} - (2.6687)\frac{20,045}{10} = -5,328.57$$

Equation of the Trend Line: $\hat{y} = -5,328.57 + 2.6687x$

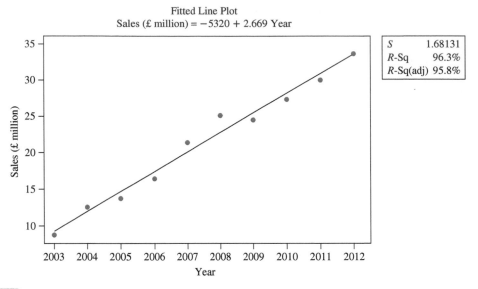

Fitted Line Plot
Sales (£ million) = −5320 + 2.669 Year

FIGURE 12.18

Minitab Graph of DWD Sales Data with a Fitted Trend Line

every yearly increase in time, sales increase by an average of £2.6687 (million). The intercept would represent company sales in the year 0 which, of course, in this problem has no meaning since the DWD Chemicals Company was not in existence in the year 0. Figure 12.18 is a Minitab display of the DWD sales data with the fitted trend line. Note that the output contains the equation of the trend line along with the values of s (standard error of the estimate) and R-Sq (r^2). As is typical with data that have a relatively strong trend, the r^2 value (0.963) is quite high.

Forecasting Using the Equation of the Trend Line

The main use of the equation of a trend line by business analysts is for forecasting outcomes for time periods in the future. Recall the caution from Section 12.8 that using a regression model to predict y values for x values outside the domain of those used to develop the model may not be valid. Despite this caution and understanding the potential drawbacks, business forecasters nevertheless extrapolate trend lines beyond the most current time periods of the data and attempt to predict outcomes for time periods in the future. To forecast for future time periods using a trend line, insert the time period of interest into the equation of the trend line and solve for. For example, suppose forecasters for the DWD Chemicals Company want to predict sales for the year 2015 using the equation of the trend line developed from their historical time series data. Replacing x in the equation of the sales trend line with 2015, results in a forecast of £40.85 (million):

$$\hat{y}(2012) = -5,328.57 + 2.6687(2012) = 4085$$

Figure 12.19 shows Minitab output for the DWD Chemicals Company data with the trend line through the data and graphical forecasts for the next three periods (2013, 2014, and 2015). Observe from the graph that the forecast for 2015 is about £41 (million).

Alternative Coding for Time Periods

If you manually calculate the equation of a trend line when the time periods are years, you notice that the calculations can get quite large and cumbersome (observe Table 12.9). However, if the years are consecutive, they can be recoded using many different possible schemes and still produce a meaningful trend line equation (albeit a different y intercept value). For example, instead of using the years 2003–2012, suppose we use the years 1 to 10. That is, 2003 = 1 (first year), 2004 = 2, 2005 = 3, and so on, to 2012 = 10. This recoding scheme produces the trend line equation of: $\hat{y} = -6.1632 + 2.6687x$ as shown in Table 12.10. Notice that the slope of the trend line is the same whether the years 2003 through 2012 are used or

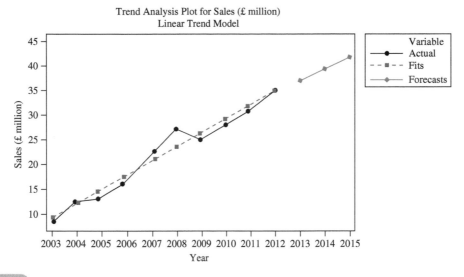

FIGURE 12.19

Minitab Output for Trend Line and Forecasts

the recoded years of 1 through 10, but the y intercept (6.1632) is different. This needs to be taken into consideration when using the equation of the trend line for forecasting. Since the new trend equation was derived from recoded data, forecasts will also need to be made using recoded data. For example, using the recoded system of 1 through 10 to represent 'years', the year 2015 is recoded as 13 (2012 = 10, 2013 = 11, 2014 = 12, and 2015 = 13). Inserting this value into the trend line equation results in a forecast of $40.86, the same as the value obtained using raw years as time.

$$\hat{y} = 6.1632 + 2.6687x = 6.1632 + 2.6687(13) = £40.86 \text{ (million)}$$

TABLE 12.10

Using Recoded Data to Calculate the Trend Line Equation

Year x	Sales y	x^2	xy
1	7.84	1	7.84
2	12.26	4	24.52
3	13.11	9	39.33
4	15.78	16	63.12
5	21.29	25	106.45
6	25.68	36	154.08
7	23.80	49	166.60
8	26.43	64	211.44
9	29.16	81	262.44
10	33.06	100	330.60
$\Sigma x = 55$	$\Sigma y = 208.41$	$\Sigma x^2 = 385$	$\Sigma xy = 1{,}366.42$

$$b_1 = \frac{\Sigma xy - \dfrac{(\Sigma x)(\Sigma y)}{n}}{\Sigma x^2 - \dfrac{(\Sigma x)^2}{n}} = \frac{(1{,}366.42) - \dfrac{(55)(208.41)}{10}}{385 - \dfrac{(55)^2}{10}} = \frac{220.165}{82.5} = 2.6687$$

$$b_0 = \frac{\Sigma y}{n} - b_1 \frac{\Sigma x}{n} = \frac{208.41}{10} - (2.6687)\frac{55}{10} = 6.1632$$

Equation of the Trend Line: $\hat{y} = -6.1632 + 2.6687x$

Similar time recoding schemes can be used in the calculating of trend line equations when the time variable is something other than years. For example, in the case of monthly time series data, the time periods can be recoded as:

January = 1, February = 2, March = 3, ..., December = 12

In the case of quarterly data over a two-year period, the time periods can be recoded with a scheme such as:

	Time Period	Recoded Time Period
Year 1:	Quarter 1	1
	Quarter 2	2
	Quarter 3	3
	Quarter 4	4
Year 2:	Quarter 1	5
	Quarter 2	6
	Quarter 3	7
	Quarter 4	8

DEMONSTRATION PROBLEM 12.7

Shown below are monthly food and beverage sales in a European country during a recent year over an eight-month period. Develop the equation of a trend line through these data and use the equation to forecast sales for October.

Month	Sales (€ million)
January	32,569
February	32,274
March	32,583
April	32,304
May	32,149
June	32,077
July	31,989
August	31,977

Solution

Shown here is a Minitab-produced scatter diagram of these time series data:

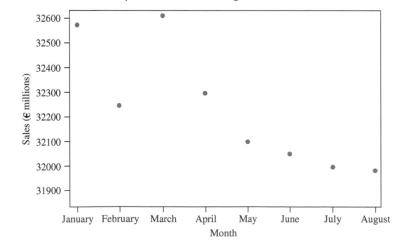

The months of January through August can be coded using the numbers of 1 through 8, respectively. Using these numbers as the time period values (x) and sales as the dependent variable (y), the following output was obtained from Minitab:

Regression Analysis: Sales versus Month

The regression equation is
Sales = 32628 − 86.2 Month

Predictor	Coef	SE Coef	T	P
Constant	32628.2	93.3	349.80	0.000
Month	-86.21	18.47	-4.67	0.003

S = 119.708   R-Sq = 78.4%   R-Sq(adj) = 74.8%

The equation of the trend line is: $\hat{y} = 32{,}628.2 - 86.21x$. A slope of -86.21 indicates that there is a downward trend of food and beverage sales over this period of time at a rate of €86.21 (million) per month. The y intercept of 32,628.2 represents what the trend line would estimate the sales to have been in period 0 or December of the previous year. The sales figure for October can be forecast by inserting $x = 10$ into this model and obtaining:

$$\hat{y}(10) = 32{,}628.2 - 86.21(10) = 31{,}766.1.$$

12.9 PROBLEMS

12.48 Determine the equation of the trend line for the data shown below on the EU area's trade balance with China over a nine-year period (provided by Eurostat). Using the trend line equation, forecast the value for the year 2012.

Year	EU Area Trade Balance with China (€ millions)
2000	−32,771
2001	−32,550
2002	−32,708
2003	−40,386
2004	−53,000
2005	−75,355
2006	−91,184
2007	−112,335
2008	−119,749
2009	−88,652
2010	−114,027

12.49 Shown below are the average attendance figures for Manchester United, the UK football team with the largest average match attendance over an eight-football-season period (according to PremierLeague.com). Use these data to construct a trend line and forecast the average attendance for the 2012/13 football season.

Football Season	Average attendance
2003/04	67,641
2004/05	67,871
2005/06	68,765
2006/07	75,826
2007/08	75,691
2008/09	75,304
2009/10	74,864
2010/11	75,109

12.50   After a slight decrease following the 2007–2008 global financial crisis, e-commerce sales in the United States have continued their strong growth for the past several years. Shown below are quarterly e-commerce sales figures released by the Census Bureau for the United States over a three-year period. Use these data to determine the equation of a trend line for e-commerce sales during this time and use the trend 'model' to forecast e-commerce sales for the third quarter of the year 2012.

Year	Quarter	Sales ($ millions)
2008	1	36,244
	2	36,691
	3	36,180
	4	33,393
2009	1	34,151
	2	35,133
	3	36,909
	4	38,269
2010	1	39,159
	2	40,472
	3	42,381
	4	44,517

Regression Analysis: Cost (€1,000) versus Number of Passengers

```
The regression equation is
Cost (€1,000) = 1.57 + 0.0407 Number of Passengers

Predictor                     Coef      SE Coef        T         P
Constant                    1.5698       0.3381     4.64     0.001
Number of Passengers      0.040702     0.004312     9.44     0.000

S = 0.177217    R-Sq = 89.9%    R-Sq(adj) = 88.9%

Analysis of Variance

Source              DF          SS          MS          F         P
Regression           1      2.7980      2.7980      89.09     0.000
Residual Error      10      0.3141      0.0314
Total               11      3.1121

           Number of       Cost
Obs        Passengers     (€1,000)      Fit      SE Fit    Residual
 1            61.0         4.2800      4.0526     0.0876      0.2274
 2            63.0         4.0800      4.1340     0.0808     -0.0540
 3            67.0         4.4200      4.2968     0.0683      0.1232
 4            69.0         4.1700      4.3782     0.0629     -0.2082
 5            70.0         4.4800      4.4189     0.0605      0.0611
 6            74.0         4.3000      4.5817     0.0533     -0.2817
 7            76.0         4.8200      4.6631     0.0516      0.1569
 8            81.0         4.7000      4.8666     0.0533     -0.1666
 9            86.0         5.1100      5.0701     0.0629      0.0399
10            91.0         5.1300      5.2736     0.0775     -0.1436
11            95.0         5.6400      5.4364     0.0912      0.2036
12            97.0         5.5600      5.5178     0.0984      0.0422
```

FIGURE 12.20

Minitab Regression Analysis of the Airline Cost Example

12.10 INTERPRETING THE OUTPUT

Although manual computations can be done, most regression problems are analysed by using a computer. In this section, computer output from both Minitab and Excel will be presented and discussed.

At the top of the Minitab regression output, shown in Figure 12.20, is the regression equation. Next is a table that describes the model in more detail. 'Coef' stands for coefficient of the regression terms. The coefficient of Number of Passengers, the x variable, is 0.040702. This value is equal to the slope of the regression line and is reflected in the regression equation. The coefficient shown next to the constant term (1.5698) is the value of the constant, which is the y intercept and also a part of the regression equation. The 'T' values are a t test for the slope and a t test for the intercept or constant. (We generally do not interpret the t test for the constant.) The t value for the slope, $t = 9.44$ with an associated probability of 0.000, is the same as the value obtained manually in section 12.7. Because the probability of the t value is given, the p-value method can be used to interpret the t value.

The next row of output is the standard error of the estimate s_e, $S = 0.177217$; the coefficient of determination, r^2, R-Sq = 89.9%; and the adjusted value of r^2, R-Sq(adj) = 88.9%. (Adjusted r^2 will be discussed in Chapter 13.) Following these items is the analysis of variance table. Note that the value of $F = 89.09$ is used to test the overall model of the regression line. The final item of the output is the predicted value and the corresponding residual for each pair of points.

Although the Excel regression output, shown in Figure 12.21 for Demonstration Problem 12.1, is somewhat different from the Minitab output, the same essential regression features are present. The regression equation is found

SUMMARY OUTPUT

Regression Statistics

Multiple R	0.942
R Square	0.886
Adjusted R Square	0.875
Standard Error	15.6491
Observations	12

ANOVA

	df	SS	MS	F	Significance F
Regression	1	19115.0632	19115.0632	78.05	0.000005
Residual	10	2448.9368	244.8937		
Total	11	21564			

	Coefficients	Standard Error	t Stat	P-value
Intercept	30.9125	13.2542	2.33	0.041888
Number of Beds	2.2315	0.2526	8.83	0.000005

RESIDUAL OUTPUT

Observation	Predicted FTEs	Residuals
1	82.237	−13.237
2	95.626	−0.626
3	95.626	6.374
4	109.015	8.985
5	124.636	1.364
6	133.562	−8.562
7	142.488	−4.488
8	151.414	26.586
9	173.729	−17.729
10	178.192	5.808
11	200.507	−24.507
12	204.970	20.030

FIGURE 12.21

Excel Regression Output for Demonstration Problem 12.1

under Coefficients at the bottom of ANOVA. The slope or coefficient of x is 2.2315 and the y-intercept is 30.9125. The standard error of the estimate for the hospital problem is given as the fourth statistic under Regression Statistics at the top of the output, Standard Error = 15.6491. The r^2 value is given as 0.886 on the second line. The t test for the slope is found under t Stat near the bottom of the ANOVA section on the 'Number of Beds' (x variable) row, $t = 8.83$. Adjacent to the t Stat is the P-value, which is the probability of the t statistic occurring by chance if the null hypothesis is true. For this slope, the probability shown is 0.000005. The ANOVA table is in the middle of the output with the F value having the same probability as the t statistic, 0.000005, and equalling t^2. The predicted values and the residuals are shown in the Residual Output section.

Decision Dilemma SOLVED

Predicting International Hourly Wages by the Price of a Big Mac

In the Decision Dilemma, questions were raised about the relationship between the price of a Big Mac hamburger and net hourly wages around the world and if a model could be developed to predict net hourly wages by the price of a Big Mac. Data were given for a sample of 27 countries. In exploring the possibility that there is a relationship between these two variables, a Pearson product-moment correlation coefficient, r, was computed to be 0.812. This r value indicates that there is a relatively high correlation between the two variables and that developing a regression model to predict one variable by the other has potential. Designating net hourly wages as the y or dependent variable and the price of a Big Mac as the x or predictor variable, the following regression output was obtained for these data using Excel.

Regression Statistics

Multiple R	0.812
R Square	0.660
Adjusted R Square	0.646
Standard Error	2.934
Observations	27

ANOVA

	df	SS	MS	F	Significance F
Regression	1	416.929	416.929	48.45	0.0000003
Residual	25	215.142	8.606		
Total	26	632.071			

	Coefficients	Standard Error	t Stat	P-Value
Intercept	−4.545	1.626	−2.79	0.009828805
Big Mac Price	4.741	0.681	6.96	0.0000003

Taken from this output, the regression model is:

Net Hourly Wage = $-4.545 + 4.741 = $ (Price of Big Mac)

While the y-intercept has virtually no practical meaning in this analysis, the slope indicates that for every dollar increase in the price of a Big Mac, there is an incremental increase of \$4.741 in net hourly wages for a country. It is worth underscoring here that just because there is a relationship between two variables, it does not mean there is a cause-and-effect relationship. That is, McDonald's cannot raise net hourly wages in a country just by increasing the cost of a Big Mac!

Using this regression model, the net hourly wage for a country with a \$3.00 Big Mac can be predicted by substituting $x = 3$ into the model:

Net Hourly Wage = $-4.545 + 4.741(3) = 9.68

That is, the model predicts that the net hourly wage for a country is \$9.68 when the price of a Big Mac is \$3.00. How good a fit is the regression model to the data? Observe from the Excel output that the F value for testing the overall significance of the model (48.45) is highly significant with a p-value of 0.0000003, and that the t statistic for testing to determine if the slope is significantly different from zero is 6.96 with a p-value of 0.0000003. In simple regression, the t statistic is the square root of the F value and these statistics relate essentially the same information – that there are significant regression effects in the model. The r^2 value is 66.0%, indicating that the model has moderate predictability. The standard error of the model, $s = 2.934$, indicates that if the error terms are approximately normally distributed, about 68% of the predicted net hourly wages would fall within ±$2.93.

Shown right is an Excel-produced line fit plot. Note from the plot that there generally appears be a linear relationship between the variables but that many of the data points fall considerably away from the fitted regression line, indicating that the price of a Big Mac only partially accounts for net hourly wages.

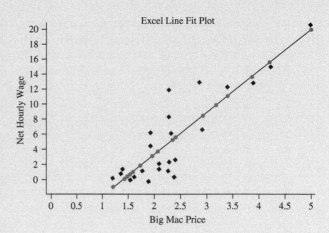

ETHICAL CONSIDERATIONS

Regression analysis offers several opportunities for unethical behaviour. One way is to present a regression model in isolation from information about the fit of the model. That is, the regression model is represented as a valid tool for prediction without any regard for how well it actually fits the data. While it is true that least squares analysis can produce a line of best fit through virtually any set of points, it does not necessarily follow that the regression model is a good predictor of the dependent variable. For example, sometimes business consultants sell regression models to companies as forecasting tools or market predictors without disclosing to the client that the r^2 value is very low, the slope of the regression line is not significant, the residuals are large, and the standard error of the estimate is large. This is unethical behaviour.

Another unethical use of simple regression analysis is stating or implying a cause-and-effect relationship between two variables just because they are highly correlated and produce a high r^2 in regression. The Decision Dilemma presents a good example of this with the regression analysis of the price of a Big Mac hamburger and the net hourly wages in a country. While the coefficient of determination is 66.0% and there appears to be a modest fit of the regression line to the data, that does not mean that increasing the price of a Big Mac in a given country will increase the country's net hourly wages. Often, two correlated variables are related to a third variable that drives the two of them but is not included in the regression analysis. In the Decision Dilemma example, both Big Mac prices and

net hourly wages may be related to exchange rates or a country's economic condition.

A third way that business analysts can act unethically in using regression analysis is to knowingly violate the assumptions underlying regression. Regression analysis requires equal error variance, independent error terms, and error terms that are normally distributed. Through the use of residual plots and other statistical techniques, a business researcher can test these assumptions. To present a regression model as fact when the assumptions underlying it are being grossly violated is unethical behaviour.

It is important to remember that since regression models are developed from sample data, when an x value is entered into a simple regression model, the resulting prediction is only a point estimate. While business people do often use regression models as predicting tools, it should be kept in mind that the prediction value is an estimate not a guaranteed outcome. By utilizing or at least pointing out confidence intervals and prediction intervals, such as those presented in Section 12.8, the business researcher places the predicted point estimate within the context of inferential estimation and is thereby acting more ethically.

And lastly, another ethical problem that arises in regression analysis is using the regression model to predict values of the independent variable that are outside the domain of values used to develop the model. The airline cost model used in this chapter was built with between 61 and 97 passengers. A linear relationship appeared to be evident between flight costs and number of passengers over this domain. This model is not guaranteed to fit values outside the domain of 61 to 97 passengers, however. In fact, either a non-linear relationship or no relationship may be present between flight costs and number of passengers if values from outside this domain are included in the model-building process. It is a mistake and probably unethical behaviour to make claims for a regression model outside the purview of the domain of values for which the model was developed.

SUMMARY

Correlation measures the degree of relatedness of variables. The most well-known measure of correlation is the Pearson product-moment coefficient of correlation, r. This value ranges from -1 to 0 to $+1$. An r value of $+1$ is perfect positive correlation and an r value of -1 is perfect negative correlation. Positive correlation means that as one variable increases in value, the other variable tends to increase. Negative correlation means that as one variable increases in value, the other variable tends to decrease. For r values near zero, little or no correlation is present.

Regression is a procedure that produces a mathematical model (function) that can be used to predict one variable by other variables. Simple regression is bivariate (two variables) and linear (only a line fit is attempted). Simple regression analysis produces a model that attempts to predict a y variable, referred to as the dependent variable, by an x variable, referred to as the independent variable. The general form of the equation of the simple regression line is the slope-intercept equation of a line. The equation of the simple regression model consists of a slope of the line as a coefficient of x and a y-intercept value as a constant.

After the equation of the line has been developed, several statistics are available that can be used to determine how well the line fits the data. Using the historical data values of x, predicted values of y (denoted as $\hat{y}$) can be calculated by inserting values of x into the regression equation. The predicted values can then be compared to the actual values of y to determine how well the regression equation fits the known data. The difference between a specific y value and its associated predicted y value is called the residual or error of prediction. Examination of the residuals can offer insight into the magnitude of the errors produced by a model. In addition, residual analysis can be used to help determine whether the assumptions underlying the regression analysis have been met. Specifically, graphs of the residuals can reveal (1) lack of linearity, (2) lack of homogeneity of error variance, and (3) independence of error terms. Geometrically, the residuals are the vertical distances from the y values to the regression line. Because the equation that yields the regression line is derived in such a way that the line is in the geometric middle of the points, the sum of the residuals is zero.

A single value of error measurement called the standard error of the estimate, s_e, can be computed. The standard error of the estimate is the standard deviation of error of a model. The value of s_e can be used as a single guide to the magnitude of the error produced by the regression model as opposed to examining all the residuals.

Another widely used statistic for testing the strength of a regression model is r^2, or the coefficient of determination. The coefficient of determination is the proportion of total variance of the y variable accounted for or predicted by x. The coefficient of determination ranges from 0 to 1. The higher the r^2 is, the stronger is the predictability of the model.

Testing to determine whether the slope of the regression line is different from zero is another way to judge the fit of the regression model to the data. If the population slope of the regression line is not different from zero, the regression model is not adding significant predictability to the dependent variable. A t statistic is used to test the significance of the slope. The overall significance of the regression model can be tested using an F statistic. In simple regression, because only one predictor is present, this test accomplishes the same thing as the t test of the slope and $F = t^2$.

One of the most prevalent uses of a regression model is to predict the values of y for given values of x. Recognizing that the predicted value is often not the same as the actual value, a confidence interval has been developed to yield a range within which the mean y value for a given x should fall. A prediction interval for a single y value for a given x value also is specified. This second interval is wider because it allows for the wide diversity of individual values, whereas the confidence interval for the mean y value reflects only the range of average y values for a given x.

Time-series data are data that are gathered over a period of time at regular intervals. Developing the equation of a forecasting trend line for time-series data is a special case of simple regression analysis where the time factor is the predictor variable. The time variable can be in units of years, months, weeks, quarters, and others.

KEY TERMS

coefficient of determination (r^2)
confidence interval
dependent variable
deterministic model
heteroscedasticity
homoscedasticity
independent variable
least squares analysis
outliers

prediction interval
probabilistic model
regression analysis
residual
residual plot
scatter plot
simple regression
standard error of the estimate (s_e)
sum of squares of error (SSE)

GO ONLINE TO DISCOVER THE EXTRA FEATURES FOR THIS CHAPTER

The Student Study Guide containing solutions to the odd-numbered questions, additional Quizzes and Concept Review Activities, Excel and Minitab databases, additional data files in Excel and Minitab, and more worked examples.
www.wiley.com/college/cortinhas

FORMULAS

Pearson's product-moment correlation coefficient

$$r = \frac{\Sigma(x-\bar{x})(y-\bar{y})}{\sqrt{\Sigma(x-\bar{x})^2 \Sigma(y-\bar{y})^2}}$$

$$= \frac{\Sigma xy - \frac{(\Sigma x \Sigma y)}{n}}{\sqrt{\left[\Sigma x^2 - \frac{(\Sigma x)^2}{n}\right]\left[\Sigma y^2 - \frac{(\Sigma y)^2}{n}\right]}}$$

Equation of the simple regression line

$$\hat{y} = \beta_0 + \beta_1 x$$

Sum of squares

$$SS_{xx} = \Sigma x^2 - \frac{(\Sigma x)^2}{n}$$

$$SS_{yy} = \Sigma y^2 - \frac{(\Sigma y)^2}{n}$$

$$SS_{xy} = \Sigma xy - \frac{\Sigma x \Sigma y}{n}$$

Slope of the regression line

$$b_1 = \frac{\Sigma(x-\bar{x})(y-\bar{y})}{\Sigma(x-\bar{x})^2} = \frac{\Sigma xy - n\bar{x}\,\bar{y}}{\Sigma x^2 - n\bar{x}^2}$$

$$= \frac{\Sigma xy - \frac{(\Sigma x)(\Sigma y)}{n}}{\Sigma x^2 - \frac{(\Sigma x)^2}{n}}$$

y-intercept of the regression line

$$b_0 = y^2 - b_1\bar{x} = \frac{\Sigma y}{n} - b_1\frac{(\Sigma x)}{n}$$

Sum of squares of error

$$SSE = \Sigma(y-\hat{y})^2 = \Sigma y^2 - b_0\Sigma y - b_1\Sigma xy$$

Standard error of the estimate

$$s_e = \sqrt{\frac{SSE}{n-2}}$$

Coefficient of determination

$$r^2 = 1 - \frac{SSE}{SS_{yy}} = 1 - \frac{SSE}{\Sigma y^2 - \frac{(\Sigma y)^2}{n}}$$

Computational formula for r^2

$$r^2 = \frac{b_1^2 SS_{xx}}{SS_{yy}}$$

t test of slope

$$t = \frac{b_1 - \beta_1}{s_b}$$

$$s_b = \frac{s_e}{\sqrt{SS_{xx}}}$$

Confidence interval to estimate $E(y_x)$ for a given value of x

$$\hat{y} \pm t_{\alpha/2, n-2} s_e \sqrt{\frac{1}{n} + \frac{(x_0 - \bar{x})^2}{SS_{xx}}}$$

Prediction interval to estimate y for a given value of x

$$\hat{y} \pm t_{\alpha/2, n-2} s_e \sqrt{1 + \frac{1}{n} + \frac{(x_0 - \bar{x})^2}{SS_{xx}}}$$

SUPPLEMENTARY PROBLEMS

CALCULATING THE STATISTICS

12.51 Determine the Pearson product-moment correlation coefficient for the following data:

x	1	10	9	6	5	3	2
y	8	4	4	5	7	7	9

12.52 Use the following data for parts (a) to (f).

x	5	7	3	16	12	9
y	8	9	11	27	15	13

a. Determine the equation of the least squares regression line to predict y by x.
b. Using the x values, solve for the predicted values of y and the residuals.
c. Solve for s_e.
d. Solve for r^2.
e. Test the slope of the regression line. Use $\sigma = 0.01$.
f. Comment on the results determined in parts (b) to (e), and make a statement about the fit of the line.

12.53 Use the following data for parts (a) to (g):

x	53	47	41	50	58	62	45	60
y	5	5	7	4	10	12	3	11

a. Determine the equation of the simple regression line to predict y from x.
b. Using the x values, solve for the predicted values of y and the residuals.
c. Solve for SSE.
d. Calculate the standard error of the estimate.
e. Determine the coefficient of determination.
f. Test the slope of the regression line. Assume $\alpha = 0.05$. What do you conclude about the slope?
g. Comment on parts (d) and (e).

12.54 If you were to develop a regression line to predict y by x, what value would the coefficient of determination have?

x	213	196	184	202	221	247
y	76	65	62	68	71	75

12.55 Determine the equation of the least squares regression line to predict y from the following data:

x	47	94	68	73	80	49	52	61
y	14	40	34	31	36	19	20	21

a. Construct a 95% confidence interval to estimate the mean y value for $x = 60$.
b. Construct a 95% prediction interval to estimate an individual y value for $x = 70$.
c. Interpret the results obtained in parts (a) and (b).

12.56 Determine the equation of the trend line through the cost data below. Use the equation of the line to forecast cost for year 7.

Year	Cost (€ millions)
1	56
2	54
3	49
4	46
5	45

TESTING YOUR UNDERSTANDING

12.57 A manager of a car dealership believes there is a relationship between the number of salespeople on duty and the number of cars sold. Suppose the sample below is used to develop a simple regression model to predict the number of cars sold by the number of salespeople. Solve for r^2 and explain what r^2 means in this problem.

Week	Number of Cars Sold	Number of Sales people
1	79	6
2	64	6
3	49	4
4	23	2
5	52	3

12.58 Executives of a video rental chain want to predict the success of a potential new store. The company's researcher begins by gathering information on number of rentals and average family income from several of the chain's present outlets.

Rentals	Average Family Income (£000s)
710	65
529	43
314	29
504	47
619	52
428	50
317	46
205	29
468	31
545	43
607	49
694	64

Develop a regression model to predict the number of rentals per day by the average family income. Comment on the output.

12.59 It seems logical to assume that banks with more capital would have greater assets. This assumption is mitigated, however, by several possibilities: some banks were more affected than others by the 2007–2008 global financial crisis, some banks may have different reserve requirements, and so on. The data shown here were published by Bankersalmanac.com. Perform a simple regression analysis to predict a bank's capital by its assets. How strong is the relationship?

Bank	Assets ($ millions)	Capital ($ millions)
BNP Paribas SA	2,675,627	34,428
Deutsche Bank AG	2,551,727	3,187
Barclays PLC	2,326,004	22,626

(continued)

Bank	Assets ($ millions)	Capital ($ millions)
Crédit Agricole SA	2,133,810	9,648
Industrial & Commercial Bank of China Limited	2,043,861	53,003
The Royal Bank of Scotland Group PLC	2,040,790	10,317
The Bank of Tokyo-Mitsubishi UFJ Ltd	1,644,768	18,293
China Construction Bank Corporation	1,641,683	37,967
JPMorgan Chase Bank National Association	1,631,621	1,785
Banco Santander SA	1,590,560	5,892
Bank of China Limited	1,588,462	42,392
Lloyds Banking Group PLC	1,574,667	2,457
Agricultural Bank of China Limited	1,569,865	49,324
Société Générale	1,515,897	1,249
Bank of America	1,482,278	3,020
UBS AG	1,409,574	410
Sumitomo Mitsui Banking Corporation	1,282,706	18,924
ING Bank NV	1,263,419	752

12.60 Shown here are the total employment labour force figures for the country of Albania over a 15-year period published by the International Monetary Fund. Develop the equation of a trend line through these data and use the equation to predict the total employment labour force of Albania for the year 2014.

Year	Total Employment (000s)
1996	158.16
1997	193.53
1998	235.04
1999	239.79
2000	215.09
2001	180.51
2002	172.39
2003	166.00
2004	159.00
2005	154.95
2006	150.332
2007	144.458
2008	140.599
2009	142.012
2010	143.877

12.61 How strong is the correlation between the inflation rate and 30-year treasury yields? The following data published by Fuji Securities are given as pairs of inflation rates and treasury yields for selected years over a 35-year period:

Inflation Rate	30-Year Treasure Yield
1.57%	3.05%
2.23	3.93
2.17	4.68
4.53	6.57
7.25	8.27
9.25	12.01
5.00	10.27
4.62	8.45

Compute the Pearson product-moment correlation coefficient to determine the strength of the correlation between these two variables. Comment on the strength and direction of the correlation.

12.62 According to a 2010 report by the Food and Agriculture Organization, the amount of inland capture fisheries production in tonnes has been increasing steadily over the years. However, the growth in the amount of inland capture fisheries production has varied greatly by country as shown in the data below. Some countries have remained relatively constant; some have dropped in terms of tonnes of fish caught, while others have grown significantly.

Inland Capture Fisheries	2004 (tonnes)	2008 (tonnes)
China	2,097,167	2,248,177
Bangladesh	732,067	1,060,181
India	527,290	953,106
Myanmar	454,260	814,740
Uganda	371,789	450,000
Cambodia	250,000	365,000
Indonesia	330,879	323,150
Nigeria	182,264	304,413
United Republic of Tanzania	312,040	281,690
Brazil	246,101	243,000
Egypt	282,099	237,572
Thailand	203,200	231,100
Democratic Republic of Kongo	231,772	23,000
Russian Federation	178,403	216,841

Develop a simple regression model to predict the 2008 inland capture fisheries by the 2004 inland

capture fisheries. According to the model, if a country had captured 270,000 tonnes in 2004, what would the predicted number be for 2008? Construct a confidence interval for the average y value for the 270,000 landings. Use the t statistic to test to determine whether the slope is significantly different from zero. Use $\alpha = 0.05$.

12.63 People in the aerospace industry believe the cost of a space project is a function of the weight of the major object being sent into space. Use the following data to develop a regression model to predict the cost of a space project by the weight of the space object. Determine r^2 and s_e.

Weight (tons)	Cost (€ millions)
1.897	53.6
3.019	184.9
0.453	6.4
0.988	23.5
1.058	33.4
2.100	110.4
2.387	104.6

12.64 The following data represent the CO_2 emissions per inhabitant and the real GDP per capita (in euros) of the European Union formed of 27 countries (EU-27) for a period of 14 years.

Time Period	CO_2 emissions	Real per capita GDP (€)
1	8.7	16,700
2	8.9	17,000
3	8.7	17,400
4	8.6	17,900
5	8.5	18,400
6	8.5	19,100
7	8.7	19,400
8	8.6	19,600
9	8.7	19,800
10	8.7	20,200
11	8.6	20,500
12	8.6	21,100
13	8.4	21,600
14	8.2	21,600

Develop a regression model to predict the CO_2 emissions in the EU-27 by the real per capita GDP. Comment on the result and on the strength of the model. Develop a time-series trend line for per capita GDP using the time periods given.

Forecast per capita GDP for period 16 using this equation.

12.65 Is the price of a digital camera a function of its megapixel resolution? Shown here are the price and megapixels for seven cameras as advertised by the comparison site *PriceRunner.co.uk*.

Model	Price (£)	Megapixels
Fujifilm FinePix S2950HD Black	132	14
Nikon D3100 + 18–55 mm VR Lens Kit	400	14.2
Nikon D3x + 8192MB	5,067	24.5
Canon EOS 5D Mark II + 24 – 105 mm Lens Kit	2,280	21.1
Sony Cyber-shot DSC-HX9V Black	287	16.2
Panasonic Lumix DMC-FT3 Red	281	12.1
Samsung PL80 Black	71	12.4

Use the data to develop a regression line to predict the price of a camera by the amount of megapixels incorporated in it. Compute s_e and r^2. Assuming $\alpha = 0.05$, test the slope of the regression line. Comment on the strength of the regression model.

12.66 Can the consumption of water in a city be predicted by temperature? The following data represent a sample of a day's water consumption and the high temperature for that day.

Water Use (millions of litres)	Temperature (degrees Celsius)
829	39.44
212	3.89
405	25.00
488	25.56
257	10.00
697	35.56
568	32.22
424	23.89

Develop a least squares regression line to predict the amount of water used in a day in a city by the high temperature for that day. What would be the predicted water usage for a temperature of 37.78°C? Evaluate the regression model by calculating s_e, by calculating r^2, and by testing the slope. Let $\alpha = .01$.

INTERPRETING THE OUTPUT

12.67 Study the following Minitab output from a regression analysis to predict y from x:
  a. What is the equation of the regression model?
  b. What is the meaning of the coefficient of x?
  c. What is the result of the test of the slope of the regression model? Let $\alpha = 0.10$. Why is the t ratio negative?
  d. Comment on r^2 and the standard error of the estimate.
  e. Comment on the relationship of the F value to the t ratio for x.
  f. The correlation coefficient for these two variables is –0.7918. Is this result surprising to you? Why or why not?

```
Regression Analysis: Y versus X
The regression equation is
Y = 67.2 - 0.0565 X

Predictor   Coef        SE Coef   T       p
Constant    67.231      5.0461    3.32    0.000
X           -0.05650    0.01027   -5.50   0.000

S=10.32  R-Sq=62.7%  R-Sq(adj)=60.6%
Analysis of Variance
Source          DF  SS      MS      F       P
Regression      1   3222.9  3222.9  30.25   0.000
Residual Error  18  1918.0  106.6
Total           19  5141.0
```

12.68 Study the following Excel regression output for an analysis attempting to predict the number of union members in the United Kingdom by the size of the labour force for selected years over a 16-year period from data published by the Office for National Statistics and the Department for Business Innovation and Skills. Analyse the computer output. Discuss the strength of the model in terms of proportion of variation accounted for, slope, and overall predictability. Using the equation of the regression line, attempt to predict the number of union members when the labour force is 110,000. Note that the model was developed with data already recoded in 1,000 units. Use the data in the model as is.

SUMMARY OUTPUT

Regression Statistics	
Multiple R	0.525255047
R Square	0.275892865
Adjusted R Square	0.224170926
Standard Error	137.2140794
Observations	16

ANOVA

	df	SS	MS	F	Significance F
Regression	1	100429.9	100429.8998	5.334156	0.036678
Residual	14	263587.9	18827.70359		
Total	15	364017.8			

	Coefficients	Standard Error	t Stat	P-value	Lower 95%	Upper 95%	Lower 95.0%	Upper 95.0%
Intercept	9351.80367	1032.762	9.055141909	3.15E-07	7136.75	11566.86	7136.75	11566.86
X Variable 1	−0.081703125	0.035376	−2.309579087	0.036678	−0.15758	−0.00583	−0.15758	−0.00583

12.69 Study the following Minitab residual diagnostic graphs. Comment on any possible violations of regression assumptions.

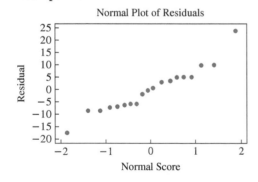

Normal Plot of Residuals

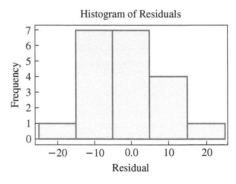

Histogram of Residuals

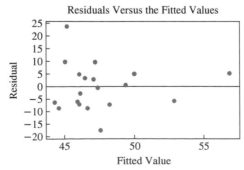

Residuals Versus the Fitted Values

ANALYSING THE DATABASES

1. Develop a regression model from the Manufacturing database to predict the Number of Production Workers by the Number of Employees. Discuss the model and its strength on the basis of statistics presented in this chapter. Now develop a regression model to predict New Capital Expenditures by the Number of Employees. Discuss this model and its strengths. Compare the two models. Does it make sense that the Number of Production Workers and New Capital Expenditures could each be predicted by the Number of Employees? Why or why not?

2. Using the Hospital database, develop a regression model to predict the number of Personnel by the number of Births. Now develop a regression model to predict number of Personnel by number of Beds. Examine the regression output. Which model is stronger in predicting number of Personnel? Explain why, using techniques presented in this chapter. Use the second regression model to predict the number of Personnel in a hospital that has 110 beds. Construct a 95% confidence interval around this prediction for the average value of y.

3. Analyse all the variables except Type in the Financial database by using a correlation matrix. The seven variables in this database are capable of producing 21 pairs of correlations. Which are most highly correlated? Select the variable that is most highly correlated with P/E ratio and use it as a predictor to develop a regression model to predict P/E ratio. How did the model do?

4. Construct a correlation matrix for the seven stock market indexes around the world. Describe what you find. That is, what indicators seem to be most strongly related to other indicators? Now focus on the three American stock market indicators. Which pair of these stock indicators is most correlated? Develop a regression model to predict the DJIA by the Nikkei 225. How strong is the model? Develop a regression model to predict the DJIA by the Hang Seng. How strong is the model? Develop a regression model to predict the DJIA by the IPC. How strong is the model? Compare the three models.

CASE

THE NEW ZEALAND WINE INDUSTRY

The wine industry in New Zealand is growing strongly. In 2009 (to year end June), total sales of New Zealand wines were a staggering 172 million litres, the highest the industry has ever seen. The same year witnessed the opening of the 640th winery. Export sales of 112 million litres accounted for over 65% of total sales for the first time ever.

However, the industry has not been without its challenges. A strong New Zealand dollar, rising costs, difficult market conditions, and economic uncertainty have all taken their toll. Notwithstanding these current problems, according to the Wine Institute of New Zealand's New Zealand Winegrowers Annual Report, 2009, the future is bright with expectations of continued growth in sales in the foreseeable years.

DISCUSSION

1. Members of the New Zealand wine industry are interested in trying to predict both domestic and export sales. It is probable that domestic consumption of New Zealand wine is affected by the availability of wines produced in other countries. In particular, given the high quality and affordability of Australian wines, imports from Australia could have an adverse effect on domestic consumption of New Zealand wine. The data below relate to consumption per capita (in litres) of New Zealand wine and wine imports from Australia (in millions of litres) for the years ended June 1995 to 2005. Use the techniques learned in this chapter to analyse the data. Include both regression and correlation

techniques. Discuss the strength of the relationship and any models that are developed.

	Consumption per capita (litres NZ wine)	Wine imports from Australia (million litres for year ended in June)
1995	8.5	15.462
1996	9.9	13.391
1997	10.4	16.201
1998	10	21.772
1999	10.1	20.762
2000	10.6	20.655
2001	9.3	21.331
2002	8.2	25.091
2003	8.8	32.363
2004	8.8	27.636
2005	11.2	24.34
2006	12.1	27.25
2007	12.2	36.497
2008	11.1	22.412
2009	13.9	20.019

Source: Wine Institute of New Zealand, *New Zealand Winegrowers Annual Reports*, available at www.nzwine.com.

2. Suppose that the New Zealand government and wine industry are keen to predict export sales of New Zealand wine so they can try to increase future export sales. Many of the factors that determine exports will be out of the hands of both the wine industry and the government. However, one major factor that will help determine this will be the total productive capacity of the New Zealand wine industry. The following data relate to export volumes of New Zealand wine (millions of litres) and the number of wineries in New Zealand over the years (ending June) 1994 to 2010. Using these data and techniques from this chapter, write a brief

report to the New Zealand government about the predictability of export sales from the number of wineries. In this report, in light of your results, state how it might be possible to increase export sales in the future.

	Number of wineries	Export volume (million of litres)
1994	190	7.9
1995	204	7.8
1996	238	11
1997	262	13.1
1998	293	15.2
1999	334	16.6
2000	358	19.2
2001	382	19.2
2002	398	23
2003	421	27.1
2004	463	31.1
2005	516	51.4
2006	530	57.8
2007	543	76
2008	585	88.6
2009	643	112.6
2010	672	142

Source: Wine Institute of New Zealand, *New Zealand Winegrowers Annual Reports*, available at www.nzwine.com.

3. Suppose that, at least in the short run, the number of wineries is fixed. It might be possible to increase total production of New Zealand wine by improving the average grape yield of the existing wineries. The Excel output on the next page displays the results of a regression predicting total wine production (in millions of litres) by average yield (in tonnes per hectare). Suppose you were asked by the Wine Institute of New Zealand to analyse the data and summarise the results. What would you find?

MINITAB OUTPUT

```
Regression Analysis: Sales versus Satisfaction

The regression equation is
Sales = 1.73 + 0.162 CustSat

Predictor      Coef       StDev        T         P
Constant      1.7332      0.4364      3.97      0.002
CustSat       0.16245     0.01490    10.90      0.000

S = 0.4113    R-Sq = 90.1%    R-Sq(adj) = 89.4%

Analysis of Variance

Source           DF        SS         MS         F         P
Regression        1      20.098     20.098    118.80     0.000
Residual Error   13       2.199      0.169
Total            14      22.297
```

EXCEL OUTPUT

SUMMARY OUTPUT

Regression Statistics	
Multiple R	0.949
R Square	0.901
Adjusted R Square	0.894
Standard Error	0.411
Observations	15

ANOVA

	df	SS	MS	F	Significance F
Regression	1	20.098	20.098	118.8	0.000
Residual	13	2.199	0.169		
Total	14	22.297			

	Coefficients	Standard Error	t Stat	P-value
Intercept	1.733	0.436	3.97	0.0016
Sick Days	0.162	0.015	10.90	0.0000

4. Delta Wire increased productivity from 70,000 to 90,000 pounds per week during a time when it instituted a basic skills training programme. Suppose this programme was implemented over an 18-month period and that the data right are the number of total cumulative basic skills hours of training and the per week productivity figures taken once a month over this time. Use techniques from this chapter to analyse the data and make a brief report to Delta about the predictability of productivity from cumulative hours of training.

Cumulative Hours of Training	Productivity (in pounds per week)
0	70,000
100	70,350
250	70,500
375	72,600
525	74,000
750	76,500
875	77,000
1,100	77,400
1,300	77,900
1,450	77,200
1,660	78,900
1,900	81,000
2,300	82,500
2,600	84,000
2,850	86,500
3,150	87,000
3,500	88,600
4,000	90,000

Source: Adapted from 'Delta Wire Corporation', *Strengthening America's Competitiveness: Resource Management Insights for Small Business Success.* Published by Warner Books on behalf of Connecticut Mutual Life Insurance Company and the US Chamber of Commerce in association with the Blue Chip Enterprise Initiative, 1991, International Monetary Fund; Terri Bergman, 'Training: The Case for Increased Investment', *Employment Relations Today,* Winter 1994–1995, pp. 381–391, available at www.ed.psu.edu/nwac/document/train/invest.html.

USING THE COMPUTER

EXCEL

- Excel has the capability of doing simple regression analysis. For a more inclusive analysis, use the **Data Analysis** tool. For a more 'a la carte' approach, use Excel's **Insert Function**.

- To use the **Data Analysis** tool for a more inclusive analysis, begin by selecting the **Data** tab on the Excel worksheet. From the **Analysis** panel at the right top of the **Data** tab worksheet, click on **Data Analysis**. If your Excel worksheet does not show the **Data Analysis** option, then you can load it as an add-in following directions given in Chapter 2. From the **Data Analysis** pulldown menu, select **Regression**.

In the **Regression** dialog box, input the location of the *y* values in **Input Y Range**. Input the location of the *x* values in **Input X Range**. Input **Labels** and input **Confidence Level**. To pass the line through the origin, check **Constant is Zero**. To print out the raw residuals, check **Residuals**. To print out residuals converted to *z* scores, check **Standardized Residuals**. For a plot of the residuals, check **Residual Plots**. For a plot of the line through the points, check **Line Fit Plots**. Standard output includes r, r^2, s_e, and an ANOVA table with the F test, the slope and intercept, t statistics with associated p-values, and any optionally requested output such as graphs or residuals.

- To use the **Insert Function** (f_x) go to the **Formulas** tab on an Excel worksheet (top centre tab). The **Insert Function** is on the far left of the menu bar. In the **Insert Function** dialog box at the top, there is a pulldown menu where it says **Or select a category**. From the pulldown menu associated with this command, select **Statistical**. Select **INTERCEPT** from the **Insert Function's Statistical** menu to solve for the y-intercept, **RSQ** to solve for r^2, **SLOPE** to solve for the slope, and **STEYX** to solve for the standard error of the estimate.

MINITAB

- Minitab has a relatively thorough capability to perform regression analysis. To begin, select **Stat** from the menu bar. Select **Regression** from the **Stat** pulldown menu. Select **Regression** from the **Regression** pulldown menu. Place the column name or column location of the y variable in **Response**. Place the column name or column location of the x variable in **Predictors**. Select **Graphs** for options relating to residual plots. Use this option and check **Four in one** to produce the residual diagnostic plots shown in the chapter. Select **Options** for confidence intervals and prediction intervals. Select **Results** for controlling the regression analysis output. Select **Storage** to store fits and/or residuals.

- To obtain a fitted-line plot, select **Stat** from the menu bar. Select **Regression** from the **Stat** pulldown menu. Select **Fitted Line Plot** from the **Regression** pulldown menu. In the Fitted Line Plot dialog box, place the column name or column location of the y variable in **Response(Y)**.

- Place the column name or column location of the x variable in **Response(X)**. Check **Type of Regression Model** as **Linear** (Chapter 12), **Quadratic**, or **Cubic**.

- Select **Graphs** for options relating to residual plots. Use this option and check **Four in one** to produce the residual diagnostic plots shown in the chapter. Select **Options** for confidence intervals and prediction intervals. Select **Storage** to store fits and/or residuals.

CHAPTER 13

Multiple Regression Analysis

LEARNING OBJECTIVES

This chapter presents the potential of multiple regression analysis as a tool in business decision making and its applications, thereby enabling you to:

1. Explain how, by extending the simple regression model to a multiple regression model with two independent variables, it is possible to determine the multiple regression equation for any number of unknowns

2. Examine significance tests of both the overall regression model and the regression coefficients

3. Calculate the residual, standard error of the estimate, coefficient of multiple determination, and adjusted coefficient of multiple determination of a regression model

4. Use a computer to find and interpret multiple regression outputs

Decision Dilemma

Are You Going to Hate Your New Job?

Getting a new job can be an exciting and energizing event in your life.

But what if you discover after a short time on the job that you hate your job? Is there any way to determine ahead of time whether you will love or hate your job? Sue Shellenbarger of the *Wall Street Journal* discusses some of the things to look for when interviewing for a position that may provide clues as to whether you will be happy on that job.

Among other things, work cultures vary from hip, freewheeling start-ups to old-school organizational-driven domains. Some organizations place pressure on workers to feel tense and to work long hours while others place more emphasis on creativity and the bottom line. Shellenbarger suggests that job interviewees pay close attention to how they are treated in an interview. Are they just another cog in the wheel or are they valued as an individual? Is a work–life balance apparent within the company? Ask what a typical workday is like at that firm. Inquire about the values that undergird the management by asking questions such as 'What is your proudest accomplishment?' Ask about flexible schedules and how job's training is managed. For example, does the worker have to go to job training on their own time?

A recent paper on the determinants of job satisfaction across the EU-15 countries identified a long list of variables that seem to affect the level of job satisfaction for EU-15 workers. A number of variables were found to have a positive relation with job satisfaction, including the level of education of the worker, the industry, self-employment status, the smaller the size of the firm, lower working hours, job's supervisionary role, amongst others.

Suppose another researcher gathered survey data asking 19 employees to rate their job satisfaction on a scale from 0 to 100 (with 100 being perfectly satisfied) and on a number of questions that are thought to affect the level of job satisfaction. These include, relationship with supervisor, overall quality of the work environment, total hours worked per week and opportunities for advancement. Suppose the data below represent the results of this survey. Assume that relationship with supervisor is rated on a scale from 1 to 5 (1 represents poor relationship and 5 represents an excellent relationship), overall quality of the work environment is rated on a scale from 0 to 10 (0 represents poor work environment and 10 represents an excellent work environment), and opportunities for advancement is rated on a scale from 1 to 5 (1 represents no opportunities and 5 represents excellent opportunities).

Job Satisfaction	Relationship With Supervisor	Overall Quality of Work Environment	Total Hours Worked per Week	Opportunities for Advancement
55	3	6	55	4
20	1	1	60	3
85	4	8	45	1
65	4	5	65	5
45	3	4	40	3
70	4	6	50	4
35	2	2	75	2
60	4	7	40	3
95	5	8	45	5
65	3	7	60	1
85	3	7	55	3
10	1	1	50	2
75	4	6	45	4
80	4	8	40	5
50	3	5	60	5
90	5	10	55	3
75	3	8	70	4
45	2	4	40	2
65	3	7	55	1

Managerial and Statistical Questions

1. Several variables are presented that may be related to job satisfaction. Which variables are stronger predictors of job satisfaction? Might other variables not mentioned here be related to job satisfaction?

2. Is it possible to develop a mathematical model to predict job satisfaction using the data given? If so, how strong is the model? With four independent variables, will we need to develop four different simple regression models and compare their results?

Source: Adapted from Sue Shellenbarger, 'How to Find Out if You're Going to Hate a New Job Before You Agree to Take It', *Wall Street Journal*, 13 June 2002, p. D1 and José María Millán, S. Jolanda A. Hessels, Roy Thurik, and Rafael Aguado, 'Determinants of Job Satisfaction Across the EU-15: A Comparison of Self-Employed and Paid Employees', 22 February 2011, Tinbergen Institute Discussion Paper 11-043/3.

Simple regression analysis (discussed in Chapter 12) is bivariate linear regression in which one **dependent variable**, y, is predicted by one **independent variable**, x. Examples of simple regression applications include models to predict retail sales by population density, FTSE 100 averages by prime interest rates, crude oil production by energy consumption, and CEO compensation by quarterly sales. However, in many cases, other independent variables, taken in conjunction with these variables, can make the regression model a better fit in predicting the dependent variable. For example, sales could be predicted by the size of store and number of competitors in addition to population density. A model to predict the FTSE 100 average could include, in addition to the prime interest rate, such predictors as yesterday's volume, the bond interest rate, and the producer price index. A model to predict CEO compensation could be developed by using variables such as company earnings per share, age of CEO, and size of company in addition to quarterly sales. A model could perhaps be developed to predict the cost of outsourcing by such variables as unit price, export taxes, cost of money, damage in transit, and other factors. Each of these examples contains only one dependent variable, y, as with simple regression analysis. However, multiple independent variables, x (predictors) are involved. *Regression analysis with two or more independent variables or with at least one non-linear predictor* is called **multiple regression** analysis.

13.1 THE MULTIPLE REGRESSION MODEL

Multiple regression analysis is similar in principle to simple regression analysis. However, it is more complex conceptually and computationally. Recall from Chapter 12 that the equation of the probabilistic simple regression model is

$$y = \beta_0 + \beta_1 x + \varepsilon$$

where:

y = the value of the dependent variable
β_0 = the propulation y intercept
β_1 = the population slope
ε = the error of prediction

Extending this notion to multiple regression gives the general equation for the probabilistic multiple regression model.

$$y = \beta_0 + \beta_1 x_1 + \beta_2 x_2 + \beta_3 x_3 + \cdots + \beta_k x_k + \varepsilon$$

where:

y = the value of the dependent variable
β_0 = the regression constant
β_1 = the partial regression coefficient for independent variable 1
β_2 = the partial regression coefficient for independent variable 2
β_3 = the partial regression coefficient for independent variable 3
β_k = the partial regression coefficient for independent variable k
k = the number of independent variables

In multiple regression analysis, the dependent variable, y, is sometimes referred to as the **response variable**. The **partial regression coefficient** of an independent variable, β_i, *represents the increase that will occur in the value of y from a one-unit increase in that independent variable if all other variables are held constant*. The 'full' (versus partial) regression coefficient of an independent variable is a coefficient obtained from the bivariate model (simple regression) in which the independent variable is the sole predictor of y. The partial regression coefficients occur because more than one predictor is included in a model. The partial regression coefficients are analogous to β_1, the slope of the simple regression model in Chapter 12.

In actuality, the partial regression coefficients and the regression constant of a multiple regression model are population values and are unknown. In virtually all research, these values are estimated by using sample information. Shown here is the form of the equation for estimating y with sample information.

$$\hat{y} = b_0 + b_1 x_1 + b_2 x_2 + b_3 x_3 + \cdots + b_k x_k$$

where:

$\hat{y}$ = the predicted value of y
b_0 = the estimate of the regression constant
b_1 = the estimate of regression coefficient 1
b_2 = the estimate of regression coefficient 2
b_3 = the estimate of regression coefficient 3
b_k = the estimate of regression coefficient k
k = the number of independent variables

Multiple Regression Model with Two Independent Variables (First-Order)

The simplest multiple regression model is one constructed with two independent variables, where the highest power of either variable is 1 (first-order regression model). The regression model is

$$y = \beta_0 + \beta_1 x_1 + \beta_2 x_2 + \varepsilon$$

The constant and coefficients are estimated from sample information, resulting in the following model:

$$\hat{y} = b_0 + b_1 x_1 + b_2 x_2$$

Figure 13.1 is a three-dimensional graph of a series of points (x_1, x_2, y) representing values from three variables used in a multiple regression model to predict the market price of a house by the number of square metres in the house and its age. Simple regression models yield a line that is fit through data points in the xy plane. In multiple regression analysis, the resulting model produces a **response surface**. In the multiple regression model shown here with two independent first-order variables, the response surface is a **response plane**. The response plane for such a model is fit in a three-dimensional space (x_1, x_2, y).

If such a response plane is fit into the points shown in Figure 13.1, the result is the graph in Figure 13.2. Notice that most of the points are not on the plane. As in simple regression, an error in the fit of the model in multiple regression is

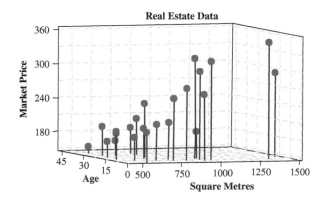

FIGURE 13.1

Points in a Sample Space

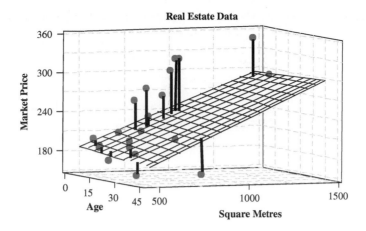

Response Plane for a First-Order Two-Predictor Multiple Regression Model

usually present. The distances shown in the graph from the points to the response plane are the errors of fit, or residuals $(y - \hat{y})$. Multiple regression models with three or more independent variables involve more than three dimensions and are difficult to depict geometrically.

Observe in Figure 13.2 that the regression model attempts to fit a plane into the three dimensional plot of points. Notice that the plane intercepts the y axis. Figure 13.2 depicts some values of y for various values of x_1 and x_2. The error of the response plane (ε) in predicting or determining the y values is the distance from the points to the plane.

Determining the Multiple Regression Equation

The simple regression equations for determining the sample slope and intercept given in Chapter 12 are the result of using methods of calculus to minimize the sum of squares of error for the regression model. The procedure for developing these equations involves solving two simultaneous equations with two unknowns, b_0 and b_1. Finding the sample slope and intercept from these formulas requires the values of Σx, Σy, Σxy, and Σx^2.

The procedure for determining formulas to solve for multiple regression coefficients is similar. The formulas are established to meet an objective of *minimizing the sum of squares of error for the model*. Hence, the regression analysis shown here is referred to as **least squares analysis**. Methods of calculus are applied, resulting in $k + 1$ equations with $k + 1$ unknowns (b_0 and k values of b_i) for multiple regression analyses with k independent variables. Thus, a regression model with six independent variables will generate seven simultaneous equations with seven unknowns (b_0, b_1, b_2, b_3, b_4, b_5, b_6).

For multiple regression models with two independent variables, the result is three simultaneous equations with three unknowns (b_0, b_1, and b_2).

$$b_0 n + b_1 \Sigma x_1 + b_2 \Sigma x_2 = \Sigma y$$
$$b_0 \Sigma x_1 + b_1 \Sigma x_1^2 + b_2 \Sigma x_1 x_2 = \Sigma x_1 y$$
$$b_0 \Sigma x_2 + b_1 \Sigma x_1 x_2 + b_2 \Sigma x_2^2 = \Sigma x_2 y$$

The process of solving these equations by hand is tedious and time-consuming. Solving for the regression coefficients and regression constant in a multiple regression model with two independent variables requires Σx_1, Σx_2, Σy, Σx_1^2, Σx_2^2, $\Sigma x_1 x_2$, $\Sigma x_1 y$, and $\Sigma x_2 y$. In actuality, virtually all business researchers use computer statistical software packages to solve for the regression coefficients, the regression constant, and other pertinent information. In this chapter, we will discuss computer output and assume little or no hand calculation. The emphasis will be on the interpretation of the computer output.

A Multiple Regression Model

TABLE 13.1

Real Estate Data

Market Price (€000s)	Total Number of Square Metres	Age of House (Years)
y	x_1	x_2
157.5	489	35
162.8	759	45
174.8	473	20
192.0	733	32
184.8	574	25
194.8	475	14
187.3	533	8
195.0	946	10
197.5	513	28
208.5	753	30
198.8	555	2
209.8	653	6
199.3	646	14
211.3	757	9
240.0	701	19
273.8	827	4
256.3	751	5
302.5	938	7
262.3	929	3
320.0	996	6
322.5	935	10
294.8	1,452	11
350.0	1,384	8

A real estate study was conducted in a small Dutch city to determine what variables, if any, are related to the market price of a home. Several variables were explored, including the number of bedrooms, the number of bathrooms, the age of the house, the number of square metres of living space, the total number of square metres of space, and the number of garages. Suppose the researcher wants to develop a regression model to predict the market price of a home by two variables, 'total number of square metres in the house' and 'the age of the house'. Listed in Table 13.1 are the data for these three variables.

A number of statistical software packages can perform multiple regression analysis, including Excel and Minitab. The output for the Minitab multiple regression analysis on the real estate data is given in Figure 13.3. (Excel output for this type of analysis is shown in Demonstration Problem 13.1.)

The Minitab output for regression analysis begins with 'The regression equation is'. From Figure 13.3, the regression equation for the real estate data in Table 13.1 is

$$\hat{y} = 143 + 0.14534x_1 - 1.6668x_2$$

The regression constant, 143, is the y-intercept. The y-intercept is the value of $\hat{y}$ if both x_1 (number of square metres) and x_2 (age) are zero. In this example, a practical understanding of the y-intercept is meaningless. It makes little sense to say that a house containing no square metres ($x_1 = 0$) and no years of age ($x_2 = 0$) would cost €143,000. Note in Figure 13.2 that the response plane crosses the y-axis (price) at 143.

```
Regression Analysis: Market Price versus Square Metres, Age

The regression equation is
Market Price = 143 + 0.145 Square Metres - 1.67 Age

Predictor            Coef      SE Coef        T          P
Constant           143.41        24.99      5.74      0.000
Square Metres      0.14534      0.02578     5.64      0.000
Age               -1.6668        0.5695    -2.93      0.008

S = 29.8743     R-Sq = 74.1%        R-Sq(adj) = 71.6%

Analysis of Variance

Source             DF        SS          MS        F         P
Regression          2      51199       25600     28.68     0.000
Residual Error     20      17850         892
Total              22      69049
```

FIGURE 13.3

Minitab Output of Regression for the Real Estate Example

The coefficient of x_1 (total number of square metres in the house) is 0.14534, which means that a one-unit increase in square metres would result in a predicted increase of $0.14534 \cdot (\text{€}1,000) = \text{€}145.34$ in the price of the home if age were held constant. All other variables being held constant, the addition of one square metre of space in the house results in a predicted increase of €145.34 in the price of the home.

The coefficient of x_2 (age) is -1.6668. The negative sign on the coefficient denotes an inverse relationship between the age of a house and the price of the house: the older the house, the lower the price. In this case, if the total number of square metres in the house is kept constant, a one-unit increase in the age of the house (one year) will result in $-1.6668 \cdot (\text{€}1,000) = -\text{€}1,666.8$, a predicted €1,666.8 drop in the price.

In examining the regression coefficients, it is important to remember that the independent variables are often measured in different units. It is usually not wise to compare the regression coefficients of predictors in a multiple regression model and decide that the variable with the largest regression coefficient is the best predictor. In this example, the two variables are in different units, square metres and years. Just because x_2 has the larger coefficient (1.6668) does not necessarily make x_2 the strongest predictor of y.

This regression model can be used to predict the price of a house in this small Dutch city. If the house has 750 square metres total and is 12 years old, $x_1 = 750$ and $x_2 = 12$. Substituting these values into the regression model yields

$$\hat{y} = 143 + 0.14534x_1 - 1.6668x_2$$
$$= 143 + 0.14534(750) - 1.6668(12) = 232.0034$$

The predicted price of the house is €232,003.40. Figure 13.2 is a graph of these data with the response plane and the residual distances.

DEMONSTRATION PROBLEM 13.1

Since 1993, the government bond yield in the United Kingdom has varied from over 8% to less than 4%. What factor in the UK economy seems to be related to the government bond yield? Two possible predictors of the government bond yield are the annual unemployment rate and the household savings rate in the United Kingdom. Shown below are data for the annual government bond yield over an 18-year period in the United Kingdom along with the annual unemployment rate and the annual average household saving ratio (as a percentage of disposable personal income). Use these data to develop a multiple regression model to predict the annual government bond yield by the unemployment rate and the average household saving ratio. Determine the predicted government bond yield if the unemployment rate is 6.5 and the average personal saving is 5.0.

Year	Government Bond Yield	Unemployment Rate	Household Saving Ratio
1993	7.5467	10.40	10.8
1994	8.1525	9.53	9.3
1995	8.3192	8.65	10.3
1996	7.9392	8.13	9.4
1997	7.1342	7.00	9.6
1998	5.5958	6.28	7.4
1999	5.0150	5.98	5.2
2000	5.3267	5.45	4.7
2001	5.0092	5.08	6.0
2002	4.9150	5.18	4.8
2003	4.5783	5.03	5.1
2004	4.9258	4.75	3.7
2005	4.4583	4.88	3.9

(continued)

Year	Government Bond Yield	Unemployment Rate	Household Saving Ratio
2006	4.3742	5.43	3.4
2007	5.0383	5.35	2.6
2008	4.5842	5.70	2.0
2009	3.6475	7.63	6.0
2010	3.6125	7.88	5.3

Solution

The following output shows the results of analysing the data by using the regression portion of Excel.

SUMMARY OUTPUT

Regression Statistics	
Multiple R	0.8382
R Square	0.7026
Adjusted R Square	0.6630
Standard Error	0.8922
Observations	18

ANOVA

	df	SS	MS	F	Significance F
Regression	2	28.2121	14.1060	17.72	0.000112135
Residual	15	11.9401	0.7960		
Total	17	40.1522			

	Coefficients	Standard Error	t Stat	P-value
Intercept	2.8034	0.9073	3.09	0.0075
Unemployment Rates	−0.0277	0.2155	−0.13	0.8996
Household Savings Ratio	0.4839	0.1363	3.55	0.0029

The regression equation is

$$\hat{y} = 2.8034 - 0.277x_1 + 0.4839x_2$$

where:

$\hat{y}$ = government bond yield
x_1 = unemployment rate
x_2 = household saving ratio

The model indicates that for every one-unit (1%) increase in the unemployment rate, the predicted prime interest rate decreases by 0.0277%, if the household saving ratio is held constant. The model also indicates that for every one-unit (1%) increase in the household saving ratio, the predicted government bond yield increases by 0.48%, if unemployment is held constant.

If the unemployment rate is 6.5 and the household saving ratio is 5.0, the predicted government bond yield is 3.422%:

$$\hat{y} = 2.8034 - (0.277)(6.5) + (0.4839)(5.0) = 3.422$$

13.1   PROBLEMS

13.1   Use a computer to develop the equation of the regression model for the data below. Comment on the regression coefficients. Determine the predicted value of y for $x_1 = 200$ and $x_2 = 7$.

y	x_1	x_2
12	174	3
18	281	9
31	189	4
28	202	8
52	149	9
47	188	12
38	215	5
22	150	11
36	167	8
17	135	5

13.2   Use a computer to develop the equation of the regression model for the data below. Comment on the regression coefficients. Determine the predicted value of y for $x_1 = 33$, $x_2 = 29$, and $x_3 = 13$.

y	x_1	x_2	x_3
114	21	6	5
94	43	25	8
87	56	42	25
98	19	27	9
101	29	20	12
85	34	45	21
94	40	33	14
107	32	14	11
119	16	4	7
93	18	31	16
108	27	12	10
117	31	3	8

13.3   Using the data below, determine the equation of the regression model. How many independent variables are there? Comment on the meaning of these regression coefficients.

Predictor	Coefficient
Constant	121.62
x_1	−.174
x_2	6.02
x_3	.00026
x_4	.0041

13.4   Use the data below to determine the equation of the multiple regression model. Comment on the regression coefficients.

Predictor	Coefficient
Constant	31,409.5
x_1	.08425
x_2	289.62
x_3	−.0947

13.5 Is there a particular product that is an indicator of per capita personal consumption for countries around the world? Shown below are data on per capita personal consumption, paper consumption, fish consumption, and petrol consumption for 11 countries. Use the data to develop a multiple regression model to predict per capita personal consumption by paper consumption, fish consumption, and petrol consumption. Discuss the meaning of the partial regression weights.

Country	Per Capita Personal Consumption (in US$)	Paper Consumption (kg per person)	Fish Consumption (lbs per person)	Petrol Consumption (litres per person)
Bangladesh	836	1	23	2
Greece	3,145	85	53	394
Italy	21,785	204	48	368
Japan	37,931	250	141	447
Kenya	276	4	12	16
Norway	1,913	156	113	477
Philippines	2,195	19	65	43
Portugal	3,154	116	133	257
United Kingdom	19,539	207	44	460
United States	109,521	308	47	1,624
Venezuela	622	27	40	528

13.6 Jensen, Solberg, and Zorn investigated the relationship of insider ownership, debt, and dividend policies in companies. One of their findings was that firms with high insider ownership choose lower levels of both debt and dividends. Shown here is a sample of data of these three variables for 11 different industries. Use the data to develop the equation of the regression model to predict insider ownership by debt ratio and dividend payout. Comment on the regression coefficients.

Industry	Insider Ownership	Debt Ratio	Dividend Payout
Mining	8.2	14.2	10.4
Food and beverage	18.4	20.8	14.3
Furniture	11.8	18.6	12.1
Publishing	28.0	18.5	11.8
Petroleum refining	7.4	28.2	10.6
Glass and cement	15.4	24.7	12.6
Motor vehicle	15.7	15.6	12.6
Department store	18.4	21.7	7.2
Restaurant	13.4	23.0	11.3
Amusement	18.1	46.7	4.1
Hospital	10.0	35.8	9.0

13.2 SIGNIFICANCE TESTS OF THE REGRESSION MODEL AND ITS COEFFICIENTS

Multiple regression models can be developed to fit almost any data set if the level of measurement is adequate and enough data points are available. Once a model has been constructed, it is important to test the model to determine whether it fits the data well and whether the assumptions underlying regression analysis are met. Assessing the adequacy of the regression model can be done in several ways, including testing the overall significance of the model, studying the significance tests of the regression coefficients, computing the residuals, examining the standard error of the estimate, and observing the coefficient of determination. In this section, we examine significance tests of the regression model and of its coefficients.

Testing the Overall Model

With simple regression, a t test of the slope of the regression line is used to determine whether the population slope of the regression line is different from zero – that is, whether the independent variable contributes significantly in linearly predicting the dependent variable.

The hypotheses for this test, presented in Chapter 12 are

$$H_0: \beta_1 = 0$$
$$H_a: \beta_1 \neq 0$$

For multiple regression, an analogous test makes use of the F statistic. The overall significance of the multiple regression model is tested with the following hypotheses:

$$H_0: \beta_1 = \beta_2 = \beta_3 = \ldots = \beta_k = 0$$
$$H_a: \text{At least one of the regression coefficients is} \neq 0.$$

If we fail to reject the null hypothesis, we are stating that the regression model has no significant predictability for the dependent variable. A rejection of the null hypothesis indicates that at least one of the independent variables is adding significant predictability for y.

This F test of overall significance is often given as a part of the standard multiple regression output from statistical software packages. The output appears as an analysis of variance (ANOVA) table. Shown here is the ANOVA table for the real estate example taken from the Minitab output in Figure 13.3.

```
Analysis of Variance
```

Source	DF	SS	MS	F	p
Regression	2	51199	25600	28.68	0.000
Residual Error	20	17850	892		
Total	22	69049			

The F value is 28.68; because $p = 0.000$, the F value is significant at $\alpha = 0.001$. The null hypothesis is rejected, and there is at least one significant predictor of house price in this analysis.

The F value is calculated by the following equation.

$$F = \frac{MS_{reg}}{MS_{err}} = \frac{SS_{reg}/df_{reg}}{SS_{err}/df_{err}} = \frac{SSR/k}{SSE/(N-k-1)}$$

where:
$MS = $ mean square
$SS = $ sum of squares
$df = $ degrees of freedom
$k = $ number of independent variables
$N = $ number of observations

Note that in the ANOVA table for the real estate example, $df_{reg} = 2$. The degrees of freedom formula for regression is the number of regression coefficients plus the regression constant minus 1. The net result is the number of regression coefficients, which equals the number of independent variables, k. The real estate example uses two independent variables, so $k = 2$. Degrees of freedom error in multiple regression equals the total number of observations minus the number of regression coefficients minus the regression constant, or $N - k - 1$. For the real estate example, $N = 23$; thus, $df_{err} = 23 - 2 - 1 = 20$.

As shown in Chapter 11, $MS = SS/df$. The F ratio is formed by dividing MS_{reg} by MS_{err}. In using the F distribution table to determine a critical value against which to test the observed F value, the degrees of freedom numerator is df_{reg}

and the degrees of freedom denominator is df_{err}. The table F value is obtained in the usual manner, as presented in Chapter 11. With $\alpha = 0.01$ for the real estate example, the table value is

$$F_{0.01,2,20} = 5.85$$

Comparing the observed F of 28.68 to this table value shows that the decision is to reject the null hypothesis. This same conclusion was reached using the p-value method from the computer output.

If a regression model has only one linear independent variable, it is a simple regression model. In that case, the F test for the overall model is the same as the t test for significance of the population slope. The F value displayed in the regression ANOVA table is related to the t test for the slope in the simple regression case as follows:

$$F = t^2$$

In simple regression, the F value and the t value give redundant information about the overall test of the model.

Most researchers who use multiple regression analysis will observe the value of F and its p-value rather early in the process. If F is not significant, then no population regression coefficient is significantly different from zero, and the regression model has no predictability for the dependent variable.

Significance Tests of the Regression Coefficients

In multiple regression, individual significance tests can be computed for each regression coefficient using a t test. Each of these t tests is analogous to the t test for the slope used in Chapter 12 for simple regression analysis. The hypotheses for testing the regression coefficient of each independent variable take the following form:

$$H_0: \beta_1 = 0$$
$$H_a: \beta_1 \neq 0$$

$$H_0: \beta_2 = 0$$
$$H_a: \beta_2 \neq 0$$

$$\vdots$$

$$H_0: \beta_k = 0$$
$$H_a: \beta_k \neq 0$$

Most multiple regression computer packages yield observed t values to test the individual regression coefficients as standard output. Shown here are the t values and their associated probabilities for the real estate example as displayed with the multiple regression output in Figure 13.3.

Variable	T	P
Square metres	5.64	.000
Age	−2.93	.008

At $\alpha = 0.05$, the null hypothesis is rejected for both variables because the probabilities (p) associated with their t values are less than 0.05. If the t ratios for any predictor variables are not significant (fail to reject the null hypothesis), the researcher might decide to drop that variable(s) from the analysis as a non-significant predictor(s). Other factors can enter into this decision. In Chapter 14, we will explore techniques for model building in which some variable sorting is required.

The degrees of freedom for each of these individual tests of regression coefficients are $n - k - 1$. In this particular example because there are $k = 2$ predictor variables, the degrees of freedom are $23 - 2 - 1 = 20$. With $\alpha = 0.50$ and a two-tailed test, the critical table t value is

$$t_{0.025,25} = \pm 2.086$$

Notice from the t ratios shown here that if this critical table t value had been used as the hypothesis test criterion instead of the p-value method, the results would have been the same. Testing the regression coefficients not only gives the researcher some insight into the fit of the regression model, but it also helps in the evaluation of how worthwhile individual independent variables are in predicting y.

13.2 PROBLEMS

13.7 Examine the Minitab output shown here for a multiple regression analysis. How many predictors were there in this model? Comment on the overall significance of the regression model. Discuss the t ratios of the variables and their significance.

```
The regression equation is
   Y = 4.096 − 5.111X₁ + 2.662X₂ + 1.557X₃ + 1.141X₄ + 1.650X₅
       − 1.248X₆ + 0.436X₇ + 0.962X₈ + 1.289X₉
```

Predictor	Coef	Stdev	T	p
Constant	4.096	1.2884	3.24	.006
X_1	−5.111	1.8700	2.73	.011
X_2	2.662	2.0796	1.28	.212
X_3	1.557	1.2811	1.22	.235
X_4	1.141	1.4712	0.78	.445
X_5	1.650	1.4994	1.10	.281
X_6	−1.248	1.2735	0.98	.336
X_7	0.436	0.3617	1.21	.239
X_8	0.962	1.1896	0.81	.426
X_9	1.289	1.9182	0.67	.508

$S = 3.503$ $R\text{-sq} = 40.8\%$ $R\text{-sq(adj.)} = 20.3\%$

Analysis of Variance

Source	DF	SS	MS	F	p
Regression	9	219.746	24.416	1.99	.0825
Error	26	319.004	12.269		
Total	35	538.750			

13.8 Displayed here is the Minitab output for a multiple regression analysis. Study the ANOVA table and the t ratios and use these to discuss the strengths of the regression model and the predictors. Does this model appear to fit the data well? From the information here, what recommendations would you make about the predictor variables in the model?

```
The regression equation is
   Y = 34.7 + 0.0763X₁ + 0.00026X₂ − 1.12X₃
```

Predictor	Coef	Stdev	T	p
Constant	34.672	5.256	6.60	.000
X_1	0.07629	0.02234	3.41	.005
X_2	0.000259	0.001031	0.25	.805
X_3	−1.1212	0.9955	−1.13	.230

$S = 9.722$ $R\text{-sq} = 51.5\%$ $R\text{-sq(adj)} = 40.4\%$

Analysis of Variance

Source	DF	SS	MS	F	p
Regression	3	1306.99	435.66	4.61	.021
Error	13	1228.78	94.52		
Total	16	2535.77			

13.9 Using the data in Problem 13.5, develop a multiple regression model to predict per capita personal consumption by the consumption of paper, fish, and petrol. Discuss the output and pay particular attention to the F test and the t tests.

13.10 Using the data from Problem 13.6, develop a multiple regression model to predict insider ownership from debt ratio and dividend payout. Comment on the strength of the model and the predictors by examining the ANOVA table and the t tests.

13.11 Develop a multiple regression model to predict y from x_1, x_2, and x_3 using the data below. Discuss the values of F and t.

y	x_1	x_2	x_3
5.3	44	11	401
3.6	24	40	219
5.1	46	13	394
4.9	38	18	362
7.0	61	3	453
6.4	58	5	468
5.2	47	14	386
4.6	36	24	357
2.9	19	52	206
4.0	31	29	301
3.8	24	37	243
3.8	27	36	228
4.8	36	21	342
5.4	50	11	421
5.8	55	9	445

13.12 Use the data below to develop a regression model to predict y from x_1 and x_2. Comment on the output. Develop a regression model to predict y from x_1 only. Compare the results of this model with those of the model using both predictors. What might you conclude by examining the output from both regression models?

y	x_1	x_2
28	12.6	134
43	11.4	126
45	11.5	143
49	11.1	152
57	10.4	143
68	9.6	147
74	9.8	128
81	8.4	119
82	8.8	130
86	8.9	135
101	8.1	141
112	7.6	123
114	7.8	121
119	7.4	129
124	6.4	135

13.13 Study the Excel multiple regression output below. How many predictors are in this model? How many observations? What is the equation of the regression line? Discuss the strength of the model in terms F. Which predictors, if any, are significant? Why or why not? Comment on the overall effectiveness of the model.

SUMMARY OUTPUT

Regression Statistics	
Multiple R	0.842
R Square	0.710
Adjusted R Square	0.630
Standard Error	109.430
Observations	15

ANOVA

	df	SS	MS	F	Significance F
Regression	3	321946.82	107315.6	8.96	0.0027
Residual	11	131723.20	11974.8		
Total	14	453670.00			

	Coefficients	Standard Error	t Stat	P-value
Intercept	657.0530	167.460	3.92	0.0024
X Variable 1	5.7103	1.792	3.19	0.0087
X Variable 2	−0.4169	0.322	−1.29	0.2222
X Variable 3	−3.4715	1.443	−2.41	0.0349

13.3   RESIDUALS, STANDARD ERROR OF THE ESTIMATE, AND R^2

Three more statistical tools for examining the strength of a regression model are the residuals, the standard error of the estimate, and the coefficient of multiple determination.

Residuals

The **residual**, or error, of the regression model is *the difference between the y value and the predicted value, $\hat{y}$.*

$$\text{Residual} = y - \hat{y}$$

The residuals for a multiple regression model are solved for in the same manner as they are with simple regression. First, a predicted value, $\hat{y}$, is determined by entering the value for each independent variable for a given set of observations into the multiple regression equation and solving for $\hat{y}$. Next, the value of $y - \hat{y}$ is computed for each set of observations. Shown here are the calculations for the residuals of the first set of observations from Table 13.1. The predicted value of y for $x_1 = 1605$ and $x_2 = 35$ is

$$\hat{y} = 143 + 0.14534(489) - 1.6668(35) = 155.733$$

Actual value of $y = 157.5$
Residual $= y - \hat{y} = 157.5 - 155.733 = 1.767$

All residuals for the real estate data and the regression model displayed in Table 13.1 and Figure 13.3 are displayed in Table 13.2.

TABLE 13.2

Residuals for the Real Estate Regression Model

y	$\hat{y}$	$y - \hat{y}$
157.5	155.733	1.767
162.8	178.307	−15.507
174.8	178.410	−3.610
192	196.197	−4.197
184.8	184.755	0.045
194.8	188.701	6.099
187.3	207.132	−19.832
195	263.824	−68.824
197.5	170.889	26.611
208.5	202.437	6.063
198.8	220.330	−21.530
209.8	227.906	−18.106
199.3	213.554	−14.254
211.3	238.021	−26.721
240	213.214	26.786
273.8	256.529	17.271
256.3	243.816	12.484
302.5	267.661	34.839
262.3	273.020	−10.720
320	277.758	42.242
322.5	262.225	60.275
294.8	335.699	−40.899
350	330.816	19.184

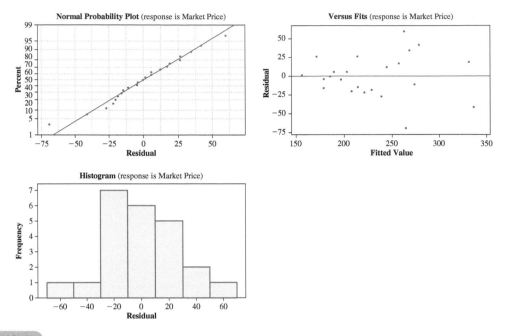

FIGURE 13.4

Minitab Residual Diagnosis for the Real Estate Example

An examination of the residuals in Table 13.2 can reveal some information about the fit of the real estate regression model. The business researcher can observe the residuals and decide whether the errors are small enough to support the accuracy of the model. The house price figures are in units of €1,000. Two of the 23 residuals are more than 60.00, or more than €60,000 off in their prediction. On the other hand, one residual is less than 1, or €1,000 off in its prediction.

Residuals are also helpful in locating outliers. **Outliers** are *data points that are apart, or far, from the mainstream of the other data.* They are sometimes data points that were mistakenly recorded or measured. Because every data point influences the regression model, outliers can exert an overly important influence on the model based on their distance from other points. In examining the residuals in Table 13.2 for outliers, the eighth residual listed is –68.824. This error indicates that the regression model was not nearly as successful in predicting house price on this particular house as it was with others (an error of more than €68,000). For whatever reason, this data point stands somewhat apart from other data points and may be considered an outlier.

Residuals are also useful in testing the assumptions underlying regression analysis. Figure 13.4 displays Minitab diagnostic techniques for the real estate example. In the top right is a graph of the residuals. Notice that residual variance seems to increase in the right half of the plot, indicating potential heteroscedasticity. As discussed in Chapter 12, one of the assumptions underlying regression analysis is that the error terms have homoscedasticity or homogeneous variance. That assumption might be violated in this example. The normal plot of residuals is nearly a straight line, indicating that the assumption of normally distributed error terms probably has not been violated.

SSE and Standard Error of the Estimate

One of the properties of a regression model is that the residuals sum to zero. As pointed out in Chapter 12, this property precludes the possibility of computing an 'average' residual as a single measure of error. In an effort to compute a single statistic that can represent the error in a regression analysis, the zero-sum property can be overcome by *squaring the residuals and then summing the squares.* Such an operation produces the sum of squares of error (SSE).

The formula for computing the sum of squares error (SSE) for multiple regression is the same as it is for simple regression.

$$SSE = \Sigma(y - \hat{y})^2$$

For the real estate example, SSE can be computed by squaring and summing the residuals displayed in Table 13.2.

$$
\begin{aligned}
SSE = [&(1.767)^2 + (-15.507)^2 + (-3.610)^2 + (-4.197)^2 \\
+ &(0.045)^2 + (6.099)^2 + (-19.832)^2 + (-68.824)^2 \\
+ &(26.611)^2 + (6.063)^2 + (-21.530)^2 + (-18.106)^2 \\
+ &(-14.254)^2 + (-26.721)^2 + (26.786)^2 + (17.271)^2 \\
+ &(12.484)^2 + (34.839)^2 + (-10.720)^2 + (42.242)^2 \\
+ &(60.275)^2 + (-40.899)^2 + (19.184)^2] \\
= &\, 17,850
\end{aligned}
$$

SSE can also be obtained directly from the multiple regression computer output by selecting the value of SS (sum of squares) listed beside error. Shown here is the ANOVA portion of the output displayed in Figure 13.3, which is the result of a multiple regression analysis model developed to predict house prices. Note that the SS for error shown in the ANOVA table equals the value of $\Sigma(y - \hat{y})^2$ just computed (2,861.0).

SSE

```
Analysis of Variance
```

Source	DF	SS	MS	F	P
Regression	2	51199	25600	28.68	.000
Error	20	17850	892		
Total	22	69049			

SSE has limited usage as a measure of error. However, it is a tool used to solve for other, more useful measures. One of those is the **standard error of the estimate**, s_e, which is essentially *the standard deviation of residuals (error) for the regression model.* As explained in Chapter 12, an assumption underlying regression analysis is that the error terms are approximately normally distributed with a mean of zero. With this information and by the empirical rule, approximately 68% of the residuals should be within $\pm 1 s_e$ and 95% should be within $\pm 2 s_e$. This property makes the standard error of the estimate a useful tool in estimating how accurately a regression model is fitting the data.

The standard error of the estimate is computed by dividing SSE by the degrees of freedom of error for the model and taking the square root.

$$s_e = \sqrt{\frac{SSE}{n - k - 1}}$$

where:

n = number of observations
k = number of independent variables

The value of s_e can be computed for the real estate example as follows:

$$s_e = \sqrt{\frac{SSE}{n - k - 1}} = \sqrt{\frac{17850}{23 - 2 - 1}} = 29.874$$

The standard error of the estimate, s_e, is usually given as standard output from regression analysis by computer software packages. The Minitab output displayed in Figure 13.3 contains the standard error of the estimate for the real estate example.

$$S = 29.874$$

By the empirical rule, approximately 68% of the residuals should be within $\pm 1 s_e = \pm 1(11.96) = \pm 11.96$. Because house prices are in units of €1,000, approximately 68% of the predictions are within ± 29.874(€1,000), or $\pm$€29,874. Examining the output displayed in Table 13.2, 18/23, or about 78%, of the residuals are within this span. According to the empirical rule, approximately 95% of the residuals should be within $\pm 2 s_e$, or $\pm 2(29.874) = \pm 59.75$. Further examination of the residual values in Table 13.2 shows that 21 of 23, or 91%, fall within this range. The business researcher can study the standard error of the estimate and these empirical rule-related ranges and decide whether the error of the regression model is sufficiently small to justify further use of the model.

Coefficient of Multiple Determination (R²)

The **coefficient of multiple determination (R²)** is analogous to the coefficient of determination (r^2) discussed in Chapter 12. R^2 represents *the proportion of variation of the dependent variable, y, accounted for by the independent variables in the regression model.* As with r^2, the range of possible values for R^2 is from 0 to 1. An R^2 of 0 indicates no relationship between the predictor variables in the model and y. An R^2 of 1 indicates that 100% of the variability of y has been accounted for by the predictors. Of course, it is desirable for R^2 to be high, indicating the strong predictability of a regression model. The coefficient of multiple determination can be calculated by the following formula:

$$R^2 = \frac{SSR}{SS_{yy}} = 1 - \frac{SSE}{SS_{yy}}$$

R^2 can be calculated in the real estate example by using the sum of squares regression (SSR), the sum of squares error (SSE), and sum of squares total (SS_{yy}) from the ANOVA portion of Figure 13.3.

Analysis of Variance

Source	DF	SS	MS	F	p
Regression	2	51199	25600	28.68	.000
Error	20	17850	892		
Total	22	69049			

$$R^2 = \frac{SSR}{SS_{yy}} = \frac{51199}{69049} = 0.741$$

or

$$R^2 = 1 - \frac{SSE}{SS_{yy}} = 1 - \frac{17850}{69049} = 0.741$$

In addition, virtually all statistical software packages print out R^2 as standard output with multiple regression analysis. A re-examination of Figure 13.3 reveals that R^2 is given as

$$R\text{-sq} = 74.1\%$$

This result indicates that a relatively high proportion of the variation of the dependent variable, house price, is accounted for by the independent variables in this regression model.

Adjusted R^2

As additional independent variables are added to a regression model, the value of R^2 cannot decrease, and in most cases it will increase. In the formulas for determining R^2,

$$R^2 = \frac{\text{SSR}}{\text{SS}_{yy}} = 1 - \frac{\text{SSE}}{\text{SS}_{yy}}$$

The value of SS_{yy} for a given set of observations will remain the same as independent variables are added to the regression analysis because SS_{yy} is the sum of squares for the dependent variable. Because additional independent variables are likely to increase SSR at least by some amount, the value of R^2 will probably increase for any additional independent variables.

However, sometimes additional independent variables add no *significant* information to the regression model, yet R^2 increases. R^2 therefore may yield an inflated figure. Statisticians have developed an **adjusted R^2** *to take into consideration both the additional information each new independent variable brings to the regression model and the changed degrees of freedom of regression.* Many standard statistical computer packages now compute and report adjusted R^2 as part of the output. The formula for computing adjusted R^2 is

$$\text{Adjusted } R^2 = 1 - \frac{\text{SSE}/(n-k-1)}{\text{SS}_{yy}/(n-1)}$$

The value of adjusted R^2 for the real estate example can be solved by using information from the ANOVA portion of the computer output in Figure 13.3.

```
                       n - 1      SS_YY
             n - k - 1       SSE

   Analysis of Variance

   Source        DF      SS       MS        F        p

   Regression    2      51199    25600    28.68     .000
   Error         20     17850      892
   Total         22     69049

   SSE = 17850      SS_YY = 69049      n - k - 1 = 20    n - 1 = 22
```

$$\text{Adj. } R^2 = 1 - \left[\frac{17850/20}{69049/22}\right] = 1 - 0.284 = 0.716$$

The standard Minitab regression output in Figure 13.3 contains the value of the adjusted R^2 already computed. For the real estate example, this value is shown as

$$R\text{-sq(adj.)} = 71.6\%$$

A comparison of R^2 (0.741) with the adjusted R^2 (0.716) for this example shows that the adjusted R^2 reduces the overall proportion of variation of the dependent variable accounted for by the independent variables by a factor of 0.025, or 2.5%. The gap between the R^2 and adjusted R^2 tends to increase as non-significant independent variables are added to the regression model. As n increases, the difference between R^2 and adjusted R^2 becomes less.

STATISTICS IN BUSINESS TODAY

Using Regression Analysis to Help Select a Robot

Several factors contribute to the success of a manufacturing firm in the world markets. Some examples are creating more efficient plants, lowering labour costs, increasing the quality of products, improving the standards of supplier materials, and learning more about international markets. Basically, success boils down to producing a better product for less cost.

One way to achieve that goal is to improve the technology of manufacturing facilities. Many companies are now using robots in plants to increase productivity and reduce labour costs. The use of such technology is relatively new. The science of selecting and purchasing robots is imperfect and often involves considerable subjectivity.

Two researchers, Moutaz Khouja and David Booth, devised a way to use multiple regression to assist decision makers in robot selection. After sorting through 20 of the more promising variables, they found that the most important variables related to robot performance are repeatability, accuracy, load capacity, and velocity. Accuracy is measured by the distance between where the robot goes on a single trial and the centre of all points to which it goes on repeated trials. Repeatability is the radius of the circle that just includes all points to which the robot goes on repeated trials. Repeatability is of most concern to decision makers because it is hardest to correct. Accuracy can be viewed as bias and is easier to correct. Load capacity is the maximum weight that the robot can handle, and velocity is the maximum tip velocity of the robot arm.

Khouja and Booth used data gathered from 27 robots and regression analysis to develop a multiple regression model that attempts to predict repeatability of robots (the variable of most concern for decision makers) by the velocity and load capacity of robots. Using the resulting regression model and the residuals of the fit, they developed a ranking system for selecting robots that takes into account repeatability, load capacity, velocity, and cost.

Source: Adapted from Moutaz Khouja and David E. Booth, 'A Decision Model for the Robot Selection Problem Using Robust Regression', *Decision Sciences* 22, no. 3 (July/August 1991), pp. 656–62. The *Decision Sciences* journal is published by the Decision Sciences Institute, located at Georgia State University.

13.3   PROBLEMS

13.14   Study the Minitab output shown in Problem 13.7. Comment on the overall strength of the regression model in light of S, R^2, and adjusted R^2.

13.15   Study the Minitab output shown in Problem 13.8. Comment on the overall strength of the regression model in light of S, R^2, and adjusted R^2.

13.16   Using the regression output obtained by working Problem 13.5, comment on the overall strength of the regression model using S, R^2, and adjusted R^2.

13.17   Using the regression output obtained by working Problem 13.6, comment on the overall strength of the regression model using S, R^2, and adjusted R^2.

13.18   Using the regression output obtained by working Problem 13.11, comment on the overall strength of the regression model using S, R^2, and adjusted R^2.

13.19 Using the regression output obtained by working Problem 13.12, comment on the overall strength of the regression model using S, R^2, and adjusted R^2.

13.20 Study the Excel output shown in Problem 13.13. Comment on the overall strength of the regression model in light of S, R^2, and adjusted R^2.

13.21 Study the Minitab residual diagnostic output that follows. Discuss any potential problems with meeting the regression assumptions for this regression analysis based on the residual graphics.

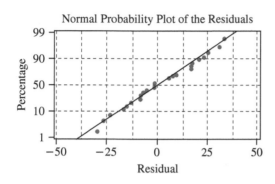

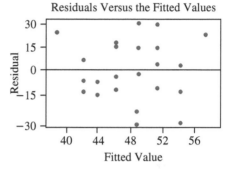

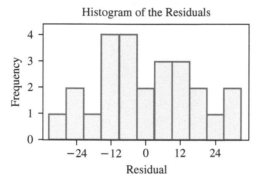

13.4   INTERPRETING MULTIPLE REGRESSION COMPUTER OUTPUT

A Re-examination of the Multiple Regression Output

Figure 13.5 shows again the Minitab multiple regression output for the real estate example. Many of the concepts discussed thus far in the chapter are highlighted. Note the following items:

1. The equation of the regression model
2. The ANOVA table with the F value for the overall test of the model
3. The t ratios, which test the significance of the regression coefficients
4. The value of SSE
5. The value of s_e
6. The value of R^2
7. The value of adjusted R^2

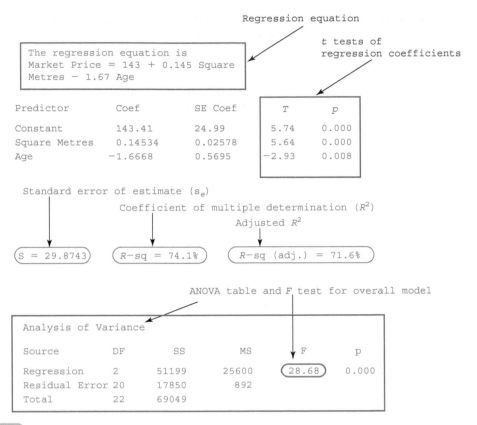

Regression equation

t tests of
regression coefficients

The regression equation is
Market Price = 143 + 0.145 Square
Metres − 1.67 Age

Predictor	Coef	SE Coef	T	p
Constant	143.41	24.99	5.74	0.000
Square Metres	0.14534	0.02578	5.64	0.000
Age	−1.6668	0.5695	−2.93	0.008

Standard error of estimate (s_e)

Coefficient of multiple determination (R^2)

Adjusted R^2

S = 29.8743   R−sq = 74.1%   R−sq (adj.) = 71.6%

ANOVA table and F test for overall model

Analysis of Variance

Source	DF	SS	MS	F	p
Regression	2	51199	25600	28.68	0.000
Residual Error	20	17850	892		
Total	22	69049			

FIGURE 13.5

Annotated Version of the Minitab Output of Regression for the Real Estate Example

DEMONSTRATION PROBLEM 13.2

Discuss the Excel multiple regression output for Demonstration Problem 13.1. Comment on the F test for the overall significance of the model, the t tests of the regression coefficients, and the values of s_e, R^2, and adjusted R^2.

Solution

This multiple regression analysis was done to predict the government bond yield using the predictors of unemployment and household saving ratio. The equation of the regression model was presented in the solution of Demonstration Problem 13.1. Shown here is the complete multiple regression output from the Excel analysis of the data.

The value of F for this problem is 17.72, with a p-value of 0.0001, which is significant at $\alpha = 0.01$. On the basis of this information, the null hypothesis is rejected for the overall test of significance. At least one of the predictor variables is statistically significant from zero, and there is significant predictability of the prime interest rate by this model.

An examination of the t ratios supports this conclusion using $\alpha = 0.01$. The t ratio for unemployment is −0.13 with an associated p-value of 0.8996. This information indicates that unemployment is not a significant predictor of prime interest rates. On the other hand, the t ratio for household savings ratio is 3.55 with an associated p-value of 0.0029, indicating that personal saving is a significant predictor of prime interest rates at $\alpha = 0.01$.

The standard error of the estimate is $s_e = 0.8922$, indicating that approximately 68% of the residuals are within ± 0.8922. An examination of the Excel-produced residuals shows that actually 13 out of 18, or 72%, fall in this interval. Approximately 95% of the residuals should be within $\pm 2(0.8922) = \pm 1.7844$, and an examination of the Excel-produced residuals shows that 17 out of 18, or 94%, of the residuals are within this interval.

R^2 for this regression analysis is 0.7026, or 70.26%. However, the adjusted R^2 is only 66.30%, indicating that there is some inflation in the R^2 value. Overall, there is modest predictability in this model, and only the one predictor variable, household savings ratio, is a significant predictor.

SUMMARY OUTPUT

Regression Statistics	
Multiple R	0.8382
R Square	0.7026
Adjusted R Square	0.6630
Standard Error	0.8922
Observations	18

ANOVA

	df	SS	MS	F	Significance F
Regression	2	28.2121	14.1060	17.7210	0.0001
Residual	15	11.9401	0.7960		
Total	17	40.1522			

	Coefficients	Standard Error	t Stat	P-value
Intercept	2.8034	0.9073	3.09	0.0075
Unemployment Rates	−0.0277	0.2155	−0.13	0.8996
Household Savings Ratio	0.4839	0.1363	3.55	0.0029

RESIDUAL OUTPUT

Observation	Predicted Prime Interest Rate	Residuals
1	7.7416	−0.1949
2	7.0398	1.1127
3	7.5481	0.7711
4	7.1270	0.8122
5	7.2550	−0.1208
6	6.2104	−0.6146
7	5.1541	−0.1391
8	4.9269	0.3998
9	5.5661	−0.5569
10	4.9827	−0.0677
11	5.1320	−0.5537
12	4.4624	0.4634
13	4.5555	−0.0972
14	4.2984	0.0758
15	3.9135	1.1248
16	3.6135	0.9707
17	5.4956	−1.8481
18	5.1500	−1.5375

13.22 Study the Minitab regression output that follows. How many predictors are there? What is the equation of the regression model? Using the key statistics discussed in this chapter, discuss the strength of the model and the predictors.

```
Regression Analysis: Y versus X₁, X₂, X₃, X₄

The regression equation is
Y = -55.9 + 0.0105 X₁ - 0.107 X₂ + 0.579 X₃ - 0.870 X₄
Predictor         Coef     SE Coef        T         P
Constant        -55.93       24.22     -2.31     0.025
       X₁       0.01049     0.02100      0.50     0.619
       X₂      -0.10720     0.03503     -3.06     0.003
       X₃       0.57922     0.07633      7.59     0.000
       X₄       -0.8695      0.1498     -5.81     0.000
S = 9.025    R-Sq = 80.2%     R-Sq(adj) = 78.7%
```

```
Analysis of Variance

Source             DF       SS       MS        F        p

Regression          2    18088.5   4522.1    55.52    0.000
Residual Error     55     4479.7     81.4
Total              59    22568.2
```

13.23 Study the Excel regression output that follows. How many predictors are there? What is the equation of the regression model? Using the key statistics discussed in this chapter, discuss the strength of the model and its predictors.

SUMMARY OUTPUT

Regression Statistics	
Multiple R	0.814
R Square	0.663
Adjusted R Square	0.636
Standard Error	51.761
Observations	28

ANOVA

	df	SS	MS	F	Significance F
Regression	2	131567.02	65783.51	24.55	0.0000013
Residual	25	66979.65	2679.19		
Total	27	198546.68			

	Coefficients	Standard Error	t Stat	P-value
Intercept	203.3937	67.518	3.01	0.0059
X₁	1.1151	0.528	2.11	0.0448
X₂	2.2115	0.567	-3.90	0.0006

Decision Dilemma SOLVED

Are You Going to Hate Your New Job?

In the Decision Dilemma, several variables are considered in attempting to determine whether a person will like his or her new job. Four predictor (independent) variables are given with the data set: relationship with supervisor, overall quality of work environment, total hours worked per week, and opportunities for advancement. Other possible variables might include level of education of the worker, the industry, self-employment status, the smaller the size of the firm, lower working hours, jobs supervisionary role, work culture, amount of pressure, how the interviewee is treated during the interview, availability of flexible scheduling, size of office, amount of time allotted for lunch, availability of management, and many others.

Using the data that are given, a multiple regression model can be developed to predict job satisfaction from the four independent variables. Such an analysis allows the business researcher to study the entire data set in one model rather than constructing four different simple regression models, one for each independent variable.

In the multiple regression model, job satisfaction is the dependent variable. There are 19 observations. The Excel regression output for this problem follows.

The test for overall significance of the model produced an F of 33.89 with a p-value of 0.00000046 (significant at $\alpha = 0.000001$). The R^2 of 0.906 and adjusted R^2 of 0.880 indicate strong predictability in the model. The standard error of the estimate, 8.03, can be viewed in light of the job satisfaction values that ranged from 10 to 95 and the residuals, which are not shown here. Sixteen of the 19 residuals (over 84%) are within the standard error of the estimate. Examining the t statistics and their associated p-values reveals that only one independent variable, 'overall quality of work environment' ($t = 3.92$, p-value $= 0.0015$), is significant at $\alpha = 0.01$. Using a more generous α of 0.10, one could argue that 'relationship with supervisor' is also a significant predictor of job satisfaction. Judging by their large p-values, it appears that 'total hours worked per week' and 'opportunities for advancement' are not good predictors of job satisfaction.

SUMMARY OUTPUT

Regression Statistics

Multiple R	0.952
R Square	0.906
Adjusted R Square	0.880
Standard Error	8.03
Observations	19

ANOVA

	df	SS	MS	F	Significance F
Regression	4	8748.967	2187.242	33.89	0.00000046
Residual	14	903.664	64.547		
Total	18	9652.632			

	Coefficients	Standard Error	t Stat	P-value
Intercept	2.6961	13.005	0.21	0.8387
Relationship with Supervisor	6.9211	3.774	1.83	0.0880
Overall Quality of Work Environment	6.0814	1.550	3.92	0.0015
Total Hours Worked per Week	0.1063	0.1925	0.55	0.5895
Opportunities for Advancement	0.3881	1.6322	0.24	0.8155

ETHICAL CONSIDERATIONS

Multiple regression analysis can be used either intentionally or unintentionally in questionable or unethical ways. When degrees of freedom are small, an inflated value of R^2 can be obtained, leading to overenthusiastic expectations about the predictability of a regression model. To prevent this type of reliance, a researcher should take into account the nature of the data, the variables, and the value of the adjusted R^2.

Another misleading aspect of multiple regression can be the tendency of researchers to assume cause-and-effect relationships between the dependent variable and predictors. Just because independent variables produce a significant R^2 does not necessarily mean those variables are causing the deviation of the y values. Indeed, some other force not in the model may be driving both the independent variables and the dependent variable over the range of values being studied.

Some people use the estimates of the regression coefficients to compare the worth of the predictor variables; the larger the coefficient, the greater is its worth. At least two problems can be found in this approach. The first is that most variables are measured in different units. Thus, regression coefficient weights are partly a function of the unit of measurement of the variable. Second, if multicollinearity (discussed in Chapter 14) is present, the interpretation of the regression coefficients is questionable. In addition, the presence of multicollinearity raises several issues about the interpretation of other regression output. Researchers who ignore this problem are at risk of presenting spurious results.

Another danger in using regression analysis is in the extrapolation of the model to values beyond the range of values used to derive the model. A regression model that fits data within a given range does not necessarily fit data outside that range. One of the uses of regression analysis is in the area of forecasting. Users need to be aware that what has occurred in the past is not guaranteed to continue to occur in the future. Unscrupulous and sometimes even well-intentioned business decision makers can use regression models to project conclusions about the future that have little or no basis. The receiver of such messages should be cautioned that regression models may lack validity outside the range of values in which the models were developed.

SUMMARY

Multiple regression analysis is a statistical tool in which a mathematical model is developed in an attempt to predict a dependent variable by two or more independent variables or in which at least one predictor is non-linear. Because doing multiple regression analysis by hand is extremely tedious and time-consuming, it is almost always done on a computer.

The standard output from a multiple regression analysis is similar to that of simple regression analysis. A regression equation is produced with a constant that is analogous to the y-intercept in simple regression and with estimates of the regression coefficients that are analogous to the estimate of the slope in simple regression. An F test for the overall model is computed to determine whether at least one of the regression coefficients is significantly different from zero. This F value is usually displayed in an ANOVA table, which is part of the regression output. The ANOVA table also contains the sum of squares of error and sum of squares of regression, which are used to compute other statistics in the model.

Most multiple regression computer output contains t values, which are used to determine the significance of the regression coefficients. Using these t values, statisticians can make decisions about including or excluding variables from the model.

Residuals, standard error of the estimate, and R^2 are also standard computer regression output with multiple regression. The coefficient of determination for simple regression models is denoted r^2, whereas for multiple regression it is R^2. The interpretation of residuals, standard error of the estimate, and R^2 in multiple regression is similar to that in simple regression. Because R^2 can be inflated with non-significant variables in the mix, an adjusted R^2 is often computed. Unlike R^2, adjusted R^2 takes into account the degrees of freedom and the number of observations.

GO ONLINE TO DISCOVER THE EXTRA FEATURES FOR THIS CHAPTER

The Student Study Guide containing solutions to the odd-numbered questions, additional Quizzes and Concept Review Activities, Excel and Minitab databases, additional data files in Excel and Minitab, and more worked examples.
www.wiley.com/college/cortinhas

FORMULAS

The F value

$$F = \frac{\text{MS}_{\text{reg}}}{\text{MS}_{\text{err}}} = \frac{\text{SS}_{\text{reg}}/\text{df}_{\text{reg}}}{\text{SS}_{\text{err}}/\text{df}_{\text{err}}} = \frac{\text{SSR}/k}{\text{SSE}/(N-k-1)}$$

Sum of squares of error

$$\text{SSE} = \Sigma(y - \hat{y})^2$$

Standard error of the estimate

$$s_e = \sqrt{\frac{\text{SSE}}{n-k-1}}$$

Coefficient of multiple determination

$$R^2 = \frac{\text{SSR}}{\text{SS}_{yy}} = 1 - \frac{\text{SSE}}{\text{SS}_{yy}}$$

Adjusted R^2

$$\text{Adjusted } R^2 = 1 - \frac{\text{SSE}/(n-k-1)}{\text{SS}_{yy}/(n-1)}$$

SUPPLEMENTARY PROBLEMS

CALCULATING THE STATISTICS

13.24 Use the data right to develop a multiple regression model to predict y from x_1 and x_2. Discuss the output, including comments about the overall strength of the model, the significance of the regression coefficients, and other indicators of model fit.

y	x_1	x_2
198	29	1.64
214	71	2.81
211	54	2.22
219	73	2.70

(continued)

y	x_1	x_2
184	67	1.57
167	32	1.63
201	47	1.99
204	43	2.14
190	60	2.04
222	32	2.93
197	34	2.15

13.25 Given here are the data for a dependent variable, y, and independent variables. Use these data to develop a regression model to predict y. Discuss the output.

y	x_1	x_2	x_3
14	51	16.4	56
17	48	17.1	64
29	29	18.2	53
32	36	17.9	41
54	40	16.5	60
86	27	17.1	55
117	14	17.8	71
120	17	18.2	48
194	16	16.9	60
203	9	18.0	77
217	14	18.9	90
235	11	18.5	67

TESTING YOUR UNDERSTANDING

13.26 The table below presents the average price per year for several minerals over a decade. Use these data and multiple regression to produce a model to predict the average price of gold from the other variables. Comment on the results of the process.

Gold ($ per oz)	Copper (cents per lb)	Silver ($ per oz)	Aluminium (cents per lb)
161.1	64.2	4.4	39.8
308.0	93.3	11.1	61.0
613.0	101.3	20.6	71.6
460.0	84.2	10.5	76.0
376.0	72.8	8.0	76.0
424.0	76.5	11.4	77.8
361.0	66.8	8.1	81.0
318.0	67.0	6.1	81.0
368.0	66.1	5.5	81.0
448.0	82.5	7.0	72.3
438.0	120.5	6.5	110.1
382.6	130.9	5.5	87.8

13.27 The World Bank Indicators dataset includes data on 198 indicators from 209 countries from 1960 to present. Among the variables reported by this dataset are the per capita GDP, the population growth rate, the percentage of rural population and the total tax rate (as a percentage of profit). Shown here are the data for these five variables for the most recent year available for the European Union countries for which data was available. Use the data to develop a regression model to predict per capita GDP from the population growth rate, the percentage of rural population and the total tax rate. Comment on the regression model and its strengths and its weaknesses.

Country	GDP per capita (constant 2000 US$)	Population growth (% annual)	Rural population (% of total)	Total Tax Rate (% of profit)
Austria	26,106.2	0.33	32.6	55.5
Belgium	24,176.4	0.75	2.6	57.3
Bulgaria	2,542.5	−0.50	28.6	31.4
Czech Republic	7,225.4	0.63	26.5	47.2
Denmark	30,547.9	0.65	13.1	29.2
Estonia	6,113.3	−0.02	30.5	49.1
Finland	26,495.9	0.47	36.4	47.7
France	22,820.1	0.54	22.4	65.8
Germany	24,409.5	−0.28	26.3	44.9
Greece	14,843.7	0.41	38.8	47.4
Hungary	5,833.5	−0.16	32.1	57.5
Ireland	28,502.5	0.56	38.4	26.5
Italy	18,479.2	0.65	31.8	68.4
Lithuania	5,153.8	−0.55	32.9	42.7
Luxembourg	52,388.1	1.87	17.7	20.9
Netherlands	26,094.0	0.52	17.6	39.3
Poland	6,330.6	0.06	38.7	42.5
Portugal	11,588.1	0.09	39.9	42.9
Romania	2,606.6	−0.15	45.6	46.4
Slovak Republic	8,041.6	0.21	43.3	48.6
Slovenia	12,576.9	1.08	51.7	37.5
Spain	15,533.8	0.88	22.7	56.9
Sweden	30,899.3	0.89	15.4	54.6
United Kingdom	27,259.2	0.70	10.0	35.9

13.28 The International Monetary Fund produces primary commodity indexes for several different categories. Shown here are the percentage changes in the price indexes over a period of 20 years for food, beverages, industrial materials, and agricultural raw materials. Also displayed are the percentage changes in index for all commodities.

Use these data and multiple regression to develop a model that attempts to predict all commodities by the other four variables. Comment on the result of this analysis.

All Commodities	Food	Beverages	Industrial Materials	Agricultural Raw Materials
55.1	97.6	70.8	74.6	92.3
52.5	96.3	75.9	75.0	105.9
55.1	99.2	128.3	86.2	117.5
59.2	104.4	126.5	94.6	120.3
62.1	112.0	106.1	87.2	116.0
59.5	102.6	130.5	86.3	111.7
48.3	92.4	120.5	72.2	94.6
49.3	80.5	93.9	71.0	93.0
62.9	81.9	77.0	77.7	98.5
58.5	80.5	65.7	72.8	95.2
58.3	83.3	81.6	71.4	95.0
65.0	88.6	85.5	75.3	95.6
79.5	100.5	84.6	88.6	99.4
98.6	99.6	99.6	99.3	99.6
120.8	110.5	108.4	136.3	108.8
135.1	127.3	123.3	154.3	114.2
172.3	157.0	152.0	145.7	113.3
120.7	134.0	154.4	118.7	94.1
152.2	149.4	176.2	170.0	125.4

13.29 The United Nations has published detailed imports and exports annual statistics of close to 200 countries from 1962 to the most recent year. Shown here are the Russain Federation imports (in thousands of US dollars) for three farm products for the last 12 years. Use these data and multiple regression analysis to predict the imports of sweetcorn by the imports of wheat and soya beans. Comment on the results.

Sweetcorn	Wheat	Soya Beans
3,335	18,835	49,000
3,778	9,617	8,020
376	5,016	4,723
588	3,351	12,848
843	9,133	4,509
1,372	25,699	340
3,259	11,897	10,418
3,905	16,616	1,158
4,140	16,112	53,218
5,089	31,720	326,601
7,155	14,916	442,909
7,857	7,062	487,486

13.30 The consulting firm Mercer compiles cost-of-living indexes for selected cities around the world. Shown here are cost-of-living indexes for 21 different cities on five different items for a recent year. Use the data to develop a regression model to predict the rent of a luxury two-bedroom apartment by the price of a cinema ticket, a music CD, a hamburger meal, and a litre of milk. Discuss the results, highlighting both the significant and non-significant predictors.

Expenditure, (£)	Rent of a luxury two-bedroom apartment (per month)	One Cinema Ticket (international release)	One Music CD	One Hamburger Meal	One litre of Milk
Luanda	4,480.46	8.32	18.66	11.2	2.11
London	2,500	10.8	14.99	3.97	0.76
Dublin	1,050.83	9.2	14.01	6.26	0.85
Paris	1,926.52	8.67	14.01	5.17	1.18
Rome	1,576.25	6.57	19.18	5.17	1.44
Amsterdam	1,050.83	8.76	15.75	5.12	0.87
Berlin	1,050.83	7	13.13	4.81	0.83
Athens	1,225.97	7.88	16.55	4.99	0.88
Brussels	1,225.97	7.8	16.55	5.21	0.8
Madrid	1,138.4	7	18.35	4.95	0.84
Prague	1,007.05	5.7	19.51	3.54	0.57
Warsaw	1,088.68	5.01	15.24	2.85	0.56
Zagreb	1,019.02	3.96	16.66	4.2	0.75
Tokyo	2,839.6	12.78	13.49	4.61	1.69
Beijing	1,874.45	7.5	13.12	2.2	1.86
Sydney	1,418.2	8.51	13.59	3.66	1.56
New York	2,560.26	7.68	9.74	3.81	0.77
Buenos Aires	960.1	4.16	6.99	2.95	0.84
Johannesburg	833.47	3.75	12.5	2.25	0.91
Vancouver	1,393.71	7.58	12.72	4.7	1.08
Moscow	2,304.23	8.49	13.79	3.62	3.33

INTERPRETING THE OUTPUT

13.31 Shown here are the data for y and three predictors, x_1, x_2, and x_3. A multiple regression analysis has been done on these data; the Minitab results are given. Comment on the outcome of the analysis in light of the data.

y	x_1	x_2	x_3
94	21	1	204
97	25	0	198
93	22	1	184
95	27	0	200
90	29	1	182
91	20	1	159
91	18	1	147
94	25	0	196
98	26	0	228
99	24	0	242
90	28	1	162
92	23	1	180
96	25	0	219

```
Regression Analysis: Y versus X₁,X₂,X₃
The regression equation is
Y = 87.9 − 0.256X₁ − 2.71X₂ + 0.0706X₃
Predictor       Coef     SE Coef       T        P
Constant      87.890       3.445   25.51    0.000
X₁           -0.25612     0.08317   -3.08    0.013
X₂            -2.7137      0.7306   -3.71    0.005
X₃            0.07061     0.01353    5.22    0.001

S = 0.850311 R-Sq = 94.1% R-Sq(adj) = 92.1%
Analysis of Variance
Source          DF      SS      MS       F       p
Regression       3  103.185  34.395   47.57   0.000
Residual Error   9    6.507   0.723
Total           12  109.692
```

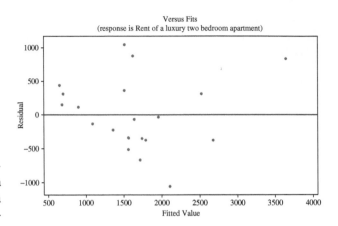

13.32 Minitab residual diagnostic output from the multiple regression analysis for the data given in Problem 13.30 follows. Discuss any potential problems with meeting the regression assumptions for this regression analysis based on the residual graphics.

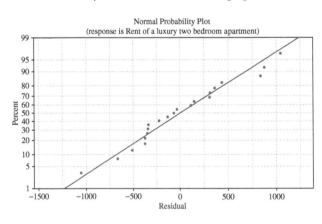

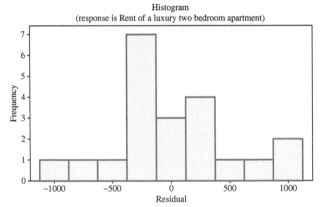

 ANALYSING THE DATABASES

1. Use the Manufacturing database to develop a multiple regression model to predict Cost of Materials by Number of Employees, New Capital Expenditures, Value Added by Manufacture, and End-of-Year Inventories. Discuss the results of the analysis.

2. Develop a regression model using the Financial database. Use Total Revenues, Total Assets, Return on Equity, Earnings Per Share, Average Yield, and Dividends Per Share to predict the average P/E ratio for a company. How strong is the model? Which variables seem to be the best predictors?

3. Using the International Stock Market database, develop a multiple regression model to predict the Nikkei by the DJIA, the Nasdaq, the S&P 500, the Hang Seng, the FTSE 100, and the IPC. Discuss the outcome, including the model, the strength of the model, and the strength of the predictors.

4. Develop a multiple regression model to predict the price of Electricity by the price of Natural Gas and the price of Steam Coal using the Energy database. How strong is the model? Which of the two predictors seems to be stronger? Why?

CASE

AUDIENCE DEMAND FOR UK FILMS

The movie industry is a multibillion pound worldwide industry. The United Kingdom makes a major contribution through local productions and providing locations for foreign productions. According to the British Film Institute's latest Statistical Yearbook, the United Kingdom is the third largest market in the world and the industry makes a substantial contribution to GDP. In 2009, UK film trade surplus was a record £929 million as the industry exported £1,476 million-worth of services, of which £935 million came from royalties and £541 million from film productions. The film and video industries employ more than 48,500 people, of whom more than 30,000 work in film and video production. UK films now attract a sizeable 24% share of total box office takings.

The top films generate huge box office takings. In 2010 the top film of the year at the UK box office was *Toy Story 3*. This film grossed £74 million to become the second highest grossing film of all time, behind 2009's *Avatar*. *Harry Potter and the Deathly Hallows: Part 1*, the seventh in the Harry Potter franchise and a joint UK/US production, grossed £52 million to claim the runner-up spot in the 2010 chart. Nine UK titles featured in the top 20 for that year. Eight of these were UK/US collaborations, produced, at least partly, in the UK and mainly financed by inward investment from the major US studios. The highest placed independent UK film was *StreetDance 3D* which appeared at number 18 in the top 20 chart.

DISCUSSION

1. The following data correspond to the annual box office gross takings in the United Kingdom between the years 2001 and 2010. It also contains the number of admissions (in millions), the number of film releases, the number of foreign language films released in the UK, the number of UK award winners (including Academy Awards®, the BAFTA film awards, and Berlin, Cannes, Sundance, Toronto, and Venice festivals), the number of cinema screens in the UK and the total production budget for UK films. Using these data, develop a multiple regression model to study how well the box office gross takings can be explained by the other variables. Which variables seem to be more promising predictors? What implications for film production and distributor executives might be evident from this analysis?

Year	Box Office Gross (£ million)	Total Admissions (million)	Film Releases	Foreign Language Films Released	Number of UK Award Winners	Number of Cinema Screens	Total Production Budget of UK films (£ million)
2001	645	155.9	352	96	25	3,164	816.1
2002	755	175.9	369	123	24	3,258	1,047.5
2003	742	167.3	423	142	22	3,318	757.6
2004	770	171.3	451	159	22	3,342	1,821.7
2005	770	164.7	467	193	23	3,357	1,634.8
2006	762	156.6	505	165	25	3,440	861.6
2007	821	162.4	516	161	32	3,514	918.1
2008	850	164.2	527	179	32	3,610	980.6
2009	944	173.5	503	148	36	3,651	651.8
2010	988	169.2	557	190	24	3,671	1,139.1

2. Think of other variables, not already mentioned, that may also be important predictors in determining the annual gross box office earnings in the UK.

3. Some film distributors offer discount cinema tickets via email. Suppose that a random sample of 25 movie-goers is undertaken, and suppose that the number of movies the person has seen in the last year,

their age and income, and the number of discount cinema tickets they have received via email in the last year, is recorded. Use the data to develop a multiple regression model to predict the number of times an individual goes to the movies per year from their age, income, and number of discount cinema tickets received. Which particular independent variables seem to have more promise in predicting the number of times a person goes to the movies? What marketing implications might be evident from this analysis?

Age	Number of movies seen at cinema last year	Number of emailed discount cinema tickets received last year	Income (£000s)
25	4	1	20
30	12	5	35
27	10	4	30
42	8	5	30
29	11	8	25
25	12	5	60
50	8	3	30
45	6	5	35

(continued)

Age	Number of movies seen at cinema last year	Number of emailed discount cinema tickets received last year	Income (£000s)
32	16	7	25
23	10	1	20
40	18	5	40
35	12	3	40
28	10	3	50
33	12	2	30
40	15	5	80
37	3	1	30
51	10	8	35
20	8	4	25
26	15	5	35
38	19	10	45
27	12	3	35
29	14	6	35
34	10	4	45
30	6	3	55
22	8	5	30

Source: Adapted from the British Film Institute Statistics Yearbook, 2011, available at www.bfi.org.uk/filmtvinfo/stats/BFI-Statistical-Yearbook-2011.pdf.

USING THE COMPUTER

EXCEL

- Excel has the capability of doing multiple regression analysis. The commands are essentially the same as those for simple regression except that the x range of data may include several columns. Excel will determine the number of predictor variables from the number of columns entered in to **Input X Range**.

- Begin by selecting the **Data** tab on the Excel worksheet. From the **Analysis** panel at the right top of the **Data** tab worksheet, click on **Data Analysis**. If your Excel worksheet does not show the **Data Analysis** option, then you can load it as an add-in following directions given in Chapter 2. From the **Data Analysis** pulldown menu, select **Regression**. In the **Regression** dialog box, input the location of the y values in **Input Y Range**. Input the location of the x values in **Input X Range**. Input **Labels** and input **Confidence Level**.

To pass the line through the origin, check **Constant is Zero**. To print out the raw residuals, check **Residuals**. To print out residuals converted to z scores, check **Standardized Residuals**. For a plot of the residuals, check **Residual Plots**. For a plot of the line through the points check **Line Fit Plots**.

- Standard output includes R, R^2, s_e, and an ANOVA table with the F test, the slope and intercept, t statistics with associated p-values, and any optionally requested output, such as graphs or residuals.

MINITAB

- Multiple regression analysis can be performed using Minitab using the following commands: select **Stat** from the menu bar. Select **Regression** from the **Stat** pulldown menu. Select **Regression** from the **Regression** pulldown menu. Place the column name or column location of the y variable in **Response**. Place the column name(s) or column location(s) of the x variable(s) in **Predictors**. Select **Graphs** for options relating to residual plots. Use this option and check **Four in one** to produce the residual diagnostic plots shown in the chapter. Select **Options** for confidence intervals and prediction intervals. Select **Results** for controlling the regression analysis output. Select **Storage** to store fits and/or residuals.

- To obtain a fitted-line plot, select **<u>S</u>tat** from the menu bar. Select **<u>R</u>egression** from the **<u>S</u>tat** pulldown menu. Select **<u>F</u>itted Line Plot** from the **<u>R</u>egression** pulldown menu. In the Fitted Line Plot dialog box, place the column name or column location of the y variable in **Response(Y)**. Place the column name(s) or column location(s) of the x variable(s) in **Response(X)**. Check **Type of Regression Model** as **Linear**, **Quadratic**, or **Cubic**. Select **Graphs** for options relating to residual plots. Use this option and check **Four in one** to produce the residual diagnostic plots shown in the chapter. Select **Options** for confidence intervals and prediction intervals. Select **Storage** to store fits and/or residuals.

CHAPTER 14

Building Multiple Regression Models

LEARNING OBJECTIVES

This chapter presents several advanced topics in multiple regression analysis, enabling you to:

1. Generalize linear regression models as polynomial regression models using model transformation and Tukey's ladder of transformation, accounting for possible interaction among the independent variables

2. Examine the role of indicator, or dummy, variables as predictors or independent variables in multiple regression analysis

3. Use all possible regressions, stepwise regression, forward selection, and backward elimination search procedures to develop regression models that account for the most variation in the dependent variable and are parsimonious

4. Recognize when multicollinearity is present, understanding general techniques for preventing and controlling it

Decision Dilemma

Determining Compensation for CEOs

The debate on whether chief executive officers deserve their pay packages has increased in intensity since the advent of the global financial crisis. The critics, who include politicians, investors and the general public, claim that excessive executive pay is both a symptom and a cause of the financial crisis.

As pointed out by Adrian Wooldridge in *The Economist*, it is not hard to understand this public outrage: executive pay has exploded since the 1980s. For most of the post-war era, executives earned a few multiples of their company's median pay. Since then, starting in America and slowly spreading to the rest of the world, the multiples have increased dramatically. Today, many workers earn in a year what their boss takes home in an evening.

Chief executive officers for large companies receive widely varying amounts of compensation for their work. Why is the range so wide? What are some of the variables that seem to contribute to the diversity of CEO compensation packages?

As a starting place, one might examine the role of company size as measured by sales volume, number of employees, number of plants, and so on in driving CEO compensation. It could be argued that CEOs of larger companies carry larger responsibilities and hence should receive higher compensation. Some researchers believe CEO compensation is related to such things as industry performance of the firm, percentage of stock that has outside ownership, and proportion of insiders on the board. At least a significant proportion of CEOs are likely to be compensated according to the performance of their companies during the fiscal period preceding compensation. Company performance can be measured by such variables as earnings per share, percentage change in profit, sales, and profit. In addition, some theorize that companies with outside ownership are more orientated towards declaring dividends to stockholders than towards large CEO compensation packages.

Do CEOs' individual and family characteristics play a role in their compensation? Do such things as CEO age, degrees obtained, marital status, military experience, and number of children matter in compensation? Do type of industry and geographic location of the company

matter? What are the significant factors in determining CEO compensation?

Suppose a researcher trying to answer these questions collected the data presented below from 20 randomly selected European companies listed in the stock markets. In the first column on the left are the annual compensation figures (in €000s) for 20 CEOs. Those figures represent salary, bonuses, and any other cash remuneration given to the CEO as part of compensation. The four columns to the right contain data on four variables associated with each CEO's company: sales, number of employees, capital investment, and whether the company is in manufacturing. Sales figures and capital investment figures are given in € millions.

Cash Compensation	Sales	Number of Employees	Capital Investment	Manufacturing
212	35.0	248.00	10.5	1
226	27.2	156.00	3.8	0
237	49.5	348.00	14.9	1
239	34.0	196.00	5.0	0
242	52.8	371.00	15.9	1
245	37.6	216.00	5.7	0
253	60.7	425.00	18.3	1
262	49.2	285.00	8.0	0
271	75.1	524.00	22.6	1
285	69.0	401.00	12.3	0
329	137.2	947.00	41.4	1
340	140.1	825.00	30.3	0
353	162.9	961.00	36.7	0
384	221.7	1517.00	67.1	1
405	261.6	1784.00	79.2	1
411	300.1	1788.00	79.8	0
456	455.5	2733.00	135.7	0
478	437.6	2957.00	132.7	1
525	802.1	4857.00	278.4	0
564	731.5	4896.00	222.2	1

Managerial and Statistical Questions

1. Can a model be developed to predict CEO compensation?
2. If a model is developed, how can the model be evaluated to determine whether it is valid?

(continued)

3. Is it possible to sort out variables that appear to be related to CEO compensation and determine which variables are more significant predictors?

4. Are some of the variables related to CEO compensation but in a non-linear manner?

5. Are some variables highly interrelated and redundant in their potential for determining CEO compensation?

Sources: Adapted from Jeffrey L. Kerr and Leslie Kren, 'Effect of Relative Decision Monitoring on Chief Executive Compensation', *Academy of Management Journal,* vol. 35, no. 2 (June 1992). Used with permission. Robin L. Bartlett, James H. Grant, and Timothy I. Miller, 'The Earnings of Top Executives: Compensating Differentials for Risky Business', *Quarterly Reviews of Economics and Finance,* vol. 32, no. 1 (Spring 1992). Used with permission. Executive Pay, Economist Debates, *The Economist,* at www.economist.com/debate/days/view/402.

14.1   NON-LINEAR MODELS: MATHEMATICAL TRANSFORMATION

The regression models presented thus far are based on the general linear regression model, which has the form

$$y = \beta_0 + \beta_1 x_1 + \beta_2 x_2 + \cdots + \beta_k x_k + \varepsilon \tag{14.1}$$

where:

$\beta_0 =$ the regression constant
$\beta_1, \beta_2 \ldots, \beta_k$ are the partial regression coefficients for the k independent variables
$x_1, \ldots, x_k$ are the independent variables
$k =$ the number of independent variables

In this general linear model, the parameters, β_i, are linear. It does not mean, however, that the dependent variable, y, is necessarily linearly related to the predictor variables. Scatter plots sometimes reveal a curvilinear relationship between x and y. Multiple regression response surfaces are not restricted to linear surfaces and may be curvilinear.

To this point, the variables, x_i, have represented different predictors. For example, in the real estate example presented in Chapter 13, the variables x_1 and x_2 represented two predictors: number of square metres in the house and the age of the house, respectively. Certainly, regression models can be developed for more than two predictors. For example, a marketing site location model could be developed in which sales, as the response variable, is predicted by population density, number of competitors, size of the store, and number of salespeople. Such a model could take the form

$$y = \beta_0 + \beta_1 x_1 + \beta_2 x_2 + \beta_3 x_3 + \beta_4 x_4 + \varepsilon$$

This regression model has four x_i variables, each of which represents a different predictor.

The general linear model also applies to situations in which some x_i represent recoded data from a predictor variable already represented in the model by another independent variable. In some models, x_i represents variables that have undergone a mathematical transformation to allow the model to follow the form of the general linear model.

In this section of this chapter, we explore some of these other models, including polynomial regression models, regression models with interaction, and models with transformed variables.

Polynomial Regression

Regression models in which the highest power of any predictor variable is 1 and in which there are no interaction terms – cross products $(x_i \cdot x_j)$ – are referred to as *first-order models*. Simple regression models like those presented in Chapter 12 are *first-order models with one independent variable.* The general model for simple regression is

$$y = \beta_0 + \beta_1 x_1 + \varepsilon$$

If a second independent variable is added, the model is referred to as a first-order model with two independent variables and appears as

$$y = \beta_0 + \beta_1 x_1 + \beta_2 x_2 + \varepsilon$$

Polynomial regression models are regression models that are second-order or higher-order models. They contain squared, cubed, or higher powers of the predictor variable(s) and contain response surfaces that are curvilinear. Yet, they are still special cases of the general linear model given in Formula 14.1.

Consider a regression model with one independent variable where the model includes a second predictor, which is the independent variable squared. Such a model is referred to as a second-order model with one independent variable because the highest power among the predictors is 2, but there is still only one independent variable. This model takes the following form:

$$y = \beta_0 + \beta_1 x_1 + \beta_2 x_1^2 + \varepsilon$$

This model can be used to explore the possible fit of a quadratic model in predicting a dependent variable. A **quadratic regression model** is *a multiple regression model in which the predictors are a variable and the square of the variable.* How can this be a special case of the general linear model? Let, x_2 of the general linear model be equal to x_1^2 then $y = \beta_0 + \beta_1 x_1 + \beta_2 x_1^2 + \varepsilon$, becomes $y = \beta_0 + \beta_1 x_1 + \beta_2 x_2 + \varepsilon$. Through what process does a researcher go to develop the regression constant and coefficients for a curvilinear model such as this one?

Multiple regression analysis assumes a linear fit of the regression coefficients and regression constant, but not necessarily a linear relationship of the independent variable values (x). Hence, a researcher can often accomplish curvilinear regression by recoding the data before the multiple regression analysis is attempted.

As an example, consider the data given in Table 14.1. This table contains sales volumes (in £ millions) for 13 manufacturing companies along with the number of manufacturer's representatives associated with each firm. A simple regression analysis to predict sales by the number of manufacturer's representatives results in the Excel output in Figure 14.1.

TABLE 14.1

Sales Data for 13 Manufacturing Companies

Manufacturer	Sales (£ millions)	Number of Manufacturing Representatives
1	2.1	2
2	3.6	1
3	6.2	2
4	10.4	3
5	22.8	4
6	35.6	4
7	57.1	5
8	83.5	5
9	109.4	6
10	128.6	7
11	196.8	8
12	280.0	10
13	462.3	11

SUMMARY OUTPUT

Regression Statistics	
Multiple R	0.933
R Square	0.870
Adjusted R Square	0.858
Standard Error	51.098
Observations	13

ANOVA

	df	SS	MS	F	Significance F
Regression	1	192395.416	192395.416	73.69	0.0000033
Residual	11	28721.452	2611.041		
Total	12	221116.868			

	Coefficients	Standard Error	t Stat	P-value
Intercept	−107.029	28.737	−3.72	0.0033561
Reps	41.026	4.779	8.58	0.0000033

FIGURE 14.1

Excel Simple Regression Output for Manufacturing Example

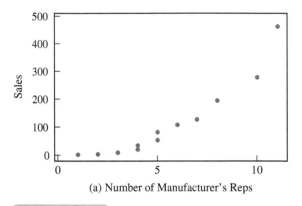
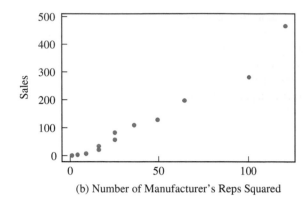

(a) Number of Manufacturer's Reps   (b) Number of Manufacturer's Reps Squared

FIGURE 14.2

Minitab Scatter Plots of Manufacturing Data

This regression output shows a regression model with an r^2 of 87.0%, a standard error of the estimate equal to 51.10, a significant overall F test for the model, and a significant t ratio for the predictor number of manufacturer's representatives.

Figure 14.2(a) is a scatter plot for the data in Table 14.1. Notice that the plot of number of representatives and sales is not a straight line and is an indication that the relationship between the two variables may be curvilinear. To explore the possibility that a quadratic relationship may exist between sales and number of representatives, the business researcher creates a second predictor variable (number of manufacturer's representatives)^2, to use in the regression analysis to predict sales along with number of manufacturer's representatives, as shown in Table 14.2. Thus, a variable can be created to explore second-order parabolic relationships by squaring the data from the independent variable of the linear model and entering it into the analysis. Figure 14.2(b) is a scatter plot of sales with (number of manufacturer's reps)^2. Note that this graph, with the squared term, more closely approaches a straight line than does the graph in Figure 14.2(a). By recoding the predictor variable, the researcher creates a potentially better regression fit.

With these data, a multiple regression model can be developed. Figure 14.3 shows the Excel output for the regression analysis to predict sales by number of manufacturer's representatives and (number of manufacturer's representatives)^2.

TABLE 14.2

Display of Manufacturing Data with Newly Created Variable

Manufacturer	Sales (£ millions) y	Number of Reps x_1	(Number of Reps)^2 $x_2 = x_1(^2)$
1	2.1	2	4
2	3.6	1	1
3	6.2	2	4
4	10.4	3	9
5	22.8	4	16
6	35.6	4	16
7	57.1	5	25
8	83.5	5	25
9	109.4	6	36
10	128.6	7	49
11	196.8	8	64
12	280.0	10	100
13	462.3	11	121

Examine the output in Figure 14.3 and compare it with the output in Figure 14.1 for the simple regression model. The R^2 for this model is 97.3%, which is an increase from the r^2 of 87.0% for the single linear predictor model. The standard error of the estimate for this model is 24.59, which is considerably lower than the 51.10 value obtained from the simple regression model. Remember, the sales figures were £ millions. The quadratic model reduced the standard error of the estimate by 26.51(£1,000,000), or £26,510,000. It appears that the quadratic model is a better model for predicting sales.

An examination of the t statistic for the squared term and its associated probability in Figure 14.3 shows that it is statistically significant at $\alpha = 0.001$ ($t = 6.12$ with a probability of 0.0001). If this t statistic were not significant, the researcher would most likely drop the squared term and revert to the first-order model (simple regression model).

In theory, third- and higher-order models can be explored. Generally, business researchers tend to utilize

SUMMARY OUTPUT

Regression Statistics	
Multiple R	0.986
R Square	0.973
Adjusted R Square	0.967
Standard Error	24.593
Observations	13

ANOVA

	df	SS	MS	F	Significance F
Regression	2	215068.6001	107534.3000	177.79	0.000000015
Residual	10	6048.2676	604.8268		
Total	12	221116.8677			

	Coefficients	Standard Error	t Stat	P-value
Intercept	18.067	24.673	0.73	0.4808194
Reps	−15.723	9.550	−1.65	0.1307046
Reps Squared	4.750	0.776	6.12	0.0001123

FIGURE 14.3

Excel Output for Quadratic Model of Manufacturing Example

first- and second-order regression models more than higher-order models. Remember that most regression analysis is used in business to aid decision making. Higher-power models (third, fourth, etc.) become difficult to interpret and difficult to explain to decision makers. In addition, the business researcher is usually looking for trends and general directions. The higher the order in regression modelling, the more the model tends to follow irregular fluctuations rather than meaningful directions.

Tukey's Ladder of Transformations

As just shown with the manufacturing example, recoding data can be a useful tool in improving the regression model fit. Many other ways of recoding data can be explored in this process. John W. Tukey* presented a 'ladder of expressions' that can be explored to straighten out a plot of x and y, thereby offering potential improvement in the predictability of the regression model. **Tukey's ladder of transformations** gives the following expressions for both x and y.

Ladder for x
$\leftarrow$ Up Ladder $\downarrow$ Neutral   Down Ladder $\rightarrow$

$$\ldots, x^4,\ x^3,\ x^2,\ x,\ \sqrt{x},\ x,\ \log x, -\frac{1}{\sqrt{x}}, -\frac{1}{x}, -\frac{1}{x^2}, -\frac{1}{x^3}, -\frac{1}{x^4}, \ldots$$

Ladder for y
$\leftarrow$ Up Ladder $\downarrow$ Neutral   Down Ladder $\rightarrow$

$$\ldots,\ y^4,\ y^3,\ y^2,\ y,\ \sqrt{y},\ y,\ \log y, -\frac{1}{\sqrt{y}}, -\frac{1}{y}, -\frac{1}{y^2}, -\frac{1}{y^3}, -\frac{1}{y^4}, \ldots$$

*John W. Tukey, *Exploratory Data Analysis.* Reading, MA: Addison-Wesley, 1977.

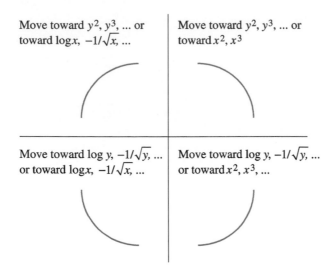

Move toward y^2, y^3, ... or toward $\log x$, $-1/\sqrt{x}$, ...

Move toward y^2, y^3, ... or toward x^2, x^3

Move toward $\log y$, $-1/\sqrt{y}$, ... or toward $\log x$, $-1/\sqrt{x}$, ...

Move toward $\log y$, $-1/\sqrt{y}$, ... or toward x^2, x^3, ...

FIGURE 14.4

Tukey's Four-Quadrant Approach

These ladders suggest to the user potential ways to recode the data. Tukey published a 'four-quadrant approach' to determining which expressions on the ladder are more appropriate for a given situation. This approach is based on the shape of the scatter plot of x and y. Figure 14.4 shows the four quadrants and the associated recoding expressions. For example, if the scatter plot of x and y indicates a shape like that shown in the upper left quadrant, recoding should move 'down the ladder' for the x variable towards

$$\log x, -\frac{1}{\sqrt{x}}, -\frac{1}{x}, -\frac{1}{x^2}, -\frac{1}{x^3}, -\frac{1}{x^4}, \ldots$$

or 'up the ladder' for the y variable towards

$$y^2, y^3, y^4, \ldots$$

Or, if the scatter plot of x and y indicates a shape like that of the lower right quadrant, the recoding should move 'up the ladder' for the x variable towards

$$x^2, x^3, x^4, \ldots$$

or 'down the ladder' for the y variable towards

$$\log y, -\frac{1}{\sqrt{y}}, -\frac{1}{y}, -\frac{1}{y^2}, -\frac{1}{y^3}, -\frac{1}{y^4}, \ldots$$

In the manufacturing example, the graph in Figure 14.2(a) is shaped like the curve in the lower right quadrant of Tukey's four-quadrant approach. His approach suggests that the business researcher move 'up the ladder' on x as was done by using the squared term. The researcher could have explored other options such as continuing on up the ladder of x or going down the ladder of y. Tukey's ladder is a continuum and leaves open other recoding possibilities between the expressions. For example, between x^2 and x^3 are many possible powers of x that can be explored, such as $x^{2.1}$, $x^{2.5}$, or $x^{2.86}$.

Regression Models with Interaction

Often when two different independent variables are used in a regression analysis, an *interaction* occurs between the two variables. This interaction was discussed in Chapter 11 in two-way analysis of variance, where one variable will act differently over a given range of values for the second variable than it does over another range of values for the second variable. For example, in a manufacturing plant, temperature and humidity might interact in such a way as to have an effect on the hardness of the raw material. The air humidity may affect the raw material differently at different temperatures.

In regression analysis, interaction can be examined as a separate independent variable. An interaction predictor variable can be designed by multiplying the data values of one variable by the values of another variable, thereby creating a new variable. A model that includes an interaction variable is

$$y = \beta_0 + \beta_1 x_1 + \beta_2 x_2 + \beta_3 x_1 x_2 + \varepsilon$$

The $x_1 x_2$ term is the interaction term. Even though this model has 1 as the highest power of any one variable, it is considered to be a second-order equation because of the $x_1 x_2$ term.

Suppose the data in Table 14.3 represent the closing stock prices for three corporations over a period of 15 months. An investment firm wants to use the prices for stocks 2 and 3 to develop a regression model to predict the price of stock 1. The form of the general linear regression equation for this model is

$$y = \beta_0 + \beta_1 x_1 + \beta_2 x_2 + \varepsilon$$

where

y = price of stock 1
x_1 = price of stock 2
x_2 = price of stock 3

TABLE 14.3

Prices of Three Stocks over a 15-Month Period

Stock 1	Stock 2	Stock 3
41	36	35
39	36	35
38	38	32
45	51	41
41	52	39
43	55	55
47	57	52
49	58	54
41	62	65
35	70	77
36	72	75
39	74	74
33	83	81
28	101	92
31	107	91

Using Minitab to develop this regression model, the firm's researcher obtains the first output, as displayed in Figure 14.5. This regression model is a first-order model with two predictors, x_1 and x_2. This model produced a modest R^2 of 0.472. Both of the t ratios are small and statistically non-significant ($t = -0.62$ with a p-value of 0.549 and $t = -0.36$ with a p-value of 0.728). Although the overall model is statistically significant, $F = 5.37$ with probability of 0.022, neither predictor is significant.

Sometimes the effects of two variables are not additive because of the interacting effects between the two variables. In such a case, the researcher can use multiple regression analysis to explore the interaction effects by including an interaction term in the equation.

$$y = \beta_0 + \beta_1 x_1 + \beta_2 x_2 + \beta_3 x_1 x_2 + \varepsilon$$

The equation fits the form of the general linear model

$$y = \beta_0 + \beta_1 x_1 + \beta_2 x_2 + \beta_3 x_3 + \varepsilon$$

where $x_3 = x_1 x_2$. Each individual observation of x_3 is obtained through a recoding process by multiplying the associated observations of x_1 and x_2.

Applying this procedure to the stock example, the researcher uses the interaction term and Minitab to obtain the second regression output shown in Figure 14.5. This output contains x_1, x_2, and the interaction term $x_1 x_2$. Observe the R^2, which equals 0.804 for this model. The introduction of the interaction term caused the R^2 to increase from 47.2% to 80.4%. In addition, the standard error of the estimate decreased from 4.570 in the first model to 2.909 in the second model. The t ratios for both the x_1 term and the interaction term are statistically significant in the second model ($t = 3.36$ with a p-value of 0.006 for x_1 and $t = -4.31$ with a probability of 0.001 for $x_1 x_2$). The inclusion of the interaction term helped the regression model account for a substantially greater amount of the dependent variable and is a significant contributor to the model.

```
Regression Analysis: Stock 1 versus Stock 2, Stock 3
─────────────────────────────────────────────────────────────
The regression equation is
Stock 1 = 50.9 - 0.119 Stock 2 - 0.071 Stock 3

Predictor        Coef    SE Coef       T        P
Constant       50.855      3.791    13.41    0.000
Stock 2       -0.1190     0.1931    -0.62    0.549
Stock 3       -0.0708     0.1990    -0.36    0.728

S = 4.57020    R-Sq = 47.2%      R-Sq(adj) = 38.4%

Analysis of Variance

Source              DF        SS       MS       F       P
Regression           2    224.29   112.15    5.37   0.022
Residual Error      12    250.64    20.89
Total               14    474.93

Regression Analysis: Stock 1 versus Stock 2, Stock 3, Interaction
─────────────────────────────────────────────────────────────
The regression equation is
Stock 1 = 12.0 + 0.879 Stock 2 + 0.220 Stock 3 - 0.00998 Interaction

Predictor          Coef    SE Coef       T        P
Constant         12.046      9.312     1.29    0.222
Stock 2          0.8788     0.2619     3.36    0.006
Stock 3          0.2205     0.1435     1.54    0.153
Interaction   -0.009985   0.002314    -4.31    0.001

S = 2.90902    R-Sq = 80.4%      R-Sq(adj) = 75.1%

Analysis of Variance

Source              DF        SS       MS       F       P
Regression           3    381.85   127.28   15.04   0.000
Residual Error      11     93.09     8.46
Total               14    474.93
```

FIGURE 14.5

Two Minitab Regression Outputs, Without and With Interaction

Figure 14.6(a) is the response surface for the first regression model presented in Figure 14.5 (the model without interaction). As you observe the response plane with stock 3 as the point of reference, you see the plane moving upwards with increasing values of stock 1 as the plane moves away from you towards smaller values of stock 2. Now examine Figure 14.6(b), the response surface for the second regression model presented in Figure 14.5 (the model with interaction). Note how the response plane is twisted, with its slope changing as it moves along stock 2. This pattern is caused by the interaction effects of stock 2 prices and stock 3 prices. A cross-section of the plane taken from left to right at any given stock 2 price produces a line that attempts to predict the price of stock 3 from the price of stock 1. As you move back through different prices of stock 2, the slope of that line changes, indicating that the relationship between stock 1 and stock 3 varies according to stock 2.

A researcher also could develop a model using two independent variables with their squares and interaction. Such a model would be a second-order model with two independent variables. The model would look like this.

$$y = \beta_0 + \beta_1 x_1 + \beta_2 x_2 + \beta_3 x_1^2 + \beta_4 x_2^2 + \beta_5 x_1 x_2 + \varepsilon$$

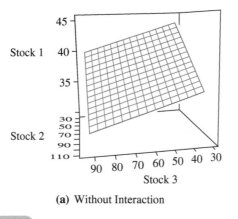
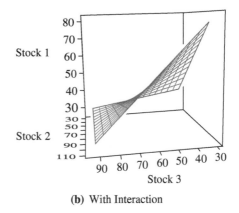

(a) Without Interaction   **(b)** With Interaction

FIGURE 14.6

Response Surfaces for the Stock Example, Without and With Interaction

Model Transformation

To this point in examining polynomial and interaction models, the focus has been on recoding values of x variables. Some multiple regression situations require that the dependent variable, y, be recoded. To examine different relationships between x and y, Tukey's four-quadrant analysis and ladder of transformations can be used to explore ways to recode x or y in attempting to construct regression models with more predictability. Included on the ladder are such y transformations as log y and $1/y$.

Suppose the following data represent the annual sales and annual advertising expenditures for seven companies. Can a regression model be developed from these figures that can be used to predict annual sales by annual advertising expenditures?

Company	Sales (€ million/year)	Advertising (€ million/year)
1	2,580	1.2
2	11,942	2.6
3	9,845	2.2
4	27,800	3.2
5	18,926	2.9
6	4,800	1.5
7	14,550	2.7

One mathematical model that is a good candidate for fitting these data is an exponential model of the form

$$y = \beta_0 \beta_1^x \varepsilon$$

This model can be transformed (by taking the log of each side) so that it is in the form of the general linear equation.

$$\log y = \log \beta_0 + x \log \beta_1$$

This transformed model requires a recoding of the y data through the use of logarithms. Notice that x is not recoded but that the regression constant and coefficient are in logarithmic scale. If we let $y' = \log y$, $\beta_0' = \log \beta_0$, and $\beta_1' = \log \beta_1$, the exponential model is in the form of the general linear model.

$$y' = \beta_0' + \beta_1' x$$

The process begins by taking the log of the y values. The data used to build the regression model and the Excel regression output for these data follow.

Log Sales (y)	Advertising (x)
3.4116	1.2
4.0771	2.6
3.9932	2.2
4.4440	3.2
4.2771	2.9
3.6812	1.5
4.1629	2.7

SUMMARY OUTPUT

Regression Statistics	
Multiple R	0.990
R Square	0.980
Adjusted R Square	0.977
Standard Error	0.0543
Observations	7

ANOVA

	df	SS	MS	F	Significance F
Regression	1	0.739215	0.739215	250.36	0.000018
Residual	5	0.014763	0.002953		
Total	6	0.753979			

	Coefficients	Standard Error	t Stat	P-value
Intercept	2.9003	0.0729	39.80	0.00000019
Advertising	0.4751	0.0300	15.82	0.00001834

A simple regression model (without the log recoding of the y variable) yields an R^2 of 87%, whereas the exponential model R^2 is 98%. The t statistic for advertising is 15.82 with a p-value of 0.00001834 in the exponential model and 5.77 with a p-value of 0.00219 in the simple regression model. Thus the exponential model gives a better fit than does the simple regression model. An examination of (x^2, y) and (x^3, y) models reveals R^2 of 0.930 and 0.969, respectively, which are quite high but still not as good as the R^2 yielded by the exponential model (the output for these models is not shown here).

The resulting equation of the exponential regression model is

$$y = 2.9003 + 0.4751x$$

In using this regression equation to determine predicted values of y for x, remember that the resulting predicted y value is in logarithmic form and the antilog of the predicted y must be taken to get the predicted y value in raw units. For example, to get the predicted y value (sales) for an advertising figure of 2.0 (€ million), substitute $x = 2.0$ into the regression equation.

$$y = 2.9003 + 0.4751x = 2.9003 + 0.4751(2.0) = 3.8505$$

The log of sales is 3.8505. Taking the antilog of 3.8505 results in the predicted sales in raw units.

$$\text{antilog}(3.8505) = 7{,}087.61(\text{€ million})$$

Thus, the exponential regression model predicts that €2.0 million of advertising will result in €7,087.61 million of sales.

Other ways can be used to transform mathematical models so that they can be treated like the general linear model. One example is an inverse model such as

$$y = \frac{1}{\beta_0 + \beta_1 x_1 + \beta_2 x_2 + \varepsilon}$$

Such a model can be manipulated algebraically into the form

$$\frac{1}{y} = \beta_0 + \beta_1 x_1 + \beta_2 x_2 + \varepsilon$$

Substituting $y' = 1/y$ into this equation results in an equation that is in the form of the general linear model.

$$y' = \beta_0 + \beta_1 x_1 + \beta_2 x_2 + \varepsilon$$

To use this 'inverse' model, recode the data values for y by using $1/y$. The regression analysis is done on the $1/y$, x_1, and x_2 data. To get predicted values of y from this model, enter the raw values of x_1 and x_2. The resulting predicted value of y from the regression equation will be the inverse of the actual predicted y value.

DEMONSTRATION PROBLEM 14.1

In the aerospace and defence industry, some cost estimators predict the cost of new space projects by using mathematical models that take the form

$$y = \beta_0 x^{\beta_1} \varepsilon$$

These cost estimators often use the weight of the object being sent into space as the predictor (x) and the cost of the object as the dependent variable (y). Quite often β_1 turns out to be a value between 0 and 1, resulting in the predicted value of y equalling some root of x.

Use the sample cost data given here to develop a cost regression model in the form just shown to determine the equation for the predicted value of y. Use this regression equation to predict the value of y for $x = 3,000$.

y (cost in billions)	x (weight in tonnes)
1.2	450
9.0	20,200
4.5	9,060
3.2	3,500
13.0	75,600
0.6	175
1.8	800
2.7	2,100

Solution

The equation

$$y = \beta_0 x^{\beta_1} \varepsilon$$

is not in the form of the general linear model, but it can be transformed by using logarithms:

$$\log y = \log \beta_0 + \beta_1 \log x + \varepsilon$$

which takes on the general linear form

$$y' = \beta_0' + \beta_1 x'$$

where:
$$y' = \log y$$
$$\beta_0' = \log \beta_0$$
$$x' = \log x$$

This equation requires that both x and y be recoded by taking the logarithm of each.

log y	log x
.0792	2.6532
.9542	4.3054
.6532	3.9571
.5051	3.5441
1.1139	4.8785
−.2218	2.2430
.2553	2.9031
.4314	3.3222

Using these data, the computer produces the following regression constant and coefficient:

$$b_0' = -1.25292 \qquad b_1 = 0.49606$$

From these values, the equation of the predicted y value is determined to be

$$\log \hat{y} = -1.25292 + 0.49606 \log x$$

If $x = 3{,}000$, $\log x = 3.47712$, and

$$\log \hat{y} = -1.25292 + 0.49606(3.47712) = 0.47194$$

then

$$\hat{y} = \text{antilog}(\log \hat{y}) = \text{antilog}(0.47194) = 2.964$$

The predicted value of y is €2.9644 billion for $x = 3{,}000$ tonnes of weight.

Taking the antilog of $b_0' = -1.25292$ yields 0.055857. From this and $b_1 = 0.49606$, the model can be written in the original form:

$$y = (0.055857)x^{0.49606}$$

Substituting $x = 3{,}000$ into this formula also yields €2.9645 billion for the predicted value of y.

14.1 PROBLEMS

14.1 Use the data below to develop a quadratic model to predict y from x. Develop a simple regression model from the data and compare the results of the two models. Does the quadratic model seem to provide any better predictability? Why or why not?

x	y	x	y
14	200	15	247
9	74	8	82
6	29	5	21
21	456	10	94
17	320		

14.2 Develop a multiple regression model of the form

$$y = b_0 b_1^x \varepsilon$$

using the following data to predict y from x. From a scatter plot and Tukey's ladder of transformation, explore ways to recode the data and develop an alternative regression model. Compare the results.

y	x	y	x
2,485	3.87	740	2.83
1,790	3.22	4010	3.62
874	2.91	3629	3.52
2,190	3.42	8010	3.92
3,610	3.55	7047	3.86
2,847	3.61	5680	3.75
1,350	3.13	1740	3.19

14.3 The World Development Indicators database compiles data on household final consumption expenditure and on the number of fixed broadband Internet subscribers for virtually every country and region in the world. Shown here are data on household final consumption expenditure (in millions of constant 2,000 US$) and the number of fixed broadband subscribers (in thousands) for India for the most recent eight years of available data.

Using these data, develop a regression model to predict total household final consumption expenditure by the number of fixed broadband Internet subscribers and by (number of fixed broadband Internet subscribers)^2. Compare this model with a regression model to predict total household final consumption by only the number of fixed broadband Internet subscribers. Construct a scatter plot of the data. Does the shape of the plot suggest some alternative models in light of Tukey's four-quadrant approach? If so, develop at least one other model and compare the model with the other two previously developed.

Household Final Consumption Expenditure (constant 2,000 US$, millions)	Fixed Broadband Internet Subscribers (thousands)
320,154	82
339,014	140
357,785	180
387,900	1,348
420,758	2,300
459,053	3,130
494,084	5,280
530,856	7,746

14.4 Dun & Bradstreet reports, among other things, information about new business incorporations and number of business failures over several years. Shown here are data on business failures and current liabilities of the failing companies over several years. Use these data and the following model to predict current liabilities of the failing companies by the number of business failures. Discuss the strength of the model.

$$y = b_0 b_1^x \varepsilon$$

Now develop a different regression model by recoding x. Use Tukey's four-quadrant approach as a resource. Compare your models.

Rate of Business Failures (10,000s)	Current Liabilities of Failing Companies (£ millions)
54	1,888
43	4,380
42	4,635
61	6,955
88	15,611
110	16,073
107	29,269
115	36,937
120	44,724
102	34,724
98	39,126
65	44,261

14.5 Use the data below to develop a curvilinear model to predict y. Include both x_1 and x_2 in the model in addition to x_1^2 and x_2^2, and the interaction term $x_1 x_2$. Comment on the overall strength of the model and the significance of each predictor. Develop a regression model with the same independent variables as the first model but without the interaction variable. Compare this model with the model with interaction.

y	x_1	x_2
47.8	6	7.1
29.1	1	4.2
81.8	11	10.0
54.3	5	8.0
29.7	3	5.7
64.0	9	8.8
37.4	3	7.1
44.5	4	5.4
42.1	4	6.5
31.6	2	4.9
78.4	11	9.1
71.9	9	8.5
17.4	2	4.2
28.8	1	5.8
34.7	2	5.9
57.6	6	7.8
84.2	12	10.2
63.2	8	9.4
39.0	3	5.7
47.3	5	7.0

14.6 What follows is Excel output from a regression model to predict y using x_1, x_2, x_1^2, x_2^2, and the interaction term $x_1 x_2$. Comment on the overall strength of the model and the significance of each predictor. The data follow the Excel output. Develop a regression model with the same independent variables as the first model but without the interaction variable. Compare this model with the model with interaction.

SUMMARY OUTPUT

Regression Statistics

Multiple R	0.954
R-Square	0.910
Adjusted R-Square	0.878
Standard Error	7.544
Observations	20

ANOVA

	df	SS	MS	F	Significance F
Regression	5	8089.274577	1617.855	28.43	0.00000073
Residual	14	796.725	56.909		
Total	19	8886			

	Coefficients	Standard Error	t Stat	P-value
Intercept	464.4433	503.0955	0.92	0.3716
X_1	−10.5101	6.0074	−1.75	0.1021
X_2	−1.2212	1.9791	−0.62	0.5471
X_1Sq	0.0357	0.0195	1.84	0.0876
X_2Sq	−0.0002	0.0021	−0.08	0.9394
$X_1{}^*X_2$	0.0243	0.0107	2.28	0.0390

y	x_1	x_2	y	x_1	x_2
34	120	190	45	96	245
56	105	240	34	79	288
78	108	238	23	66	312
90	110	250	89	88	315
23	78	255	76	80	320
34	98	230	56	73	335
45	89	266	43	69	335
67	92	270	23	75	250
78	95	272	45	63	372
65	85	288	56	74	360

14.2 INDICATOR (DUMMY) VARIABLES

Some variables are referred to as **qualitative variables** (as opposed to *quantitative* variables) because qualitative variables do not yield quantifiable outcomes. Instead, *qualitative variables yield nominal- or ordinal-level information*, which is used more to categorize items. These variables have a role in multiple regression and are referred to as **indicator**, or **dummy variables**. In this section, we will examine the role of indicator, or dummy, variables as predictors or independent variables in multiple regression analysis.

Indicator variables arise in many ways in business research. Mail questionnaire or personal interview demographic questions are prime candidates because they tend to generate qualitative measures on such items as sex, geographic region, occupation, marital status, level of education, economic class, political affiliation, religion, management/non-management status, buying/leasing a home, method of transportation, or type of broker. In one political economy study, business researchers were attempting to develop a multiple regression model to predict whether the level of economic development has a positive effect on the level of democracy. One independent variable was whether the country was a former British colony. In a second study, a site location model for pizza restaurants included indicator variables for (1) whether the restaurant served beer and (2) whether the restaurant had a salad bar.

These indicator variables are qualitative in that no interval or ratio level measurement is assigned to a response. For example, if a country is a former British colony, awarding it a score of 20 or 30 or 75 because of its location makes no sense. In terms of sex, what value would you assign to a man or a woman in a regression study? Yet these types of indicator, or dummy, variables are often useful in multiple regression studies and can be included if they are coded in the proper format.

Most researchers code indicator variables by using 0 or 1. For example, in the level of democracy study, countries which are former British colonies could be assigned a 1, and all other countries would then be assigned a 0. The assignment of 0 or 1 is arbitrary, with the number merely holding a place for the category. For this reason, the coding is referred to as 'dummy' coding; the number represents a category by holding a place and is not a measurement.

Many indicator, or dummy, variables are dichotomous, such as male/female, salad bar/no salad bar, employed/not employed, and rent/own. For these variables, a value of 1 is arbitrarily assigned to one category and a value of 0 is assigned to the other category. Some qualitative variables contain several categories, such as the variable 'type of job', which might have the categories assembler, painter, and inspector. In this case, using a coding of 1, 2, and 3, respectively, is tempting. However, that type of coding creates problems for multiple regression analysis. For one thing, the category 'inspector' would receive a value that is three times that of 'painter'. In addition, the values of 1, 2, and 3 indicate a hierarchy of job types: assembler $<$ painter $<$ inspector. The proper way to code such indicator variables is with the 0, 1 coding. Two separate independent variables should be used to code the three categories of type of job. The first variable is assembler, where a 1 is recorded if the person's job is assembler and a 0 is recorded if it is not. The second variable is painter, where a 1 is recorded if the person's job is painter and a 0 is recorded if it is not. A variable should not be assigned to inspector, because all workers in the study for whom a 1 was not recorded either for the assembler variable or the painter variable must be inspectors. Thus, coding the inspector variable would result in redundant information and is not necessary. This reasoning holds for all indicator variables with more than two categories. If an indicator variable has c categories, then $c - 1$ dummy variables must be created and inserted into the regression analysis in order to include the indicator variable in the multiple regression.*

An example of an indicator variable with more than two categories is the result of the following question taken from a typical questionnaire.

Your office is located in which region of the country?

_____ North _____ East_____ South _____ West

Suppose a researcher is using a multiple regression analysis to predict the cost of doing business and believes geographic location of the office is a potential predictor. How does the researcher insert this qualitative variable into the analysis? Because $c = 4$ for this question, three dummy variables are inserted into the analysis. Table 14.4 shows one possible way this process works with 13 respondents. Note that rows 2, 7, and 11 contain all zeros, which indicate that those respondents have offices in the West. Thus, a fourth dummy variable for the West region is not necessary and, indeed, should not be included because the information contained in such a fourth variable is contained in the other three variables.

A word of caution is in order. Because of degrees of freedom and interpretation considerations, it is important that a multiple regression analysis have enough observations to handle adequately the number of independent variables entered. Some researchers recommend as a rule of thumb at least three

TABLE 14.4

Coding for the Indicator Variable of Geographic Location for Regression Analysis

North x_1	East x_2	South x_3
1	0	0
0	0	0
1	0	0
0	0	1
0	1	0
0	1	0
0	0	0
0	0	1
1	0	0
1	0	0
0	0	0
0	1	0
0	0	1

*If c indicator variables are included in the analysis, no unique estimator of the regression coefficients can be found. J. Neter, M. H. Kuter, W. Wasserman, and C. Nachtsheim, *Applied Linear Regression Models,* 3rd edn. Chicago: Richard D. Irwin, 1996.

observations per independent variable. If a qualitative variable has multiple categories, resulting in several dummy independent variables, and if several qualitative variables are being included in an analysis, the number of predictors can rather quickly exceed the limit of recommended number of variables per number of observations. Nevertheless, dummy variables can be useful and are a way in which nominal or ordinal information can be recoded and incorporated into a multiple regression model.

As an example, consider the issue of sex discrimination in the salary earnings of workers in some industries. In examining this issue, suppose a random sample of 15 workers is drawn from a pool of employed workers in a particular industry and the workers' average monthly salaries are determined, along with their age and gender. The data are shown in Table 14.5. As sex can be only male or female, this variable is a dummy variable requiring 0, 1 coding. Suppose we arbitrarily let 1 denote male and 0 denote female. Figure 14.7 is the multiple regression model developed from the data of Table 14.5 by using Minitab to predict the dependent variable, monthly salary, by two independent variables, age and sex.

The computer output in Figure 14.7 contains the regression equation for this model.

TABLE 14.5

Data for the Monthly Salary Example

Monthly Salary (£000s)	Age (10 years)	Sex (1 = male, 0 = female)
1.548	3.2	1
1.629	3.8	1
1.011	2.7	0
1.229	3.4	0
1.746	3.6	1
1.528	4.1	1
1.018	3.8	0
1.190	3.4	0
1.551	3.3	1
0.985	3.2	0
1.610	3.5	1
1.432	2.9	1
1.215	3.3	0
0.990	2.8	0
1.585	3.5	1

$$\text{Salary} = 0.732 + 0.111 \text{ Age} + 0.459 \text{ Sex}$$

An examination of the t ratios reveals that the dummy variable 'sex' has a regression coefficient that is significant at $\alpha = 0.001$ ($t = 8.58$, $p = 0.000$). The overall model is significant at $\alpha = 0.001$ ($F = 48.54$, $p = 0.000$). The standard error of the estimate, $s_e = 0.09679$, indicates that approximately 68% of the errors of prediction are within $\pm£96.79$ ($0.09679 \cdot £1,000$). The R^2 is relatively high at 89.0%, and the adjusted R^2 is 87.2%.

```
Regression Analysis: Salary versus Age, Sex

The regression equation is
Salary = 0.732 + 0.111 Age + 0.459 Sex
Predictor      Coef     SE Coef      T       P
Constant     0.7321     0.2356     3.11    0.009
Age          0.11122    0.07208    1.54    0.149
Sex          0.45868    0.05346    8.58    0.000

S = 0.0967916    R-Sq = 89.0%    R-Sq(adj) = 87.2%


Analysis of Variance

Source          DF        SS         MS        F        P
Regression       2      0.90949    0.45474    48.54    0.000
Residual Error  12      0.11242    0.00937
Total           14      1.02191
```

FIGURE 14.7

Minitab Regression Output for the Monthly Salary Example

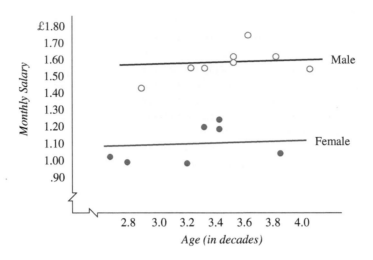

FIGURE 14.8

Regression Model for Male and Female Sex

The t value for sex indicates that it is a significant predictor of monthly salary in this model. This significance is apparent when one looks at the effects of this dummy variable another way. Figure 14.8 shows the graph of the regression equation when sex $= 1$ (male) and the graph of the regression equation when sex $= 0$ (female). When sex $= 1$ (male), the regression equation becomes

$$0.732 + 0.111(\text{Age}) + 0.459(1) = 1.191 + 0.111(\text{Age})$$

When sex $= 0$ (female), the regression equation becomes

$$732 + 0.111(\text{Age}) + 0.459(0) = 0.732 + 0.111(\text{Age}).$$

The full regression model (with both predictors) has a response surface that is a plane in a three-dimensional space. However, if a value of 1 is entered for sex into the full regression model, as just shown, the regression model is reduced to a line passing through the plane formed by monthly salary and age. If a value of 0 is entered for sex, as shown, the full regression model also reduces to a line passing through the plane formed by monthly salary and age. Figure 14.8 displays these two lines. Notice that the only difference in the two lines is the y-intercept. Observe the monthly salary with male sex, as depicted by $\circ$, versus the monthly salary with female sex, depicted by $\bullet$. The difference in the y-intercepts of these two lines is 0.459, which is the value of the regression coefficient for sex. This intercept figure signifies that, on average, men earn £459 per month more than women for this population.

STATISTICS IN BUSINESS TODAY

Predicting Export Intensity of Chinese Manufacturing Firms Using Multiple Regression Analysis

According to business researchers Hongxin Zhao and Shaoming Zou, little research has been done on the impact of external or uncontrollable variables on the export performance of a company. These two researchers conducted a study of Chinese manufacturing firms and used multiple regression to determine whether both domestic market concentration and firm location are good predictors of a firm's export intensity. The study included 999 Chinese manufacturing firms that exported. The dependent variable was 'export intensity', which was defined to be the proportion of production output that is exported and was computed by dividing the firm's export value by its production output value. The higher the proportion was, the higher the export intensity. Zhao and Zou used covariate techniques (beyond the scope of this text) to control for the fact that companies in the study varied by size, capital intensity, innovativeness, and industry. The independent variables were industry concentration and location. Industry concentration was computed as a ratio, with higher values indicating more concentration in the industry. The location variable was a composite index taking into account total freight volume, available modes of transportation, number of telephones, and size of geographic area.

The multiple regression model produced an R^2 of approximately 52%. Industry concentration was a statistically significant predictor at $\alpha = 0.01$, and the sign

on the regression coefficient indicated that a negative relationship may exist between industry concentration and export intensity. It means export intensity is lower in highly concentrated industries and higher in lower concentrated industries. The researchers believe that in a more highly concentrated industry, the handful of firms dominating the industry will stifle the export competitiveness of firms. In the absence of dominating firms in a more fragmented setting, more competition and an increasing tendency to export are noted. The location variable was also a significant predictor at $\alpha = 0.01$. Firms located in coastal areas had higher export intensities than did those located in inland areas.

Source: Hongxin Zhao and Shaoming Zou, 'The Impact of Industry Concentration and Firm Location on Export Propensity and Intensity: An Empirical Analysis of Chinese Manufacturing Firms', *Journal of International Marketing*, vol. 10, no. 1 (2002), pp. 52–71.

14.2 PROBLEMS

14.7 Analyse the data below by using a multiple regression computer software package to predict y using x_1 and x_2. Notice that x_2 is a dummy variable. Discuss the output from the regression analysis; in particular, comment on the predictability of the dummy variable.

y	x_1	x_2
16.8	27	1
13.2	16	0
14.7	13	0
15.4	11	1
11.1	17	0
16.2	19	1
14.9	24	1
13.3	21	0
17.8	16	1
17.1	23	1
14.3	18	0
13.9	16	0

14.8 Given here are the data from a dependent variable and two independent variables. The second independent variable is an indicator variable with several categories. Hence, this variable is represented by x_2, x_3, and x_4. How many categories are needed in total for this independent variable? Use a computer to perform a multiple regression analysis on this data to predict y from the x values. Discuss the output and pay particular attention to the dummy variables.

y	x_1	x_2	x_3	x_4
11	1.9	1	0	0
3	1.6	0	1	0
2	2.3	0	1	0
5	2.0	0	0	1
9	1.8	0	0	0
14	1.9	1	0	0
10	2.4	1	0	0
8	2.6	0	0	0
4	2.0	0	1	0
9	1.4	0	0	0
11	1.7	1	0	0
4	2.5	0	0	1
6	1.0	1	0	0
10	1.4	0	0	0
3	1.9	0	1	0
4	2.3	0	1	0
9	2.2	0	0	0
6	1.7	0	0	1

14.9 The Minitab output displayed here is the result of a multiple regression analysis with three independent variables. Variable x_1 is a dummy variable. Discuss the computer output and the role x_1 plays in this regression model.

```
The regression equation is
y = 121 + 13.4 x₁ - 0.632 x₂ + 1.42 x₃

Predictor        Coef        Stdev          T          p
Constant        121.31       11.56       10.50       .000
X₁               13.355       4.714        2.83       .014
X₂              -0.6322       0.2270      -2.79       .015
X₃                1.421       3.342        0.43       .678

S = 7.041     R-sq = 79.5%     R-sq(adj) = 74.7%


Analysis of Variance

Source           df          SS          MS          F          p
Regression        3       2491.98      830.66      16.76       .000
Error            13        644.49       49.58
Total            16       3136.47
```

14.10 Given here is Excel output for a multiple regression model that was developed to predict y from two independent variables, x_1 and x_2. Variable x_2 is a dummy variable. Discuss the strength of the multiple regression model on the basis of the output. Focus on the contribution of the dummy variable. Plot x_1 and y with x_2 as 0, and then plot x_1 and y with x_2 as 1. Compare the two lines and discuss the differences.

SUMMARY OUTPUT

Regression Statistics	
Multiple R	0.623
R Square	0.388
Adjusted R Square	0.341
Standard Error	11.744
Observations	29

ANOVA

	df	SS	MS	F	Significance F
Regression	2	2270.11	1135.05	8.23	0.0017
Residual	26	3585.75	137.91		
Total	28	5855.86			

	Coefficients	Standard Error	t Stat	P-value
Intercept	41.225	6.3800	6.46	0.00000076
X_1	1.081	1.3530	0.80	0.4316
X_2	−18.404	4.5470	−4.05	0.0004

14.11 *Which?* is a UK-based product-testing and campaigning charity that carries out systematic testing of consumer the results in reports in its magazine and website. The table below presents data relative to a comparison of nine cameras priced above £250.00, and includes data such as price of the camera, the resolution (in megapixels), the weight (in grams) and whether the camera has manual focus setting. Use the data below and a computer to develop a model to predict the price of a digital camera from the resolution, weight, and whether it has manual focus setting. Comment on the output.

Model	Price (£)	Resolution (in megapixels)	Weight (grams)	Manual Focus setting
Canon Digital lxus i7	414.95	7.1	123	0
Canon Powershot S90	431.78	10.0	200	1
Casio Exilim EX-FC100	257.99	9.0	175	1
Leika V-Lux 20	489.95	12.0	219	0
Ricoh Caplio GX200	299.95	12.0	244	1
Sony Cyber-shot DSC-TX9	352.33	12.0	151	0
Samsung ST550	259.00	12.0	168	0
Pentax Optio W90	254.99	12.0	163	1
Panasonic Lumix DMC-TZ3	270.49	7.1	270	0

14.12 A researcher gathered 155 observations on four variables: job satisfaction, occupation, industry, and marital status. She wants to develop a multiple regression model to predict job satisfaction by the other three variables. All three predictor variables are qualitative variables with the following categories.

1. Occupation: accounting, management, marketing, finance
2. Industry: manufacturing, healthcare, transportation
3. Marital status: married, single

How many variables will be in the regression model? Delineate the number of predictors needed in each category and discuss the total number of predictors.

14.3   MODEL-BUILDING: SEARCH PROCEDURES

To this point in the chapter, we have explored various types of multiple regression models. We evaluated the strengths of regression models and learned how to understand more about the output from multiple regression computer packages. In this section we examine procedures for developing several multiple regression model options to aid in the decision-making process.

Suppose a researcher wants to develop a multiple regression model to predict the retail index of energy prices in Germany. The researcher believes that much of the variation of the German retail energy prices is driven by variables related to production of electricity in Germany. The researcher decides to use as predictors the following five largest sources of electricity production in Germany (values in thousands of gigawatt hours).

1.   Electricity output from coal and coal products
2.   Electricity output from natural gas
3.   Electricity output from nuclear
4.   Electricity output from hydro
5.   Electricity output from solar/wind/other

The researcher collected data for each of these variables for a period between 1996 and 2010, figuring that the retail index of energy prices in Germany is driven by the same year's electricity production in Germany.

Table 14.6 shows data for the five independent variables along with the dependent variable, retail index of energy prices. Using the data presented in Table 14.6, the researcher attempted to develop a multiple regression model using five different independent variables. The result of this process was the Minitab output in Figure 14.9. Examining the output, the researcher can reach some conclusions about that particular model and its variables.

The output contains an R^2 value of 99.0%, a standard error of the estimate of 2.101, and an overall significant F value of 184.91. Notice from Figure 14.9 that the t ratios indicate that the regression coefficients of three of the predictor variables, natural gas, nuclear and hydro are not significant at $\alpha = 0.05$. If the researcher were to drop the three variables out of the regression analysis and rerun the model with the other two significant predictors only, what would

TABLE 14.6

Data for Multiple Regression Model to Predict the Retail Index of Energy Prices in Germany

Year	Retail Index of Energy Prices	Electricity Output from Coal and Coal Products	Electricity Output from Natural Gas	Electricity Output from Nuclear	Electricity Output from Hydro	Electricity Output from Solar/Wind/Other
1996	80.4	303	48	160	22	2
1997	80.8	292	50	170	17	3
1998	81.7	299	54	162	17	5
1999	85.0	288	55	170	20	6
2000	80.8	304	52	170	22	9
2001	84.1	302	58	171	23	11
2002	87.8	307	55	165	23	16
2003	92.2	320	59	165	19	19
2004	95.9	305	63	167	21	26
2005	100.0	306	69	163	20	29
2006	103.9	302	76	167	20	33
2007	111.1	310	77	141	21	43
2008	118.7	291	88	148	21	45
2009	126.1	257	79	135	19	52
2010	130.1	271	85	141	19	52

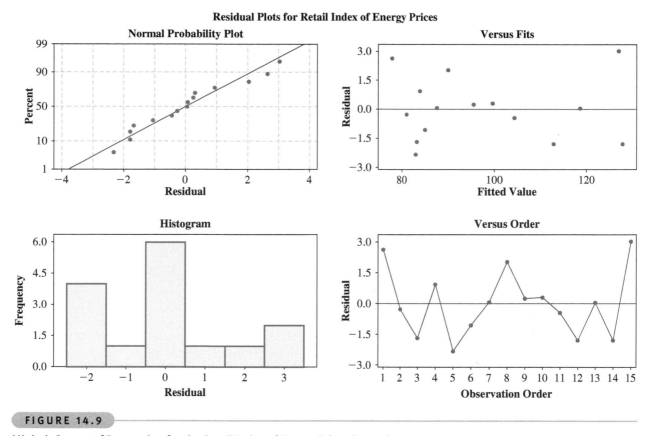

FIGURE 14.9

Minitab Output of Regression for the Retail Index of Energy Prices Example

happen to the model? What if the researcher ran a regression model with only three predictors? How would these models compare to the full model with all five predictors? Are all the predictors necessary?

Developing regression models for business decision making involves at least two considerations. The first is to develop a regression model that accounts for the most variation of the dependent variable – that is, develop models that maximize the explained proportion of the deviation of the y values. At the same time, the regression model should be as parsimonious (simple and economical) as possible. The more complicated a quantitative model becomes, the harder it is for managers to understand and implement the model. In addition, as more variables are included in a model, it becomes more expensive to gather historical data or update present data for the model. These two considerations (dependent variable explanation and parsimony of the model) are quite often in opposition to each other. Hence the business researcher, as the model builder, often needs to explore many model options.

In the retail index of energy prices regression model, if two variables explain the deviation of retail index of energy prices nearly as well as five variables, the simpler model is more attractive. How might researchers conduct regression analysis so that they can examine several models and then choose the most attractive one? The answer is to use search procedures.

Search Procedures

Search procedures are processes *whereby more than one multiple regression model is developed for a given database, and the models are compared and sorted by different criteria*, depending on the given procedure. Virtually all search procedures are done on a computer. Several search procedures are discussed in this section, including all possible regressions, stepwise regression, forward selection, and backward elimination.

TABLE 14.7

Predictors for All Possible Regressions with Five Independent Variables

Single Predictor	Two Predictors	Three Predictors	Four Predictors	Five Predictors
x_1	x_1, x_2	x_1, x_2, x_3	x_1, x_2, x_3, x_4	x_1, x_2, x_3, x_4, x_5
x_2	x_1, x_3	x_1, x_2, x_4	x_1, x_2, x_3, x_5	
x_4	x_1, x_4	x_1, x_2, x_5	x_1, x_2, x_4, x_5	
x_3	x_1, x_5	x_1, x_3, x_4	x_1, x_3, x_4, x_5	
x_5	x_2, x_3	x_1, x_3, x_5	x_2, x_3, x_4, x_5	
	x_2, x_4	x_1, x_4, x_5		
	x_2, x_5	x_2, x_3, x_4		
	x_3, x_4	x_2, x_3, x_5		
	x_3, x_5	x_2, x_4, x_5		
	x_4, x_5	x_3, x_4, x_5		

All Possible Regressions

The **all possible regressions** search procedure *computes all possible linear multiple regression models from the data using all variables*. If a data set contains k independent variables, all possible regressions will determine $2^k - 1$ different models.

For the retail index of energy prices example, the procedure of all possible regressions would produce $2^5 - 1 = 31$ different models from the $k = 5$ independent variables. With $k = 5$ predictors, the procedure produces all single-predictor models, all models with two predictors, all models with three predictors, all models with four predictors, and all models with five predictors, as shown in Table 14.7.

The all possible regressions procedure enables the business researcher to examine every model. In theory, this method eliminates the chance that the business researcher will never consider some models, as can be the case with other search procedures. On the other hand, the search through all possible models can be tedious, time-consuming, inefficient, and perhaps overwhelming.

Stepwise Regression

Perhaps the most widely known and used of the search procedures is stepwise regression. **Stepwise regression** is *a step-by-step process that begins by developing a regression model with a single predictor variable and adds and deletes predictors one step at a time*, examining the fit of the model at each step until no more significant predictors remain outside the model.

STEP 1: In step 1 of a stepwise regression procedure, the k independent variables are examined one at a time by developing a simple regression model for each independent variable to predict the dependent variable. The model containing the largest absolute value of t for an independent variable is selected, and the independent variable associated with the model is selected as the 'best' single predictor of y at the first step. Some computer software packages use an F value instead of a t value to make this determination. Most of these computer programs allow the researcher to predetermine critical values for t or F, but also contain a default value as an option. If the first independent variable selected at step 1 is denoted x_1, the model appears in the form

$$\hat{y} = b_0 + b_1 x_1$$

If, after examining all possible single-predictor models, it is concluded that none of the independent variables produces a t value that is significant at α, then the search procedure stops at step 1 and recommends no model.

STEP 2: In step 2, the stepwise procedure examines all possible two-predictor regression models with x_1 as one of the independent variables in the model and determines which of the other $k - 1$ independent variables in conjunction with x_1 produces the highest absolute t value in the model. If this other variable selected from the remaining independent variables is denoted x_2 and is included in the model selected at step 2 along with x_1, the model appears in the form

$$\hat{y} = b_0 + b_1 x_1 + b_2 x_2$$

At this point, stepwise regression pauses and examines the t value of the regression coefficient for x_1. Occasionally, the regression coefficient for x_1 will become statistically non-significant when x_2 is entered into the model. In that case, stepwise regression will drop x_1 out of the model and go back and examine which of the other $k - 2$ independent variables, if any, will produce the largest significant absolute t value when that variable is included in the model along with x_2. If no other variables show significant t values, the procedure halts. It is worth noting that the regression coefficients are likely to change from step to step to account for the new predictor being added in the process. Thus, if x_1 stays in the model at step 2, the value of b_1 at step 1 will probably be different from the value of at b_1 at step 2.

STEP 3: Step 3 begins with independent variables, x_1 and x_2 (the variables that were finally selected at step 2), in the model. At this step, a search is made to determine which of the $k - 2$ remaining independent variables in conjunction with x_1 and x_2 produces the largest significant absolute t value in the regression model. Let us denote the one that is selected as x_3. If no significant t values are acknowledged at this step, the process stops here and the model determined in step 2 is the final model. At step 3, the model appears in the form

$$\hat{y} = b_0 + b_1 x_1 + b_2 x_2 + b_3 x_3$$

In a manner similar to step 2, stepwise regression now goes back and examines the t values of the regression coefficients of x_1 and x_2 in this step 3 model. If either or both of the t values are now non-significant, the variables are dropped out of the model and the process calls for a search through the remaining $k - 3$ independent variables to determine which, if any, in conjunction with x_3 produce the largest significant t values in this model. The stepwise regression process continues step by step until no significant independent variables remain that are not in the model.

In the retail index of energy prices example, recall that Table 14.6 contained data that can be used to develop a regression model to predict retail index of energy prices in Germany from as many as five different independent variables. Figure 14.9 displayed the results of a multiple regression analysis to produce a model using all five predictors. Suppose the researcher were to use a stepwise regression search procedure on these data to find a regression model. Recall that the following independent variables were being considered:

1. Electricity output from coal and coal products
2. Electricity output from natural gas
3. Electricity output from nuclear
4. Electricity output from hydro
5. Electricity output from solar/wind/other

STEP 1. Each of the independent variables is examined one at a time to determine the strength of each predictor in a simple regression model. The results are reported in Table 14.8.

Note that the independent variable 'Solar/Wind/Other' was selected as the predictor variable, x_1, in step 1. An examination of Table 14.8 reveals that Solar/Wind/Other produced the largest absolute value (18.35) of the single

Step 1: Results of Simple Regression Using Each Independent Variable to Predict the Retail Index of Energy Prices

Dependent Variable	Independent Variable	t Ratio	R^2
Retail index of energy prices	Solar/Wind/Other	18.35	96.3%
Retail index of energy prices	Natural Gas	11.26	90.7
Retail index of energy prices	Nuclear	−6.02	73.6
Retail index of energy prices	Coal and Coal Products	−2.65	35.0
Retail index of energy prices	Hydro	−0.63	3.0
	→ Variable selected to serve as x_1		

predictors. By itself, Solar/Wind/Other accounted for 96.3% of the variation of the values (retail index of energy prices). The regression equation taken from the computer output for this model is

$$y = 75.4 + 0.937x_1$$

where:

y = Retail index of energy prices
x_1 = Solar/Wind/Other

STEP 2: In step 2, x_1 was retained initially in the model and a search was conducted among the four remaining independent variables to determine which of those variables in conjunction with x_1 produced the largest significant t value. Table 14.9 reports the results of this search.

The information in Table 14.9 shows that the model selected in step 2 includes the independent variables 'Solar/Wind/Other' and 'Coal and Coal Products'. Coal and coal products has the largest absolute t value (−4.32), and it is significant at $\alpha = 0.05$. Other variables produced varying sizes of t values. The model produced at step 2 has an R^2 of 98.5%. These two variables taken together account for almost 99% of the variation of retail index of energy prices in this sample.

Step 2: Regression Results with Two Predictors

Dependent Variable y	Independent Variable x_1	Independent Variable x_2	t Ratio of x_2	R^2
Retail index of energy prices	Solar/Wind/Other	Natural Gas	−0.69	96.4
Retail index of energy prices	Solar/Wind/Other	Nuclear	−1.88	97.1
Retail index of energy prices	Solar/Wind/Other	Coal and Coal Products	−4.32	98.5
Retail index of energy prices	Solar/Wind/Other	Hydro	−1.93	97.2
	→ Variables selected at step 2			

From other computer information, it is ascertained that the t value for the x_1 variable in this model is 22.89, which is even higher than in step 1. Therefore, x_1 will not be dropped from the model by the stepwise regression procedure. The step 2 regression model from the computer output is

$$y = 132 + 0.861x_1 - 0.184x_2$$

where:

y = Retail index of energy prices
x_1 = Solar/Wind/Other
x_2 = Coal and Coal Products

Note that the regression coefficient for x_1 changed from 0.937 at step 1 in the model to 0.961 at step 2.

The R^2 for the model in step 1 was 96.3%. Notice that none of the R^2 values produced from step 2 models is less than 96.3%. The reason is that x_1 is still in the model, so the R at this step must be at least as high as it was in step 1, when only x_1 was in the model. In addition, by examining the R^2 values in Table 14.9, you can get a feel for how much the prospective new predictor adds to the model by seeing how much R^2 increases from 96.3%. For example, with x_2 (Coal and Coal Products) added to the model, the R^2 goes up to 98.5%. However, adding the variable 'Nuclear' to x_1 increases R^2 very little (it goes up 97.1%).

STEP 3: In step 3, the search procedure continues to look for an additional predictor variable from the three independent variables remaining out of the solution. Variables x_1 and x_2 are retained in the model. Table 14.10 reports the result of this search.

In this step, regression models are explored that contain x_1 (Solar/Wind/Other) and x_2 (Coal and Coal Products) in addition to one of the three remaining variables. None of the three models produces t ratios that are significant at $\alpha = 0.05$. No new variables are added to the model produced in step 2. The stepwise regression process ends.

Figure 14.10 shows the Minitab stepwise regression output for the world crude oil production example. The results printed in the table are virtually identical to the step-by-step results discussed in this section but are in a different format.

Each column in Figure 14.10 contains information about the regression model at each step. Thus, column 1 contains data on the regression model for step 1. In each column at each step you can see the variables in the model. As an example, at step 2, Solar/Wind/Other and Coal and Coal Products are in the model. The numbers above the t ratios are the regression coefficients. The coefficients and the constant in column 2, for example, yield the regression model equation values for step 2.

$$y = 132 + 0.861x_1 - 0.184x_2$$

TABLE 14.10

Step 3: Regression Results with Three Predictors

Dependent Variable y	Independent Variable x_1	Independent Variable x_2	Independent Variable x_3	t Ratio of x_3	R^2
Retail index of energy prices	Solar/Wind/Other	Coal and Coal Products	Natural Gas	1.24	98.7%
Retail index of energy prices	Solar/Wind/Other	Coal and Coal Products	Nuclear	−0.97	98.7
Retail index of energy prices	Solar/Wind/Other	Coal and Coal Products	Hydro	−1.13	98.7
		No t ratio is significant at $\alpha = 0.05$.			
		No new variables are added to the model.			

The values of R^2 (R-Sq) and the standard error of the estimate (S) are displayed on the bottom row of the output along with the adjusted value of R^2.

Forward Selection

Another search procedure is forward selection. **Forward selection** is essentially the same as stepwise regression, but once a variable is entered into the process, it is never dropped out. Forward selection begins by finding the independent variable that will produce the largest absolute value of t (and largest R^2) in predicting y. The selected variable is denoted here as x_1 and is part of the model

$$\hat{y} = b_0 + b_1 x_1$$

Forward selection proceeds to step 2. While retaining x_1, it examines the other $k-1$ independent variables and determines which variable in the model with x_1 produces the highest absolute value of t that is significant. To this point, forward selection is the same as stepwise regression. If this second variable is designated x_2, the model is

$$\hat{y} = b_0 + b_1 x_1 + b_2 x_2$$

At this point, forward selection does not re-examine the t value of x_1. Both x_1 and x_2 remain in the model as other variables are examined and included. When independent variables are correlated in forward selection, the overlapping of information can limit the potential predictability of two or more variables in combination. Stepwise regression takes this into account, in part, when it goes back to re-examine the t values of predictors already in the model to determine whether they are still significant predictors of y given the variables that have now entered the process. In other words, stepwise regression acknowledges that the strongest single predictor of y that is selected at step 1 may not be a significant predictor of y when taken in conjunction with other variables.

```
Stepwise Regression: Retail Index versus Coal and coa, Natural gas, . . .

Alpha-to-Enter: 0.15   Alpha-to-Remove: 0.15

Response is Retail Index of Energy Prices on 5 predictors, with N = 15
Step                                     1            2
Constant                              75.45       131.95

Solar/wind/other                      0.937        0.861
T-Value                               18.35        22.89
p-Value                               0.000        0.000

Coal and coal products                            -0.184
T-Value                                            -4.32
p-Value                                            0.001

S                                      3.43         2.24
R-Sq                                  96.28        98.54
R-Sq(adj)                             96.00        98.30

Mallows Cp                            23.7          4.6
```

FIGURE 14.10

Minitab Stepwise Regression Output for the Retail Index of Energy Prices Example

Using a forward selection procedure to develop multiple regression models for the retail index of energy prices example would result in the same outcome as that provided by stepwise regression because neither x_1 nor x_2 were removed from the model in that particular stepwise regression. The difference in the two procedures is more apparent in examples where variables selected at earlier steps in the process are removed during later steps in stepwise regression.

Backward Elimination

The **backward elimination** search procedure is *a step-by-step process that begins with the 'full' model (all k predictors).* Using the *t* values, a search is made to determine whether any non-significant independent variables are in the model. If no non-significant predictors are found, the backward process ends with the full model. If non-significant predictors are found, the predictor with the smallest absolute value of *t* is eliminated and a new model is developed with $k - 1$ independent variables.

This model is then examined to determine whether it contains any independent variables with non-significant *t* values. If it does, the predictor with the smallest absolute *t* value is eliminated from the process and a new model is developed for the next step.

This procedure of identifying the smallest non-significant *t* value and eliminating that variable continues until all variables left in the model have significant *t* values. Sometimes this process yields results similar to those obtained from forward selection and other times it does not. A word of caution is in order. Backward elimination always begins with all possible predictors in the model. Sometimes the sample data do not provide enough observations to justify the use of all possible predictors at the same time in the model. In this case, backward elimination is not a suitable option with which to build regression models.

The following steps show how the backward elimination process can be used to develop multiple regression models to predict world crude oil production using the data and five predictors displayed in Table 14.6:

STEP 1: A full model is developed with all predictors. The results are shown in Table 14.11. The R^2 for this model is 99.20%. A study of Table 14.11 reveals that the predictor 'Hydro' has the smallest absolute value of a non-significant t ($t = -0.92$, $p = 0.380$). In step 2, this variable will be dropped from the model.

STEP 2: A second regression model is developed with $k - 1 = 4$ predictors. Hydro has been eliminated from consideration. The results of this multiple regression analysis are presented in Table 14.12. The computer results in Table 14.12 indicate that the variable 'Nuclear' has the smallest absolute value of a non-significant t of the variables remaining in the model ($t = -1.45$, $p = 0.178$). In step 3, this variable will be dropped from the model.

STEP 3: A third regression model is developed with $k - 2 = 3$ predictors. Both Nuclear and Hydro variables have been removed from the model. The results of this multiple regression analysis are reported in Table 14.13. The

TABLE 14.11

Step 1: Backward Elimination, Full Model

Predictor	Coefficient	t Ratio	p
Solar/Wind/Other	.63820	4.66	.000
Natural Gas	.23510	1.49	.171
Nuclear	−.12351	−1.35	.209
Coal and Coal Products	−.14549	−3.12	.001
Hydro	−.31880	−0.92	.380

Variable to be dropped from the model

TABLE 14.12

Step 2: Backward Elimination, Four Predictors

Predictor	Coefficient	t Ratio	p
Solar/Wind/Other	.61630	4.60	.001
Natural Gas	.25540	1.65	.131
Nuclear	−.13085	−1.45	.178
Coal and Coal Products	−.15980	−3.66	.004

Variable to be dropped from the model

TABLE 14.13

Step 3: Backward Elimination, Three Predictors

Predictor	Coefficient	t Ratio	p
Solar/Wind/Other	.72180	6.13	.000
Natural Gas	.19490	1.24	.240
Coal and Coal Products	−.18592	−4.45	.001

Variable to be dropped from the model

TABLE 14.14

Step 4: Backward Elimination, Two Predictors

Predictor	Coefficient	t Ratio	p
Solar/Wind/Other.	.86092	22.89	.000
Coal and Coal Products	−.18422	−4.32	.001

All variables are significant at $\alpha = 0.05$.
No variables will be dropped from this model.
The process stops.

computer results in Table 14.13 indicate that the variable 'Natural Gas' has the smallest absolute value of a nonsignificant t of the variables remaining in the model ($t = 1.24$, $p = 0.240$). In step 4, this variable will be dropped from the model.

STEP 4: A fourth regression model is developed with $k − 3 = 2$ predictors. Natural Gas, Nuclear, and Hydro variables have been removed from the model. The results of this multiple regression analysis are reported in Table 14.14. Observe that all p-values are less than $\alpha = 0.05$, indicating that all t values are significant, so no additional independent variables need to be removed. The backward elimination process ends with two predictors in the model. The final model obtained from this backward elimination process is the same model as that obtained by using stepwise regression.

14.3   PROBLEMS

14.13   Use a stepwise regression procedure and the data below to develop a multiple regression model to predict y. Discuss the variables that enter at each step, commenting on their t values and on the value of R^2.

y	x_1	x_2	x_3	y	x_1	x_2	x_3
21	5	108	57	22	13	105	51
17	11	135	34	20	10	111	43
14	14	113	21	16	20	140	20
13	9	160	25	13	19	150	14
19	16	122	43	18	14	126	29
15	18	142	40	12	21	175	22
24	7	93	52	23	6	98	38
17	9	128	38	18	15	129	40

14.14   Given here are data for a dependent variable and four potential predictors. Use these data and a stepwise regression procedure to develop a multiple regression model to predict y. Examine the values of t and R^2 at each step and comment on those values. How many steps did the procedure use? Why do you think the process stopped?

y	x_1	x_2	x_3	x_4
101	2	77	1.2	42
127	4	72	1.7	26
98	9	69	2.4	47

(*continued*)

y	x_1	x_2	x_3	x_4
79	5	53	2.6	65
118	3	88	2.9	37
114	1	53	2.7	28
110	3	82	2.8	29
94	2	61	2.6	22
96	8	60	2.4	48
73	6	64	2.1	42
108	2	76	1.8	34
124	5	74	2.2	11
82	6	50	1.5	61
89	9	57	1.6	53
76	1	72	2.0	72
109	3	74	2.8	36
123	2	99	2.6	17
125	6	81	2.5	48

14.15 The computer output given here is the result of a stepwise multiple regression analysis to predict a dependent variable by using six predictor variables. The number of observations was 108. Study the output and discuss the results. How many predictors ended up in the model? Which predictors, if any, did not enter the model?

```
STEPWISE REGRESSION OF Y ON 6 PREDICTORS, WITH N = 108
STEP               1         2         3         4
CONSTANT        8.71      6.82      6.57      5.96
X3             -2.85     -4.92     -4.97     -5.00
T-RATIO         2.11      2.94      3.04      3.07
X1                        4.42      3.72      3.22
T-RATIO                   2.64      2.20      2.05
X2                                  1.91      1.78
T-RATIO                             2.07      2.02
X6                                            1.56
T-RATIO                                       1.98
S               3.81      3.51      3.43      3.36
R-SQ           29.20     49.45     54.72     59.29
```

14.16 Study the output given here from a stepwise multiple regression analysis to predict y from four variables. Comment on the output at each step.

```
STEPWISE REGRESSION OF Y ON 4 PREDICTORS, WITH N = 63
STEP               1         2
CONSTANT       27.88     22.30
X3              0.89
T-RATIO         2.26
X2                       12.38
T-RATIO                   2.64
X4                       0.0047
T-RATIO                   2.01
S              16.52      9.47
R-SQ           42.39     68.20
```

14.17 ThyssenKrupp is an integrated materials and technology group with some 177,000 employees in around 80 countries. The data below are an extract of the company's annual financial statements and include information on sales (€ millions), return on equity (%), dividend per share (€) and earnings per share (%). Use the data and stepwise regression to predict the variable sales earned from the other three variables.

Year	Sales	Return on Equity	Dividend per Share	Earnings per Share
1998/99	32,378	7.4	0.71581	0.52
1999/00	37,209	12.4	0.75	1.02
2000/01	38,008	10	0.6	1.29
2001/02	36,698	9.2	0.4	0.42
2002/03	35,327	10.1	0.5	1.12
2003/04	39,342	17.7	0.6	1.77
2004/05	42,927	20.9	0.8	2.08
2005/06	47,125	21.1	1	3.24
2006/07	51,723	31.9	1.3	4.3
2007/08	53,426	27.2	1.3	4.59
2008/09	40,463	−24.4	0.3	−4.01
2009/10	42,621	10.92	0.45	1.77

14.18 The Department of Energy and Climate Change releases figures on the carbon emissions by various fuels and total energy consumption in the United Kingdom. Shown here are the figures for the carbon emissions by four different fuel sources, Gas, Oil, Coal and Other Solid Fuels, and Non-fuel, as well as data on total energy consumption in the UK over a 15-year period. Use the data and stepwise regression to predict the total energy consumption in the UK from the carbon emissions by gas, oil, coal and other solid fuels and non-fuel. Examine the data and discuss the output.

Year	Total Energy Consumption	Carbon Emissions by Fuel			
		Gas	Oil	Coal and Other Solid Fuels	Non-fuel
1995	150.8	41.6	54.4	48.9	6.0
1996	156.9	48.3	55.7	46.8	6.1
1997	150.4	49.5	53.5	41.8	5.6
1998	151.0	51.4	52.5	41.7	5.4
1999	148.1	55.0	51.2	36.9	5.0
2000	150.3	56.4	50.4	38.7	4.7
2001	153.4	56.5	51.0	41.4	4.6
2002	148.6	55.9	50.2	38.3	4.3
2003	151.8	56.2	50.1	41.2	4.3
2004	151.7	57.1	50.8	39.7	4.1
2005	151.1	55.4	51.7	39.7	4.2
2006	150.4	52.1	51.0	43.2	4.0
2007	148.3	53.1	50.5	40.5	4.2
2008	145.3	54.9	49.1	37.6	3.8
2009	131.1	50.0	46.4	31.5	3.2

14.4 MULTICOLLINEARITY

One problem that can arise in multiple regression analysis is multicollinearity. **Multicollinearity** is *when two or more of the independent variables of a multiple regression model are highly correlated*. Technically, if two of the independent variables are correlated, we have collinearity; when three or more independent variables are correlated, we have multicollinearity. However, the two terms are frequently used interchangeably.

The reality of business research is that most of the time some correlation between predictors (independent variables) will be present. The problem of multicollinearity arises when the intercorrelation between predictor variables is high. This relationship causes several other problems, particularly in the interpretation of the analysis.

1. It is difficult, if not impossible, to interpret the estimates of the regression coefficients.
2. Inordinately small t values for the regression coefficients may result.
3. The standard deviations of regression coefficients are overestimated.
4. The algebraic sign of estimated regression coefficients may be the opposite of what would be expected for a particular predictor variable.

The problem of multicollinearity can arise in regression analysis in a variety of business and economics research situations. For example, suppose a model is being developed to predict salaries in a given industry. Independent variables such as years of education, age, years in management, experience on the job, and years of tenure with the firm might be considered as predictors. It is obvious that several of these variables are correlated (virtually all of these variables have something to do with number of years, or time) and yield redundant information. Suppose a financial regression model is being developed to predict bond market rates by such independent variables as FTSE average, London Interbank Offered Rate (LIBOR), GNP, producer price index, and consumer price index. Several of these predictors are likely to be intercorrelated.

In problem 14.17, several of the independent variables are intercorrelated, leading to the potential of multicollinearity problems. Table 14.15 gives the correlations of the predictor variables for this problem. Note that the r value is quite high ($r > 0.78$) for dividend per share and return on equity (0.79), earnings per share and return on equity (0.98) and earnings per share and dividend per share (0.83).

Table 14.15 shows that earnings per share and return on equity are highly correlated. Earnings per share as a single predictor of sales produces the following simple regression model:

$$\hat{y} = 38{,}208 + 251(\text{return on equity})$$

Notice that the estimate of the regression coefficient, 251, is positive, indicating that as return on equity increases, sales increase. Using earnings per share as a single predictor of sales yields the following simple regression model:

$$\hat{y} = 38{,}610 + 1{,}873(\text{earnings per share})$$

The multiple regression model developed using both return on equity and earnings per share to predict sales is

$$\hat{y} = 41{,}306 - 770(\text{return on equity}) + 6{,}649(\text{earnings per share})$$

Observe that this regression model indicates a *negative* relationship between return on equity and sales (-770), which is in opposition to the *positive* relationship shown in the regression equation for earnings per share as a single predictor. Because of the multicollinearity between coal return on equity and earnings per share, these two independent variables interact in the regression analysis in such a way as to produce regression coefficient estimates that are difficult to interpret. Extreme caution should be exercised before interpreting these regression coefficient estimates.

TABLE 14.15

Correlations Among ThyssenKrupp Group Sales Predictor Variables

	Return on Equity	Dividend per Share	Earnings per Share
Return on Equity	1		
Dividend per Share	0.79	1	
Earnings per Share	0.98	0.83	1

The problem of multicollinearity can also affect the t values that are used to evaluate the regression coefficients. Because the problems of multicollinearity among predictors can result in an overestimation of the standard deviation of the regression coefficients, the t values tend to be under-representative when multicollinearity is present. In some regression models containing multicollinearity in which all t values are non-significant, the overall F value for the model is highly significant. In Section 14.1, an example was given of how including interaction when it is significant strengthens a regression model. The computer output for the regression models both with and without the interaction term was shown in Figure 14.5. The model without interaction produced a statistically significant F value but neither predictor variable was significant. Further investigation of this model reveals that the correlation between the two predictors, x_1 and x_2, is 0.945. This extremely high correlation indicates a strong collinearity between the two predictor variables.

This collinearity may explain the fact that the overall model is significant but neither predictor is significant. It also underscores one of the problems with multicollinearity: under-represented t values. The t values test the strength of the predictor given the other variables in the model. If a predictor is highly correlated with other independent variables, it will appear not to add much to the explanation of y and produce a low t value. However, had the predictor not been in the presence of these other correlated variables, the predictor might have explained a high proportion of variation of y.

Many of the problems created by multicollinearity are interpretation problems. The business researcher should be alert to and aware of multicollinearity potential with the predictors in the model and view the model outcome in light of such potential.

The problem of multicollinearity is not a simple one to overcome. However, several methods offer an approach to the problem. One way is to examine a correlation matrix like the one in Table 14.15 to search for possible intercorrelations among potential predictor variables. If several variables are highly correlated, the researcher can select the variable that is most correlated to the dependent variable and use that variable to represent the others in the analysis. One problem with this idea is that correlations can be more complex than simple correlation among variables. In other words, simple correlation values do not always reveal multiple correlation between variables. In some instances, variables may not appear to be correlated as pairs, but one variable is a linear combination of several other variables. This situation is also an example of multicollinearity, and a cursory observation of the correlation matrix will probably not reveal the problem.

Stepwise regression is another way to prevent the problem of multicollinearity. The search process enters the variables one at a time and compares the new variable to those in solution. If a new variable is entered and the t values on old variables become non-significant, the old variables are dropped out of solution. In this manner, it is more difficult for the problem of multicollinearity to affect the regression analysis. Of course, because of multicollinearity, some important predictors may not enter in to the analysis.

Other techniques are available to attempt to control for the problem of multicollinearity. One is called a **variance inflation factor**, in which a regression analysis is conducted to predict an independent variable by the other independent variables. In this case, the independent variable being predicted becomes the dependent variable. As this process is done for each of the independent variables, it is possible to determine whether any of the independent variables are a function of the other independent variables, yielding evidence of multicollinearity. By using the results from such a model, a variance inflation factor (VIF) can be computed to determine whether the standard errors of the estimates are inflated:

$$\text{VIF} = \frac{1}{1 - R_i^2}$$

where R_i^2 is the coefficient of determination for any of the models, used to predict an independent variable by the other $k - 1$ independent variables. Some researchers follow a guideline that any variance inflation factor greater than 10 or R_i^2 value more than 0.90 for the largest variance inflation factors indicates a severe multicollinearity problem.*

---

*William Mendenhall and Terry Sincich, *A Second Course in Business Statistics: Regression Analysis*. San Francisco: Dellen Publishing Company, 1989; John Neter, William Wasserman, and Michael H. Kutner, *Applied Linear Regression Models*, 2nd edn. Homewood, IL: Richard D. Irwin, 1989.

14.4 PROBLEMS

14.19 Develop a correlation matrix for the independent variables in Problem 14.13. Study the matrix and make a judgement as to whether substantial multicollinearity is present among the predictors. Why or why not?

14.20 Construct a correlation matrix for the four independent variables for Problem 14.14 and search for possible multicollinearity. What did you find and why?

14.21 In the retail index of energy prices example, you were asked to use stepwise regression to predict retail index of energy prices by the electricity output from Coal and Coal Products, electricity output from Natural Gas, electricity output from Nuclear, electricity output from Hydro, and electricity output from Solar/Wind/Other. Study the stepwise results, including the regression coefficients, to determine whether there may be a problem with multicollinearity. Construct a correlation matrix of the three variables to aid you in this task.

14.22 Study the three predictor variables in Problem 14.18 and attempt to determine whether substantial multicollinearity is present between the predictor variables. If there is a problem of multicollinearity, how might it affect the outcome of the multiple regression analysis?

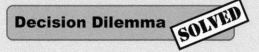

Determining Compensation for CEOs

One statistical tool that can be used to study CEO compensation is multiple regression analysis. Regression models can be developed using predictor variables, such as age, years of experience, worth of company, or others, for analysing CEO compensation. Search procedures such as stepwise regression can be used to sort out the more significant predictors of CEO compensation.

The researcher prepares for the multiple regression analysis by conducting a study of CEOs and gathering data on several variables. The data presented in the Decision Dilemma could be used for such an analysis. It seems reasonable to believe that CEO compensation is related to the size and worth of a company, therefore it makes sense to attempt to develop a regression model or models to predict CEO compensation by the variables company sales, number of employees in the company, and the capital investment of a company. Qualitative or dummy variables can also be used in such an analysis. In the database given in the Decision Dilemma, one variable indicates whether a company is a manufacturing company. One way to recode this variable for regression analysis is to assign a 1 to companies that are manufacturers and a 0 to others.

A stepwise regression procedure can sort out the variables that seem to be more important predictors of CEO compensation. A stepwise regression analysis was conducted on the Decision Dilemma database using sales, number of employees, capital investment, and whether a

company is in manufacturing as the four independent variables. The result of this analysis follows.

```
Stepwise Regression: Cash Compen versus Sales,
No. of Emp.,...
Alpha-to-Enter: 0.15 Alpha-to-Remove: 0.15
Response is Cash Com on 4 predictors, with N = 20
```

Step	1	2	3	4
Constant	243.9	232.2	223.8	223.3
No. of E	0.0696	0.1552	0.0498	
T-Value	13.67	4.97	0.98	
P-Value	0.000	0.000	0.343	
Cap. Inv		-1.66	-2.92	-3.06
T-Value		-2.77	-3.97	-4.27
P-Value		0.013	0.001	0.001
Sales			1.08	1.45
T-Value			2.46	6.10
P-Value			0.026	0.000
S	32.6	27.9	24.5	24.5
R-Sq	91.22	93.95	95.61	95.34
R-Sq(adj)	90.73	93.24	94.78	94.80

The stepwise regression analysis produces a single predictor model at step 1 with a high R^2 value of 0.9122. The number of employees variable used in a simple regression model accounts for over 91.2% of the variation of CEO compensation data. An examination of the regression coefficient of number of employees at the first step (0.0696) indicates that a one-employee increase results in

(continued)

a predicted increase of (0.0696 · €1,000) about €70 in the CEO's compensation.

At step 2, the company's capital investment enters the model. Notice that the R^2 increases only by 0.0273 and that the regression coefficient on capital investment is negative. This result seems counter-intuitive because we would expect that the more capital investment a company has, the more the CEO should be compensated for the responsibility. A Minitab simple regression analysis using only capital investment produces the following model:

```
The regression equation is
CashCompen = 257 + 1.29 CapInv
```

Notice that the regression coefficient in this model is positive as we would suppose. Multicollinearity is likely. In fact, multicollinearity is evident among sales, number of employees, and capital investment. Each is a function or determiner of company size. Examine the following correlation coefficient:

	Correlations	
	Sales	**No. Employees**
No. Employees	0.997	1
Cap. Invest	0.995	0.999

Notice that these three predictors are highly interrelated. Therefore, the interpretation of the regression coefficients and the order of entry of these variables in the stepwise regression become more difficult. Nevertheless, number of employees is most highly related to CEO compensation in these data. Observe also in the stepwise regression output that number of employees actually drops out of the model at step 4. The t ratio for number of employees is not significant ($t = 0.98$) at step 3. However, the R^2 actually drops slightly when number of employees is removed. In searching for a model that is both parsimonious and explanatory, the researcher could do worse than to merely select the model at step 1.

Researchers might want to explore more complicated non-linear models. Some of the independent variables might be related to CEO compensation but in some non-linear manner.

A brief study of the predictor variables in the Decision Dilemma database reveals that as compensation increases, the values of the data in the independent variables do not increase at a linear rate. Scatter plots of sales, number of employees, and capital investment with CEO compensation confirm this suspicion. Shown here is a scatter plot of sales with cash compensation.

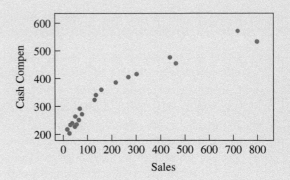

Observe that the graph suggests more of a logarithmic fit than a linear one. We can use recoding techniques presented in the chapter to conduct a multiple regression analysis to predict compensation using the log of each of these variables. In the analysis, the compensation figures remain the same, but each of the three quantitative independent variables are recoded by taking the log of each value and entering the resultant variable in the model. A second stepwise regression analysis is undertaken with the log variables in the mix along with the original variables. The results follow:

```
Stepwise Regression: Cash Compen versus Sales,
No. of Emp.,...
Alpha-to-Enter: 0.1 Alpha-to-Remove: 0.1
Response is Cash Com on 7 predictors, with N = 20
```

Step	1	2	3	4	5
Constant	−129.61	−13.23	−122.53	−147.22	−120.74
Log sale	224.3	152.2	281.4	307.8	280.8
T-Value	22.22	8.75	11.08	32.75	26.81
P-Value	0.000	0.000	0.000	0.000	0.000
No. Emp		0.0251	0.0233	0.0903	0.0828
T-Value		4.53	6.97	13.94	15.52
P-Value		0.000	0.000	0.000	0.000
Log cap			−106.4	−126.0	−109.8
T-Value			−5.58	−17.87	−15.56
P-Value			0.000	0.000	0.000
Sales				−0.434	−0.250
T-Value				−10.52	−4.11
P-Value				0.000	0.001
Cap. Inv					−0.37
T-Value					−3.51
P-Value					0.003
S	20.7	14.3	8.59	3.07	2.32
R-Sq	96.48	98.41	99.46	99.94	99.97
R-Sq(adj)	96.29	98.22	99.36	99.92	99.95

(continued)

Note that in this stepwise regression analysis, the variable log sales has the highest single predictability of compensation producing an R^2 of 0.9648, which is higher than the value at step 1 in the first stepwise regression analysis. Number of employees enters at step 2 and log of capital investment at step 3. However, such a high R^2 at step 1 leaves little room for improved predictability. Our search through the variables may well end with the decision to use the log of sales as the efficient, predictable model of compensation. The final model might be:

$$CEO\ Compensation = -129.61 + 224.3\ Log\ sales$$

Human resource managers sometimes use compensation tables to assist them in determining ranges and ballparks for salary offers. Company boards of directors can use such models as the one developed here to assist them in negotiations with possible candidates for CEO positions or to aid them in determining whether a presently employed CEO is over- or undercompensated. In addition, candidates who are searching for new CEO opportunities can use models like these to determine the potential compensation for a new position and to help them be more adequately prepared for salary negotiations should they be offered a CEO position.

Some of the variables in this study will undoubtedly produce redundant information. The use of a correlation matrix and a stepwise regression process can protect the analysis from some of the problems of multicollinearity. The use of multiple regression analysis on a large sample of CEO compensation data with many independent variables could provide some interesting and exciting results.

ETHICAL CONSIDERATIONS

Some business researchers misuse the results of search procedures by using the order in which variables come into a model (on stepwise and forward selection) to rank the variables in importance. They state that the variable entered at step 1 is the most important predictor of y, the variable entering at step 2 is second most important, and so on. In actuality, variables entering the analysis after step 1 are being analysed by how much of the unaccounted-for variation (residual variation) they are explaining, not how much they are related to y by themselves. A variable that comes into the model at the fourth step is the variable that most greatly accounts for the variation of the y values left over after the first three variables have explained the rest. However, the fourth variable taken by itself might explain more variation of y than the second or third variable when seen as single predictors.

Some people use the estimates of the regression coefficients to compare the worth of the predictor variables; the larger the coefficient is, the greater its worth. At least two problems plague this approach. The first is that most variables are measured in different units. Thus, regression coefficient weights are partly a function of the unit of measurement of the variable. Second, if multicollinearity is present, the interpretation of the regression coefficients is questionable. In addition, the presence of multicollinearity raises several issues about the interpretation of other regression output. Researchers who ignore this problem are at risk of presenting spurious results.

SUMMARY

Multiple regression analysis can handle non-linear independent variables. One way to accommodate this issue is to recode the data and enter the variables into the analysis in the normal way. Other non-linear regression models, such as exponential models, require that the entire model be transformed. Often the transformation involves the use of logarithms. In some cases, the resulting value of the regression model is in logarithmic form and the antilogarithm of the answer must be taken to determine the predicted value of y.

Indicator, or dummy, variables are qualitative variables used to represent categorical data in the multiple regression model. These variables are coded as 0, 1 and are often used to represent nominal or ordinal classification data that the researcher wants to include in the regression analysis. If a qualitative variable contains more than two categories, it generates multiple dummy variables. In general, if a qualitative variable contains c categories, $c - 1$ dummy variables should be created.

Search procedures are used to help sort through the independent variables as predictors in the examination of various possible models. Several search procedures are available, including all possible regressions, stepwise regression, forward selection, and backward elimination. The all possible regressions procedure computes every possible regression model for a set of data. The drawbacks of this procedure include the time and energy required to compute all possible regressions and the difficulty of deciding which models are most appropriate. The stepwise regression procedure involves selecting and adding one independent variable at a time to the regression process after beginning with a one-predictor model. Variables are added to the model at each step if they contain the most significant t value associated with the remaining variables. If no additional t value is statistically significant at any given step, the procedure stops. With stepwise regression, at each step the process examines the variables already in the model to determine whether their t values are still significant. If not, they are dropped from the model, and the process searches for other independent variables with large, significant t values to replace the variable(s) dropped. The forward selection procedure is the same as stepwise regression but does not drop variables out of the model once they have been included. The backward elimination procedure begins with a 'full' model, a model that contains all the independent variables. The sample size must be large enough to justify a full model, which can be a limiting factor. Backward elimination drops out the least important predictors one at a time until only significant predictors are left in the regression model. The variable with the smallest absolute t value of the statistically nonsignificant t values is the independent variable that is dropped out of the model at each step.

One of the problems in using multiple regression is multicollinearity, or correlations among the predictor variables. This problem can cause overinflated estimates of the standard deviations of regression coefficients, misinterpretation of regression coefficients, undersized t values, and misleading signs on the regression coefficients. It can be lessened by using an intercorrelation matrix of independent variables to help recognize bivariate correlation, by using stepwise regression to sort the variables one at a time, or by using statistics such as a variance inflation factor.

KEY TERMS

all possible regressions	search procedures
backward elimination	stepwise regression
dummy variable	Tukey's four-quadrant
forward selection	approach
indicator variable	Tukey's ladder of
multicollinearity	transformations
quadratic model	variance inflation
qualitative variable	factor

GO ONLINE TO DISCOVER THE EXTRA FEATURES FOR THIS CHAPTER

The Student Study Guide containing solutions to the odd-numbered questions, additional Quizzes and Concept Review Activities, Excel and Minitab databases, additional data files in Excel and Minitab, and more worked examples.
www.wiley.com/college/cortinhas

FORMULAS

Variance inflation factor

$$\text{VIF} = \frac{1}{1 - R_i^2}$$

SUPPLEMENTARY PROBLEMS

CALCULATING THE STATISTICS

14.23 Given here are the data for a dependent variable, y, and independent variables. Use these data to develop a regression model to predict y. Discuss the output. Which variable is an indicator variable? Was it a significant predictor of y?

x_1	x_2	x_3	y
0	51	16.4	14
0	48	17.1	17
1	29	18.2	29
0	36	17.9	32
0	40	16.5	54
1	27	17.1	86
1	14	17.8	117
0	17	18.2	120
1	16	16.9	194
1	9	18.0	203
1	14	18.9	217
0	11	18.5	235

14.24 Use the following data and a stepwise regression analysis to predict y. In addition to the two independent variables given here, include three other predictors in your analysis: the square of each x as a predictor and an interaction predictor. Discuss the results of the process.

x_1	x_2	y	x_1	x_2	y
10	3	2,002	5	12	1,750
5	14	1,747	6	8	1,832
8	4	1,980	5	18	1,795
7	4	1,902	7	4	1,917
6	7	1,842	8	5	1,943
7	6	1,883	6	9	1,830
4	21	1,697	5	12	1,786
11	4	2,021			

14.25 Use the x_1 values and the log of the x_1 values given here to predict the y values by using a stepwise regression procedure. Discuss the output. Were either or both of the predictors significant?

y	x_1	y	x_1
20.4	850	13.2	204
11.6	146	17.5	487
17.8	521	12.4	192
15.3	304	10.6	98
22.4	1029	19.8	703
21.9	910	17.4	394
16.4	242	19.4	647

TESTING YOUR UNDERSTANDING

14.26 The table below presents the average price per year for several minerals over a decade. Use these data and a stepwise regression procedure to produce a model to predict the average price of gold from the other variables. Comment on the results of the process.

Gold ($ per oz)	Copper (cents per lb)	Silver ($ per oz)	Aluminium (cents per lb)
161.1	64.2	4.4	39.8
308.0	93.3	11.1	61.0
613.0	101.3	20.6	71.6
460.0	84.2	10.5	76.0
376.0	72.8	8.0	76.0
424.0	76.5	11.4	77.8
361.0	66.8	8.1	81.0
318.0	67.0	6.1	81.0
368.0	66.1	5.5	81.0
448.0	82.5	7.0	72.3
438.0	120.5	6.5	110.1
382.6	130.9	5.5	87.8

14.27 The World Bank Indicators dataset includes data on 198 indicators from 209 countries from 1960 to the present. Among the variables reported by this dataset are the per capita GDP, the population growth rate, the percentage of rural population and the total tax rate (as a percentage of profit). Shown here are the data for these five variables for the most recent year available for the European Union countries for which data was available. Use the data to develop a regression model to predict per capita GDP from the population growth rate, the percentage of rural population and the total tax rate. Graph each of these predictors separately with the response variable and use Tukey's four-quadrant approach to explore possible recoding schemes for non-linear relationships. Include any of these in the regression model. Comment on the regression model and its strengths and its weaknesses.

Country	GDP per capita (constant 2000 US$)	Population growth (% annual)	Rural population (% of total)	Total Tax Rate (% of profit)
Austria	26,106.2	0.33	32.6	55.5
Belgium	24,176.4	0.75	2.6	57.3
Bulgaria	2,542.5	−0.50	28.6	31.4
Czech Republic	7,225.4	0.63	26.5	47.2
Denmark	30,547.9	0.65	13.1	29.2
Estonia	6,113.3	−0.02	30.5	49.1
Finland	26,495.9	0.47	36.4	47.7
France	22,820.1	0.54	22.4	65.8
Germany	24,409.5	−0.28	26.3	44.9
Greece	14,843.7	0.41	38.8	47.4
Hungary	5,833.5	−0.16	32.1	57.5
Ireland	28,502.5	0.56	38.4	26.5
Italy	18,479.2	0.65	31.8	68.4
Lithuania	5,153.8	−0.55	32.9	42.7
Luxembourg	52,388.1	1.87	17.7	20.9
Netherlands	26,094.0	0.52	17.6	39.3
Poland	6,330.6	0.06	38.7	42.5
Portugal	11,588.1	0.09	39.9	42.9
Romania	2,606.6	−0.15	45.6	46.4
Slovak Republic	8,041.6	0.21	43.3	48.6
Slovenia	12,576.9	1.08	51.7	37.5
Spain	15,533.8	0.88	22.7	56.9
Sweden	30,899.3	0.89	15.4	54.6
United Kingdom	27,259.2	0.70	10.0	35.9

14.28 The International Monetary Fund produces primary commodity indexes for several different categories. Shown here are the percentage changes in the price indexes over a period of 20 years for food, beverages, industrial materials, and agricultural raw materials. Also displayed are the percentage changes in index for all commodities.

Use these data and a stepwise regression procedure to develop a model that attempts to predict all commodities by the other four variables. Construct scatter plots of each of these variables with all commodities. Examine the graphs in light of Tukey's four-quadrant approach. Develop any other appropriate predictor variables by recoding data and include them in the analysis. Comment on the result of this analysis.

All Commodities	Food	Beverages	Industrial Materials	Agricultural Raw Materials
55.1	97.6	70.8	74.6	92.3
52.5	96.3	75.9	75.0	105.9
55.1	99.2	128.3	86.2	117.5
59.2	104.4	126.5	94.6	120.3
62.1	112.0	106.1	87.2	116.0
59.5	102.6	130.5	86.3	111.7
48.3	92.4	120.5	72.2	94.6
49.3	80.5	93.9	71.0	93.0
62.9	81.9	77.0	77.7	98.5
58.5	80.5	65.7	72.8	95.2
58.3	83.3	81.6	71.4	95.0
65.0	88.6	85.5	75.3	95.6
79.5	100.5	84.6	88.6	99.4
98.6	99.6	99.6	99.3	99.6
120.8	110.5	108.4	136.3	108.8
135.1	127.3	123.3	154.3	114.2
172.3	157.0	152.0	145.7	113.3
120.7	134.0	154.4	118.7	94.1
152.2	149.4	176.2	170.0	125.4

14.29 The United Nations has published detailed imports and exports annual statistics of close to 200 countries from 1962 to the most recent year. Shown here are the Russian Federation imports (in thousands of US dollars) for three farm products for the last 12 years. Use these data and a stepwise regression analysis to predict the imports of sweetcorn by the imports of wheat and soya beans. Comment on the results.

Sweetcorn	Wheat	Soya Beans
3,335	18,835	49,000
3,778	9,617	8,020
376	5,016	4,723
588	3,351	12,848
843	9,133	4,509
1,372	25,699	340
3,259	11,897	10,418
3,905	16,616	1,158
4,140	16,112	53,218
5,089	31,720	326,601
7,155	14,916	442,909
7,857	7,062	487,486

14.30 The consulting firm Mercer compiles cost-of-living indexes for selected cities around the world. Shown here are cost-of-living indexes for 21 different cities on five different items for a recent year. Use the data to develop a regression model to predict the rent of a luxury two-bedroom apartment by the price of a cinema ticket, a music CD, a hamburger meal, and a litre of milk. Discuss the results, highlighting both the significant and non-significant predictors.

Expenditure (£)	Rent of a Luxury Two-bedroom Apartment (per month)	One Cinema Ticket (international release)	One Music CD	One Hamburger Meal	One litre of Milk
Luanda	4,480.46	8.32	18.66	11.2	2.11
London	2,500.00	10.8	14.99	3.97	0.76
Dublin	1,050.83	9.2	14.01	6.26	0.85
Paris	1,926.52	8.67	14.01	5.17	1.18
Rome	1,576.25	6.57	19.18	5.17	1.44
Amsterdam	1,050.83	8.76	15.75	5.12	0.87
Berlin	1,050.83	7	13.13	4.81	0.83
Athens	1,225.97	7.88	16.55	4.99	0.88
Brussels	1,225.97	7.8	16.55	5.21	0.8
Madrid	1,138.40	7	18.35	4.95	0.84
Prague	1,007.05	5.7	19.51	3.54	0.57
Warsaw	1,088.68	5.01	15.24	2.85	0.56
Zagreb	1,019.02	3.96	16.66	4.2	0.75
Tokyo	2,839.60	12.78	13.49	4.61	1.69
Beijing	1,874.45	7.5	13.12	2.2	1.86
Sydney	1,418.20	8.51	13.59	3.66	1.56
New York	2,560.26	7.68	9.74	3.81	0.77
Buenos Aires	960.10	4.16	6.99	2.95	0.8
Johannesburg	833.47	3.75	12.5	2.25	0.91
Vancouver	1,393.71	7.58	12.72	4.7	1.08
Moscow	2,304.23	8.49	13.79	3.62	3.33

INTERPRETING THE OUTPUT

14.31 A stepwise regression procedure was used to analyse a set of 20 observations taken on four predictor variables to predict a dependent variable. The results of this procedure are given next. Discuss the results.

```
STEPWISE REGRESSION OF Y ON 4
PREDICTORS,
WITH N = 20

STEP              1          2
CONSTANT       152.2      124.5

X₁            -50.6      -43.4
T-RATIO        7.42       6.13

X₂                        1.36
T-RATIO                   2.13

S              15.2       13.9
R-SQ          75.39      80.59
```

14.32 Shown here are the data for y and three predictors, x_1, x_2, and x_3. A stepwise regression procedure has been done on these data; the results are also given. Comment on the outcome of the stepwise analysis in light of the data.

y	x_1	x_2	x_3
94	21	1	204
97	25	0	198
93	22	1	184
95	27	0	200
90	29	1	182
91	20	1	159
91	18	1	147
94	25	0	196
98	26	0	228
99	24	0	242
90	28	1	162
92	23	1	180
96	25	0	219

```
Step           1        2        3
Constant     74.81    82.18    87.89


x₃           0.099    0.067    0.071
T-Value       6.90     3.65     5.22
P-Value      0.000    0.004    0.001


x₂                    -2.26    -2.71
T-Value               -2.32    -3.71
P-Value               0.043    0.005


x₁                             -0.256
T-Value                        -3.08
P-Value                        0.013


S            1.37     1.16     0.850
R-Sq        81.24    87.82    94.07
R-Sq(adj)   79.53    85.38    92.09
```

14.33 Shown on the following page is output from two Excel regression analyses on the same problem. The first output was done on a 'full' model. In the second output, the variable with the smallest absolute t value has been removed, and the regression has been rerun like a second step of a backward elimination process. Examine the two outputs. Explain what happened, what the results mean, and what might happen in a third step.

FULL MODEL

Regression Statistics

Multiple R	0.567
R Square	0.321
Adjusted R Square	0.208
Standard Error	159.681
Observations	29

ANOVA

	df	SS	MS	F	Significance F
Regression	4	289856.08	72464.02	2.84	0.046
Residual	24	611955.23	25498.13		
Total	28	901811.31			

	Coefficients	Standard Error	t Stat	P-value
Intercept	336.79	124.0800	2.71	0.012
x_1	1.65	1.7800	0.93	0.363
x_2	−5.63	13.4700	−0.42	0.680
x_3	0.26	1.6800	0.16	0.878
x_4	185.50	66.2200	2.80	0.010

SECOND MODEL

Regression Statistics

Multiple R	0.566
R Square	0.321
Adjusted R Square	0.239
Standard Error	156.534
Observations	29

ANOVA

	df	SS	MS	F	Significance F
Regression	3	289238.1	96412.70	3.93	0.020
Residual	25	612573.20	24502.90		
Total	28	901811.3			

	Coefficients	Standard Error	t Stat	P-value
Intercept	342.92	11.34	2.97	0.006
x_1	1.83	1.31	1.40	0.174
x_2	−5.75	13.18	−0.44	0.667
x_4	181.22	59.05	3.07	0.005

ANALYSING THE DATABASES

1. Use the Manufacturing database to develop a multiple regression model to predict Cost of Materials by Number of Employees, New Capital Expenditures, Value Added by Manufacture, Value of Industry Shipments, and End-of-Year Inventories. Create indicator variables for values of industry shipments that have been coded from 1 to 4. Use a stepwise regression procedure. Does multicollinearity appear to be a problem in this analysis? Discuss the results of the analysis.

2. Construct a correlation matrix for the Hospital database variables. Are some of the variables highly correlated? Which ones and why? Perform a stepwise multiple regression analysis to predict Personnel by Control, Service, Beds, Admissions, Census, Outpatients, and Births. The variables Control and Service will need to be coded as indicator variables. Control has four subcategories, and Service has two.

3. Develop a regression model using the Financial database. Use Total Revenues, Total Assets, Return on Equity, Earnings per Share, Average Yield, and Dividends per Share to predict the average P/E ratio for a company. How strong is the model? Use stepwise regression to help sort out the variables. Several of these variables may be measuring similar things. Construct a correlation matrix to explore the possibility of multicollinearity among the predictors.

4. Using the International Stock Market database, conduct stepwise a multiple regression procedure to predict the DJIA by the Nasdaq, the S&P 500, the Nikkei, the Hang Seng, the FTSE 100, and the IPC. Discuss the outcome of the analysis including the model, the strength of the model, and the predictors.

CASE

MOBILE PHONE QUALITY AND PRICE

Worldwide sales of mobile phones are a multibillion pound business. There is severe competition among the major manufacturers to attract higher sales and greater market shares. To achieve this, companies compete with each other on price. However, for many customers price may not be as important as the perceived quality of the phone, especially as many phones are marketed at 'zero price' under a wide variety of plans and packages being offered by service providers.

Today, the range of mobile phones available is huge, as are the functions offered. Often viewed as more of a fashion accessory or a 'must have' item rather than simply a phone, today's consumers are demanding a range of attributes such as in-built camera, GPS navigation, compatibility with 3G network, a colour display, and inbuilt MP3 player. Of course, it should also be able to make or receive phone calls.

The independent consumer advice charity *Which?* regularly conducts extensive surveys of numerous makes and models of mobile phones on offer and gives them an overall rating score, as well as rating scores for a variety of other categories. Some qualitative information on the phones is also given. The ratings given are then ranked and a *Which?* 'Best Buy' list is produced, of which the table below is an extract.

Decision makers and marketers at mobile phone manufacturers would like to know what features of a mobile phone are important for consumers. This would be especially important in helping to design effective marketing and advertising campaigns. They need to know not only what the important factors are, but also which are the most important ones. For example, more consumers are now looking for 3G network compatibility. Out of the 26 mobile phones that were included in the table, all had Bluetooth (this variable was therefore excluded as it did not provide any additional information), 18 were compatible with the 3G network, 19 had Wi-Fi and 23 had a touch screen. In purchasing a mobile phone with a camera, an obvious factor consumers would consider would be the quality of the picture so the resolution of the camera (in megapixels) is included. The battery standby time, the rating of the music player and the weight of the phone are some of the other factors that may influence a consumer's choice of mobile phone. The following table shows *Which?* Best Buy overall ratings for 26 phones, along with their music player scores. The table also shows the resolution of the camera, the battery standby time and whether the phone is 3G compatible, whether it has Wi-Fi and a touch screen.

Make and Model	Overall score (%)	Battery Standby Time (hours)	Music Player Score (%)	Camera Resolution (megapixels)	Weight (grams)	3G	Wi-Fi	Touch screen
Nokia 5800 Xpressmusic	80	406	80	3.2	114	1	1	1
Nokia 5230	79	432	80	2.0	114	1	0	1
Samsung S7070 DIVA	77	800	80	3.2	93	0	0	1
Sony Ericsson U5i Vivaz	77	440	80	8.1	97	1	1	1
Nokia 6303 Classic	77	450	60	3.2	98	0	0	0
Nokia 5530 Xpressmusic	77	351	80	3.2	108	0	1	1
LG Pop GD510	76	360	60	3.0	87	0	0	1
LG KP500 Cookie	76	350	60	3.0	90	0	0	1
Samsung Galaxy S II	76	750	80	8.0	117	1	1	1
BlackBerry 8520 Curve	75	408	80	2.0	105	0	1	0
Samsung Tocco Lite	75	800	80	3.2	93	0	0	1
Samsung S8500 Wave 8GB	74	550	80	5.0	117	1	1	1
Nokia X3	74	380	80	3.2	104	0	0	0
HTC Desire HD	72	420	80	8.0	164	1	1	1
LG Optimus 2X	72	400	60	8.0	146	1	1	1
HTC Incredible S	70	370	80	8.0	136	1	1	1
HTC Desire S	70	455	80	5.0	132	1	1	1
Nokia X6	68	450	80	5.0	122	1	1	1
Sony Ericsson Xperia Play	68	413	80	5.0	173	1	1	1
Samsung Omnia 7	66	330	60	5.0	139	1	1	1
Nokia N8	66	400	80	12.0	134	1	1	1
Motorola Defy	66	240	60	5.0	118	1	1	1
LG Optimus 7	66	330	60	5.0	158	1	1	1
BlackBerry Torch 9800	65	336	80	5.0	161	1	1	1
Samsung Galaxy S	65	576	60	5.0	117	1	1	1
Nokia N97 Mini	65	310	60	5.0	139	1	1	1

Source: Adapted from *Which?* Mobile Phone Reviews, *Which?* Best Buy Phones, available at www.which.co.uk/mobile/phones/, accessed August 2011, and www.amazon.co.uk, accessed August 2011.

DISCUSSION

1. Develop a multiple regression model to predict the overall score using the four explanatory (predictor) variables corresponding to battery standby time, music player score, camera resolution and weight. Discuss the significance of the estimated model as well as the significance of the explanatory variables in predicting the overall score. Use Excel to produce a correlation matrix for the predictor variables to check whether multicollinearity is present among the variables or not. Comment on these results and any implications they may have for your modelling strategy.

2. Produce scatter plots of each explanatory variable in turn against the overall score. In light of these results and Tukey's four quadrant approach,

re-estimate the model using any appropriate data transformations.

3. Compatibility with the 3G network, Wi-Fi and a touch screen are other recent features of mobile phones. The previous table also contains data on 3G compatibility and whether the phone has Wi-Fi and touch screen (a '1' indicates the phone has 3G compatibility, Wi-Fi, and touch screen, respectively and a '0' that is does not). Develop another multiple regression model using the four explanatory variables (battery standby time, music player score, camera resolution, and weight) and the three dummy variables 3G, Wi-Fi, and touch screen. Summarize the strength of your model (including the dummy variables) and any new variables generated by recoding. What advice would you give to mobile phone manufacturers with regard to the development, release and marketing of new mobile phones?

4. We would expect price to be an important factor when choosing a mobile phone. Mobile phone manufacturers would like us to believe that phone quality is closely reflected by its price, with the assumption that better quality phones are more expensive. Using the following data on mobile phone prices (corresponding to Amazon.co.uk's price for each model) and overall quality (score), test this assertion by developing a model to predict phone price by its overall rating score. Experiment with this basic specification by trying non-linear models, such as polynomial regressions, using Tukey's four quadrant

approach to help you. What would you conclude from this analysis?

Make and Model	Amazon.co.uk Best Price (£)	Overall score (%)
Nokia 5800 Xpressmusic	174.99	80
Nokia 5230	103.49	79
Samsung S7070 DIVA	127.99	77
Sony Ericsson U5i Vivaz	317.03	77
Nokia 6303 Classic	89.99	77
Nokia 5530 Xpressmusic	225.91	77
LG Pop GD510	64.99	76
LG KP500 Cookie	54.89	76
Samsung Galaxy S II	473.33	76
BlackBerry 8520 Curve	148.99	75
Samsung Tocco Lite	71.69	75
Samsung S8500 Wave 8GB	205.99	74
Nokia X3	101.98	74
HTC Desire HD	347.17	72
LG Optimus 2X	299.99	72
HTC Incredible S	375.49	70
HTC Desire S	331.04	70
Nokia X6	199.99	68
Sony Ericsson Xperia Play	318.99	68
Samsung Omnia 7	279.99	66
Nokia N8	299.99	66
Motorola Defy	223.26	66
LG Optimus 7	299.99	66
BlackBerry Torch 9800	349.99	65
Samsung Galaxy S	359.99	65
Nokia N97 Mini	174.99	65

USING THE COMPUTER

EXCEL

- Excel does not have Model-Building Search Procedure capability. However, Excel can perform multiple regression analysis. The commands are essentially the same as those for simple regression except that the x range of data may include several columns. Excel will determine the number of predictor variables from the number of columns entered in to **Input X Range**.

- Begin by selecting the **Data** tab on the Excel worksheet. From the **Analysis** panel at the right top of the **Data** tab worksheet, click on **Data Analysis**.

If your Excel worksheet does not show the **Data Analysis** option, then you can load it as an add-in following directions given in Chapter 2. From the **Data Analysis** pulldown menu, select **Regression**. In the **Regression** dialog box, input the location of the y values in **Input Y Range**. Input the location of the **X** values in **Input X Range**. Input **Labels** and input **Confidence Level**. To pass the line through the origin, check **Constant is Zero**. To printout the raw residuals, check **Residuals**. To printout residuals converted to z scores, check **Standardized Residuals**. For a plot of the residuals, check **Residual**

Plots. For a plot of the line through the points check **Line Fit Plots**.

- Standard output includes R, R^2, s_e, and an ANOVA table with the F test, the slope and intercept, t statistics with associated p-values, and any optionally requested output, such as graphs or residuals.

MINITAB

- Minitab does have Model-Building Search Procedure capability procedures including both forward and backward Stepwise regression, Forward Selection, and Backward Elimination.

- To begin, select **Stat** from the menu bar. Select **Regression** from the **Stat** pulldown menu. Select **Stepwise** from the **Regression** pulldown menu. Place the column name or column location of the y variable in **Response**. Place the column name or column location of the x variable(s) in **Predictors**. If you want to guarantee inclusion of particular variables in the model, place the column name or column locations of such variables in **Predictors to include in every model**. This is optional. Select **Methods** for Model Building options and selection of criterion for adding or removing a variable.

- In the **Methods** dialog box, Check **Use alpha values** to use alpha as the criterion for adding or removing a variable. Check **Use F values** to use F values as the criterion for adding or removing a variable. Check **Stepwise (forward and backward)** to run a standard forward or backward stepwise regression procedure. To specify particular variables to be included in the initial model, place the column name or column location of such variables in the box labelled **Predictors in initial model**. Check **Forward selection** to run a forward selection regression. Check **Backward elimination** to run a backward elimination regression. In each of these model-building procedures, you have the option of setting particular values of alpha or F for the entering and/or removing variables from the model. Minitab defaults to an alpha of 0.15 and an F of 4.

CHAPTER 15

Time-Series Forecasting and Index Numbers

LEARNING OBJECTIVES

This chapter discusses the general use of forecasting in business, several tools that are available for making business forecasts, the nature of time-series data, and the role of index numbers in business, thereby enabling you to:

1. Differentiate among various measurements of forecasting error, including mean absolute deviation and mean square error, in order to assess which forecasting method to use

2. Describe smoothing techniques for forecasting models, including naive, simple average, moving average, weighted moving average, and exponential smoothing

3. Determine trend in time-series data by using linear regression trend analysis, quadratic model trend analysis, and Holt's two-parameter exponential smoothing method

4. Account for seasonal effects of time-series data by using decomposition and Winters' three-parameter exponential smoothing method

5. Test for autocorrelation using the Durbin-Watson test, overcoming it by adding independent variables and transforming variables and taking advantage of it with autoregression

6. Differentiate among simple index numbers, unweighted aggregate price index numbers, weighted aggregate price index numbers, Laspeyres price index numbers, and Paasche price index numbers by defining and calculating each

Decision Dilemma

Forecasting Air Pollution

The Kyoto Protocol is a protocol linked to the United Nations Framework Convention on Climate Change (UNFCCC), aimed at fighting global warming. The UNFCCC is an international environmental treaty with the goal of achieving the 'stabilization of greenhouse gas concentrations in the atmosphere at a level that would prevent dangerous anthropogenic interference with the climate system'. The Protocol was initially adopted on 11 December 1997 in Kyoto, Japan, and entered into force on 16 February 2005. As of April 2010, 191 nation states have signed and ratified the protocol, and 37 countries have committed themselves to reducing their collective greenhouse gas emissions by 5.2% from the 1990 level.

The United Kingdom is one of the countries that has committed to reducing its 1990 level of greenhouse gas emission. The Department of Energy and Climate Change (DECC) regularly collects, monitors, and verifies statistics on the UK's greenhouse gas emissions. Shown below are emission data for two air pollution variables, carbon dioxide (CO_2) and other greenhouse gases (methane, nitrous oxide, hydrofluorocarbons, perluorocarbons, and sulphur hexafluoride) over a 21-year period reported by the DECC in million tonnes carbon dioxide equivalent (data for 2010 is provisional).

Year	Net CO_2 Emissions	Other greenhouse gases
1990	589.7	191.9
1991	596.8	189.6
1992	579.8	179.7

(continued)

Year	Net CO_2 Emissions	Other greenhouse gases
1993	564.9	171.3
1994	559.2	165.0
1995	550.8	163.5
1996	572.8	161.7
1997	548.7	159.4
1998	551.6	151.9
1999	542.3	129.3
2000	549.4	122.6
2001	561.3	114.5
2002	543.7	110.0
2003	553.4	104.1
2004	552.6	101.8
2005	549.7	99.7
2006	546.3	96.6
2007	537.8	94.4
2008	525.1	92.6
2009	473.7	89.9
2010	491.7	88.0

Managerial and Statistical Questions

1. Is it possible to forecast the emissions of carbon dioxide or other greenhouse gases for the year 2015, 2020, or even 2025 using these data?
2. What techniques best forecast the emissions of carbon dioxide or other greenhouse gases for future years from these data?

Source: Adapted from statistics published by the Department of Energy and Climate Change, at www.decc.gov.uk/en/content/cms/statistics/climate_stats/data/data.aspx; 'Kyoto Protocol', Wikipedia entry at http://en.wikipedia.org/wiki/Kyoto_Protocol.

Every day, **forecasting** – *the art or science of predicting the future* – is used in the decision-making process to help business people reach conclusions about buying, selling, producing, hiring, and many other actions. As an example, consider the following items:

- Market watchers predict a resurgence of stock values next year.
- City planners forecast a water crisis in Southern Europe.
- Future brightens for solar power.
- Energy secretary sees rising demand for oil.
- CEO says difficult times won't be ending soon for airline industry.

- Life insurance outlook fades.
- Increased competition from Asian businesses will result in significant redundancies in European consumer goods industries.

How are these and other conclusions reached? What forecasting techniques are used? Are the forecasts accurate? In this chapter we discuss several forecasting techniques, how to measure the error of a forecast, and some of the problems that can occur in forecasting. In addition, this chapter will focus only on data that occur over time, time-series data.

Time-series data are *data gathered on a given characteristic over a period of time at regular intervals*. Time-series forecasting techniques attempt to account for changes over time by examining patterns, cycles, or trends, or using information about previous time periods to predict the outcome for a future time period. Time-series methods include naïve methods, averaging, smoothing, regression trend analysis, and the decomposition of the possible time-series factors, all of which are discussed in subsequent sections.

15.1   INTRODUCTION TO FORECASTING

Virtually all areas of business, including production, sales, employment, transportation, distribution, and inventory, produce and maintain time-series data. Table 15.1 provides an example of time-series data released by the International Monetary Fund's International Financial Statistics database. The table contains the average yield rates of French government bonds for a 20-year period.

Why does the average yield differ from year to year? Is it possible to use these time series data to predict average yields for year 21 or ensuing years? Figure 15.1 is a graph of these data over time. Often graphical depiction of time-series data can give a clue about any trends, cycles, or relationships that might exist there. Does the graph in Figure 15.1 show that bond yields are decreasing? Will next year's yield rate be lower or is a cycle occurring in these data that will result in an increase? To answer such questions, it is sometimes helpful to determine which of the four components of time-series data exist in the data being studied.

Time-Series Components

It is generally believed that time-series data are composed of four elements: trend, cyclicality, seasonality, and irregularity. Not all time-series data have all these elements. Consider Figure 15.2, which shows the effects of these time-series elements on data over a period of 13 years.

TABLE 15.1

Average Yields of French Government Bonds

Year	Average Yield
1	9.03667
2	8.5875
3	6.775
4	7.21583
5	7.535
6	6.31083
7	5.58167
8	4.64
9	4.60833
10	5.39417
11	4.93917
12	4.86
13	4.13
14	4.09833
15	3.41
16	3.79667
17	4.30417
18	4.23417
19	3.64919
20	3.1171

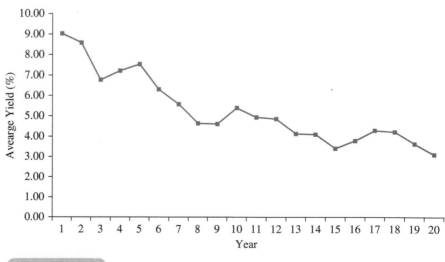

FIGURE 15.1

Excel Graph of Bond Yield Time-Series Data

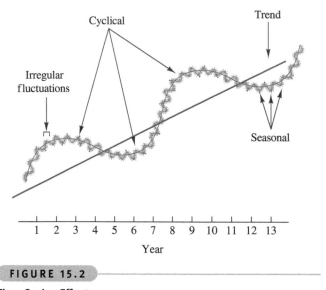

FIGURE 15.2

Time-Series Effects

The long-term general direction of data is referred to as **trend**. Notice that even though the data depicted in Figure 15.2 move through upward and downward periods, the general direction or trend is increasing (denoted in Figure 15.2 by the line). **Cycles** are *patterns of highs and lows through which data move over time periods usually of more than a year.* Notice that the data in Figure 15.2 seemingly move through two periods or cycles of highs and lows over a 13-year period. Time-series data that do not extend over a long period of time may not have enough 'history' to show **cyclical effects**. **Seasonal effects**, on the other hand, are *shorter cycles, which usually occur in time periods of less than one year.* Often seasonal effects are measured by the month, but they may occur by quarter, or may be measured in as small a time frame as a week or even a day. Note the seasonal effects shown in Figure 15.2 as up and down cycles, many of which occur during a one-year period. **Irregular fluctuations** are *rapid changes or 'bleeps' in the data, which occur in even shorter time frames than seasonal effects.* Irregular fluctuations can happen as often as day to day. They are subject to momentary change and are often unexplained. Note the irregular fluctuations in the data of Figure 15.2.

Observe again the bond yield data depicted in Figure 15.1. The general trend seems to move downward and contain three cycles. Each of the cycles traverses approximately five to eight years. It is possible, although not displayed here, that seasonal periods of highs and lows within each year result in seasonal bond yields. In addition, irregular daily fluctuations of bond yield rates may occur but are unexplainable.

Time-series data that contain no trend, cyclical, or seasonal effects are said to be **stationary**. Techniques used to forecast stationary data analyse only the irregular fluctuation effects.

The Measurement of Forecasting Error

In this chapter, several forecasting techniques will be introduced that typically produce different forecasts. How does a decision maker know which forecasting technique is doing the best job in predicting the future? One way is to compare forecast values with actual values and determine the amount of **forecasting error** a technique produces. An examination of individual errors gives some insight into the accuracy of the forecasts. However, this process can be tedious, especially for large data sets, and often a single measurement of overall forecasting error is needed for the entire set of data under consideration. Any of several methods can be used to compute error in forecasting. The choice depends on the forecaster's objective, the forecaster's familiarity with the technique, and the method of error measurement used by the computer forecasting software. Several techniques can be used to measure overall error, including mean error (ME), mean absolute deviation (MAD), mean square error (MSE), mean percentage error (MPE), and mean absolute percentage error (MAPE). Here we will consider the mean absolute deviation (MAD) and the mean square error (MSE).

Error

The **error of an individual forecast** is *the difference between the actual value and the forecast of that value.*

$$e_t = x_t - F_t$$

ERROR OF AN INDIVIDUAL FORECAST

where:
e_t = the error of the forecast
x_t = the actual value
F_t = the forecast value

Mean Absolute Deviation (MAD)

One measure of overall error in forecasting is the mean absolute deviation, MAD. The **mean absolute deviation (MAD)** is *the mean, or average, of the absolute values of the errors.*

Table 15.2 presents the number of new car registrations in the United Kingdom over an 11-year period along with the forecast for each year and the error of the forecast. An examination of these data reveals that some of the forecast errors are positive and some are negative. In summing these errors in an attempt to compute an overall measure of error, the negative and positive values offset each other resulting in an underestimation of the total error. The mean absolute deviation overcomes this problem by taking the absolute value of the error measurement, thereby analysing the magnitude of the forecast errors without regard to direction.

TABLE 15.2

New Car Registrations

Year	New Car Registrations (thousands)	Forecast	Error
1	2,222	2,311	−89
2	2,459	2,403	56
3	2,564	2,452	112
4	2,579	2,486	93
5	2,567	2,476	91
6	2,440	2,333	107
7	2,345	2,261	84
8	2,404	2,362	42
9	2,132	2,646	−514
10	1,995	2,065	−70
11	2,031	2,064	−33

MEAN ABSOLUTE DEVIATION

$$\text{MAD} = \frac{\sum |e_t|}{\text{Number of Forecasts}}$$

The mean absolute error can be computed for the forecast errors in Table 15.2 as follows.

$$\text{MAD} = \frac{|-89| + |56| + |112| + |93| + |91| + |107| + |84| + |42| + |-514| + |-70| + |-33|}{11} = 117.36$$

Mean Square Error (MSE)

The **mean square error (MSE)** is another way to circumvent the problem of the cancelling effects of positive and negative forecast errors. The MSE is *computed by squaring each error (thus creating a positive number) and averaging the squared errors.* The following formula states it more formally:

MEAN SQUARE ERROR

$$\text{MSE} = \frac{\sum e_t^2}{\text{Number of Forecasts}}$$

The mean square error can be computed for the errors shown in Table 15.2 as follows.

$$\text{MSE} = \frac{(-89)^2 + (56)^2 + (112)^2 + (93)^2 + (91)^2 + (107)^2 + (84)^2 + (42)^2 + (-514)^2 + (-70)^2 + (-33)^2}{11} = 30,089.55$$

Selection of a particular mechanism for computing error is up to the forecaster. It is important to understand that different error techniques will yield different information. The business researcher should be informed enough about the various error measurement techniques to make an educated evaluation of the forecasting results.

15.1 Use the forecast errors given here to compute MAD and MSE. Discuss the information yielded by each type of error measurement.

Period	e
1	2.3
2	1.6
3	−1.4
4	1.1
5	.3
6	−.9
7	−1.9
8	−2.1
9	.7

15.2 Determine the error for each of the forecasts below. Compute MAD and MSE.

Period	Value	Forecast	Error
1	202	–	–
2	191	202	
3	173	192	
4	169	181	
5	171	174	
6	175	172	
7	182	174	
8	196	179	
9	204	189	
10	219	198	
11	227	211	

15.3 Using the following data, determine the values of MAD and MSE. Which of these measurements of error seems to yield the best information about the forecasts? Why?

Period	Value	Forecast
1	19.4	16.6
2	23.6	19.1
3	24.0	22.0
4	26.8	24.8
5	29.2	25.9
6	35.5	28.6

15.4 Figures for hectares of organic crop area in Belgium from a 10-year period follow. The data are published by Eurostat's New Cronos Database. With these data, forecasts have been made by using techniques presented later in this chapter. Compute MAD and MSE on these forecasts. Comment on the errors.

Year	Organic crop area (ha)	Forecast
1	13,036	–
2	15,437	13,036.0
3	24,820	14,716.7
4	16,176	21,789.0
5	19,853	17,859.9
6	19,764	19,255.1
7	21,754	19,611.3
8	23,842	21,111.2
9	27,376	23,022.8
10	29,778	26,070.0

15.2 SMOOTHING TECHNIQUES

Several techniques are available to forecast time-series data that are stationary or that include no significant trend, cyclical, or seasonal effects. These techniques are often referred to as **smoothing techniques** because they *produce forecasts based on 'smoothing out' the irregular fluctuation effects in the time-series data*. Three general categories of smoothing techniques are presented here: (1) naive forecasting models, (2) averaging models, and (3) exponential smoothing.

Naive Forecasting Models

Naive forecasting models are *simple models in which it is assumed that the more recent time periods of data represent the best predictions or forecasts for future outcomes*. Naive models do not take into account data trend, cyclical effects, or seasonality. For this reason, naive models seem to work better with data that are reported on a daily or weekly basis or in situations that show no trend or seasonality. The simplest of the naive forecasting methods is the model in which the forecast for a given time period is the value for the previous time period.

$$F_t = x_{t-1}$$

where:

$$F_t = \text{the forecast value for time period } t$$
$$x_{t-1} = \text{the value for time period } t-1$$

As an example, if 532 pairs of shoes were sold by a retailer last week, this naive forecasting model would predict that the retailer will sell 532 pairs of shoes this week. With this naive model, the actual sales for this week will be the forecast for next week.

Observe the data in Table 15.3 representing the total reported number of arrivals in hotels and similar establishments in Norway for the past 12 months as reported by Eurostat. Figure 15.3 presents an Excel graph of these total arrivals over the 12-month period. From these data, we can make a naive forecast of the total number of arrivals in hotels and similar establishments in Norway for January of the next year by using the figure for December, which is 209,770.

Another version of the naive forecast might be to use the number of arrivals in January of the previous year as the forecast for January of next year, because the business researcher may believe a relationship exists between number of arrivals and the month of the year. In this case, the naive forecast for next January from Table 15.3 is 169,889 (January of the previous year). The forecaster is free to be creative with the naive forecast model method and search for other relationships or rationales within the limits of the time-series data that would seemingly produce a valid forecast.

TABLE 15.3

Total Number of Arrivals in Hotels and Similar Establishments in Norway

Month	Total Arrivals
January	169,889
February	180,802
March	228,430
April	239,828
May	323,266
June	349,055
July	480,739
August	427,902
September	332,118
October	313,023
November	281,171
December	209,770

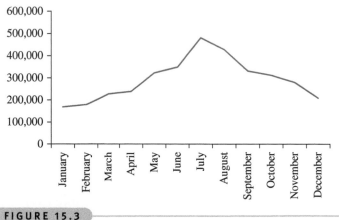

FIGURE 15.3

Excel Graph of Arrivals in Hotels and Similar Establishments in Norway over a 12-Month Period

Averaging Models

Many naive model forecasts are based on the value of one time period. Often such forecasts become a function of irregular fluctuations of the data; as a result, the forecasts are 'oversteered'. Using averaging models, a forecaster enters information from several time periods into the forecast and 'smoothes' the data. **Averaging models** are computed by *averaging data from several time periods and using the average as the forecast for the next time period.*

Simple Averages

The most elementary of the averaging models is the **simple average model**. With this model, *the forecast for time period t is the average of the values for a given number of previous time periods*, as shown in the following equation:

$$F_t = \frac{X_{t-1} + X_{t-2} + X_{t-3} + \cdots + X_{t-m}}{n}$$

The data in Table 15.4 provide the costs of standard grade burning oil in the United Kingdom for three years. Figure 15.4 displays a Minitab graph of these data.

A simple 12-month average could be used to forecast the cost of standard grade burning oil for July of year 3 from the data in Table 15.4 by averaging the values for July of year 2 through June of year 3 (the preceding 12 months).

$$F_{July, year 3} = \frac{44.5 + 44.2 + 42.9 + 45.3 + 46.6 + 50.2 + 55.1 + 55.6 + 57.6 + 61.2 + 60.4 + 58.8}{12} = 51.9$$

With this **simple average**, the forecast for year 3 July standard grade burning oil cost is 51.9 pence. Note that none of the previous 12-month figures equal this value and that this average is not necessarily more closely related to values early in the period than to those late in the period. The use of the simple average over 12 months tends to smooth the variations, or fluctuations, that occur during this time.

Moving Averages

Suppose we were to attempt to forecast the standard grade burning oil cost for August of year 3 by using averages as the forecasting method. Would we still use the simple average for July of year 2

TABLE 15.4

Cost of Standard Grade Burning Oil (pence per litre)

Time Frame	Cost of Burning Oil
January (year 1)	36.0
February	33.9
March	31.8
April	33.2
May	34.5
June	36.1
July	36.0
August	37.1
September	37.4
October	38.0
November	39.8
December	40.1
January (year 2)	42.5
February	43.2
March	45.1
April	46.7
May	47.4
June	46.8
July	44.5
August	44.2
September	42.9
October	45.3
November	46.6
December	50.2
January (year 3)	55.1
February	55.6
March	57.6
April	61.2
May	60.4
June	58.8

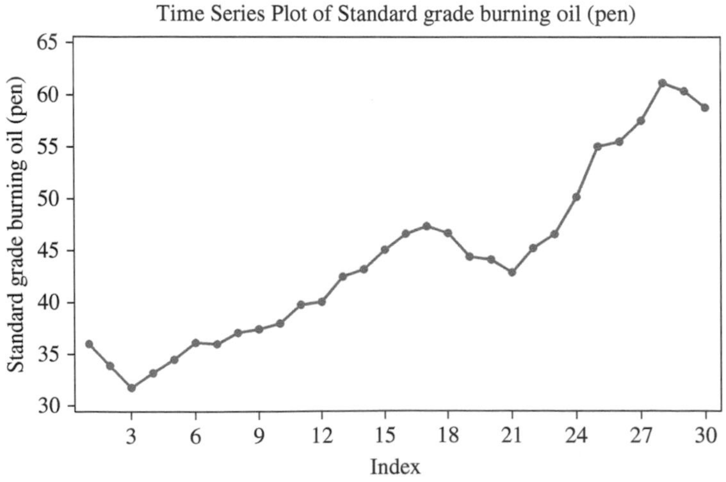

Time Series Plot of Standard grade burning oil (pen)

FIGURE 15.4

Minitab Graph of Standard Grade Burning Oil Data

through to August of year 3 as we did to forecast for July of year 3? Instead of using the same 12 months' average used to forecast July of year 3, it would seem to make sense to use the 12 months prior to August of year 3 (August of year 2 through to July of year 3) to average for the new forecast. Suppose in July of year 3 the cost of burning oil is 59.3 pence. We could forecast August of year 3 with a new average that includes the same months used to forecast July of year 3, but without the value for July of year 2 and with the value of July of year 3 added.

$$F_{Aug,year3} = \frac{44.2 + 42.9 + 45.3 + 46.6 + 50.2 + 55.1 + 55.6 + 57.6 + 61.2 + 60.4 + 58.8 + 59.3}{12} = 53.12$$

Computing an average of the values from August of year 2 through to July of year 3 produces a moving average, which can be used to forecast the cost of burning oil for August of year 3. In computing this moving average, the earliest of the previous 12 values, July of year 2, is dropped and the most recent value, July of year 3, is included.

A **moving average** is *an average that is updated or recomputed for every new time period being considered.* The most recent information is utilized in each new moving average. This advantage is offset by the disadvantages that (1) it is difficult to choose the optimal length of time for which to compute the moving average, and (2) moving averages do not usually adjust for such time-series effects as trend, cycles, or seasonality. To determine the more optimal lengths for which to compute the moving averages, we would need to forecast with several different average lengths and compare the errors produced by them.

DEMONSTRATION PROBLEM 15.1

Shown here are shipments (in millions of dollars) for electric lighting and wiring equipment over a 12-month period. Use these data to compute a four-month moving average for all available months.

Month	Shipments
January	1,056
February	1,345
March	1,381
April	1,191
May	1,259
June	1,361
July	1,110
August	1,334
September	1,416
October	1,282
November	1,341
December	1,382

Solution

The first moving average is

$$\text{4-Month Moving Average} = \frac{1056 + 1345 + 1381 + 1191}{4} = 1243.25$$

This first four-month moving average can be used to forecast the shipments in May. Because 1,259 shipments were actually made in May, the error of the forecast is

$$\text{Error}_{May} = 1,259 - 1,243.25 = 15.75$$

Shown next, along with the monthly shipments, are the four-month moving averages and the errors of forecast when using the four-month moving averages to predict the next month's shipments. The first moving average is displayed beside the month of May because it is computed by using January, February,

March, and April and because it is being used to forecast the shipments for May. The rest of the four-month moving averages and errors of forecast are as shown.

	Four-Month Moving Forecast		
Month	Shipments	Average	Error
January	1,056	–	–
February	1,345	–	–
March	1,381	–	–
April	1,191	–	–
May	1,259	1,243.25	15.75
June	1,361	1,294.00	67.00
July	1,110	1,298.00	−188.00
August	1,334	1,230.25	103.75
September	1,416	1,266.00	150.00
October	1,282	1,305.25	−23.25
November	1,341	1,285.50	55.50
December	1,382	1,343.25	38.75

The following Minitab graph shows the actual shipment values and the forecast shipment values based on the four-month moving averages. Notice that the moving averages are 'smoothed' in comparison with the individual data values. They appear to be less volatile and seem to be attempting to follow the general trend of the data.

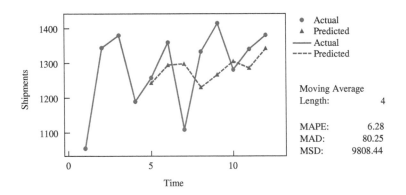

Weighted Moving Averages

A forecaster may want to place more weight on certain periods of time than on others. For example, a forecaster might believe that the previous month's value is three times as important in forecasting as other months. *A moving average in which some time periods are weighted differently than others* is called a **weighted moving average.**

As an example, suppose a three-month weighted average is computed by weighting last month's value by 3, the value for the previous month by 2, and the value for the month before that by 1. This weighted average is computed as

$$\bar{x}_{\text{weighted}} = \frac{3(M_{t-1}) + 2(M_{t-2}) + 1(M_{t-3})}{6}$$

where:

M_{t-1} = last month's value
M_{t-2} = value for the previous month
M_{t-3} = value for the month before the previous month

Notice that the divisor is 6. With a weighted average, the divisor always equals the total number of weights. In this example, the value of M_{t-1} counts three times as much as the value for M_{t-3}.

DEMONSTRATION PROBLEM 15.2

Compute a four-month weighted moving average for the electric lighting and wiring data from Demonstration Problem 15.1, using weights of 4 for last month's value, 2 for the previous month's value, and 1 for each of the values from the two months prior to that.

Solution

The first weighted average is

$$\frac{4(1191)+2(1381)+1(1345)+1(1056)}{8}=1240.875$$

This moving average is recomputed for each ensuing month. Displayed next are the monthly values, the weighted moving averages, and the forecast error for the data.

Month	Shipments	Four-Month Weighted Moving Average Forecast	Error
January	1,056	–	–
February	1,345	–	–
March	1,381	–	–
April	1,191	–	–
May	1,259	1,240.9	18.1
June	1,361	1,268.0	93.0
July	1,110	1,316.8	−206.8
August	1,334	1,201.5	132.5
September	1,416	1,272.0	144.0
October	1,282	1,350.4	−68.4
November	1,341	1,300.5	40.5
December	1,382	1,334.8	47.2

Note that in this problem the errors obtained by using the four-month weighted moving average were greater than most of the errors obtained by using an unweighted four-month moving average, as shown here.

Forecast Error, Unweighted Four-Month Moving Average	Forecast Error, Weighted Four-Month Moving Average
–	–
–	–
–	–
–	–
15.8	18.1
67.0	93.0
−188.0	−206.8
103.8	132.5
150.0	144.0
−23.3	−68.4
55.5	40.5
38.8	47.2

Larger errors with weighted moving averages are not always the case. The forecaster can experiment with different weights in using the weighted moving average as a technique. Many possible weighting schemes can be used.

Exponential Smoothing

Another forecasting technique, **exponential smoothing**, is *used to weight data from previous time periods with exponentially decreasing importance in the forecast.* Exponential smoothing is accomplished by multiplying the actual value for the present time period, X_t, by a value between 0 and 1 (the exponential smoothing constant) referred to as α (not the same α used for a Type I error) and adding that result to the product of the present time period's forecast F_t and $(1 - \alpha)$. The following is a more formalized version:

$$F_{t+1} = \alpha \cdot x_t + (1 - \alpha) \cdot F_t$$

EXPONENTIAL
SMOOTHING

where:

F_{t+1} = the forecast for the next time period $(t + 1)$
F_t = the forecast for the present time period (t)
X_t = the actual value for the present time period
α = a value between 0 and 1 referred to as the
exponential smoothing constant.

The value of α is determined by the forecaster. The essence of this procedure is that the new forecast is a combination of the present forecast and the present actual value. If α is chosen to be less than 0.5, less weight is placed on the actual value than on the forecast of that value. If α is chosen to be greater than 0.5, more weight is being put on the actual value than on the forecast value.

As an example, suppose the prime interest rate for a time period is 5% and the forecast of the prime interest rate for this time period was 6%. If the forecast of the prime interest rate for the next period is determined by exponential smoothing with $\alpha = 3$, the forecast is

$$F_{t+1} = (0.3)(5\%) + (1.0 - 0.3)(6\%) = 5.7\%$$

Notice that the forecast value of 5.7% for the next period is weighted more towards the previous forecast of 6% than toward the actual value of 5% because α is 0.3. Suppose we use $\alpha = 0.7$ as the exponential smoothing constant. Then,

$$F_{t+1} = (0.7)(5\%) + (1.0 - 0.7)(6\%) = 5.3\%$$

This value is closer to the actual value of 5% than the previous forecast of 6% because the exponential smoothing constant, α, is greater than 0.5.

To see why this procedure is called exponential smoothing, examine the formula for exponential smoothing again.

$$F_{t+1} = \alpha \cdot X_t + (1 - \alpha) \cdot F_t$$

If exponential smoothing has been used over a period of time, the forecast for F_t will have been obtained by

$$F_t = \alpha \cdot X_{t-1} + (1 - \alpha) \cdot F_{t-1}$$

Substituting this forecast value, F_t, into the preceding equation for F_{t-1} produces

$$F_{t+1} = \alpha \cdot X_t + (1 - \alpha)[\alpha \cdot X_{t-1} + (1 - \alpha) \cdot F_{t-1}]$$
$$= \alpha \cdot X_t + \alpha (1 - \alpha) \cdot X_{t-1} + (1 - \alpha)^2 F_{t-1}$$

but

$$F_{t-1} = \alpha \cdot X_{t-2} + (1 - \alpha) \cdot F_{t-2}$$

Substituting this value of F_{t-1} into the preceding equation for F_{t+1} produces

$$F_{t+1} = \alpha \cdot X_t + \alpha (1 - \alpha) \cdot X_{t-1} + (1 - \alpha)^2 F_{t-1}$$

$$= \alpha \cdot X_t + \alpha (1 - \alpha) \cdot X_{t-1} + (1 - \alpha)^2 [\alpha \cdot X_{t-2} + (1 - \alpha)F_{t-2}]$$

$$= \alpha \cdot X_t + \alpha (1 - \alpha) \cdot X_{t-1} + \alpha(1 - \alpha)^2 \cdot X_{t-2} + (1 - \alpha)^3 F_{t-2}$$

Continuing this process shows that the weights on previous-period values and forecasts include $(1 - \alpha)^n$ (exponential values). The following chart shows the values of α, $(1 - \alpha)$, $(1 - \alpha)^2$, and $(1 - \alpha)^3$ for three different values of alpha. Included is the value of $\alpha(1 - \alpha)^3$, which is the weight of the actual value for three time periods back. Notice the rapidly decreasing emphasis on values for earlier time periods. The impact of exponential smoothing on time-series data is to place much more emphasis on recent time periods. The choice of α determines the amount of emphasis.

α	$1 - \alpha$	$(1 - \alpha)^2$	$(1 - \alpha)^3$	$\alpha(1 - \alpha)^3$
.2	.8	.64	.512	.1024
.5	.5	.25	.125	.0625
.8	.2	.04	.008	.0064

Some forecasters use the computer to analyse time-series data for various values of α. By setting up criteria with which to judge the forecasting errors, forecasters can select the value of α that best fits the data.

The exponential smoothing formula

$$F_{t+1} = \alpha \cdot X_t + (1 - \alpha) \cdot F_t$$

can be rearranged algebraically as

$$F_{t+1} = F_t + \alpha(X_t - F_t)$$

This form of the equation shows that the new forecast, F_{t-1}, equals the old forecast, F_t, plus an adjustment based on α times the error of the old forecast $(X_t - F_t)$. The smaller α is, the less impact the error has on the new forecast and the more the new forecast is like the old. It demonstrates the dampening effect of α on the forecasts.

DEMONSTRATION PROBLEM 15.3

The Department for Communities and Local Government reports the total units of new houses started. The total units of new houses over a 20-year recent period in Scotland are given here. Use exponential smoothing to forecast the values for each ensuing time period. Work the problem using $\alpha = 0.2$, 0.5, and 0.8.

Year	Total Units	Year	Total Units
1	21,340	11	22,320
2	17,560	12	23,180
3	19,930	13	22,270
4	23,870	14	27,050
5	25,470	15	27,000
6	23,470	16	26,370
7	22,010	17	28,420
8	21,680	18	26,590
9	20,510	19	19,590
10	22,650	20	15,130

(Continued)

Solution

An Excel graph of these data is shown below.

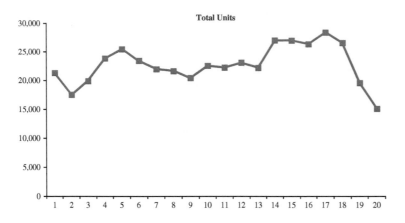

The following table provides the forecasts with each of the three values of alpha. Note that because no forecast is given for the first time period, we cannot compute a forecast based on exponential smoothing for the second period. Instead, we use the actual value for the first period as the forecast for the second period to get started. As examples, the forecasts for the third, fourth, and fifth periods are computed for $\alpha = 0.2$ as follows.

$$F_3 = 0.2(17,560) + 0.8(21,340) = 20,584$$
$$F_4 = 0.2(19,930) + 0.8(20,584) = 20,453$$
$$F_5 = 0.2(23,870) + 0.8(20,453) = 21,137$$

Year	Total Units	$\alpha = 0.2$ F	e	$\alpha = 0.5$ F	e	$\alpha = 0.8$ F	e
1	21,340	–	–	–	–	–	–
2	17,560	21,340	−3,780.0	21,340	−3,780.0	21,340	−3,780.0
3	19,930	20,584	−654.0	19,450	480.0	18,316	1,614.0
4	23,870	20,453	3,416.8	19,690	4,180.0	19,607	4,262.8
5	25,470	21,137	4,333.4	21,780	3,690.0	23,017	2,452.6
6	23,470	22,003	1,466.8	23,625	−155.0	24,979	−1,509.5
7	22,010	22,297	−286.6	23,548	−1,537.5	23,772	−1,761.9
8	21,680	22,239	−559.3	22,779	−1,098.8	22,362	−682.4
9	20,510	22,127	−1,617.4	22,229	−1,719.4	21,816	−1,306.5
10	22,650	21,804	846.1	21,370	1,280.3	20,771	1,878.7
11	22,320	21,973	346.8	22,010	310.2	22,274	45.7
12	23,180	22,043	1,137.5	22,165	1,015.1	22,311	869.1
13	22,270	22,270	0.0	22,672	−402.5	23,006	−736.2
14	27,050	22,270	4,780.0	22,471	4,578.8	22,417	4,632.8
15	27,000	23,226	3,774.0	24,761	2,239.4	26,123	876.6
16	26,370	23,981	2,389.2	25,880	489.7	26,825	−454.7
17	28,420	24,459	3,961.4	26,125	2,294.8	26,461	1,959.1
18	26,590	25,251	1,339.1	27,273	−682.6	28,028	−1,438.2
19	19,590	25,519	−5,928.7	26,931	−7,341.3	26,878	−7,287.6
20	15,130	24,333	−9,203.0	23,261	−8,130.6	21,048	−5,917.5

	$\alpha = 0.2$	$\alpha = 0.5$	$\alpha = 0.8$
MAD:	2,622.1	2,389.8	2,287.7
MSE:	12,241,658.0	10,896,862.3	8,943,945.6

Which value of alpha works best on the data? At the bottom of the preceding analysis are the values of two different measurements of error for each of the three different values of alpha. With each measurement of error, $\alpha = 0.8$ produces the smallest measurement of error. Observe from the Excel graph of the original data that the data vary up and down considerably. In exponential smoothing, the value of alpha is multiplied by the actual value and $1 - \alpha$ is multiplied by the forecast value to get the next forecast. Because the actual values are varying considerably, the exponential smoothing value with the largest alpha seems to be forecasting the best. By placing the greatest weight on the actual values, the new forecast seems to predict the new value better.

STATISTICS IN BUSINESS TODAY

Inflation Targeting in the Euro Zone

The Euro system's primary objective is to maintain price stability in the Euro zone. The Euro system consists of the European Central Bank (ECB), which decides the monetary policy, and the central banks of the member states that belong to the Eurozone, which apply the monetary policy decided by the ECB.

The ECB controls overheating of the economy through excessive borrowing and spending by raising (lowering) the interest rate on main refinancing operations, also known as the benchmark interest rate (The ECB also has a number of other instruments of monetary policy at its disposal but the weekly main refinancing operations are undoubtedly the most important). The ECB aims to confine inflation close to but below 2%. The following graph shows the movements of inflation and the ECB's interest rate for main refinancing operations over the turbulent period between October 2008 and July 2011.

The global financial crisis that started in 2007 induced the deepest economic downturn since World War II and the fastest deceleration in inflation since the launch of the Euro. The ECB confronted this situation by rapidly lowering interest rates by a staggering 2.75% in about six months (from 3.75% in Oct 2008 to 1% in May 2009). The slow improvement of economic conditions in Europe combined with high energy prices slowly boosted inflation to levels above the target of 2%. This in turn led to ECB raising interest rates first to 1.25% (in April 2011) and then to 1.5% (July 2011).

The ECB has a forecasting model to enable it to predict the likely value of inflation in a given month. Two popular economic models used to forecast inflation are the Traditional Phillips Curve (TPC) model and the New Keynesian Phillips Curve (NKPC) model. The TPC model essentially takes a 'backward-looking'

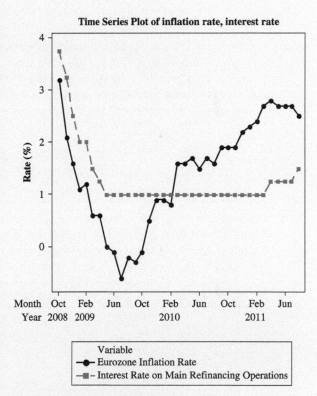

Time Series Plot of inflation rate, interest rate

Variable
—●— Eurozone Inflation Rate
—■- Interest Rate on Main Refinancing Operations

approach by modelling inflation as a function of lagged inflation and the output gap (i.e., excess demand). In contrast, the NKPC model is 'forward-looking' and gives prominence to the role of expected inflation and changes in marginal costs. As has been emphasized in this chapter, the performance of the various models can be evaluated by comparing the mean square errors and the mean absolute deviations.

Source: Adapted from data from the European Central Bank's website, available at http://sdw.ecb.europa.eu/quickview.do?SERIES_KEY=122.ICP.M.U2.N.000000.4.ANR and http://sdw.ecb.europa.eu/browse.do?node=2018801.

15.2   PROBLEMS

15.5   Use the following time-series data to answer the given questions:

Time Period	Value	Time Period	Value
1	27	6	66
2	31	7	71
3	58	8	86
4	63	9	101
5	59	10	97

    a. Develop forecasts for periods 5 through 10 using four-month moving averages.

    b. Develop forecasts for periods 5 through 10 using four-month weighted moving averages. Weight the most recent month by a factor of 4, the previous month by 2, and the other months by 1.

    c. Compute the errors of the forecasts in parts (a) and (b) and observe the differences in the errors forecast by the two different techniques.

15.6   Following are time-series data for eight different periods. Use exponential smoothing to forecast the values for periods 3 through 8. Use the value for the first period as the forecast for the second period. Compute forecasts using two different values of alpha, $\alpha = 0.1$ and $\alpha = 0.8$. Compute the errors for each forecast and compare the errors produced by using the two different exponential smoothing constants.

Time Period	Value	Time Period	Value
1	211	5	242
2	228	6	227
3	236	7	217
4	241	8	203

15.7   Following are time-series data for nine time periods. Use exponential smoothing with constants of 0.3 and 0.7 to forecast time periods 3 through 9. Let the value for time period 1 be the forecast for time period 2. Compute additional forecasts for time periods 4 through 9 using a three-month moving average. Compute the errors for the forecasts and discuss the size of errors under each method.

Time Period	Value	Time Period	Value
1	9.4	6	11.0
2	8.2	7	10.3
3	7.9	8	9.5
4	9.0	9	9.1
5	9.8		

15.8   The Federal Statistical Office of Germany publishes data relating to sustainable development in Germany, including data on resource protection, climate protection, renewable energies, and land use. Shown here are data on housing and transport area in Germany over a 17-year period (in hectares).

    a. Use these data to develop forecasts for the years 6 through 17 using a five-year moving average.

    b. Use these data to develop forecasts for the years 6 through 17 using a five-year weighted moving average. Weight the most recent year by 6, the previous year by 4, the year before that by 2, and the other years by 1.

    c. Compute the errors of the forecasts in parts (a) and (b) and observe the differences in the errors of the forecasts.

Year	Housing and Transport Area (ha)
1	40,305
2	40,742
3	41,179
4	41,615
5	42,052
6	42,506
7	42,982
8	43,459
9	43,939
10	44,381
11	44,780
12	45,141
13	45,621
14	46,050
15	46,436
16	46,789
17	47,137

15.9 The data below show the global number of issues from initial public offerings (IPOs) for a 15-year period released by the study Global IPO Trends by Ernst & Young. Use these data to develop forecasts for the years 3 through 15 using exponential smoothing techniques with alpha values of 0.2 and 0.9. Let the forecast for year 2 be the value for year 1. Compare the results by examining the errors of the forecasts.

Year	Number of Issues
1	1,837
2	1,748
3	1,042
4	1,372
5	1,883
6	876
7	847
8	812
9	1,520
10	1,552
11	1,796
12	2,014
13	769
14	577
15	1,393

15.3 TREND ANALYSIS

There are several ways to determine trend in time-series data and one of the more prominent is regression analysis. In Section 12.9, we explored the use of simple regression analysis in determining the equation of a trend line. In time-series regression trend analysis, the response variable, Y, is the variable being forecast, and the independent variable, X, represents time.

Many possible trend fits can be explored with time-series data. In this section we examine only the linear model and the quadratic model because they are the easiest to understand and simplest to compute. Because seasonal effects can confound trend analysis, it is assumed here that no seasonal effects occur in the data or they were removed prior to determining the trend.

Linear Regression Trend Analysis

The data in Table 15.5 represent 35 years of data on the average length of the working week in Canada for manufacturing workers. A regression line can be fit to these data by using the time periods as the independent variable and length of workweek as the dependent variable. Because the time periods are consecutive, they can be entered as X along with the time-series data (Y) into a regression analysis. The linear model explored in this example is

$$Y_i = \beta_0 + \beta_1 X_{ti} + \varepsilon_i$$

where:

  Y_i = data value for period i

  X_{ti} = ith time period

TABLE 15.5

Average Hours per Week in Manufacturing by Canadian Workers

Time Period	Hours	Time Period	Hours
1	37.2	19	36.0
2	37.0	20	35.7
3	37.4	21	35.6
4	37.5	22	35.2
5	37.7	23	34.8
6	37.7	24	35.3
7	37.4	25	35.6
8	37.2	26	35.6
9	37.3	27	35.6
10	37.2	28	35.9
11	36.9	29	36.0
12	36.7	30	35.7
13	36.7	31	35.7
14	36.5	32	35.5
15	36.3	33	35.6
16	35.9	34	36.3
17	35.8	35	36.5
18	35.9		

Source: Data prepared by the US Bureau of Labor Statistics, Office of Productivity and Technology.

SUMMARY OUTPUT

Regression Statistics	
Multiple R	0.782
R Square	0.611
Adjusted R Square	0.600
Standard Error	0.5090
Observations	35

ANOVA

	df	SS	MS	F	Significance F
Regression	1	13.4467	13.4467	51.91	0.000000029
Residual	33	8.5487	0.2591		
Total	34	21.9954			

	Coefficients	Standard Error	t Stat	P-value
Intercept	37.4161	0.1758	212.81	0.000000000
Year	−0.0614	0.0085	−7.20	0.000000029

FIGURE 15.5

Excel Regression Output for Hours Worked Using Linear Trend

Figure 15.5 shows the Excel regression output for this example. By using the coefficients of the X variable and intercept, the equation of the trend line can be determined to be

$$\hat{Y} = 37.4161 - 0.0614X_t$$

The slope indicates that for every unit increase in time period, X_t, a predicted decrease of 0.0614 occurs in the length of the average workweek in manufacturing. Because the working week is measured in hours, the length of the average work-week decreases by an average of $(0.0614)(60 \text{ minutes}) = 3.7$ minutes each year in Canada in manufacturing. The Y intercept, 37.4161, indicates that in the year prior to the first period of these data the average work-week was 37.4161 hours.

The probability of the t ratio (0.00000003) indicates that significant linear trend is present in the data. In addition, $R^2 = 0.611$ indicates considerable predictability in the model. Inserting the various period values (1, 2, 3, ... , 35) into the preceding regression equation produces the predicted values of Y that are the trend. For example, for period 23 the predicted value is

$$\hat{Y} = 37.4161 - 0.0614(23) = 36.0 \text{ hours}$$

The model was developed with 35 periods (years). From this model, the average work-week in Canada in manufacturing for period 41 (the 41st year) can be forecast:

$$\hat{Y} = 37.4161 - 0.0614(41) = 34.9 \text{ hours}$$

Figure 15.6 presents an Excel scatter plot of the average workweek lengths over the 35 periods (years). In this Excel plot, the trend line has been fitted through the points. Observe the general downward trend of the data, but also note the somewhat cyclical nature of the points. Because of this pattern, a forecaster might want to determine whether a quadratic model is a better fit for trend.

Regression Trend Analysis Using Quadratic Models

In addition to linear regression, forecasters can explore using quadratic regression models to predict data by using the time-series periods. The quadratic regression model is

$$Y_i = \beta_0 + \beta_1 X_{ti} + \beta_2 X_{ti}^2 + \varepsilon_i$$

where:

Y_i = the time-series data value for period i

X_{ti} = the ith period

X_{ti}^2 = the square of the ith period

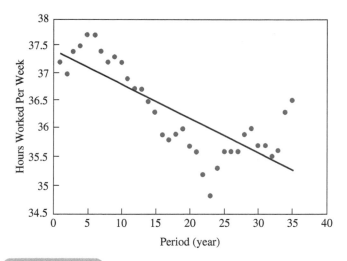

FIGURE 15.6

Excel Graph of Canadian Manufacturing Data with Trend Line

This model can be implemented in time-series trend analysis by using the time periods squared as an additional predictor. Thus, in the hours worked example, besides using $X_t = 1, 2, 3, 4, \ldots, 35$ as a predictor, we would also use $X_t^2 = 1, 4, 9, 16, \ldots, 1{,}225$ as a predictor.

Table 15.6 provides the data needed to compute a quadratic regression trend model on the manufacturing working week data. Note that the table includes the original data, the time periods, and the time periods squared.

TABLE 15.6

Data for Quadratic Fit of Manufacturing Working Week Example

Time Period	(Time Period)²	Hours	Time Period	(Time Period)²	Hours
1	1	37.2	19	361	36.0
2	4	37.0	20	400	35.7
3	9	37.4	21	441	35.6
4	16	37.5	22	484	35.2
5	25	37.7	23	529	34.8
6	36	37.7	24	576	35.3
7	49	37.4	25	625	35.6
8	64	37.2	26	676	35.6
9	81	37.3	27	729	35.6
10	100	37.2	28	784	35.9
11	121	36.9	29	841	36.0
12	144	36.7	30	900	35.7
13	169	36.7	31	961	35.7
14	196	36.5	32	1,024	35.5
15	225	36.3	33	1,089	35.6
16	256	35.9	34	1,156	36.3
17	289	35.8	35	1,225	36.5
18	324	35.9			

Source: Data prepared by the US Bureau of Labor Statistics, Office of Productivity and Technology.

SUMMARY OUTPUT

Regression Statistics	
Multiple R	0.873
R Square	0.761
Adjusted R Square	0.747
Standard Error	0.4049
Observations	35

ANOVA

	df	SS	MS	F	Significance F
Regression	2	16.7483	8.3741	51.07	0.0000000001
Residual	32	5.2472	0.1640		
Total	34	21.9954			

	Coefficients	Standard Error	t Stat	P-value
Intercept	38.1644	0.2177	175.34	0.0000000
Time Period	−0.1827	0.0279	−6.55	0.0000002
(Time Period)Sq	0.0034	0.0008	4.49	0.0000876

FIGURE 15.7

Excel Regression Output for Canadian Manufacturing Example with Quadratic Trend

The Excel computer output for this quadratic trend regression analysis is shown in Figure 15.7. We see that the quadratic regression model produces an R^2 of 0.761 with both of 0.611 with X_t and X_t^2 in the model. The linear model produced an R^2 of 0.611 with X_t alone. The quadratic regression seems to add some predictability to the trend model. Figure 15.8 displays an Excel scatter plot of the working week data with a second-degree polynomial fit through the data.

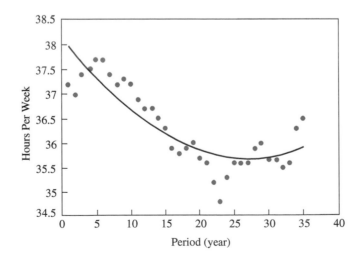

FIGURE 15.8

Excel Graph of Canadian Manufacturing Data with a Second-Degree Polynomial Fit

DEMONSTRATION PROBLEM 15.4

Following are data on the employed labour force of Greece for 1991 through 2010, obtained from the Organization for Economic Cooperation and Development's Main Economic Indicators Database. Use regression analysis to fit a trend line through the data. Explore a quadratic regression trend also. Does either model do well? Compare the two models.

Year	Labour Force (thousands)
1991	3,934.0
1992	4,034.0
1993	4,112.4
1994	4,188.9
1995	4,244.5
1996	4,314.0
1997	4,292.7
1998	4,525.8
1999	4,586.2
2000	4,612.0
2001	4,580.3
2002	4,655.9
2003	4,734.4
2004	4,818.8
2005	4,846.3
2006	4,886.8
2007	4,916.7
2008	4,937.3
2009	4,979.8
2010	5,017.4

Solution

Recode the time periods as 1 through 20 and let that be X. Run the regression analysis with the labour force members as Y, the dependent variable, and the time period as the independent variable. Now square all the X values, resulting in 1, 4, 9, . . ., 324, 361, 400 and let those formulate a second predictor (X^2). Run the regression analysis to predict the number in the labour force with both the time period variable (X) and the (time period)^2 variable. The Minitab output for each of these regression analyses follows.

```
Regression Analysis: Labour Force versus Year
The regression equation is
Labour Force = 3967 + 56.6 Year

Predictor           Coef      SE Coef          T          P
Constant         3966.77        27.76     142.90      0.000
Year              56.584         2.317      24.42      0.000
S = 59.7571    R-Sq = 97.1%    R-Sq(adj) =96.9%
```

```
Analysis of Variance

Source              DF          SS          MS          F          P
Regression           1     2129190     2129190     596.26      0.000
Residual Error      18       64276        3571
Total               19     2193467
```

```
Regression Analysis: Labour Force versus Year, Year Sq

The regression equation is
Labour Force = 3853 + 87.5 Year - 1.47 Year sq

Predictor           Coef      SE Coef           T          P
Constant         3853.47         2924      131.78      0.000
Year              87.487        6.413       13.64      0.000
Year Sq          -1.4715       0.2966       -4.96      0.000
S = 39.3031   R-Sq = 98.8%     R-Sq(adj) =98.7%
Analysis of Variance
Source              DF          SS          MS          F          P
Regression           2     2167206     1083603      701.48      0.000
Residual Error      14       26261        1545
Total               16     2193467
```

A comparison of the models shows that the linear model accounts for over 97% of the variability in the labour force figures, and the quadratic model only increases that predictability to 98.8%. Shown next are Minitab scatter plots of the data. First is the linear model, and then the quadratic model is presented.

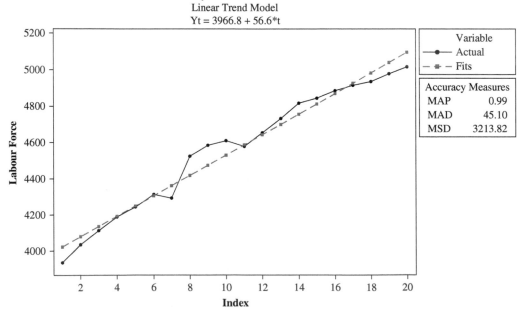

Trend Analysis Plot for Labour Force
Linear Trend Model
$Yt = 3966.8 + 56.6*t$

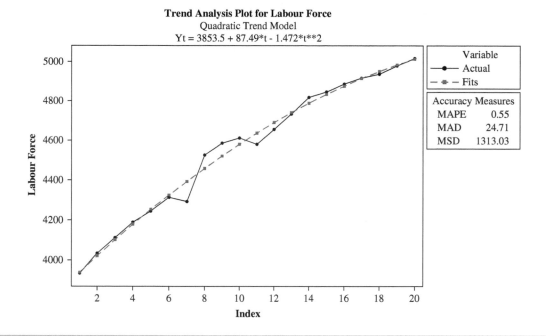

Trend Analysis Plot for Labour Force
Quadratic Trend Model
Yt = 3853.5 + 87.49*t - 1.472*t**2

Holt's Two-Parameter Exponential Smoothing Method

The exponential smoothing technique presented in Section 15.2 (single exponential smoothing) is appropriate to use in forecasting stationary time-series data but is ineffective in forecasting time-series data with a trend because the forecasts will lag behind the trend. However, another exponential smoothing technique, Holt's two-parameter exponential smoothing method, can be used for trend analysis. Holt's technique uses weights (β) to smooth the trend in a manner similar to the smoothing used in single exponential smoothing (α). Using these two weights and several equations, Holt's method is able to develop forecasts that include both a smoothing value and a trend value.

15.3 PROBLEMS

15.10 The Office for National Statistics compiles data on household expenditure on tobacco in the United Kingdom. Shown here are the household expenditures on tobacco at current prices in the UK over the past 22 years. Use a computer to develop a regression model to fit the trend effects for these data. Use a linear model and then try a quadratic model. How well does either model fit the data?

Year	Household Expenditure on Tobacco (£ millions)	Year	Household Expenditure on Tobacco (£ millions)
1	8,170	12	14,222
2	8,649	13	14,458
3	9,648	14	14,622
4	10,280	15	15,266
5	10,759	16	15,305
6	10,933	17	15,377
7	11,519	18	15,649
8	12,265	19	15,653
9	12,648	20	15,650
10	13,363	21	16,356
11	14,292	22	17,668

15.11   The data below on the number of people employed in Agriculture in Finland for the years 1985 through 2010 are provided by the Organization for Economic Cooperation and Development. Using regression techniques discussed in this section, analyse the data for trend. Develop a scatter plot of the data and fit the trend line through the data. Discuss the strength of the model.

Year	Employment in Agriculture (000s)	Year	Employment in Agriculture (000s)
1985	279.4	1998	143.7
1986	265.7	1999	144
1987	251.5	2000	142.1
1988	237.6	2001	135
1989	217.8	2002	126.7
1990	206.8	2003	120.4
1991	197.7	2004	116.2
1992	186.8	2005	115.5
1993	173.5	2006	114
1994	167.4	2007	112.8
1995	158.1	2008	114.6
1996	147.7	2009	118.6
1997	152.7	2010	115.3

15.12   Shown below are the average number of construction permits granted for buildings and construction work (residential and non-residential buildings) in Germany for a recent nine-year period and published by the German Federal Statistical Office. Plot the data, fit a trend line, and discuss the strength of the regression model. In addition, explore a quadratic trend and compare the results of the two models.

Year	Permits Granted
1	24,898
2	22,641
3	20,175
4	20,653
5	15,685
6	15,337
7	15,377
8	15,942
9	17,018

15.4   SEASONAL EFFECTS

Earlier in the chapter, we discussed the notion that time-series data consist of four elements: trend, cyclical effects, seasonality, and irregularity. In this section, we examine techniques for identifying seasonal effects. **Seasonal effects** are *patterns of data behaviour that occur in periods of time of less than one year.* How can we separate out the seasonal effects?

Decomposition

One of the main techniques for isolating the effects of seasonality is **decomposition**. The decomposition methodology presented here uses the multiplicative model as its basis. The multiplicative model is:

$$T \cdot C \cdot S \cdot I$$

where:
   T = trend
   C = cyclicality
   S = seasonality
   I = irregularity

EU27 Exports (€ millions)

Year	Quarter	Shipments
1	1	623,855
	2	636,003
	3	592,653
	4	644,932
2	1	668,380
	2	664,167
	3	641,195
	4	687,222
3	1	704,570
	2	716,329
	3	674,936
	4	622,189
4	1	545,629
	2	533,511
	3	540,789
	4	576,979
5	1	596,438
	2	637,083
	3	627,934
	4	676,929

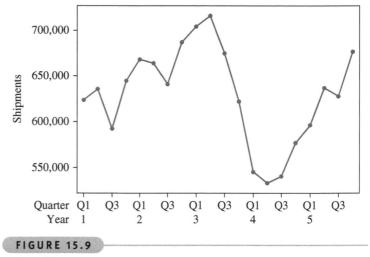

FIGURE 15.9

Minitab Time-Series Graph of EU27 Total Exports Data

To illustrate the decomposition process, we will use the five-year quarterly time-series data on the 27 country European Union (EU27) total exports in millions of euros given in Table 15.7. Figure 15.9 provides a graph of these data.

According to the multiplicative time-series model, $T \cdot C \cdot S \cdot I$, the data can contain the elements of trend, cyclical effects, seasonal effects, and irregular fluctuations. The process of isolating the seasonal effects begins by determining $T \cdot C$ for each value and dividing the time-series data $(T \cdot C \cdot S \cdot I)$ by $T \cdot C$. The result is

$$\frac{T \cdot C \cdot S \cdot I}{T \cdot C} = S \cdot I$$

The resulting expression contains seasonal effects along with irregular fluctuations. After reducing the time-series data to the effects of *SI* (seasonality and irregularity), a method for eliminating the irregular fluctuations can be applied, leaving only the seasonal effects.

Suppose we start with time-series data that cover several years and are measured in quarterly increments. If we average the data over four quarters, we will have 'dampened' the seasonal effects of the data because the rise and fall of values during the quarterly periods will have been averaged out over the year.

We begin by computing a four-quarter moving average for quarter 1 through quarter 4 of year 1, using the data from Table 15.7.

$$\text{Four-quarter average} = \frac{623{,}855 + 636{,}003 + 592{,}653 + 644{,}932}{4} = \frac{2{,}497{,}443}{4} = 624{,}360.8$$

The four-quarter moving average for quarter 1 through quarter 4 of year 1 is €624,360.8 million-worth of exports. Because the four-quarter average is in the middle of the four quarters, it would be placed in the decomposition table between quarter 2 and quarter 3.

Quarter 1

Quarter 2

— 624,360.8

Quarter 3

Quarter 4

TABLE 15.8

Development of Four-Quarter Moving Averages for the EU27 Total Exports Data

Quarter	Actual Values $(T \times C \times S \times I)$	Four-Quarter Moving Total	Eight-Quarter (Two-Year) Moving Total	Four-Quarter Centred Moving Average $(T \times C)$	Ratios of Actual Values to Moving Averages $(S \times I) \times (100)$
1 (year 1)	623,855				
2	636,003				
3	592,653	2,497,443	5,039,412	629,926	94.08
4	644,932	2,541,968	5,112,101	639,013	100.93
1 (year 2)	668,380	2,570,132	5,188,807	648,601	103.05
2	664,167	2,618,675	5,279,639	659,955	100.64
3	641,195	2,660,964	5,358,118	669,765	95.73
4	687,222	2,697,154	5,446,469	680,809	100.94
1 (year 3)	704,570	2,749,315	5,532,371	691,546	101.88
2	716,329	2,783,056	5,501,078	687,635	104.17
3	674,936	2,718,023	5,277,104	659,638	102.32
4	622,189	2,559,082	4,935,345	616,918	100.85
1 (year 4)	545,629	2,376,264	4,618,381	577,298	94.51
2	533,511	2,242,117	4,439,024	554,878	96.15
3	540,789	2,196,907	4,444,623	555,578	97.34
4	576,979	2,247,716	4,599,004	574,875	100.37
1 (year 5)	596,438	2,351,288	4,789,721	598,715	99.62
2	637,083	2,438,433	4,976,816	622,102	102.41
3	627,934	2,538,383			
4	676,929				

To remove seasonal effects, we need to determine a value that is 'centred' with each month. To find this value, instead of using a four-quarter moving average, we use four-quarter moving totals and then sum two consecutive moving totals. This eight-quarter total value is divided by 8 to produce a 'centred' four-quarter moving average that lines up across from a quarter. Using this method is analogous to computing two consecutive four-quarter moving averages and averaging them, thus producing a value that falls on line with a quarter, in between the two averages. The results of using this procedure on the data from Table 15.7 are shown in Table 15.8 in column 5.

A four-quarter moving total can be computed on these data starting with quarter 1 of year 1 through quarter 4 of year 1 as follows:

$$First\ Moving\ Total = 623,855 + 636,033 + 592,653 + 644,932 = 2,497,443$$

In Table 15.8, 16,498 is between quarter 2 and quarter 3 of year 1. The four-month moving total for quarter 2 of year 1 through quarter 1 of year 2 is

$$Second\ Moving\ Total = 636,033 + 592,653 + 644,932 + 668,380 = 2,541,968$$

In Table 15.8, this value is between quarter 3 and quarter 4 of year 1. The eight-quarter (two-year) moving total is computed for quarter 3 of year 1 as

$$Eight-Quarter\ Moving\ Total = 2,497,443 + 2,541,968 = 5,039,412$$

Notice that in Table 15.8 this value is centred with quarter 3 of year 1 because it is between the two adjacent four-quarter moving totals. Dividing this total by 8 produces the four-quarter moving average for quarter 3 of year 1 shown in column 5 of Table 15.8.

$$\frac{5,039,412}{8} = 629,926$$

Column 3 contains the uncentred four-quarter moving totals, column 4 contains the two-year centred moving totals, and column 5 contains the four-quarter centred moving averages.

The four-quarter centred moving averages shown in column 5 of Table 15.8 represent $T \cdot C$. Seasonal effects have been removed from the original data (actual values) by summing across the four-quarter periods. Seasonal effects are removed when the data are summed across the time periods that include the seasonal periods and the irregular effects are smoothed, leaving only trend and cycle.

Column 2 of Table 15.8 contains the original data (actual values), which include all effects ($T \cdot C \cdot S \cdot I$). Column 5 contains only the trend and cyclical effects, $T \cdot C$. If column 2 is divided by column 5, the result is $S \cdot I$, which is displayed in column 6 of Table 15.8.

The values in column 6, sometimes called ratios of actuals to moving average, have been multiplied by 100 to index the values. These values are thus seasonal indexes.

TABLE 15.9

Seasonal Indexes for the EU27 Total Exports Data

Quarter	Year 1	Year 2	Year 3	Year 4	Year 5
1	–	103.05	101.88	94.51	99.62
2	–	100.64	104.17	96.15	102.41
3	94.08	95.73	102.32	97.34	–
4	100.93	100.94	100.85	100.37	–

TABLE 15.10

Final Seasonal Indexes for the EU27 Total Exports Data

Quarter	Index
1	100.83
2	101.60
3	96.61
4	100.97

An **index number** is *a ratio of a measure taken during one time frame to that same measure taken during another time frame, usually denoted as the time period.* Often the ratio is multiplied by 100 and expressed as a percentage. Index numbers will be discussed more fully in section 15.6. Column 6 contains the effects of seasonality and irregular fluctuations. Now we must remove the irregular effects.

Table 15.9 contains the values from column 6 of Table 15.8 organized by quarter and year. Each quarter in these data has four seasonal indexes. Throwing out the high and low index for each quarter eliminates the extreme values. The remaining two indexes are averaged as follows for quarter 1:

Quarter 1:	103.05	101.88	94.51	99.62
Eliminate:		94.51	and	103.05
Average the Remaining Indexes:				

$$\bar{X}_{Q1index} = \frac{101.88 + 99.62}{2} = 100.75$$

Table 15.10 gives the final seasonal indexes for all the quarters of these data, after adjusting each seasonal index by 1.000747 (the small adjustment of each index is required because the sum of the original indexes was not 400).

After the final adjusted seasonal indexes are determined, the original data can be **deseasonalized.** The deseasonalization of actual values is relatively common with data published by the government and other agencies. Data can be deseasonalized by dividing the actual values, which consist of $T \cdot C \cdot S \cdot I$, by the final adjusted seasonal effects.

$$\text{Deseasonalized Data} = \frac{T \cdot C \cdot S \cdot I}{S} = T \cdot C \cdot I$$

Because the seasonal effects are in terms of index numbers, the seasonal indexes must be divided by 100 before deseasonalization. Shown here are the computations for deseasonalizing the EU27 total exports data from Table 15.7 for quarter 1 of year 1.

$$\text{Year 1 Quarter 1 Actual} = 623{,}855$$

$$\text{Year 1 Quarter 1 Seasonal Index} = 100.83$$

$$\text{Year 1 Quarter 1 Deseasonalised Value} = \frac{623{,}855}{1.0083} = 618{,}719.6$$

TABLE 15.11

Deseasonalized EU27 Total Exports Data

Year	Quarter	Shipments Actual Values $(T \times C \times S \times I)$	Seasonal Indexes S	Deseasonalized Data $T \times C \times I$
1	1	623,855	100.83	618,720
	2	636,003	101.60	625,987
	3	592,653	96.61	613,449
	4	644,932	100.97	638,736
2	1	668,380	100.83	662,878
	2	664,167	101.60	653,708
	3	641,195	96.61	663,694
	4	687,222	100.97	680,620
3	1	704,570	100.83	698,770
	2	716,329	101.60	705,048
	3	674,936	96.61	698,619
	4	622,189	100.97	616,212
4	1	545,629	100.83	541,138
	2	533,511	101.60	525,109
	3	540,789	96.61	559,765
	4	576,979	100.97	571,436
5	1	596,438	100.83	591,528
	2	637,083	101.60	627,050
	3	627,934	96.61	649,968
	4	676,929	100.97	670,426

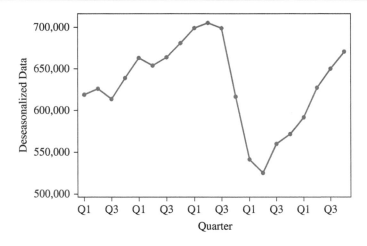

FIGURE 15.10

Graph of the Deseasonalized EU27 Total Exports Data

Table 15.11 gives the deseasonalized data for this example for all years. Figure 15.10 is a graph of the deseasonalized data.

Finding Seasonal Effects with the Computer

Through Minitab, decomposition can be performed on the computer with relative ease. The commands for this procedure are given at the end of the chapter in the Using the Computer section. Figure 15.11 displays Minitab output for seasonal decomposition of the EU27 total exports example. Note that the seasonal indexes are virtually identical to those shown in Table 15.10 computed by hand.

Winters' Three-Parameter Exponential Smoothing Method

Holt's two-parameter exponential smoothing method can be extended to include seasonal analysis. This technique, referred to as Winters' method, not only smoothes observations and trend but also smoothes the seasonal effects. In addition to the single exponential smoothing weight of α and the trend weight of β, Winters' method introduces γ, a weight for seasonality. Using these three weights and several equations, Winters' method is able to develop forecasts that include a smoothing value for observations, a trend value, and a seasonal value.

```
Time-Series Decomposition for EU27 Exports
Multiplicative Model
Data            EU27 Exports
Length          20
NMissing        0
Seasonal        Indices

Period          Index
1               1.00827
2               1.01599
3               0.96608
4               1.00966
```

FIGURE 15.11

Minitab Output for Seasonal Decomposition of the EU27 Total Exports Data

15.4 PROBLEMS

15.13 The Organization for Economic Cooperation and Development publishes statistics on the unemployed level of various countries by month. Shown here are the unemployment figures of Sweden for January of a recent year through December of the next year. Use these data to compute 12-month centred moving averages $(T \cdot C)$. Using these computed values, determine the seasonal effects $(S \cdot I)$.

Month	Number of Unemployed	Month	Number of Unemployed
January (1st year)	350,700	January (2nd year)	451,600
February	387,300	February	446,700
March	403,600	March	440,100
April	402,800	April	472,100
May	445,600	May	434,200
June	502,600	June	488,100
July	398,100	July	414,600
August	392,800	August	370,100
September	404,800	September	385,000
October	394,700	October	368,800
November	390,100	November	349,000
December	416,600	December	363,500

15.14 Eurostat publishes monthly data on cows' milk collection and the production of other agricultural products for the European Union. The cows' milk collection figures are given in thousands of tonnes. Use the data to analyse the effects of seasonality, trend, and cycle. Develop the trend model with a linear model only.

Month	Cows' Milk Collection	Month	Cows' Milk Collection
January (year 1)	897.581	January (year 4)	951.649
February	808.805	February	897.729
March	861.762	March	940.504
April	891.831	April	912.388

(continued)

Month	Cows' Milk Collection	Month	Cows' Milk Collection
May	910.296	May	951.719
June	873.726	June	900.677
July	880.656	July	894.520
August	867.262	August	876.301
September	838.819	September	869.568
October	859.484	October	902.843
November	826.092	November	885.840
December	891.113	December	952.246

Month	Cows' Milk Collection	Month	Cows' Milk Collection
January (year 2)	914.921	January (year 5)	1,004.200
February	835.796	February	918.000
March	913.167	March	964.100
April	890.999	April	986.800
May	935.610	May	1,007.600
June	891.537	June	969.200
July	882.065	July	957.200
August	875.634	August	940.300
September	839.468	September	907.500
October	877.451	October	931.000
November	848.680	November	912.300
December	920.244	December	970.400

Month	Cows' Milk Collection	Month	Cows' Milk Collection
January (year 3)	941.600	January (year 6)	982.700
February	850.400	February	916.900
March	933.900	March	1,016.300
April	916.400	April	998.200
May	937.100	May	1,023.800
June	876.800	June	979.500
July	874.300	July	982.300
August	873.400	August	972.000
September	842.800	September	923.300
October	888.300	October	945.200
November	872.200	November	923.700
December	930.000	December	969.600

15.5   AUTOCORRELATION AND AUTOREGRESSION

Data values gathered over time are often correlated with values from past time periods. This characteristic can cause problems in the use of regression in forecasting and at the same time can open some opportunities. One of the problems that can occur in regressing data over time is autocorrelation.

Autocorrelation

Autocorrelation, or **serial correlation**, occurs in data *when the error terms of a regression forecasting model are correlated.* The likelihood of this occurring with business data increases over time, particularly with economic variables. Autocorrelation can be a problem in using regression analysis as the forecasting method because one of the assumptions underlying regression analysis is that the error terms are independent or random (not correlated). In most business analysis situations, the correlation of error terms is likely to occur as positive autocorrelation (positive errors are associated with positive errors of comparable magnitude and negative errors are associated with negative errors of comparable magnitude).

When autocorrelation occurs in a regression analysis, several possible problems might arise. First, the estimates of the regression coefficients no longer have the minimum variance property and may be inefficient. Second, the variance of the error terms may be greatly underestimated by the mean square error value. Third, the true standard deviation of the estimated regression coefficient may be seriously underestimated. Fourth, the confidence intervals and tests using the t and F distributions are no longer strictly applicable.

First-order autocorrelation results from correlation between the error terms of adjacent time periods (as opposed to two or more previous periods). If first-order autocorrelation is present, the error for one time period, e_t, is a function of the error of the previous time period, e_{t-1}, as follows:

$$e_t = \rho e_{t-1} + v_t$$

The first-order autocorrelation coefficient, ρ, measures the correlation between the error terms. It is a value that lies between -1 and 0 and $+1$, as does the coefficient of correlation discussed in Chapter 12. v_t is a normally distributed independent error term. If positive autocorrelation is present, the value of ρ is between 0 and $+1$. If the value of ρ is 0, $e_t = v_t$, which means there is no autocorrelation and e_t is just a random, independent error term.

One way to *test to determine whether autocorrelation is present in a time-series regression analysis* is by using the **Durbin–Watson test** for autocorrelation. Shown next is the formula for computing a Durbin–Watson test for autocorrelation.

DURBIN–WATSON TEST	$$D = \dfrac{\displaystyle\sum_{t=2}^{n}(e_t - e_{t-1})^2}{\displaystyle\sum_{t=1}^{n} e_i^2}$$

where:

n = the number of observations

Note from the formula that the Durbin–Watson test involves finding the difference between successive values of error $(e_t - e_{t-1})$. If errors are positively correlated, this difference will be smaller than with random or independent errors. Squaring this term eliminates the cancellation effects of positive and negative terms.

The null hypothesis for this test is that there is *no* autocorrelation. For a two-tailed test, the alternative hypothesis is that there *is* autocorrelation.

$$H_0: \rho = 0$$
$$H_a: \rho \neq 0$$

As mentioned before, most business forecasting autocorrelation is positive autocorrelation. In most cases, a one-tailed test is used.

$$H_0: \rho = 0$$
$$H_a: \rho > 0$$

In the Durbin–Watson test, D is the observed value of the Durbin–Watson statistic using the residuals from the regression analysis. A critical value for D can be obtained from the values of α, n, and k by using Table A.9 in Appendix A, where α is the level of significance, n is the number of data items, and k is the number of predictors. Two Durbin–Watson tables are given in the appendix. One table contains values for $\alpha = 0.01$ and the other for $\alpha = 0.05$. The Durbin–Watson tables in Appendix A include values for d_U and d_L. These values range from 0 to 4. If the observed value of D is above d_U, we fail to reject the null hypothesis and there is no significant autocorrelation. If the observed value of D is below d_L, the null hypothesis is rejected and there is autocorrelation. Sometimes the observed statistic, D, is between the values of d_U and d_L. In this case, the Durbin–Watson test is inconclusive.

As an example, consider Table 15.12, which contains estimates published each year by the Department of Trade & Industry, of the level of oil and gas reserves in the UK over a 21-year period. A regression line can be fit through these data to determine whether the amount of gas reserves can be predicted by the amount of oil reserves. The resulting errors of prediction can be tested by the Durbin–Watson statistic for the presence of significant positive autocorrelation by using $\alpha = 0.05$. The hypotheses are

$$H_0: \rho = 0$$
$$H_a: \rho > 0$$

TABLE 15.12

Estimates of UK level of recoverable Oil and Gas over a 21-Year Period

Year	Oil (million tonnes)	Gas (billion cubic metres)
1	8.597	1,770
2	8.572	1,780
3	8.649	1,805
4	8.688	1,865
5	8.879	1,915
6	8.971	1,910
7	8.680	1,915
8	8.349	1,960
9	8.140	1,980
10	7.613	1,795
11	7.355	1,750
12	7.417	1,630
13	7.171	1,535
14	6.847	1,329
15	6.662	1,241
16	6.560	1,169
17	6.465	1,006
18	6.452	967
19	6.252	940
20	5.881	907
21	5.822	840

The following regression equation was obtained by means of a Minitab computer analysis:

$$\text{Gas Reserves} = -231.124 + 1.0834 \text{ Oil Reserves}$$

Using the values for oil reserves (X) from Table 15.12 and the regression equation shown here, predicted values of Y (gas reserves) can be computed. From the predicted values and the actual values, the errors of prediction for each time interval, e_t, can be calculated. Table 15.13 shows the values of $\hat{Y}$, e_t, e_t^2 ($e_t - e_{t-1}$), and $(e_t - e_{t-1})^2$ for this example. Note that the first predicted value of Y is

$$\hat{Y}_1 = -231.124 + 1.0834(1,810) = 1,730,92$$

The error for year 1 is

$$\text{Actual}_1 - \text{Predicted}_1 = 1,770 - 1,730.92 = 39.08$$

The value of $e_t - e_t - 1$ for year 1 and year 2 is computed by subtracting the error for year 1 from the error of year 2.

$$e_{year2} - e_{year1} = 43.66 - 39.08 = 4.58$$

The Durbin–Watson statistic can now be computed:

$$D = \frac{\sum_{t=2}^{n}(e_t - e_{t-1})^2}{\sum_{t=1}^{n} e_t^2} = \frac{140{,}921.12}{316{,}270.16} = 0.4456$$

TABLE 15.13

Predicted Values and Error Terms for the Oil and Gas Reserves Data

Year	$\hat{Y}$	e_t	e_t^2	$e_t - e_{t-1}$	$(e_t - e_{t-1})^2$
1	1,730.92	39.08	1,527.56	–	–
2	1,736.34	43.66	1,906.54	4.58	20.98
3	1,893.52	−88.52	7,835.08	−132.18	17,471.55
4	2,018.18	−153.18	23,462.89	−64.66	4,180.92
5	2,039.86	−124.86	15,589.02	28.32	802.02
6	2,018.18	−108.18	11,702.05	16.68	278.22
7	1,817.64	97.36	9,479.75	205.54	42,246.69
8	1,963.98	−3.98	15.81	−101.34	10,269.80
9	1,953.14	26.86	721.67	30.84	951.11
10	1,720.08	74.92	5,613.61	48.06	2,309.76
11	1,573.74	176.26	31,069.00	101.34	10,269.80
12	1,384.04	245.96	60,498.29	69.70	4,858.09
13	1,319.00	216.00	46,657.73	−29.96	897.60
14	1,225.77	103.23	10,656.02	−112.78	12,718.43
15	1,142.30	98.70	9,740.90	−4.53	20.54
16	1,208.43	−39.43	1,554.57	−138.12	19,078.24
17	1,142.30	−136.30	18,578.78	−96.88	9,384.96
18	1,128.21	−161.21	25,989.31	−24.91	620.41
19	1,046.91	−106.91	11,430.18	54.30	2,948.49
20	993.80	−86.80	7,533.55	20.12	404.65
21	961.28	−121.28	14,707.87	−34.48	1,188.87
			$\sum e_t^2 = 316{,}270.16$		$\sum(e_t - e_{t-1})^2 = 140{,}921.12$

Because we used a simple linear regression, the value of k is 1. The sample size, n, is 21, and $\alpha = 0.05$. The critical values in Table A.9 are

$$d_U = 1.42 \text{ and } d_L = 1.22$$

Because the computed D statistic, 0.4456, is less than the value of $d_L = 1.22$, the null hypothesis is rejected. A positive autocorrelation is present in this example.

Ways to Overcome the Autocorrelation Problem

Several approaches to data analysis can be used when autocorrelation is present. One uses additional independent variables and another transforms the independent variable.

Addition of Independent Variables

Often the reason autocorrelation occurs in regression analyses is that one or more important predictor variables have been left out of the analysis. For example, suppose a researcher develops a regression forecasting model that attempts to predict sales of new homes by sales of used homes over some period of time. Such a model might contain significant autocorrelation. The exclusion of the variable 'prime mortgage interest rate' might be a factor driving the autocorrelation between the other two variables. Adding this variable to the regression model might significantly reduce the autocorrelation.

Transforming Variables

When the inclusion of additional variables is not helpful in reducing autocorrelation to an acceptable level, transforming the data in the variables may help to solve the problem. One such method is the **first-differences approach.** With the first-differences approach, *each value of X is subtracted from each succeeding time period value of X*; these 'differences' become the new and transformed X variable. The same process is used to transform the Y variable. The regression analysis is then computed on the transformed X and transformed Y variables to compute a new model that is hopefully free of significant autocorrelation effects.

Another way is to generate new variables by using the percentage changes from period to period and regressing these new variables. A third way is to use autoregression models.

Autoregression

A forecasting technique that takes advantage of the relationship of values (Y_t) to previous-period values $(Y_{t-1}, Y_{t-2}, Y_{t-3}, \ldots)$ is called autoregression. **Autoregression** is *a multiple regression technique in which the independent variables are time-lagged versions of the dependent variable,* which means we try to predict a value of Y from values of Y from previous time periods. The independent variable can be lagged for one, two, three, or more time periods. An autoregressive model containing independent variables for three time periods looks like this:

$$\hat{Y} = b_0 + b_1 Y_{t-1} + b_2 Y_{t-2} + b_3 Y_{t-3}$$

As an example, we shall attempt to predict the volume of gas reserves, displayed in Table 15.12, by using data lagged for both one and two time periods. The data used in this analysis are displayed in Table 15.14. Using Excel, a multiple regression model is developed to predict the values of Y_t by the values of Y_{t-1} and Y_{t-2}. The results appear in Figure 15.12. Note that the regression analysis does not use data from years 1 and 2 of Table 15.14 because there are no values for the two lagged variables for one or both of those years.

The autoregression model is

$$Y_t = 794.5946 + 0.6225 Y_{t-1} - 0.1348 Y_{t-2}$$

The relatively low value of R^2 (41.1) and relatively high value of s_e(328.867) indicate that this regression model has only weak predictability. Interestingly, the one-period lagged variable is quite significant ($t = 2.89$ with a p-value of 0.009857), but the two-period lagged variable is not significant ($t = -0.73$ with a p-value of 0.476278) indicating the presence of first-order autocorrelation.

TABLE 15.14

Time-Lagged Gas Reserves Data

Year	Gas Reserves Y_t	One-Period Lagged $Y_{t-1}(X_1)$	Two-Period Lagged $Y_{t-2}(X_2)$
1	1,770	–	–
2	1,780	1,770	–
3	1,805	1,780	1,770
4	1,865	1,805	1,780
5	1,915	1,865	1,805
6	1,910	1,915	1,865
7	1,915	1,910	1,915
8	1,960	1,915	1,910
9	1,980	1,960	1,915
10	1,795	1,980	1,960
11	1,750	1,795	1,980
12	1,630	1,750	1,795
13	1,535	1,630	1,750
14	1,329	1,535	1,630
15	1,241	1,329	1,535
16	1,169	1,241	1,329
17	1,006	1,169	1,241
18	967	1,006	1,169
19	940	967	1,006
20	907	940	967
21	840	907	940

SUMMARY OUTPUT

Regression Statistics	
Multiple R	0.643
R Square	0.414
Adjusted R Square	0.349
Standard Error	328.867
Observations	21.000

ANOVA

	df	SS	MS	F	Significance F
Regression	2	1375505	687752.7	6.36	0.00815
Residual	18	1946764	108153.6		
Total	20	3322270			

	Coefficients	Standard Error	t Stat	P-value
Intercept	794.5946	228.8297	3.47	0.002718
Lagged 1	0.6225	0.2158	2.89	0.009857
Lagged 2	−0.1348	0.1853	−0.73	0.476278

FIGURE 15.12

Excel Autoregression Results for Gas Reserves Data

Autoregression can be a useful tool in locating seasonal or cyclical effects in time series data. For example, if the data are given in monthly increments, autoregression using variables lagged by as much as 12 months can search for the predictability of previous monthly time periods. If data are given in quarterly time periods, autoregression of up to four periods removed can be a useful tool in locating the predictability of data from previous quarters. When the time periods are in years, lagging the data by yearly periods and using autoregression can help in locating cyclical predictability.

15.5 PROBLEMS

15.15 Eurostat publishes consumer price indexes (CPIs) on many commodities. Following are the CPIs for food and for wine over the period February 2009 to July 2011 for the European Union (27 countries). Use these data to develop a linear regression model to forecast the percentage change in food CPIs by the percentage change in wine CPIs. Compute a Durbin–Watson statistic to determine whether significant autocorrelation is present in the model. Let $\alpha = 0.05$.

Month and Year	Food	Wine	Month and Year	Food	Wine
Feb. 2009	115.34	108.62	May 2010	115.30	110.63
Mar. 2009	115.22	108.55	Jun. 2010	115.35	110.68
Apr. 2009	115.15	109.00	Jul. 2010	115.38	110.90
May 2009	115.26	109.67	Aug. 2010	114.99	110.99
Jun. 2009	114.72	109.42	Sept. 2010	115.19	111.15
Jul. 2009	113.82	109.60	Oct. 2010	115.51	111.65
Aug. 2009	112.93	110.04	Nov. 2010	116.22	111.04
Sept. 2009	112.72	110.11	Dec. 2010	116.92	110.44
Oct. 2009	112.88	110.26	Jan. 2011	117.46	111.96
Nov. 2009	113.26	109.95	Feb. 2011	118.28	111.90
Dec. 2009	113.55	109.63	Mar. 2011	118.80	112.52
Jan. 2010	114.36	110.61	Apr. 2011	119.14	112.95
Feb. 2010	114.60	110.12	May 2011	119.91	113.74
Mar. 2010	115.10	110.31	Jun. 2011	119.67	113.44
Apr. 2010	115.49	110.82	Jul. 2011	119.10	113.58

15.16 Use the data from Problem 15.15 to create a regression forecasting model using the first-differences data transformation. How do the results from this model differ from those obtained in Problem 15.15?

15.17 The OECD publishes trade, prices, and macroeconomic data of its member countries. Below are recent data on Austria's real GDP and female unemployment. Use these data to develop a simple regression forecasting model that attempts to predict the number of female unemployed by the real GDP. Compute a Durbin–Watson statistic for this regression model and determine whether significant autocorrelation is present. Let $\alpha = 0.05$.

Year	Real GDP	Female Unemployment	Year	Real GDP	Female Unemployment
1	132,924.4	1,218,000	16	187,023.7	1,590,000
2	136,849.4	1,213,000	17	193,747.6	1,598,000
3	136,932.4	1,286,000	18	200,218.8	1,623,000
4	140,295.9	1,277,000	19	207,528.8	1,631,000
5	143,546.0	1,301,000	20	208,608.2	1,639,000
6	145,478.7	1,319,000	21	212,044.8	1,682,000
7	149,652.4	1,335,000	22	213,743.5	1,705,000
8	155,251.7	1,355,000	23	219,182.4	1,682,000
9	161,727.1	1,393,000	24	224,573.9	1,729,200
10	167,125.9	1,424,000	25	232,655.2	1,780,675

(continued)

Year	Real GDP	Female Unemployment	Year	Real GDP	Female Unemployment
11	170,280.9	1,475,000	26	241,332.0	1,819,325
12	170,918.3	1,494,000	27	246,590.1	1,867,825
13	174,700.6	1,581,000	28	237,002.1	1,892,050
14	179,136.5	1,596,000	29	241,650.8	1,899,000
15	183,131.1	1,579,000			

15.18 Use the data in Problem 15.17 to compute a regression model after recoding the data by the first-differences approach. Compute a Durbin–Watson statistic to determine whether significant autocorrelation is present in this first-differences model. Compare this model with the model determined in Problem 15.17, and compare the significance of the Durbin–Watson statistics for the two problems. $\alpha = 0.05$.

15.19 The World Bank publishes data on arable land for about 240 countries or groups of countries. Data on arable land in China (millions of hectares) between 1980 and 2008 follow. Use these time-series data to develop an autoregression model with a one-period lag. Now try an autoregression model with a two-period lag. Discuss the results and compare the two models.

Year	Arable Land (millions of hectares)	Year	Arable Land (millions of hectares)
1980	97.0	1995	122.1
1981	97.6	1996	121.4
1982	97.8	1997	120.2
1983	111.3	1998	121.6
1984	112.0	1999	121.1
1985	120.9	2000	121.0
1986	120.9	2001	120.1
1987	121.6	2002	119.0
1988	122.2	2003	116.7
1989	123.3	2004	122.2
1990	123.7	2005	118.4
1991	124.2	2006	118.2
1992	124.4	2007	109.4
1993	122.0	2008	108.6
1994	121.8		

15.20 The European Commission publishes data on the number of outgoing Erasmus students in the European Union. Shown here are the number of outgoing Italian Erasmus students for a 23-year period. Use these data to develop an autoregression forecasting model with a two-period lag. Discuss the results of this analysis.

Year	Number of Italian Erasmus Students	Year	Number of Italian Erasmus Students
1	220	13	12,421
2	1,365	14	13,253
3	2,295	15	13,950
4	3,355	16	15,225
5	4,202	17	16,829
6	5,308	18	16,440
7	6,808	19	16,389
8	7,217	20	17,195
9	8,969	21	17,562
10	8,907	22	17,754
11	9,271	23	19,118
12	10,875		

15.6   INDEX NUMBERS

One particular type of descriptive measure that is useful in allowing comparisons of data over time is the index number. An index number is, in part, a ratio of a measure taken during one time frame to that same measure taken during another time frame, usually denoted as the base period. Often the ratio is multiplied by 100 and is expressed as a percentage. When expressed as a percentage, index numbers serve as an alternative to comparing raw numbers. Index number users become accustomed to interpreting measures for a given time period in light of a base period on a scale in which the base period has an index of 100(%). Index numbers are used to compare phenomena from one time period to another and are especially helpful in highlighting inter-period differences.

Index numbers are widely used around the world to relate information about stock markets, inflation, sales, exports and imports, agriculture, and many other things. Some examples of specific indexes are the employment cost index, price index for construction, index of manufacturing capacity, producer price index, consumer price index, Euro Stoxx 50, index of output, and Nikkei 225 average. This section, although recognizing the importance of stock indexes and others, will focus on price indexes.

The motivation for using an index number is to reduce data to an easier-to-use, more convenient form. As an example, examine the raw data on number of company liquidations in England and Wales from 1990 through 2010 shown in Table 15.15. An analyst can describe these data by observing that, in general, the number of company liquidations has had a general decreasing trend since the early 1990s but that that tendency has not been regular. How do the number of company liquidations in 2008 compare with 1998? How do the number of company liquidations in 2000 compare to 1990 with 1992? To answer these questions without index numbers, a business researcher would probably resort to subtracting the number of company liquidations for the years of interest and comparing the corresponding increases or decreases. This process can be tedious and frustrating for decision makers who must maximize their effort in minimal time. Using simple index numbers, the business researcher can transform these data into values that are more usable and make it easier to compare other years to one particular key year.

TABLE 15.15

Company Liquidations in England and Wales

Year	Company Liquidations
1990	15,051
1991	21,827
1992	24,425
1993	20,708
1994	16,728
1995	14,536
1996	13,461
1997	12,610
1998	13,203
1999	14,280
2000	14,317
2001	14,972
2002	16,306
2003	14,184
2004	12,192
2005	12,893
2006	13,137
2007	12,507
2008	15,535
2009	19,077
2010	16,045

Simple Index Numbers

How are index numbers computed? The equation for computing a **simple index number** follows.

SIMPLE INDEX NUMBER

$$I_i = \frac{X_i}{X_0}(100)$$

where:

$X_0 = $ the quantity, price, or cost in the base year
$X_i = $ the quantity, price, or cost in the year of interest
$I_i = $ the index number for the year of interest

Suppose bankruptcy researchers examining the data from Table 15.15 decide to compute index numbers using 1990 as the base year. The index number for the year 2000 is

$$I_{2000} = \frac{X_{2000}}{X_{1990}}(100) = \frac{14,317}{15,051}(100) = 95.1$$

TABLE 15.16

TABLE 15.16

Index Numbers for Company Liquidations in England and Wales

Year	Company Liquidations	Index Number
1990	15,051	100.0
1991	21,827	145.0
1992	24,425	162.3
1993	20,708	137.6
1994	16,728	111.1
1995	14,536	96.6
1996	13,461	89.4
1997	12,610	83.8
1998	13,203	87.7
1999	14,280	94.9
2000	14,317	95.1
2001	14,972	99.5
2002	16,306	108.3
2003	14,184	94.2
2004	12,192	81.0
2005	12,893	85.7
2006	13,137	87.3
2007	12,507	83.1
2008	15,535	103.2
2009	19,077	126.7
2010	16,045	106.6

Table 15.16 displays all the index numbers for the data in Table 15.15, with 1990 as the base year, along with the raw data. A cursory glance at these index numbers reveals a general decrease in the number of bankruptcies for most of the years since 1990 (because the index has been going down). In particular, the greatest drop in number seems to have occurred between 1993 and 1994 – a drop of 26 in the index. There have also been a number of years where the number of company liquidations were higher than those in the base year (index numbers over 100), the years between 1991 and 1994, 2002 and the years after 2008. Because most people are easily able to understand the concept of 100%, it is likely that decision makers can make quick judgements on the number of business bankruptcies in England and Wales from one year relative to another by examining the index numbers over this period.

Unweighted Aggregate Price Index Numbers

The use of simple index numbers makes possible the conversion of prices, costs, quantities, and so on for different time periods into a number scale with the base year equalling 100%. One of the drawbacks of simple index numbers, however, is that each time period is represented by only one item or commodity. When multiple items are involved, multiple sets of index numbers are possible. Suppose a decision maker is interested in combining or pooling the prices of several items, creating a 'market basket' in order to compare the prices for several years. Fortunately, a technique does exist for combining several items and determining index numbers for the total (aggregate). Because this technique is used mostly in determining price indexes, the focus in this section is on developing aggregate price indexes. The formula for constructing the **unweighted aggregate price index number** follows.

UNWEIGHTED AGGREGATE PRICE INDEX NUMBER

$$I_i = \frac{\sum P_i}{\sum P_0}(100)$$

where:

P_i = the price of an item in the year of interest (i)
P_0 = the price of an item in the base year (0)
I_i = the index number for the year of interest (i)

Suppose a Ministry of Labour wants to compare the cost of family food buying over the years. Ministry officials decide that instead of using a single food item to do this comparison, they will use a food basket that consists of five items: eggs, milk, bananas, potatoes, and sugar. They gathered price information on these five items for the years 1995, 2000, and 2012. The items and the prices are listed in Table 15.17.

From the data in Table 15.17 and the formula, the unweighted aggregate price indexes for the years 1995, 2000, and 2012 can be computed by using 1995 as the base year. The first step is to add together, or aggregate, the prices for all the food basket items in a given year. These totals are shown in the last row of Table 15.17. The index numbers are constructed by using these totals (not individual item prices): $\sum P_{1995} = 3.56$, $\sum P_{2000} = 4.22$, $\sum P_{2012} = 5.20$. From these figures, the unweighted aggregate price index for 2000 is computed as follows:

$$\text{For 2000: } I_{2000} = \frac{\sum P_{2000}}{\sum P_{1995}}(100) = \frac{4.22}{3.56}(100) = 118.32$$

Weighted Aggregate Price Index Numbers

A major drawback to unweighted aggregate price indexes is that they are *unweighted* – that is, equal weight is put on each item by assuming the market basket contains only one of each item. This assumption may or may not be true. For example, a household may consume 5 kilograms of bananas per year but drink 50 litres of milk. In addition, unweighted aggregate index numbers are dependent on the units selected for various items. For example, if milk is measured in pints instead of litres, the price of milk used in determining the index numbers is considerably lower. A class of index numbers that can be used to avoid these problems is weighted aggregate price index numbers.

TABLE 15.17

Prices for a Basket of Food Items

Item	1995	Year 2000	2012
Eggs (dozen)	0.78	0.86	1.15
Milk (per litre)	0.60	0.73	0.95
Bananas (per kg.)	0.79	1.01	1.12
Potatoes (per kg.)	0.62	0.68	0.88
Sugar (per kg.)	0.77	0.93	1.10
Total of Items	3.56	4.22	5.20

Weighted aggregate price index numbers are *computed by multiplying quantity weights and item prices in determining the market basket worth for a given year.* Sometimes when price and quantity are multiplied to construct index numbers, the index numbers are referred to as *value indexes.* Thus, weighted aggregate price index numbers are also value indexes.

Including quantities eliminates the problems caused by how many of each item are consumed per time period and the units of items. If 50 litres of milk but only 5 kilograms of bananas are consumed, weighted aggregate price index numbers will reflect those weights. If the business researcher switches from litres of milk to pints, the prices will change downward but the quantity will increase (1.7598 pints in a litre).

In general, weighted aggregate price indexes are constructed by multiplying the price of each item by its quantity and then summing these products for the market basket over a given time period (often a year). The ratio of this sum for one time period of interest (year) to a base time period of interest (base year) is multiplied by 100. The following formula reflects a weighted aggregate price index computed by using quantity weights from each time period (year):

$$I_i = \frac{\Sigma P_i Q_i}{\Sigma P_0 Q_0}(100)$$

One of the problems with this formula is the implication that new and possibly different quantities apply for each time period. However, business researchers expend much time and money ascertaining the quantities used in a market basket. Re-determining quantity weights for each year is therefore often prohibitive for most organizations (even the government). Two particular types of weighted aggregate price indexes offer a solution to the problem of which quantity weights to use. The first and most widely used is the Laspeyres price index. The second and less widely used is the Paasche price index.

Laspeyres Price Index

The **Laspeyres price index** is *a weighted aggregate price index computed by using the quantities of the base period (year) for all other years.* The advantages of this technique are that the price indexes for all years can be compared, and new quantities do not have to be determined for each year. The formula for constructing the Laspeyres price index follows.

LASPEYRES PRICE INDEX

$$I_L = \frac{\Sigma P_i Q_0}{\Sigma P_0 Q_0}(100)$$

Notice that the formula requires the base period quantities (Q_0) in both the numerator and the denominator.

TABLE 15.18

Food Basket Items with Quantity Weights

		Price	
Item	Quantity	1995	2012
Eggs (dozen)	45	0.78	1.15
Milk (per litre)	60	0.60	0.95
Bananas (per kg.)	12	0.79	1.12
Potatoes (per kg.)	55	0.62	0.88
Sugar (per kg.)	36	0.77	1.10

In Table 15.17, a food basket is presented in which aggregate price indexes are computed. This food basket consisted of eggs, milk, bananas, potatoes, and sugar. The prices of these items were combined (aggregated) for a given year and the price indexes were computed from these aggregate figures. The unweighted aggregate price indexes computed on these data gave all items equal importance. Suppose that the business researchers realize that applying equal weight to these five items is probably not a representative way to construct this food basket and consequently ascertain quantity weights on each food item for one year's consumption. Table 15.18 lists these five items, their prices, and their quantity usage weights for the base year (1995). From these data, the business researchers can compute Laspeyres price indexes.

The Laspeyres price index for 2012 with 1995 as the base year is:

$$\sum P_i Q_0 = \sum P_{2012} Q_{1995}$$
$$= \sum [(1.15)(45) + (0.95)(60) + (1.12)(12) + (0.88)(55) + (1.10)(36)]$$
$$= 51.75 + 57.00 + 13.44 + 48.40 + 39.60 = 210.19$$
$$\sum P_0 Q_0 = \sum P_{1995} Q_{1995}$$
$$= \sum [(0.78)(45) + (0.60)(60) + (0.79)(12) + (0.62)(55) + (0.77)(36)]$$
$$= 35.10 + 36.14 + 9.52 + 33.95 + 27.78 = 142.49$$

Paasche Price Index

The **Paasche price index** is *a weighted aggregate price index computed by using the quantities for the year of interest in computations for a given year.* The advantage of this technique is that it incorporates current quantity figures in the calculations. One disadvantage is that ascertaining quantity figures for each time period is expensive. The formula for computing Paasche price indexes follows.

PAASCHE PRICE INDEX	$I_p = \dfrac{\sum P_i Q_i}{\sum P_0 Q_i}(100)$

Suppose the yearly quantities for the basket of food items listed in Table 15.18 are determined. The result is the quantities and prices shown in Table 15.19 for the years 1995 and 2012 that can be used to compute Paasche price index numbers.

TABLE 15.19

Food Basket Items with Yearly Quantity Weights for 1995 and 2012

Item	P_{1995}	Q_{1995}	P_{2012}	Q_{2012}
Eggs (dozen)	0.78	45	1.15	42
Milk (per litre)	0.60	60	0.95	57
Bananas (per kg)	0.79	12	1.12	13
Potatoes (per kg)	0.62	55	0.88	52
Sugar (per kg)	0.77	36	1.1	36

The Paasche price index numbers can be determined for 2012 by using a base year of 1995 as follows.

For 2012:

$$\sum P_{2012}Q_{2012} = \sum[(1.15)(42) + (0.95)(57) + (1.12)(13) + (0.88)(52) + (1.10)(36)]$$

$$= 48.30 + 54.15 + 14.56 + 45.76 + 39.60 = 202.37$$

$$\sum P_{1995}Q_{2012} = \sum[(0.78)(42) + (0.60)(57) + (0.79)(13) + (0.62)(52) + (0.77)(36)]$$

$$= 32.76 + 34.33 + 10.32 + 32.10 + 27.78$$

$$= 137.29$$

$$I_{2012} = \frac{\sum P_{2012}Q_{2012}}{\sum P_{1995}Q_{2012}}(100) = \frac{202.37}{137.29}(100) = 147.41$$

DEMONSTRATION PROBLEM 15.5

The Southwest Paediatrics Clinic has been in business for 18 years. The office manager noticed that prices of clinic materials and office supplies fluctuate over time. To get a handle on the price trends for running the clinic, the office manager examined prices of six items the clinic uses as part of its operation. Shown here are the items, their prices, and the quantities for the years 2011 and 2012. Use these data to develop unweighted aggregate price indexes for 2012 with a base year of 2011. Compute the Laspeyres price index for the year 2012 using 2011 as the base year. Compute the Paasche index number for 2012 using 2011 as the base year.

	2011		**2012**	
Item	**Price**	**Quantity**	**Price**	**Quantity**
Syringes (dozen)	6.70	150	6.95	135
Cotton swabs (box)	1.35	60	1.45	65
Patient record forms (pad)	5.10	8	6.25	12
Children's Tylenol (bottle)	4.50	25	4.95	30
Computer paper (box)	11.95	6	13.20	8
Thermometers	7.90	4	9.00	2
Totals	37.50		41.80	

Solution

Unweighted Aggregate Index for 2012:

$$I_{2012} = \frac{\sum P_{2012}}{\sum P_{2011}}(100) = \frac{41.80}{37.50}(100) = 111.5$$

Laspeyres Index for 2012:

$$\sum P_{2012}Q_{2011} = [(6.95)(150) + (1.45)(60) + (6.25)(8) + (4.95)(25) + (13.20)(6) + (9.00)(4)]$$

$$= 1{,}042.50 + 87.00 + 50.00 + 123.75 + 79.20 + 36.00$$

$$= 1{,}418.45$$

$$\sum P_{2011}Q_{2011} = [(6.70)(150) + (1.35)(60) + (5.10)(8) + (4.50)(25) + (11.95)(6) + (7.90)(4)]$$

$$= 1{,}005.00 + 81.00 + 40.80 + 1125.50 + 71.70 + 31.60$$

$$= 1{,}342.60$$

$$I_{2012} = \frac{\sum P_{2012}Q_{2011}}{\sum P_{2011}Q_{2011}}(100) = \frac{1{,}418.45}{1{,}342.60}(100) = 105.6$$

Paasche Index for 2012:

$$\sum P_{2012}Q_{2012} = [(6.95)(135)+(1.45)(65)+(6.25)(12)+(4.95)(30)+(13.20)(8)+(9.00)(2)]$$

$$= 938.25 + 94.25 + 75.00 + 148.50 + 105.60 + 18.00$$

$$= 1{,}379.60$$

$$\sum P_{2011}Q_{2012} = [(6.70)(135)+(1.35)(65)+(5.10)(12)+(4.50)(30)+(11.95)(8)+(7.90)(2)]$$

$$= 904.50 + 87.75 + 61.20 + 135.00 + 95.60 + 15.80$$

$$= 1{,}299.85$$

$$I_{2012} = \frac{\sum P_{2012}Q_{2012}}{\sum P_{2011}Q_{2012}}(100) = \frac{1{,}379.60}{1{,}299.85}(100) = 106.1$$

15.6   PROBLEMS

15.21 Suppose the data below represent the price of 20 reams of office paper over a 50-year time frame. Find the simple index numbers for the data.

a. Let 1950 be the base year.

b. Let 1980 be the base year.

Year	Price	Year	Price
1950	22.45	1980	69.75
1955	31.40	1985	73.44
1960	32.33	1990	80.05
1965	36.50	1995	84.61
1970	44.90	2000	87.28
1975	61.24	2005	89.56

15.22 The World Intellectual Property Organization reports yearly figures for patents issued around the world. Following are the numbers of patents applications for the years 1980 through 2009 at the European Patent Office. Using these data and a base year of 2000, determine the simple index numbers for each year.

Year	Number of Patents (000s)	Year	Number of Patents (000s)
1980	18,596	1995	60,559
1981	24,119	1996	64,035
1982	27,422	1997	72,904
1983	30,664	1998	82,087
1984	35,982	1999	89,359
1985	36,916	2000	100,692
1986	41,342	2001	110,027
1987	45,069	2002	106,243
1988	49,774	2003	116,604
1989	55,774	2004	123,701
1990	60,754	2005	128,713
1991	55,984	2006	135,231
1992	58,896	2007	140,763
1993	56,974	2008	146,150
1994	57,842	2009	134,580

15.23 Using the data that follow, compute the aggregate index numbers for the four types of meat. Let 1995 be the base year for this market basket of goods.

Items	Year		
	1995	**2002**	**2012**
Sirloin Steaks (per kg.)	11.53	13.4	15.17
Sausage (per kg.)	2.21	2.15	2.51
Bacon (per kg.)	5.92	6.68	8.35
Ribeye steak (per kg.)	13.38	13.1	14.49

15.24 Suppose the data below are prices of market goods involved in household transportation for the years 2004 through 2012. Using 2010 as a base year, compute aggregate transportation price indexes for this data.

Items	Year								
	2004	**2005**	**2006**	**2007**	**2008**	**2009**	**2010**	**2011**	**2012**
Petrol (per litre)	0.815	0.859	0.911	0.911	0.800	0.867	1.028	1.295	1.445
Oil (per litre)	7.98	8.01	8.18	8.17	8.01	8.11	8.3	8.45	8.48
Transmission fluid (per litre)	8.8	8.82	8.98	8.96	8.94	8.9	8.92	8.94	8.96
Radiator coolant (per litre)	5.95	5.96	6.24	6.21	6.19	6.05	6.12	6.1	6.24

15.25 Calculate Laspeyres price indexes for 2010–2012 from the data below. Use 2000 as the base year.

	Quantity	Price			
Item	**2000**	**2000**	**2010**	**2011**	**2012**
1	21	0.50	0.67	0.68	0.71
2	6	1.23	1.85	1.90	1.91
3	17	0.84	0.75	0.75	0.80
4	43	0.15	0.21	0.25	0.25

15.26 Calculate Paasche price indexes for 2011 and 2012 using the following data and 2010 as the base year:

Item	2010 Price	2011		2012	
		Price	**Quantity**	**Price**	**Quantity**
1	22.50	27.80	13	28.11	12
2	10.90	13.10	5	13.25	8
3	1.85	2.25	41	2.35	44

Decision Dilemma SOLVED

Forecasting Air Pollution

In searching for the most effective forecasting technique to use to forecast either the net CO_2 or other greenhouse gas emissions, it is useful to determine whether a trend is evident in either set of time-series data. Minitab's trend analysis output is presented here for other greenhouse gas emissions.

(continued)

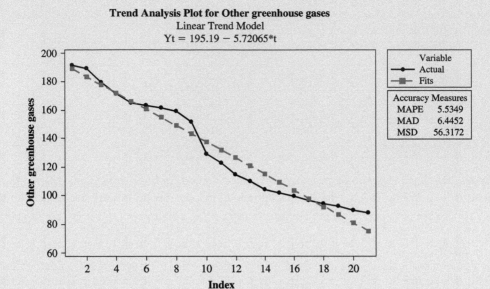

In observing the fit of this trend line and the time-series plot, it is evident that there appears to be more of a quadratic trend than a linear trend. Therefore, a Minitab-produced quadratic trend model was run and the results are presented below. Note that the error measures are all smaller for the quadratic model and that the curve fits the data much better than does the linear model.

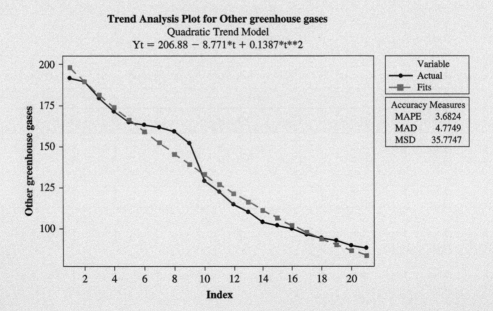

Various smoothing techniques can be used to forecast time-series data. After exploring several moving average models to predict other greenhouse gas emissions, it was determined that a one-year moving average fits the data relatively well. The results of a Minitab moving average graphical analysis of greenhouse gas emissions using a one-year moving average is shown below. Note that the forecasts shadow the actual values quite well.

(*continued*)

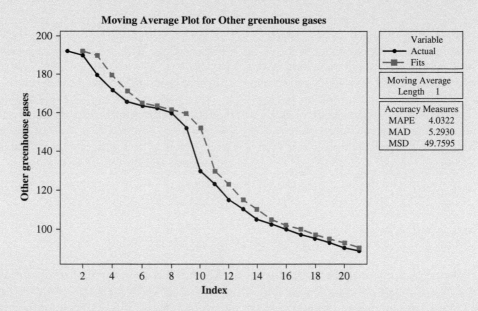

The effectiveness of exponential smoothing as a forecasting tool for nitrogen oxide emissions was tested using Minitab for several values of α. Through this analysis, it was determined that the best forecasts were obtained for values of α near 1, indicating that the actual value for the previous time period was a much stronger contributor to the forecast than the previous time period's forecast. Shown below is a Minitab-produced graphical analysis of an exponential smoothing forecast of the other greenhouse gas data using an alpha of 0.95. You are encouraged to explore other methods for forecasting net CO_2 and other greenhouse gas emissions.

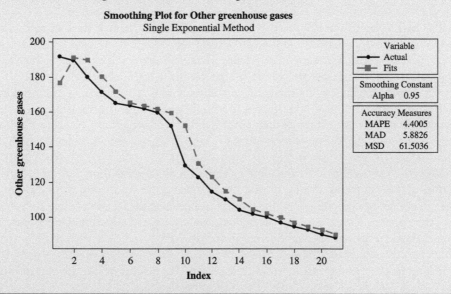

ETHICAL CONSIDERATIONS

The true test of a forecast is the accuracy of the prediction. Until the actual value is obtained for a given time period, the accuracy of the forecast is unknown. Many forecasters make predictions in society, including card readers, religious leaders, and self-proclaimed prophets. The proof of the forecast is in the outcome. The same holds true in the business world. Forecasts are made about everything from market share to interest rates to number of international air travellers. Many businesses fail because of faulty forecasts.

Forecasting is perhaps as much an art as a science. To keep forecasting ethical, the consumer of the forecast should be given the caveats and limitations of the forecast. The forecaster should be honestly cautious in selling the predictions to a client. In addition, the forecaster should be constantly on the lookout for changes in the business setting being modelled and quickly translate and incorporate those changes into the forecasting model.

Unethical behaviour can occur in forecasting when particular data are selected to develop a model that has been predetermined to produce certain results. As mentioned previously, statistics can be used to 'prove' almost anything. The ethical forecaster lets the data drive the model and is constantly seeking honest input from new variables to revise the forecast. He or she strives to communicate the limitations of both the forecasts and the models to clients.

SUMMARY

Time-series data are data that have been gathered at regular intervals over a period of time. It is generally believed that time-series data are composed of four elements – trend, cyclical effects, seasonality, and irregularity. Trend is the long-term general direction of the time-series data. Cyclical effects are the business and economic cycles that occur over periods of more than one year. Seasonal effects are patterns or cycles of data behaviour that occur over time periods of less than one year. Irregular fluctuations are unaccounted-for 'blips' or variations that occur over short periods of time.

One way to establish the validity of a forecast is to examine the forecasting error. The error of a forecast is the difference between the actual value and the forecast value. Computing a value to measure forecasting error can be done in several different ways. This chapter presents mean absolute deviation and mean square error for this task.

Regression analysis with either linear or quadratic models can be used to explore trend. Regression trend analysis is a special case of regression analysis in which the dependent variable is the data to be forecast and the independent variable is the time periods numbered consecutively from 1 to k, where k is the number of time periods. For the quadratic model, a second independent variable is constructed by squaring the values in the first independent variable, and both independent variables are included in the analysis.

One group of time-series forecasting methods contains smoothing techniques. Among these techniques are naive models, averaging techniques, and simple exponential smoothing. These techniques do much better if the time-series data are stationary or show no significant trend or seasonal effects. Naive forecasting models are models in which it is assumed that the more recent time periods of data represent the best predictions or forecasts for future outcomes.

Simple averages use the average value for some given length of previous time periods to forecast the value for the next period. Moving averages are time period averages that are revised for each time period by including the most recent value(s) in the computation of the average and deleting the value or values that are farthest away from the present time period. A special case of the moving average is the weighted moving average, in which different weights are placed on the values from different time periods.

Simple (single) exponential smoothing is a technique in which data from previous time periods are weighted exponentially to forecast the value for the present time period. The forecaster has the option of selecting how much to weight more recent values versus those of previous time periods.

Decomposition is a method for isolating the four possible effects in time-series data, trend, cyclical effects, seasonality, and irregular fluctuations.

Autocorrelation or serial correlation occurs when the error terms from forecasts are correlated over time. In regression analysis, this effect is particularly disturbing because one of the assumptions is that the error terms are independent. One way to test for autocorrelation is to use the Durbin–Watson test.

A number of methods attempt to overcome the effects of autocorrelation on the data. One way is to determine whether at least one independent variable is missing and, if so, include it or them in the model. Another way is to transform the variables. One transformation technique is the first-differences approach, in which each value of X is subtracted from the succeeding time period value of X and the differences are used as the values of the *X* variable. The same approach is used to transform the *Y* variable. The forecasting model is then developed from the transformed variables.

Autoregression is a forecasting technique in which time-series data are predicted by independent variables that are lagged versions of the original dependent variable data. A variable that is lagged one period is derived from values of the previous time period. Other variables can be lagged two or more periods.

Index numbers can be used to translate raw data into numbers that are more readily comparable. Simple index numbers are constructed by creating the ratio of the raw data value for a given time period to the raw data value for the base period and multiplying the ratio by 100. The index number for the base time period is designated to be 100.

Unweighted aggregate price index numbers are constructed by summing the prices of several items for a time period and comparing that sum to the sum of the prices of the same items during a base time period and multiplying the ratio by 100. Weighted aggregate price indexes are index numbers utilizing the prices of several items, and the items are weighted by their quantity usage.

The Laspeyres price index uses the quantity weights from the base year in all calculations. The Paasche price index uses the quantity weights for the current time period for both the current time period and the base time period in calculations.

KEY TERMS

autocorrelation	mean square error (MSE)
autoregression	moving average
averaging models	naive forecasting models
cycles	Paasche price index
cyclical effects	seasonal effects
decomposition	serial correlation
deseasonalized data	simple average
Durbin–Watson test	simple average model
error of an individual forecast	simple index number
exponential smoothing	smoothing techniques
first-differences approach	stationary
forecasting	time-series datatrend
forecasting error	unweighted aggregate price
index number	index number
irregular fluctuations	weighted aggregate price
Laspeyres price index	index numbers
mean absolute deviation (MAD)	weighted moving average

GO ONLINE TO DISCOVER THE EXTRA FEATURES FOR THIS CHAPTER

FORMULAS

Individual forecast error

$$e_t = X_t - F_t$$

Mean absolute deviation

$$\text{MAD} = \frac{\sum |e_i|}{\text{Number of Forecasts}}$$

Mean square error

$$\text{MSE} = \frac{\sum e_i^2}{\text{Number of Forecasts}}$$

Exponential smoothing

$$F_{t+1} = \alpha \cdot X_t + (1 - \alpha) \cdot F_t$$

Durbin–Watson test

$$D = \frac{\sum\limits_{t=2}^{n} (e_t - e_{t-1})^2}{\sum\limits_{t=1}^{n} e_t^2}$$

SUPPLEMENTARY PROBLEMS

CALCULATING THE STATISTICS

15.27 Following are the average three-month interest rates in Poland over a 25-month period published by Eurostat.

Month	Yield	Month	Yield
1	5.49	13	7.91
2	4.69	14	4.17
3	4.3	15	4.13
4	4.2	16	3.69
5	4.52	17	3.85
6	4.6	18	3.86
7	4.26	19	7.35
8	4.16	20	3.82
9	4.18	21	3.82
10	4.18	22	3.83
11	4.19	23	3.86
12	4.23	24	3.92

a. Explore trends in these data by using regression trend analysis. How strong are the models? Is the quadratic model significantly stronger than the linear trend model?

b. Use a four-month moving average to forecast values for each of the ensuing months.

c. Use simple exponential smoothing to forecast values for each of the ensuing months. Let $\alpha = 0.3$ and then let $\alpha = 0.7$. Which weight produces better forecasts?

d. Compute MAD for the forecasts obtained in parts (b) and (c) and compare the results.

e. Determine seasonal effects using decomposition on these data. Let the seasonal effects have four periods. After determining the seasonal indexes, deseasonalize the data.

15.28 Compute index numbers for the following data using 2000 as the base year:

Year	Quantity	Year	Quantity
1997	2,073	2005	2,520
1998	2,290	2006	2,529
1999	2,349	2007	2,483
2000	2,313	2008	2,467
2001	2,456	2009	2,397
2002	2,508	2010	2,351
2003	2,463	2011	2,308
2004	2,499		

15.29 Compute unweighted aggregate price index numbers for each of the given years using 2010 as the base year.

Item	2008	2009	2010	2011	2012
1	3.21	3.37	3.80	3.73	3.65
2	.51	.55	.68	.62	.59
3	.83	.90	.91	1.02	1.06
4	1.30	1.32	1.33	1.32	1.30
5	1.67	1.72	1.90	1.99	1.98
6	.62	.67	.70	.72	.71

15.30 Using the following data and 2009 as the base year, compute the Laspeyres price index for 2012 and the Paasche price index for 2011:

Item	2009		2010	
	Price	Quantity	Price	Quantity
1	2.75	12	2.98	9
2	0.85	47	0.89	52
3	1.33	20	1.32	28

Item	2011		2012	
	Price	Quantity	Price	Quantity
1	3.10	9	3.21	11
2	0.95	61	0.98	66
3	1.36	25	1.40	32

TESTING YOUR UNDERSTANDING

15.31 The following data contain the annual quantity (tonnes) of total fish caught by Portugal over a 30-year period as published by the Food and Agriculture Organization of the United Nations.

a. Use a three-year moving average to forecast the quantity of fish for the years 1983 through 2009 for these data. Compute the error of each forecast and then determine the mean absolute deviation of error for the forecast.

b. Use exponential smoothing and $\alpha = 0.2$ to forecast the data from 1983 through 2009. Let the forecast for 1981 equal the actual value for 1980. Compute the error of each forecast and then determine the mean absolute deviation of error for the forecast.

c. Compare the results obtained in parts (a) and (b) using MAD. Which technique seems to perform better? Why?

Year	Quantity	Year	Quantity
1980	276,121	1995	272,795
1981	269,469	1996	270,798
1982	259,938	1997	234,992
1983	255,798	1998	236,328
1984	311,486	1999	218,875
1985	327,919	2000	199,657
1986	413,184	2001	202,673

(continued)

Year	Quantity	Year	Quantity
1987	397,710	2002	211,706
1988	353,693	2003	220,481
1989	340,032	2004	228,269
1990	337,660	2005	219,077
1991	337,132	2006	237,743
1992	306,884	2007	262,770
1993	301,751	2008	231,816
1994	276,668	2009	207,058

15.32 The Organization for Economic Cooperation and Development publishes a series of statistics on trade in goods and services. Included in these statistics are the monthly value of total exports of goods in national currencies. Displayed here is a portion of these data representing the United Kingdom's total exports of goods from January of year 1 through to December of year 5. Use time-series decomposition methods to develop the seasonal indexes for these data.

Time Period	Total Exports of Goods (£ million)	Time Period	Total Exports of Goods (£ million)
January (year 1)	18,433	January (year 2)	16,746
February	20,783	February	16,921
March	25,952	March	19,426
April	22,742	April	17,914
May	24,427	May	18,543
June	23,535	June	18,966
July	18,321	July	18,315
August	17,405	August	17,373
September	18,552	September	18,121
October	18,821	October	19,964
November	19,163	November	20,460
December	17,102	December	17,790
January (year 3)	18,532	January (year 4)	16,978
February	20,147	February	17,896
March	20,207	March	19,219
April	21,500	April	18,448
May	21,339	May	17,258
June	22,207	June	18,474
July	22,879	July	19,321
August	18,997	August	16,274
September	22,073	September	19,749
October	22,276	October	20,963
November	19,483	November	20,206
December	19,041	December	20,254

Time Period	Total Exports of Goods (£ million)
January (year 5)	17,551
February	20,579
March	22,934
April	21,423
May	20,799
June	23,354
July	22,147
August	19,943
September	22,796
October	23,308
November	24,085
December	23,253

15.33 Use the seasonal indexes computed to deseasonalize the data in Problem 15.32.

15.34 Determine the trend for the data in Problem 15.32 using the deseasonalized data from Problem 15.33. Explore both a linear and a quadratic model in an attempt to develop the better trend model.

15.35 Shown here are retail price figures and quantity estimates for five different food commodities over three years. Use these data and a base year of 2010 to compute unweighted aggregate price indexes for this market basket of food. Using a base year of 2010, calculate Laspeyres price indexes and Paasche price indexes for 2011 and 2012.

Item	2010 Price	2010 Quantity	2011 Price	2011 Quantity	2012 Price	2012 Quantity
Margarine (500 g)	1.26	21	1.32	23	1.39	22
Granulated Sugar (1 kg)	0.94	5	0.97	3	1.12	4
Fresh Milk (1 litre)	1.43	70	1.56	68	1.62	65
Cola (2 litres)	1.05	12	1.02	13	1.25	11
Frozen Pizza (340 g)	2.81	27	2.86	29	2.99	28

15.36 Given below are data on inward flows of foreign direct investment (FDI) from countries belonging to the Organization for Economic Cooperation and Development (OECD) and on the ratio of total trade to GDP released by the OECD and the Penn World Tables, respectively, for a 25-year period for Spain. Develop a regression model to predict the inward flows of FDI by the ratio of total trade to GDP. Use this model to predict the FDI inward flows for a year in which there are 7.0 (million) business establishments. Discuss the strength of the regression model. Use these data and the regression model to compute a Durbin–Watson test to determine whether significant autocorrelation is present. Let alpha be 0.05.

Inward FDI (€ millions)	Ratio of Total Trade to GDP (%)
1,318.4	40.91
1,728.4	35.41
3,086.1	36.33
3,474.0	36.70
5,134.7	37.32
7,644.9	35.55
6,768.0	35.42
6,963.4	35.91
7,071.0	36.93
7,268.0	41.63
4,489.0	44.77
5,031.0	46.71
5,499.0	51.74
10,330.0	53.55
17,127.0	55.19
42,524.0	61.20
31,450.0	59.57
39,717.0	56.78
22,913.0	55.02
19,441.0	55.88
19,922.0	56.69
22,562.0	59.04
48,574.0	60.51
43,106.0	58.69

15.37 Shown here are the nominal effective exchange rates (NEERs) for the Euro area for the years 1990 through 2007 from the Eurostat Data website. Use the data to answer the following questions.

Year	NEER	Year	NEER
1990	111.77	1999	95.64
1991	107.83	2000	85.58
1992	110.84	2001	86.89
1993	105.2	2002	89.54
1994	103.82	2003	99.95
1995	108.94	2004	103.49
1996	108.77	2005	103.27
1997	99.56	2006	104.01
1998	101.32	2007	109.27

a. Compute the four-year moving average to forecast the NEERs from 1994 through 2007.

b. Compute the four-year weighted moving average to forecast the NEERs from 1994 through 2007. Weight the most recent year by 4, the next most recent year by 3, the next year by 2, and the last year of the four by 1.

c. Determine the errors for parts (a) and (b). Compute MSE for parts (a) and (b). Compare the MSE values and comment on the effectiveness of the moving average versus the weighted moving average for these data.

15.38 Yahoo! UK & Ireland Finance web page publishes data on the FTSE 100 index prices. Given are the FTSE 100 index prices from May of 2009 through August of 2011. The FTSE is a share index of the 100 most highly capitalized UK companies listed on the London Stock Exchange. Use these data to develop autoregression models for a one-month lag and a four-month lag. Compare the results of these two models. Which model seems to yield better predictions? Why?

Time Period	FTSE 100 Index Price (close)
May (2009)	4,417.9
June	4,249.2
July	4,608.4
August	4,908.9
September	5,133.9
October	5,044.5
November	5,190.7
December	5,412.9
January (2010)	5,188.5
February	5,354.5
March	5,679.6
April	5,553.3
May	5,188.4
June	4,916.9
July	5,258.0
August	5,225.2
September	5,548.6
October	5,675.2
November	5,528.3
December	5,899.9
January (2011)	5,862.9
February	5,994.0
March	5,908.8
April	6,069.9

(continued)

Time Period	FTSE 100 Index Price (close)
May	5,990.0
June	5,945.7
July	5,815.2
August	5,131.1

15.39 Eurostat publishes data on the number of tourist nights for a large number of countries. Shown here are the number of tourist nights (in thousands) of tourists staying in Romania from the first quarter of year 1 through the fourth quarter of year 6. Use these data to determine the seasonal indexes for the data through time-series decomposition methods. Use the four-quarter centred moving average in the computations.

Time Period	Number of Tourist Nights (thousands)
1st quarter (year 1)	6,817
2nd quarter	8,795
3rd quarter	25,424
4th quarter	10,999
1st quarter (year 2)	7,735
2nd quarter	9,130
3rd quarter	19,969
4th quarter	9,284
1st quarter (year 3)	6,414
2nd quarter	7,682
3rd quarter	16,530
4th quarter	6,721
1st quarter (year 4)	5,524
2nd quarter	7,229
3rd quarter	33,038
4th quarter	12,147
1st quarter (year 5)	9,092
2nd quarter	10,783
3rd quarter	24,095
4th quarter	10,795
1st quarter (year 6)	9,132
2nd quarter	11,609
3rd quarter	24,553
4th quarter	10,735

15.40 Use the seasonal indexes computed to deseasonalize the data in Problem 15.39.

15.41 Use both a linear and quadratic model to explore trends in the deseasonalized data from Problem 15.40. Which model seems to produce a better fit of the data?

15.42 The European Central Bank publishes data on official reserve assets of the European central bank vis-à-vis extra euro area. The data below give the average yearly amounts of official gold reserves (in € millions) held by the European central bank over a 10-year period. Use these data to develop an autoregression model with a one-period lag. Discuss the strength of the model.

Year	Official Gold Reserves of European Central Bank (€ millions)
1	50,463
2	47,833
3	47,618
4	40,513
5	37,319
6	38,865
7	40,897
8	40,734
9	43,595
10	55,307

15.43 The data shown here, from Statistics Sweden, show gross savings in equity and money market funds (millions of Swedish Krona, SEK). Use these data to develop a regression model to forecast the equity fund assets by money market funds. Conduct a Durbin–Watson test on the data and the regression model to determine whether significant autocorrelation is present. Let $\alpha = 0.01$.

Time Period	Equity Funds	Money Market Funds
1	38,958	10,145
2	32,166	12,286
3	32,552	8,551
4	50,336	18,598
5	67,289	11,316
6	46,786	30,700
7	34,970	8,654
8	51,545	20,013
9	55,076	19,697
10	58,115	13,373
11	38,288	19,181
12	74,290	25,917
13	35,116	20,732
14	42,793	13,432
15	29,745	20,003

(continued)

Time Period	Equity Funds	Money Market Funds
16	52,049	26,312
17	34,541	13,937
18	63,903	15,045
19	47,270	10,994
20	82,518	19,318
21	73,643	17,899
22	159,770	28,080
23	51,692	14,608
24	98,095	14,648
25	65,945	23,395
26	58,853	17,227

15.44 The gross domestic product per capita in purchasing power parity for China for a period of 18 years, as reported by the United Nations' World Development Indicators, are shown here. Use these data and exponential smoothing to develop forecasts for the years 2 through to 18. Try $\alpha = 0.1$, 0.5 and 0.8, and compare the results using MAD. Discuss your findings. Select the value of alpha that worked best and use your exponential smoothing results to predict the figure for 19.

Year	GDP per capita (PPP, current international $)	Year	GDP per capita (PPP, current international $)
1	1,024	10	2,600
2	1,179	11	2,863
3	1,346	12	3,197
4	1,507	13	3,599
5	1,672	14	4,115
6	1,841	15	4,761
7	1,988	16	5,594
8	2,152	17	6,234
9	2,364	18	6,828

INTERPRETING THE OUTPUT

15.45 Shown below is the Excel output for a regression analysis to predict the number of business bankruptcy filings over a 16-year period by the number of consumer bankruptcy filings. How strong is the model? Note the residuals. Compute a Durbin–Watson statistic from the data and discuss the presence of autocorrelation in this model.

SUMMARY OUTPUT

Regression Statistics	
Multiple R	0.529
R Square	0.280
Adjusted R Square	0.228
Standard Error	8179.84
Observations	16

ANOVA

	df	SS	MS	F	Significance F
Regression	1	364069877.4	364069877.4	5.44	0.0351
Residual	14	936737379.6	66909812.8		
Total	15	1300807257.0			

	Coefficients	Standard Error	t Stat	P-value
Intercept	75532.43621	4980.08791	15.17	0.0000
Year	−0.01574	0.00675	−2.33	0.0351

RESIDUAL OUTPUT

Observation	Predicted Bus. Bankruptcies	Residuals
1	70638.58	−1338.6
2	71024.28	−8588.3
3	71054.61	−7050.6
4	70161.99	1115.0
5	68462.72	12772.3
6	67733.25	14712.8
7	66882.45	−3029.4
8	65834.05	−2599.1
9	64230.61	622.4
10	61801.70	9747.3
11	61354.16	9288.8
12	62738.76	−434.8
13	63249.36	−10875.4
14	61767.01	−9808.0
15	57826.69	−4277.7
16	54283.80	−256.8

ANALYSING THE DATABASES

1. Use the Agricultural time-series database and the variable Green Beans to forecast the number of green beans for period 169 by using the following techniques:
   a. Five-period moving average
   b. Simple exponential smoothing with $\alpha = 0.6$
   c. Time-series linear trend model
   d. Decomposition

2. Use decomposition on Carrots in the Agricultural database to determine the seasonal indexes. These data actually represent 14 years of 12-month data. Do the seasonal indexes indicate the presence of some seasonal effects? Run an autoregression model to predict Carrots by a 1-month lag and another by a 12-month lag. Compare the two models. Because vegetables are somewhat seasonal, is the 12-month lag model significant?

3. Use the Energy database to forecast the 2010 price of automotive diesel by using simple exponential smoothing of previous automotive diesel data. Let $\alpha = 0.2$ and $\alpha = 0.8$. Compare the forecast with the actual figure. Which of the two models produces the forecast with the least error?

4. Use the International Labour database to develop a regression model to predict the unemployment rate for Germany by the unemployment rate of Italy. Test for autocorrelation and discuss its presence or absence in this regression analysis.

CASE

NANDO'S

Nando's began in 1987 when Robert Brozin and Fernando Duarte bought a restaurant called Chickenland in Rosettenville, southern Johannesburg, South Africa. They renamed the restaurant Nando's after Fernando. The restaurant incorporated influences from former Portuguese colonists from Mozambique, many of whom had settled on the south-eastern side of Johannesburg, after their homeland's independence in 1975. Since day one, expansion was a vital part of their vision. Growing from a single restaurant in 1987, Nando's had already expanded to four by 1990. To finance the rapid expansion plans, Nando's brought in the South African group Hollard as partners. New outlets were kept as a separate company which eventually made decision making inefficient and the implementation of a common strategy all but impossible. This led to Nando's buying out the joint venture partners in 1995. By this time there were 45 outlets in South Africa and another 17 in other African countries (Namibia, Botswana, Zimbabwe, and Swaziland).

By early 1997, another 46 outlets had been established in eight countries outside the African continent, including Portugal (1989), Australia (1991), the United Kingdom (1992), Israel (1993), and Canada (1994). The majority of these outlets were established by bringing in local partners, mostly by the use of franchising. When exchange controls began to relax in the mid-1990s, Nando's began to take a direct equity holding in those countries. This was seen as essential to facilitating Nando's international ambitions and providing ideal support to maintaining the 'Nando's Way' of doing business.

By 1995, managing this increasingly complex international structure had become very difficult and a new international holding was formed – Nando's International Holdings (NIH). On 27 April 1997, Nando's Group Holdings (NGH) – the South African operation – was successfully listed on the Johannesburg Stock Exchange. NIH held 54% of NGH – the remainder being made available to former joint venture partners and members of the public. The main purposes of the share offer and listing were to facilitate restructuring of the group and to increase its capital base. Brozin, however, was certain that he wanted to keep full control of all operations and keep his company independent. By the end of 2006, Nando's operated more than 400 restaurants located in 30 countries worldwide and the company was set to continue its growth into the future.

DISCUSSION

1. After the foundation of Nando's International Holdings in the mid-1990s, Nando's continued its upward climb of record sales around the world. Suppose the figures shown here are Nando's monthly sales figures from January 2004 through December 2012 (in €000s). Are any trends evident in the data? Does Nando's have a seasonal component to its sales? Shown after the sales figures is a Minitab output from a decomposition analysis of the sales figures using 12-month seasonality. Next an Excel graph displays the data with a trend line. Examine the data, the output, and any additional analysis you feel is helpful, and write a short report on Nando's sales. Include a discussion of the general direction of sales and any seasonal tendencies that might be occurring.

Month	2004	2005	2006	2007	2008	2009	2010	2011	2012
January	139.7	165.1	177.8	228.6	266.7	431.8	381.0	431.8	495.3
February	114.3	177.8	203.2	254.0	317.5	457.2	406.4	444.5	533.4
March	101.6	177.8	228.6	266.7	368.3	457.2	431.8	495.3	635.0
April	152.4	203.2	279.4	342.9	431.8	482.6	457.2	533.4	673.1
May	215.9	241.3	317.5	355.6	457.2	533.4	495.3	558.8	749.3
June	228.6	279.4	330.2	406.4	571.5	622.3	584.2	647.7	812.8
July	215.9	292.1	368.3	444.5	546.1	660.4	609.6	673.1	800.1
August	190.5	317.5	355.6	431.8	482.6	520.7	558.8	660.4	736.6
September	177.8	203.2	241.3	330.2	431.8	508.0	508.0	609.6	685.8
October	139.7	177.8	215.9	330.2	406.4	482.6	495.3	584.2	635.0
November	139.7	165.1	215.9	304.8	393.7	457.2	444.5	520.7	622.3
December	152.4	177.8	203.2	292.1	406.4	431.8	419.1	482.6	622.3

```
Time-Series Decomposition
for Sales Multiplicative
Model

Data:        Sales
Length:      108
NMissing:    0

Fitted Trend Equation

Yt = 121.481 + 5.12862*t

Seasonal Indices

Period        Index
  1          0.79487
  2          0.85125
  3          0.92600
  4          1.02227
  5          1.11591
  6          1.24281
  7          1.31791
  8          1.16422
  9          0.99201
 10          0.91524
 11          0.85071
 12          0.80679
Accuracy Measures
MAPE:           8.04
MAD:           29.51
MSD:         1407.55
```

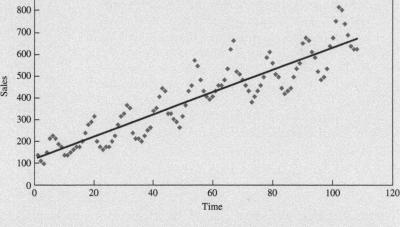

Calculate the error of the forecasts and determine which forecasting method seems to do the best job of minimizing error. Study the data and explain the behaviour of the per-unit window cleaning costs since 1999.

2. Suppose Nando's accountants computed the average cost of window cleaning per restaurant for each year since 1999, as reported here. Use techniques in this chapter to analyse the data. Forecast the per-unit window cleaning costs through the year 2012. Use smoothing techniques, moving averages, trend analysis, and any others that seem appropriate.

Year	Per-Unit Labour Cost	Year	Per-Unit Labour Cost
1999	80.15	2006	59.84
2000	85.29	2007	57.29
2001	85.75	2008	58.74
2002	64.23	2009	55.01
2003	63.70	2010	56.20
2004	62.54	2011	55.93
2005	60.19	2012	55.60

Source: Adapted from 'Nando's spice for success: wit, integrity and passion for the family', available at www.btimes.co.za/97/0504/comp/comp2.htm. See also 'Nando's Story' and 'Nando's Worldwide', available at www. nandos.com.

EXCEL

- Excel has the capability of forecasting using several of the techniques presented in this chapter. Two of the forecasting techniques are accessed using the **Data Analysis** tool, and two other forecasting techniques are accessed using the **Insert Function**.

- To use the **Data Analysis** tool, begin by selecting the **Data** tab on the Excel worksheet. From the **Analysis** panel at the top right of the **Data** tab worksheet, click on **Data Analysis**. If your Excel worksheet does not show the **Data Analysis** option, then you can load it as an add-in following directions given in Chapter 2.

- To do exponential smoothing, select **Exponential Smoothing** from the **Data Analysis** pulldown menu. In the dialog box, input the location of the data to be smoothed in **Input Range**. Input the value of the dampening factor in **Damping factor**. Excel will default to 0.3. Input the location of the upper left cell of the output table in the **Output Range** space. The output consists of forecast values of the data. If you check **Standard Errors**, a second column of output will be given of standard errors.

- To compute moving averages, select **Moving Average** from the **Data Analysis** pulldown menu. In the dialog box, input the location of the data for which the moving averages are to be computed in **Input Range**. Record how many values you want to include in computing the moving average in **Interval**. The default number is three values. Input the location of the upper left cell of the output table in **Output Range**. The output consists of the moving averages. If you check **Standard Errors**, a second column of output will be given of standard errors.

- To use the **Insert Function** (f_x) to compute forecasts and/or to fit a trend line, go to the **Formulas** tab on an Excel worksheet (top centre tab). The **Insert Function** is on the far left of the menu bar. In the **Insert Function** dialog box at the top, there is a pulldown menu where it says **Or select a category**. From the pulldown menu associated with this command, select **Statistical.**

- To compute forecasts using linear regression, select **FORECAST** from the **Insert Function's Statistical** menu. In the first line of the **FORECAST** dialog box, place the value of x for which you want a predicted value in **X**. An entry here is required. On the second line, place the location of the y values to be used in the development of the regression model in **Known_y's**. On the third line, place the location of the x values to be used in the development of the regression model in **Known_x's**. The output consists of the predicted value.

- To fit a trend line to data, select **TREND** from the **Insert Function's Statistical** menu. On the first line of the **TREND** dialog box, place the location of the y values to be used in the development of the regression model in **Known_y's**. On the second line, place the location of the x values to be used in the development of the regression model in **Known_x's**. Note that the x values can consist of more than one column if you want to fit a polynomial curve. To accomplish this, place squared values of x, cubed values of x, and so on as desired in other columns, and include those columns in **Known_x**. On the third line, place the values for which you want to return corresponding y values in **New_x's**. In the fourth line, place **TRUE** in **Const** if you want to get a value for the constant as usual (default option). Place **FALSE** if you want to set b_0 to zero.

MINITAB

- There are several forecasting techniques available through Minitab. These techniques are accessed in the following way: select **Stat** from the menu bar, and from the ensuing pulldown menu, select **Time Series**. From this pulldown menu select one of several forecasting techniques as detailed below.

- To begin a **Time Series Plot**, select which of the four types of plots you want from **Simple, Multiple, With Groups, or Multiple with Groups**. Enter the column containing the values that you want to plot in **Series**. Other options include **Time/Scale**, where you can determine what time frame you want to use along the x-axis; **Labels**, where you input titles and data labels; **Data View**, where you can choose how you want the graph to appear with options of symbols, connect line, or project lines; **Multiple Graphs**; and **Data Options**.

- To begin a **Trend Analysis**, place the location of the time-series data in the **Variables** slot. Under **Model**

Type, select the type of model you want to create from **Linear**, **Quadratic**, **Exponential growth**, or **S-Curve**. You can generate forecasts from your model by checking **Generate forecasts** and inserting how many forecasts you want and the starting point. Other options include **Time**, where you can determine what time frame you want to use along the x-axis; **Options**, where you input titles and data weights; **Storage**, where you can choose to store fits and/or residuals; **Graphs**, where you can choose from several graphical display options; and **Results,** which offers you three different ways to display the results.

- To begin a **Decomposition**, place the location of the time-series data in the **Variables** slot. Choose the **Model Type** by selecting from **Multiplicative, Additive, Trend plus seasonal**, or **Seasonal only**. You can generate forecasts from your model by checking **Generate forecasts** and inserting how many forecasts you want and the starting point. Other options include **Time**, where you can determine what time frame you want to use along the x-axis; **Options,** where you input the title and the seasonal location of the first observation; **Storage,** where you can choose to store trend line, detrended data, seasonals, seasonally adjusted data, fits, and residuals; **Graphs,** where you can choose from several graphical display options; and **Results**, which offers you three different ways to display the results.

- To begin a **Moving Average**, place the location of the time-series data in the **Variables** slot. Enter a positive integer to indicate desired length for the moving average in the **MA Length** slot. Check the **Center the moving averages** box if you want to place the moving average values at the period that is in the centre of the range rather than at the end of the range. You can generate forecasts from your model by checking **Generate forecasts** and inserting how many forecasts you want and the starting point. Other options include **Time**, where you can determine what time frame you want to use along the x-axis; **Options**, where you input the title; **Storage**, where you can choose to store moving averages, fits, and residuals; **Graphs**, where you can choose from several graphical display options; and

Results, which offers you three different ways to display the results.

- To begin **Single Exp Smoothing**, place the location of the time-series data in the **Variables** slot. Under **Weight to Use in Smoothing**, if you choose **Optimal ARIMA**, the forecasts will use the default weight, which Minitab computes by fitting an ARIMA $(0, 1, 1)$ model to the data. With this option, Minitab calculates the initial smoothed value by backcasting. If you choose **Use**, you can enter a specific weight that is between 0 and 2. You can generate forecasts from your model by checking **Generate forecasts** and inserting how many forecasts you want and the starting point. Other options include **Time**, where you can determine what time frame you want to use along the x-axis; **Options**, where you input the title; **Storage**, where you can choose to store smoothed data, fits, and residuals; **Graphs**, where you can choose from several graphical display options; and **Results**, which offers you three different ways to display the results.

- To begin **Differences**, enter the column containing the variable for which you want to compute differences in **Series**. Enter a storage column for the differences in the box beside **Store differences in**. In the box beside **Lag**, enter the value for the lag. The default lag value is 1.

- To begin **Lag**, enter the column containing the variable that you want to lag in **Series**. Enter the storage column for the lags in **Store lags in**. Enter the value for the lag in **Lag**. The default lag value is 1.

- To begin **Autocorrelation**, enter the column containing the time series in **Series**. If you want to use the default number of lags, choose **Default number of lags**. This number is $n/4$ for a series with less than or equal to 240 observations or $\sqrt{x} + 45.\text{eps}\$\$$ for a series with more than 240 observations, where n is the number of observations in the series. By selecting Number of lags, you can enter the number of lags to use instead of the default. The maximum number of lags is $n - 1$. Check **Store ACF** to store the autocorrelation values in the next available column. Check **Store t Statistics** to store the t statistics. Check **Store Ljung-Box Q Statistics** to store the Ljung–Box Q statistics.

Categorical Data and Non-parametric Statistics

THIS FIFTH AND FINAL UNIT OF THE TEXT INCLUDES chapters 16 and 17. Chapter 16, Analysis of Categorical Data, presents two well-known chi-square techniques for analysing frequency data that have been organized into nominal or ordinal categories – the chi-square goodness-of-fit test and the chi-square test of independence. Chapter 17, Non-parametric Statistics, contains some of the more well-known non-parametric statistics for analysing nominal and ordinal data. These techniques include the runs test, the Mann–Whitney U test, the Wilcoxon matched-pairs signed rank test, the Kruskal–Wallis test, the Friedman test, and Spearman's rank correlation.

CHAPTER 16

Analysis of Categorical Data

LEARNING OBJECTIVES

The overall objective of this chapter is to give you an understanding of two statistical techniques used to analyse categorical data, thereby enabling you to:

1. Use the chi-square goodness-of-fit test to analyse probabilities of multinomial distribution trials along a single dimension and test a population proportion

2. Use the chi-square test of independence to perform contingency analysis

Decision Dilemma

Business Planning in Small Firms

Do the characteristics of the owner-managers of small firms influence whether or not those firms have a formal business plan? A recent study by Richbell, Watts, and Wardle that draws on data from a survey of the owners/managers of small metalworking firms in Sheffield, UK, tried to answer that question.

Richbell, Watts, and Wardle proposed to study this question firstly because an effective business plan is frequently seen as one of the most important factors in business success; secondly, because external agencies providing funding for either start-up or expansion usually require a plan; thirdly, because the need to plan is emphasized by numerous national and local business advice agencies; and, finally, because the business plan can be used by the small firm to serve as a strategic planning document for the entrepreneurs and serve as a subsequent monitoring device.

The data used in the study relate to a random sample that was drawn from a sample of 214 small independent Sheffield-based firms listed in Dun and Bradsheet's directory of businesses in South Yorkshire. Seventy firms participated in the survey based on face-to-face interviews and these firms represented a response rate of 75%.

Survey questions were stated in such a way as to generate frequencies. The respondents were asked a large number of questions on their own characteristics and previous work experience. One of the issues researched was the relationship between the level of education of the owner-manager and the existence of a business plan. Another area of research was determining the presence/absence of a business plan among those who had previously been employed by another firm and whether or not they had worked for a medium/large firm immediately before setting up their own business. The tables right report the answers of the owners/managers on business planning and full-time education and business planning and sector experience.

In addition, the study found that owner-managers in 53 firms in the sample had previous work experience. Of these, 15 were moving into a new sector and the rest were staying within the sector in which they had experience.

Furthermore, the study found that 66.7% of the 15 owner-managers that were moving into a new sector were business planners while this was only true of just over one-third (36.8%) of those staying within the sector in which they had experience.

Business Planning and Full-Time Education

	Minimum Education	Above Minimum Education
With Plan	16	16
Without Plan	27	11

Business Planning and Previous Size

	Small (50 or fewer employees)	Medium/Large
With Plan	9	15
Without Plan	18	9

Managerial and Statistical Questions

1. Is there a difference between owner-managers with minimum education and those with above minimum education in having a business plan?

2. Is there a difference between owner-managers who had previously been employed by a small firm or a medium/large firm in having a business plan?

3. In comparing the incidence of business planning among managers that started a business in a new sector and those who stayed with the sector in which they had experience, the researchers reported percentages. Are the differences in percentages merely chance differences from samples or is there a significant difference between business planning among managers that started a business in a new sector and those that stayed in the sector in which they had experience? What statistical technique is appropriate for analysing these data?

Source: Adapted from Richbell, S. M., H. D. Watts, and P. Wardle, 'Owner-managers and business planning in the small firm', *International Small Business Journal*, vol. 24, no. 5 (October 2006), pp. 496–514.

In this chapter, we explore techniques for analysing categorical data. **Categorical data** are *non-numerical data that are frequency counts of categories from one or more variables.* For example, it is determined that of the 790 people attending a convention, 240 are engineers, 160 are managers, 310 are sales representatives, and 80 are information technologists. The variable is 'position in company' with four categories: engineers, managers, sales representatives, and information technologists. The data are not ratings or sales figures but rather frequency counts of how many of each position attended. Research questions producing this type of data are often analysed using chi-square techniques. The chi-square distribution was introduced in Chapters 8 and 9. The techniques presented here for analysing categorical data, the *chi-square goodness-of-fit test* and *the chi-square test of independence,* are an outgrowth of the binomial distribution and the inferential techniques for analysing population proportions.

16.1 CHI-SQUARE GOODNESS-OF-FIT TEST

In Chapter 5, we studied the binomial distribution in which only two possible outcomes could occur on a single trial in an experiment. An extension of the binomial distribution is a multinomial distribution in which more than two possible outcomes can occur in a single trial. **The chi-square goodness-of-fit test** is *used to analyse probabilities of multinomial distribution trials along a single dimension.* For example, if the variable being studied is economic class with three possible outcomes of lower income class, middle income class, and upper income class, the single dimension is economic class and the three possible outcomes are the three classes. On each trial, one and only one of the outcomes can occur. In other words, a family unit must be classified either as lower income class, middle income class, or upper income class and cannot be in more than one class.

The chi-square goodness-of-fit test compares the *expected*, or theoretical, *frequencies* of categories from a population distribution to the *observed*, or actual, *frequencies* from a distribution to determine whether there is a difference between what was expected and what was observed. For example, airline industry officials might theorize that the ages of airline ticket purchasers are distributed in a particular way. To validate or reject this expected distribution, an actual sample of ticket purchaser ages can be gathered randomly, and the observed results can be compared to the expected results with the chi-square goodness-of-fit test. This test also can be used to determine whether the observed arrivals at teller windows at a bank are Poisson distributed, as might be expected. In the paper industry, manufacturers can use the chi-square goodness-of-fit test to determine whether the demand for paper follows a uniform distribution throughout the year.

Formula 16.1 is used to compute a chi-square goodness-of-fit test.

CHI-SQUARE GOODNESS-OF-FIT TEST (16.1)

$$\chi^2 = \sum \frac{(f_o - f_e)^2}{f_e}$$

$$df = k - 1 - c$$

where:

f_o = frequency of observed values
f_e = frequency of expected values
k = number of categories
c = number of parameters being estimated from the sample data

This formula compares the frequency of observed values to the frequency of the expected values across the distribution. The test loses one degree of freedom because the total number of expected frequencies must equal the number of observed frequencies; that is, the observed total taken from the sample is used as the total for the expected frequencies. In addition, in some instances a population parameter, such as λ, μ, or σ, is estimated from the sample data to determine the frequency distribution of expected values. Each time this estimation occurs, an additional degree of freedom is lost. As a rule, if a uniform distribution is being used as the expected distribution or if an expected distribution of values is given, $k - 1$ degrees of freedom are used in the test. In testing to determine whether an observed distribution is Poisson, the degrees of freedom are $k - 2$ because an additional degree of freedom is lost in estimating λ. In testing to determine whether an observed distribution is normal, the degrees of freedom are $k - 3$ because two additional degrees of freedom are lost in estimating both μ and σ from the observed sample data.

TABLE 16.1

Results of a Local Survey of Consumer Satisfaction

Response	Frequency (f_o)
Excellent	21
Pretty good	109
Only fair	62
Poor	15

Karl Pearson introduced the chi-square test in 1900. The **chi-square distribution** is *the sum of the squares of k independent random variables* and therefore can never be less than zero; it extends indefinitely in the positive direction. Actually the chi-square distributions constitute a family, with each distribution defined by the degrees of freedom (df) associated with it. For small df values the chi-square distribution is skewed considerably to the right (positive values). As the df increase, the chi-square distribution begins to approach the normal curve. Table values for the chi-square distribution are given in Appendix A. Because of space limitations, chi-square values are listed only for certain probabilities.

How can the chi-square goodness-of-fit test be applied to business situations? One survey of US consumers conducted by the *Wall Street Journal* and NBC News asked the question: 'In general, how would you rate the level of service that American businesses provide?' The distribution of responses to this question was as follows:

Excellent	8%
Pretty good	47%
Only fair	34%
Poor	11%

Suppose a store manager wants to find out whether the results of this consumer survey apply to customers of supermarkets in her city. To do so, she interviews 207 randomly selected consumers as they leave supermarkets in various parts of the city. She asks the customers how they would rate the level of service at the supermarket from which they had just exited. Mirroring the *WSJ* and NBC survey, the response categories are excellent, pretty good, only fair, and poor. The observed responses from this study are given in Table 16.1. Now the manager can use a chi-square goodness-of-fit test to determine whether the observed frequencies of responses from this survey are the same as the frequencies that would be expected on the basis of the national survey.

HYPOTHESIZE

STEP 1: The hypotheses for this example follows.

> H_o: The observed distribution is the same as the expected distribution.
>
> H_a: The observed distribution is not the same as the expected distribution.

TEST

STEP 2: The statistical test being used is

$$\chi^2 = \Sigma \frac{(f_o - f_e)^2}{f_e}$$

STEP 3: Let $\alpha = 0.05$.

STEP 4: Chi-square goodness-of-fit tests are one tailed because a chi-square of zero indicates perfect agreement between distributions. Any deviation from zero difference occurs in the positive direction only because chi-square is determined by a sum of squared values and can never be negative. With four categories in this example (excellent, pretty good, only fair, and poor), $k = 4$. The degrees of freedom are $k - 1$ because the expected distribution is given: $k - 1 = 4 - 1 = 3$. For $\alpha = 0.05$ and df $= 3$, the critical chi-square value is

$$\chi^2_{0.05,3} = 7.8417$$

After the data are analysed, an observed chi-square greater than 7.8147 must be computed in order to reject the null hypothesis.

TABLE 16.2

Construction of Expected Values for Service Satisfaction Study

Response	Expected Proportion	Expected Frequency (f_e) (proportion × sample total)
Excellent	.08	(.08)(207) = 16.56
Pretty good	.47	(.47)(207) = 97.29
Only fair	.34	(.34)(207) = 70.38
Poor	.11	(.11)(207) = 22.77
		207.00

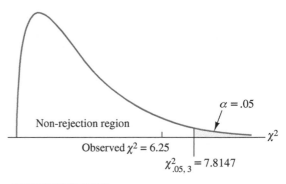

FIGURE 16. 1

Minitab Graph of Chi-Square Distribution for Service Satisfaction Example

TABLE 16.3

Calculation of Chi-Square for Service Satisfaction Example

Response	f_o	f_e	$\dfrac{(f_o - f_e)^2}{f_e}$
Excellent	21	16.56	1.19
Pretty good	109	97.29	1.41
Only fair	62	70.38	1.00
Poor	15	22.77	2.65
	207	207.00	6.25

STEP 5: The observed values gathered in the sample data from Table 16.1 sum to 207. Thus $n = 207$. The expected proportions are given, but the expected frequencies must be calculated by multiplying the expected proportions by the sample total of the observed frequencies, as shown in Table 16.2.

STEP 6: The chi-square goodness-of-fit can then be calculated, as shown in Table 16.3.

ACTION

STEP 7: Because the observed value of chi-square of 6.25 is not greater than the critical table value of 7.8147, the store manager will not reject the null hypothesis.

BUSINESS IMPLICATIONS

STEP 8: Thus the data gathered in the sample of 207 supermarket shoppers indicate that the distribution of responses of supermarket shoppers in the manager's city is not significantly different from the distribution of responses to the national survey.

The store manager may conclude that her customers do not appear to have attitudes different from those people who took the survey. Figure 16.1 depicts the chi-square distribution produced by using Minitab for this example, along with the observed and critical values.

DEMONSTRATION PROBLEM 16.1

Milk producers would like to know whether the sales of milk are distributed uniformly over a year so they can plan for milk production and storage. A uniform distribution means that the frequencies are the same in all categories. In this situation, the producers are attempting to determine whether the amounts of milk sold are the same for each month of the year. They ascertain the number of litres of milk sold by sampling one supermarket each month during a year, obtaining the data on the following page. Use $\alpha = 0.01$ to test whether the data fit a uniform distribution.

Month	Litres	Month	Litres
January	1,610	August	1,350
February	1,585	September	1,495
March	1,649	October	1,564
April	1,590	November	1,602
May	1,540	December	1,655
June	1,397	Total	18,447
July	1,410		

Solution

HYPOTHESIZE

STEP 1: The hypotheses follow.

H_0: The monthly figures for milk sales are uniformly distributed.
H_a: The monthly figures for milk sales are not uniformly distributed.

TEST

STEP 2: The statistical test used is

$$\chi^2 = \Sigma \frac{(f_0 - f_e)^2}{f_e}$$

STEP 3: Alpha is 0.01.

STEP 4: There are 12 categories and a uniform distribution is the expected distribution, so the degrees of freedom are $k - 1 = 12 - 1 = 11$. For $\alpha = 0.01$, the critical value is $\chi^2_{0.01,11} = 24.725$. An observed chi-square value of more than 24.725 must be obtained to reject the null hypothesis.

STEP 5: The data are given in the preceding table.

STEP 6: The first step in calculating the test statistic is to determine the expected frequencies. The total for the expected frequencies must equal the total for the observed frequencies (18,447). If the frequencies are uniformly distributed, the same number of litres of milk is expected to be sold each month. The expected monthly figure is

$$\frac{18,447}{12} = 1,537.25 \text{ litres}$$

The following table shows the observed frequencies, the expected frequencies, and the chi-square calculations for this problem:

Month	f_0	f_e	$\frac{(f_0 - f_e)^2}{f_e}$
January	1,610	1,537.25	3.44
February	1,585	1,537.25	1.48
March	1,649	1,537.25	8.12
April	1,590	1,537.25	1.81
May	1,540	1,537.25	0.00
June	1,397	1,537.25	12.80
July	1,410	1,537.25	10.53
August	1,350	1,537.25	22.81
September	1,495	1,537.25	1.16
October	1,564	1,537.25	0.47
November	1,602	1,537.25	2.73
December	1,655	1,537.25	9.02
Total	18,447	18,447.00	$\chi^2 = 74.37$

ACTION

STEP 7: The observed χ^2 value of 74.37 is greater than the critical table value of $\chi^2_{0.01,11} = 24.725$, so the decision is to reject the null hypothesis. This problem provides enough evidence to indicate that the distribution of milk sales is not uniform.

BUSINESS IMPLICATIONS

STEP 8: Because retail milk demand is not uniformly distributed, sales and production managers need to generate a production plan to cope with uneven demand. In times of heavy demand, more milk will need to be processed or on reserve; in times of less demand, provision for milk storage or for a reduction in the purchase of milk from dairy farmers will be necessary.

The following Minitab graph depicts the chi-square distribution, critical chi-square value, and observed chi-square value:

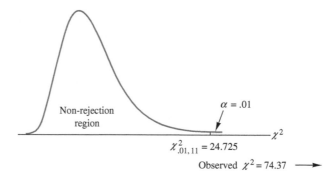

Caution: When the expected value of a category is small, a large chi-square value can be obtained erroneously, leading to a Type I error. To control for this potential error, the chi-square goodness-of-fit test should not be used when any of the expected frequencies is less than 5. If the observed data produce expected values of less than 5, combining adjacent categories (when meaningful) to create larger frequencies may be possible.

DEMONSTRATION PROBLEM 16.2

Chapter 5 indicated that, quite often in the business world, random arrivals are Poisson distributed. This distribution is characterized by an average arrival rate, λ, per some interval. Suppose a teller supervisor believes the distribution of random arrivals at a local bank is Poisson and sets out to test this hypothesis by gathering information. The data below represent a distribution of frequency of arrivals during one-minute intervals at the bank. Use $\alpha = 0.05$ to test these data in an effort to determine whether they are Poisson distributed.

Number of Arrivals	Observed Frequencies
0	7
1	18
2	25
3	17
4	12
≥ 5	5

Solution

HYPOTHESIZE

STEP 1: The hypotheses follow.

H_0: The frequency distribution is Poisson.
H_a: The frequency distribution is not Poisson.

TEST

STEP 2: The appropriate statistical test for this problem is

$$\chi^2 = \Sigma \frac{(f_0 - f_e)^2}{f_e}$$

STEP 3: Alpha is 0.05.

STEP 4: The degrees of freedom are $k - 2 = 6 - 1 - 1 = 4$ because the expected distribution is Poisson. An extra degree of freedom is lost, because the value of lambda must be calculated by using the observed sample data. For $\alpha = 0.05$, the critical table value is $\chi^2_{0.05,4} = 9.4877$. The decision rule is to reject the null hypothesis if the observed chi-square is greater than $\chi^2_{0.05,4} = 9.4877$.

STEP 5: To determine the expected frequencies, the supervisor must obtain the probability of each category of arrivals and then multiply each by the total of the observed frequencies. These probabilities are obtained by determining lambda and then using the Poisson table. As it is the mean of a Poisson distribution, lambda can be determined from the observed data by computing the mean of the data. In this case, the supervisor computes a weighted average by summing the product of number of arrivals and frequency of those arrivals and dividing that sum by the total number of observed frequencies.

Number of Arrivals	Observed Frequencies	Arrival × Observed
0	7	0
1	18	18
2	25	50
3	17	51
4	12	48
≥5	5	25
	84	192

$$\lambda = \frac{192}{84} - 2.3$$

With this value of lambda and the Poisson distribution table in Appendix A, the supervisor can determine the probabilities of the number of arrivals in each category. The expected probabilities are determined from Table A.3 by looking up the values of $x = 0, 1, 2, 3$, and 4 in the column under $\lambda = 2.3$, shown in the table below as expected probabilities. The probability for $x \geq 5$ is determined by summing the probabilities for the values of $x = 5, 6, 7, 8$, and so on. Using these probabilities and the total of 84 from the observed data, the supervisor computes the expected frequencies by multiplying each expected probability by the total (84).

Arrivals	Expected Probabilities	Expected Frequencies
0	.1003	8.42
1	.2306	19.37
2	.2652	22.28

3	.2033	17.08
4	.1169	9.82
≥ 5	.0837	7.03
		84.00

STEP 6: The supervisor uses these expected frequencies and the observed frequencies to compute the observed value of chi-square.

Arrivals	Observed Frequencies	Expected Frequencies	$\frac{(f_o - f_e)^2}{f_e}$
0	7	8.42	.24
1	18	19.37	.10
2	25	22.28	.33
3	17	17.08	.00
4	12	9.82	.48
≥ 5	5	7.03	.59
	84	84.00	$\chi^2 = 1.74$

ACTION

STEP 7: The observed value of 1.74 is not greater than the critical chi-square value of 9.4877, so the supervisor's decision is to not reject the null hypothesis. In other words, he fails to reject the hypothesis that the distribution of bank arrivals is Poisson.

BUSINESS IMPLICATIONS

STEP 8: The supervisor can use the Poisson distribution as the basis for other types of analysis, such as queuing modelling.

The following Minitab graph depicts the chi-square distribution, critical value, and computed value:

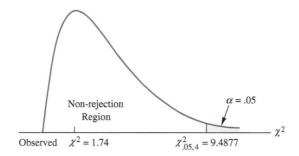

Testing a Population Proportion by Using the Chi-Square Goodness-of-Fit Test as an Alternative Technique to the z Test

In Chapter 9 we discussed a technique for testing the value of a population proportion. When sample size is large enough ($n \cdot p \geq 5$ and $n \cdot q \geq 5$), sample proportions are normally distributed and the following formula can be used to test hypotheses about p:

$$z = \frac{\hat{p} - p}{\sqrt{\dfrac{p \cdot q}{n}}}$$

The chi-square goodness-of-fit test can also be used to conduct tests about p; this situation can be viewed as a special case of the chi-square goodness-of-fit test where the number of classifications equals two (binomial distribution situation). The observed chi-square is computed in the same way as in any other chi-square goodness-of-fit test, but because the test contains only two classifications (success or failure), $k = 2$ and the degrees of freedom are $k - 1 = 2 - 1 = 1$.

As an example, we will work two problems from Section 9.4 by using chi-square methodology. The first example in Section 9.4 tests the hypothesis that exactly 8% of a manufacturer's products are defective. The following hypotheses are being tested:

$$H_o: p = 0.08$$
$$H_a: p \neq 0.08$$

The value of alpha was given to be 0.10. To test these hypotheses, a business researcher randomly selected a sample of 200 items and determined that 33 of the items had at least one flaw.

Working this problem by the chi-square goodness-of-fit test, we view it as a two-category expected distribution in which we expect 0.08 defects and 0.92 non-defects. The observed categories are 33 defects and $200 - 33 = 167$ non-defects. Using the total observed items (200), we can determine an expected distribution as $0.08(200) = 16$ and $0.92(200) = 184$. Shown here are the observed and expected frequencies.

	f_o	f_e
Defects	33	16
Non-defects	167	184

Alpha is 0.10 and this test is two tailed, so $\alpha/2 = 0.05$. The degrees of freedom are 1. The critical table chi-square value is

$$\chi^2_{0.05,1} = 3.8415$$

An observed chi-square value greater than this value must be obtained to reject the null hypothesis. The chi-square for this problem is calculated as follows:

$$\chi^2 = \sum \frac{(f_o - f_e)^2}{f_e} = \frac{(33 - 16)^2}{16} + \frac{(167 - 184)^2}{184} = 18.06 + 1.57 = 19.63$$

Notice that this observed value of chi-square, 19.63, is greater than the critical table value, 3.8415. The decision is to reject the null hypotheses. The manufacturer does not produce 8% defects according to this analysis. Observing the actual sample result, in which 0.165 of the sample was defective, indicates that the proportion of the population that is defective might be greater than 8%.

The results obtained are approximately the same as those computed in Chapter 9, in which an observed z value of 4.43 was determined and compared to a critical z value of 1.645, causing us to reject the null hypothesis. This result is not surprising to researchers who understand that when the degrees of freedom equal 1, the value of χ^2 equals z^2.

DEMONSTRATION PROBLEM 16.3

Rework Demonstration Problem 9.3 using the chi-square goodness-of-fit technique.

Solution

In this problem, we tested to determine whether the French consume a significantly higher proportion of breakfast cereal as their primary breakfast food than the 0.34 figure for Europe. The hypotheses were

$$H_o: p = 0.34$$
$$H_a: p > 0.34$$

The value of alpha was 0.05, and it is a one-tailed test. The degrees of freedom are $k-1 = 2-1 = 1$, as there are $k = 2$ categories (cereal or not cereal). The critical table value for chi-square is

$$\chi^2_{0.05,1} = 3.8415$$

To test these hypotheses, a sample of 550 people were contacted. Of these, 212 declared that breakfast cereal was their primary breakfast food. The observed categories are 212 and $550 - 212 = 338$. The expected categories are determined by multiplying 0.34 and 0.66 by the observed total number (550). Thus, the expected categories are $0.34(550) = 187$ and $0.66(550) = 363$. These frequencies follow.

	f_o	f_e
Cereal	212	187
Not cereal	338	363

The observed chi-square is determined by

$$\chi^2 = \Sigma \frac{(f_o - f_e)^2}{f_e} = \frac{(212 - 187)^2}{187} + \frac{(338 - 363)^2}{363} = 3.34 + 1.72 = 5.06$$

This observed chi-square, 5.06, is greater than the critical chi-square value of 3.8415. The decision is to reject the null hypothesis. The proportion of residents who eat breakfast cereal as their primary breakfast food is higher in France than in Europe as a whole. In Demonstration Problem 9.3, an observed z value of 2.23 was obtained, which was greater than the critical value of 1.645, allowing us to reject the null hypothesis. The results obtained by the two different methods (χ^2 and z) are essentially the same, with the observed value of χ^2 approximately equal to z^2 ($z = 2.23$, $z^2 = 4.97 \approx \chi^2$).

16.1 PROBLEMS

16.1 Use a chi-square goodness-of-fit test to determine whether the observed frequencies are distributed the same as the expected frequencies ($\alpha = 0.05$).

Category	f_o	f_e
1	53	68
2	37	42
3	32	33
4	28	22
5	18	10
6	15	8

16.2 Use the following data and $\alpha = 0.01$ to determine whether the observed frequencies represent a uniform distribution:

Category	f_o
1	19
2	17
3	14
4	18
5	19
6	21
7	18
8	18

16.3   Are the data below Poisson distributed? Use $\alpha = 0.05$ and the chi-square goodness-of-fit test to answer this question. What is your estimated lambda?

Number of Arrivals	f_o
0	28
1	17
2	11
3	5

16.4   Use the chi-square goodness-of-fit test to determine if the observed data below are normally distributed. Let $\alpha = 0.05$. What are your estimated mean and standard deviation?

Category	Observed
10–under 20	6
20–under 30	14
30–under 40	29
40–under 50	38
50–under 60	25
60–under 70	10
70–under 80	7

16.5   In one survey, successful female entrepreneurs were asked to state their personal definition of success in terms of several categories from which they could select. Thirty-nine per cent responded that happiness was their definition of success, 12% said that sales/profit was their definition, 18% responded that helping others was their definition, and 31% responded that achievements/challenge was their definition. Suppose you wanted to determine whether male entrepreneurs felt the same way and took a random sample of men, resulting in the data below. Use the chi-square goodness-of-fit test to determine whether the observed frequency distribution of data for men is the same as the distribution for women. Let $\alpha = 0.05$.

Definition	f_o
Happiness	42
Sales/profit	95
Helping others	27
Achievements/challenge	63

16.6   The percentages below come from a national survey of the ages of online music shoppers. A local survey produced the observed values. Does the evidence in the observed data indicate that we should reject the national survey distribution for local online music shoppers? Use $\alpha = 0.01$.

Age	Percentage from Survey	f_o
10–14	9	22
15–19	23	50
20–24	22	43
25–29	14	29
30–34	10	19
≥ 35	22	49

16.7   The general manager of a European football team believes the ages of purchasers of match tickets are normally distributed. The data on the following page represent the distribution of ages for a sample of observed purchasers of football match tickets. Use the chi-square goodness-of-fit test to determine whether this distribution is significantly different from the normal distribution. Assume that $\alpha = 0.05$.

Age of Purchaser	Frequency
10–under 20	16
20–under 30	44
30–under 40	61
40–under 50	56
50–under 60	35
60–under 70	19

16.8   An emergency call service keeps records of emergency telephone calls. A study of 150 five-minute time intervals resulted in the distribution of number of calls below. For example, during 18 of the five-minute intervals, no calls occurred. Use the chi-square goodness-of-fit test and $\alpha = 0.01$ to determine whether this distribution is Poisson.

Number of Calls (per five-minute interval)	Frequency
0	18
1	28
2	47
3	21
4	16
5	11
6 or more	9

16.9   According to an extensive survey conducted for *Business Marketing* by Leo J. Shapiro & Associates, 66% of all computer companies are going to spend more on marketing this year than in previous years. Only 33% of other information technology companies and 28% of non-information technology companies are going to spend more. Suppose a researcher wanted to conduct a survey of her own to test the claim that 28% of all non-information technology companies are spending more on marketing next year than this year. She randomly selects 270 companies and determines that 62 of the companies do plan to spend more on marketing next year. Use $\alpha = 0.05$, the chi-square goodness-of-fit test, and the sample data to test to determine whether the 28% figure holds for all non-information technology companies.

16.10   Cross-cultural training is rapidly becoming a popular way to prepare executives for foreign management positions within their company. This training includes such aspects as foreign language, pre-visit orientations, meetings with former expatriates, and cultural background information on the country. According to Runzheimer International, 30% of all major companies provide formal cross-cultural programmes to their executives being relocated in foreign countries. Suppose a researcher wants to test this figure for companies in the communications industry to determine whether the figure is too high for that industry. In a random sample, 180 communications firms are contacted; 42 provide such a programme. Let $\alpha = 0.05$ and use the chi-square goodness-of-fit test to determine whether the 0.30 proportion for all major companies is too high for this industry.

16.2   CONTINGENCY ANALYSIS: CHI-SQUARE TEST OF INDEPENDENCE

The chi-square goodness-of-fit test is used to analyse the distribution of frequencies for categories of *one* variable, such as age or number of bank arrivals, to determine whether the distribution of these frequencies is the same as some hypothesized or expected distribution. However, the goodness-of-fit test cannot be used to analyse *two* variables simultaneously. A different chi-square test, the **chi-square test of independence**, can be *used to analyse the frequencies of two variables with multiple categories to determine whether the two variables are independent.* Many times this type of analysis is desirable. For example, a market researcher might want to determine whether the type of soft drink preferred by a consumer is independent of the consumer's age. An organizational behaviourist might want to know whether absenteeism is independent of job classification. Financial investors might want to determine whether type of preferred stock investment is independent of the region where the investor resides.

The chi-square test of independence can be used to analyse any level of data measurement, but it is particularly useful in analysing nominal data. Suppose a business researcher is interested in determining whether geographic region is independent of type of financial investment. On a questionnaire, the following two questions might be used to measure geographic region and type of financial investment:

In which region of the country do you reside?

A. North   B. East   C. South   D. West

Which type of financial investment are you most likely to make today?

E. Stocks   F. Bonds   G. Sovereign Bonds

The business researcher would *tally the frequencies of responses* to these two questions into a two-way table called a **contingency table**. Because the chi-square test of independence uses a contingency table, this test is sometimes referred to as **contingency analysis**.

Depicted in Table 16.4 is a contingency table for these two variables. Variable 1, geographic region, uses four categories: A, B, C, and D. Variable 2, type of financial investment, uses three categories: E, F, and G. The observed frequency for each cell is denoted as O_{ij}, where i is the row and j is the column. Thus, O_{13} is the observed frequency for the cell in the first row and third column. The expected frequencies are denoted in a similar manner.

If the two variables are independent, they are not related. In a sense, the chi-square test of independence is a test of whether the variables are related. The null hypothesis for a chi-square test of independence is that the two variables are independent (not related). If the null hypothesis is rejected, the conclusion is that the two variables are not independent and are related.

Assume at the beginning that variable 1 and variable 2 are independent. The probability of the intersection of two of their respective categories, A and F, can be found by using the multiplicative law for independent events presented in Chapter 4:

$$P(A \cap F) = P(A) \cdot P(F)$$

If A and F are independent, then

$$P(A) = \frac{n_A}{N}, P(F) = \frac{n_F}{N}, \text{ and } P(A \cap F) = \frac{n_A}{N} \cdot \frac{n_F}{N}$$

If $P(A \cap F)$ is multiplied by the total number of frequencies, N, the expected frequency for the cell of A and F can be determined.

TABLE 16.4

Contingency Table for the Investment Example

	Type of Financial Investment			
	E	F	G	
A			O_{13}	n_A
B				n_B
C				n_C
D				n_D
	n_E	n_F	n_G	N

Geographic Region: A, B, C, D

$$e_{AF} = \frac{n_A}{N} \cdot \frac{n_F}{N}(N) = \frac{n_A \cdot n_F}{N}$$

In general, if the two variables are independent, the expected frequency values of each cell can be determined by

$$e_{ij} = \frac{n_i \cdot n_j}{N}$$

where:

i = the row
j = the column
n_i = the total of row i
n_j = the total of column j
N = the total of all frequencies

Using these expected frequency values and the observed frequency values, we can compute a chi-square test of independence to determine whether the variables are independent. Formula 16.2 is the formula for accomplishing this test.

CHI-SQUARE TEST OF INDEPENDENCE (16.2)

$$\chi^2 = \sum\sum \frac{(f_o - f_e)^2}{f_e}$$

where:

$df = (r-1)(c-1)$
r = number of rows
c = number of columns

The null hypothesis for a chi-square test of independence is that the two variables are independent. The alternative hypothesis is that the variables are not independent. This test is one tailed. The degrees of freedom are $(r-1)(c-1)$. Note that Formula 16.2 is similar to Formula 16.1, with the exception that the values are summed across both rows and columns and the degrees of freedom are different.

Suppose a business researcher wants to determine whether type of petrol preferred is independent of a person's income. She takes a random survey of petrol purchasers, asking them one question about petrol preference and a second question about income. The respondent is to check whether he or she prefers (1) regular petrol, (2) premium petrol, or (3) extra premium petrol. The respondent also is to check his or her income brackets as being (1) less than €30,000, (2) €30,000 to €49,999, (3) €50,000 to €99,999, or (4) more than €100,000. The business researcher tallies the responses and obtains the results in Table 16.5. Using $\alpha = 0.01$, she can use the chi-square test of independence to determine whether type of petrol preferred is independent of income level.

HYPOTHESIZE

STEP 1: The hypotheses follow.

H_0: Type of petrol is independent of income.
H_a: Type of petrol is not independent of income.

TEST

STEP 2: The appropriate statistical test is

$$\chi^2 = \sum\sum \frac{(f_o - f_e)^2}{f_e}$$

STEP 3: Alpha is 0.01.

STEP 4: Here, there are four rows ($r = 4$) and three columns ($c = 3$). The degrees of freedom are $(4-1)(3-1) = 6$. The critical value of chi-square for $\alpha = 0.01$ is $\chi^2_{0.01,6} = 16.8119$. The decision rule is to reject the null hypothesis if the observed chi-square is greater than 16.8119.

STEP 5: The observed data appear in Table 16.5.

STEP 6: To determine the observed value of chi-square, the researcher must compute the expected frequencies. The

TABLE 16.5

Contingency Table for the Petrol Consumer Example

Income	Type of Petrol			
	Regular	Premium	Extra Premium	
Less than €30,000	85	16	6	107
€30,000 to €49,999	102	27	13	142
€50,000 to €99,999	36	22	15	73
More than €100,000	15	23	25	63
	238	88	59	385

expected values for this example are calculated as follows, with the first term in the subscript (and numerator) representing the row and the second term in the subscript (and numerator) representing the column:

$$e_{11} = \frac{(n_{1.})(n_{.1})}{N} = \frac{(107)(238)}{385} = 66.15$$

$$e_{21} = \frac{(n_{2.})(n_{.1})}{N} = \frac{(142)(238)}{385} = 87.78$$

$$e_{12} = \frac{(n_{1.})(n_{.2})}{N} = \frac{(107)(88)}{385} = 24.46$$

$$e_{22} = \frac{(n_{2.})(n_{.2})}{N} = \frac{(142)(88)}{385} = 32.46$$

$$e_{13} = \frac{(n_{1.})(n_{.3})}{N} = \frac{(107)(59)}{385} = 16.40$$

$$e_{23} = \frac{(n_{2.})(n_{.3})}{N} = \frac{(142)(59)}{385} = 21.76$$

$$e_{31} = \frac{(n_{3.})(n_{.1})}{N} = \frac{(73)(238)}{385} = 45.13$$

$$e_{41} = \frac{(n_{4.})(n_{.1})}{N} = \frac{(63)(238)}{385} = 38.95$$

$$e_{32} = \frac{(n_{3.})(n_{.2})}{N} = \frac{(73)(88)}{385} = 16.69$$

$$e_{42} = \frac{(n_{4.})(n_{.2})}{N} = \frac{(63)(88)}{385} = 14.40$$

$$e_{33} = \frac{(n_{3.})(n_{.3})}{N} = \frac{(73)(59)}{385} = 11.19$$

$$e_{43} = \frac{(n_{4.})(n_{.3})}{N} = \frac{(63)(59)}{385} = 9.65$$

The researcher then lists the expected frequencies in the cells of the contingency tables along with observed frequencies. In this text, expected frequencies are enclosed in parentheses. Table 16.6 provides the contingency table for this example.

TABLE 16.6

Contingency Table of Observed and Expected Frequencies for Petrol Consumer Example

		Type of Petrol		
		Regular	*Premium*	*Extra Premium*	
	Less than €30,000	(66.15) 85	(24.46) 16	(16.40) 6	107
	€30,000 to €49,999	(87.78) 102	(32.46) 27	(21.76) 13	142
Income	€50,000 to €99,999	(45.13) 36	(16.69) 22	(11.19) 15	73
	More than €100,000	(38.95) 15	(14.40) 23	(9.65) 25	63
		238	88	59	385

Next, the researcher computes the chi-square value by summing $(f_o - f_e)^2 / f_e$ for all cells.

$$\chi^2 = \frac{(85 - 66.15)^2}{66.15} + \frac{(16 - 24.46)^2}{24.46} + \frac{(6 - 16.40)^2}{16.40} + \frac{(102 - 87.78)^2}{87.78} + \frac{(27 - 32.46)^2}{32.46}$$

$$+ \frac{(13 - 21.76)^2}{21.76} + \frac{(36 - 45.13)^2}{45.13} + \frac{(22 - 16.69)^2}{16.69} + \frac{(15 - 11.19)^2}{11.19}$$

$$+ \frac{(15 - 38.95)^2}{38.95} + \frac{(23 - 14.40)^2}{14.40} + \frac{(25 - 9.65)^2}{9.65} = 5.37 + 2.93 + 6.60 + 2.30$$

$$+ 0.92 + 3.53 + 1.85 + 1.69 + 1.30 + 14.73 + 5.14 + 24.42 = 70.78$$

ACTION

step 7: The observed value of chi-square, 70.78, is greater than the critical value of chi-square, 16.8119, obtained from Table A.8 in Appendix A. The business researcher's decision is to reject the null hypothesis; that is, type of petrol preferred is not independent of income.

BUSINESS IMPLICATION

STEP 8: Having established that conclusion, the business researcher can then examine the outcome to determine which people, by income brackets, tend to purchase which type of petrol and use this information in market decisions.

Figure 16.2 is the Minitab output for calculating the chi-square value. Figure 16.3 is the Minitab chi-square graph with the critical value, the rejection region, and the observed χ^2.

```
Chi-square contributions are printed
below expected counts

                                    Extra
           Regular   Premium   Premium   Total
   1            85        16         6      107
             66.15     24.46     16.40
             5.374     2.924     6.593

   2           102        27        13      142
             87.78     32.46     21.76
             2.303     0.918     3.527

   3            36        22        15       73
             45.13     16.69     11.19
             1.846     1.693     1.300

   4            15        23        25       63
             38.95     14.40      9.65
            14.723     5.136    24.391

Total          238        88        59      385
Chi-Sq=70.727, DF=6, P-Value=0.000
```

FIGURE 16.2

Minitab Output for Petrol Consumer Example

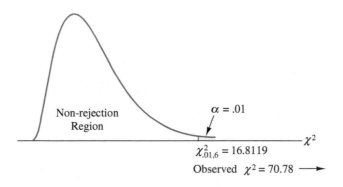

FIGURE 16.3

Minitab Graph of Chi-Square Distribution for Petrol Consumer Example

DEMONSTRATION PROBLEM 16.4

Is the type of beverage ordered with lunch at a restaurant independent of the age of the consumer? A random poll of 309 lunch customers is taken, resulting in the contingency table below of observed values. Use $\alpha = 0.01$ to determine whether the two variables are independent.

Preferred Beverage

		Coffee/Tea	Soft Drink	Other (Milk, etc.)	
	21−34	26	95	18	139
Age	35−55	41	40	20	101
	>55	24	13	32	69
		91	148	70	309

Solution

HYPOTHESIZE

STEP 1: The hypotheses follow.

H_o: Type of beverage preferred is independent of age.
H_a: Type of beverage preferred is not independent of age.

TEST

STEP 2: The appropriate statistical test is

$$\chi^2 = \Sigma\Sigma \frac{(f_o - f_e)^2}{f_e}$$

STEP 3: Alpha is 0.01.
STEP 4: The degrees of freedom are $(3-1)(3-1) = 4$, and the critical value is $\chi^2_{0.05,4} = 13.2767$. The decision rule is to reject the null hypothesis if the observed value of chi-square is greater than 13.2767.
STEP 5: The sample data were shown previously.
STEP 6: The expected frequencies are the product of the row and column totals divided by the grand total. The contingency table, with expected frequencies, follows.

Preferred Beverage

		Coffee/Tea	Soft Drink	Other (Milk, etc.)	
	21−34	(40.94) 26	(66.58) 95	(31.49) 18	139
Age	35−55	(29.74) 41	(48.38) 40	(22.88) 20	101
	>55	(20.32) 24	(33.05) 13	(15.63) 32	69
		91	148	70	309

For these values, the observed χ^2 is

$$\chi^2 = \frac{(26-40.94)^2}{40.94} + \frac{(95-66.58)^2}{66.58} + \frac{(18-31.49)^2}{31.49} + \frac{(41-29.74)^2}{29.4}$$

$$+ \frac{(40-48.38)^2}{48.38} + \frac{(20-22.88)^2}{22.88} + \frac{(24-20.32)^2}{20.32} + \frac{(13-33.05)^2}{33.05}$$

$$+ \frac{(32-15.63)^2}{15.63} = 5.45 + 12.13 + 5.78 + 4.26 + 1.45 + 0.36 + 0.67$$

$$+ 12.16 + 17.15 = 59.41$$

ACTION

STEP 7: The observed value of chi-square, 59.41, is greater than the critical value, 13.2767, so the null hypothesis is rejected.

BUSINESS IMPLICATIONS

STEP 8: The two variables – preferred beverage and age – are not independent. The type of beverage that a customer orders with lunch is related to or dependent on age. Examination of the categories reveals that younger people tend to prefer soft drinks and older people prefer other types of beverages. Managers of eating establishments and marketers of beverage products can utilize such information in targeting their market and in providing appropriate products.

Caution: *As with the chi-square goodness-of-fit test, small expected frequencies can lead to inordinately large chi-square values with the chi-square test of independence. Hence contingency tables should not be used with expected cell values of less than 5. One way to avoid small expected values is to collapse (combine) columns or rows whenever possible and whenever doing so makes sense.*

STATISTICS IN BUSINESS TODAY

Risk Taking by Ad Agencies

A study was conducted by Douglas C. West to examine under what conditions, if any, advertising agencies take more risk. The primary statistical technique used to analyse the data in the study was the chi-square test of independence. Although several studies have previously examined risk on the part of advertisers, little research addresses the willingness of advertising agencies to take risks on behalf of their clients. West theorized that people are more likely to take risks when the outcome affects someone else's livelihood and income rather than their own; consequently, advertising agencies might be more willing to take risks than advertisers. In addition, he theorized that advertising agencies might be more willing to take risks with smaller clients rather than large clients and that newer agencies might tend to be more willing to take risks than older agencies.

The study involved 64 account directors and/or heads of creative departments of advertising agencies selected from a standard directory. Respondents were presented with two advertising options under a plan to launch a new product. Plan A was a standard one with an average rate of return (risk averse), and Plan B was an uncertain one in which there is a 50% chance of getting a lower rate of return than the client's worst forecast and a 50% chance of getting a better rate of return than the client's highest forecast (risk seeking). Using a chi-square test of independence, the percentages of respondents selecting each plan were compared to percentages produced in a similar study with advertisers. The result was that the proportions of agency respondents that were risk averse/risk seeking were not significantly different from the proportions of advertisers that were risk averse/risk seeking ($\chi^2 = 3.165$, $p = 0.076$). Agencies and advertisers were also compared on four degrees of risk in light of the risk taken with their most recent client. The result showed no significant difference between agencies and advertisers on the amount of risk taken with their most recent client ($\chi^2 = 3.165$, $p = 0.076$, $\alpha = 0.05$). Thus, on two questions, there was no difference in the risk taking between agencies and advertisers.

Are there circumstances under which an advertising agency might be more willing to take risks? Respondents were asked to what degree they were willing to take risks if the client is their smallest client versus if the client is their largest client. A 4×2 contingency table was constructed with four degrees of risk and the two client sizes. Analysis of these data produced a significant chi-square of 9.819 ($p = 0.021$)

showing that agencies tended to be more risk taking with smaller clients than with large clients.

The effect of agency age on participant selection of Plan A versus Plan B was analysed using a 2×2 contingency table. Agencies were separated into two age categories (3−19 years versus 20−135 years) with Plan A and Plan B as the risk options. Using a chi-square test of independence, it was determined that a significantly higher proportion of the younger agencies were more risk seeking than older agencies ($\chi^2 = 6.75$, $p = 0.01$).

In this study, the chi-square test of independence allowed for categorical comparisons between agencies and advertisers, between small clients and large clients, and between young and old agencies. In many other studies, chi-square categorical statistics can be used to study business phenomena.

Source: Adapted from Douglas C. West, '360° of Creative Risk', *Journal of Advertising Research*, vol. 39, no. 1 (January/February 1999), pp. 39–50.

16.2   PROBLEMS

16.11   Use the contingency table below to test whether variable 1 is independent of variable 2. Let $\alpha = 0.01$.

	Variable 2	
Variable 1	203	326
	68	110

16.12   Use the contingency table below to determine whether variable 1 is independent of variable 2. Let $\alpha = 0.01$.

	Variable 2			
Variable 1	24	13	47	58
	93	59	187	244

16.13   Use the contingency table below and the chi-square test of independence to determine whether social class is independent of number of children in a family. Let $\alpha = 0.05$.

		Social Class		
		Lower	Middle	Upper
	0	7	18	6
Number of	1	9	38	23
Children	2 or 3	34	97	58
	More than 3	47	31	30

16.14 A group of 30-year-olds is interviewed to determine whether the type of music most listened to by people in their age category is independent of the geographic location of their residence. Use the chi-square test of independence with alpha = 0.01 and the following contingency table to determine whether music preference is independent of geographic location:

		Type of Music Preferred			
		Rock	R & B	Country	Classical
	North-east	140	32	5	18
Geographic Region	South-west	134	41	52	8
	Centre	154	27	8	13

16.15 Is the transportation mode used to ship goods independent of type of industry? Suppose the following contingency table represents frequency counts of types of transportation used by the publishing and the computer hardware industries. Analyse the data by using the chi-square test of independence to determine whether type of industry is independent of transportation mode. Let alpha = 0.05.

		Transportation Mode		
		Air	Train	Truck
Industry	Publishing	32	12	41
	Computer Hardware	5	6	24

16.16 The Department for Communities and Local Government regularly releases Housing and Planning statistics on the United Kingdom dwelling stock. The contingency table below shows the distribution of the dwelling stock by tenure and three regions of the United Kingdom. Is the type of tenure of the dwelling independent of the region where the dwelling is located? Use alpha = 0.01 and conduct a chi-square test of independence to determine whether, in fact, the type of tenure is independent of region where the dwelling is located.

		Dwelling Stock by Tenure and Region (thousands of dwellings)			
		Owner-Occupied	Rented Privately with a Job or Business	Rented from Housing Association	Rented from Local Authorities
	North-west	2,180	309	355	207
Region	London	1,820	654	295	451
	South-west	1,703	321	173	120

16.17 A market research company conducted a survey of CEOs to investigate their perceptions about planned government changes to employer-sponsored pension schemes. CEOs were told that the employer contributions to the fund were going to increase by 2% whilst the employee contribution was going to decrease by 1%. They were then asked: 'Do you agree or disagree with the proposed changes to employer-sponsored pension schemes?' Suppose the table below shows the CEO's responses. Can you conclude that the perceptions of CEOs are independent of age? Use $\alpha = 0.01$.

	Age Group				
	<30	30−39	40−49	50−59	>60
Agree	12	21	42	31	37
Disagree	55	72	92	96	86

Decision Dilemma SOLVED

Business Planning in Small Firms

Richbell, Watts, and Wardle examined how small firm owner-manager characteristics influenced whether or not those firms have a business plan. These researchers chose to analyse this problem by using frequency counts for various characteristics of the owner-managers. The two tables of data displayed in the Decision Dilemma contain frequencies for whether or not owner-managers have business plans on two criteria: their level of education and whether or not they have sector experience. Because each of these tables contains categorical data with two variables, a contingency analysis (chi-square test of independence) is an appropriate statistical technique for analysing the data. The null hypothesis is that business planning is independent of each of the criteria being analysed. The alternative hypothesis is that business planning is not independent of level of full-time education (level of full-time education makes a difference to whether or not a business plan exists) or previous size (if owner-manager had previously been employed by another firm and whether or not they had worked for a medium/large firm immediately before setting up their own business makes a difference to whether business planning take place). Minitab chi-square analysis produced the output below. For full-time education:

$$\text{Chi-Sq} = 3.249, \text{ DF} = 1, \text{ } P\text{-Value} = 0.071$$

The p-value of 0.071 indicates that we fail to reject the null hypothesis. On full-time education as a criterion, business planning is independent of respondents with above minimum education levels. There appears to be no difference between respondents that have business plans and those who do not on the full-time level of education criterion.

For previous size:

$$\text{Chi-Sq} = 4.339, \text{ DF} = 1, \text{ } P\text{-Value} = 0.037$$

The p-value of 0.037 indicates the rejection of the null hypothesis. Business planning is not independent of whether or not owner-managers had worked for a medium/large firm immediately before setting up their own business. In perusing the raw data, it is evident that whereas about two-thirds of the respondents with recent experience in working in a medium/large firm were business planners, this was true of only a third of those formerly working in small firms.

Richbell, Watts, and Wardle also found that of the respondents that had previous work experience, 66.7% of owners-managers that were moving into a new sector were business planners, while this was only true of just over one-third (36.8%) of those staying with the sector in which they had experience. Because sample statistics are used, is enough evidence provided to declare that there is a significant difference in proportions between owners-managers moving into a new sector and those staying with the sector they had previous experience in? Techniques presented in Chapter 10 can be used to statistically test this hypothesis. However, the chi-square test of independence can also be used. The two variables are business planning (with plan, without plan) and whether or not the previous work experience of the owner-managers is in the same sector as their new business (same sector, different sector) producing a 2×2 table:

	Sector Experience	
	Same Sector	Different Sector
With plan	14	10
Without plan	24	15

Recall that 15 owner-managers were moving into a new sector and the rest (38) were staying with the sector in which they had experience. The raw numbers are obtained by multiplying the samples sizes by the percentages (e.g., sample size for same sector = 38 and 38(0.368) = 14). Minitab gives a chi-square value of 3.861 with a p-value of 0.049. Based on this result, there is a significant difference between owner-managers with previous experience in the same sector and in a different sector on whether or not the company has a business plan.

ETHICAL CONSIDERATIONS

The usage of chi-square goodness-of-fit tests and chi-square tests of independence becomes an issue when the expected frequencies are too small. Considerable debate surrounds the discussion of how small is too small. In this chapter, we used an expected frequency of less than 5 as too small. As an example, suppose an expected frequency is 2. If the observed value is 6, then the calculation of $(f_o - f_e)^2/f_e$ results in $(6 - 2)^2/2 = 8$ just for this pair of observed and expected frequencies. Such a contribution to the overall computed chi-square can inordinately affect the total chi-square value and skew the analysis. Researchers should exercise caution in using small expected frequencies with chi-square tests lest they arrive at an incorrect statistical outcome.

SUMMARY

The chapter presented two techniques for analysing categorical data. Categorical data are *non-numerical data that are frequency counts of categories from one or more variables.* Categorical data producing this type of data are often analysed using chi-square techniques. The two techniques presented for analysing categorical data are the chi-square goodness-of-fit test and the chi-square test of independence. These techniques are an outgrowth of the binomial distribution and the inferential techniques for analysing population proportions.

The chi-square goodness-of-fit test is used to compare a theoretical or expected distribution of measurements for several categories of a variable with the actual or observed distribution of measurements. It can be used to determine whether a distribution of values fits a given distribution, such as the Poisson or normal distribution. If only two categories are used, the test offers the equivalent of a z test for a single proportion.

The chi-square test of independence is used to analyse frequencies for categories of two variables to determine whether the two variables are independent. The data used in analysis by a chi-square test of independence are arranged in a two-dimensional table called a contingency table. For this reason, the test is sometimes referred to as contingency analysis. A chi-square test of independence is computed in a manner similar to that used with the chi-square goodness-of-fit test. Expected values are computed for each cell of the contingency table and then compared to observed values with the chi-square statistic. Both the chi-square test of independence and the chi-square goodness-of-fit test require that expected values be greater than or equal to 5.

KEY TERMS

categorical data

chi-square distribution

chi-square goodness-of-fit test

chi-square test of independence

contingency analysis

contingency table

GO ONLINE TO DISCOVER THE EXTRA FEATURES FOR THIS CHAPTER

The Student Study Guide containing solutions to the odd-numbered questions, additional Quizzes and Concept Review Activities, Excel and Minitab databases, additional data files in Excel and Minitab, and more worked examples.

www.wiley.com/college/cortinhas

FORMULAS

χ^2 goodness-of-fit test

$$\chi^2 = \sum \frac{(f_o - f_e)^2}{f_e}$$

$$df = k - 1 - c$$

χ^2 test of independence

$$\chi^2 = \sum\sum \frac{(f_o - f_e)^2}{f_e}$$

$$df = (r - 1)(c - 1)$$

SUPPLEMENTARY PROBLEMS

CALCULATING THE STATISTICS

16.18 Use a chi-square goodness-of-fit test to determine whether the following observed frequencies are distributed the same as the expected frequencies. Let $\alpha = 0.01$.

Category	f_o	f_e
1	214	206
2	235	232
3	279	268
4	281	284
5	264	268
6	254	232
7	211	206

16.19 Use the chi-square contingency analysis to test to determine whether variable 1 is independent of variable 2. Use 5% level of significance.

	Variable 2		
	12	23	21
Variable 1	8	17	20
	7	11	18

TESTING YOUR UNDERSTANDING

16.20 Is a manufacturer's geographic location independent of type of customer? Use the data below for companies with primarily industrial customers and companies with primarily retail customers to test this question. Let $\alpha = 0.10$.

		Geographic Location		
		North	West	South
Customer	Industrial Customer	230	115	68
Type	Retail Customer	185	143	89

16.21 A local market sells six different kinds of home-made biscuits during its monthly market in a number of different stalls. The market manager is curious about whether sales of the six kinds of biscuits are uniformly distributed. He randomly selects the amounts of each kind of biscuit sold from five stalls and combines them into the observed data below. Use $\alpha = 0.05$ to determine whether the data indicate that sales for these six kinds of biscuits are uniformly distributed.

Type of Biscuits	Observed Frequency
Chocolate chip	189
Peanut butter	168
Cheese cracker	155
Lemon flavoured	161
Chocolate mint	216
Vanilla filled	165

16.22 A researcher interviewed 2,067 people and asked whether they were the primary decision makers in the household when buying a new car last year. Two hundred and seven were men and had bought a new car last year. Sixty-five were women and had bought a new car last year. Eight hundred and eleven of the responses were from men who did not buy a car last year. Nine hundred and eighty-four were from women who did not buy a car last year. Use these data to determine whether gender is independent of being a major decision maker in purchasing a car last year. Let $\alpha = 0.05$.

16.23 Are random arrivals at a shoe store at the local shopping centre Poisson distributed? Suppose a shopping centre employee researches this question by gathering data for arrivals during one-minute intervals on a weekday between 6.30 P.M. and 8.00 P.M. The data obtained follow. Use $\alpha = 0.05$ to determine whether the observed data seem to be from a Poisson distribution.

Arrivals per Minute	Observed Frequency
0	26
1	40
2	57
3	32
4	17
5	12
6	8

16.24 According to *Wikipedia*, the distribution of market share for the top seven TV channels in the United Kingdom in a recent year was: BBC1 20.7%, ITV 16.4%, BBC2 6.9%, Channel Four 6.4%, Channel Five 4.6%, ITV3 2.0%, and ITV2 1.9%. Suppose a marketing analyst wants to determine whether this distribution fits that of her geographic region. She randomly surveys 1,726 people and asks them to name their favourite TV channel. The responses are: BBC1 314, ITV 259, BBC2 122, Channel Four 121, Channel Five 98, ITV3 49, ITV2 48, and others 715. She then tests to determine whether the local distribution of TV channel preferences is the same or different from the national figures, using $\alpha = 0.05$. What does she find?

16.25 Are the types of professional jobs held in the computing industry independent of the number of years a person has worked in the industry? Suppose 246 workers are interviewed. Use the results obtained to determine whether type of professional job held in the computer industry is independent of years worked in the industry. Let $\alpha = 0.01$.

	Professional Position			
Years	Manager	Programmer	Operator	Systems Analyst
0–3	6	37	11	13
4–8	28	16	23	24
More than 8	47	10	12	19

16.26 A study by the Organization for Economic Cooperation and Development (OECD) found that the average annual hours actually worked per worker has decreased in most of Western Europe in the first decade of the 2000s. Suppose an OECD-wide survey is conducted that found

that 43% of the responding workers in the survey cited 'less business, less work' as the number one reason for this reduction in the annual working hours. Suppose you want to test this figure in the Netherlands to determine whether Dutch workers feel the same way. A random sample of 315 Dutch full-time workers whose work-week has been getting shorter is chosen. They are offered a selection of possible reasons for this reduction and 120 pick 'less business, less work'. Use techniques presented in this chapter and an alpha of 0.05 to test to determine whether the 43% figure for Western Europe for this reason holds true in the Netherlands.

16.27 Is the number of sports that a student currently regularly practices independent of the type of college or university being attended? Suppose students were randomly selected from three types of colleges and universities and the data shown represent the results of a survey of those students. Use a chi-square test of independence to answer the question. Let $\alpha = 0.05$.

		Type of College or University		
		Community College	Large University	Small University
Number of Sports Regularly Practised	0	25	178	31
	1	49	141	12
	2	31	54	8
	3 or more	22	14	6

INTERPRETING THE OUTPUT

16.28 A survey by Ipsos-Reid reported in American Demographics showed that if a person was given a $1,000 wind-fall, 36% would spend the money on home improvement, 24% on leisure travel/vacation, 15% on clothing, 15% on home entertainment or electronic products, and 10% on local entertainment including restaurants and movies. Suppose a researcher believes that these results would not be the same if posed to Europeans. The researcher conducts a new survey interviewing 200 randomly selected European adults asking these same questions. A chi-square goodness-of-fit test is conducted to compare the results of the new survey to the one taken by Ipsos-Reid. The Excel results follow. The observed and expected values are for the categories as already listed and appear in the same order. Discuss the findings. How did the distribution of results from the new survey compare to the old? Discuss the business implications of this outcome.

Europeans	Americans
Observed	*Expected*
36	72
64	48
42	30
38	30
20	20

The *p*-value for the chi-square goodness-of-fit test is: **0.0000043**. The observed chi-square for the goodness-of-fit test is: **30.18**.

16.29 Do men and women prefer the same colours of cars? That is, is sex independent of colour preference for cars? Suppose a study is undertaken to address this question. A random sample of men and women are asked which of five colours (silver, white, black, green, blue) they prefer in a car. The results as analysed using Minitab are shown here. Discuss the test used, the hypotheses, the findings, and the business implications.

```
       Chi-Square Test: Men, Women
Expected counts are printed below
observed counts. Chi-square contributions
are printed below expected counts.

              Men    Women   Total
    1          90       52     142
            85.20    56.80
            0.270    0.406

    2          75       58     133
            79.80    53.20
            0.289    0.433

    3          63       30      93
            55.80    37.20
            0.929    1.394

    4          39       33      72
            43.20    28.80
            0.408    0.612

    5          33       27      60
            36.00    24.00
            0.250    0.375

Total         300      200     500
Chi-Sq=5.366, DF=4, P-Value=0.252
```

ANALYSING THE DATABASE

1. The Financial database contains seven different types of companies. These seven are denoted by the variable Type. Use a chi-square goodness-of-fit test to determine whether the seven types of companies are uniformly distributed in this database.

2. In the Manufacturing database, is the Value of Industrial Shipments (a four-category variable) uniformly distributed across the database?

3. Use a chi-square test of independence to determine whether Control is independent of Service in the Hospital database. Comment on the results of this test.

4. In the Financial database, is Average Yield independent of Type? Create a new dummy variable by assigning a 0 if a company's average yield is less than 2 and a 1 otherwise. Use a chi-square test of independence to answer the question.

SPORTS ADVERTISING AND THE OLYMPICS

Why do companies use sports themes and sports celebrities in advertisements? The answer to this question partly lies in the universal appeal of sports and the fact that it is itself a multi-billion dollar industry. Companies are prepared to invest millions of dollars to connect their names and brands to major sporting events and famous athletes. For example, soccer player Cristiano Ronaldo is estimated to have had a €24 million sponsorship deal with Nike alone. Roger Federer, no. 3 2011 ATP ranking tennis player, is reported to earn around US$40 million annually endorsing products from Nike, Gillette, Jura, Mercedes-Benz, and NetJets. Companies pursue high-profile athletes because of their 'stopping power', which can be defined as the ability to draw attention to advertising messages. Stopping power has led many athletes to register their nicknames as trademarks. For example, Olympic gold medallist swimmer Kieran Perkins registered his nickname, *Superfish*. Olympic gold medallist and world record holder Ian Thorpe registered the name *Thorpedo* and launched a range of products bearing that name. More recently, Olympic gold medallist Usain Bolt registered his name and now trades products bearing his name.

Kambitsis *et al.* (2002) conducted a study to examine issues related to the involvement of sports celebrities in advertising. Specifically, they wanted to examine the motives for using sports or sporting celebrities in adverts, and to analyse the appearance of sporting celebrities in print advertisements prior to and during the Sydney Olympic Games. The research was conducted in Sydney five months before the Olympics and one month after them. Three monthly Australian magazines with high circulation were reviewed during this period in order to identify the different level of sport celebrities' involvement in advertisements before and during the Olympics. The same procedure was followed for two major daily newspapers published in Sydney for a period of 19 weeks. In addition, semi-structured interviews were conducted with two elite athletes, one sports manager, and one advertising executive.

The study found that, in general, advertisements endorsed by sports celebrities appeared simultaneously with sports-related news clips or articles. However, such advertisements were rare when there were no major sporting events taking place concurrently. For magazines, it was found that advertising companies appeared to favour swimmers as sports celebrities, using a sport-specific magazine as the chosen media. Track and field athletes, tennis players, and golfers followed in that order. Contingency tables for a comparison of pre-Olympic and Olympic editions of a sports magazine and a health/lifestyle magazine follow.

| | Sports magazine | | Health-lifestyle magazine | | |
	Sports products & services	Non-sport products & services	Sports products & services	Non-sport products & services	Total
Pre-Olympic edition	15	41	5	37	98
Olympic Edition	15	51	5	59	130
Total	30	92	10	96	228

| | Sports magazine | | Health-lifestyle magazine | | |
	Sports celebrities endorsing sports/ non-sports products	Unknown athletes endorsing sports/ non-sports products	Sports celebrities endorsing sports/ non-sports products	Unknown athletes endorsing sports/ non-sports products	Total
Pre-Olympic edition	15	41	5	37	98
Olympic Edition	15	51	5	59	130
Total	30	92	10	96	228

The authors found that, in more than 30 endorsements involving famous athletes, most featured swimmers, track and field athletes, and tennis players, and only two used footballers.

DISCUSSION

1. What could be a possible explanation for this observation?

2. Using techniques introduced in this chapter, test whether the number of advertisements for sports and non-sports products and services was affected by the Olympics.

3. Determine whether the appearance of sportspeople in print advertisements was affected by the Olympics.

4. Explain the implications of your results in each question.

Source: Extracts from Kambitsis, C. Y. Harahousou, N. Theodorakis, and G. Chatzibeis, 'Sports advertising in print media: the case of 2000 Olympic Games', *Corporate Communications*, vol. 7, no. 3 (2002), pp. 155–61; ABI/INFORM Global; Wikipedia.

USING THE COMPUTER

EXCEL

- Excel can compute a chi-square goodness-of-fit test but not a chi-square test of independence.

- To compute a chi-square goodness-of-fit test using Excel, begin with the **Insert Function** (f_x). To access the **Insert Function**, go to the **Formulas** tab on an Excel worksheet (top centre tab). The **Insert Function** is on the far left of the menu bar. In the **Insert Function** dialog box at the top, there is a pulldown menu where it says **Or select a category**. From the pulldown menu associated with this command, select **Statistical**. Select **CHITEST** from the **Insert Function's Statistical** menu. In the **CHITEST** dialog box, place the location of the observed values in **Actual_range**. Place the location of the expected values in **Expected_range**. The output will consist of a *p*-value. To determine

the observed chi-square from this *p*-value, go back to the **Insert Function** (f_x) and select **Statistical** and then **CHIINV**. In the **CHIINV** dialog box, place the *p*-value in **Probability** and the degrees of freedom in **Deg_freedom**. The output is the chi-square value.

MINITAB

- Minitab can compute the chi-square test of independence but does not directly compute the chi-square goodness-of-fit test. To compute the chi-square test of independence, from the **Stat** pulldown menu, select **Tables**. From the **Tables** pull-down menu, select **Chi-Square Test (table in work-sheet)**. Enter the columns containing the tabular data in the space beside **Columns containing the data**.

CHAPTER 17

Non-parametric Statistics

LEARNING OBJECTIVES

This chapter presents several non-parametric statistics that can be used to analyse data specifically, thereby enabling you to:

1. Use both the small-sample and large-sample runs tests to determine whether the order of observations in a sample is random
2. Use both the small-sample and large-sample cases of the Mann–Whitney U test to determine if there is a difference in two independent populations
3. Use both the small-sample and large-sample cases of the Wilcoxon matched-pairs signed rank test to compare the difference in two related samples
4. Use the Kruskal–Wallis test to determine whether samples come from the same or different populations
5. Use the Friedman test to determine whether different treatment levels come from the same population when a blocking variable is available
6. Use Spearman's rank correlation to analyse the degree of association of two variables

Decision Dilemma

How Is the Doughnut Business?

By investing US$5,000, William Rosenberg founded the Industrial Luncheon Services company in 1946 to deliver meals and coffee break snacks to customers in Massachusetts, USA. Building on his success in this venture, Rosenberg opened his first coffee and doughnut shop called the 'Open Kettle' in 1948. In 1950, Rosenberg changed the name of his shop to Dunkin' Donuts, and thus, the first Dunkin' Donuts shop was established. The first Dunkin' Donuts franchise was awarded in 1955, and by 1963, there were 100 Dunkin' Donuts shops. In 1970, the first overseas Dunkin' Donuts shop was opened in Japan, and by 1979, there were 1,000 Dunkin' Donuts shops worldwide.

Today, there are over 9,700 Dunkin' Donuts worldwide in 31 countries. In the United States alone, there are over 6,700 Dunkin' Donuts locations. Dunkin' Donuts is one of the world's largest coffee and baked goods chains, serving more than three million customers per day. Dunkin' Donuts sells 52 varieties of doughnuts and more than a dozen coffee beverages as well as an array of bagels, breakfast sandwiches, and other baked goods. Presently, Dunkin' Donuts is a brand of Dunkin' Brands, Inc., based in Canton, Massachusetts.

Suppose researchers at Dunkin' Donuts are studying several manufacturing and marketing questions in an effort to improve the consistency of their products and understand their market. Manufacturing engineers are concerned that the various machines produce a consistent doughnut size. In an effort to test this issue, four machines are selected for a study. Each machine is set to produce a doughnut that is supposed to be about 7.62 centimetres (3 inches) in diameter. A random sample of doughnuts is taken from each machine and the diameters of the doughnuts are measured. The result is the data shown as follows:

Machine 1	Machine 2	Machine 3	Machine 4
7.58	7.41	7.56	7.72
7.52	7.44	7.55	7.65
7.50	7.42	7.50	7.67
7.52	7.38	7.58	7.70
7.48	7.45	7.53	7.69
	7.40		7.71
			7.73

Suppose Dunkin' Donuts implements a national advertising campaign in Malaysia. Marketing researchers want to determine whether the campaign has increased the number of doughnuts sold at various outlets around the country. Ten stores are randomly selected and the number of doughnuts sold between 8 and 9 a.m. on a Tuesday is measured both before and after the campaign is implemented. The data follow:

Outlet	Before	After
1	301	374
2	198	187
3	278	332
4	205	212
5	249	243
6	410	478
7	360	386
8	124	141
9	253	251
10	190	264

Do bigger stores have greater sales? To test this question, suppose sales data were gathered from seven Dunkin' Donuts stores along with store size. These figures are used to rank the seven stores on each variable. The ranked data follow.

Store	Sales Rank	Size Rank
1	6	7
2	2	2
3	3	6
4	7	5
5	5	4
6	1	1
7	4	3

Managerial and Statistical Questions

1. The manufacturing researchers who are testing to determine whether there is a difference in the size of doughnuts by machine want to run a one-way ANOVA, but they have serious doubts that the ANOVA assumptions can be met by these data. Is it still possible to analyse the data using statistics?

2. The market researchers are uncertain that normal distribution assumptions underlying the matched-pairs *t* test can be met with the number of doughnuts data. How can the before-and-after data still be used to test the effectiveness of the advertisements?

3. If the sales and store size data are given as ranks, how do we compute a correlation to answer the research question about the relationship of sales and store size?

The Pearson product-moment correlation coefficient requires at least interval-level data, and these data are given as ordinal level.

Source: Adapted from information presented on the Dunkin' Donuts website at www.dunkindonuts.com/aboutus./company/. Please note that the data discussed in the problem is fictional, was not supplied by Dunkin' Donuts, and does not necessarily represent Dunkin' Donuts' experience.

Except for the chi-square analyses presented in Chapter 16, all statistical techniques presented in the text thus far are parametric techniques. **Parametric statistics** are *statistical techniques based on assumptions about the population from which the sample data are selected*. For example, if a *t* statistic is being used to conduct a hypothesis test about a population mean, the assumption is that the data being analysed are randomly selected from a *normally distributed* population. The name *parametric statistics* refers to the fact that an assumption (here, normally distributed data) is being made about the data used to test or estimate the parameter (in this case, the population mean). In addition, the use of parametric statistics requires quantitative measurements that yield interval- or ratio-level data.

For data that do not meet the assumptions made about the population, or when the level of data being measured is qualitative, statistical techniques called non-parametric, or distribution-free, techniques are used. **Non-parametric statistics** *are based on fewer assumptions about the population and the parameters than are parametric statistics*. Sometimes they are referred to as *distribution-free* statistics because many of them can be used regardless of the shape of the population distribution. A variety of non-parametric statistics are available for use with nominal or ordinal data. Some require at least ordinal-level data, but others can be specifically targeted for use with nominal-level data.

Non-parametric techniques have the following advantages:

1. Sometimes there is no parametric alternative to the use of non-parametric statistics.
2. Certain non-parametric tests can be used to analyse nominal data.
3. Certain non-parametric tests can be used to analyse ordinal data.
4. The computations on non-parametric statistics are usually less complicated than those for parametric statistics, particularly for small samples.
5. Probability statements obtained from most non-parametric tests are exact probabilities.

Using non-parametric statistics also has some disadvantages:

1. Non-parametric tests can be wasteful of data if parametric tests are available for use with the data.
2. Non-parametric tests are usually not as widely available and well known as parametric tests.
3. For large samples, the calculations for many non-parametric statistics can be tedious.

Entire courses and texts are dedicated to the study of non-parametric statistics. This text presents only some of the more important techniques: runs test, Mann–Whitney *U* test, Wilcoxon matched-pairs signed ranks test, Kruskal–Wallis test, Friedman test, Spearman's rank correlation coefficient, chi-square test of goodness-of-fit, and chi-square test of independence. The chi-square goodness-of-fit test and the chi-square test of independence were presented in Chapter 16. The others are presented in this chapter.

Figure 17.1 contains a tree diagram that displays all of the non-parametric techniques presented in this chapter, with the exception of Spearman's Rank Correlation, which is used to analyse the degree of association of two variables. As you peruse the tree diagram, you will see that there is a test of randomness, the runs test, two tests of the differences of two populations, the Mann–Whitney *U* test and the Wilcoxon matched-pairs signed rank test, and two tests of the differences of three or more populations – the Kruskal–Wallis test and the Friedman test.

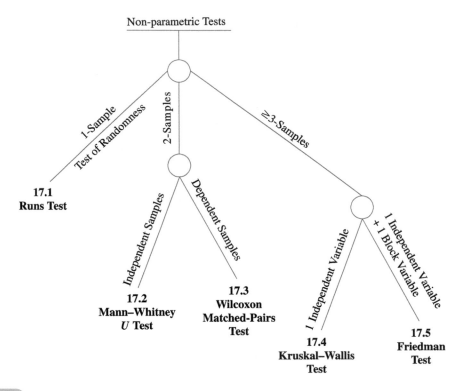

FIGURE 17.1

Tree Diagram Taxonomy of Non-parametric Inferential Techniques

17.1 RUNS TEST

The one-sample **runs test** is *a non-parametric test of randomness*. The runs test is *used to determine whether the order or sequence of observations in a sample is random*. The runs test examines the number of 'runs' of each of two possible characteristics that sample items may have. A *run* is a succession of observations that have a particular trait of one of the characteristics under study. For example, if a sample of people contains both men and women, one run could be a continuous succession of women. In tossing coins, the outcome of three heads in a row would constitute a run, as would a succession of seven tails.

Suppose a researcher takes a random sample of 15 people who arrive at a Carrefour to shop. Eight of the people are women and seven are men. If these people arrive randomly at the store, it makes sense that the sequence of arrivals would have some mix of men and women, but not probably a perfect mix. That is, it seems unlikely (although possible) that the sequence of a random sample of such shoppers would be first eight women and then seven men. In such a case, there are two runs. Suppose, however, the sequence of shoppers is woman, man, woman, man, woman, and so on all the way through the sample. This would result in 15 'runs'. Each of these cases is possible, but neither is highly likely in a random scenario. In fact, if there are just two runs, it seems possible that a group of women came shopping together followed by a group of men who did likewise. In that case, the observations would not be random. Similarly, a pattern of woman–man all the way through may make the business researcher suspicious that what has been observed is not really individual random arrivals, but actually random arrivals of couples.

In a random sample, the number of runs is likely to be somewhere between these extremes. What number of runs is reasonable? The one-sample runs test takes into consideration the size of the sample, n, the number of observations in the sample having each characteristic, n_1, n_2 (man, woman, etc.), and the number of runs in the sample, R, to reach conclusions about hypotheses of randomness. The following hypotheses are tested by the one-sample runs test:

H$_0$: The observations in the sample are randomly generated.
H$_a$: The observations in the sample are not randomly generated.

The one-sample runs test is conducted differently for small samples than it is for large samples. Each test is presented here. First, we consider the small-sample case.

Small-Sample Runs Test

If both n_1 and n_2 are less than or equal to 20, the small-sample runs test is appropriate. In the example of shoppers with $n_1 = 7$ men and $n_2 = 8$ women, the small-sample runs test could be used to test for randomness. The test is carried out by comparing the observed number of runs, R, to critical values of runs for the given values of n_1 and n_2. The critical values of R are given in Tables A.11 and A.12 in Appendix A for $\alpha = 0.05$. Table A.11 contains critical values of R for the lower tail of the distribution in which so few runs occur that the probability of that many runs or fewer runs occurring is less than 0.025 ($\alpha/2$). Table A.12 contains critical values of R for the upper tail of the distribution in which so many runs occur that the probability of that many runs or more occurring is less than 0.025 ($\alpha/2$). Any observed value of R that is less than or equal to the critical value of the lower tail (Table A.11) results in the rejection of the null hypothesis and the conclusion that the sample data are not random. Any observed value of R that is equal to or greater than the critical value in the upper tail (Table A.12) also results in the rejection of the null hypothesis and the conclusion that the sample data are not random.

As an example, suppose 26 cola drinkers are sampled randomly to determine whether they prefer regular cola or diet cola. The random sample contains 18 regular cola drinkers and eight diet cola drinkers. Let C denote regular cola drinkers and D denote diet cola drinkers. Suppose the sequence of sampled cola drinkers is DCCCCCDCCDCCCCDCDCCCDDDCCC. Is this sequence of cola drinkers evidence that the sample is not random? Applying the HTAB system of hypothesis testing to this problem results in:

HYPOTHESIZE

STEP 1: The hypotheses tested follow.

H_0: The observations in the sample were generated randomly.

H_a: The observations in the sample were not generated randomly.

TEST

STEP 2: Let n_1 denote the number of regular cola drinkers and n_2 denote the number of diet cola drinkers. Because $n_1 = 18$ and $n_2 = 8$, the small-sample runs test is the appropriate test.

STEP 3: Alpha is 0.05.

STEP 4: With $n_1 = 18$ and $n_2 = 18$, Table A.11 yields a critical value of 7 and Table A.12 yields a critical value of 17. If there are seven or fewer runs or 17 or more runs, the decision rule is to reject the null hypothesis.

STEP 5: The sample data are given as

DCCCCCDCCDCCCCDCDCCCDDDCCC

STEP 6: Tally the number of runs in this sample.

1	2	3	4	5	6	7	8	9	10	11	12
D	CCCCC	D	CC	D	CCCC	D	C	D	CCC	DDD	CCC

The number of runs, R, is 12.

ACTION

STEP 7: Because the value of R falls between the critical values of 7 and 17, the decision is to not reject the null hypothesis. Not enough evidence is provided to declare that the data are not random.

BUSINESS IMPLICATIONS

STEP 8: The cola researcher can proceed with the study under the assumption that the sample represents randomly selected cola drinkers.

Minitab has the capability of analysing data by using the runs test. Figure 17.2 is the Minitab output for the cola example runs test. Notice that the output includes the number of runs, 12, and the significance level of the test. For this analysis, diet cola was coded as a 1 and regular cola coded as a 2. The Minitab runs test is a two-tailed test, and the reported significance of the test is equivalent to a p-value. Because the significance is 0.9710, the decision is to not reject the null hypothesis.

```
Runs Test: Cola
Runs test for Cola
Runs above and below K = 1.69231

The observed number of runs = 12
The expected number of runs = 12.0769
18 observations above K, 8 below
* N is small, so the following approximation may be invalid.
P-value = 0.971
```

FIGURE 17.2

Minitab Output for the Cola Example

Large-Sample Runs Test

Tables A.11 and A.12 in Appendix A do not contain critical values for n_1 and n_2 greater than 20. Fortunately, the sampling distribution of R is approximately normal with a mean and standard deviation of

$$\mu_R = \frac{2n_1n_2}{n_1+n_2}+1 \text{ and } \sigma_R = \sqrt{\frac{2n_1n_2(2n_1n_2-n_1-n_2)}{(n_1+n_2)^2(n_1+n_2-1)}}$$

The test statistic is a z statistic computed as

$$z = \frac{R-\mu_R}{\sigma_R} = \frac{R-\left(\dfrac{2n_1n_2}{n_1+n_2}+1\right)}{\sqrt{\dfrac{2n_1n_2(2n_1n_2-n_1-n_2)}{(n_1+n_2)^2(n_1+n_2-1)}}}$$

The following hypotheses are being tested.

H_0: The observations in the sample were randomly generated.
H_a: The observations in the sample were not randomly generated.

The critical z values are obtained in the usual way by using alpha and Table A.5.

Consider the following manufacturing example. A machine produces parts that are occasionally flawed. When the machine is working in adjustment, flaws still occur but seem to happen randomly. A quality-control person randomly selects 50 of the parts produced by the machine today and examines them one at a time in the order in which they were made. The result is 40 parts with no flaws and 10 parts with flaws. The sequence of no flaws (denoted by N) and flaws (denoted by F) is shown below. Using an alpha of 0.05, the quality controller tests to determine whether the machine is producing randomly (the flaws are occurring randomly).

NNN F NNNNNNN F NN FF NNNNNN F NNNN F NNNNNN

FFFF NNNNNNNNNNNN

HYPOTHESIZE

STEP 1: The hypotheses follow.

H_0: The observations in the sample were generated randomly.
H_a: The observations in the sample were not generated randomly.

TEST

STEP 2: The appropriate statistical test is the large-sample runs test. The test statistic is

$$z = \frac{R - \mu_R}{\sigma_R} = \frac{R - \left(\dfrac{2n_1 n_2}{n_1 + n_2} + 1\right)}{\sqrt{\dfrac{2n_1 n_2 (2n_1 n_2 - n_1 - n_2)}{(n_1 + n_2)^2 (n_1 + n_2 - 1)}}}$$

STEP 3: The value of alpha is 0.05.

STEP 4: This test is two tailed. Too few or too many runs could indicate that the machine is not producing flaws randomly. With $\alpha = 0.05$ and $\alpha/2 = 0.025$, the critical values are $z_{0.025} = \pm 1.96$. The decision rule is to reject the null hypothesis if the observed value of the test statistic is greater than 1.96 or less than -1.96.

STEP 5: The preceding sequence provides the sample data. The value of n_1 is 40 and the value of n_2 is 10. The number of runs (R) is 13.

STEP 6:

$$\mu_R = \frac{2(40)(10)}{40 + 10} + 1 = 17$$

$$\sigma_R = \sqrt{\frac{2(40)(10)[2(40)(10) - 40 - 10]}{(40 + 10)^2 (40 + 10 - 1)}} = 2.213$$

$$z = \frac{13 - 17}{2.213} = -1.81$$

ACTION

STEP 7: Because the observed value of the test statistic, $z = -1.81$, is greater than the lower-tail critical value, $z = -1.96$, the decision is to not reject the null hypothesis.

BUSINESS IMPLICATIONS

STEP 8: There is no evidence that the machine is not producing flaws randomly. If the null hypothesis had been rejected, there might be concern that the machine is producing flaws systematically and thereby is in need of inspection or repair.

Figure 17.3 is the Minitab output for this example. The value of K is the average of the observations. The data were entered into Minitab with a non-flaw coded as a 0 and a flaw as a 1. The value $K = 0.20$ is merely the average of these coded values. In Minitab, a run is a sequence of observations above or below this mean, which effectively yields the same thing as the number of 0s in a row (non-flaws) or number of 1s in a row (flaws). The non-flaws and flaws could have been coded as any two different numbers and the same results would have been achieved. The output shows the number of runs as 13 (the same number obtained manually) and a test significance (p-value) equal to 0.071. The test statistic is not significant at $\alpha = 0.05$ because the p-value is greater than 0.05.

```
Runs Test: Flaws

Runs Test for Flaws
Runs above and below K = 0.2

The observed number of runs = 13
The expected number of runs = 17
10 Observations above K, 40 below
* N is small, so the following approximation may be invalid.
P-value = 0.071
```

FIGURE 17.3

Minitab Output for the Flawed Parts Example

17.1 Test the following sequence of observations by using the runs test and $\alpha = 0.05$ to determine whether the process produced random results.

X X X Y X X Y Y Y X Y X Y X X Y Y Y Y X

17.2 Test the following sequence of observations by using the runs test and $\alpha = 0.05$ to determine whether the process produced random results.

M M N N N N M M M M M M N N M M M M M N M M
N N N N N N N N N N N N M M M M M M M M M M

17.3 A process produced good parts and defective parts. A sample of 60 parts was taken and inspected. Eight defective parts were found. The sequence of good and defective parts was analysed by using Minitab. The output is given here. With a two-tailed test and $\alpha = 0.05$, what conclusions can be reached about the randomness of the sample?

```
Runs Test: Defects

Runs test for Defects
Runs above and below K = 0.1333

The observed number of runs = 11
The expected number of runs = 14.8667
8 Observations above K, 52 below
P-value = 0.0264
```

17.4 A China Central Television survey showed that 45% of all Chinese people consider themselves happy. Suppose a researcher randomly samples 27 Chinese people and asks whether they are happy and that 15 say 'Yes'. The sequence of Yes and No responses is recorded and tested for randomness by means of Minitab. The output follows. Using an alpha of 0.05 and a two-tailed test, what could you conclude about the randomness of the sample?

```
Runs Test: happy

Runs test for happy
Runs above and below K = 0.518519

The observed number of runs = 18
The expected number of runs = 14.4815
14 observations above K, 13 below
P-value = 0.167
```

17.5 The 2001 census found that more than 70% of the UK population had a religion. Suppose a researcher in your region conducts a similar poll and asks the same question with the result that of 64 people interviewed, 40 declare they have a religion. The sequence of responses to this question is given below with Y denoting Yes and N denoting No. Use the runs test and $\alpha = 0.05$ to test this sequence and determine whether the responses are random.

Y Y N Y Y N N Y Y Y N N Y N N Y Y Y Y Y N Y Y Y Y N N Y Y N N N
Y Y Y N N Y Y Y Y N Y N Y Y Y N N N Y N N Y Y Y Y Y N N Y Y Y Y

17.6 Deloitte Touche Tohmatsu Limited's (DTTL) Global Manufacturing Industry group conducted a global survey to explore consumer adoption of electric vehicles (EVs). The survey found that 69% of European consumers either identify themselves as potential first movers, or at least willing to consider purchasing an electric vehicle. Suppose a survey is conducted in your country to ask the same question of 13 randomly selected nationals. The results are that five of the sample say they are willing to consider purchasing an electric vehicle and eight say they are not willing to consider such purchase. The sequence of responses follows. (Y denotes a Yes answer and N denotes a No answer.) Use $\alpha = 0.05$ to test to determine whether this sequence represents a random sample.

N N N N Y Y Y N N N N Y Y

17.2   MANN–WHITNEY *U* TEST

The **Mann–Whitney *U* test** is a *non-parametric counterpart of the t test used to compare the means of two independent populations*. This test was developed by Henry B. Mann and D. R. Whitney in 1947. Recall that the *t* test for independent samples presented in Chapter 10 can be used when data are at least interval in measurement and the populations are normally distributed. However, if the assumption of a normally distributed population is invalid or if the data are only ordinal in measurement, the *t* test should not be used. In such cases, the Mann–Whitney *U* test is an acceptable option for analysing the data. The following assumptions underlie the use of the Mann–Whitney *U* test:

1. The samples are independent.
2. The level of data is at least ordinal.

The two-tailed hypotheses being tested with the Mann–Whitney *U* test are as follows.

$$H_0: \text{The two populations are identical.}$$
$$H_a: \text{The two populations are not identical.}$$

Computation of the *U* test begins by arbitrarily designating two samples as group 1 and group 2. The data from the two groups are combined into one group, with each data value retaining a group identifier of its original group. The pooled values are then ranked from 1 to *n*, with the smallest value being assigned a rank of 1. The sum of the ranks of values from group 1 is computed and designated as W_1 and the sum of the ranks of values from group 2 is designated as W_2.

The Mann–Whitney *U* test is implemented differently for small samples than for large samples. If both $n_1, n_2 \leq 10$, the samples are considered small. If either n_1 or n_2 is greater than 10, the samples are considered large.

Small-Sample Case

With small samples, the next step is to calculate a *U* statistic for W_1 and for W_2 as

$$U_1 = n_1 n_2 + \frac{n_1(n_1+1)}{2} - W_1 \text{ and } U_2 = n_1 n_2 + \frac{n_2(n_2+1)}{2} - W_2$$

The test statistic is the smallest of these two *U* values. Both values do not need to be calculated; instead, one value of *U* can be calculated and the other can be found by using the transformation

$$U' = n_1 \cdot n_2 - U$$

Table A.13 in Appendix A contains *p*-values for U. To determine the *p*-value for a *U* from the table, let n_1 denote the size of the smaller sample and n_2 the size of the larger sample. Using the particular table in Table A.13 for n_1, n_2, locate the value of *U* in the left column. At the intersection of the *U* and n_1 is the *p*-value for a one-tailed test. For a two-tailed test, double the *p*-value shown in the table.

DEMONSTRATION PROBLEM 17.1

Is there a difference between health service workers and educational service workers in the amount of compensation employers pay them per hour? Suppose a random sample of seven health service workers is taken along with a random sample of eight educational service workers from different parts of the United Kingdom. Each of their employers is interviewed and figures are obtained on the amount paid per hour for employee compensation for these workers. The data below indicate total compensation per hour. Use a Mann–Whitney U test to determine whether these two populations are different in employee compensation.

Health Service Worker	Educational Service Worker
£20.10	£26.19
19.80	23.88
22.36	25.50
18.75	21.64
21.90	24.85
22.96	25.30
20.75	24.12
	23.45

Solution

HYPOTHESIZE

STEP 1: The hypotheses are as follows:

H_0: The health service population is identical to the educational service population on employee compensation.

H_a: The health service population is not identical to the educational service population on employee compensation.

TEST

STEP 2: Because we cannot be certain the populations are normally distributed, we chose a non-parametric alternative to the t test for independent populations: the small-sample Mann–Whitney U test.

STEP 3: Let alpha be 0.05.

STEP 4: If the final p-value from Table A.13 (after doubling for a two-tailed test here) is less than 0.05, the decision is to reject the null hypothesis.

STEP 5: The sample data were already provided.

STEP 6: We combine scores from the two groups and rank them from smallest to largest while retaining group identifier information.

Total Employee Compensation	Rank	Group
£18.75	1	H
19.80	2	H
20.10	3	H
20.75	4	H
21.64	5	E
21.90	6	H
22.36	7	H
22.96	8	H
23.45	9	E
23.88	10	E
24.12	11	E
24.85	12	E
25.30	13	E
25.50	14	E

(continued)

Total Employee Compensation	Rank	Group
26.19	15	E

$W_1 = 1 + 2 + 3 + 4 + 6 + 7 + 8 = 31$
$W_2 = 5 + 9 + 10 + 11 + 12 + 13 + 14 + 15 = 89$

$$U_1 = (7)(8) + \frac{(7)(8)}{2} - 31 = 53$$

$$U_2 = (7)(8) + \frac{(8)(9)}{2} - 89 = 3$$

Because U_2 is the smaller value of U, we use $U = 3$ as the test statistic for Table A.13. Because it is the smallest size, let $n_1 = 7$; $n_2 = 8$.

ACTION

STEP 7: Table A.13 yields a p-value of 0.0011. Because this test is two tailed, we double the table p-value, producing a final p-value of 0.0022. Because the p-value is less than $\alpha = 0.05$, the null hypothesis is rejected. The statistical conclusion is that the populations are not identical.

BUSINESS IMPLICATIONS

STEP 8: An examination of the total compensation figures from the samples indicates that employers pay educational service workers more per hour than they pay health service workers.

As shown in Figure 17.4, Minitab has the capability of computing a Mann–Whitney U test. The output includes a p-value of 0.0046 for the two-tailed test for Demonstration Problem 17.1. The decision based on the computer output is to reject the null hypothesis, which is consistent with what we computed. The difference in p-value is due to rounding error in the table.

Large-Sample Case

For large sample sizes, the value of U is approximately normally distributed. Using an average expected U value for groups of this size and a standard deviation of U's allows computation of a z score for the U value. The probability of yielding a z score of this magnitude, given no difference between the groups, is computed. A decision is then made whether to reject the null hypothesis. A z score can be calculated from U by the following formulas.

LARGE-SAMPLE FORMULAS MANN–WHITNEY U TEST (17.1)	$\mu_U = \dfrac{n_1 \cdot n_2}{2}, \quad \sigma_U = \sqrt{\dfrac{n_1 \cdot n_2 (n_1 + n_2 + 1)}{12}}, \quad z = \dfrac{U - \mu_U}{\sigma_U}$

```
Mann-Whitney Test and CI: HS Worker, EdS Worker
            N    Median
HS Worker   7    20.750
EdS Worker  8    24.485
Point estimate for ETA1-ETA2 is -3.385
95.7 Percent CI for ETA1-ETA2 is (-5.370, -1.551)
W = 31.0
Test of ETA1 = ETA2 vs ETA1 not = ETA2 is significant at 0.0046
```

FIGURE 17.4

Minitab Output for Demonstration Problem 17.1

For example, the Mann–Whitney U test can be used to determine whether there is a difference in the average income of European families who subscribe to cable television and families who do not subscribe to cable television. Suppose a sample of 14 families that have identified themselves as cable television subscribers and a sample of 13 families that have identified themselves as non-cable television subscribers are selected randomly.

HYPOTHESIZE

STEP 1: The hypotheses for this example are as follows.

H_0: The incomes of cable television subscribers and non-cable television subscribers are identical.
H_a: The incomes of cable television subscribers and non-cable television subscribers are not identical.

TABLE 17.1

Income of Cable TV Subscribers and Non-Cable TV Subscribers

Cable TV Subscribers (CTV)	Non-Cable TV Subscribers (Non-CTV)
24,500	41,000
39,400	32,500
36,800	33,000
43,000	21,000
57,960	40,500
32,000	32,400
61,000	16,000
34,000	21,500
43,500	39,500
55,000	27,600
39,000	43,500
62,500	51,900
61,400	27,800
53,000	
$n_1 = 14$	$n_2 = 13$

TEST

STEP 2: Use the Mann–Whitney U test for large samples.

STEP 3: Let $\alpha = 0.05$.

STEP 4: Because this test is two tailed with $\alpha = 0.05$, the critical values are $z_{0.025} = +1.96$. If the test statistic is greater than 1.96 or less than –1.96, the decision is to reject the null hypothesis.

STEP 5: The average annual reported income in euros for each family in the two samples is given in Table 17.1.

STEP 6: The first step towards computing a Mann–Whitney U test is to combine these two columns of data into one group and rank the data from lowest to highest, while maintaining the identification of each original group. Table 17.2 shows the results of this step.

Note that in the case of a tie, the ranks associated with the tie are averaged across the values that tie. For example, two incomes of €43,500 appear in the sample. These incomes represent ranks 19 and 20. Each value therefore is awarded a ranking of 19.5, or the average of 19 and 20.

If cable TV subscribers are designated as group 1, W_1 can be computed by summing the ranks of all the incomes of cable TV subscribers in the sample.

$$W_1 = 4 + 7 + 11 + 12 + 13 + 14 + 18 + 19.5 + 22 + 23 + 24 + 25 + 26 + 27$$
$$= 245.5$$

TABLE 17.2

Ranks of Incomes from Combined Groups of Cable TV Subscribers and Non-Cable TV Subscribers

Income	Rank	Group	Income	Rank	Group
16,000	1	Non-CTV	39,500	15	Non-CTV
21,000	2	Non-CTV	40,500	16	Non-CTV
21,500	3	Non-CTV	41,000	17	Non-CTV
24,500	4	CTV	43,000	18	CTV
27,600	5	Non-CTV	43,500	19.5	CTV
27,800	6	Non-CTV	43,500	19.5	Non-CTV
32,000	7	CTV	51,900	21	Non-CTV
32,400	8	Non-CTV	53,000	22	CTV
32,500	9	Non-CTV	55,000	23	CTV
33,000	10	Non-CTV	57,960	24	CTV
34,000	11	CTV	61,000	25	CTV
36,800	12	CTV	61,400	26	CTV
39,000	13	CTV	62,500	27	CTV
39,400	14	CTV			

Then W_1 is used to compute the U value. Because $n_1 = 14$ and $n_2 = 13$, then

$$U = n_1 n_2 + \frac{n_1(n_1+1)}{2} - W_1 = (14)(13) + \frac{(14)(15)}{2} - 245.5 = 41.5$$

Because $n_1, n_2 > 10$, U is approximately normally distributed, with a mean of

$$\mu_U = \frac{n_1 \cdot n_2}{2} = \frac{(14)(13)}{2} = 91$$

and a standard deviation of

$$\sigma_U = \sqrt{\frac{n_1 \cdot n_2(n_1 + n_2 + 1)}{12}} = \sqrt{\frac{(14)(13)(28)}{28}} = 20.6$$

A z value now can be computed to determine the probability of the sample U value coming from the distribution with $\mu_U = 91$ and $\sigma_U = 20.6$ if there is no difference in the populations.

$$z = \frac{U - \mu_U}{\sigma_U} = \frac{41.5 - 91}{20.6} = \frac{-49.5}{20.6} = -2.40$$

ACTION

STEP 7: The observed value of z is –2.40, which is less than $z_{\alpha/2} = -1.96$ so the results are in the rejection region. That is, there is a difference between the income of cable TV subscribers and that of non-cable TV subscribers. Examination of the sample data confirms that in general, the income of cable TV subscribers is higher than that of non-cable TV subscribers.

BUSINESS IMPLICATIONS

STEP 8: The fact that cable TV subscribers have higher average income can affect the type of programming on cable TV in terms of both trying to please present viewers and offering programmes that might attract viewers of other income levels. In addition, fund-raising drives can be made to appeal to the viewers with higher incomes.

Assignment of cable TV subscribers to group 1 was arbitrary. If non-cable TV subscriber viewers had been designated as group 1, the results would have been the same but the observed z value would have been positive.

Figure 17.5 is the Minitab output for this example. Note that Minitab does not produce a z value but rather yields the value of W and the probability of the test results occurring by chance (0.0174). Because the p-value (0.0174) is less than $\alpha = 0.05$, the decision based on the computer output is to reject the null hypothesis. The p-value of the observed test statistic ($z = -2.40$) is 0.0164. The difference is likely to be due to a rounding error.

```
Mann-Whitney Test and CI: CTV, Non-PBS

           N    Median
CTV       14    43250
Non-CTV   13    32500
Point estimate for ETA1-ETA2 is 12500
95.1 Percent CI for ETA1-ETA2 is (3000,22000)
W = 245.5
Test of ETA1 = ETA2 vs ETA1 not = ETA2 is significant at 0.0174
The test is significant at 0.0174 (adjusted for ties)
```

FIGURE 17.5

Minitab Output for the Cable TV Subscribers Example

DEMONSTRATION PROBLEM 17.2

Do construction workers who purchase lunch from street vendors spend less per meal than construction workers who go to restaurants for lunch? To test this question, a researcher selects two random samples of construction workers, one group that purchases lunch from street vendors and one group that purchases lunch from restaurants. Workers are asked to record how much they spend on lunch that day. The data follow. Use the data and a Mann–Whitney U test to analyse the data to determine whether street-vendor lunches are significantly cheaper than restaurant lunches. Let $\alpha = 0.01$.

Vendor	Restaurant
€2.75	€4.10
3.29	4.75
4.53	3.95
3.61	3.50
3.10	4.25
4.29	4.98
2.25	5.75
2.97	4.10
4.01	2.70
3.68	3.65
3.15	5.11
2.97	4.80
4.05	6.25
3.60	3.89
	4.80
	5.50
$n_1 = 14$	$n_2 = 16$

Solution

HYPOTHESIZE

STEP 1: The hypotheses follow.

> H_0: The populations of construction-worker spending for lunch at vendors and restaurants are the same.
>
> H_a: The population of construction-worker spending at vendors is shifted to the left of the population of construction-worker spending at restaurants.

TEST

STEP 2: The large-sample Mann–Whitney U test is appropriate. The test statistic is the z.

STEP 3: Alpha is 0.01.

STEP 4: If the p-value of the sample statistic is less than 0.01, the decision is to reject the null hypothesis.

STEP 5: The sample data are given.

STEP 6: Determine the value of W_1 by combining the groups, while retaining group identification and ranking all the values from 1 to 30 (14 + 16), with 1 representing the smallest value.

Value	Rank	Group	Value	Rank	Group
€2.25	1	V	€4.01	16	V
2.70	2	R	4.05	17	V
2.75	3	V	4.10	18.5	R

(continued)

Value	Rank	Group	Value	Rank	Group
2.97	4.5	V	4.10	18.5	R
2.97	4.5	V	4.25	20	R
3.10	6	V	4.29	21	V
3.15	7	V	4.53	22	V
3.29	8	V	4.75	23	R
3.50	9	R	4.80	24.5	R
3.60	10	V	4.80	24.5	R
3.61	11	V	4.98	26	R
3.65	12	R	5.11	27	R
3.68	13	V	5.50	28	R
3.89	14	R	5.75	29	R
3.95	15	R	6.25	30	R

Summing the ranks for the vendor sample gives

$$W_1 = 1 + 3 + 4.5 + 4.5 + 6 + 7 + 8 + 10 + 11 + 13 + 16 + 17 + 21 + 22 = 144$$

Solving for U, μ_U, and σ_U yields

$$U = (14)(16) + \frac{(14)(15)}{2} - 144 = 185 \qquad \mu_U = \frac{(14)(16)}{2} = 112$$

$$\sigma_U = \sqrt{\frac{(14)(16)(31)}{12}} = 24.1$$

Solving for the observed z value gives

$$z = \frac{185 - 112}{24.1} = 3.03$$

ACTION

STEP 7: The p-value associated with $z = 3.03$ is 0.0012. The null hypothesis is rejected.

BUSINESS IMPLICATIONS

STEP 8: The business researcher concludes that construction-worker spending at vendors is less than the spending at restaurants for lunches.

17.2 PROBLEMS

17.7 Use the Mann–Whitney U test and the data below to determine whether there is a significant difference between the values of group 1 and group 2. Let $\alpha = 0.05$.

Group 1	Group 2
15	23
17	14
26	24
11	13
18	22
21	23
13	18
29	21

17.8 The data shown represent two random samples gathered from two populations. Is there sufficient evidence in the data to determine whether the values of population 1 are significantly larger than the values of population 2? Use the Mann–Whitney U test and $\alpha = 0.01$.

Sample 1	Sample 2
224	203
256	218
231	229
222	230
248	211
283	230
241	209
217	223
240	219
255	236
216	227
	208
	214

17.9 According to a Rapleaf study of 30 million social networkers, more women are on social networks and on average have slightly more friends than men. Suppose you want to validate these results by taking your own sample in a new social networking site. The following data represent the number of friends people have in the new social network. The samples are independent. Use a Mann–Whitney U test to determine whether the number of friends females have is greater than the number of friends males have. Let $\alpha = 0.01$.

Males	Females
12	16
13	15
8	10
11	17
9	13
6	12
11	14
	9
	13

17.10 Suppose 12 urban households and 12 rural households are selected randomly and each family is asked to report the amount spent on food at home annually. The results follow. Use a Mann–Whitney U test to determine whether there is a significant difference between urban and rural households in the amounts spent for food at home. Use $\alpha = 0.05$.

Urban	Rural	Urban	Rural
£2,110	£2,050	£1,950	£2,770
2,655	2,800	2,480	3,100
2,710	2,975	2,630	2,685
2,540	2,075	2,750	2,790
2,200	2,490	2,850	2,995
2,175	2,585	2,850	2,995

17.11 Does the male stock market investor earn significantly more than the female stock market investor? One study by the New York Stock Exchange showed that the male investor has an income of $46,400 and that the female investor has an income of $39,400. Suppose an analyst wanted to 'prove' that the male investor earns more than the female investor. The following data represent random samples of male and female investors from across the United States. The analyst uses the Mann–Whitney U test to determine whether the male investor earns significantly more than the female investor for $\alpha = 0.01$. What does the analyst find?

Male	Female
$50,100	$41,200
47,800	36,600
45,000	44,500
51,500	47,800
55,000	42,500
53,850	47,500
51,500	40,500
63,900	28,900
57,800	48,000
61,100	42,300
51,000	40,000
	31,400

17.12 The Land Registry of England and Wales reports that the median price of an existing single-family home in Teignbridge is £224,160 and the median price of an existing single-family home in East Devon is £248,493. Suppose a survey of 13 randomly selected single-family homes is taken in Teignbridge and a survey of 15 randomly selected single-family homes is taken in East Devon with the resulting prices shown here. Use a Mann–Whitney U test to determine whether there is a significant difference in the price of a single-family home in these two locations. Let $\alpha = 0.05$.

Teignbridge	East Devon
234,157	243,947
238,057	234,127
235,062	235,238
237,016	237,359
235,940	240,031
236,981	239,114
240,479	242,012
240,102	244,500
239,638	236,419
241,861	237,867
241,408	237,741
232,405	234,514
241,730	242,136
	236,333
	243,968

17.3 WILCOXON MATCHED-PAIRS SIGNED RANK TEST

The Mann–Whitney U test presented in Section 17.2 is a non-parametric alternative to the t test for two *independent* samples. If the two samples are *related*, the U test is not applicable. A test that does handle related data is the **Wilcoxon matched-pairs signed rank test**, which serves as *a non-parametric alternative to the t test for two related samples*. Developed by Frank Wilcoxon in 1945, the Wilcoxon test, like the t test for two related samples, is used to analyse several different types of studies when the data of one group are related to the data in the other group, including before-and-after studies, studies in which measures are taken on the same person or object under two different conditions, and studies of twins or other relatives.

The Wilcoxon test utilizes the differences of the scores of the two matched groups in a manner similar to that of the t test for two related samples. After the difference scores have been computed, the Wilcoxon test ranks all differences regardless of whether the difference is positive or negative. The values are ranked from smallest to largest, with a rank of 1 assigned to the smallest difference. If a difference is negative, the rank is given a negative sign. The sum of the positive ranks is tallied along with the sum of the negative ranks. Zero differences representing ties between scores from the two groups are ignored, and the value of n is reduced accordingly. When ties occur between ranks, the ranks are averaged over the values. The smallest sum of ranks (either $+$ or $-$) is used in the analysis and is represented by T. The Wilcoxon matched-pairs signed rank test procedure for determining statistical significance differs with sample size. When the number of matched pairs, n, is greater than 15, the value of T is approximately normally distributed and a z score is computed to test the null hypothesis. When sample size is small, $n \leq 15$, a different procedure is followed.

Two assumptions underlie the use of this technique:

1. The paired data are selected randomly.
2. The underlying distributions are symmetrical.

The following hypotheses are being tested. For two-tailed tests:

$$H_0: M_d = 0 \qquad H_a: M_d \neq 0$$

For one-tailed tests:

$$H_0: M_d = 0 \qquad H_a: M_d > 0$$

or

$$H_0: M_d = 0 \qquad H_a: M_d < 0$$

where M_d is the median.

Small-Sample Case ($n \leq 15$)

When sample size is small, a critical value against which to compare T can be found in Table A.14 to determine whether the null hypothesis should be rejected. The critical value is located by using n and α. Critical values are given in the table for $\alpha = 0.05$, 0.025, 0.01, and 0.005 for two-tailed tests and $\alpha = 0.10$, 0.05, 0.02, and 0.01 for one-tailed tests. If the observed value of T is less than or equal to the critical value of T, the decision is to reject the null hypothesis.

As an example, consider the survey by Visa Europe that estimated the average point-of-sale spend per card in Europe. The European average was €2,524. Suppose six points of sale in Germany are matched with six identical points of sale in France, and their amounts of transactions using Visa cards for last year are obtained. The data are as follows:

Points-of-Sale Pair	Germany	France
1	€1,950	€1,760
2	1,840	1,870
3	2,015	1,810
4	1,580	1,660
5	1,790	1,340
6	1,925	1,765

A credit card analyst uses $\alpha = 0.05$ to test to determine whether there is a significant difference in annual average point-of-sale spend per card spending between these two locations.

HYPOTHESIZE

STEP 1: The following hypotheses are being tested:

$$H_0: M_d = 0$$
$$H_a: M_d \neq 0$$

TEST

STEP 2: Because the sample size of pairs is six, the small-sample Wilcoxon matched-pairs signed ranks test is appropriate if the underlying distributions are assumed to be symmetrical.

STEP 3: Alpha is 0.05.

STEP 4: From Table A.14, if the observed value of T is less than or equal to 1, the decision is to reject the null hypothesis.

STEP 5: The sample data were listed earlier.

STEP 6:

POS Pair	Germany	France	d	Rank
1	€1,950	€1,760	+190	+4
2	1,840	1,870	−30	−1
3	2,015	1,810	+205	+5
4	1,580	1,660	−80	−2
5	1,790	1,340	+450	+6
6	1,925	1,765	+160	+3

$$T = \text{minimum of } (T_+, T_-)$$
$$T_+ = 4 + 5 + 6 + 3 = 18$$
$$T_- = 1 + 2 = 3$$
$$T = \text{minimum of } (18, 3) = 3$$

ACTION

STEP 7: Because $T = 3$ is greater than critical $T = 1$, the decision is not to reject the null hypothesis.

BUSINESS IMPLICATIONS

STEP 8: Not enough evidence is provided to declare that Germany and France differ in average point-of-sale spend per card. This information may be useful to credit and debit card providers and employers in the two areas and particularly to businesses that either operate in both countries or are planning to move from one to the other.

Large-Sample Case ($n > 15$)

For large samples, the T statistic is approximately normally distributed and a z score can be used as the test statistic. Formula 17.2 contains the necessary equations to complete this procedure.

WILCOXON MATCHED-PAIRS SIGNED RANK TEST

$$\mu_T = \frac{(n)(n+1)}{4}$$

$$\sigma_T = \sqrt{\frac{(n)(n+1)(2n+1)}{24}}$$

$$z = \frac{T - \mu_T}{\sigma_T}$$

where:

n = number of pairs
T = total ranks for either $+$ or $-$ differences, whichever is less in magnitude

This technique can be applied to the airline industry, where an analyst might want to determine whether there is a difference in the cost per mile of airfares in Europe between 1982 and 2012 for various cities. The data in Table 17.3 represent the costs per mile of airline tickets for a sample of 17 cities for both 1982 and 2012.

TABLE 17.3

Airline Ticket Costs for Various Cities

City	1982	2012	d	Rank
1	20.3	22.8	−2.5	−8
2	19.5	12.7	+6.8	+17
3	18.6	14.1	+4.5	+13
4	20.9	16.1	+4.8	+15
5	19.9	25.2	−5.3	−16
6	18.6	20.2	−1.6	−4
7	19.6	14.9	+4.7	+14
8	23.2	21.3	+1.9	+6.5
9	21.8	18.7	+3.1	+10
10	20.3	20.9	−0.6	−1
11	19.2	22.6	−3.4	−11.5
12	19.5	16.9	+2.6	+9
13	18.7	20.6	−1.9	−6.5
14	17.7	18.5	−0.8	−2
15	21.6	23.4	−1.8	−5
16	22.4	21.3	+1.1	+3
17	20.8	17.4	+3.4	+11.5

HYPOTHESIZE

STEP 1: The analyst states the hypotheses as follows.

$$H_0: M_d = 0$$
$$H_a: M_d \neq 0$$

TEST

STEP 2: The analyst applies a Wilcoxon matched-pairs signed rank test to the data to test the difference in cents per mile for the two periods of time. She assumes the underlying distributions are symmetrical.

STEP 3: Use $\alpha = 0.05$.

STEP 4: Because this test is two tailed, $\alpha/2 = 0.025$ and the critical values are $z = \pm 1.96$. If the observed value of the test statistic is greater than 1.96 or less than −1.96, the null hypothesis is rejected.

STEP 5: The sample data are given in Table 17.3.

STEP 6: The analyst begins the process by computing a difference score, d. Which year's data are subtracted from the other does not matter as long as consistency in direction is maintained. For the data in Table 17.3, the analyst subtracted the 2012 figures from the 1982 figures. The sign of the difference is left on the difference score. Next, she ranks the differences without regard to sign, but the sign is left on the rank as an identifier. Note the tie for ranks 6 and 7; each is given a rank of 6.5, the average of the two ranks. The same applies to ranks 11 and 12.

After the analyst ranks all difference values regardless of sign, she sums the positive ranks (T_1) and the negative ranks (T_2). She then determines the T value from these two sums as the smallest T_1 or T_2.

$$T = \text{minimum of } (T_+, T_-)$$
$$T_+ = 17 + 13 + 15 + 14 + 6.5 + 10 + 9 + 3 + 11.5 = 99$$
$$T_- = 8 + 16 + 4 + 1 + 11.5 + 6.5 + 2 + 5 = 54$$
$$T = \text{minimum of } (99, 54) = 54$$

The T value is normally distributed for large sample sizes, with a mean and standard deviation of

$$\mu_T = \frac{(n)(n+1)}{4} = \frac{(17)(18)}{4} = 76.5$$

$$\sigma_T = \sqrt{\frac{(n)(n+1)(2n+1)}{24}} = \sqrt{\frac{(17)(18)(35)}{24}} = 21.1$$

The observed z value is

$$z = \frac{T - \mu_T}{\sigma_T} = \frac{54 - 76.5}{21.1} = -1.07$$

ACTION

STEP 7: The critical z value for this two-tailed test is $z_{0.025} = \pm 1.96$. The observed $z = -1.07$, so the analyst fails to reject the null hypothesis. There is no significant difference in the cost of airline tickets between 1982 and 2012.

BUSINESS IMPLICATIONS

STEP 8: Promoters in the airline industry can use this type of information (the fact that ticket prices have not increased significantly in 30 years) to sell their product as a good buy. In addition, industry managers could use it as an argument for raising prices.

DEMONSTRATION PROBLEM 17.3

During the 1980s and 1990s, worldwide businesses increasingly emphasized quality control. One of the arguments in favour of quality-control programmes is that quality control can increase productivity. Suppose a company implemented a quality-control programme and has been operating under it for two years. The company's president wants to determine whether worker productivity significantly increased since installation of the programme. Company records contain the figures for items produced per worker during a sample of production runs two years ago. Productivity figures on the same workers are gathered now and compared to the previous figures. The following data represent items produced per hour. The company's statistical analyst uses the Wilcoxon matched-pairs signed rank test to determine whether there is a significant increase in per worker production for $\alpha = 0.01$.

Worker	Before	After	Worker	Before	After
1	5	11	11	2	6
2	4	9	12	5	10
3	9	9	13	4	9
4	6	8	14	5	7
5	3	5	15	8	9
6	8	7	16	7	6
7	7	9	17	9	10
8	10	9	18	5	8
9	3	7	19	4	5
10	7	9	20	3	6

Solution

HYPOTHESIZE

STEP 1: The hypotheses are as follows:

$$H_o: M_d = 0$$
$$H_a: M_d < 0$$

TEST

STEP 2: The analyst applies a Wilcoxon matched-pairs signed rank test to the data to test the difference in productivity from before to after. He assumes the underlying distributions are symmetrical.

STEP 3: Use $\alpha = 0.01$.

STEP 4: This test is one tailed. The critical value is $z = -2.33$. If the observed value of the test statistic is less than -2.33, the null hypothesis is rejected.

STEP 5: The sample data are as already given.

STEP 6: The analyst computes the difference values, and, because zero differences are to be eliminated, deletes worker 3 from the study. This reduces n from 20 to 19. He then ranks the differences regardless of sign. The differences that are the same (ties) receive the average rank for those values. For example, the differences for workers 4, 5, 7, 10, and 14 are the same. The ranks for these five are 7, 8, 9, 10, and 11, so each worker receives the rank of 9, the average of these five ranks.

Worker	Before	After	d	Rank
1	5	11	−6	−19
2	4	9	−5	−17
3	9	9	0	delete
4	6	8	−2	−9
5	3	5	−2	−9
6	8	7	+1	+3.5
7	7	9	−2	−9
8	10	9	+1	+3.5
9	3	7	−4	−14.5
10	7	9	−2	−9
11	2	6	−4	−14.5
12	5	10	−5	−17
13	4	9	−5	−17
14	5	7	−2	−9
15	8	9	−1	−3.5
16	7	6	+1	+3.5
17	9	10	−1	−3.5
18	5	8	−3	−12.5
19	4	5	−1	−3.5
20	3	6	−3	−12.5

The analyst determines the values of T_+, T_-, and T to be

$$T_+ = 3.5 + 3.5 + 3.5 = 10.5$$
$$T_- = 19 + 17 + 9 + 9 + 9 + 14.5 + 9 + 14.5 + 17 + 17 + 9 + 3.5 + 3.5 + 12.5 + 3.5 + 12.5$$
$$= 179.5$$
$$T = \text{minimum of } (10.5, 179.5) = 10.5$$

The mean and standard deviation of T are

$$\mu_T = \frac{(n)(n+1)}{4} = \frac{(19)(20)}{4} = 95$$

$$\sigma_T = \sqrt{\frac{(n)(n+1)(2n+1)}{24}} = \sqrt{\frac{(19)(20)(39)}{24}} = 24.8$$

The observed z value is

$$Z = \frac{T - \mu_T}{\sigma_T} = \frac{10.5 - 95}{24.8} = -3.41$$

ACTION

STEP 7: The observed z value (−3.41) is in the rejection region, so the analyst rejects the null hypothesis. The productivity is significantly greater after the implementation of quality control at this company.

BUSINESS IMPLICATIONS

STEP 8: Managers, the quality team, and any consultants can point to the figures as validation of the efficacy of the quality programme. Such results could be used to justify further activity in the area of quality.

```
Wilcoxon Signed Rank Test: difference
Test of median = 0.000000 versus median < 0.000000
                         N for    Wilcoxon                Estimated
                  N      Test    Statistic        P        Median
difference 20     19        10.5   0.0000         -2.000
```

FIGURE 17.6

Minitab Output for Demonstration Problem 17.3

Figure 17.6 is Minitab output for Demonstration Problem 17.3. Minitab does not produce a z test statistic for the Wilcoxon matched-pairs signed rank test. Instead, it calculates a Wilcoxon statistic that is equivalent to T. A p-value of 0.000 is produced for this T value. The p-value of the observed $z = -3.41$ determined in Demonstration Problem 17.3 is 0.0003.

17.3 PROBLEMS

17.13 Use the Wilcoxon matched-pairs signed rank test to determine whether there is a significant difference between the two groups of related data given. Use $\alpha = 0.10$. Assume the underlying distributions are symmetrical.

1	2	1	2
212	179	220	223
234	184	218	217
219	213	234	208
199	167	212	215
194	189	219	187
206	200	196	198
234	212	178	189
225	221	213	201

17.14 Use the Wilcoxon matched-pairs signed rank test and $\alpha = 0.05$ to analyse the before-and-after measurements given. Assume the underlying distributions are symmetrical.

Before	After
49	43
41	29
47	30
39	38
53	40
51	43
51	46
49	40
38	42
54	50
46	47
50	47
44	39
49	49
45	47

17.15 A corporation owns a chain of several hundred petrol stations across the European Union. The marketing director wants to test a proposed marketing campaign by running ads on some local television stations and determining whether petrol sales at a sample of the company's stations increase after the advertising. The following data represent petrol sales for a day before and a day after the advertising campaign. Use the Wilcoxon matched-pairs signed rank test to determine whether sales increased significantly after the advertising campaign. Let $\alpha = 0.05$. Assume the underlying distributions are symmetrical.

Station	Before	After
1	€10,500	€12,600
2	8,870	10,660
3	12,300	11,890
4	10,510	14,630
5	5,570	8,580
6	9,150	10,115
7	11,980	14,350
8	6,740	6,900
9	7,340	8,890
10	13,400	16,540
11	12,200	11,300
12	10,570	13,330
13	9,880	9,990
14	12,100	14,050
15	9,000	9,500
16	11,800	12,450
17	10,500	13,450

17.16 Most supermarkets have invested heavily in optical scanner systems to expedite customer checkout, increase checkout productivity, and improve product accountability. These systems are not 100% effective, and items often have to be scanned several times. Sometimes items are entered into the manual cash register because the scanner cannot read the item number. In general, do optical scanners register significantly more items than manual entry systems do? The following data are from an experiment in which a supermarket selected 14 of its best checkers and measured their productivity both when using a scanner and when working manually. The data show the number of items checked per hour by each method. Use a Wilcoxon matched-pairs signed rank test and $\alpha = 0.05$ to test the difference. Assume the underlying distributions are symmetrical.

Checker	Manual	Scanner
1	426	473
2	387	446
3	410	421
4	506	510
5	411	465
6	398	409
7	427	414
8	449	459
9	407	502
10	438	439
11	418	456
12	482	499
13	512	517
14	402	437

17.17 People's attitudes towards big multinational companies change over time and probably are cyclical. Suppose the data below represent a survey of 20 adults taken in 2000 and again in 2012 in which each adult was asked to rate big multinational companies overall on a scale from 1 to 100 in terms of positive opinion. A response of 1 indicates a low opinion and a response of 100 indicates a high opinion. Use a Wilcoxon matched-pairs signed rank test to determine whether the scores from 2012 are significantly higher than the scores from 2000. Use $\alpha = 0.10$. Assume the underlying distributions are symmetrical.

Person	2000	2012
1	49	54
2	27	38
3	39	38
4	75	80
5	59	53
6	67	68
7	22	43
8	61	67
9	58	73
10	60	55
11	72	58
12	62	57
13	49	63
14	48	49
15	19	39
16	32	34
17	60	66
18	80	90
19	55	57
20	68	58

17.18 Suppose 16 people in various industries are contacted in 2011 and asked to rate business conditions on several factors. The ratings of each person are tallied into a 'business optimism' score. The same people are contacted in 2012 and asked to do the same thing. The higher the score, the more optimistic the person is. Shown here are the 2011 and 2012 scores for the 16 people. Use a Wilcoxon matched-pairs signed rank test to determine whether people were less optimistic in 2012 than in 2011. Assume the underlying distributions are symmetrical and that alpha is 0.05.

Industry	April 2011	April 2012
1	63.1	57.4
2	67.1	66.4
3	65.5	61.8
4	68.0	65.3
5	66.6	63.5
6	65.7	66.4
7	69.2	64.9
8	67.0	65.2
9	65.2	65.1
10	60.7	62.2
11	63.4	60.3
12	59.2	57.4
13	62.9	58.2
14	69.4	65.3
15	67.3	67.2
16	66.8	64.1

17.4   KRUSKAL–WALLIS TEST

The *non-parametric alternative to the one-way analysis of variance* is the **Kruskal–Wallis test**, developed in 1952 by William H. Kruskal and W. Allen Wallis. Like the one-way analysis of variance, the Kruskal–Wallis test is used to determine whether $c \geq 3$ samples come from the same or different populations. Whereas the one-way ANOVA is based on the assumptions of normally distributed populations, independent groups, at least interval level data, and equal population variances, the Kruskal–Wallis test can be used to analyse ordinal data and is not based on any assumption about population shape. The Kruskal–Wallis test is based on the assumption that the c groups are independent and that individual items are selected randomly.

The hypotheses tested by the Kruskal–Wallis test follow.

H_0: The c populations are identical.
H_a: At least one of the c populations is different.

This test determines whether all of the groups come from the same or equal populations or whether at least one group comes from a different population.

The process of computing a Kruskal–Wallis K statistic begins with ranking the data in all the groups together, as though they were from one group. The smallest value is awarded a 1. As usual, for ties, each value is given the average rank for those tied values. Unlike one-way ANOVA, in which the raw data are analysed, the Kruskal–Wallis test analyses the ranks of the data.

Formula 17.3 is used to compute a Kruskal–Wallis K statistic.

KRUSKAL–WALLIS TEST (17.3)

$$K = \frac{12}{n(n+1)} \left(\sum_{j=1}^{c} \frac{T_j^2}{n_j} \right) - 3(n+1)$$

where:

c = number of groups
n = total number of items
T_j = total of ranks in a group
n_j = number of items in a group
$K \approx \chi^2$, with df $= c - 1$

The K value is approximately chi-square distributed, with $c - 1$ degrees of freedom as long as n_j is not less than 5 for any group.

Suppose a researcher wants to determine whether the size of the clinic produces significant differences in the number of patients seen by each doctor per day. She takes a random sample of doctors from (1) small, (2) medium-size, or (3) large clinics. Table 17.4 shows the data she obtained.

Three groups are targeted in this study, so $c = 3$, and $n = 18$ physicians, with the numbers of patients ranked for these doctors. The researcher sums the ranks within each column to obtain T_j, as shown in Table 17.5.

The Kruskal–Wallis K is

$$K = \frac{12}{18(18+1)}(1,897) - 3(18+1) = 9.56$$

The critical chi-square value is $\chi^2_{\alpha,df}$. If $\alpha = 0.05$ and df for $c - 1 = 3 - 1 = 2$, $\chi^2_{0.05,2} = 5.9915$. This test is always one tailed, and the rejection region is always in the right tail of the distribution. Because $K = 9.56$ is larger than the critical χ^2 value, the researcher rejects the null hypothesis. The number of patients seen per doctor is not the same in these three sizes of clinics. Examination of the values in each group reveals that doctors in small clinics see fewer patients per doctor, and that large clinic doctors see more patients per doctor.

Figure 17.7 is the Minitab computer output for this example. The statistic H printed in the output is equivalent to the K statistic calculated here (both K and H are 9.56).

TABLE 17.4

Number of Clinic Patients per Doctor

Small Clinic	Medium-size Clinic	Large Clinic
13	24	26
15	16	22
20	19	31
18	22	27
23	25	28
	14	33
	17	

TABLE 17.5

Kruskal–Wallis Analysis of Doctors' Patients

Small Clinic	Medium-size Clinic	Large Clinic
1	12	14
3	4	9.5
8	7	17
6	9.5	15
11	13	16
	2	18
	5	
$T_1 = 29$	$T_2 = 52.5$	$T_3 = 89.5$
$n_1 = 5$	$n_2 = 7$	$n_3 = 6$

$n = 18$

$$\sum_{j=1}^{3} \frac{T_j^2}{n_j} = \frac{(29)^2}{5} + \frac{(52.5)^2}{7} + \frac{(89.5)^2}{6} = 1{,}897$$

```
Kruskal-Wallis Test on Patients

Group     N   Median   Ave Rank      Z
1         5   18.00        5.8    -1.82
2         7   19.00        7.5    -1.27
3         6   27.50       14.9     3.04
Overall  18                9.5

H = 9.56   DF = 2   P = 0.008
H = 9.57   DF = 2   P = 0.008 (adjusted for ties)
```

FIGURE 17.7

Minitab Output for the Doctors' Patients Example

DEMONSTRATION PROBLEM 17.4

Agribusiness researchers are interested in determining the conditions under which Christmas trees grow fastest. A random sample of equivalent-size seedlings is divided into four groups. The trees are all grown in the same field. One group is left to grow naturally, one group is given extra water, one group is given fertilizer spikes, and one group is given fertilizer spikes and extra water. At the end of one year, the seedlings are measured for growth (in height). These measurements are shown for each group. Use the Kruskal–Wallis test to determine whether there is a significant difference in the growth of trees in these groups. Use $\alpha = 0.01$.

Group 1 (native)	Group 2 (+ water)	Group 3 (+ fertilizer)	Group 4 (+ water and fertilizer)
8	10	11	18
5	12	14	20
7	11	10	16
11	9	16	15
9	13	17	14
6	12	12	22

Solution

Here, $n = 24$, and $n_j = 6$ in each group.

HYPOTHESIZE

STEP 1: The hypotheses follow.

$$H_o\text{: group 1} = \text{group 2} = \text{group 3} = \text{group 4}$$
$$H_a\text{: At least one group is different.}$$

TEST

STEP 2: The Kruskal–Wallis K is the appropriate test statistic.

STEP 3: Alpha is 0.01.

STEP 4: The degrees of freedom are $c - 1 = 4 - 1 = 3$. The critical value of chi-square is $\chi^2_{0.01,3} = 11.3449$. If the observed value of K is greater than 11.3449, the decision is to reject the null hypothesis.

STEP 5: The data are as shown previously.

STEP 6: Ranking all group values yields the following.

1	2	3	4	
4	7.5	10	22	
1	13	16.5	23	
3	10	7.5	19.5	
10	5.5	19.5	18	
5.5	15	21	16.5	
2	13	13	24	
$T_1 = 25.5$	$T_2 = 64.0$	$T_3 = 87.5$	$T_4 = 123.0$	
$n_1 = 6$	$n_2 = 6$	$n_3 = 6$	$n_4 = 6$	$n = 24$

$$\sum_{j=1}^{c} \frac{T_j^2}{n_j} = \frac{(25.5)^2}{6} + \frac{(64)^2}{6} + \frac{(87.5)^2}{6} + \frac{(123)^2}{6} = 4{,}588.6$$

$$K = \frac{12}{24(24+1)}(4{,}588.6) - 3(24+1) = 16.77$$

ACTION

STEP 7: The observed K value is 16.77 and the critical $\chi^2_{0.01,3} = 11.3449$. Because the observed value is greater than the table value, the null hypothesis is rejected. There is a significant difference in the way the trees grow.

BUSINESS IMPLICATIONS

STEP 8: From the increased heights in the original data, the trees with both water and fertilizer seem to be doing the best. However, these are sample data; without analysing the pairs of samples with non-parametric multiple comparisons (not included in this text), it is difficult to conclude whether the water/fertilizer group is actually growing faster than the others. It appears that the trees under natural conditions are growing more slowly than the others. The following diagram shows the relationship of the observed K value and the critical chi-square value:

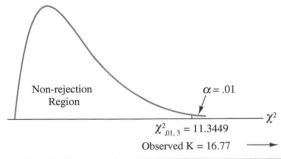

17.4 **PROBLEMS**

17.19 Use the Kruskal–Wallis test to determine whether groups 1 through 5 come from different populations. Let $\alpha = 0.01$.

1	2	3	4	5
157	165	219	286	197
188	197	257	243	215
175	204	243	259	235
174	214	231	250	217
201	183	217	279	240
203		203		233
				213

17.20 Use the Kruskal–Wallis test to determine whether there is a significant difference in the groups below. Use $\alpha = 0.05$.

Group 1	19	21	29	22	37	42	
Group 2	30	38	35	24	29		
Group 3	39	32	41	44	30	27	33

17.21 Is there a difference in the amount of customers' initial deposits when they open savings accounts according to geographic region of Italy? To test this question, an analyst selects savings and loan offices of equal size from four regions of Italy. The offices selected are located in areas having similar economic and population characteristics. The analyst randomly selects adult customers who are opening their first savings account and obtains the following euro amounts. Use the Kruskal–Wallis test to determine whether there is a significant difference between geographic regions. Use $\alpha = 0.05$.

Region 1	Region 2	Region 3	Region 4
1,200	225	675	1,075
450	950	500	1,050
110	100	1,100	750
800	350	310	180
375	275	660	330
200			680
			425

17.22 Does the asking price of a new car vary according to whether the dealership is in a small town, a city, or a suburban area? To test this question, a researcher randomly selects dealerships selling Skodas in the Netherlands. The researcher goes to these dealerships posing as a prospective buyer and makes a serious inquiry as to the asking price of a new Skoda Octavia Classic (each having the same equipment). The following data represent the results of this sample. Is there a significant difference between prices according to the area in which the dealership is located? Use the Kruskal–Wallis test and $\alpha = 0.05$.

Small Town	City	Suburb
€21,800	€22,300	€22,000
22,500	21,900	22,600
21,750	21,900	22,800
22,200	22,650	22,050
21,600	21,800	21,250
		22,550

17.23 Europeans travel to various destinations during public holidays and for long weekend breaks. How long do people stay at these destinations? Does the length of stay differ according to location? Suppose the following data were taken from a survey of people who were asked how many nights they stay at their chosen destinations. Use a Kruskal–Wallis test to determine whether there is a significant difference in the duration of stay by type of holiday destination. Let $\alpha = 0.05$.

Amusement Park	Seaside	City	National Park
0	3	2	2
1	2	2	4
1	3	3	3
0	5	2	4
2	4	3	3
1	4	2	5
0	3	3	4
	5	3	4
	2	1	
		3	

17.24 Do workers on different shifts get different amounts of sleep per week? Some people believe that shift workers who regularly work the night shift (12.00 a.m. to 8.00 a.m.) or afternoon shift (4.00 p.m. to 12.00 a.m.) are unable to get the same amount of sleep as day workers because of family schedules, noise, amount of daylight, and other factors. To test this theory, a researcher samples workers from day, afternoon, and night shifts and asks each worker to keep a sleep journal for one week. The following data represent the number of hours of sleep per week per worker for the different shifts. Use the Kruskal–Wallis test to determine whether there is a significant difference in the number of hours of sleep per week for workers on these shifts. Use $\alpha = 0.05$.

Day Shift	Afternoon Shift	Night Shift
52	45	41
57	48	46
53	44	39
56	51	49
55	48	42
50	54	35
51	49	52
	43	

STATISTICS IN BUSINESS TODAY

Job Search for the Class of 2011

The UK Graduate Careers Survey 2011 was conducted by High Fliers Research, an independent research company that specializes in student and graduate research. The survey was based on face-to-face interviews with 17,851 final-year students from 30 leading universities.

This sample includes more than a fifth of the finalists who were due to graduate from these universities in the 2010/11 academic year.

The survey revealed that the average starting salary expected by graduates was £22,600 and that investment banking was the most popular career choice for the year's university leavers. Other popular choices for graduates were expected to be careers in the media, teaching, and marketing. Additionally, the survey found that: a further 25% of the 'class of 2011' planned to remain at university for postgraduate study; 8% expected to take temporary or voluntary work; 13% were preparing to take time off or go travelling; and 14% of finalists had yet to decide what to do next.

In total, 59% of finalists had made graduate job applications by the time the survey took place in March, compared with the 55% who had applied by the same point in the previous year. The average number of graduate job applications made by finalists had increased too, from 5.7 applications per student in 2009/10, to 6.8 applications per student in this year. Fewer finalists had applied for graduate positions in the police and the armed forces, while the number of graduate job hunters keen to work for the government or elsewhere in the public sector had dropped by one-fifth, perhaps a reflection of the dramatic cost-cutting measures across the public sector being implemented by the coalition government. The survey also found that, after five years in work, graduates expected to be earning an average of £39,900 and one-sixth of this year's university-leavers believed that their salary will be £100,000 or more by the age of 30.

Non-parametric techniques can be used in studies similar to this one. For example, one study published in January 2011 found an increase of 8.9% in graduate job offers in the UK. If the distributions of job offers are unknown, the Mann–Whitney U test could be used to test to determine if there is a significant difference in the number of job offers between 2010 and 2011. Furthermore, suppose the demographics of graduate job starters can be broken down by gender. A Kruskal–Wallis test could be used to determine if there is a significant difference in income levels of male and female graduates. In addition, a Wilcoxon matched-pairs signed rank test could in future be used to determine if the wages of 2011 graduates significantly increased from one year to the next for the same set of graduates.

Source: Adapted from 'The UK Graduate Careers Survey 2011', available at www.highfliers.co.uk/download/Release2011.pdf.

17.5   FRIEDMAN TEST

The **Friedman test**, developed by M. Friedman in 1937, is *a non-parametric alternative to the randomized block design* discussed in Chapter 11. The randomized block design has the same assumptions as other ANOVA procedures, including observations are drawn from normally distributed populations. When this assumption cannot be met or when the researcher has ranked data, the Friedman test provides a non-parametric alternative.

Three assumptions underlie the Friedman test:

1. The blocks are independent.
2. No interaction is present between blocks and treatments.
3. Observations within each block can be ranked.

The hypotheses being tested are as follows:

H_0: The treatment populations are equal.
H_a: At least one treatment population yields larger values than at least one other treatment population.

The first step in computing a Friedman test is to convert all raw data to ranks (unless the data are already ranked). However, unlike the Kruskal–Wallis test where all data are ranked together, the data in a Friedman test are ranked *within* each block from smallest (1) to largest (c). Each block contains c ranks, where c is the number of treatment levels. Using these ranks, the Friedman test will test to determine whether it is likely that the different treatment levels (columns) came from the same population. Formula 17.4 is used to calculate the test statistic, which is approximately chi-square distributed with $df = c - 1$ if $c > 4$, or when $c = 3$ and $b > 9$, or when $c = 4$ and $b > 4$.

FRIEDMAN TEST (17.4)

$$\chi_r^2 = \frac{12}{bc(c+1)} \sum_{j=1}^{c} R_j^2 - 3b(c+1)$$

where:

c = number of treatment levels (columns)
b = number of blocks (rows)
R_j = total of ranks for a particular treatment level (column)
j = particular treatment level (column)
$\chi_r^2 \approx \chi^2$, with $df = c - 1$

As an example, suppose a manufacturing company assembles microcircuits that contain a plastic housing. Managers are concerned about an unacceptably high number of the products that sustained housing damage during shipment. The housing component is made by four different suppliers. Managers have decided to conduct a study of the plastic housing by randomly selecting five housings made by each of the four suppliers. To determine whether a supplier is consistent during the production week, one housing is selected for each day of the week. That is, for each supplier, a housing made on Monday is selected, one made on Tuesday is selected, and so on.

In analysing the data, the treatment variable is supplier and the treatment levels are the four suppliers. The blocking effect is day of the week with each day representing a block level. The quality-control team wants to determine whether there is any significant difference in the tensile strength of the plastic housing by supplier. The data are given here (in pounds per inch).

Day	Supplier 1	Supplier 2	Supplier 3	Supplier 4
Monday	62	63	57	61
Tuesday	63	61	59	65
Wednesday	61	62	56	63
Thursday	62	60	57	64
Friday	64	63	58	66

HYPOTHESIZE

STEP 1: The hypotheses follow.

H_0: The supplier populations are equal.

H_a: At least one supplier population yields larger values than at least one other supplier population.

TEST

STEP 2: The quality researchers do not feel they have enough evidence to conclude that the observations come from normally distributed populations. Because they are analysing a randomized block design, the Friedman test is appropriate.

STEP 3: Let $\alpha = 0.05$.

STEP 4: For four treatment levels (suppliers), $c = 4$ and $df = 4 - 1 = 3$. The critical value is $\chi^2_{0.05,3} = 7.8147$. If the observed chi-square is greater than 7.8147, the decision is to reject the null hypothesis.

STEP 5: The sample data are as given.

STEP 6: The calculations begin by ranking the observations in each row with 1 designating the rank of the smallest observation. The ranks are then summed for each column, producing R_j. The values of R_j are squared and then summed.

Because the study is concerned with five days of the week, five blocking levels are used and $b = 5$. The value of R_j is computed as shown in the following table.

Day	Supplier 1	Supplier 2	Supplier 3	Supplier 4
Monday	3	4	1	2
Tuesday	3	2	1	4
Wednesday	2	3	1	4
Thursday	3	2	1	4
Friday	3	2	1	4
R_j	14	13	5	18
R_j^2	196	169	25	324

$$\sum_{j=1}^{4} R_j^2 = (196 + 169 + 25 + 324) = 714$$

$$\chi_r^2 = \frac{12}{bc(c+1)} \sum_{j=1}^{c} R_j^2 - 3b(c+1) = \frac{12}{5(4)(4+1)}(714) - 3(5)(4+1) = 10.68$$

ACTION

STEP 7: Because the observed value of $\chi_r^2 = 10.68$ is greater than the critical value, $\chi_{0.05,3}^2 = 7.8147$, the decision is to reject the null hypothesis.

BUSINESS IMPLICATIONS

STEP 8: Statistically, there is a significant difference in the tensile strength of housings made by different suppliers. The sample data indicate that supplier 3 is producing housings with a lower tensile strength than those made by other suppliers and that supplier 4 is producing housings with higher tensile strength. Further study by managers and a quality team may result in attempts to bring supplier 3 up to standard on tensile strength or perhaps cancellation of the contract.

Figure 17.8 displays the chi-square distribution for df = 3 along with the critical value, the observed value of the test statistic, and the rejection region. Figure 17.9 is the Minitab output for the Friedman test. The computer output contains the value of χ_r^2 referred to as S along with the p-value of 0.014, which informs the researcher that the null hypothesis is rejected at an alpha of 0.05. Additional information is given about the medians and the column sum totals of ranks.

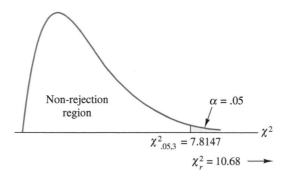

FIGURE 17.8

Distribution for Tensile Strength Example

```
Friedman Test: Tensile Strength Versus Supplier Blocked by Day
S = 10.68     DF = 3     P = 0.014
                                Sum
                                 of
Supplier   N   Est Median     Ranks
1          5       62.125      14.0
2          5       61.375      13.0
3          5       56.875       5.0
4          5       64.125      18.0
Grand median = 61.125
```

FIGURE 17.9

Minitab Output for the Tensile Strength Example

DEMONSTRATION PROBLEM 17.5

A market research company wants to determine brand preference for refrigerators. Five companies contracted with the research company to have their products be included in the study. As part of the study, the research company randomly selects 10 potential refrigerator buyers and shows them one of each of the five brands. Each survey participant is then asked to rank the refrigerator brands from 1 to 5. The results of these rankings are given in the table. Use the Friedman test and $\alpha = 0.01$ to determine whether there are any significant differences between the rankings of these brands.

Solution

HYPOTHESIZE

STEP 1: The hypotheses are as follows:

H_0: The brand populations are equal.

H_a: At least one brand population yields larger values than at least one other brand population.

TEST

STEP 2: The market researchers collected ranked data that are ordinal in level. The Friedman test is the appropriate test.

STEP 3: Let $\alpha = 0.01$.

STEP 4: Because the study uses five treatment levels (brands), $c = 5$ and $df = 5 - 1 = 4$. The critical value is $\chi^2_{0.01,4} = 13.2767$. If the observed chi-square is greater than 13.2767, the decision is to reject the null hypothesis.

STEP 5: The sample data follow.

STEP 6: The ranks are totalled for each column, squared, and then summed across the column totals. The results are shown in the table.

Individual	Brand A	Brand B	Brand C	Brand D	Brand E
1	3	5	2	4	1
2	1	3	2	4	5
3	3	4	5	2	1
4	2	3	1	4	5
5	5	4	2	1	3
6	1	5	3	4	2
7	4	1	3	2	5

(continued)

Individual	Brand A	Brand B	Brand C	Brand D	Brand E
8	2	3	4	5	1
9	2	4	5	3	1
10	3	5	4	2	1
R_j	26	37	31	31	25
R_j^2	676	1,369	961	961	625

$$\sum R_j^2 = 4{,}592$$

The value of χ_r^2 is

$$\chi_r^2 = \frac{12}{bc(c+1)} \sum_{j=1}^{c} R_j^2 - 3b(c+1) = \frac{12}{10(5)(5+1)}(4{,}592) - 3(10)(5+1) = 3.68$$

ACTION

STEP 7: Because the observed value of $\chi_r^2 = 3.68$ is not greater than the critical value, $\chi_{0.01,4}^2 = 13.2767$, the researchers fail to reject the null hypothesis.

BUSINESS IMPLICATIONS

STEP 8: Potential refrigerator purchasers appear to have no significant brand preference. Marketing managers for the various companies might want to develop strategies for positively distinguishing their product from the others.

The chi-square distribution for four degrees of freedom, produced by Minitab, is shown with the observed test statistic and the critical value. In addition, Minitab output for the Friedman test is shown. Note that the p-value is 0.451, which underscores the decision not to reject the null hypothesis at $\alpha = 0.01$.

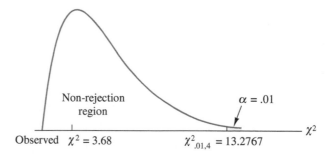

Minitab Friedman Output:

```
Friedman Test: Rank versus Brand Blocked by Individual
S = 3.68   DF = 4   P = 0.451

                     Sum
              Est     of
Brand   N   Median  Ranks
1      10   2.300   26.0
2      10   4.000   37.0
3      10   3.000   31.0
4      10   3.000   31.0
5      10   1.700   25.0
Grand median = 2.800
```

17.25 Use the data below to test to determine whether there are any differences between treatment levels. Let $\alpha = 0.05$.

		\multicolumn{5}{c}{Treatment}				
		1	2	3	4	5
	1	200	214	212	215	208
	2	198	211	214	217	206
Block	3	207	206	213	216	207
	4	213	210	215	219	204
	5	211	209	210	221	205

17.26 Use the Friedman test and $\alpha = 0.05$ to test the following data to determine whether there is a significant difference between treatment levels:

		\multicolumn{6}{c}{Treatment}					
		1	2	3	4	5	6
	1	29	32	31	38	35	33
	2	33	35	30	42	34	31
	3	26	34	32	39	36	35
	4	30	33	35	41	37	32
Block	5	33	31	32	35	37	36
	6	31	34	33	37	36	35
	7	26	32	35	43	36	34
	8	32	29	31	38	37	35
	9	30	31	34	41	39	35

17.27 An experiment is undertaken to study the effects of four different medical treatments on the recovery time for a medical disorder. Six doctors are involved in the study. One patient with the disorder is sampled for each doctor under each treatment, resulting in 24 patients in the study. Recovery time in days is the observed measurement. The data are given here. Use the Friedman test and $\alpha = 0.01$ to determine whether there is a significant difference in recovery times for the four different medical treatments.

		\multicolumn{4}{c}{Treatment}			
		1	2	3	4
	1	3	7	5	4
	2	4	5	6	3
Doctor	3	3	6	5	4
	4	3	6	7	4
	5	2	6	7	3
	6	4	5	7	3

17.28 Does the configuration of the working week have any impact on productivity? This question is raised by a researcher who wants to compare the traditional five-day working week with a four-day working week and a working week with three 12-hour days and one four-hour day. The researcher conducts the experiment in a factory making small electronic parts. He selects 10 workers who spend a month working under each type of work-week configuration. The researcher randomly selects one day from each of the three months (three work-week configurations) for each of the 10 workers. The observed measurement is the number of parts produced per day by each worker. Use the Friedman test to determine whether there is a difference in productivity by work-week configuration.

		Work-week Configuration		
		5 Days	4 Days	$3^1/_2$ Days
	1	37	33	28
	2	44	38	36
	3	35	29	31
	4	41	40	36
Worker	5	38	39	35
	6	34	27	23
	7	43	38	39
	8	39	35	32
	9	41	38	37
	10	36	30	31

17.29 Shown here is Minitab output from a Friedman test. What is the size of the experimental design in terms of treatment levels and blocks? Discuss the outcome of the experiment in terms of any statistical conclusions.

```
Friedman Test: Observations versus Treatment Blocked by Block

S = 2.04    DF = 3    P = 0.564

                        Sum
                Est      of
Treatment   N   Median   Ranks
1           5   3.250    15.0
2           5   2.000    10.0
3           5   2.750    11.0
4           5   4.000    14.0

Grand median = 3.000
```

17.30 Shown here is Minitab output for a Friedman test. Discuss the experimental design and the outcome of the experiment.

```
Friedman Test: Observations versus Treatment Blocked by Block

S = 13.71    DF = 4    P = 0.009

                        Sum
                Est      of
Treatment   N   Median   Ranks
1           7   21.000   12.0
2           7   24.000   14.0
3           7   29.800   30.0
4           7   27.600   26.0
5           7   27.600   23.0

Grand median = 26.000
```

17.6 SPEARMAN'S RANK CORRELATION

In Chapter 12, the Pearson product-moment correlation coefficient, r, was presented and discussed as a technique to measure the amount or degree of association between two variables. The Pearson r requires at least interval level of measurement for the data. When only ordinal-level data or ranked data are available, **Spearman's rank correlation**, r_s, can be used to analyse the degree of association of two variables. Charles E. Spearman (1863–1945) developed this correlation coefficient.

The formula for calculating a Spearman's rank correlation is as follows:

SPEARMAN'S RANK
CORRELATION (17.7)

$$r_s = 1 - \frac{6\sum d^2}{n(n^2 - 1)}$$

where:

n = number of pairs being correlated
d = the difference in the ranks of each pair

The Spearman's rank correlation formula is derived from the Pearson product-moment formula and utilizes the ranks of the n pairs instead of the raw data. The value of d is the difference in the ranks of each pair.

The process begins by the assignment of ranks within each group. The difference in ranks between each group (d) is calculated by subtracting the rank of a member of one group from the rank of its associated member of the other group. The differences (d) are then squared and summed. The number of pairs in the groups is represented by n.

The interpretation of r_s values is similar to the interpretation of r values. Positive correlations indicate that high values of one variable tend to be associated with high values of the other variable, and low values of one variable tend to be associated with low values of the other variable. Correlations near $+1$ indicate high positive correlations, and correlations near -1 indicate high negative correlations. Negative correlations indicate that high values of one variable tend to be associated with low values of the other, and vice versa. Correlations near zero indicate little or no association between variables.

Listed in Table 17.6 are the average government bond yield and consumer price index over a 10-year period for Greece. The data were published by the International Monetary Fund. Suppose we want to determine the strength of association between the Greek government bond yield and consumer price index by using Spearman's rank correlation.

The government bond yields are ranked and the consumer price index are ranked. The difference in ranks is computed for each year. The differences are squared and summed, producing $\sum d^2 = 190$. The number of pairs, n, is 10. The value of $r_s = -0.152$ indicates that there is a very modest negative correlation between government bond yields and the consumer price index in Greece. The calculations of this Spearman's rank correlation are enumerated in Table 17.7.

TABLE 17.6

Government Bond Yield and Consumer Price Index over a 10-Year Period in Greece

Year	Government Bond Yield (%)	Consumer Price Index (%)
1	6.10	3.17
2	5.30	3.37
3	5.12	3.63
4	4.27	3.53
5	4.26	2.90
6	3.59	3.55
7	4.07	3.20
8	4.50	2.90
9	4.80	4.15
10	5.17	1.21

TABLE 17.7

Calculations of Spearman's Rank Correlation for Government Bond Yield and Consumer Price Index over a 10-Year Period in Greece

Year	Rank: Government Bond Yield	Rank: Consumer Price Index	d	d^2
1	10	4	6	36
2	9	6	3	9
3	7	9	-2	4
4	4	7	-3	9
5	3	3	0	0
6	1	8	-7	49
7	2	5	-3	9
8	5	2	3	9
9	6	10	-4	16
10	8	1	7	49

$$r_s = 1 - \frac{6\sum d^2}{n(n^2 - 1)} = 1 - \frac{6(190)}{10(10^2 - 1)} = -0.152$$

DEMONSTRATION PROBLEM 17.6

How strong is the correlation between the price of silver and the price of gold? In an effort to estimate this association, a commodities analyst gathered the data shown over a selected number of years from *Wikipedia*. She computes a Spearman's rank correlation for these data.

Year	Silver Price (US$ per troy ounce)	Gold Price (US$ per troy ounce)
1940	0.34	33
1960	0.91	35
1970	1.63	35
1980	16.39	612
1990	4.06	383
2000	4.95	279
2005	7.31	444
2009	14.67	972
2010	20.19	1,225

Solution

Here, $n = 9$. When the analyst ranks the values within each group and computes the values of d and d^2, she obtains the following:

Year	Silver	Gold	d	d^2
1940	1	1	0	0
1960	2	2.5	−0.5	0.25
1970	3	2.5	0.5	0.25
1980	8	7	1	1
1990	4	5	−1	1
2000	5	4	1	1
2005	6	6	0	0
2009	7	8	−1	1
2010	9	9	0	0
				$\Sigma d^2 = 4.5$

$$r_s = 1 - \frac{6\Sigma d^2}{n(n^2 - 1)} = 1 - \frac{6(4.5)}{9(9 - 1)} = +.963$$

A high positive correlation is computed between the price of silver and the price of gold.

17.31 Compute a Spearman's rank correlation for the following variables to determine the degree of association between the two variables:

x	y
23	201
41	259
37	234
29	240
25	231
17	209
33	229
41	246
40	248
28	227
19	200

17.32 The data below are the ranks for values of the two variables, x and y. Compute a Spearman's rank correlation to determine the degree of relation between the two variables.

x	y	x	y
4	6	3	2
5	8	1	3
8	7	2	1
11	10	9	11
10	9	6	4
7	5		

17.33 Compute a Spearman's rank correlation for the following data:

x	y	x	y
99	108	80	124
67	139	57	162
82	117	49	145
46	168	91	102

17.34 Over a period of a few months, is there a strong correlation between the value of the British pound and the UK Government Bond Yield? The following data represent a sample of these quantities over a period of time. Compute a Spearman's rank correlation to determine the strength of the relationship between the government bond yield rates and the value of the pound:

British Pound Value (US$ per Pound)	Government Bond Yield (%)	British Pound Value (US$ per Pound)	Government Bond Yield (%)
1.592	1.78	1.609	2.56
1.553	2.00	1.666	2.59
1.566	2.32	1.650	2.28
1.589	2.51	1.601	2.09
1.623	2.74		

17.35 Shown here are the average annual central bank discount rates over a 14-year period for the Euro-area and the United States, according to the International Monetary Fund. Compute a Spearman's rank correlation to determine the degree of association between these two variables.

Year	Euro Area	United States
1999	4.00	5.50
2000	5.75	6.50
2001	4.25	1.75
2002	3.75	1.25
2003	3.00	1.00
2004	3.00	2.25
2005	3.25	4.25
2006	4.50	5.25
2007	5.00	4.25
2008	3.00	0.13
2009	1.75	0.13
2010	1.75	0.13

17.36 Shown here are the net inward and outward foreign direct investment (from non-EU countries) as a percentage of GDP figures in Bulgaria as reported by Eurostat over an 11-year period. Use these data to calculate a Spearman's rank correlation to determine the degree of association between inward and outward FDI in Bulgaria over this period. Was the association strong? Comment.

Year	Inward FDI from non-EU Countries (% of GDP)	Outward FDI from non-EU Countries (% of GDP)
1	5.8	0.1
2	7.7	0
3	5	0.1
4	3.8	0.2
5	10.1	0.1
6	13.4	−0.8
7	13.6	1.1
8	23.5	0.5
9	29.4	0.7
10	18.9	1.4
11	9.4	−0.3

17.37 Is there a correlation between the FTSE100 share index and the NASDAQ composite index? Shown below are the values for these two variables over a recent 11-month period. Compute a Spearman's rank correlation to determine the degree of association between these two variables.

Month	FTSE100 Index	NASDAQ Index
1	5,528.3	14.04
2	5,899.9	14.1
3	5,862.9	14.6
4	5,994.0	14.95
5	5,908.8	14.65
6	6,069.9	15.05
7	5,990.0	14.85
8	5,945.7	14.38
9	5,815.2	14.23
10	5,394.5	13.66
11	5,066.8	13.08

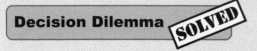

Decision Dilemma SOLVED

How Is the Doughnut Business?

The Dunkin' Donuts researchers' dilemma is that in each of the three studies presented, the assumptions underlying the use of parametric statistics are in question or have not been met. The distribution of the data is unknown bringing into question the normal distribution assumption or the level of data is only ordinal. For each study, a nonparametric technique presented in this chapter could be appropriately used to analyse the data.

The differences in doughnut sizes according to machine can be analysed using the Kruskal–Wallis test. The independent variable is machine with four levels of classification. The dependent variable is size of doughnut in centimetres. The Kruskal–Wallis test is not based on any assumption about population shape. The following Minitab output is from a Kruskal–Wallis test on the machine data presented in the Decision Dilemma.

```
Kruskal-Wallis Test: Diameter versus Machine
Kruskal-Wallis Test on Diameter
Machine   N    Median   Ave Rank      Z
1         5    7.520      10.4      -0.60
2         6    7.415       3.5      -3.57
3         5    7.550      12.6       0.22
4         7    7.700      20.0       3.74
Overall   23              12.0
H = 19.48  DF = 3  P = 0.000
H = 19.51  DF = 3  P = 0.000 (adjusted for ties)
```

Because the H statistic (Minitab's equivalent to the K statistic) has a p-value of 0.000, there is a significant difference in the diameter of the doughnut according to machine at $\alpha = 0.001$. An examination of median values reveals that machine 4 is producing the largest doughnuts and machine 2 the smallest.

How well did the advertising work? One way to address this question is to perform a before-and-after test of the number of doughnuts sold. The non-parametric alternative to the matched-pairs t test is the Wilcoxon matched-pairs signed rank test. The analysis for these data is:

Before	After	d	Rank
301	374	−73	−9
198	187	11	4
278	332	−54	−7

(continued)

Before	After	d	Rank
205	212	−7	−3
249	243	6	2
410	478	−68	−8
360	386	−26	−6
124	141	−17	−5
253	251	2	1
190	264	−74	−10

$T_+ = 4 + 2 + 1 = 7$
$T_- = 9 + 7 + 3 + 8 + 6 + 5 + 10 = 48$
observed $T = \min(T_+, T_-) = 7$
critical T for .025 and $n = 10$ is 8

Using a two-sided test and $\alpha = 0.05$, the critical T value is 8. Because the observed T is 7, the decision is to reject the null hypothesis. There is a significant difference between the before and after number of donuts sold. An observation of the ranks and raw data reveals that a majority of the stores experienced an increase in sales after the advertising campaign.

Do bigger stores have greater sales? Because the data are given as ranks, it is appropriate to use Spearman's Rank Correlation to determine the extent of the correlation between these two variables. Shown below are the calculations of a Spearman's Rank correlation for this problem.

Sales	Size	d	d^2
6	7	−1	1
2	2	0	0
3	6	−3	9
7	5	2	4
5	4	1	1
1	1	0	0
4	3	1	1
		$\sum d^2 = 16$	

$$r_s = 1 - \frac{6\sum d^2}{n(n^2 - 1)} = 1 - \frac{6(16)}{7(49 - 1)} = 0.714$$

There is a relatively strong correlation (0.714) between sales and size of store. It is not, however, a perfect correlation, which leaves room for other factors that may determine a store's sales such as location, attractiveness of store, population density, number of employees, management style, and others.

ETHICAL CONSIDERATIONS

The researcher should be aware of all assumptions underlying the usage of statistical techniques. Many parametric techniques have level-of-data requirements and assumptions about the distribution of the population or assumptions about the parameters. Inasmuch as these assumptions and requirements are not met, the researcher sets herself or himself up for misuse of statistical analysis. Spurious results can follow, and misguided conclusions can be reached. Non-parametric statistics can be used in many cases to avoid such pitfalls. In addition, some non-parametric statistics require at least ordinal-level data.

SUMMARY

Non-parametric statistics are a group of techniques that can be used for statistical analysis when the data are less than interval in measurement or when assumptions about population parameters, such as shape of the distribution, cannot be met. Non-parametric tests offer several advantages. Sometimes the non-parametric test is the only technique available, with no parametric alternative. Non-parametric tests can be used to analyse nominal- or ordinal-level data. Computations from non-parametric tests are usually simpler than those used with parametric tests. Probability statements obtained from most non-parametric tests are exact probabilities. Non-parametric techniques also have some disadvantages. They are wasteful of data whenever a parametric technique can be used. Non-parametric tests are not as widely available as parametric tests. For large sample sizes, the calculations of non-parametric statistics can be tedious.

Many of the parametric techniques presented in this text have corresponding non-parametric techniques. The six non-parametric statistical techniques presented here are the runs test, the Mann–Whitney U test, the Wilcoxon matched-pairs signed rank test, the Kruskal–Wallis test, the Friedman test, and Spearman's rank correlation.

The runs test is a non-parametric test of randomness. It is used to determine whether the order of sequence of observations in a sample is random. A run is a succession of observations that have a particular characteristic. If data are truly random, neither a very high number of runs nor a very small number of runs is likely to be present.

The Mann–Whitney U test is a non-parametric version of the t test of the means from two independent samples. When the assumption of normally distributed data cannot be met or if the data are only ordinal in level of measurement, the Mann–Whitney U test can be used in place of the t test. The Mann–Whitney U test – like many non-parametric tests – works with the ranks of data rather than the raw data.

The Wilcoxon matched-pairs signed rank test is used as an alternative to the t test for related measures when assumptions cannot be met or if the data are ordinal in measurement. In contrast to the Mann–Whitney U test, the Wilcoxon test is used when the data are related in some way. The Wilcoxon test is used to analyse the data by ranks of the differences of the raw data.

The Kruskal–Wallis test is a non-parametric one-way analysis of variance technique. It is particularly useful when the assumptions underlying the F test of the parametric one-way ANOVA cannot be met. The Kruskal–Wallis test is usually used when the researcher wants to determine whether three or more groups or samples are from the same or equivalent populations. This test is based on the assumption that the sample items are selected randomly and that the groups are independent. The raw data are converted to ranks and the Kruskal–Wallis test is used to analyse the ranks with the equivalent of a chi-square statistic.

The Friedman test is a non-parametric alternative to the randomized block design. Friedman's test is computed by ranking the observations within each block and then summing the ranks for each treatment level. The resulting test statistic χ^2 is approximately chi-square distributed.

If two variables contain data that are ordinal in level of measurement, a Spearman's rank correlation can be used to determine the amount of relationship or association between the variables. Spearman's rank correlation coefficient is a non-parametric alternative to Pearson's product-moment correlation coefficient. Spearman's rank correlation coefficient is interpreted in a manner similar to the Pearson r.

Friedman test
Kruskal–Wallis test
Mann–Whitney U test
non-parametric statistics

parametric statistics
runs test
Spearman's rank correlation
Wilcoxon matched-pairs signed rank test

GO ONLINE TO DISCOVER THE EXTRA FEATURES FOR THIS CHAPTER

The Student Study Guide containing solutions to the odd-numbered questions, additional Quizzes and Concept Review Activities, Excel and Minitab databases, additional data files in Excel and Minitab, and more worked examples.
www.wiley.com/college/cortinhas

FORMULAS

Large-sample runs test

$$\mu_R = \frac{2n_1 n_2}{n_1 + n_2} + 1$$

$$\sigma_R = \sqrt{\frac{2n_1 n_2 (2n_1 n_2 - n_1 - n_2)}{(n_1 + n_2)^2 (n_1 + n_2 - 1)}}$$

$$z = \frac{R - \mu_R}{\sigma_R} = \frac{R - \left(\dfrac{2n_1 n_2}{n_1 + n_2} + 1\right)}{\sqrt{\dfrac{2n_1 n_2 (2n_1 n_2 - n_1 - n_2)}{(n_1 + n_2)^2 (n_1 + n_2 - 1)}}}$$

Mann–Whitney U test small sample:

$$U_1 = n_1 n_2 + \frac{n_1 (n_2 + 1)}{2} - W_1$$

$$U_2 = n_1 n_2 + \frac{n_2 (n_2 + 1)}{2} - W_2$$

$$U' = n_1 \cdot n_2 - U$$

Large sample:

$$\mu_U = \frac{n_1 \cdot n_2}{2}$$

$$\sigma_U = \sqrt{\frac{n_1 \cdot n_2 (n_1 + n_2 + 1)}{12}}$$

$$z = \frac{U - \mu_U}{\sigma_U}$$

Wilcoxon matched-pair signed rank test

$$\mu_T = \frac{(n)(n+1)}{4}$$

$$\sigma_T = \sqrt{\frac{(n)(n+1)(2n+1)}{24}}$$

$$z = \frac{T - \mu_T}{\sigma_T}$$

Kruskal–Wallis test

$$K = \frac{12}{n(n+1)} \left(\sum_{j=1}^{c} \frac{T_j^2}{n_j} \right) - 3(n+1)$$

Friedman test

$$\chi^2 = \frac{12}{bc(c+1)} \sum_{j=1}^{c} R_j^2 - 3b(c+1)$$

Spearman's rank correlation

$$r_s = 1 - \frac{6 \sum d_2}{n(n^2 - 1)}$$

CALCULATING THE STATISTICS

17.38 Use the runs test to determine whether the sample is random. Let alpha be 0.05.

1	1	1	1	2	2	2	2	2	2	2	1	1	1	2	2	2	
2	2	2	2	2	1	2	1	2	2	1	1	1	1	2	2	2	

17.39 Use the Mann–Whitney U test and $\alpha = 0.01$ to determine whether there is a significant difference between the populations represented by the two samples given here.

Sample 1	Sample 2
573	547
532	566
544	551
565	538
540	557
548	560
536	557
523	547

17.40 Use the Wilcoxon matched-pairs signed rank test to determine whether there is a significant difference between the related populations represented by the matched pairs given here. Assume $\alpha = 0.05$.

Group 1	Group 2
5.6	6.4
1.3	1.5
4.7	4.6
3.8	4.3
2.4	2.1
5.5	6.0
5.1	5.2
4.6	4.5
3.7	4.5

17.41 Use the Kruskal–Wallis test and $\alpha = 0.01$ to determine whether the four groups come from different populations.

Group 1	Group 2	Group 3	Group 4
6	4	3	1
11	13	7	4
8	6	7	5
10	8	5	6
13	12	10	9
7	9	8	6
10	8	5	7

17.42 Use the Friedman test to determine whether the treatment groups come from different populations. Let alpha be 0.05.

Block	Group 1	Group 2	Group 3	Group 4
1	16	14	15	17
2	8	6	5	9
3	19	17	13	18
4	24	26	25	21
5	13	10	9	11
6	19	11	18	13
7	21	16	14	15

17.43 Compute a Spearman's rank correlation to determine the degree of association between the two variables.

Variable 1	Variable 2
101	87
129	89
133	84
147	79
156	70
179	64
183	67
190	71

TESTING YOUR UNDERSTANDING

17.44 Commercial fish farming is a growing industry in Europe. What makes fish raised commercially grow faster and larger? Suppose that a fish industry study is conducted over the three summer months in an effort to determine whether the amount of water allotted per fish makes any difference in the speed with which the fish grow. The data below represent the inches of growth of salmon in fish farms for different volumes of water per fish. Use $\alpha = 0.01$ to test whether there is a significant difference in fish growth by volume of allotted water.

1 Litre per Fish	5 Litres per Fish	10 Litres per Fish
1.1	2.9	3.1
1.4	2.5	2.4
1.7	2.6	3.0
1.3	2.2	2.3
1.9	2.1	2.9
1.4	2.0	1.9
2.1	2.7	

17.45 Manchester Partners International claims that 60% of the banking executives who lose their job stay in banking, whereas 40% leave banking. Suppose 40 people who have lost their job as a banking executive are contacted and are asked whether they are still in banking. The results follow. Test to determine whether this sample appears to be random on the basis of the sequence of those who have left banking and those who have not. Let L denote left banking and S denote stayed in banking. Let $\alpha = 0.05$.

S S L S L L S S S S S L S S L L L S S L
L L L S S L S S S S S S L L S L S S L S

17.46 Three machines produce the same part. Ten different machine operators work these machines. A quality team wants to determine whether the machines are producing parts that are significantly different from each other in weight. The team devises an experimental design in which a random part is selected from each of the 10 machine operators on each machine. The results follow. Using alpha of 0.05, test to determine whether there is a difference in machines.

Operator	Machine 1	Machine 2	Machine 3
1	231	229	234
2	233	232	231
3	229	233	230
4	232	235	231
5	235	228	232
6	234	237	231
7	236	233	230
8	230	229	227
9	228	230	229
10	237	238	234

17.47 In some firefighting organizations, you must serve as a firefighter for some period of time before you can become part of the emergency medical service arm of the organization. Does that mean EMS workers are older, on average, than traditional firefighters? Use the data shown and $\alpha = 0.05$ to test whether EMS workers are significantly older than firefighters. Assume the two groups are independent and you do not want to use a t test to analyse the data.

Firefighters	EMS Workers	Firefighters	EMS Workers
23	27	32	39
37	29	24	33
28	30	21	30

(continued)

Firefighters	EMS Workers	Firefighters	EMS Workers
25	33	27	28
41	28		27
36	36		30

17.48 Automobile dealers usually advertise in the yellow pages of the telephone book. Sometimes they have to pay to be listed in the white pages, and some dealerships opt to save money by omitting that listing, assuming most people will use the yellow pages to find the telephone number. A two-year study is conducted with 20 car dealerships where in one year the dealer is listed in the white pages and the other year it is not. Ten of the dealerships are listed in the white pages the first year and the other 10 are listed there in the second year in an attempt to control for economic cycles. The data below represent the numbers of units sold per year. Is there a significant difference between the number of units sold when the dealership is listed in the white pages and the number sold when it is not listed? Assume all companies are continuously listed in the yellow pages, that the t test is not appropriate, and that $\alpha = 0.01$.

Dealer	With Listing	Without Listing
1	1,180	1,209
2	874	902
3	1,071	862
4	668	503
5	889	974
6	724	675
7	880	821
8	482	567
9	796	602
10	1,207	1,097
11	968	962
12	1,027	1,045
13	1,158	896
14	670	708
15	849	642
16	559	327
17	449	483
18	992	978
19	1,046	973
20	852	841

17.49 Suppose you want to take a random sample of statistics test scores to determine whether there is any significant difference between the statistics scores for the test given in March and the scores for the test given in June. You gather the data on the following page from a sample of persons who took each test. Use the Mann–Whitney U test to determine whether there is a significant difference in the two test results. Let $\alpha = 0.10$.

March	June
83.1	53.8
87.7	72.3
92.3	96.9
66.2	90.8
76.9	93.8
78.5	80
81.5	70.8
86.2	84.6
84.6	81.5
75.4	87.7

17.50 Does impulse buying really increase sales? A market researcher is curious to find out whether the location of packages of chewing gum in a grocery store really has anything to do with volume of gum sales. As a test, gum is moved to a different location in the store every Monday for four weeks (four locations). To control the experiment for type of gum, six different brands are moved around. Sales representatives keep track of how many packs of each type of gum are sold every Monday for the four weeks. The results follow. Test to determine whether there are any differences in the volume of gum sold at the various locations. Let $\alpha = 0.05$.

		Location			
		1	2	3	4
	A	176	58	111	120
	B	156	62	98	117
Brand	C	203	89	117	105
	D	183	73	118	113
	E	147	46	101	114
	F	190	83	113	115

17.51 Does deodorant sell better in a box or without additional packaging? An experiment in a large store is designed in which, for one month, all deodorants are sold packaged in a box and, during a second month, all deodorants are removed from the box and sold without packaging. Is there a significant difference in the number of units of deodorant sold with and without the additional packaging? Let $\alpha = 0.05$.

Deodorant	Box	No Box
1	185	170
2	109	112

(continued)

Deodorant	Box	No Box
3	92	90
4	105	87
5	60	51
6	45	49
7	25	11
8	58	40
9	161	165
10	108	82
11	89	94
12	123	139
13	34	21
14	68	55
15	59	60
16	78	52

17.52 Some people drink coffee to relieve stress on the job. Is there a correlation between the number of cups of coffee consumed on the job and perceived job stress? Suppose the data shown represent the number of cups of coffee consumed per week and a stress rating for the job on a scale of 0 to 100 for nine managers in the same industry. Determine the correlation between these two variables, assuming you do not want to use the Pearson product-moment correlation coefficient.

Cups of Coffee per Week	Job Stress
25	80
41	85
16	35
0	45
11	30
28	50
34	65
18	40
5	20

17.53 A Gallup/Air Transport Association survey showed that in a recent year 52% of all air trips were for pleasure/personal and 48% were for business. Suppose the organization randomly samples 30 air travellers and asks them to state the purpose of their trip. The results are shown here with B denoting business and P denoting personal. Test the sequence of these data to determine whether the data are random. Let $\alpha = 0.05$.

B	P	B	P	B	B	P	B	P	P	P	B	P	B	P	P
P	B	P	B	B	B	P	B	P	P	B	B	P	P	B	B

17.54 Does a statistics course improve a student's mathematics skills, as measured by a national test? Suppose a random sample of 13 students takes the same national mathematics examination just prior to enrolling in a statistics course and just after completing the course. Listed are the students' quantitative scores from both examinations. Use $\alpha = 0.01$ to determine whether the scores after the statistics course are significantly higher than the scores before.

Student	Before	After
1	71.7	77.5
2	80.8	79.2
3	86.7	89.2
4	60.0	68.3
5	73.3	70.8
6	83.3	84.2
7	70.8	75.0
8	78.3	80.0
9	85.8	86.7
10	71.7	71.7
11	75.0	76.7
12	82.5	83.3
13	90.0	88.3

17.55 Should male managers wear a tie during the workday to command respect and demonstrate professionalism? Suppose a measurement scale has been developed that generates a management professionalism score. A random sample of managers in a high-tech industry is selected for the study, some of whom wear ties at work and others of whom do not. One subordinate is selected randomly from each manager's department and asked to complete the scale on their boss's professionalism. Analyse the data taken from these independent groups to determine whether the managers with the ties received significantly higher professionalism scores. Let $\alpha = 0.05$.

With Tie	Without Tie
27	22
23	16
25	25
22	19
25	21
26	24
21	20
25	19
26	23

(continued)

With Tie	Without Tie
28	26
22	17

17.56 Many fast-food restaurants have soft drink dispensers with preset amounts, so that when the operator merely pushes a button for the desired drink the cup is automatically filled. This method apparently saves time and seems to increase worker productivity. To test this conclusion, a researcher randomly selects 18 workers from the fast-food industry, 9 from a restaurant with automatic soft drink dispensers and 9 from a comparable restaurant with manual soft drink dispensers. The samples are independent. During a comparable hour, the amount of sales rung up by the worker is recorded. Assume that $\alpha = 0.01$ and that a t test is not appropriate. Test whether workers with automatic dispensers are significantly more productive (higher sales per hour).

Automatic Dispenser	Manual Dispenser
€153	€105
128	118
143	129
110	114
152	125
168	117
144	106
137	92
118	126

17.57 A particular metal part can be produced at different temperatures. All other variables being equal, a company would like to determine whether the strength of the metal part is significantly different for different temperatures. Given are the strengths of random samples of parts produced under different temperatures. Use $\alpha = 0.01$ and determine whether there is a significant difference in the strength of the part for different temperatures.

45°	55°	70°	85°
216	228	219	218
215	224	220	216
218	225	221	217
216	222	223	221
219	226	224	218
214	225		217

NON-PARAMETRIC STATISTICS

17.58 Is there a strong correlation between the number of miles driven by a salesperson and sales volume achieved? Data were gathered from nine salespeople who worked territories of similar size and potential. Determine the correlation coefficient for these data. Assume the data are ordinal in level of measurement.

Sales	Miles per Month
150,000	1,500
210,000	2,100
285,000	3,200
301,000	2,400
335,000	2,200
390,000	2,500
400,000	3,300
425,000	3,100
440,000	3,600

17.59 Workers in three different but comparable companies were asked to rate the use of quality-control techniques in their firms on a 50-point scale. A score of 50 represents nearly perfect implementation of quality-control techniques and 0 represents no implementation. Workers are divided into three independent groups. One group worked in a company that had required all its workers to attend a three-day seminar on quality control one year ago. A second group worked in a company in which each worker was part of a quality circle group that had been meeting at least once a month for a year. The third group of workers was employed by a company in which management had been actively involved in the quality-control process for more than a year. Use $\alpha = 0.10$ to determine whether there is a significant difference between the three groups, as measured by the ratings.

Attended 3-Day Seminar	Quality Circles	Management Involved
9	27	16
11	38	21
17	25	18
10	40	28
22	31	29
15	19	20
6	35	31

17.60 The scores given are husband–wife scores on a marketing measure. Use the Wilcoxon matched-pairs signed rank test to determine whether the

wives' scores are significantly higher on the marketing measure than the husbands'. Assume that $\alpha = 0.01$.

Husbands	Wives
27	35
22	29
28	30
19	20
28	27
29	31
18	22
21	19
25	29
18	28
20	21
24	22
23	33
25	38
22	34
16	31
23	36
30	31

INTERPRETING THE OUTPUT

17.61 Study the following Minitab output. What statistical test was run? What type of design was it? What was the result of the test?

```
Friedman  Test:  Observations  versus
Treatment Blocked by Block

   S = 11.31    DF = 3    P = 0.010

                               Sum
                        Est     Of
   Treatment    N    Median   Ranks
   1           10    20.125    17.0
   2           10    25.875    33.0
   3           10    24.500    30.5
   4           10    22.500    19.5
Grand median = 23.250
```

17.62 Examine the Minitab output below. Discuss the statistical test, its intent, and its outcome.

```
Runs Test

Runs above and below K =  1.4200
The observed number of runs = 28
The expected number of runs = 25.3600
21 Observations above K, 29 below
P-value = 0.439
```

17.63 Study the Minitab output below. What statistical test was being computed by Minitab? What are the results of this analysis?

```
Mann-Whitney Test and CI

             N    Median
1st Group   16    37.000
2nd Group   16    46.500

Point estimate for ETA1-ETA2 is -8.000
95.7 Percent CI for ETA1-ETA2 is
(-13.999, -2.997)
W = 191.5
Test of ETA1 = ETA2 vs ETA1 not = ETA2
is significant at 0.0067
```

17.64 Study the following Minitab output. What type of statistical test was done? What were the hypotheses, and what was the outcome? Discuss.

```
Kruskal-Wallis Test on Observations

Machine   N   Median   Ave Rank      Z
1         5    35.00      14.8     0.82
2         6    25.50       4.2    -3.33
3         7    35.00      15.0     1.11
4         6    35.00      16.0     1.40
Overall  24              12.5

H = 11.21    DF = 3    P = 0.011
```

ANALYSING THE DATABASES

1. Compute a Spearman's rank correlation between New Capital Expenditures and End-of-Year Inventories in the Manufacture database. Is the amount spent annually on New Capital Expenditures related to the End-of-Year Inventories? Are these two variables highly correlated? Explain.

2. Use a Kruskal–Wallis test to determine whether there is a significant difference between the four levels of Value of Industry Shipments on Number of Employees for the Manufacture database. Discuss the results.

3. Use a Mann–Whitney U test to determine whether there is a significant difference between hospitals that are general hospitals and those that are psychiatric (Service variable) on Personnel for the Hospital database. Discuss the results.

4. Use the Kruskal–Wallis test to determine if there is a significant difference in Total Revenues by Type of Company in the Financial database.

CASE

RALEIGH

Raleigh's history started in 1887, in Raleigh Street, Nottingham. Frank Bowden, a prosperous 38-year-old, bought a bicycle made by Messrs. Woodhead, Angois, and Ellis, because his doctor had told him to ride a bicycle for his health. Bowden was impressed by his bicycle and went to Raleigh Street to find the makers. Woodhead, Angois, and Ellis were making three bikes, a week. Bowden made them an offer and bought the business. Production rose and three years later he needed a bigger workshop, which he found in a four-storey building in Russell Street. He changed the company's name to Raleigh Cycles to commemorate the original address. In six years Bowden created a business which became the biggest bicycle manufacturing company in the world, occupying seven and a half acres in Faraday Road, Lenton, Nottingham.

While bicycle production had steadily risen through the mid-1950s, the British market began to decline with the increasing affordability and popularity of the automobile. So, for much of the post-War era, British bicycle manufacturers largely competed with each other in the export market. The 1950s saw the creation of the British Cycle Corporation under the Tube Investments Group which owned Phillips, Hercules, Sun, Armstrong, and Norman. In 1957, Raleigh bought BSA Cycles Ltd, BSA's bicycle division, from the parent group. In 1960, Tube Investments acquired Raleigh and merged the British Cycle Corporation with Raleigh to form TI-Raleigh which had 75% of the UK market.

In 1982, rights to the Raleigh USA name were purchased by the Huffy Corporation. Under the terms of the agreement, Raleigh of England licensed Huffy to design and distribute Raleigh bicycles in the USA, and Huffy was given instant access to a nationwide network of bike shops. The renamed Raleigh Cycle Company of America sold bikes in the US while the rest of the world, including Canada, received Raleigh of England bikes. At that time, production of some Raleigh models was shifted to Japan, with Bridgestone Corporation of Japan manufacturing most of these bikes. By 1984, all Raleighs for the American market, except the top-of-the range Team Professional and Prestige road bikes, were produced in the Far East.

In 1987, the leading German bicycle manufacturer Derby Cycle bought Raleigh USA from Huffy. Today, Raleigh Cycle Company of America parts and frames are mass-produced in China and Taiwan for Derby Cycle and assembled in other plants. In 2000, Derby Cycle controlled Raleigh USA, Raleigh UK, Raleigh Canada, and Raleigh Ireland. In the latter three markets, Raleigh was the number-one manufacturer of bicycles. Derby Cycle began a series of divestitures, selling Sturmey-Archer to SunRace of Taiwan, and selling Brooks to Selle Royal of Italy. In 2001, following financial problems, there was a management buy-out of all the remaining Raleigh companies led by Alan Finden-Crofts who was advised by Paul Revill, an employee of the company at the time who remains employed by Raleigh UK at the Nottingham Head Office. By 2003, assembly of bicycles had ended in the UK with 280 assembly and factory staff made redundant, and all bicycles coming from East Asia.

DISCUSSION

1. What is the age of the target market for Raleigh bikes in the UK? One theory is that in counties where mountain bikes are more popular, the mean age of their customers is older than in counties where relatively little mountain biking is done. In an attempt to test this theory, a random sample of Cumbrian customers is taken along with a random sample of customers in Essex. The ages for these customers are given here. The customer is defined as 'the person for whom the bike is primarily purchased'. The shape of the population distribution of bicycle customer ages is unknown. Analyse the data and discuss the implications for Raleigh manufacturing and sales.

Cumbria	Essex
29	11
38	14
31	15
17	12
36	14
28	25
44	14
9	11
32	8
23	
35	

2. Suppose for a particular model of bike, the specified weight of a handle bar is 200 grams and Raleigh uses three different suppliers of handle bars. Suppose Raleigh conducts a quality-control study in which handle bars are randomly selected from each supplier and weighed. The results (in grams) are shown next. It is uncertain whether handle bar weight is normally distributed in the population. Analyse the data and discuss what the business implications are to Raleigh.

Supplier 1	Supplier 2	Supplier 3
200.76	197.38	192.63
202.63	207.24	199.68
198.03	201.56	203.07
201.24	194.53	195.18
202.88	197.21	189.11
194.62	198.94	
203.58		
205.41		

3. Quality technicians at Raleigh's manufacturing plant examine their finished products for paint flaws. Paint inspections are done on a production run of 75 bicycles. The inspection data are coded

and the data analysed using Minitab. If a bicycle's paint job contained no flaws, a 0 is recorded; and if it contained at least one flaw, the code used is a 1. Inspectors want to determine whether the flawed bikes occur in a random fashion or in a non-random pattern. Study the Minitab output. Determine whether the flaws occur randomly. Report on the proportion of flawed bikes and discuss the implications of these results to Raleigh's production management.

```
Runs Test: Paint Flaw

Runs above and below K = 0.2533

The observed number of runs = 29
The expected number of runs = 29.3733
19 Observations above K, 56 below
P-value = 0.908
```

Source: Adapted from the *Wikipedia* entry for Raleigh Bicycle Company, available at http://en.wikipedia.org/wiki/Raleigh_Bicycle_Company, and Raleigh website, available at www.raleigh.co.uk.

USING THE COMPUTER

MINITAB

- Five of the non-parametric statistics presented in this chapter can be accessed using Minitab. For each non-parametric technique: select **Stat** from the menu bar. From the **Stat** pull-down menu, select **Nonparametrics**. From **Nonparametrics** pulldown menu, select the appropriate non-parametric technique from **Runs Test**, **Mann-Whitney**, **Kruskal-Wallis**, **Friedman**, and **1-Sample Wilcoxon**.

- To begin Runs test select **Runs Test** from the **Nonparametrics** pulldown menu. Supply the location of the column with the data in the **Variables** space: Check either **Above and below the mean** or **Above and below**. Minitab will default to **Above and below the mean** and will use the mean of the numbers to determine when the runs stop. Select **Above and below** if you want to supply your own value.

- To begin a Mann–Whitney U test, select **Mann-Whitney** from the **Nonparametrics** pulldown menu. Place the column location of the data from the first sample in **First Sample**. Place the column location of the data from the second sample in **Second Sample**. Insert the level of confidence in **Confidence level**. In the slot provided beside **Alternative**, select the form of the alternative hypothesis. Choose from **not equal**, **less than**, and **greater than**.

- To begin a Kruskal–Wallis test, all observations from all treatment levels must be located in one column. Place the location of the column containing these observations in the space provided beside **Response**. Enter the treatment levels to match the observations in a second column. Place the location of the column containing these treatment levels in the space provided beside **Factor**.

- To begin a Friedman test, all observations must be located in one column. Place the location of the column containing these observations in the space provided beside **Response**. Enter the treatment levels to match the observations in a second column. Place the location of the column containing these treatment levels in the space provided beside **Treatment**. Enter the block levels to match the observations in a third column. Place the location of the column containing these block levels in the space provided beside **Blocks**. You have the option of storing residuals and storing fits by clicking on **Store residuals** and **Store fits**. You are not allowed to store the fits without storing the residuals.

- There is no Wilcoxin matched-pairs signed rank test in Minitab. You must manipulate Minitab to perform this test. Either enter the data from the two related samples in two columns and use the calculator under **Calc** on the main menu bar to subtract the two columns, thereby creating a third column of differences; or enter the differences in a column to begin with. Select **1-sample Wilcoxon**. In the space provided next to **Variables**, enter the location of the column containing the differences. Place the level of confidence in the box beside **Confidence interval**. Minitab will default to 95%. Minitab will default a hypothesis test of a 0.0 median. To test any other value, check **Test median** and enter the new test value for the median. In the slot provided beside **Alternative**, select the form of the alternative hypothesis. Choose from **not equal**, **less than**, and **greater than**.

Tables

TABLE A.1

Random Numbers

12651	61646	11769	75109	86996	97669	25757	32535	07122	76763
81769	74436	02630	72310	45049	18029	07469	42341	98173	79260
36737	98863	77240	76251	00654	64688	09343	70278	67331	98729
82861	54371	76610	94934	72748	44124	05610	53750	95938	01485
21325	15732	24127	37431	09723	63529	73977	95218	96074	42138
74146	47887	62463	23045	41490	07954	22597	60012	98866	90959
90759	64410	54179	66075	61051	75385	51378	08360	95946	95547
55683	98078	02238	91540	21219	17720	87817	41705	95785	12563
79686	17969	76061	83748	55920	83612	41540	86492	06447	60568
70333	00201	86201	69716	78185	62154	77930	67663	29529	75116
14042	53536	07779	04157	41172	36473	42123	43929	50533	33437
59911	08256	06596	48416	69770	68797	56080	14223	59199	30162
62368	62623	62742	14891	39247	52242	98832	69533	91174	57979
57529	97751	54976	48957	74599	08759	78494	52785	68526	64618
15469	90574	78033	66885	13936	42117	71831	22961	94225	31816
18625	23674	53850	32827	81647	80820	00420	63555	74489	80141
74626	68394	88562	70745	23701	45630	65891	58220	35442	60414
11119	16519	27384	90199	79210	76965	99546	30323	31664	22845
41101	17336	48951	53674	17880	45260	08575	49321	36191	17095
32123	91576	84221	78902	82010	30847	62329	63898	23268	74283
26091	68409	69704	82267	14751	13151	93115	01437	56945	89661
67680	79790	48462	59278	44185	29616	76531	19589	83139	28454
15184	19260	14073	07026	25264	08388	27182	22557	61501	67481
58010	45039	57181	10238	36874	28546	37444	80824	63981	39942
56425	53996	86245	32623	78858	08143	60377	42925	42815	11159
82630	84066	13592	60642	17904	99718	63432	88642	37858	25431
14927	40909	23900	48761	44860	92467	31742	87142	03607	32059
23740	22505	07489	85986	74420	21744	97711	36648	35620	97949
32990	97446	03711	63824	07953	85965	87089	11687	92414	67257
05310	24058	91946	78437	34365	82469	12430	84754	19354	72745
21839	39937	27534	88913	49055	19218	47712	67677	51889	70926
08833	42549	93981	94051	28382	83725	72643	64233	97252	17133
58336	11139	47479	00931	91560	95372	97642	33856	54825	55680
62032	91144	75478	47431	52726	30289	42411	91886	51818	78292
45171	30557	53116	04118	58301	24375	65609	85810	18620	49198
91611	62656	60128	35609	63698	78356	50682	22505	01692	36291
55472	63819	86314	49174	93582	73604	78614	78849	23096	72825
18573	09729	74091	53994	10970	86557	65661	41854	26037	53296
60866	02955	90288	82136	83644	94455	06560	78029	98768	71296
45043	55608	82767	60890	74646	79485	13619	98868	40857	19415
17831	09737	79473	75945	28394	79334	70577	38048	03607	06932
40137	03981	07585	18128	11178	32601	27994	05641	22600	86064
77776	31343	14576	97706	16039	47517	43300	59080	80392	63189
69605	44104	40103	95635	05635	81673	68657	09559	23510	95875
19916	52934	26499	09821	97331	80993	61299	36979	73599	35055
02606	58552	07678	56619	65325	30705	99582	53390	46357	13244
65183	73160	87131	35530	47946	09854	18080	02321	05809	04893
10740	98914	44916	11322	89717	88189	30143	52687	19420	60061
98642	89822	71691	51573	83666	61642	46683	33761	47542	23551
60139	25601	93663	25547	02654	94829	48672	28736	84994	13071

Reprinted with permission from *A Million Random Digits with 100,000 Normal Deviates* by the RAND Corporation. New York: The Free Press, 1955. Copyright 1955 and 1983 by RAND.

TABLE A.2

Binomial Probability Distribution

					$n=1$				
					Probability				
x	.1	.2	.3	.4	.5	.6	.7	.8	.9
0	.900	.800	.700	.600	.500	.400	.300	.200	.100
1	.100	.200	.300	.400	.500	.600	.700	.800	.900

					$n=2$				
					Probability				
x	.1	.2	.3	.4	.5	.6	.7	.8	.9
0	.810	.640	.490	.360	.250	.160	.090	.040	.010
1	.180	.320	.420	.480	.500	.480	.420	.320	.180
2	.010	.040	.090	.160	.250	.360	.490	.640	.810

					$n=3$				
					Probability				
x	.1	.2	.3	.4	.5	.6	.7	.8	.9
0	.729	.512	.343	.216	.125	.064	.027	.008	.001
1	.243	.384	.441	.432	.375	.288	.189	.096	.027
2	.027	.096	.189	.288	.375	.432	.441	.384	.243
3	.001	.008	.027	.064	.125	.216	.343	.512	.729

					$n=4$				
					Probability				
x	.1	.2	.3	.4	.5	.6	.7	.8	.9
0	.656	.410	.240	.130	.063	.026	.008	.002	.000
1	.292	.410	.412	.346	.250	.154	.076	.026	.004
2	.049	.154	.265	.346	.375	.346	.265	.154	.049
3	.004	.026	.076	.154	.250	.346	.412	.410	.292
4	.000	.002	.008	.026	.063	.130	.240	.410	.656

					$n=5$				
					Probability				
x	.1	.2	.3	.4	.5	.6	.7	.8	.9
0	.590	.328	.168	.078	.031	.010	.002	.000	.000
1	.328	.410	.360	.259	.156	.077	.028	.006	.000
2	.073	.205	.309	.346	.313	.230	.132	.051	.008
3	.008	.051	.132	.230	.313	.346	.309	.205	.073
4	.000	.006	.028	.077	.156	.259	.360	.410	.328
5	.000	.000	.002	.010	.031	.078	.168	.328	.590

(*Continued*)

TABLE A.2

Binomial Probability Distribution (*Continued*)

					$n = 6$				
					Probability				
x	.1	.2	.3	.4	.5	.6	.7	.8	.9
0	.531	.262	.118	.047	.016	.004	.001	.000	.000
1	.354	.393	.303	.187	.094	.037	.010	.002	.000
2	.098	.246	.324	.311	.234	.138	.060	.015	.001
3	.015	.082	.185	.276	.313	.276	.185	.082	.015
4	.001	.015	.060	.138	.234	.311	.324	.246	.098
5	.000	.002	.010	.037	.094	.187	.303	.393	.354
6	.000	.000	.001	.004	.016	.047	.118	.262	.531

					$n = 7$				
					Probability				
x	.1	.2	.3	.4	.5	.6	.7	.8	.9
0	.478	.210	.082	.028	.008	.002	.000	.000	.000
1	.372	.367	.247	.131	.055	.017	.004	.000	.000
2	.124	.275	.318	.261	.164	.077	.025	.004	.000
3	.023	.115	.227	.290	.273	.194	.097	.029	.003
4	.003	.029	.097	.194	.273	.290	.227	.115	.023
5	.000	.004	.025	.077	.164	.261	.318	.275	.124
6	.000	.000	.004	.017	.055	.131	.247	.367	.372
7	.000	.000	.000	.002	.008	.028	.082	.210	.478

					$n = 8$				
					Probability				
x	.1	.2	.3	.4	.5	.6	.7	.8	.9
0	.430	.168	.058	.017	.004	.001	.000	.000	.000
1	.383	.336	.198	.090	.031	.008	.001	.000	.000
2	.149	.294	.296	.209	.109	.041	.010	.001	.000
3	.033	.147	.254	.279	.219	.124	.047	.009	.000
4	.005	.046	.136	.232	.273	.232	.136	.046	.005
5	.000	.009	.047	.124	.219	.279	.254	.147	.033
6	.000	.001	.010	.041	.109	.209	.296	.294	.149
7	.000	.000	.001	.008	.031	.090	.198	.336	.383
8	.000	.000	.000	.001	.004	.017	.058	.168	.430

					$n = 9$				
					Probability				
x	.1	.2	.3	.4	.5	.6	.7	.8	.9
0	.387	.134	.040	.010	.002	.000	.000	.000	.000
1	.387	.302	.156	.060	.018	.004	.000	.000	.000
2	.172	.302	.267	.161	.070	.021	.004	.000	.000
3	.045	.176	.267	.251	.164	.074	.021	.003	.000
4	.007	.066	.172	.251	.246	.167	.074	.017	.001
5	.001	.017	.074	.167	.246	.251	.172	.066	.007
6	.000	.003	.021	.074	.164	.251	.267	.176	.045
7	.000	.000	.004	.021	.070	.161	.267	.302	.172
8	.000	.000	.000	.004	.018	.060	.156	.302	.387
9	.000	.000	.000	.000	.002	.010	.040	.134	.387

TABLE A.2

Binomial Probability Distribution (*Continued*)

n = 10

x	Probability								
	.1	.2	.3	.4	.5	.6	.7	.8	.9
0	.349	.107	.028	.006	.001	.000	.000	.000	.000
1	.387	.268	.121	.040	.010	.002	.000	.000	.000
2	.194	.302	.233	.121	.044	.011	.001	.000	.000
3	.057	.201	.267	.215	.117	.042	.009	.001	.000
4	.011	.088	.200	.251	.205	.111	.037	.006	.000
5	.001	.026	.103	.201	.246	.201	.103	.026	.001
6	.000	.006	.037	.111	.205	.251	.200	.088	.011
7	.000	.001	.009	.042	.117	.215	.267	.201	.057
8	.000	.000	.001	.011	.044	.121	.233	.302	.194
9	.000	.000	.000	.002	.010	.040	.121	.268	.387
10	.000	.000	.000	.000	.001	.006	.028	.107	.349

n = 11

x	Probability								
	.1	.2	.3	.4	.5	.6	.7	.8	.9
0	.314	.086	.020	.004	.000	.000	.000	.000	.000
1	.384	.236	.093	.027	.005	.001	.000	.000	.000
2	.213	.295	.200	.089	.027	.005	.001	.000	.000
3	.071	.221	.257	.177	.081	.023	.004	.000	.000
4	.016	.111	.220	.236	.161	.070	.017	.002	.000
5	.002	.039	.132	.221	.226	.147	.057	.010	.000
6	.000	.010	.057	.147	.226	.221	.132	.039	.002
7	.000	.002	.017	.070	.161	.236	.220	.111	.016
8	.000	.000	.004	.023	.081	.177	.257	.221	.071
9	.000	.000	.001	.005	.027	.089	.200	.295	.213
10	.000	.000	.000	.001	.005	.027	.093	.236	.384
11	.000	.000	.000	.000	.000	.004	.020	.086	.314

n = 12

x	Probability								
	.1	.2	.3	.4	.5	.6	.7	.8	.9
0	.282	.069	.014	.002	.000	.000	.000	.000	.000
1	.377	.206	.071	.017	.003	.000	.000	.000	.000
2	.230	.283	.168	.064	.016	.002	.000	.000	.000
3	.085	.236	.240	.142	.054	.012	.001	.000	.000
4	.021	.133	.231	.213	.121	.042	.008	.001	.000
5	.004	.053	.158	.227	.193	.101	.029	.003	.000
6	.000	.016	.079	.177	.226	.177	.079	.016	.000
7	.000	.003	.029	.101	.193	.227	.158	.053	.004
8	.000	.001	.008	.042	.121	.213	.231	.133	.021
9	.000	.000	.001	.012	.054	.142	.240	.236	.085
10	.000	.000	.000	.002	.016	.064	.168	.283	.230
11	.000	.000	.000	.000	.003	.017	.071	.206	.377
12	.000	.000	.000	.000	.000	.002	.014	.069	.282

(*Continued*)

TABLE A.2

Binomial Probability Distribution (*Continued*)

					$n = 13$					
					Probability					
x	.1	.2	.3	.4	.5	.6	.7	.8	.9
0	.254	.055	.010	.001	.000	.000	.000	.000	.000
1	.367	.179	.054	.011	.002	.000	.000	.000	.000
2	.245	.268	.139	.045	.010	.001	.000	.000	.000
3	.100	.246	.218	.111	.035	.006	.001	.000	.000
4	.028	.154	.234	.184	.087	.024	.003	.000	.000
5	.006	.069	.180	.221	.157	.066	.014	.001	.000
6	.001	.023	.103	.197	.209	.131	.044	.006	.000
7	.000	.006	.044	.131	.209	.197	.103	.023	.001
8	.000	.001	.014	.066	.157	.221	.180	.069	.006
9	.000	.000	.003	.024	.087	.184	.234	.154	.028
10	.000	.000	.001	.006	.035	.111	.218	.246	.100
11	.000	.000	.000	.001	.010	.045	.139	.268	.245
12	.000	.000	.000	.000	.002	.011	.054	.179	.367
13	.000	.000	.000	.000	.000	.001	.010	.055	.254

					$n = 14$					
					Probability					
x	.1	.2	.3	.4	.5	.6	.7	.8	.9
0	.229	.044	.007	.001	.000	.000	.000	.000	.000
1	.356	.154	.041	.007	.001	.000	.000	.000	.000
2	.257	.250	.113	.032	.006	.001	.000	.000	.000
3	.114	.250	.194	.085	.022	.003	.000	.000	.000
4	.035	.172	.229	.155	.061	.014	.001	.000	.000
5	.008	.086	.196	.207	.122	.041	.007	.000	.000
6	.001	.032	.126	.207	.183	.092	.023	.002	.000
7	.000	.009	.062	.157	.209	.157	.062	.009	.000
8	.000	.002	.023	.092	.183	.207	.126	.032	.001
9	.000	.000	.007	.041	.122	.207	.196	.086	.008
10	.000	.000	.001	.014	.061	.155	.229	.172	.035
11	.000	.000	.000	.003	.022	.085	.194	.250	.114
12	.000	.000	.000	.001	.006	.032	.113	.250	.257
13	.000	.000	.000	.000	.001	.007	.041	.154	.356
14	.000	.000	.000	.000	.000	.001	.007	.044	.229

TABLE A.2

Binomial Probability Distribution (*Continued*)

n = 15

					Probability				
x	.1	.2	.3	.4	.5	.6	.7	.8	.9
0	.206	.035	.005	.000	.000	.000	.000	.000	.000
1	.343	.132	.031	.005	.000	.000	.000	.000	.000
2	.267	.231	.092	.022	.003	.000	.000	.000	.000
3	.129	.250	.170	.063	.014	.002	.000	.000	.000
4	.043	.188	.219	.127	.042	.007	.001	.000	.000
5	.010	.103	.206	.186	.092	.024	.003	.000	.000
6	.002	.043	.147	.207	.153	.061	.012	.001	.000
7	.000	.014	.081	.177	.196	.118	.035	.003	.000
8	.000	.003	.035	.118	.196	.177	.081	.014	.000
9	.000	.001	.012	.061	.153	.207	.147	.043	.002
10	.000	.000	.003	.024	.092	.186	.206	.103	.010
11	.000	.000	.001	.007	.042	.127	.219	.188	.043
12	.000	.000	.000	.002	.014	.063	.170	.250	.129
13	.000	.000	.000	.000	.003	.022	.092	.231	.267
14	.000	.000	.000	.000	.000	.005	.031	.132	.343
15	.000	.000	.000	.000	.000	.000	.005	.035	.206

n = 16

					Probability				
x	.1	.2	.3	.4	.5	.6	.7	.8	.9
0	.185	.028	.003	.000	.000	.000	.000	.000	.000
1	.329	.113	.023	.003	.000	.000	.000	.000	.000
2	.275	.211	.073	.015	.002	.000	.000	.000	.000
3	.142	.246	.146	.047	.009	.001	.000	.000	.000
4	.051	.200	.204	.101	.028	.004	.000	.000	.000
5	.014	.120	.210	.162	.067	.014	.001	.000	.000
6	.003	.055	.165	.198	.122	.039	.006	.000	.000
7	.000	.020	.101	.189	.175	.084	.019	.001	.000
8	.000	.006	.049	.142	.196	.142	.049	.006	.000
9	.000	.001	.019	.084	.175	.189	.101	.020	.000
10	.000	.000	.006	.039	.122	.198	.165	.055	.003
11	.000	.000	.001	.014	.067	.162	.210	.120	.014
12	.000	.000	.000	.004	.028	.101	.204	.200	.051
13	.000	.000	.000	.001	.009	.047	.146	.246	.142
14	.000	.000	.000	.000	.002	.015	.073	.211	.275
15	.000	.000	.000	.000	.000	.003	.023	.113	.329
16	.000	.000	.000	.000	.000	.000	.003	.028	.185

(*Continued*)

TABLE A.2

Binomial Probability Distribution (*Continued*)

					$n = 17$				
					Probability				
x	.1	.2	.3	.4	.5	.6	.7	.8	.9
0	.167	.023	.002	.000	.000	.000	.000	.000	.000
1	.315	.096	.017	.002	.000	.000	.000	.000	.000
2	.280	.191	.058	.010	.001	.000	.000	.000	.000
3	.156	.239	.125	.034	.005	.000	.000	.000	.000
4	.060	.209	.187	.080	.018	.002	.000	.000	.000
5	.017	.136	.208	.138	.047	.008	.001	.000	.000
6	.004	.068	.178	.184	.094	.024	.003	.000	.000
7	.001	.027	.120	.193	.148	.057	.009	.000	.000
8	.000	.008	.064	.161	.185	.107	.028	.002	.000
9	.000	.002	.028	.107	.185	.161	.064	.008	.000
10	.000	.000	.009	.057	.148	.193	.120	.027	.001
11	.000	.000	.003	.024	.094	.184	.178	.068	.004
12	.000	.000	.001	.008	.047	.138	.208	.136	.017
13	.000	.000	.000	.002	.018	.080	.187	.209	.060
14	.000	.000	.000	.000	.005	.034	.125	.239	.156
15	.000	.000	.000	.000	.001	.010	.058	.191	.280
16	.000	.000	.000	.000	.000	.002	.017	.096	.315
17	.000	.000	.000	.000	.000	.000	.002	.023	.167

					$n = 18$				
					Probability				
x	.1	.2	.3	.4	.5	.6	.7	.8	.9
0	.150	.018	.002	.000	.000	.000	.000	.000	.000
1	.300	.081	.013	.001	.000	.000	.000	.000	.000
2	.284	.172	.046	.007	.001	.000	.000	.000	.000
3	.168	.230	.105	.025	.003	.000	.000	.000	.000
4	.070	.215	.168	.061	.012	.001	.000	.000	.000
5	.022	.151	.202	.115	.033	.004	.000	.000	.000
6	.005	.082	.187	.166	.071	.015	.001	.000	.000
7	.001	.035	.138	.189	.121	.037	.005	.000	.000
8	.000	.012	.081	.173	.167	.077	.015	.001	.000
9	.000	.003	.039	.128	.185	.128	.039	.003	.000
10	.000	.001	.015	.077	.167	.173	.081	.012	.000
11	.000	.000	.005	.037	.121	.189	.138	.035	.001
12	.000	.000	.001	.015	.071	.166	.187	.082	.005
13	.000	.000	.000	.004	.033	.115	.202	.151	.022
14	.000	.000	.000	.001	.012	.061	.168	.215	.070
15	.000	.000	.000	.000	.003	.025	.105	.230	.168
16	.000	.000	.000	.000	.001	.007	.046	.172	.284
17	.000	.000	.000	.000	.000	.001	.013	.081	.300
18	.000	.000	.000	.000	.000	.000	.002	.018	.150

TABLE A.2

Binomial Probability Distribution (*Continued*)

n = 19

	Probability								
x	.1	.2	.3	.4	.5	.6	.7	.8	.9
0	.135	.014	.001	.000	.000	.000	.000	.000	.000
1	.285	.068	.009	.001	.000	.000	.000	.000	.000
2	.285	.154	.036	.005	.000	.000	.000	.000	.000
3	.180	.218	.087	.017	.002	.000	.000	.000	.000
4	.080	.218	.149	.047	.007	.001	.000	.000	.000
5	.027	.164	.192	.093	.022	.002	.000	.000	.000
6	.007	.095	.192	.145	.052	.008	.001	.000	.000
7	.001	.044	.153	.180	.096	.024	.002	.000	.000
8	.000	.017	.098	.180	.144	.053	.008	.000	.000
9	.000	.005	.051	.146	.176	.098	.022	.001	.000
10	.000	.001	.022	.098	.176	.146	.051	.005	.000
11	.000	.000	.008	.053	.144	.180	.098	.017	.000
12	.000	.000	.002	.024	.096	.180	.153	.044	.001
13	.000	.000	.001	.008	.052	.145	.192	.095	.007
14	.000	.000	.000	.002	.022	.093	.192	.164	.027
15	.000	.000	.000	.001	.007	.047	.149	.218	.080
16	.000	.000	.000	.000	.002	.017	.087	.218	.180
17	.000	.000	.000	.000	.000	.005	.036	.154	.285
18	.000	.000	.000	.000	.000	.001	.009	.068	.285
19	.000	.000	.000	.000	.000	.000	.001	.014	.135

n = 20

	Probability								
x	.1	.2	.3	.4	.5	.6	.7	.8	.9
0	.122	.012	.001	.000	.000	.000	.000	.000	.000
1	.270	.058	.007	.000	.000	.000	.000	.000	.000
2	.285	.137	.028	.003	.000	.000	.000	.000	.000
3	.190	.205	.072	.012	.001	.000	.000	.000	.000
4	.090	.218	.130	.035	.005	.000	.000	.000	.000
5	.032	.175	.179	.075	.015	.001	.000	.000	.000
6	.009	.109	.192	.124	.037	.005	.000	.000	.000
7	.002	.055	.164	.166	.074	.015	.001	.000	.000
8	.000	.022	.114	.180	.120	.035	.004	.000	.000
9	.000	.007	.065	.160	.160	.071	.012	.000	.000
10	.000	.002	.031	.117	.176	.117	.031	.002	.000
11	.000	.000	.012	.071	.160	.160	.065	.007	.000
12	.000	.000	.004	.035	.120	.180	.114	.022	.000
13	.000	.000	.001	.015	.074	.166	.164	.055	.002
14	.000	.000	.000	.005	.037	.124	.192	.109	.009
15	.000	.000	.000	.001	.015	.075	.179	.175	.032
16	.000	.000	.000	.000	.005	.035	.130	.218	.090
17	.000	.000	.000	.000	.001	.012	.072	.205	.190
18	.000	.000	.000	.000	.000	.003	.028	.137	.285
19	.000	.000	.000	.000	.000	.000	.007	.058	.270
20	.000	.000	.000	.000	.000	.000	.001	.012	.122

(Continued)

TABLE A.2

Binomial Probability Distribution (*Continued*)

					n = 25				
					Probability				
x	.1	.2	.3	.4	.5	.6	.7	.8	.9
0	.072	.004	.000	.000	.000	.000	.000	.000	.000
1	.199	.024	.001	.000	.000	.000	.000	.000	.000
2	.266	.071	.007	.000	.000	.000	.000	.000	.000
3	.226	.136	.024	.002	.000	.000	.000	.000	.000
4	.138	.187	.057	.007	.000	.000	.000	.000	.000
5	.065	.196	.103	.020	.002	.000	.000	.000	.000
6	.024	.163	.147	.044	.005	.000	.000	.000	.000
7	.007	.111	.171	.080	.014	.001	.000	.000	.000
8	.002	.062	.165	.120	.032	.003	.000	.000	.000
9	.000	.029	.134	.151	.061	.009	.000	.000	.000
10	.000	.012	.092	.161	.097	.021	.001	.000	.000
11	.000	.004	.054	.147	.133	.043	.004	.000	.000
12	.000	.001	.027	.114	.155	.076	.011	.000	.000
13	.000	.000	.011	.076	.155	.114	.027	.001	.000
14	.000	.000	.004	.043	.133	.147	.054	.004	.000
15	.000	.000	.001	.021	.097	.161	.092	.012	.000
16	.000	.000	.000	.009	.061	.151	.134	.029	.000
17	.000	.000	.000	.003	.032	.120	.165	.062	.002
18	.000	.000	.000	.001	.014	.080	.171	.111	.007
19	.000	.000	.000	.000	.005	.044	.147	.163	.024
20	.000	.000	.000	.000	.002	.020	.103	.196	.065
21	.000	.000	.000	.000	.000	.007	.057	.187	.138
22	.000	.000	.000	.000	.000	.002	.024	.136	.226
23	.000	.000	.000	.000	.000	.000	.007	.071	.266
24	.000	.000	.000	.000	.000	.000	.001	.024	.199
25	.000	.000	.000	.000	.000	.000	.000	.004	.072

TABLE A.3

Poisson Probabilities

					λ					
x	.005	.01	.02	.03	.04	.05	.06	.07	.08	.09
0	.9950	.9900	.9802	.9704	.9608	.9512	.9418	.9324	.9231	.9139
1	.0050	.0099	.0196	.0291	.0384	.0476	.0565	.0653	.0738	.0823
2	.0000	.0000	.0002	.0004	.0008	.0012	.0017	.0023	.0030	.0037
3	.0000	.0000	.0000	.0000	.0000	.0000	.0000	.0001	.0001	.0001

x	.1	.2	.3	.4	.5	.6	.7	.8	.9	1.0
0	.9048	.8187	.7408	.6703	.6065	.5488	.4966	.4493	.4066	.3679
1	.0905	.1637	.2222	.2681	.3033	.3293	.3476	.3595	.3659	.3679
2	.0045	.0164	.0333	.0536	.0758	.0988	.1217	.1438	.1647	.1839
3	.0002	.0011	.0033	.0072	.0126	.0198	.0284	.0383	.0494	.0613
4	.0000	.0001	.0003	.0007	.0016	.0030	.0050	.0077	.0111	.0153
5	.0000	.0000	.0000	.0001	.0002	.0004	.0007	.0012	.0020	.0031
6	.0000	.0000	.0000	.0000	.0000	.0000	.0001	.0002	.0003	.0005
7	.0000	.0000	.0000	.0000	.0000	.0000	.0000	.0000	.0000	.0001

x	1.1	1.2	1.3	1.4	1.5	1.6	1.7	1.8	1.9	2.0
0	.3329	.3012	.2725	.2466	.2231	.2019	.1827	.1653	.1496	.1353
1	.3662	.3614	.3543	.3452	.3347	.3230	.3106	.2975	.2842	.2707
2	.2014	.2169	.2303	.2417	.2510	.2584	.2640	.2678	.2700	.2707
3	.0738	.0867	.0998	.1128	.1255	.1378	.1496	.1607	.1710	.1804
4	.0203	.0260	.0324	.0395	.0471	.0551	.0636	.0723	.0812	.0902
5	.0045	.0062	.0084	.0111	.0141	.0176	.0216	.0260	.0309	.0361
6	.0008	.0012	.0018	.0026	.0035	.0047	.0061	.0078	.0098	.0120
7	.0001	.0002	.0003	.0005	.0008	.0011	.0015	.0020	.0027	.0034
8	.0000	.0000	.0001	.0001	.0001	.0002	.0003	.0005	.0006	.0009
9	.0000	.0000	.0000	.0000	.0000	.0000	.0001	.0001	.0001	.0002

x	2.1	2.2	2.3	2.4	2.5	2.6	2.7	2.8	2.9	3.0
0	.1225	.1108	.1003	.0907	.0821	.0743	.0672	.0608	.0550	.0498
1	.2572	.2438	.2306	.2177	.2052	.1931	.1815	.1703	.1596	.1494
2	.2700	.2681	.2652	.2613	.2565	.2510	.2450	.2384	.2314	.2240
3	.1890	.1966	.2033	.2090	.2138	.2176	.2205	.2225	.2237	.2240
4	.0992	.1082	.1169	.1254	.1336	.1414	.1488	.1557	.1622	.1680
5	.0417	.0476	.0538	.0602	.0668	.0735	.0804	.0872	.0940	.1008
6	.0146	.0174	.0206	.0241	.0278	.0319	.0362	.0407	.0455	.0504
7	.0044	.0055	.0068	.0083	.0099	.0118	.0139	.0163	.0188	.0216
8	.0011	.0015	.0019	.0025	.0031	.0038	.0047	.0057	.0068	.0081
9	.0003	.0004	.0005	.0007	.0009	.0011	.0014	.0018	.0022	.0027
10	.0001	.0001	.0001	.0002	.0002	.0003	.0004	.0005	.0006	.0008
11	.0000	.0000	.0000	.0000	.0000	.0001	.0001	.0001	.0002	.0002
12	.0000	.0000	.0000	.0000	.0000	.0000	.0000	.0000	.0000	.0001

(*Continued*)

TABLE A.3

Poisson Probabilities (*Continued*)

					λ					
x	3.1	3.2	3.3	3.4	3.5	3.6	3.7	3.8	3.9	4.0
0	.0450	.0408	.0369	.0334	.0302	.0273	.0247	.0224	.0202	.0183
1	.1397	.1304	.1217	.1135	.1057	.0984	.0915	.0850	.0789	.0733
2	.2165	.2087	.2008	.1929	.1850	.1771	.1692	.1615	.1539	.1465
3	.2237	.2226	.2209	.2186	.2158	.2125	.2087	.2046	.2001	.1954
4	.1733	.1781	.1823	.1858	.1888	.1912	.1931	.1944	.1951	.1954
5	.1075	.1140	.1203	.1264	.1322	.1377	.1429	.1477	.1522	.1563
6	.0555	.0608	.0662	.0716	.0771	.0826	.0881	.0936	.0989	.1042
7	.0246	.0278	.0312	.0348	.0385	.0425	.0466	.0508	.0551	.0595
8	.0095	.0111	.0129	.0148	.0169	.0191	.0215	.0241	.0269	.0298
9	.0033	.0040	.0047	.0056	.0066	.0076	.0089	.0102	.0116	.0132
10	.0010	.0013	.0016	.0019	.0023	.0028	.0033	.0039	.0045	.0053
11	.0003	.0004	.0005	.0006	.0007	.0009	.0011	.0013	.0016	.0019
12	.0001	.0001	.0001	.0002	.0002	.0003	.0003	.0004	.0005	.0006
13	.0000	.0000	.0000	.0000	.0001	.0001	.0001	.0001	.0002	.0002
14	.0000	.0000	.0000	.0000	.0000	.0000	.0000	.0000	.0000	.0001

x	4.1	4.2	4.3	4.4	4.5	4.6	4.7	4.8	4.9	5.0
0	.0166	.0150	.0136	.0123	.0111	.0101	.0091	.0082	.0074	.0067
1	.0679	.0630	.0583	.0540	.0500	.0462	.0427	.0395	.0365	.0337
2	.1393	.1323	.1254	.1188	.1125	.1063	.1005	.0948	.0894	.0842
3	.1904	.1852	.1798	.1743	.1687	.1631	.1574	.1517	.1460	.1404
4	.1951	.1944	.1933	.1917	.1898	.1875	.1849	.1820	.1789	.1755
5	.1600	.1633	.1662	.1687	.1708	.1725	.1738	.1747	.1753	.1755
6	.1093	.1143	.1191	.1237	.1281	.1323	.1362	.1398	.1432	.1462
7	.0640	.0686	.0732	.0778	.0824	.0869	.0914	.0959	.1002	.1044
8	.0328	.0360	.0393	.0428	.0463	.0500	.0537	.0575	.0614	.0653
9	.0150	.0168	.0188	.0209	.0232	.0255	.0281	.0307	.0334	.0363
10	.0061	.0071	.0081	.0092	.0104	.0118	.0132	.0147	.0164	.0181
11	.0023	.0027	.0032	.0037	.0043	.0049	.0056	.0064	.0073	.0082
12	.0008	.0009	.0011	.0013	.0016	.0019	.0022	.0026	.0030	.0034
13	.0002	.0003	.0004	.0005	.0006	.0007	.0008	.0009	.0011	.0013
14	.0001	.0001	.0001	.0001	.0002	.0002	.0003	.0003	.0004	.0005
15	.0000	.0000	.0000	.0000	.0001	.0001	.0001	.0001	.0001	.0002

TABLE A.3

Poisson Probabilities (*Continued*)

					λ					
x	5.1	5.2	5.3	5.4	5.5	5.6	5.7	5.8	5.9	6.0
0	.0061	.0055	.0050	.0045	.0041	.0037	.0033	.0030	.0027	.0025
1	.0311	.0287	.0265	.0244	.0225	.0207	.0191	.0176	.0162	.0149
2	.0793	.0746	.0701	.0659	.0618	.0580	.0544	.0509	.0477	.0446
3	.1348	.1293	.1239	.1185	.1133	.1082	.1033	.0985	.0938	.0892
4	.1719	.1681	.1641	.1600	.1558	.1515	.1472	.1428	.1383	.1339
5	.1753	.1748	.1740	.1728	.1714	.1697	.1678	.1656	.1632	.1606
6	.1490	.1515	.1537	.1555	.1571	.1584	.1594	.1601	.1605	.1606
7	.1086	.1125	.1163	.1200	.1234	.1267	.1298	.1326	.1353	.1377
8	.0692	.0731	.0771	.0810	.0849	.0887	.0925	.0962	.0998	.1033
9	.0392	.0423	.0454	.0486	.0519	.0552	.0586	.0620	.0654	.0688
10	.0200	.0220	.0241	.0262	.0285	.0309	.0334	.0359	.0386	.0413
11	.0093	.0104	.0116	.0129	.0143	.0157	.0173	.0190	.0207	.0225
12	.0039	.0045	.0051	.0058	.0065	.0073	.0082	.0092	.0102	.0113
13	.0015	.0018	.0021	.0024	.0028	.0032	.0036	.0041	.0046	.0052
14	.0006	.0007	.0008	.0009	.0011	.0013	.0015	.0017	.0019	.0022
15	.0002	.0002	.0003	.0003	.0004	.0005	.0006	.0007	.0008	.0009
16	.0001	.0001	.0001	.0001	.0001	.0002	.0002	.0002	.0003	.0003
17	.0000	.0000	.0000	.0000	.0000	.0001	.0001	.0001	.0001	.0001

x	6.1	6.2	6.3	6.4	6.5	6.6	6.7	6.8	6.9	7.0
0	.0022	.0020	.0018	.0017	.0015	.0014	.0012	.0011	.0010	.0009
1	.0137	.0126	.0116	.0106	.0098	.0090	.0082	.0076	.0070	.0064
2	.0417	.0390	.0364	.0340	.0318	.0296	.0276	.0258	.0240	.0223
3	.0848	.0806	.0765	.0726	.0688	.0652	.0617	.0584	.0552	.0521
4	.1294	.1249	.1205	.1162	.1118	.1076	.1034	.0992	.0952	.0912
5	.1579	.1549	.1519	.1487	.1454	.1420	.1385	.1349	.1314	.1277
6	.1605	.1601	.1595	.1586	.1575	.1562	.1546	.1529	.1511	.1490
7	.1399	.1418	.1435	.1450	.1462	.1472	.1480	.1486	.1489	.1490
8	.1066	.1099	.1130	.1160	.1188	.1215	.1240	.1263	.1284	.1304
9	.0723	.0757	.0791	.0825	.0858	.0891	.0923	.0954	.0985	.1014
10	.0441	.0469	.0498	.0528	.0558	.0588	.0618	.0649	.0679	.0710
11	.0244	.0265	.0285	.0307	.0330	.0353	.0377	.0401	.0426	.0452
12	.0124	.0137	.0150	.0164	.0179	.0194	.0210	.0227	.0245	.0263
13	.0058	.0065	.0073	.0081	.0089	.0099	.0108	.0119	.0130	.0142
14	.0025	.0029	.0033	.0037	.0041	.0046	.0052	.0058	.0064	.0071
15	.0010	.0012	.0014	.0016	.0018	.0020	.0023	.0026	.0029	.0033
16	.0004	.0005	.0005	.0006	.0007	.0008	.0010	.0011	.0013	.0014
17	.0001	.0002	.0002	.0002	.0003	.0003	.0004	.0004	.0005	.0006
18	.0000	.0001	.0001	.0001	.0001	.0001	.0001	.0002	.0002	.0002
19	.0000	.0000	.0000	.0000	.0000	.0000	.0001	.0001	.0001	.0001

(*Continued*)

TABLE A.3

Poisson Probabilities (*Continued*)

					λ					
x	7.1	7.2	7.3	7.4	7.5	7.6	7.7	7.8	7.9	8.0
0	.0008	.0007	.0007	.0006	.0006	.0005	.0005	.0004	.0004	.0003
1	.0059	.0054	.0049	.0045	.0041	.0038	.0035	.0032	.0029	.0027
2	.0208	.0194	.0180	.0167	.0156	.0145	.0134	.0125	.0116	.0107
3	.0492	.0464	.0438	.0413	.0389	.0366	.0345	.0324	.0305	.0286
4	.0874	.0836	.0799	.0764	.0729	.0696	.0663	.0632	.0602	.0573
5	.1241	.1204	.1167	.1130	.1094	.1057	.1021	.0986	.0951	.0916
6	.1468	.1445	.1420	.1394	.1367	.1339	.1311	.1282	.1252	.1221
7	.1489	.1486	.1481	.1474	.1465	.1454	.1442	.1428	.1413	.1396
8	.1321	.1337	.1351	.1363	.1373	.1381	.1388	.1392	.1395	.1396
9	.1042	.1070	.1096	.1121	.1144	.1167	.1187	.1207	.1224	.1241
10	.0740	.0770	.0800	.0829	.0858	.0887	.0914	.0941	.0967	.0993
11	.0478	.0504	.0531	.0558	.0585	.0613	.0640	.0667	.0695	.0722
12	.0283	.0303	.0323	.0344	.0366	.0388	.0411	.0434	.0457	.0481
13	.0154	.0168	.0181	.0196	.0211	.0227	.0243	.0260	.0278	.0296
14	.0078	.0086	.0095	.0104	.0113	.0123	.0134	.0145	.0157	.0169
15	.0037	.0041	.0046	.0051	.0057	.0062	.0069	.0075	.0083	.0090
16	.0016	.0019	.0021	.0024	.0026	.0030	.0033	.0037	.0041	.0045
17	.0007	.0008	.0009	.0010	.0012	.0013	.0015	.0017	.0019	.0021
18	.0003	.0003	.0004	.0004	.0005	.0006	.0006	.0007	.0008	.0009
19	.0001	.0001	.0001	.0002	.0002	.0002	.0003	.0003	.0003	.0004
20	.0000	.0000	.0001	.0001	.0001	.0001	.0001	.0001	.0001	.0002
21	.0000	.0000	.0000	.0000	.0000	.0000	.0000	.0000	.0001	.0001

x	8.1	8.2	8.3	8.4	8.5	8.6	8.7	8.8	8.9	9.0
0	.0003	.0003	.0002	.0002	.0002	.0002	.0002	.0002	.0001	.0001
1	.0025	.0023	.0021	.0019	.0017	.0016	.0014	.0013	.0012	.0011
2	.0100	.0092	.0086	.0079	.0074	.0068	.0063	.0058	.0054	.0050
3	.0269	.0252	.0237	.0222	.0208	.0195	.0183	.0171	.0160	.0150
4	.0544	.0517	.0491	.0466	.0443	.0420	.0398	.0377	.0357	.0337
5	.0882	.0849	.0816	.0784	.0752	.0722	.0692	.0663	.0635	.0607
6	.1191	.1160	.1128	.1097	.1066	.1034	.1003	.0972	.0941	.0911
7	.1378	.1358	.1338	.1317	.1294	.1271	.1247	.1222	.1197	.1171
8	.1395	.1392	.1388	.1382	.1375	.1366	.1356	.1344	.1332	.1318
9	.1256	.1269	.1280	.1290	.1299	.1306	.1311	.1315	.1317	.1318
10	.1017	.1040	.1063	.1084	.1104	.1123	.1140	.1157	.1172	.1186
11	.0749	.0776	.0802	.0828	.0853	.0878	.0902	.0925	.0948	.0970
12	.0505	.0530	.0555	.0579	.0604	.0629	.0654	.0679	.0703	.0728
13	.0315	.0334	.0354	.0374	.0395	.0416	.0438	.0459	.0481	.0504
14	.0182	.0196	.0210	.0225	.0240	.0256	.0272	.0289	.0306	.0324
15	.0098	.0107	.0116	.0126	.0136	.0147	.0158	.0169	.0182	.0194
16	.0050	.0055	.0060	.0066	.0072	.0079	.0086	.0093	.0101	.0109
17	.0024	.0026	.0029	.0033	.0036	.0040	.0044	.0048	.0053	.0058
18	.0011	.0012	.0014	.0015	.0017	.0019	.0021	.0024	.0026	.0029
19	.0005	.0005	.0006	.0007	.0008	.0009	.0010	.0011	.0012	.0014
20	.0002	.0002	.0002	.0003	.0003	.0004	.0004	.0005	.0005	.0006
21	.0001	.0001	.0001	.0001	.0001	.0002	.0002	.0002	.0002	.0003
22	.0000	.0000	.0000	.0000	.0001	.0001	.0001	.0001	.0001	.0001

TABLE A.3

Poisson Probabilities (*Continued*)

					λ					
x	9.1	9.2	9.3	9.4	9.5	9.6	9.7	9.8	9.9	10.0
0	.0001	.0001	.0001	.0001	.0001	.0001	.0001	.0001	.0001	.0000
1	.0010	.0009	.0009	.0008	.0007	.0007	.0006	.0005	.0005	.0005
2	.0046	.0043	.0040	.0037	.0034	.0031	.0029	.0027	.0025	.0023
3	.0140	.0131	.0123	.0115	.0107	.0100	.0093	.0087	.0081	.0076
4	.0319	.0302	.0285	.0269	.0254	.0240	.0226	.0213	.0201	.0189
5	.0581	.0555	.0530	.0506	.0483	.0460	.0439	.0418	.0398	.0378
6	.0881	.0851	.0822	.0793	.0764	.0736	.0709	.0682	.0656	.0631
7	.1145	.1118	.1091	.1064	.1037	.1010	.0982	.0955	.0928	.0901
8	.1302	.1286	.1269	.1251	.1232	.1212	.1191	.1170	.1148	.1126
9	.1317	.1315	.1311	.1306	.1300	.1293	.1284	.1274	.1263	.1251
10	.1198	.1210	.1219	.1228	.1235	.1241	.1245	.1249	.1250	.1251
11	.0991	.1012	.1031	.1049	.1067	.1083	.1098	.1112	.1125	.1137
12	.0752	.0776	.0799	.0822	.0844	.0866	.0888	.0908	.0928	.0948
13	.0526	.0549	.0572	.0594	.0617	.0640	.0662	.0685	.0707	.0729
14	.0342	.0361	.0380	.0399	.0419	.0439	.0459	.0479	.0500	.0521
15	.0208	.0221	.0235	.0250	.0265	.0281	.0297	.0313	.0330	.0347
16	.0118	.0127	.0137	.0147	.0157	.0168	.0180	.0192	.0204	.0217
17	.0063	.0069	.0075	.0081	.0088	.0095	.0103	.0111	.0119	.0128
18	.0032	.0035	.0039	.0042	.0046	.0051	.0055	.0060	.0065	.0071
19	.0015	.0017	.0019	.0021	.0023	.0026	.0028	.0031	.0034	.0037
20	.0007	.0008	.0009	.0010	.0011	.0012	.0014	.0015	.0017	.0019
21	.0003	.0003	.0004	.0004	.0005	.0006	.0006	.0007	.0008	.0009
22	.0001	.0001	.0002	.0002	.0002	.0002	.0003	.0003	.0004	.0004
23	.0000	.0001	.0001	.0001	.0001	.0001	.0001	.0001	.0002	.0002
24	.0000	.0000	.0000	.0000	.0000	.0000	.0000	.0001	.0001	.0001

TABLE A.4

The e^{-x} Table

x	e^{-x}	x	e^{-x}	x	e^{-x}	x	e^{-x}
0.0	1.0000	3.0	0.0498	6.0	0.00248	9.0	0.00012
0.1	0.9048	3.1	0.0450	6.1	0.00224	9.1	0.00011
0.2	0.8187	3.2	0.0408	6.2	0.00203	9.2	0.00010
0.3	0.7408	3.3	0.0369	6.3	0.00184	9.3	0.00009
0.4	0.6703	3.4	0.0334	6.4	0.00166	9.4	0.00008
0.5	0.6065	3.5	0.0302	6.5	0.00150	9.5	0.00007
0.6	0.5488	3.6	0.0273	6.6	0.00136	9.6	0.00007
0.7	0.4966	3.7	0.0247	6.7	0.00123	9.7	0.00006
0.8	0.4493	3.8	0.0224	6.8	0.00111	9.8	0.00006
0.9	0.4066	3.9	0.0202	6.9	0.00101	9.9	0.00005
1.0	0.3679	4.0	0.0183	7.0	0.00091	10.0	0.00005
1.1	0.3329	4.1	0.0166	7.1	0.00083		
1.2	0.3012	4.2	0.0150	7.2	0.00075		
1.3	0.2725	4.3	0.0136	7.3	0.00068		
1.4	0.2466	4.4	0.0123	7.4	0.00061		
1.5	0.2231	4.5	0.0111	7.5	0.00055		
1.6	0.2019	4.6	0.0101	7.6	0.00050		
1.7	0.1827	4.7	0.0091	7.7	0.00045		
1.8	0.1653	4.8	0.0082	7.8	0.00041		
1.9	0.1496	4.9	0.0074	7.9	0.00037		
2.0	0.1353	5.0	0.0067	8.0	0.00034		
2.1	0.1225	5.1	0.0061	8.1	0.00030		
2.2	0.1108	5.2	0.0055	8.2	0.00027		
2.3	0.1003	5.3	0.0050	8.3	0.00025		
2.4	0.0907	5.4	0.0045	8.4	0.00022		
2.5	0.0821	5.5	0.0041	8.5	0.00020		
2.6	0.0743	5.6	0.0037	8.6	0.00018		
2.7	0.0672	5.7	0.0033	8.7	0.00017		
2.8	0.0608	5.8	0.0030	8.8	0.00015		
2.9	0.0550	5.9	0.0027	8.9	0.00014		

TABLE A.5

Areas of the Standard Normal Distribution

The entries in this table are the probabilities that a standard normal random variable is between 0 and z (the shaded area).

z	0.00	0.01	0.02	0.03	0.04	0.05	0.06	0.07	0.08	0.09
0.0	.0000	.0040	.0080	.0120	.0160	.0199	.0239	.0279	.0319	.0359
0.1	.0398	.0438	.0478	.0517	.0557	.0596	.0636	.0675	.0714	.0753
0.2	.0793	.0832	.0871	.0910	.0948	.0987	.1026	.1064	.1103	.1141
0.3	.1179	.1217	.1255	.1293	.1331	.1368	.1406	.1443	.1480	.1517
0.4	.1554	.1591	.1628	.1664	.1700	.1736	.1772	.1808	.1844	.1879
0.5	.1915	.1950	.1985	.2019	.2054	.2088	.2123	.2157	.2190	.2224
0.6	.2257	.2291	.2324	.2357	.2389	.2422	.2454	.2486	.2517	.2549
0.7	.2580	.2611	.2642	.2673	.2704	.2734	.2764	.2794	.2823	.2852
0.8	.2881	.2910	.2939	.2967	.2995	.3023	.3051	.3078	.3106	.3133
0.9	.3159	.3186	.3212	.3238	.3264	.3289	.3315	.3340	.3365	.3389
1.0	.3413	.3438	.3461	.3485	.3508	.3531	.3554	.3577	.3599	.3621
1.1	.3643	.3665	.3686	.3708	.3729	.3749	.3770	.3790	.3810	.3830
1.2	.3849	.3869	.3888	.3907	.3925	.3944	.3962	.3980	.3997	.4015
1.3	.4032	.4049	.4066	.4082	.4099	.4115	.4131	.4147	.4162	.4177
1.4	.4192	.4207	.4222	.4236	.4251	.4265	.4279	.4292	.4306	.4319
1.5	.4332	.4345	.4357	.4370	.4382	.4394	.4406	.4418	.4429	.4441
1.6	.4452	.4463	.4474	.4484	.4495	.4505	.4515	.4525	.4535	.4545
1.7	.4554	.4564	.4573	.4582	.4591	.4599	.4608	.4616	.4625	.4633
1.8	.4641	.4649	.4656	.4664	.4671	.4678	.4686	.4693	.4699	.4706
1.9	.4713	.4719	.4726	.4732	.4738	.4744	.4750	.4756	.4761	.4767
2.0	.4772	.4778	.4783	.4788	.4793	.4798	.4803	.4808	.4812	.4817
2.1	.4821	.4826	.4830	.4834	.4838	.4842	.4846	.4850	.4854	.4857
2.2	.4861	.4864	.4868	.4871	.4875	.4878	.4881	.4884	.4887	.4890
2.3	.4893	.4896	.4898	.4901	.4904	.4906	.4909	.4911	.4913	.4916
2.4	.4918	.4920	.4922	.4925	.4927	.4929	.4931	.4932	.4934	.4936
2.5	.4938	.4940	.4941	.4943	.4945	.4946	.4948	.4949	.4951	.4952
2.6	.4953	.4955	.4956	.4957	.4959	.4960	.4961	.4962	.4963	.4964
2.7	.4965	.4966	.4967	.4968	.4969	.4970	.4971	.4972	.4973	.4974
2.8	.4974	.4975	.4976	.4977	.4977	.4978	.4979	.4979	.4980	.4981
2.9	.4981	.4982	.4982	.4983	.4984	.4984	.4985	.4985	.4986	.4986
3.0	.4987	.4987	.4987	.4988	.4988	.4989	.4989	.4989	.4990	.4990
3.1	.4990	.4991	.4991	.4991	.4992	.4992	.4992	.4992	.4993	.4993
3.2	.4993	.4993	.4994	.4994	.4994	.4994	.4994	.4995	.4995	.4995
3.3	.4995	.4995	.4995	.4996	.4996	.4996	.4996	.4996	.4996	.4997
3.4	.4997	.4997	.4997	.4997	.4997	.4997	.4997	.4997	.4997	.4998
3.5	.4998									
4.0	.49997									
4.5	.499997									
5.0	.4999997									
6.0	.499999999									

TABLE A.6

Critical Values from the *t* Distribution

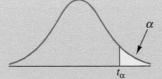

	Values of α for one-tailed test and $\alpha/2$ for two-tailed test					
df	$t_{.100}$	$t_{.050}$	$t_{.025}$	$t_{.010}$	$t_{.005}$	$t_{.001}$
1	3.078	6.314	12.706	31.821	63.656	318.289
2	1.886	2.920	4.303	6.965	9.925	22.328
3	1.638	2.353	3.182	4.541	5.841	10.214
4	1.533	2.132	2.776	3.747	4.604	7.173
5	1.476	2.015	2.571	3.365	4.032	5.894
6	1.440	1.943	2.447	3.143	3.707	5.208
7	1.415	1.895	2.365	2.998	3.499	4.785
8	1.397	1.860	2.306	2.896	3.355	4.501
9	1.383	1.833	2.262	2.821	3.250	4.297
10	1.372	1.812	2.228	2.764	3.169	4.144
11	1.363	1.796	2.201	2.718	3.106	4.025
12	1.356	1.782	2.179	2.681	3.055	3.930
13	1.350	1.771	2.160	2.650	3.012	3.852
14	1.345	1.761	2.145	2.624	2.977	3.787
15	1.341	1.753	2.131	2.602	2.947	3.733
16	1.337	1.746	2.120	2.583	2.921	3.686
17	1.333	1.740	2.110	2.567	2.898	3.646
18	1.330	1.734	2.101	2.552	2.878	3.610
19	1.328	1.729	2.093	2.539	2.861	3.579
20	1.325	1.725	2.086	2.528	2.845	3.552

TABLE A.6

Critical Values from the *t* Distribution (*Continued*)

df	Values of α for one-tailed test and $\alpha/2$ for two-tailed test					
	$t_{.100}$	$t_{.050}$	$t_{.025}$	$t_{.010}$	$t_{.005}$	$t_{.001}$
21	1.323	1.721	2.080	2.518	2.831	3.527
22	1.321	1.717	2.074	2.508	2.819	3.505
23	1.319	1.714	2.069	2.500	2.807	3.485
24	1.318	1.711	2.064	2.492	2.797	3.467
25	1.316	1.708	2.060	2.485	2.787	3.450
26	1.315	1.706	2.056	2.479	2.779	3.435
27	1.314	1.703	2.052	2.473	2.771	3.421
28	1.313	1.701	2.048	2.467	2.763	3.408
29	1.311	1.699	2.045	2.462	2.756	3.396
30	1.310	1.697	2.042	2.457	2.750	3.385
40	1.303	1.684	2.021	2.423	2.704	3.307
50	1.299	1.676	2.009	2.403	2.678	3.261
60	1.296	1.671	2.000	2.390	2.660	3.232
70	1.294	1.667	1.994	2.381	2.648	3.211
80	1.292	1.664	1.990	2.374	2.639	3.195
90	1.291	1.662	1.987	2.368	2.632	3.183
100	1.290	1.660	1.984	2.364	2.626	3.174
150	1.287	1.655	1.976	2.351	2.609	3.145
200	1.286	1.653	1.972	2.345	2.601	3.131
∞	1.282	1.645	1.960	2.326	2.576	3.090

TABLE A.7

Percentage Points of the *F* Distribution

$$\alpha = .10$$

ν_2	Numerator Degrees of Freedom								
	1	2	3	4	5	6	7	8	9
1	39.86	49.50	53.59	55.83	57.24	58.20	58.91	59.44	59.86
2	8.53	9.00	9.16	9.24	9.29	9.33	9.35	9.37	9.38
3	5.54	5.46	5.39	5.34	5.31	5.28	5.27	5.25	5.24
4	4.54	4.32	4.19	4.11	4.05	4.01	3.98	3.95	3.94
5	4.06	3.78	3.62	3.52	3.45	3.40	3.37	3.34	3.32
6	3.78	3.46	3.29	3.18	3.11	3.05	3.01	2.98	2.96
7	3.59	3.26	3.07	2.96	2.88	2.83	2.78	2.75	2.72
8	3.46	3.11	2.92	2.81	2.73	2.67	2.62	2.59	2.56
9	3.36	3.01	2.81	2.69	2.61	2.55	2.51	2.47	2.44
10	3.29	2.92	2.73	2.61	2.52	2.46	2.41	2.38	2.35
11	3.23	2.86	2.66	2.54	2.45	2.39	2.34	2.30	2.27
12	3.18	2.81	2.61	2.48	2.39	2.33	2.28	2.24	2.21
13	3.14	2.76	2.56	2.43	2.35	2.28	2.23	2.20	2.16
14	3.10	2.73	2.52	2.39	2.31	2.24	2.19	2.15	2.12
15	3.07	2.70	2.49	2.36	2.27	2.21	2.16	2.12	2.09
16	3.05	2.67	2.46	2.33	2.24	2.18	2.13	2.09	2.06
17	3.03	2.64	2.44	2.31	2.22	2.15	2.10	2.06	2.03
18	3.01	2.62	2.42	2.29	2.20	2.13	2.08	2.04	2.00
19	2.99	2.61	2.40	2.27	2.18	2.11	2.06	2.02	1.98
20	2.97	2.59	2.38	2.25	2.16	2.09	2.04	2.00	1.96
21	2.96	2.57	2.36	2.23	2.14	2.08	2.02	1.98	1.95
22	2.95	2.56	2.35	2.22	2.13	2.06	2.01	1.97	1.93
23	2.94	2.55	2.34	2.21	2.11	2.05	1.99	1.95	1.92
24	2.93	2.54	2.33	2.19	2.10	2.04	1.98	1.94	1.91
25	2.92	2.53	2.32	2.18	2.09	2.02	1.97	1.93	1.89
26	2.91	2.52	2.31	2.17	2.08	2.01	1.96	1.92	1.88
27	2.90	2.51	2.30	2.17	2.07	2.00	1.95	1.91	1.87
28	2.89	2.50	2.29	2.16	2.06	2.00	1.94	1.90	1.87
29	2.89	2.50	2.28	2.15	2.06	1.99	1.93	1.89	1.86
30	2.88	2.49	2.28	2.14	2.05	1.98	1.93	1.88	1.85
40	2.84	2.44	2.23	2.09	2.00	1.93	1.87	1.83	1.79
60	2.79	2.39	2.18	2.04	1.95	1.87	1.82	1.77	1.74
120	2.75	2.35	2.13	1.99	1.90	1.82	1.77	1.72	1.68
∞	2.71	2.30	2.08	1.94	1.85	1.77	1.72	1.67	1.63

Denominator Degrees of Freedom

TABLE A.7

Percentage Points of the *F* Distribution (*Continued*)

					$\alpha = .10$						v_1
				Numerator Degrees of Freedom							v_2
10	12	15	20	24	30	40	60	120	∞		
60.19	60.71	61.22	61.74	62.00	62.26	62.53	62.79	63.06	63.33	1	
9.39	9.41	9.42	9.44	9.45	9.46	9.47	9.47	9.48	9.49	2	
5.23	5.22	5.20	5.18	5.18	5.17	5.16	5.15	5.14	5.13	3	
3.92	3.90	3.87	3.84	3.83	3.82	3.80	3.79	3.78	3.76	4	
3.30	3.27	3.24	3.21	3.19	3.17	3.16	3.14	3.12	3.10	5	
2.94	2.90	2.87	2.84	2.82	2.80	2.78	2.76	2.74	2.72	6	
2.70	2.67	2.63	2.59	2.58	2.56	2.54	2.51	2.49	2.47	7	
2.54	2.50	2.46	2.42	2.40	2.38	2.36	2.34	2.32	2.29	8	
2.42	2.38	2.34	2.30	2.28	2.25	2.23	2.21	2.18	2.16	9	
2.32	2.28	2.24	2.20	2.18	2.16	2.13	2.11	2.08	2.06	10	
2.25	2.21	2.17	2.12	2.10	2.08	2.05	2.03	2.00	1.97	11	
2.19	2.15	2.10	2.06	2.04	2.01	1.99	1.96	1.93	1.90	12	Denominator Degrees of Freedom
2.14	2.10	2.05	2.01	1.98	1.96	1.93	1.90	1.88	1.85	13	
2.10	2.05	2.01	1.96	1.94	1.91	1.89	1.86	1.83	1.80	14	
2.06	2.02	1.97	1.92	1.90	1.87	1.85	1.82	1.79	1.76	15	
2.03	1.99	1.94	1.89	1.87	1.84	1.81	1.78	1.75	1.72	16	
2.00	1.96	1.91	1.86	1.84	1.81	1.78	1.75	1.72	1.69	17	
1.98	1.93	1.89	1.84	1.81	1.78	1.75	1.72	1.69	1.66	18	
1.96	1.91	1.86	1.81	1.79	1.76	1.73	1.70	1.67	1.63	19	
1.94	1.89	1.84	1.79	1.77	1.74	1.71	1.68	1.64	1.61	20	
1.92	1.87	1.83	1.78	1.75	1.72	1.69	1.66	1.62	1.59	21	
1.90	1.86	1.81	1.76	1.73	1.70	1.67	1.64	1.60	1.57	22	
1.89	1.84	1.80	1.74	1.72	1.69	1.66	1.62	1.59	1.55	23	
1.88	1.83	1.78	1.73	1.70	1.67	1.64	1.61	1.57	1.53	24	
1.87	1.82	1.77	1.72	1.69	1.66	1.63	1.59	1.56	1.52	25	
1.86	1.81	1.76	1.71	1.68	1.65	1.61	1.58	1.54	1.50	26	
1.85	1.80	1.75	1.70	1.67	1.64	1.60	1.57	1.53	1.49	27	
1.84	1.79	1.74	1.69	1.66	1.63	1.59	1.56	1.52	1.48	28	
1.83	1.78	1.73	1.68	1.65	1.62	1.58	1.55	1.51	1.47	29	
1.82	1.77	1.72	1.67	1.64	1.61	1.57	1.54	1.50	1.46	30	
1.76	1.71	1.66	1.61	1.57	1.54	1.51	1.47	1.42	1.38	40	
1.71	1.66	1.60	1.54	1.51	1.48	1.44	1.40	1.35	1.29	60	
1.65	1.60	1.55	1.48	1.45	1.41	1.37	1.32	1.26	1.19	120	
1.60	1.55	1.49	1.42	1.38	1.34	1.30	1.24	1.17	1.00	∞	

(*Continued*)

TABLE A.7

Percentage Points of the F Distribution (*Continued*)

v_2 \ v_1	$\alpha = .05$ Numerator Degrees of Freedom								
	1	2	3	4	5	6	7	8	9
1	161.45	199.50	215.71	224.58	230.16	233.99	236.77	238.88	240.54
2	18.51	19.00	19.16	19.25	19.30	19.33	19.35	19.37	19.38
3	10.13	9.55	9.28	9.12	9.01	8.94	8.89	8.85	8.81
4	7.71	6.94	6.59	6.39	6.26	6.16	6.09	6.04	6.00
5	6.61	5.79	5.41	5.19	5.05	4.95	4.88	4.82	4.77
6	5.99	5.14	4.76	4.53	4.39	4.28	4.21	4.15	4.10
7	5.59	4.74	4.35	4.12	3.97	3.87	3.79	3.73	3.68
8	5.32	4.46	4.07	3.84	3.69	3.58	3.50	3.44	3.39
9	5.12	4.26	3.86	3.63	3.48	3.37	3.29	3.23	3.18
10	4.96	4.10	3.71	3.48	3.33	3.22	3.14	3.07	3.02
11	4.84	3.98	3.59	3.36	3.20	3.09	3.01	2.95	2.90
12	4.75	3.89	3.49	3.26	3.11	3.00	2.91	2.85	2.80
13	4.67	3.81	3.41	3.18	3.03	2.92	2.83	2.77	2.71
14	4.60	3.74	3.34	3.11	2.96	2.85	2.76	2.70	2.65
15	4.54	3.68	3.29	3.06	2.90	2.79	2.71	2.64	2.59
16	4.49	3.63	3.24	3.01	2.85	2.74	2.66	2.59	2.54
17	4.45	3.59	3.20	2.96	2.81	2.70	2.61	2.55	2.49
18	4.41	3.55	3.16	2.93	2.77	2.66	2.58	2.51	2.46
19	4.38	3.52	3.13	2.90	2.74	2.63	2.54	2.48	2.42
20	4.35	3.49	3.10	2.87	2.71	2.60	2.51	2.45	2.39
21	4.32	3.47	3.07	2.84	2.68	2.57	2.49	2.42	2.37
22	4.30	3.44	3.05	2.82	2.66	2.55	2.46	2.40	2.34
23	4.28	3.42	3.03	2.80	2.64	2.53	2.44	2.37	2.32
24	4.26	3.40	3.01	2.78	2.62	2.51	2.42	2.36	2.30
25	4.24	3.39	2.99	2.76	2.60	2.49	2.40	2.34	2.28
26	4.23	3.37	2.98	2.74	2.59	2.47	2.39	2.32	2.27
27	4.21	3.35	2.96	2.73	2.57	2.46	2.37	2.31	2.25
28	4.20	3.34	2.95	2.71	2.56	2.45	2.36	2.29	2.24
29	4.18	3.33	2.93	2.70	2.55	2.43	2.35	2.28	2.22
30	4.17	3.32	2.92	2.69	2.53	2.42	2.33	2.27	2.21
40	4.08	3.23	2.84	2.61	2.45	2.34	2.25	2.18	2.12
60	4.00	3.15	2.76	2.53	2.37	2.25	2.17	2.10	2.04
120	3.92	3.07	2.68	2.45	2.29	2.18	2.09	2.02	1.96
∞	3.84	3.00	2.60	2.37	2.21	2.10	2.01	1.94	1.88

Denominator Degrees of Freedom

TABLE A.7

Percentage Points of the F Distribution (*Continued*)

										v_1	
				$\alpha = .05$							
				Numerator Degrees of Freedom							v_2
10	**12**	**15**	**20**	**24**	**30**	**40**	**60**	**120**	**∞**		
241.88	243.90	245.90	248.00	249.10	250.10	251.10	252.20	253.30	254.30	1	
19.40	19.41	19.43	19.45	19.45	19.46	19.47	19.48	19.49	19.50	2	
8.79	8.74	8.70	8.66	8.64	8.62	8.59	8.57	8.55	8.53	3	
5.96	5.91	5.86	5.80	5.77	5.75	5.72	5.69	5.66	5.63	4	
4.74	4.68	4.62	4.56	4.53	4.50	4.46	4.43	4.40	4.36	5	
4.06	4.00	3.94	3.87	3.84	3.81	3.77	3.74	3.70	3.67	6	
3.64	3.57	3.51	3.44	3.41	3.38	3.34	3.30	3.27	3.23	7	
3.35	3.28	3.22	3.15	3.12	3.08	3.04	3.01	2.97	2.93	8	
3.14	3.07	3.01	2.94	2.90	2.86	2.83	2.79	2.75	2.71	9	
2.98	2.91	2.85	2.77	2.74	2.70	2.66	2.62	2.58	2.54	10	
2.85	2.79	2.72	2.65	2.61	2.57	2.53	2.49	2.45	2.40	11	
2.75	2.69	2.62	2.54	2.51	2.47	2.43	2.38	2.34	2.30	12	
2.67	2.60	2.53	2.46	2.42	2.38	2.34	2.30	2.25	2.21	13	
2.60	2.53	2.46	2.39	2.35	2.31	2.27	2.22	2.18	2.13	14	
2.54	2.48	2.40	2.33	2.29	2.25	2.20	2.16	2.11	2.07	15	
2.49	2.42	2.35	2.28	2.24	2.19	2.15	2.11	2.06	2.01	16	
2.45	2.38	2.31	2.23	2.19	2.15	2.10	2.06	2.01	1.96	17	
2.41	2.34	2.27	2.19	2.15	2.11	2.06	2.02	1.97	1.92	18	
2.38	2.31	2.23	2.16	2.11	2.07	2.03	1.98	1.93	1.88	19	
2.35	2.28	2.20	2.12	2.08	2.04	1.99	1.95	1.90	1.84	20	
2.32	2.25	2.18	2.10	2.05	2.01	1.96	1.92	1.87	1.81	21	
2.30	2.23	2.15	2.07	2.03	1.98	1.94	1.89	1.84	1.78	22	
2.27	2.20	2.13	2.05	2.01	1.96	1.91	1.86	1.81	1.76	23	
2.25	2.18	2.11	2.03	1.98	1.94	1.89	1.84	1.79	1.73	24	
2.24	2.16	2.09	2.01	1.96	1.92	1.87	1.82	1.77	1.71	25	
2.22	2.15	2.07	1.99	1.95	1.90	1.85	1.80	1.75	1.69	26	
2.20	2.13	2.06	1.97	1.93	1.88	1.84	1.79	1.73	1.67	27	
2.19	2.12	2.04	1.96	1.91	1.87	1.82	1.77	1.71	1.65	28	
2.18	2.10	2.03	1.94	1.90	1.85	1.81	1.75	1.70	1.64	29	
2.16	2.09	2.01	1.93	1.89	1.84	1.79	1.74	1.68	1.62	30	
2.08	2.00	1.92	1.84	1.79	1.74	1.69	1.64	1.58	1.51	40	
1.99	1.92	1.84	1.75	1.70	1.65	1.59	1.53	1.47	1.39	60	
1.91	1.83	1.75	1.66	1.61	1.55	1.50	1.43	1.35	1.25	120	
1.83	1.75	1.67	1.57	1.52	1.46	1.39	1.32	1.22	1.00	∞	

Denominator Degrees of Freedom

(*Continued*)

TABLE A.7

Percentage Points of the *F* Distribution (*Continued*)

					$\alpha = .025$					
v_1 v_2					Numerator Degrees of Freedom					
	1	2	3	4	5	6	7	8	9
1	647.79	799.48	864.15	899.60	921.83	937.11	948.20	956.64	963.28
2	38.51	39.00	39.17	39.25	39.30	39.33	39.36	39.37	39.39
3	17.44	16.04	15.44	15.10	14.88	14.73	14.62	14.54	14.47
4	12.22	10.65	9.98	9.60	9.36	9.20	9.07	8.98	8.90
5	10.01	8.43	7.76	7.39	7.15	6.98	6.85	6.76	6.68
6	8.81	7.26	6.60	6.23	5.99	5.82	5.70	5.60	5.52
7	8.07	6.54	5.89	5.52	5.29	5.12	4.99	4.90	4.82
8	7.57	6.06	5.42	5.05	4.82	4.65	4.53	4.43	4.36
9	7.21	5.71	5.08	4.72	4.48	4.32	4.20	4.10	4.03
10	6.94	5.46	4.83	4.47	4.24	4.07	3.95	3.85	3.78
11	6.72	5.26	4.63	4.28	4.04	3.88	3.76	3.66	3.59
12	6.55	5.10	4.47	4.12	3.89	3.73	3.61	3.51	3.44
13	6.41	4.97	4.35	4.00	3.77	3.60	3.48	3.39	3.31
14	6.30	4.86	4.24	3.89	3.66	3.50	3.38	3.29	3.21
15	6.20	4.77	4.15	3.80	3.58	3.41	3.29	3.20	3.12
16	6.12	4.69	4.08	3.73	3.50	3.34	3.22	3.12	3.05
17	6.04	4.62	4.01	3.66	3.44	3.28	3.16	3.06	2.98
18	5.98	4.56	3.95	3.61	3.38	3.22	3.10	3.01	2.93
19	5.92	4.51	3.90	3.56	3.33	3.17	3.05	2.96	2.88
20	5.87	4.46	3.86	3.51	3.29	3.13	3.01	2.91	2.84
21	5.83	4.42	3.82	3.48	3.25	3.09	2.97	2.87	2.80
22	5.79	4.38	3.78	3.44	3.22	3.05	2.93	2.84	2.76
23	5.75	4.35	3.75	3.41	3.18	3.02	2.90	2.81	2.73
24	5.72	4.32	3.72	3.38	3.15	2.99	2.87	2.78	2.70
25	5.69	4.29	3.69	3.35	3.13	2.97	2.85	2.75	2.68
26	5.66	4.27	3.67	3.33	3.10	2.94	2.82	2.73	2.65
27	5.63	4.24	3.65	3.31	3.08	2.92	2.80	2.71	2.63
28	5.61	4.22	3.63	3.29	3.06	2.90	2.78	2.69	2.61
29	5.59	4.20	3.61	3.27	3.04	2.88	2.76	2.67	2.59
30	5.57	4.18	3.59	3.25	3.03	2.87	2.75	2.65	2.57
40	5.42	4.05	3.46	3.13	2.90	2.74	2.62	2.53	2.45
60	5.29	3.93	3.34	3.01	2.79	2.63	2.51	2.41	2.33
120	5.15	3.80	3.23	2.89	2.67	2.52	2.39	2.30	2.22
∞	5.02	3.69	3.12	2.79	2.57	2.41	2.29	2.19	2.11

Denominator Degrees of Freedom

TABLE A.7

Percentage Points of the *F* Distribution (*Continued*)

										v_1
				$\alpha = .025$						
				Numerator Degrees of Freedom						v_2
10	12	15	20	24	30	40	60	120	∞	
968.63	976.72	984.87	993.08	997.27	1001.40	1005.60	1009.79	1014.04	1018.00	1
39.40	39.41	39.43	39.45	39.46	39.46	39.47	39.48	39.49	39.50	2
14.42	14.34	14.25	14.17	14.12	14.08	14.04	13.99	13.95	13.90	3
8.84	8.75	8.66	8.56	8.51	8.46	8.41	8.36	8.31	8.26	4
6.62	6.52	6.43	6.33	6.28	6.23	6.18	6.12	6.07	6.02	5
5.46	5.37	5.27	5.17	5.12	5.07	5.01	4.96	4.90	4.85	6
4.76	4.67	4.57	4.47	4.41	4.36	4.31	4.25	4.20	4.14	7
4.30	4.20	4.10	4.00	3.95	3.89	3.84	3.78	3.73	3.67	8
3.96	3.87	3.77	3.67	3.61	3.56	3.51	3.45	3.39	3.33	9
3.72	3.62	3.52	3.42	3.37	3.31	3.26	3.20	3.14	3.08	10
3.53	3.43	3.33	3.23	3.17	3.12	3.06	3.00	2.94	2.88	11
3.37	3.28	3.18	3.07	3.02	2.96	2.91	2.85	2.79	2.72	12
3.25	3.15	3.05	2.95	2.89	2.84	2.78	2.72	2.66	2.60	13
3.15	3.05	2.95	2.84	2.79	2.73	2.67	2.61	2.55	2.49	14
3.06	2.96	2.86	2.76	2.70	2.64	2.59	2.52	2.46	2.40	15
2.99	2.89	2.79	2.68	2.63	2.57	2.51	2.45	2.38	2.32	16
2.92	2.82	2.72	2.62	2.56	2.50	2.44	2.38	2.32	2.25	17
2.87	2.77	2.67	2.56	2.50	2.44	2.38	2.32	2.26	2.19	18
2.82	2.72	2.62	2.51	2.45	2.39	2.33	2.27	2.20	2.13	19
2.77	2.68	2.57	2.46	2.41	2.35	2.29	2.22	2.16	2.09	20
2.73	2.64	2.53	2.42	2.37	2.31	2.25	2.18	2.11	2.04	21
2.70	2.60	2.50	2.39	2.33	2.27	2.21	2.14	2.08	2.00	22
2.67	2.57	2.47	2.36	2.30	2.24	2.18	2.11	2.04	1.97	23
2.64	2.54	2.44	2.33	2.27	2.21	2.15	2.08	2.01	1.94	24
2.61	2.51	2.41	2.30	2.24	2.18	2.12	2.05	1.98	1.91	25
2.59	2.49	2.39	2.28	2.22	2.16	2.09	2.03	1.95	1.88	26
2.57	2.47	2.36	2.25	2.19	2.13	2.07	2.00	1.93	1.85	27
2.55	2.45	2.34	2.23	2.17	2.11	2.05	1.98	1.91	1.83	28
2.53	2.43	2.32	2.21	2.15	2.09	2.03	1.96	1.89	1.81	29
2.51	2.41	2.31	2.20	2.14	2.07	2.01	1.94	1.87	1.79	30
2.39	2.29	2.18	2.07	2.01	1.94	1.88	1.80	1.72	1.64	40
2.27	2.17	2.06	1.94	1.88	1.82	1.74	1.67	1.58	1.48	60
2.16	2.05	1.94	1.82	1.76	1.69	1.61	1.53	1.43	1.31	120
2.05	1.94	1.83	1.71	1.64	1.57	1.48	1.39	1.27	1.00	∞

Denominator Degrees of Freedom

(*Continued*)

TABLE A.7

Percentage Points of the *F* Distribution (*Continued*)

v_2 \ v_1	$\alpha = .01$ Numerator Degrees of Freedom								
	1	2	3	4	5	6	7	8	9
1	4052.18	4999.34	5403.53	5624.26	5763.96	5858.95	5928.33	5980.95	6022.40
2	98.50	99.00	99.16	99.25	99.30	99.33	99.36	99.38	99.39
3	34.12	30.82	29.46	28.71	28.24	27.91	27.67	27.49	27.34
4	21.20	18.00	16.69	15.98	15.52	15.21	14.98	14.80	14.66
5	16.26	13.27	12.06	11.39	10.97	10.67	10.46	10.29	10.16
6	13.75	10.92	9.78	9.15	8.75	8.47	8.26	8.10	7.98
7	12.25	9.55	8.45	7.85	7.46	7.19	6.99	6.84	6.72
8	11.26	8.65	7.59	7.01	6.63	6.37	6.18	6.03	5.91
9	10.56	8.02	6.99	6.42	6.06	5.80	5.61	5.47	5.35
10	10.04	7.56	6.55	5.99	5.64	5.39	5.20	5.06	4.94
11	9.65	7.21	6.22	5.67	5.32	5.07	4.89	4.74	4.63
12	9.33	6.93	5.95	5.41	5.06	4.82	4.64	4.50	4.39
13	9.07	6.70	5.74	5.21	4.86	4.62	4.44	4.30	4.19
14	8.86	6.51	5.56	5.04	4.69	4.46	4.28	4.14	4.03
15	8.68	6.36	5.42	4.89	4.56	4.32	4.14	4.00	3.89
16	8.53	6.23	5.29	4.77	4.44	4.20	4.03	3.89	3.78
17	8.40	6.11	5.19	4.67	4.34	4.10	3.93	3.79	3.68
18	8.29	6.01	5.09	4.58	4.25	4.01	3.84	3.71	3.60
19	8.18	5.93	5.01	4.50	4.17	3.94	3.77	3.63	3.52
20	8.10	5.85	4.94	4.43	4.10	3.87	3.70	3.56	3.46
21	8.02	5.78	4.87	4.37	4.04	3.81	3.64	3.51	3.40
22	7.95	5.72	4.82	4.31	3.99	3.76	3.59	3.45	3.35
23	7.88	5.66	4.76	4.26	3.94	3.71	3.54	3.41	3.30
24	7.82	5.61	4.72	4.22	3.90	3.67	3.50	3.36	3.26
25	7.77	5.57	4.68	4.18	3.85	3.63	3.46	3.32	3.22
26	7.72	5.53	4.64	4.14	3.82	3.59	3.42	3.29	3.18
27	7.68	5.49	4.60	4.11	3.78	3.56	3.39	3.26	3.15
28	7.64	5.45	4.57	4.07	3.75	3.53	3.36	3.23	3.12
29	7.60	5.42	4.54	4.04	3.73	3.50	3.33	3.20	3.09
30	7.56	5.39	4.51	4.02	3.70	3.47	3.30	3.17	3.07
40	7.31	5.18	4.31	3.83	3.51	3.29	3.12	2.99	2.89
60	7.08	4.98	4.13	3.65	3.34	3.12	2.95	2.82	2.72
120	6.85	4.79	3.95	3.48	3.17	2.96	2.79	2.66	2.56
∞	6.63	4.61	3.78	3.32	3.02	2.80	2.64	2.51	2.41

Denominator Degrees of Freedom

TABLE A.7

Percentage Points of the *F* Distribution (*Continued*)

					$\alpha = .01$						v_1
				Numerator Degrees of Freedom							v_2
10	**12**	**15**	**20**	**24**	**30**	**40**	**60**	**120**	**∞**		
6055.93	6106.68	6156.97	6208.66	6234.27	6260.35	6286.43	6312.97	6339.51	6366.00	1	
99.40	99.42	99.43	99.45	99.46	99.47	99.48	99.48	99.49	99.50	2	
27.23	27.05	26.87	26.69	26.60	26.50	26.41	26.32	26.22	26.13	3	
14.55	14.37	14.20	14.02	13.93	13.84	13.75	13.65	13.56	13.46	4	
10.05	9.89	9.72	9.55	9.47	9.38	9.29	9.20	9.11	9.02	5	
7.87	7.72	7.56	7.40	7.31	7.23	7.14	7.06	6.97	6.88	6	
6.62	6.47	6.31	6.16	6.07	5.99	5.91	5.82	5.74	5.65	7	
5.81	5.67	5.52	5.36	5.28	5.20	5.12	5.03	4.95	4.86	8	
5.26	5.11	4.96	4.81	4.73	4.65	4.57	4.48	4.40	4.31	9	
4.85	4.71	4.56	4.41	4.33	4.25	4.17	4.08	4.00	3.91	10	
4.54	4.40	4.25	4.10	4.02	3.94	3.86	3.78	3.69	3.60	11	
4.30	4.16	4.01	3.86	3.78	3.70	3.62	3.54	3.45	3.36	12	
4.10	3.96	3.82	3.66	3.59	3.51	3.43	3.34	3.25	3.17	13	
3.94	3.80	3.66	3.51	3.43	3.35	3.27	3.18	3.09	3.00	14	
3.80	3.67	3.52	3.37	3.29	3.21	3.13	3.05	2.96	2.87	15	
3.69	3.55	3.41	3.26	3.18	3.10	3.02	2.93	2.84	2.75	16	
3.59	3.46	3.31	3.16	3.08	3.00	2.92	2.83	2.75	2.65	17	
3.51	3.37	3.23	3.08	3.00	2.92	2.84	2.75	2.66	2.57	18	
3.43	3.30	3.15	3.00	2.92	2.84	2.76	2.67	2.58	2.49	19	
3.37	3.23	3.09	2.94	2.86	2.78	2.69	2.61	2.52	2.42	20	
3.31	3.17	3.03	2.88	2.80	2.72	2.64	2.55	2.46	2.36	21	
3.26	3.12	2.98	2.83	2.75	2.67	2.58	2.50	2.40	2.31	22	
3.21	3.07	2.93	2.78	2.70	2.62	2.54	2.45	2.35	2.26	23	
3.17	3.03	2.89	2.74	2.66	2.58	2.49	2.40	2.31	2.21	24	
3.13	2.99	2.85	2.70	2.62	2.54	2.45	2.36	2.27	2.17	25	
3.09	2.96	2.81	2.66	2.58	2.50	2.42	2.33	2.23	2.13	26	
3.06	2.93	2.78	2.63	2.55	2.47	2.38	2.29	2.20	2.10	27	
3.03	2.90	2.75	2.60	2.52	2.44	2.35	2.26	2.17	2.06	28	
3.00	2.87	2.73	2.57	2.49	2.41	2.33	2.23	2.14	2.03	29	
2.98	2.84	2.70	2.55	2.47	2.39	2.30	2.21	2.11	2.01	30	
2.80	2.66	2.52	2.37	2.29	2.20	2.11	2.02	1.92	1.80	40	
2.63	2.50	2.35	2.20	2.12	2.03	1.94	1.84	1.73	1.60	60	
2.47	2.34	2.19	2.03	1.95	1.86	1.76	1.66	1.53	1.38	120	
2.32	2.18	2.04	1.88	1.79	1.70	1.59	1.47	1.32	1.00	∞	

Denominator Degrees of Freedom

(*Continued*)

TABLE A.7

Percentage Points of the *F* Distribution (*Continued*)

v_2	\multicolumn{9}{c}{$\alpha = .005$ — Numerator Degrees of Freedom}								
	1	2	3	4	5	6	7	8	9
1	16212.46	19997.36	21614.13	22500.75	23055.82	23439.53	23715.20	23923.81	24091.45
2	198.50	199.01	199.16	199.24	199.30	199.33	199.36	199.38	199.39
3	55.55	49.80	47.47	46.20	45.39	44.84	44.43	44.13	43.88
4	31.33	26.28	24.26	23.15	22.46	21.98	21.62	21.35	21.14
5	22.78	18.31	16.53	15.56	14.94	14.51	14.20	13.96	13.77
6	18.63	14.54	12.92	12.03	11.46	11.07	10.79	10.57	10.39
7	16.24	12.40	10.88	10.05	9.52	9.16	8.89	8.68	8.51
8	14.69	11.04	9.60	8.81	8.30	7.95	7.69	7.50	7.34
9	13.61	10.11	8.72	7.96	7.47	7.13	6.88	6.69	6.54
10	12.83	9.43	8.08	7.34	6.87	6.54	6.30	6.12	5.97
11	12.23	8.91	7.60	6.88	6.42	6.10	5.86	5.68	5.54
12	11.75	8.51	7.23	6.52	6.07	5.76	5.52	5.35	5.20
13	11.37	8.19	6.93	6.23	5.79	5.48	5.25	5.08	4.94
14	11.06	7.92	6.68	6.00	5.56	5.26	5.03	4.86	4.72
15	10.80	7.70	6.48	5.80	5.37	5.07	4.85	4.67	4.54
16	10.58	7.51	6.30	5.64	5.21	4.91	4.69	4.52	4.38
17	10.38	7.35	6.16	5.50	5.07	4.78	4.56	4.39	4.25
18	10.22	7.21	6.03	5.37	4.96	4.66	4.44	4.28	4.14
19	10.07	7.09	5.92	5.27	4.85	4.56	4.34	4.18	4.04
20	9.94	6.99	5.82	5.17	4.76	4.47	4.26	4.09	3.96
21	9.83	6.89	5.73	5.09	4.68	4.39	4.18	4.01	3.88
22	9.73	6.81	5.65	5.02	4.61	4.32	4.11	3.94	3.81
23	9.63	6.73	5.58	4.95	4.54	4.26	4.05	3.88	3.75
24	9.55	6.66	5.52	4.89	4.49	4.20	3.99	3.83	3.69
25	9.48	6.60	5.46	4.84	4.43	4.15	3.94	3.78	3.64
26	9.41	6.54	5.41	4.79	4.38	4.10	3.89	3.73	3.60
27	9.34	6.49	5.36	4.74	4.34	4.06	3.85	3.69	3.56
28	9.28	6.44	5.32	4.70	4.30	4.02	3.81	3.65	3.52
29	9.23	6.40	5.28	4.66	4.26	3.98	3.77	3.61	3.48
30	9.18	6.35	5.24	4.62	4.23	3.95	3.74	3.58	3.45
40	8.83	6.07	4.98	4.37	3.99	3.71	3.51	3.35	3.22
60	8.49	5.79	4.73	4.14	3.76	3.49	3.29	3.13	3.01
120	8.18	5.54	4.50	3.92	3.55	3.28	3.09	2.93	2.81
∞	7.88	5.30	4.28	3.72	3.35	3.09	2.90	2.74	2.62

Denominator Degrees of Freedom (v_2)

TABLE A.7

Percentage Points of the F Distribution (*Continued*)

										v_1	
				$\alpha = .005$							v_2
			Numerator Degrees of Freedom								
10	12	15	20	24	30	40	60	120	∞		
24221.84	24426.73	24631.62	24836.51	24937.09	25041.40	25145.71	25253.74	25358.05	25465.00	1	
199.39	199.42	199.43	199.45	199.45	199.48	199.48	199.48	199.49	199.50	2	
43.68	43.39	43.08	42.78	42.62	42.47	42.31	42.15	41.99	41.83	3	
20.97	20.70	20.44	20.17	20.03	19.89	19.75	19.61	19.47	19.32	4	
13.62	13.38	13.15	12.90	12.78	12.66	12.53	12.40	12.27	12.14	5	
10.25	10.03	9.81	9.59	9.47	9.36	9.24	9.12	9.00	8.88	6	
8.38	8.18	7.97	7.75	7.64	7.53	7.42	7.31	7.19	7.08	7	
7.21	7.01	6.81	6.61	6.50	6.40	6.29	6.18	6.06	5.95	8	
6.42	6.23	6.03	5.83	5.73	5.62	5.52	5.41	5.30	5.19	9	
5.85	5.66	5.47	5.27	5.17	5.07	4.97	4.86	4.75	4.64	10	
5.42	5.24	5.05	4.86	4.76	4.65	4.55	4.45	4.34	4.23	11	
5.09	4.91	4.72	4.53	4.43	4.33	4.23	4.12	4.01	3.90	12	Denominator Degrees of Freedom
4.82	4.64	4.46	4.27	4.17	4.07	3.97	3.87	3.76	3.65	13	
4.60	4.43	4.25	4.06	3.96	3.86	3.76	3.66	3.55	3.44	14	
4.42	4.25	4.07	3.88	3.79	3.69	3.59	3.48	3.37	3.26	15	
4.27	4.10	3.92	3.73	3.64	3.54	3.44	3.33	3.22	3.11	16	
4.14	3.97	3.79	3.61	3.51	3.41	3.31	3.21	3.10	2.98	17	
4.03	3.86	3.68	3.50	3.40	3.30	3.20	3.10	2.99	2.87	18	
3.93	3.76	3.59	3.40	3.31	3.21	3.11	3.00	2.89	2.78	19	
3.85	3.68	3.50	3.32	3.22	3.12	3.02	2.92	2.81	2.69	20	
3.77	3.60	3.43	3.24	3.15	3.05	2.95	2.84	2.73	2.61	21	
3.70	3.54	3.36	3.18	3.08	2.98	2.88	2.77	2.66	2.55	22	
3.64	3.47	3.30	3.12	3.02	2.92	2.82	2.71	2.60	2.48	23	
3.59	3.42	3.25	3.06	2.97	2.87	2.77	2.66	2.55	2.43	24	
3.54	3.37	3.20	3.01	2.92	2.82	2.72	2.61	2.50	2.38	25	
3.49	3.33	3.15	2.97	2.87	2.77	2.67	2.56	2.45	2.33	26	
3.45	3.28	3.11	2.93	2.83	2.73	2.63	2.52	2.41	2.29	27	
3.41	3.25	3.07	2.89	2.79	2.69	2.59	2.48	2.37	2.25	28	
3.38	3.21	3.04	2.86	2.76	2.66	2.56	2.45	2.33	2.21	29	
3.34	3.18	3.01	2.82	2.73	2.63	2.52	2.42	2.30	2.18	30	
3.12	2.95	2.78	2.60	2.50	2.40	2.30	2.18	2.06	1.93	40	
2.90	2.74	2.57	2.39	2.29	2.19	2.08	1.96	1.83	1.69	60	
2.71	2.54	2.37	2.19	2.09	1.98	1.87	1.75	1.61	1.43	120	
2.52	2.36	2.19	2.00	1.90	1.79	1.67	1.53	1.36	1.00	∞	

TABLE A.8

The Chi-Square Table

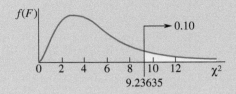

Values of χ^2 for Selected Probabilities

Example: df (Number of degrees of freedom) = 5, the tail above χ^2 = 9.23635 represents 0.10 or 10% of area under the curve.

Degrees of Freedom	Area in Upper Tail									
	.995	.99	.975	.95	.9	.1	.05	.025	.01	.005
1	0.0000393	0.0001571	0.0009821	0.0039322	0.0157907	2.7055	3.8415	5.0239	6.6349	7.8794
2	0.010025	0.020100	0.050636	0.102586	0.210721	4.6052	5.9915	7.3778	9.2104	10.5965
3	0.07172	0.11483	0.21579	0.35185	0.58438	6.2514	7.8147	9.3484	11.3449	12.8381
4	0.20698	0.29711	0.48442	0.71072	1.06362	7.7794	9.4877	11.1433	13.2767	14.8602
5	0.41175	0.55430	0.83121	1.14548	1.61031	9.2363	11.0705	12.8325	15.0863	16.7496
6	0.67573	0.87208	1.23734	1.63538	2.20413	10.6446	12.5916	14.4494	16.8119	18.5475
7	0.98925	1.23903	1.68986	2.16735	2.83311	12.0170	14.0671	16.0128	18.4753	20.2777
8	1.34440	1.64651	2.17972	2.73263	3.48954	13.3616	15.5073	17.5345	20.0902	21.9549
9	1.73491	2.08789	2.70039	3.32512	4.16816	14.6837	16.9190	19.0228	21.6660	23.5893
10	2.15585	2.55820	3.24696	3.94030	4.86518	15.9872	18.3070	20.4832	23.2093	25.1881
11	2.60320	3.05350	3.81574	4.57481	5.57779	17.2750	19.6752	21.9200	24.7250	26.7569
12	3.07379	3.57055	4.40378	5.22603	6.30380	18.5493	21.0261	23.3367	26.2170	28.2997
13	3.56504	4.10690	5.00874	5.89186	7.04150	19.8119	22.3620	24.7356	27.6882	29.8193
14	4.07466	4.66042	5.62872	6.57063	7.78954	21.0641	23.6848	26.1189	29.1412	31.3194
15	4.60087	5.22936	6.26212	7.26093	8.54675	22.3071	24.9958	27.4884	30.5780	32.8015
16	5.14216	5.81220	6.90766	7.96164	9.31224	23.5418	26.2962	28.8453	31.9999	34.2671
17	5.69727	6.40774	7.56418	8.67175	10.08518	24.7690	27.5871	30.1910	33.4087	35.7184
18	6.26477	7.01490	8.23074	9.39045	10.86494	25.9894	28.8693	31.5264	34.8052	37.1564
19	6.84392	7.63270	8.90651	10.11701	11.65091	27.2036	30.1435	32.8523	36.1908	38.5821
20	7.43381	8.26037	9.59077	10.85080	12.44260	28.4120	31.4104	34.1696	37.5663	39.9969
21	8.03360	8.89717	10.28291	11.59132	13.23960	29.6151	32.6706	35.4789	38.9322	41.4009
22	8.64268	9.54249	10.98233	12.33801	14.04149	30.8133	33.9245	36.7807	40.2894	42.7957
23	9.26038	10.19569	11.68853	13.09051	14.84795	32.0069	35.1725	38.0756	41.6383	44.1814
24	9.88620	10.85635	12.40115	13.84842	15.65868	33.1962	36.4150	39.3641	42.9798	45.5584
25	10.51965	11.52395	13.11971	14.61140	16.47341	34.3816	37.6525	40.6465	44.3140	46.9280
26	11.16022	12.19818	13.84388	15.37916	17.29188	35.5632	38.8851	41.9231	45.6416	48.2898
27	11.80765	12.87847	14.57337	16.15139	18.11389	36.7412	40.1133	43.1945	46.9628	49.6450
28	12.46128	13.56467	15.30785	16.92788	18.93924	37.9159	41.3372	44.4608	48.2782	50.9936
29	13.12107	14.25641	16.04705	17.70838	19.76774	39.0875	42.5569	45.7223	49.5878	52.3355
30	13.78668	14.95346	16.79076	18.49267	20.59924	40.2560	43.7730	46.9792	50.8922	53.6719
40	20.70658	22.16420	24.43306	26.50930	29.05052	51.8050	55.7585	59.3417	63.6908	66.7660
50	27.99082	29.70673	32.35738	34.76424	37.68864	63.1671	67.5048	71.4202	76.1538	79.4898
60	35.53440	37.48480	40.48171	43.18797	46.45888	74.3970	79.0820	83.2977	88.3794	91.9518
70	43.27531	45.44170	48.75754	51.73926	55.32894	85.5270	90.5313	95.0231	100.4251	104.2148
80	51.17193	53.53998	57.15315	60.39146	64.27784	96.5782	101.8795	106.6285	112.3288	116.3209
90	59.19633	61.75402	65.64659	69.12602	73.29108	107.5650	113.1452	118.1359	124.1162	128.2987
100	67.32753	70.06500	74.22188	77.92944	82.35813	118.4980	124.3221	129.5613	135.8069	140.1697

Critical Values for the Durbin-Watson Test

Entries in the table give the critical values for a one-tailed Durbin-Watson test for autocorrelation.
For a two-tailed test, the level of significance is doubled.

Significant Points of d_L and d_U: $\alpha = .05$
Number of Independent Variables

k	1		2		3		4		5	
n	d_L	d_U	d_L	d_U	d_L	d_U	d_L	d_U	d_L	d_U
15	1.08	1.36	0.95	1.54	0.82	1.75	0.69	1.97	0.56	2.21
16	1.10	1.37	0.98	1.54	0.86	1.73	0.74	1.93	0.62	2.15
17	1.13	1.38	1.02	1.54	0.90	1.71	0.78	1.90	0.67	2.10
18	1.16	1.39	1.05	1.53	0.93	1.69	0.82	1.87	0.71	2.06
19	1.18	1.40	1.08	1.53	0.97	1.68	0.86	1.85	0.75	2.02
20	1.20	1.41	1.10	1.54	1.00	1.68	0.90	1.83	0.79	1.99
21	1.22	1.42	1.13	1.54	1.03	1.67	0.93	1.81	0.83	1.96
22	1.24	1.43	1.15	1.54	1.05	1.66	0.96	1.80	0.86	1.94
23	1.26	1.44	1.17	1.54	1.08	1.66	0.99	1.79	0.90	1.92
24	1.27	1.45	1.19	1.55	1.10	1.66	1.01	1.78	0.93	1.90
25	1.29	1.45	1.21	1.55	1.12	1.66	1.04	1.77	0.95	1.89
26	1.30	1.46	1.22	1.55	1.14	1.65	1.06	1.76	0.98	1.88
27	1.32	1.47	1.24	1.56	1.16	1.65	1.08	1.76	1.01	1.86
28	1.33	1.48	1.26	1.56	1.18	1.65	1.10	1.75	1.03	1.85
29	1.34	1.48	1.27	1.56	1.20	1.65	1.12	1.74	1.05	1.84
30	1.35	1.49	1.28	1.57	1.21	1.65	1.14	1.74	1.07	1.83
31	1.36	1.50	1.30	1.57	1.23	1.65	1.16	1.74	1.09	1.83
32	1.37	1.50	1.31	1.57	1.24	1.65	1.18	1.73	1.11	1.82
33	1.38	1.51	1.32	1.58	1.26	1.65	1.19	1.73	1.13	1.81
34	1.39	1.51	1.33	1.58	1.27	1.65	1.21	1.73	1.15	1.81
35	1.40	1.52	1.34	1.58	1.28	1.65	1.22	1.73	1.16	1.80
36	1.41	1.52	1.35	1.59	1.29	1.65	1.24	1.73	1.18	1.80
37	1.42	1.53	1.36	1.59	1.31	1.66	1.25	1.72	1.19	1.80
38	1.43	1.54	1.37	1.59	1.32	1.66	1.26	1.72	1.21	1.79
39	1.43	1.54	1.38	1.60	1.33	1.66	1.27	1.72	1.22	1.79
40	1.44	1.54	1.39	1.60	1.34	1.66	1.29	1.72	1.23	1.79
45	1.48	1.57	1.43	1.62	1.38	1.67	1.34	1.72	1.29	1.78
50	1.50	1.59	1.46	1.63	1.42	1.67	1.38	1.72	1.34	1.77
55	1.53	1.60	1.49	1.64	1.45	1.68	1.41	1.72	1.38	1.77
60	1.55	1.62	1.51	1.65	1.48	1.69	1.44	1.73	1.41	1.77
65	1.57	1.63	1.54	1.66	1.50	1.70	1.47	1.73	1.44	1.77
70	1.58	1.64	1.55	1.67	1.52	1.70	1.49	1.74	1.46	1.77
75	1.60	1.65	1.57	1.68	1.54	1.71	1.51	1.74	1.49	1.77
80	1.61	1.66	1.59	1.69	1.56	1.72	1.53	1.74	1.51	1.77
85	1.62	1.67	1.60	1.70	1.57	1.72	1.55	1.75	1.52	1.77
90	1.63	1.68	1.61	1.70	1.59	1.73	1.57	1.75	1.54	1.78
95	1.64	1.69	1.62	1.71	1.60	1.73	1.58	1.75	1.56	1.78
100	1.65	1.69	1.63	1.72	1.61	1.74	1.59	1.76	1.57	1.78

This table is reprinted by permission of *Biometrika* trustees from J. Durbin and G. S. Watson, "Testing for Serial Correlation in Least Square Regression II," *Biometrika*, vol. 38, 1951, pp. 159–78.

(Continued)

TABLE A.9

Critical Values for the Durbin-Watson Test (*Continued*)

			Significant Points of d_L and d_U: $\alpha = .01$ Number of Independent Variables								
k		1		2		3		4		5
n	d_L	d_U	d_L	d_U	d_L	d_U	d_L	d_U	d_L	d_U
15	0.81	1.07	0.70	1.25	0.59	1.46	0.49	1.70	0.39	1.96
16	0.84	1.09	0.74	1.25	0.63	1.44	0.53	1.66	0.44	1.90
17	0.87	1.10	0.77	1.25	0.67	1.43	0.57	1.63	0.48	1.85
18	0.90	1.12	0.80	1.26	0.71	1.42	0.61	1.60	0.52	1.80
19	0.93	1.13	0.83	1.26	0.74	1.41	0.65	1.58	0.56	1.77
20	0.95	1.15	0.86	1.27	0.77	1.41	0.68	1.57	0.60	1.74
21	0.97	1.16	0.89	1.27	0.80	1.41	0.72	1.55	0.63	1.71
22	1.00	1.17	0.91	1.28	0.83	1.40	0.75	1.54	0.66	1.69
23	1.02	1.19	0.94	1.29	0.86	1.40	0.77	1.53	0.70	1.67
24	1.04	1.20	0.96	1.30	0.88	1.41	0.80	1.53	0.72	1.66
25	1.05	1.21	0.98	1.30	0.90	1.41	0.83	1.52	0.75	1.65
26	1.07	1.22	1.00	1.31	0.93	1.41	0.85	1.52	0.78	1.64
27	1.09	1.23	1.02	1.32	0.95	1.41	0.88	1.51	0.81	1.63
28	1.10	1.24	1.04	1.32	0.97	1.41	0.90	1.51	0.83	1.62
29	1.12	1.25	1.05	1.33	0.99	1.42	0.92	1.51	0.85	1.61
30	1.13	1.26	1.07	1.34	1.01	1.42	0.94	1.51	0.88	1.61
31	1.15	1.27	1.08	1.34	1.02	1.42	0.96	1.51	0.90	1.60
32	1.16	1.28	1.10	1.35	1.04	1.43	0.98	1.51	0.92	1.60
33	1.17	1.29	1.11	1.36	1.05	1.43	1.00	1.51	0.94	1.59
34	1.18	1.30	1.13	1.36	1.07	1.43	1.01	1.51	0.95	1.59
35	1.19	1.31	1.14	1.37	1.08	1.44	1.03	1.51	0.97	1.59
36	1.21	1.32	1.15	1.38	1.10	1.44	1.04	1.51	0.99	1.59
37	1.22	1.32	1.16	1.38	1.11	1.45	1.06	1.51	1.00	1.59
38	1.23	1.33	1.18	1.39	1.12	1.45	1.07	1.52	1.02	1.58
39	1.24	1.34	1.19	1.39	1.14	1.45	1.09	1.52	1.03	1.58
40	1.25	1.34	1.20	1.40	1.15	1.46	1.10	1.52	1.05	1.58
45	1.29	1.38	1.24	1.42	1.20	1.48	1.16	1.53	1.11	1.58
50	1.32	1.40	1.28	1.45	1.24	1.49	1.20	1.54	1.16	1.59
55	1.36	1.43	1.32	1.47	1.28	1.51	1.25	1.55	1.21	1.59
60	1.38	1.45	1.35	1.48	1.32	1.52	1.28	1.56	1.25	1.60
65	1.41	1.47	1.38	1.50	1.35	1.53	1.31	1.57	1.28	1.61
70	1.43	1.49	1.40	1.52	1.37	1.55	1.34	1.58	1.31	1.61
75	1.45	1.50	1.42	1.53	1.39	1.56	1.37	1.59	1.34	1.62
80	1.47	1.52	1.44	1.54	1.42	1.57	1.39	1.60	1.36	1.62
85	1.48	1.53	1.46	1.55	1.43	1.58	1.41	1.60	1.39	1.63
90	1.50	1.54	1.47	1.56	1.45	1.59	1.43	1.61	1.41	1.64
95	1.51	1.55	1.49	1.57	1.47	1.60	1.45	1.62	1.42	1.64
100	1.52	1.56	1.50	1.58	1.48	1.60	1.46	1.63	1.44	1.65

TABLE A.10

Critical Values of the Studentized Range (q) Distribution

| Degrees of Freedom | $\alpha = .05$ Number of Populations | | | | | | | | | | | | | | | | | | |
	2	3	4	5	6	7	8	9	10	11	12	13	14	15	16	17	18	19	20
1	18.0	27.0	32.8	37.1	40.4	43.1	45.4	47.4	49.1	50.6	52.0	53.2	54.3	55.4	56.3	57.2	58.0	58.8	59.6
2	6.08	8.33	9.80	10.9	11.7	12.4	13.0	13.5	14.0	14.4	14.7	15.1	15.4	15.7	15.9	16.1	16.4	16.6	16.8
3	4.50	5.91	6.82	7.50	8.04	8.48	8.85	9.18	9.46	9.72	9.95	10.2	10.3	10.5	10.7	10.8	11.0	11.1	11.2
4	3.93	5.04	5.76	6.29	6.71	7.05	7.35	7.60	7.83	8.03	8.21	8.37	8.52	8.66	8.79	8.91	9.03	9.13	9.23
5	3.64	4.60	5.22	5.67	6.03	6.33	6.58	6.80	6.99	7.17	7.32	7.47	7.60	7.72	7.83	7.93	8.03	8.12	8.21
6	3.46	4.34	4.90	5.30	5.63	5.90	6.12	6.32	6.49	6.65	6.79	6.92	7.03	7.14	7.24	7.34	7.43	7.51	7.59
7	3.34	4.16	4.68	5.06	5.36	5.61	5.82	6.00	6.16	6.30	6.43	6.55	6.66	6.76	6.85	6.94	7.02	7.10	7.17
8	3.26	4.04	4.53	4.89	5.17	5.40	5.60	5.77	5.92	6.05	6.18	6.29	6.39	6.48	6.57	6.65	6.73	6.80	6.87
9	3.20	3.95	4.41	4.76	5.02	5.24	5.43	5.59	5.74	5.87	5.98	6.09	6.19	6.28	6.36	6.44	6.51	6.58	6.64
10	3.15	3.88	4.33	4.65	4.91	5.12	5.30	5.46	5.60	5.72	5.83	5.93	6.03	6.11	6.19	6.27	6.34	6.40	6.47
11	3.11	3.82	4.26	4.57	4.82	5.03	5.20	5.35	5.49	5.61	5.71	5.81	5.90	5.98	6.06	6.13	6.20	6.27	6.33
12	3.08	3.77	4.20	4.51	4.75	4.95	5.12	5.27	5.39	5.51	5.61	5.71	5.80	5.88	5.95	6.02	6.09	6.15	6.21
13	3.06	3.73	4.15	4.45	4.69	4.88	5.05	5.19	5.32	5.43	5.53	5.63	5.71	5.79	5.86	5.93	5.99	6.05	6.11
14	3.03	3.70	4.11	4.41	4.64	4.83	4.99	5.13	5.25	5.36	5.46	5.55	5.64	5.71	5.79	5.85	5.91	5.97	6.03
15	3.01	3.67	4.08	4.37	4.59	4.78	4.94	5.08	5.20	5.31	5.40	5.49	5.57	5.65	5.72	5.78	5.85	5.90	5.96
16	3.00	3.65	4.05	4.33	4.56	4.74	4.90	5.03	5.15	5.26	5.35	5.44	5.52	5.59	5.66	5.73	5.79	5.84	5.90
17	2.98	3.63	4.02	4.30	4.52	4.70	4.86	4.99	5.11	5.21	5.31	5.39	5.47	5.54	5.61	5.67	5.73	5.79	5.84
18	2.97	3.61	4.00	4.28	4.49	4.67	4.82	4.96	5.07	5.17	5.27	5.35	5.43	5.50	5.57	5.63	5.69	5.74	5.79
19	2.96	3.59	3.98	4.25	4.47	4.65	4.79	4.92	5.04	5.14	5.23	5.31	5.39	5.46	5.53	5.59	5.65	5.70	5.75
20	2.95	3.58	3.96	4.23	4.45	4.62	4.77	4.90	5.01	5.11	5.20	5.28	5.36	5.43	5.49	5.55	5.61	5.66	5.71
24	2.92	3.53	3.90	4.17	4.37	4.54	4.68	4.81	4.92	5.01	5.10	5.18	5.25	5.32	5.38	5.44	5.49	5.55	5.59
30	2.89	3.49	3.85	4.10	4.30	4.46	4.60	4.72	4.82	4.92	5.00	5.08	5.15	5.21	5.27	5.33	5.38	5.43	5.47
40	2.86	3.44	3.79	4.04	4.23	4.39	4.52	4.63	4.73	4.82	4.90	4.98	5.04	5.11	5.16	5.22	5.27	5.31	5.36
60	2.83	3.40	3.74	3.98	4.16	4.31	4.44	4.55	4.65	4.73	4.81	4.88	4.94	5.00	5.06	5.11	5.15	5.20	5.24
120	2.80	3.36	3.68	3.92	4.10	4.24	4.36	4.47	4.56	4.64	4.71	4.78	4.84	4.90	4.95	5.00	5.04	5.09	5.13
∞	2.77	3.31	3.63	3.86	4.03	4.17	4.29	4.39	4.47	4.55	4.62	4.68	4.74	4.80	4.85	4.89	4.93	4.97	5.01

(*Continued*)

TABLE A.10

Critical Values of the Studentized Range (q) Distribution (*Continued*)

									$\alpha = .01$										
Degrees									**Number of Populations**										
of Freedom	**2**	**3**	**4**	**5**	**6**	**7**	**8**	**9**	**10**	**11**	**12**	**13**	**14**	**15**	**16**	**17**	**18**	**19**	**20**
1	90.0	135.	164.	186.	202.	216.	227.	237.	246.	253.	260.	266.	272.	277.	282.	286.	290.	294.	298.
2	14.0	19.0	22.3	24.7	26.6	28.2	29.5	30.7	31.7	32.6	33.4	34.1	34.8	35.4	36.0	36.5	37.0	37.5	37.9
3	8.26	10.6	12.2	13.3	14.2	15.0	15.6	16.2	16.7	17.1	17.5	17.9	18.2	18.5	18.8	19.1	19.3	19.5	19.8
4	6.51	8.12	9.17	9.96	10.6	11.1	11.5	11.9	12.3	12.6	12.8	13.1	13.3	13.5	13.7	13.9	14.1	14.2	14.4
5	5.70	6.97	7.80	8.42	8.91	9.32	9.67	9.97	10.2	10.5	10.7	10.9	11.1	11.2	11.4	11.6	11.7	11.8	11.9
6	5.24	6.33	7.03	7.56	7.97	8.32	8.61	8.87	9.10	9.30	9.49	9.65	9.81	9.95	10.1	10.2	10.3	10.4	10.5
7	4.95	5.92	6.54	7.01	7.37	7.68	7.94	8.17	8.37	8.55	8.71	8.86	9.00	9.12	9.24	9.35	9.46	9.55	9.65
8	4.74	5.63	6.20	6.63	6.96	7.24	7.47	7.68	7.87	8.03	8.18	8.31	8.44	8.55	8.66	8.76	8.85	8.94	9.03
9	4.60	5.43	5.96	6.35	6.66	6.91	7.13	7.32	7.49	7.65	7.78	7.91	8.03	8.13	8.23	8.32	8.41	8.49	8.57
10	4.48	5.27	5.77	6.14	6.43	6.67	6.87	7.05	7.21	7.36	7.48	7.60	7.71	7.81	7.91	7.99	8.07	8.15	8.22
11	4.39	5.14	5.62	5.97	6.25	6.48	6.67	6.84	6.99	7.13	7.25	7.36	7.46	7.56	7.65	7.73	7.81	7.88	7.95
12	4.32	5.04	5.50	5.84	6.10	6.32	6.51	6.67	6.81	6.94	7.06	7.17	7.26	7.36	7.44	7.52	7.59	7.66	7.73
13	4.26	4.96	5.40	5.73	5.98	6.19	6.37	6.53	6.67	6.79	6.90	7.01	7.10	7.19	7.27	7.34	7.42	7.48	7.55
14	4.21	4.89	5.32	5.63	5.88	6.08	6.26	6.41	6.54	6.66	6.77	6.87	6.96	7.05	7.12	7.20	7.27	7.33	7.39
15	4.17	4.83	5.25	5.56	5.80	5.99	6.16	6.31	6.44	6.55	6.66	6.76	6.84	6.93	7.00	7.07	7.14	7.20	7.26
16	4.13	4.78	5.19	5.49	5.72	5.92	6.08	6.22	6.35	6.46	6.56	6.66	6.74	6.82	6.90	6.97	7.03	7.09	7.15
17	4.10	4.74	5.14	5.43	5.66	5.85	6.01	6.15	6.27	6.38	6.48	6.57	6.66	6.73	6.80	6.87	6.94	7.00	7.05
18	4.07	4.70	5.09	5.38	5.60	5.79	5.94	6.08	6.20	6.31	6.41	6.50	6.58	6.65	6.72	6.79	6.85	6.91	6.96
19	4.05	4.67	5.05	5.33	5.55	5.73	5.89	6.02	6.14	6.25	6.34	6.43	6.51	6.58	6.65	6.72	6.78	6.84	6.89
20	4.02	4.64	5.02	5.29	5.51	5.69	5.84	5.97	6.09	6.19	6.29	6.37	6.45	6.52	6.59	6.65	6.71	6.76	6.82
24	3.96	4.54	4.91	5.17	5.37	5.54	5.69	5.81	5.92	6.02	6.11	6.19	6.26	6.33	6.39	6.45	6.51	6.56	6.61
30	3.89	4.45	4.80	5.05	5.24	5.40	5.54	5.65	5.76	5.85	5.93	6.01	6.08	6.14	6.20	6.26	6.31	6.36	6.41
40	3.82	4.37	4.70	4.93	5.11	5.27	5.39	5.50	5.60	5.69	5.77	5.84	5.90	5.96	6.02	6.07	6.12	6.17	6.21
60	3.76	4.28	4.60	4.82	4.99	5.13	5.25	5.36	5.45	5.53	5.60	5.67	5.73	5.79	5.84	5.89	5.93	5.98	6.02
120	3.70	4.20	4.50	4.71	4.87	5.01	5.12	5.21	5.30	5.38	5.44	5.51	5.56	5.61	5.66	5.71	5.75	5.79	5.83
∞	3.64	4.12	4.40	4.60	4.76	4.88	4.99	5.08	5.16	5.23	5.29	5.35	5.40	5.45	5.49	5.54	5.57	5.61	5.65

TABLE A.11

Critical Values of R for the Runs Test: Lower Tail

n_1 \ n_2	\multicolumn{19}{c}{$\alpha = .025$}

n_1 \ n_2	2	3	4	5	6	7	8	9	10	11	12	13	14	15	16	17	18	19	20
2											2	2	2	2	2	2	2	2	2
3				2	2	2	2	2	2	2	2	2	2	3	3	3	3	3	3
4			2	2	2	3	3	3	3	3	3	3	3	3	4	4	4	4	4
5			2	2	3	3	3	3	3	4	4	4	4	4	4	4	5	5	5
6		2	2	3	3	3	3	4	4	4	4	5	5	5	5	5	5	6	6
7		2	2	3	3	3	4	4	5	5	5	5	5	6	6	6	6	6	6
8		2	3	3	3	4	4	5	5	5	6	6	6	6	6	7	7	7	7
9		2	3	3	4	4	5	5	5	6	6	6	7	7	7	7	8	8	8
10		2	3	3	4	5	5	5	6	6	7	7	7	7	8	8	8	8	9
11		2	3	4	4	5	5	6	6	7	7	7	8	8	8	9	9	9	9
12	2	2	3	4	4	5	6	6	7	7	7	8	8	8	9	9	9	10	10
13	2	2	3	4	5	5	6	6	7	7	8	8	9	9	9	10	10	10	10
14	2	2	3	4	5	5	6	7	7	8	8	9	9	9	10	10	10	11	11
15	2	3	3	4	5	6	6	7	7	8	8	9	9	10	10	11	11	11	12
16	2	3	4	4	5	6	6	7	8	8	9	9	10	10	11	11	11	12	12
17	2	3	4	4	5	6	7	7	8	9	9	10	10	11	11	11	12	12	13
18	2	3	4	5	5	6	7	8	8	9	9	10	10	11	11	12	12	13	13
19	2	3	4	5	6	6	7	8	8	9	10	10	11	11	12	12	13	13	13
20	2	3	4	5	6	6	7	8	9	9	10	10	11	12	12	13	13	13	14

Source: Adapted from F. S. Swed and C. Eisenhart, *Ann. Math. Statist.*, vol. 14, 1943, pp. 83–86.

TABLE A.12

Critical Values of R for the Runs Test: Upper Tail

n_1 \ n_2	\multicolumn{19}{c}{$\alpha = .025$}

n_1 \ n_2	2	3	4	5	6	7	8	9	10	11	12	13	14	15	16	17	18	19	20
2																			
3																			
4				9	9														
5			9	10	10	11	11												
6			9	10	11	12	12	13	13	13	13								
7				11	12	13	13	14	14	14	14	15	15	15					
8				11	12	13	14	14	15	15	16	16	16	16	17	17	17	17	17
9					13	14	14	15	16	16	16	17	17	18	18	18	18	18	18
10					13	14	15	16	16	17	17	18	18	18	19	19	19	20	20
11					13	14	15	16	17	17	18	19	19	19	20	20	20	21	21
12					13	14	16	16	17	18	19	19	20	20	21	21	21	22	22
13						15	16	17	18	19	19	20	20	21	21	22	22	23	23
14						15	16	17	18	19	20	20	21	22	22	23	23	23	24
15						15	16	18	18	19	20	21	22	22	23	23	24	24	25
16							17	18	19	20	21	21	22	23	23	24	25	25	25
17							17	18	19	20	21	22	23	23	24	25	25	26	26
18							17	18	19	20	21	22	23	24	25	25	26	26	27
19							17	18	20	21	22	23	23	24	25	26	26	27	27
20							17	18	20	21	22	23	24	25	25	26	27	27	28

TABLE A.13

p-Values for Mann-Whitney *U* Statistic Small Samples ($n_1 \leq n_2$)

		n_1			
$n_2 = 3$	U_0	1	2	3	
	0	.25	.10	.05	
	1	.50	.20	.10	
	2		.40	.20	
	3		.60	.35	
	4			.50	

		n_1			
$n_2 = 4$	U_0	1	2	3	4
	0	.2000	.0667	.0286	.0143
	1	.4000	.1333	.0571	.0286
	2	.6000	.2667	.1143	.0571
	3		.4000	.2000	.1000
	4		.6000	.3143	.1714
	5			.4286	.2429
	6			.5714	.3429
	7				.4429
	8				.5571

		n_1				
$n_2 = 5$	U_0	1	2	3	4	5
	0	.1667	.0476	.0179	.0079	.0040
	1	.3333	.0952	.0357	.0159	.0079
	2	.5000	.1905	.0714	.0317	.0159
	3		.2857	.1250	.0556	.0278
	4		.4286	.1964	.0952	.0476
	5		.5714	.2857	.1429	.0754
	6			.3929	.2063	.1111
	7			.5000	.2778	.1548
	8				.3651	.2103
	9				.4524	.2738
	10				.5476	.3452
	11					.4206
	12					.5000

Computed by M. Pagano, Dept. of Statistics, University of Florida. Reprinted by permission from William Mendenhall and James E. Reinmuth, *Statistics for Management and Economics,* 5th ed. Copyright © 1986 by PWS-KENT Publishers, Boston.

TABLE A.13

p-Values for Mann-Whitney *U* Statistic Small Samples ($n_1 \leq n_2$) (*Continued*)

$n_2 = 6$	U_0	1	2	3	4	5	6
	0	.1429	.0357	.0119	.0048	.0022	.0011
	1	.2857	.0714	.0238	.0095	.0043	.0022
	2	.4286	.1429	.0476	.0190	.0087	.0043
	3	.5714	.2143	.0833	.0333	.0152	.0076
	4		.3214	.1310	.0571	.0260	.0130
	5		.4286	.1905	.0857	.0411	.0206
	6		.5714	.2738	.1286	.0628	.0325
	7			.3571	.1762	.0887	.0465
	8			.4524	.2381	.1234	.0660
	9			.5476	.3048	.1645	.0898
	10				.3810	.2143	.1201
	11				.4571	.2684	.1548
	12				.5429	.3312	.1970
	13					.3961	.2424
	14					.4654	.2944
	15					.5346	.3496
	16						.4091
	17						.4686
	18						.5314

(header: n_1)

$n_2 = 7$	U_0	1	2	3	4	5	6	7
	0	.1250	.0278	.0083	.0030	.0013	.0006	.0003
	1	.2500	.0556	.0167	.0061	.0025	.0012	.0006
	2	.3750	.1111	.0333	.0121	.0051	.0023	.0012
	3	.5000	.1667	.0583	.0212	.0088	.0041	.0020
	4		.2500	.0917	.0364	.0152	.0070	.0035
	5		.3333	.1333	.0545	.0240	.0111	.0055
	6		.4444	.1917	.0818	.0366	.0175	.0087
	7		.5556	.2583	.1152	.0530	.0256	.0131
	8			.3333	.1576	.0745	.0367	.0189
	9			.4167	.2061	.1010	.0507	.0265
	10			.5000	.2636	.1338	.0688	.0364
	11				.3242	.1717	.0903	.0487
	12				.3939	.2159	.1171	.0641
	13				.4636	.2652	.1474	.0825
	14				.5364	.3194	.1830	.1043
	15					.3775	.2226	.1297
	16					.4381	.2669	.1588
	17					.5000	.3141	.1914
	18						.3654	.2279
	19						.4178	.2675
	20						.4726	.3100
	21						.5274	.3552
	22							.4024
	23							.4508
	24							.5000

(*Continued*)

TABLE A.13

p-Values for Mann-Whitney *U* Statistic Small Samples ($n_1 \leq n_2$) (*Continued*)

$n_2 = 8$	U_0	1	2	3	4	5	6	7	8
	0	.1111	.0222	.0061	.0020	.0008	.0003	.0002	.0001
	1	.2222	.0444	.0121	.0040	.0016	.0007	.0003	.0002
	2	.3333	.0889	.0242	.0081	.0031	.0013	.0006	.0003
	3	.4444	.1333	.0424	.0141	.0054	.0023	.0011	.0005
	4	.5556	.2000	.0667	.0242	.0093	.0040	.0019	.0009
	5		.2667	.0970	.0364	.0148	.0063	.0030	.0015
	6		.3556	.1394	.0545	.0225	.0100	.0047	.0023
	7		.4444	.1879	.0768	.0326	.0147	.0070	.0035
	8		.5556	.2485	.1071	.0466	.0213	.0103	.0052
	9			.3152	.1414	.0637	.0296	.0145	.0074
	10			.3879	.1838	.0855	.0406	.0200	.0103
	11			.4606	.2303	.1111	.0539	.0270	.0141
	12			.5394	.2848	.1422	.0709	.0361	.0190
	13				.3414	.1772	.0906	.0469	.0249
	14				.4040	.2176	.1142	.0603	.0325
	15				.4667	.2618	.1412	.0760	.0415
	16				.5333	.3108	.1725	.0946	.0524
	17					.3621	.2068	.1159	.0652
	18					.4165	.2454	.1405	.0803
	19					.4716	.2864	.1678	.0974
	20					.5284	.3310	.1984	.1172
	21						.3773	.2317	.1393
	22						.4259	.2679	.1641
	23						.4749	.3063	.1911
	24						.5251	.3472	.2209
	25							.3894	.2527
	26							.4333	.2869
	27							.4775	.3227
	28							.5225	.3605
	29								.3992
	30								.4392
	31								.4796
	32								.5204

TABLE A.13

p-Values for Mann-Whitney *U* Statistic Small Samples ($n_1 \leq n_2$) (*Continued*)

$n_2 = 9$	U_0	n_1 1	2	3	4	5	6	7	8	9
	0	.1000	.0182	.0045	.0014	.0005	.0002	.0001	.0000	.0000
	1	.2000	.0364	.0091	.0028	.0010	.0004	.0002	.0001	.0000
	2	.3000	.0727	.0182	.0056	.0020	.0008	.0003	.0002	.0001
	3	.4000	.1091	.0318	.0098	.0035	.0014	.0006	.0003	.0001
	4	.5000	.1636	.0500	.0168	.0060	.0024	.0010	.0005	.0002
	5		.2182	.0727	.0252	.0095	.0038	.0017	.0008	.0004
	6		.2909	.1045	.0378	.0145	.0060	.0026	.0012	.0006
	7		.3636	.1409	.0531	.0210	.0088	.0039	.0019	.0009
	8		.4545	.1864	.0741	.0300	.0128	.0058	.0028	.0014
	9		.5455	.2409	.0993	.0415	.0180	.0082	.0039	.0020
	10			.3000	.1301	.0559	.0248	.0115	.0056	.0028
	11			.3636	.1650	.0734	.0332	.0156	.0076	.0039
	12			.4318	.2070	.0949	.0440	.0209	.0103	.0053
	13			.5000	.2517	.1199	.0567	.0274	.0137	.0071
	14				.3021	.1489	.0723	.0356	.0180	.0094
	15				.3552	.1818	.0905	.0454	.0232	.0122
	16				.4126	.2188	.1119	.0571	.0296	.0157
	17				.4699	.2592	.1361	.0708	.0372	.0200
	18				.5301	.3032	.1638	.0869	.0464	.0252
	19					.3497	.1942	.1052	.0570	.0313
	20					.3986	.2280	.1261	.0694	.0385
	21					.4491	.2643	.1496	.0836	.0470
	22					.5000	.3035	.1755	.0998	.0567
	23						.3445	.2039	.1179	.0680
	24						.3878	.2349	.1383	.0807
	25						.4320	.2680	.1606	.0951
	26						.4773	.3032	.1852	.1112
	27						.5227	.3403	.2117	.1290
	28							.3788	.2404	.1487
	29							.4185	.2707	.1701
	30							.4591	.3029	.1933
	31							.5000	.3365	.2181
	32								.3715	.2447
	33								.4074	.2729
	34								.4442	.3024
	35								.4813	.3332
	36								.5187	.3652
	37									.3981
	38									.4317
	39									.4657
	40									.5000

(*Continued*)

p-Values for Mann-Whitney *U* Statistic Small Samples ($n_1 \leq n_2$) (*Continued*)

$n_2 = 10$	U_0	1	2	3	4	5	6	7	8	9	10
	0	.0909	.0152	.0035	.0010	.0003	.0001	.0001	.0000	.0000	.0000
	1	.1818	.0303	.0070	.0020	.0007	.0002	.0001	.0000	.0000	.0000
	2	.2727	.0606	.0140	.0040	.0013	.0005	.0002	.0001	.0000	.0000
	3	.3636	.0909	.0245	.0070	.0023	.0009	.0004	.0002	.0001	.0000
	4	.4545	.1364	.0385	.0120	.0040	.0015	.0006	.0003	.0001	.0001
	5	.5455	.1818	.0559	.0180	.0063	.0024	.0010	.0004	.0002	.0001
	6		.2424	.0804	.0270	.0097	.0037	.0015	.0007	.0003	.0002
	7		.3030	.1084	.0380	.0140	.0055	.0023	.0010	.0005	.0002
	8		.3788	.1434	.0529	.0200	.0080	.0034	.0015	.0007	.0004
	9		.4545	.1853	.0709	.0276	.0112	.0048	.0022	.0011	.0005
	10		.5455	.2343	.0939	.0376	.0156	.0068	.0031	.0015	.0008
	11			.2867	.1199	.0496	.0210	.0093	.0043	.0021	.0010
	12			.3462	.1518	.0646	.0280	.0125	.0058	.0028	.0014
	13			.4056	.1868	.0823	.0363	.0165	.0078	.0038	.0019
	14			.4685	.2268	.1032	.0467	.0215	.0103	.0051	.0026
	15			.5315	.2697	.1272	.0589	.0277	.0133	.0066	.0034
	16				.3177	.1548	.0736	.0351	.0171	.0086	.0045
	17				.3666	.1855	.0903	.0439	.0217	.0110	.0057
	18				.4196	.2198	.1099	.0544	.0273	.0140	.0073
	19				.4725	.2567	.1317	.0665	.0338	.0175	.0093
	20				.5275	.2970	.1566	.0806	.0416	.0217	.0116
	21					.3393	.1838	.0966	.0506	.0267	.0144
	22					.3839	.2139	.1148	.0610	.0326	.0177
	23					.4296	.2461	.1349	.0729	.0394	.0216
	24					.4765	.2811	.1574	.0864	.0474	.0262
	25					.5235	.3177	.1819	.1015	.0564	.0315
	26						.3564	.2087	.1185	.0667	.0376
	27						.3962	.2374	.1371	.0782	.0446
	28						.4374	.2681	.1577	.0912	.0526
	29						.4789	.3004	.1800	.1055	.0615
	30						.5211	.3345	.2041	.1214	.0716
	31							.3698	.2299	.1388	.0827
	32							.4063	.2574	.1577	.0952
	33							.4434	.2863	.1781	.1088
	34							.4811	.3167	.2001	.1237
	35							.5189	.3482	.2235	.1399
	36								.3809	.2483	.1575
	37								.4143	.2745	.1763
	38								.4484	.3019	.1965
	39								.4827	.3304	.2179
	40								.5173	.3598	.2406
	41									.3901	.2644
	42									.4211	.2894
	43									.4524	.3153
	44									.4841	.3421
	45									.5159	.3697
	46										.3980
	47										.4267
	48										.4559
	49										.4853
	50										.5147

TABLE A.14

Critical Values of *T* for the Wilcoxon Matched-Pairs Signed Rank Test (Small Samples)

1-SIDED	2-SIDED	n = 5	n = 6	n = 7	n = 8	n = 9	n = 10
$\alpha = .05$	$\alpha = .10$	1	2	4	6	8	11
$\alpha = .025$	$\alpha = .05$		1	2	4	6	8
$\alpha = .01$	$\alpha = .02$			0	2	3	5
$\alpha = .005$	$\alpha = .01$				0	2	3

1-SIDED	2-SIDED	n = 11	n = 12	n = 13	n = 14	n = 15	n = 16
$\alpha = .05$	$\alpha = .10$	14	17	21	26	30	36
$\alpha = .025$	$\alpha = .05$	11	14	17	21	25	30
$\alpha = .01$	$\alpha = .02$	7	10	13	16	20	24
$\alpha = .005$	$\alpha = .01$	5	7	10	13	16	19

1-SIDED	2-SIDED	n = 17	n = 18	n = 19	n = 20	n = 21	n = 22
$\alpha = .05$	$\alpha = .10$	41	47	54	60	68	75
$\alpha = .025$	$\alpha = .05$	35	40	46	52	59	66
$\alpha = .01$	$\alpha = .02$	28	33	38	43	49	56
$\alpha = .005$	$\alpha = .01$	23	28	32	37	43	49

1-SIDED	2-SIDED	n = 23	n = 24	n = 25	n = 26	n = 27	n = 28
$\alpha = .05$	$\alpha = .10$	83	92	101	110	120	130
$\alpha = .025$	$\alpha = .05$	73	81	90	98	107	117
$\alpha = .01$	$\alpha = .02$	62	69	77	85	93	102
$\alpha = .005$	$\alpha = .01$	55	61	68	76	84	92

1-SIDED	2-SIDED	n = 29	n = 30	n = 31	n = 32	n = 33	n = 34
$\alpha = .05$	$\alpha = .10$	141	152	163	175	188	201
$\alpha = .025$	$\alpha = .05$	127	137	148	159	171	183
$\alpha = .01$	$\alpha = .02$	111	120	130	141	151	162
$\alpha = .005$	$\alpha = .01$	100	109	118	128	138	149

1-SIDED	2-SIDED	n = 35	n = 36	n = 37	n = 38	n = 39	
$\alpha = .05$	$\alpha = .10$	214	228	242	256	271	
$\alpha = .025$	$\alpha = .05$	195	208	222	235	250	
$\alpha = .01$	$\alpha = .02$	174	186	198	211	224	
$\alpha = .005$	$\alpha = .01$	160	171	183	195	208	

1-SIDED	2-SIDED	n = 40	n = 41	n = 42	n = 43	n = 44	n = 45
$\alpha = .05$	$\alpha = .10$	287	303	319	336	353	371
$\alpha = .025$	$\alpha = .05$	264	279	295	311	327	344
$\alpha = .01$	$\alpha = .02$	238	252	267	281	297	313
$\alpha = .005$	$\alpha = .01$	221	234	248	262	277	292

1-SIDED	2-SIDED	n = 46	n = 47	n = 48	n = 49	n = 50	
$\alpha = .05$	$\alpha = .10$	389	408	427	446	466	
$\alpha = .025$	$\alpha = .05$	361	379	397	415	434	
$\alpha = .01$	$\alpha = .02$	329	345	362	380	398	
$\alpha = .005$	$\alpha = .01$	307	323	339	356	373	

From E. Wilcoxon and R. A. Wilcox, "Some Rapid Approximate Statistical Procedures," 1964. Reprinted by permission of Lederle Labs, a division of the American Cyanamid Co.

TABLE A.15

Factors for Control Charts

	AVERAGES				RANGES	
Number of Items In Sample	**Factors for Control Limits**		**Factors for Central Line**		**Factors for Control Limits**	
n	A_2	A_3	d_2		D_3	D_4
2	1.880	2.659	1.128		0	3.267
3	1.023	1.954	1.693		0	2.575
4	0.729	1.628	2.059		0	2.282
5	0.577	1.427	2.326		0	2.115
6	0.483	1.287	2.534		0	2.004
7	0.419	1.182	2.704		0.076	1.924
8	0.373	1.099	2.847		0.136	1.864
9	0.337	1.032	2.970		0.184	1.816
10	0.308	0.975	3.078		0.223	1.777
11	0.285	0.927	3.173		0.256	1.744
12	0.266	0.886	3.258		0.284	1.716
13	0.249	0.850	3.336		0.308	1.692
14	0.235	0.817	3.407		0.329	1.671
15	0.223	0.789	3.472		0.348	1.652

Adapted from American Society for Testing and Materials, *Manual on Quality Control of Materials,* 1951, Table B2, p. 115. For a more detailed table and explanation, see Acheson J. Duncan, *Quality Control and Industrial Statistics,* 3d ed. Homewood, IL.: Richard D. Irwin, 1974, Table M, p. 927.

Answers to Selected Odd-Numbered Quantitative Problems

Chapter 1

1.5. a. ratio
  b. ratio
  c. ordinal
  d. nominal
  e. ratio
  f. ratio
  g. nominal
  h. ratio

1.7. a. 900 electric contractors
  b. 35 electric contractors
  c. average score for 35 participants
  d. average score for all 900 electric contractors

Chapter 2

No answers given

Chapter 3

3.1. 4

3.3. 294

3.5. −1

3.7. 107; 127; 145; 114; 127.5; 143.5

3.9. 3,895; 6,19; 3,055; 4,96; 7,545; 9,37

3.11. a. 8
  b. 2.041
  c. 6.204
  d. 2.491
  e. 4
  f. 0.69; −0.92; −0.11; 1.89; −1.32; −0.52; 0.29

3.13. a. 4.598
  b. 4.598

3.15. 58631.36; 242.139

3.17. a. 0.75
  b. 0.84
  c. 0.609
  d. 0.902

3.19. a. 2.667
  b. 11.060
  c. 3.326
  d. 5
  e. −0.85
  f. 37.65%

3.21. Between 113 and 137
  Between 101 and 149
  Between 89 and 161

3.23. 2.236

3.25. 95%; 2.5%; 0.15%; 16%

3.27. 4.64; 3.59; 1

3.29. 185.694; 13.627

3.31. a. 44.9
  b. 39
  c. 44.82
  d. 187.2
  e. 13.7

3.33. a. 41.67
  b. 45
  c. 46
  d. 40.56
  e. 6.37

3.35. skewed right

3.37. 0.726

3.39. no outliers; negatively skewed

3.41. 2.5; 2; 2; 7; 1; 3; 2

3.43. Mean=33 136,15
       Median = 26 804,5
       P30 = 13 995
       P60 = 34451
       P90 = 80 904
       Q1 = 13 661,5
       Q3 = 39 307
       Range = 80 064
       IQR = 25 645,5

3.45. a. 2621.7 and 2,146
       b. 7,580 and 1,081
       c. 4,194,334.6; 2.048
       d. −0.615; 1.925
       e. +0.6967748

3.47. a. 33.412; 32.5
       b. 58.483; 7.647

3.49. 10.78%; 6.43%

3.51. a. 392 to 446; 365 to 473; 338 to 500
       b. 79.7%
       c. −0.704
       d. The two variables are not mutually exclusive.

3.53. skewed right

3.55. 21.93; 18.14

Chapter 4

4.1.   15; 0.60
4.3.   {4, 8, 10, 14, 16, 18, 20, 22, 26, 28, 30}
4.5.   20, combinations; 0.60
4.7.   38,760
4.9.   a. 0.7167
       b. 0.5000
       c. 0.65
       d. 0.5167
4.11. not solvable
4.13. a. 0.86
       b. 0.31
       c. 0.14
       d. The two variables are not mutually exclusive.
4.15. a. 0.2807
       b. 0.0526
       c. 0.0000
       d. 0.0000
4.17. a. 0.0122
       b. 0.0144
4.19. a. 0.63
       b. 0.037

c. 0.803
d. 0.197
e. 0.963
f. 0.333

4.21. a. 0.3978
       b. 0.4522
       c. 0.0378

4.23. a. 0.2286
       b. 0.2297
       c. 0.3231
       d. 0.0000

4.25. not independent

4.27. a. 0.4054
       b. 0.3261
       c. 0.4074
       d. 0.32

4.29. a. 0.73
       b. 0.054
       c. 0.1731
       d. 0.5742

4.31. 0.0538; 0.5161; 0.4301

4.33. 0.7941; 0.2059

4.35. a. 0.4211
       b. 0.6316
       c. 0.2105
       d. 0.1250
       e. 0.5263
       f. 0.0000
       g. 0.6667
       h. 0.0000
       i. variables 1 and 2 are not independent

4.37. a. 0.135
       b. 0.105
       c. 0.533
       d. 0.391
       e. 0.505
       f. 0.365

4.39. a. 0.96
       b. 0.56
       c. 0.92
       d. 0.40
       e. 0.444

4.41. a. 0.39
       b. 0.40

c. 0.48
d. not independent
e. not mutually exclusive

4.43. a. 0.0294
b. 0.6706
c. 0.3294
d. 0.2706

4.45. a. 0.2625
b. 0.74375
c. 0.60
d. 0.25625
e. 0.0875

4.47. a. 0.20
b. 0.6429
c. 0.40
d. 0.60
e. 0.40
f. 0.3333

4.49. a. 0.469
b. 0.164
c. 0.2360
d. 0.1934
e. 0.754

4.51. a. 0.2130
b. 0.4370
c. 0.2240
d. 0.6086
e. 0.3914
f. 0.8662

4.53. a. 0.276
b. 0.686
c. 0.816
d. 0.590
e. 0.4023

Chapter 5

5.1. 2.666; 1.8364; 1.3552
5.3. 0.956; 1.1305
5.5. a. 0.0036
b. 0.1147
c. 0.3822
d. 0.5838

5.7. a. 14; 2.05
b. 24.5; 3.99
c. 50; 5

5.9. a. 0.000061
b. 0.000000333
c. 0.49892699

5.11. a. 0.969
b. 0.207
c. 0.002

5.13. a. 0.1032
b. 0.0000
c. 0.0352
d. 0.3480

5.15. a. 0.0538
b. 0.1539
c. 0.4142
d. 0.0672
e. 0.0244
f. 0.3604

5.17. a. 6.3; 2.51
b. 1.3; 1.14
c. 8.9; 2.98
d. 0.6; 0.775

5.19. 3.5
a. 0.0302
b. 0.1424
c. 0.0817
d. 0.42
e. 0.1009

5.21. a. 0.1827
b. 0.3106
c. 0.5068
d. 0.2513
e. 0.058

5.23. a. 0.3012
b. 0.0000
c. 0.0336

5.25. a. 0.0104
b. 0.0000
c. 0.1653
d. 0.9636

5.27. a. 0.5091
b. 0.2937
c. 0.4167
d. 0.0014

5.29. a. 0.051
b. 0.004
c. 0.397

5.31. a. 0.4
b. 0.6
c. 0.1143

5.33. 0.0474

5.35. a. 0.124
b. 0.849
c. 0.090
d. 0.000

5.37. a. 0.1607
b. 0.7626
c. 0.3504
d. 0.5429

5.39. a. 0.062
b. 0.002
c. 5
d. 0.035058
e. 0.734
f. 0.007
g. 0.000
h. x=4, expected number 4.4

5.41. a. 0.2644
b. 0.0694
c. 0.0029
d. 0.7521

5.43. a. 6.60
b. 0.0036054

5.45. a. 0.0687
b. 0.020
c. 0.1032
d. 2.28

5.47. 0.174

5.49. a. 0.3012
b. 0.1203
c. 0.7065

5.51. a. 0.0002
b. 0.0595
d. 0.2330

5.53. a. 0.0907
b. 0.0358
c. 0.1517
d. 0.8781

5.55. a. 0.265
b. 0.0136
c. 0.0067

5.57. a. 0.3768
b. 0.9191
c. 0.0142

5.59. a. 0.0455
b. 0.1602
c. 0.4247

Chapter 6

6.1. a. $1/40 = 0.025$
b. 220; 11.547
c. 0.25
d. 0.3750
e. 0.6250

6.3. 2.97; 0.098; 0.2941

6.5. 981.5; 0.000294; 0.2353; 0.0000; 0.2353

6.7. a. 0.7967
b. 0.0023
c. 0.312

6.9. a. 0.2758
b. 0.0392
c. 0.3718

6.11. a. 188.25
b. 244.65
c. 163.81
d. 206.11

6.13. 235.59

6.15. 7.11

6.17. a. $P(x \leq 16.5 \mid \mu = 21$ and $\sigma = 2.51)$
b. $P(10.5 \leq x \leq 20.5 \mid \mu = 12.5$ and $\sigma = 2.5)$
c. $P(21.5 \leq x \leq 22.5 \mid \mu = 24$ and $\sigma = 3.10)$
d. $P(x \geq 14.5 \mid \mu = 7.2$ and $\sigma = 1.99)$

6.19. a. 0.1170; 0.120
b. 0.4090; 0.415
c. 0.1985; 0.196
d. fails test

6.21. 0.0495

6.23. a. 0.0011
b. 0.0846
c. 0.5257
d. 0.192

6.27. a. 0.0012
b. 0.8700

c. 0.0011

d. 0.9918

6.29. **a.** 0.0000

b. 0.0000

c. 0.0872

d. 0.41 minutes

6.31. $\mu = 222.2$

a. 01054

b. 0.5934

6.33. 15; 15; 0.1248

6.35. **a.** 0.1587

b. 0.0013

c. 0.6915

d. 0.9270

e. 0.0000

6.37. **a.** 0.0202

b. 0.9817

c. 0.1849

d. 0.4449

6.39. 0.00714

6.41. **a.** 0.281

b. 0.121

c. 0.0432

6.43. 0.5319; 41.5; 0.0213

6.45. **a.** 0.1492

b. 0.4182

c. 0.3849

d. 0.0188

6.47. **a.** 0.3409

b. 0.4247

c. 0.6554

d. 0.1307

6.49. **a.** 0.0025

b. 0.8944

c. 0.3482

6.51. **a.** 0.0655

b. 0.6502

c. 0.9993

6.53. $11,428.57

6.55. **a.** 0.5488

b. 0.2592

c. 1.67 months

6.57. 853; 931.75; 1,182

6.59. 0.1446; 5.6%

Chapter 7

7.7. 825

7.13. **a.** 0.0548

b. 0.7881

c. 0.0082

d. 0.8575

e. 0.1664

7.15. 11.11

7.17. **a.** 0.9772

b. 0.2385

c. 0.1469

d. 0.1230

7.19. 0.0000

7.21. **a.** 0.0526

b. 0.0217

c. 0.0000

d. 23.57

7.23. **a.** 0.1492

b. 0.9404

c. 0.6985

d. 0.1445

e. 0.0000

7.25. 0.26

7.27. **a.** 0.1977

b. 0.2843

c. 0.9881

7.29. **a.** 0.1020

b. 0.7568

c. 0.7019

7.31. 55; 45; 90; 25; 35

7.37. **a.** 0.2546

b. 0.0028

c. 0.2141

7.41. **a.** 0.9952

b. 0.9394

c. 0.0000

7.43. **a.** 0.0838

b. 0.2451

c. 0.2517

d. 0.1539

e. 0.0197

7.45. **a.** 0.8534

b. 0.0256

c. 0.0007

7.49.   **a.**   0.7175

　　　　　b.   0.1685

　　　　　c.   0.0314

7.51.   0.2119

Chapter 8

8.1.   **a.** $24.11 \leq \mu \leq 25.89$

　　　　b. $113.17 \leq \mu \leq 126.03$

　　　　c. $3.136 \leq \mu \leq 3.702$

　　　　d. $54.55 \leq \mu \leq 58.85$

8.3. $45.92 \leq \mu \leq 48.08$

8.5. $66, 62.75 \leq \mu \leq 69.25$

8.7. $5.3, 5.13 \leq \mu \leq 5.47$

8.9. $2.852 \leq \mu \leq 3.760$

8.11. $60.818 \leq \mu \leq 98.255$

8.13. $42.18 \leq \mu \leq 49.06$

8.15. $120.6 \leq \mu \leq 136.2, 128.4$

8.17. $15.631 \leq \mu \leq 16.545, 16.088$

8.19. $2.26886 \leq \mu \leq 2.45346, 2.36116, 0.0923$

8.21. $36.77 \leq \mu \leq 62.83$

8.23. $7.53 \leq \mu \leq 14.66$

8.25.   **a.** $0.386 \leq p \leq 0.634$

　　　　b. $0.777 \leq p \leq 0.863$

　　　　c. $0.456 \leq p \leq 0.504$

　　　　d. $0.246 \leq p \leq 0.394$

8.27. $0.38 \leq p \leq 0.56$

　　　　$0.36 \leq p \leq 0.58$

　　　　$0.33 \leq p \leq 0.61$

8.29.   **a.** $0.4287 \leq p \leq 0.5113$

　　　　b. $0.2488 \leq p \leq 0.3112$

8.31.   **a.**   0.266

　　　　b. $0.247 \leq p \leq 0.285$

8.33. $0.5935 \leq p \leq 0.6665$

8.35.   **a.** $18.46 \leq \sigma^2 \leq 189.73$

　　　　b. $0.64 \leq \sigma^2 \leq 7.46$

　　　　c. $645.45 \leq \sigma^2 \leq 1{,}923.10$

　　　　d. $12.61 \leq \sigma^2 \leq 31.89$

8.37. $9.71 \leq \sigma^2 \leq 46.03; 18.49$

8.39. $14{,}084{,}038.51 \leq \sigma^2 \leq 69{,}553{,}848.45$

8.41.   **a.**   2,522

　　　　b.   601

　　　　c.   268

　　　　d.   16,577

8.43.   106

8.45.   1,083

8.47.   97

8.49. $12.03, 11.78 \leq \mu \leq 12.28, 11.72 \leq \mu \leq 12.34,$
　　　　$11.58 \leq \mu \leq 12.48$

8.51. $29.133 \leq \sigma^2 \leq 148.235, 25.911 \leq \sigma^2 \leq 182.529$

8.53. $9.19 \leq \mu \leq 12.34$

8.55. $2.307 \leq \sigma^2 \leq 15.374$

8.57. $36.231 \leq \mu \leq 38.281$

8.59. $0.542 \leq p \leq 0.596; 0.569$

8.61. $5.892 \leq \mu \leq 7.542$

8.63. $0.726 \leq p \leq 0.814$

8.65. $34.11 \leq \mu \leq 53.29, 101.44 \leq \sigma^2 \leq 821.35$

8.67. $-0.20 \leq \mu \leq 5.16, 2.48$

8.69.   543

8.71. $0.0026 \leq \sigma^2 \leq 0.0071$

Chapter 9

9.1.   **a.** $z = 2.77$, reject H_0

　　　　b. p-value $= 0.0028$, reject H_0

　　　　c.   22.115; 27.885

9.3.   **a.** $z = 1.59$, reject H_0

　　　　b.   0.0559

　　　　c.   1212.04

9.5. $z = 1.75$, fail to reject H_0

9.7. $z = 1.46$, fail to reject H_0

9.9. $z = 0.24, 0.4052$, fail to reject H_0

9.11. $t = 0.56$, fail to reject H_0

9.13. $t = 2.44$, reject H_0

9.15. $t = 1.59$, fail to reject H_0

9.17. $t = -3.31$, reject H_0

9.19. $t = -2.02$, fail to reject H_0

9.21.   reject; differs from survey

9.23. $z = -1.66$, fail to reject H_0

9.25. $z = 1.742$, fail to reject H_0

9.27. $z = 1.23$, fail to reject H_0
　　　　$z = 1.34$, fail to reject H_0

9.29. $z = -3.11$, reject H_0

9.31.   **a.** $\chi^2 = 22.4$, fail to reject H_0

　　　　b. $\chi^2 = 42$, reject H_0

　　　　c. $\chi^2 = 2.64$, fail to reject H_0

　　　　d. $\chi^2 = 2.4$, reject H_0

9.33. $\chi^2 = 21.7$, fail to reject H_0

9.35. $\chi^2 = 17.34$, reject H_0

9.37.   **a.** $\beta = 0.8159$

　　　　b. $\beta = 0.7422$

c. $\beta = 0.5636$

d. $\beta = 0.3669$

9.39. **a.** $\beta = 0.3632$

   b. $\beta = 0.0122$

   c. $\beta = 0.0000$

9.41. $z = -0.47$, fail to reject H_0; 0.6664; 0.1949; 0.0000

9.43. $t = -1.98$, reject H_0

9.45. $\chi^2 = 32.675$, fail to reject H_0

9.47. $z = -0.40$, fail to reject H_0

9.49. $z = -3.72$, reject H_0

9.51. $t = -5.70$, reject H_0

9.53. $\chi^2 = 106.47$, reject H_0

9.55. $t = -2.80$, reject H_0

9.57. $z = 3.96$, reject H_0

9.59. $t = 4.50$, reject H_0

9.61. $\chi^2 = 10.85$, fail to reject H_0

Chapter 10

10.1. **a.** $z = -1.01$, fail to reject H_0

   b. -2.41

   c. 0.1562

10.3. **a.** $z = 5.48$, reject H_0

   b. $4.04 \le \mu_1 - \mu_2 \le 10.02$

10.5. $-1.86 \le \mu_1 - \mu_2 \le -0.54$

10.7. $z = -2.32$, fail to reject H_0

10.9. $z = -2.27$, reject H_0

10.11. $t = -1.05$, fail to reject H_0

10.13. $t = -4.64$, reject H_0

10.15. **a.** $1{,}905.38 \le \mu_1 - \mu_2 \le 3{,}894.62$

   b. $t = -4.91$, reject H_0

10.17. $t = 2.06$, reject H_0

10.19. $t = 4.95$, fail to reject H_0; $10.29 \le \mu_1 - \mu_2 \le 7.89$

10.21. $t = 3.31$, reject H_0

10.23. $26.29 \le D \le 54.83$

10.25. $-3{,}415.6 \le D \le 6{,}021.2$

10.27. $6.58 \le D \le 49.60$

10.29. $63.71 \le D \le 86.29$

10.31. **a.** $z = 0.75$, fail to reject H_0

   b. $z = 4.83$, reject H_0

10.33. $z = -3.35$, reject H_0

10.35. $z = -0.94$, fail to reject H_0

10.37. $z = 2.35$, reject H_0

10.39. $F = 1.80$, fail to reject H_0

10.41. $F = 0.81$, fail to reject H_0

10.43. $F = 1.53$, fail to reject H_0

10.45. $z = -2.38$, reject H_0

10.47. $t = 0.85$, fail to reject H_0

10.49. $t = -5.26$, reject H_0

10.51. $z = -1.20$, fail to reject H_0

10.53. $F = 1.24$, fail to reject H_0

10.55. $-3.201 \le D \le 2.313$

10.57. $F = 1.31$, fail to reject H_0

10.59. $t = 2.98$, reject H_0

10.61. $z = 6.78$, reject H_0

10.63. $3.553 \le D \le 5.447$

10.65. $t = 6.71$, reject H_0

10.67. $0.142 \le p_1 - p_2 \le 0.250$

10.69. $z = 8.86$, reject H_0

10.71. $t = 4.52$, reject H_0

Chapter 11

11.5. $F = 11.07$, reject H_0

11.7. $F = 13.00$, reject H_0

11.9. 4; 50; 54; 145.8975; 19.4436; $F = 7.50$; reject H_0

11.11. $F = 13.9$, reject H_0

11.13. $F = 11.76$, reject H_0

11.15. 4 levels; sizes 18, 15, 21, and 11; $F = 2.95$, $p = 0.04$; means = 226.73, 238.79, 232.58, and 239.82

11.17. HSD = 0.896; groups 3 and 6 significantly different

11.19. HSD = 1.586; groups 1 and 2 significantly different

11.21. HSD = 10.27; groups 1 and 3 significantly different

11.23. $HSD_{1,3} = 0.0381$; groups 1 and 3 significantly different

11.25. $HSD_{1,3} = 0.0381$; $HSD_{2,3} = 1.620$; groups 1 and 3 and 2 and 3 significantly different

11.29. $F = 1.48$, fail to reject H_0

11.31. $F = 3.90$, fail to reject H_0

11.33. $F = 15.37$, reject H_0

11.37. 2; 1; 4 row levels; 3 column levels; yes $df_{row} = 3$, $df_{col.} = 2$, $df_{int.} = 6$, $df_{error} = 12$, $df_{total} = 23$

11.39. $MS_{row} = 1.047$, $MS_{col.} = 1.281$, $MS_{int.} = 0.258$, $MS_{error} = 0.436$, $F_{row} = 2.40$, $F_{col.} = 2.94$, $F_{int.} = 0.59$, fail to reject any hypothesis

11.41. $F_{row} = 87.25$, reject; $F_{col.} = 63.67$, reject; $F_{int.} = 2.07$, fail to reject H_0

11.43. $F_{row} = 34.31$, reject H_0; $F_{col.} = 14.20$, reject; $F_{int.} = 3.32$, reject H_0

11.45. no significant interaction or row effects; significant column effects

11.47. $F = 8.82$ reject; HSD = 3.33 means 1 and 2, 2 and 3, and 2 and 4 significantly different

11.49. $df_{treat.} = 5$, $MS_{treat.} = 42.0$, $df_{error} = 36$, $MS_{error} = 18.194$, $F = 2.31$

11.51. 1 treatment variable, 3 levels; 1 blocking variable, 6 levels; $df_{treat.} = 2$, $df_{block} = 5$, $df_{error} = 10$

11.53. $F_{treat.} = 31.51$, reject; $F_{blocks} = 43.20$, reject H_0; HSD = 8.757, no pairs significant

11.55. $F_{rows} = 38.21$, reject; $F_{col.} = 0.23$, fail to reject; $F_{inter} = 1.30$, fail to reject H_0

11.57. $F = 7.38$, reject H_0

11.59. $F = 0.46$, fail to reject H_0

11.61. $F_{treat.} = 13.64$, reject H_0

Chapter 12

12.1. -0.927

12.3. -0.6945

12.5. 0.944; 0.823; 0.952

12.7. $\hat{y} = 144.414 - 0.898x$

12.9. $\hat{y} = 15.460 - 0.715x$

12.11. $\hat{y} = 12,999.45 - 0.0431x$

12.13. $\hat{y} = 13.625 + 2.303x$, -1.1694, 3.9511, -1.3811, 2.7394, -4.1401

12.15. 18.6597, 37.5229, 51.8948, 62.6737, 86.0281, 118.3648, 122.8561; 6.3403, -8.5229, -5.8948, 7.3263, 1.9720, -6.3648, 5.1439

12.17. 4.0259, 11.1722, 9.7429, 12.6014, 10.4576; 0.9741, 0.8278, -0.7429, 2.3986, -3.4575

12.19. 4.7244; -0.9836; -0.3996; -6.7537; 2.7683; 0.6442; no apparent violations

12.21. The error terms appear to be non-independent

12.23. Violation of the homoscedasticity assumption

12.25. SSE = 272.0, $s_e = 7.376$, 6 out of 7 and 7 out of 7

12.27. SSE = 19.8885, $s_e = 2.575$

12.29. $s_e = 4.391$

12.31. $\hat{y} = 17.043 - 0.0000068x$; $s_e = 7.53$

12.33. $r^2 = 0.972$

12.35. $r^2 = 0.685$

12.37. $\hat{y} = 35.285 + 0.573x$; $s_e = 3.0696$; $r^2 = 0.447$

12.39. $t = -13.18$, reject H_0

12.41. $t = -2.56$, fail to reject H_0

12.43. F is significant at $\alpha = 0.05$; $t = 2.874$; reject at $\alpha = 0.05$

12.45. $38.523 \le y \le 70.705$, $10.447 \le y \le 44.901$

12.47. $0.97 \le E(y_{10}) \le 15.65$

12.49. $\hat{y} = 66,916.57 + 1,270.512x$; $\hat{y}(20,12/13) = 79,621.69$

12.51. $r = -0.940$

12.53. a. $\hat{y} = -11.335 + 0.355x$

b. 7.48, 5.35, 3.22, 6.415, 9.225, 10.675, 4.64, 9.965, -2.48, -0.35, 3.78, -2.415, 0.745, 1.325, -1.64, 1.035

c. SSE = 32.4649

d. $s_e = 2.3261$

e. $r^2 = 0.608$

f. $t = 3.05$, reject H_0

12.55. a. $20.92 \le E(y_{60}) \le 26.8$

b. $20.994 \le y \le 37.688$

12.57. $r^2 = 0.826$

12.59. $\hat{y} = 1,043.22 + 0.009287504x$; $r^2 = 0.046$

12.61. $r = 0.8998$

12.63. $\hat{y} = -39.0071 + 66.36277x$; $r^2 = 0.906$; $s_e = 21.13$

12.65. $\hat{y} = -4,864.22 + 371.77x$; $s_e = 706.33$; $r^2 = 0.88$; $t = 5.68$

Chapter 13

13.1. $\hat{y} = 25.03 - 0.0497x_1 + 1.928x_2$, 28.586

13.3. $\hat{y} = 121.62 - 0.174x_1 + 6.02x_2 + 0.00026x_3 + 0.0041x_4$, 4

13.5. Per capita consumption $= -7,629.627 + 116.2549$ paper consumption $- 120.0904$ fish consumption $+ 45.73328$ petrol consumption

13.7. 9; fail to reject null overall at $\alpha = 0.05$; only $t = 2.73$ for x_1; significant at $\alpha = 0.05$; $s_e = 3.503$; $R^2 = 0.408$, adj. $R^2 = 0.203$

13.9. Per capita consumption $= -7,629.627 + 116.2549$ paper consumption $- 120.0904$ fish consumption $+ 45.73328$ petrol consumption; $F = 14.319$ with p-value $= 0.0023$; $t = 2.67$ with p-value $= 0.032$ for petrol consumption. The p-values of the t statistics for the other two predictors are insignificant

13.11. $\hat{y} = 3.981 + 0.07322x_1 - 0.03232x_2 - 0.003886x_3$; $F = 100.47$ significant at $\alpha = 0.001$; $t = 3.50$ for x_1 significant at $\alpha = 0.001$; $s_e = 0.2331$; $R^2 = 0.965$; adj. $R^2 = 0.955$

13.13. 3 predictors; 15 observations; $\hat{y} = 657.053 + 5.710x_1 - 0.417x_2 - 3.471x_3$; $R^2 = 0.842$; adj.

$R^2 = 0.630$; $F = 8.96$ with $p = 0.0027$; x_1 significant at $\alpha = 0.01$; x_3 significant at $\alpha = 0.05$;

13.15. $s_e = 9.722$; $R^2 = 0.515$; adj. $R^2 = 0.404$

13.17. $s_e = 6.544$; $R^2 = 0.005$; adj. $R^2 = 0.000$

13.19. model with x_1, x_2: $s_e = 6.333$; $R^2 = 0.963$; adj. $R^2 = 0.957$; model with x_1: $s_e = 6.124$; $R^2 = 0.963$; adj. $R^2 = 0.960$

13.21. heterogeneity of variance

13.23. Two predictors; $\hat{y} = 203.3937 + 1.1151x_1 - 2.2115x_2$; $F = 24.55$; reject; $R^2 = 0.663$; adj. $R^2 = 0.636$

13.25. $\hat{y} = 362 - 4.75x_1 - 13.9x_2 + 1.87x_3$; $F = 16.05$; reject; $s_e = 37.07$; $R^2 = 0.858$; adj. $R^2 = 0.804$; x_1 only significant predictor

13.27. GDP per capita $= 32,612 + 13,011$ Population Growth $- 351$ Rural Population $- 207$ Total Tax Rate; $F = 13.86$ with $p = 0.000$ (significant at $\alpha = 0.01$); $R^2 = 0.676$ and adj. $R^2 = 0.627$; Population Growth and Rural Population significant at $\alpha = 0.05$; Total Tax Rate not significant

13.29. Sweetcorn $= 1,830 + 0.026$ Wheat $+ 0.011$ Soya Beans; $F = 13.32$ with $p = 0.003$ (significant at $\alpha = 0.01$); $s_e = 1,394.7$; $R^2 = 73.2\%$; adj. $R^2 = 67.3\%$; Soya Beans significant at $\alpha = 0.01$; Wheat not significant

Chapter 14

14.1. Simple model: $\hat{y} = -147.27 + 27.128x$; $F = 229.67$ with $p = 0.000$; $s_e = 27.27$; $R^2 = 0.97$; adj. $R^2 = 0.966$

Quadratic model: $\hat{y} = -22.01 + 3.385x_1 + 0.9373x_2$; $F = 578.76$ with $p = 0.000$; $s_e = 12.3$; $R^2 = 0.995$; adj. $R^2 = 0.993$; for x_1: $t = 0.75$; for x_2: $t = 5.33$

14.3. Simple regression model: $\hat{y} = 346,704 + 26.5x$; $R^2 = 0.936$; adj. $R^2 = 0.926$; $t = 9.41$ with $p = 0.000$

Quadratic regression model: $\hat{y} = 332,664 + 45.8x - 0.003x^2$; $R^2 = 0.991$; adj. $R^2 = 0.982$; for x: $t = 8.33$ with $p = 0.000$; for x^2: $t = -3.67$ with $p = 0.015$

14.5. $\hat{y} = -28.61 - 2.68x_1 + 18.25x_2 - 0.2135x_1^2 - 1.533x_2^2 + 1.22x_1x_2$; $F = 63.43$; $s_e = 4.669$; $R^2 = 0.958$; adj. $R^2 = 0.943$; no significant t ratios. Model with no interaction term: $R^2 = 0.957$

14.7. $\hat{y} = 13.619 - 0.01201x_1 + 2.988x_2$; $F = 8.43$ significant at $\alpha = 0.01$; $t = 3.88$ for x_2 (dummy

variable) significant at $\alpha = 0.01$; $s_e = 1.245$; $R^2 = 0.652$; adj. $R^2 = 0.575$

14.9. x_1 and x_2 are significant predictors at $\alpha = 0.05$

14.11. Price (£) $= 358 + 1.6$ Megapixels $- 0.092$ Weight (grams); $F = 0.14$ with $p = 0.930$ (not significant at $\alpha = 0.05$); $s_e = 108.21$; $R^2 = 0.079$; adj. $R^2 = -0.474$

14.13. Step 1: x_2 entered, $t = -7.53$, $R^2 = 0.794$
Step 2: x_3 entered, $t_2 = -4.60$, $t_3 = 2.93$, $R^2 = 0.876$

14.15. Four predictors; x_4 and x_5 are not in model

14.17. The procedure stopped at step 1; only Dividend per Share entered the process; $R^2 = 60.06$; $t = 3.38$; model was $\hat{y} = 30,328 + 15,296x_1$

14.19.

	y	x_1	x_2
x_1	-0.653		
x_2	-0.891	0.650	
x_3	0.821	-0.615	-0.688

14.21. The stepwise regression analysis of the retail index of energy prices example resulted in two of the five predictor variables being included in the model (solar/wind/other and coal and coal products). The simple regression model yielded an R^2 of .963 jumping to .985 with the two predictors. An examination of the predictor intercorrelations reveals that Solar/Wind/Other and Natural Gas have very high correlation (0.96). Fairly high correlations between Nuclear and Natural Gas (0.746), Nuclear and Solar/Wind/Other (0.82) and Nuclear and Coal and Coal Products (0.589). These might suggest multicollinearity

14.23. $\hat{y} = 564 - 27.99x_1 - 6.115x_2 - 15.90x_3$; $F = 11.32$ with $p = 0.003$; $s_e = 42.88$; $R^2 = 0.809$, adj. $R^2 = 0.738$; x_2 only significant predictor; x_1 is a non-significant indicator variable

14.25. The procedure stopped at step 1 with only log x in the model; regression model $= -13.20 + 11.64 \log x_1$; $R^2 = 0.9617$

14.27. GDP per capita $= 32,531 + 13,018$ Population Growth $- 349$ Rural Population $- 207$ Total Tax Rate; $F = 13.88$ with $p = 0.000$ (significant at $\alpha = 0.01$); $R^2 = 0.676$ and adj. $R^2 = 0.627$; Population Growth and Rural Population significant at $\alpha = 0.05$, Total Tax Rate not significant

14.29. Sweetcorn $= 2,175 + 0.0111$ Soya Beans; $R^2 = 0.7247$; t ratio for Soya Beans $= 5.13$, Wheat did not enter the analysis

14.31. $\hat{y} = 124.5 - 43.4x_1 + 1.36x_2$; $R^2 = 0.8059$; x_3 and x_4 did not enter the process

Chapter 15

15.1. MAD = 1.367, MSE = 2.27

15.3. MAD = 3.583, MSE = 15.765

15.5. **a.** 44.75, 52.75, 61.50, 64.75, 70.50, 81

b. 53.25, 56.375, 62.875, 67.25, 76.375, 89.125

c. difference in errors: 8.5, 3.626, 1.375, 2.5, 5.875, 8.125

15.7. $\alpha = 0.3$: 9.4, 9, 8.7, 8.8, 9.1, 9.7, 9.9, 9.8
$\alpha = 0.7$: 9.4, 8.6, 8.1, 8.7, 9.5, 10.6, 10.4, 9.8

15.9. MAD = 479.45 for $\alpha = 0.2$, = 453.3497 for $\alpha = 0.9$; smaller error with $\alpha = 0.2$

15.11. Quadratic model is superior, with higher R^2 (99.2 vs. 88.1) and lower standard errors of the estimate (4.85 vs. 18.03)

15.13. $T \cdot C$: 411,679.167; 418,358.333; 422,354.167; 426,762.5; 429,175; 428,095.833; 428,179.167; 427,920.833; 426,150; 424,245.833; 421,454.167; 417,529.167

$S \cdot I$: 95.79; 93.00; 94.94; 91.61; 90.03; 96.39; 104.47; 103.40; 102.30; 110.23; 102.05; 115.80;

15.15. $D = 0.33$, reject the null hypothesis; significant autocorrelation

15.17. $D = 0.65$; significant autocorrelation

15.19. 1 lag: Arable Land = 25.9 + 0.783 lag 1; $R^2 = 76.8\%$; $s_e = 3.502$
2 lags: Arable Land = 34.5 + 0.705 lag 1 + 0.007 lag 2; $R^2 = 68.9\%$; $s_e = 3.497$

15.21. **a.** 100, 139.9, 144, 162.6, 200, 272.8, 310.7, 327.1, 356.6, 376.9, 388.8, 398.9

b. 32.2, 45, 46.4, 52.3, 64.4, 87.8, 100, 105.3, 114.8, 121.3, 125.1, 128.4

15.23. 100; 106.9; 122.6

15.25. 121.6; 127.4; 131.4

15.27. **a.** Linear: Interest Rates = 4.71 − 0.043 Month; $R^2 = 64.0\%$

Quadratic: Interest Rates = 4.98 − 0.106 Month + 0.00251 (Month)^2; $R^2 = 72.1\%$.

The linear model is a strong model. The quadratic term adds some predictability but has a smaller t ratio than does the linear term.

b. MAD = 0.1129

c. MAD $(\alpha = 0.3)$ =0.2397; MAD $(\alpha = 0.7)$ =0.1513; $\alpha = 0.7$ produces better forecasts

d. MAD = 0.1129 (b), =0.2397 and 0.1513 (c); MAD from (b) produces lowest error

e. The highs and lows of each period (underlined) are eliminated and the others are averaged resulting in:

Seasonal Indexes:	1st	100.30
	2nd	100.61
	3rd	99.28
	4th	97.58
	total	397.77

Since the total is not 400, adjust each seasonal index by multiplying by = 1.005606 resulting in the final seasonal indexes of:

	1st	100.86
	2nd	101.17
	3rd	99.84
	4th	98.13

15.29. 100; 104.8; 114.5; 115.5; 114.1

15.31. $\text{MAD}_{\text{mov.avg.}} = 28{,}993$, $\text{MAD}_{\alpha=023} = 35{,}921$; three-year moving average was best

15.33. Jan. 92.13; Feb. 99.28; March 104.72; April 101.68; May 102.44; June 105.44; July 101.12; Aug. 89.59; Sept. 99.65; Oct. 106.42; Nov. 102.46; Dec. 95.28

15.35. Laspeyres: 105.2, 111.0; Paasche: 105.1, 110.8

15.37. **a.** $\text{MSE}_{\text{mov.avg.}} = 66.22$

b. $\text{MSE}_{\text{wt.mov.avg.}} = 50.39$

c. Weighted moving average is best

15.39. Adjusted seasonal indexes: $Q_1 = 60.72$, $Q_2 = 78.07$, $Q_3 = 182.25$ and $Q_4 = 78.97$.

15.43. Equity Funds = 24469 + 1.85 Money Market Funds $D = 1.71$, rejected; not positively autocorrelated

15.45. $D = 0.98$, reject H_0

Chapter 16

16.1. $\chi^2 = 18.095$, reject H_0

16.3. $\chi^2 = 2.001$, fail to reject, $\lambda = 0.9$

16.5. $\chi^2 = 198.48$, reject H_0

16.7. $\chi^2 = 2.45$, fail to reject H_0

16.9. $\chi^2 = 3.398$, fail to reject H_0

16.11. $\chi^2 = 0.00$, fail to reject H_0

16.13. $\chi^2 = 34.97$, reject H_0

16.15. $\chi^2 = 6.43$, reject H_0

16.17. $\chi^2 = 5.998$, fail to reject H_0

16.19. $\chi^2 = 1.652$, fail to reject H_0

16.21. $\chi^2 = 14.91$, reject H_0

16.23. $\chi^2 = 7.25$, fail to reject H_0

16.25. $\chi^2 = 59.63$, reject H_0

16.27. $\chi^2 = 54.63$, reject H_0

Chapter 17

17.1. $R = 11$, fail to reject H_0

17.3. $\alpha/2 = 0.025$, p-value $= 0.0264$, fail to reject H_0

17.5. $R = 27$, $z = -1.08$, fail to reject H_0

17.7. $U = 26.5$, p-value $= 0.6454$, fail to reject H_0

17.9. $U = 11$, p-value $= 0.0156$, fail to reject H_0

17.11. $z = -3.78$, reject H_0

17.13. $z = -2.59$, reject H_0

17.15. $z = -3.20$, reject H_0

17.17. $z = -1.75$, reject H_0

17.19. $K = 21.21$, reject H_0

17.21. $K = 2.75$, fail to reject H_0

17.23. $K = 18.99$, reject H_0

17.25. $\chi^2 = 13.8$, reject H_0

17.27. $\chi^2 = 13.8$, reject H_0

17.29. 4, 5, $S = 2.04$, fail to reject H_0

17.31. $r_s = -0.893$

17.33. $r_s = -0.95$

17.35. $r_s = 0.83$

17.37. $r_s = 0.936$

17.39. $U = 20$, p-value $= 0.2344$, fail to reject H_0

17.41. $K = 7.75$, fail to reject H_0

17.43. $r_s = -0.81$

17.45. $z = -0.40$, fail to reject H_0

17.47. $z = 0.96$, fail to reject H_0

17.49. $U = 45.5$, p-value $= 0.739$, fail to reject H_0

17.51. $z = -1.91$, fail to reject H_0

17.53. $R = 21$, fail to reject H_0

17.55. $z = -2.43$, reject H_0

17.57. $K = 17.21$, reject H_0

17.59. $K = 11.96$, reject H_0

Chapter 18 (On Wiley website)

18.5. $\bar{\bar{x}} = 4.51$, UCL $= 5.17$, LCL $= 3.85$
$\bar{R} = 0.90$,   UCL $= 2.05$, LCL $= 0$

18.7. $p = 0.05$, UCL $= 0.1534$, LCL $= 0.000$

18.9. $\bar{c} = 1.34375$, UCL $= 4.82136$, LCL $= 0.000$

18.11. Chart 1: nine consecutive points below centerline, four out of five points in the outer 2/3 of the lower region

Chart 2: eight consecutive points above the centerline

Chart 3: in control

18.15. $p = 0.104$, LCL $= 0.000$, UCL $= 0.234$

18.17. $\bar{c} = 2.13889$, UCL $= 6.52637$, LCL $= 0.0000$

18.19. $\bar{x} = 14.99854$, UCL $= 15.02269$,
LCL $= 14.97439$ $\bar{R} = 0.05$, UCL $= 0.1002$,
LCL $= 0.0000$

18.21. $\bar{c} = 0.64$, UCL $= 3.04$, LCL $= 0.0000$

18.23. $p = 0.06$, LCL $= 0.000$, UCL $= 0.1726$

Chapter 19 (On Wiley website)

19.1. a. 390

b. 70

c. 82, 296

d. 140

19.3. 60, 10

19.7. 31.75, 6.50

19.9. Lock in $= 85, 182.5, 97.5$

19.11. a. 75,000

b. Avoider

c. $>75,000$

19.13. 244.275, 194.275

19.15. 21,012.32, 12.32

19.17. b. 267.5, 235

c. 352.5, 85

19.19. a. 2,000, 200

b. 500

19.21. 875,650

19.23. Reduction: 0.60, 0.2333, 0.1667 Constant: 0.10, 0.6222, 0.2778

Increase: 0.0375, 0.0875, 0.8750, 21,425.55, 2,675.55

A

a posteriori After the experiment; pairwise comparisons made by the researcher *after* determining that there is a significant overall F value from ANOVA; also called post hoc.

a priori Determined before, or prior to, an experiment.

adjusted R^2 A modified value of R^2 in which the degrees of freedom are taken into account, thereby allowing the researcher to determine whether the value of R^2 is inflated for a particular multiple regression model.

after-process quality control A type of quality control in which product attributes are measured by inspection after the manufacturing process is completed to determine whether the product is acceptable.

all possible regressions A multiple regression search procedure in which all possible multiple linear regression models are determined from the data using all variables.

alpha (α) The probability of committing a Type I error; also called the level of significance.

alternative hypothesis The hypothesis that complements the null hypothesis; usually it is the hypothesis that the researcher is interested in proving.

analysis of variance (ANOVA) A technique for statistically analysing the data from a completely randomized design; uses the F test to determine whether there is a significant difference in two or more independent groups.

arithmetic mean The average of a group of numbers.

autocorrelation A problem that arises in regression analysis when the data occur over time and the error terms are correlated; also called serial correlation.

autoregression A multiple regression forecasting technique in which the independent variables are time-lagged versions of the dependent variable.

averaging models Forecasting models in which the forecast is the average of several preceding time periods.

B

backward elimination A step-by-step multiple regression search procedure that begins with a full model containing all predictors. A search is made to determine if there are any non-significant independent variables in the model. If there are no non-significant predictors, then the backward process ends with the full model. If there are non-significant predictors, then the predictor with the smallest absolute value of t is eliminated and a new model is developed with the remaining variables. This procedure continues until only variables with significant t values remain in the model.

bar graph A bar graph (or bar chart) is a chart that contains two or more categories along one axis and a series of bars, one for each category, along the other axis. Usually the length of the bar represents the magnitude of the measure for each category. A bar graph is qualitative and may be either horizontal or vertical.

Bayes' rule An extension of the conditional law of probabilities discovered by Thomas Bayes that can be used to revise probabilities.

benchmarking A quality control method in which a company attempts to develop and establish total quality management from product to process by examining and emulating the best practices and techniques used in their industry.

beta (β) The probability of committing a Type II error.

bimodal Data sets that have two modes.

binomial distribution Widely known discrete distribution in which there are only two possibilities on any one trial.

blocking variable A variable that the researcher wants to control but is not the treatment variable of interest.

bounds The error portion of the confidence interval that is added and/or subtracted from the point estimate to form the confidence interval.

box-and-whisker plot A diagram that utilizes the upper and lower quartiles along with the median and the two most extreme values to depict a distribution graphically; sometimes called a box plot.

C

c chart A quality control chart for attribute compliance that displays the number of non-conformances per item or unit.

categorical data Non-numerical data that are frequency counts of categories from one or more variables.

cause-and-effect diagram A tool for displaying possible causes for a quality problem and the inter-relationships among the causes; also called a fishbone diagram or an Ishikawa diagram.

census A process of gathering data from the whole population for a given measurement of interest.

centreline The middle horizontal line of a control chart, often determined either by a product or service specification or by computing an expected value from sample information.

central limit theorem A theorem that states that regardless of the shape of a population, the distributions of sample means and proportions are normal if sample sizes are large.

Chebyshev's theorem A theorem stating that at least $1 - 1/k^2$ values will fall within $\pm k$ standard deviations of the mean regardless of the shape of the distribution.

check sheet Simple forms consisting of multiple categories and columns for recording tallies for displaying the frequency of outcomes for some quality-related event or activity.

chi-square distribution A continuous distribution determined by the sum of the squares of k independent random variables.

chi-square goodness-of-fit test A statistical test used to analyse probabilities of multinomial distribution trials along a single dimension; compares expected, or theoretical, frequencies of categories from a population distribution to the observed, or actual, frequencies from a distribution.

chi-square test of independence A statistical test used to analyse the frequencies of two variables with multiple categories to determine whether the two variables are independent.

class mark Another name for class midpoint; the midpoint of each class interval in grouped data.

class midpoint For any given class interval of a frequency distribution, the value halfway across the class interval; the average of the two class endpoints.

classical method of assigning probabilities Probabilities assigned based on rules and laws.

classification variable The independent variable of an experimental design that was present prior to the experiment and is not the result of the researcher's manipulations or control.

classifications The subcategories of the independent variable used by the researcher in the experimental design; also called levels.

cluster (or area) sampling A type of random sampling in which the population is divided into non-overlapping areas or clusters and elements are randomly sampled from the areas or clusters.

coefficient of correlation (r) A statistic developed by Karl Pearson to measure the linear correlation of two variables. Also called the Pearson product-moment correlation coefficient.

coefficient of determination (r^2) The proportion of variability of the dependent variable accounted for or explained by the independent variable in a regression model.

coefficient of multiple determination (R^2) The proportion of variation of the dependent variable accounted for by the independent variables in the regression model.

coefficient of skewness A measure of the degree of skewness that exists in a distribution of numbers; compares the mean and the median in light of the magnitude of the standard deviation.

coefficient of variation (CV) The ratio of the standard deviation to the mean, expressed as a percentage.

collectively exhaustive events A list containing all possible elementary events for an experiment.

combinations Used to determine the number of possible ways n things can happen from N total possibilities when sampling without replacement.

complement of a union The only possible case other than the union of sets X and Y; the probability that neither X nor Y is in the outcome.

complementary events Two events, one of which comprises all the elementary events of an experiment that are not in the other event.

completely randomized design An experimental design wherein there is one treatment or independent variable with two or more treatment levels and one dependent variable. This design is analysed by analysis of variance.

concomitant variables Variables that are not being controlled by the researcher in the experiment but can have an effect on the outcome of the treatment being studied; also called confounding variables.

conditional probability The probability of the occurrence of one event given that another event has occurred.

confidence interval A range of values within which the analyst can declare, with some confidence, the population parameter lies.

confounding variables Variables that are not being controlled by the researcher in the experiment but can have an effect on the outcome of the treatment being studied; also called concomitant variables.

contingency analysis Another name for the chi-square test of independence.

contingency table A two-way table that contains the frequencies of responses to two questions; also called a raw values matrix.

continuous distributions Distributions constructed from continuous random variables.

continuous random variables Variables that take on values at every point over a given interval.

control chart A quality control graph that contains an upper control limit, a lower control limit, and a centreline; used to evaluate whether a process is or is not in a state of statistical control.

convenience sampling A non-random sampling technique in which items for the sample are selected for the convenience of the researcher.

correction for continuity A correction made when a binomial distribution problem is approximated by the normal distribution because a discrete distribution problem is being approximated by a continuous distribution.

correlation A measure of the degree of relatedness of two or more variables.

covariance The variance of x and y together.

critical value The value that divides the non-rejection region from the rejection region.

critical value method A method of testing hypotheses in which the sample statistic is compared with a critical value in order to reach a conclusion about rejecting or failing to reject the null hypothesis.

cumulative frequency A running total of frequencies through the classes of a frequency distribution.

cycles Patterns of highs and lows through which data move over time periods usually of more than a year.

cyclical effects The rise and fall of time-series data over periods longer than one year.

D

decision alternatives The various choices or options available to the decision maker in any given problem situation.

decision analysis A category of quantitative business techniques particularly targeted at clarifying and enhancing the decision-making process.

decision making under certainty A decision-making situation in which the states of nature are known.

decision making under risk A decision-making situation in which it is uncertain which states of nature will occur but the probability of each state of nature occurring has been determined.

decision making under uncertainty A decision-making situation in which the states of nature that may occur are unknown and the probability of a state of nature occurring is also unknown.

decision table A matrix that displays the decision alternatives, the states of nature, and the payoffs for a particular decision-making problem; also called a payoff table.

decision trees A flowchart-like depiction of the decision process that includes the various decision alternatives, the various states of nature, and the payoffs.

decomposition Breaking down the effects of time-series data into the four component parts of trend, cyclical, seasonal, and irregular.

degrees of freedom A mathematical adjustment made to the size of the sample; used along with α to locate values in statistical tables.

dependent samples Two or more samples selected in such a way as to be dependent or related; each item or person in one sample has a corresponding matched or related item in the other samples. Also called related samples.

dependent variable In regression analysis, the variable that is being predicted.

descriptive statistics Statistics that have been gathered on a group to describe or reach conclusions about that same group.

deseasonalized data Time-series data in which the effects of seasonality have been removed.

Design for Six Sigma A quality scheme, an offshoot of Six Sigma, that places an emphasis on designing a product or process right the first time thereby allowing organizations the opportunity to reach even higher sigma levels through Six Sigma.

deterministic model Mathematical models that produce an 'exact' output for a given input.

deviation from the mean The difference between a number and the average of the set of numbers of which the number is a part.

discrete distributions Distributions constructed from discrete random variables.

discrete random variables Random variables in which the set of all possible values is at most a finite or a countably infinite number of possible values.

disproportionate stratified random sampling A type of stratified random sampling in which the proportions of items selected from the strata for the final sample do not reflect the proportions of the strata in the population.

dot plot A dot plot is a relatively simple statistical chart used to display continuous quantitative data where each data value is plotted along the horizontal axis and is represented on the chart by a dot.

dummy variable Another name for a qualitative or indicator variable; usually coded as 0 or 1 and represents whether or not a given item or person possesses a certain characteristic.

Durbin–Watson test A statistical test for determining whether significant autocorrelation is present in a time-series regression model.

E

elementary events Events that cannot be decomposed or broken down into other events.

empirical rule A guideline that states the approximate percentage of values that fall within a given number of standard deviations of a mean of a set of data that are normally distributed.

EMV'er A decision maker who bases his or her decision on the expected monetary value of the decision alternative.

error of an individual forecast The difference between the actual value and the forecast of that value.

error of estimation The difference between the statistic computed to estimate a parameter and the parameter.

event An outcome of an experiment.

expected monetary value (EMV) A value of a decision alternative computed by multiplying the probability of each state of nature by the state's associated payoff and summing these products across the states of nature.

expected value The long-run average of occurrences; sometimes referred to as the mean value.

expected value of perfect information The difference between the payoff that would occur if the decision maker knew which states of nature would occur and the expected monetary payoff from the best decision alternative when there is no information about the occurrence of the states of nature.

expected value of sample information The difference between the expected monetary value with information and the expected monetary value without information.

experiment A process that produces outcomes.

experimental design A plan and a structure to test hypotheses in which the researcher either controls or manipulates one or more variables.

exponential distribution A continuous distribution closely related to the Poisson distribution that describes the times between random occurrences.

exponential smoothing A forecasting technique in which a weighting system is used to determine the importance of previous time periods in the forecast.

F

F **distribution** A distribution based on the ratio of two random variances; used in testing two variances and in analysis of variance.

F **value** The ratio of two sample variances, used to reach statistical conclusions regarding the null hypothesis; in ANOVA, the ratio of the treatment variance to the error variance.

factorial design An experimental design in which two or more independent variables are studied simultaneously and every level of each treatment is studied under the conditions of every level of all other treatments. Also called a factorial experiment.

factors Another name for the independent variables of an experimental design.

Failure Mode and Effects Analysis (FMEA) A systematic way for identifying the effects of potential product or process failure. It includes methodology for eliminating or reducing the chance of a failure occurring.

finite correction factor A statistical adjustment made to the *z* formula for sample means; adjusts for the fact that a population is finite and the size is known.

first-differences approach A method of transforming data in an attempt to reduce or remove autocorrelation from a time-series regression model; results in each data value being subtracted from each succeeding time period data value, producing a new, transformed value.

fishbone diagram A display of possible causes of a quality problem and the inter-relationships among the causes. The problem is diagrammed along the main line of the 'fish' and possible causes are diagrammed as line segments angled off in such a way as to give the appearance of a fish skeleton. Also called an Ishikawa diagram or a cause-and-effect diagram.

flowchart A schematic representation of all the activities and interactions that occur in a process.

forecasting The art or science of predicting the future.

forecasting error A single measure of the overall error of a forecast for an entire set of data.

forward selection A multiple regression search procedure that is essentially the same as stepwise regression analysis except that once a variable is entered into the process, it is never deleted.

frame A list, map, directory, or some other source that is being used to represent the population in the process of sampling.

frequency distribution A summary of data presented in the form of class intervals and frequencies.

frequency polygon A graph constructed by plotting a dot for the frequencies at the class midpoints and connecting the dots.

Friedman test A non-parametric alternative to the randomized block design.

G

grouped data Data that have been organized into a frequency distribution.

H

heteroscedasticity The condition that occurs when the error variances produced by a regression model are not constant.

histogram A type of vertical bar chart constructed by graphing line segments for the frequencies of classes across the class intervals and connecting each to the *x* axis to form a series of rectangles.

homoscedasticity The condition that occurs when the error variances produced by a regression model are constant.

Hurwicz criterion An approach to decision making in which the maximum and minimum payoffs selected from each decision alternative are used with a weight, α, between 0 and 1 to determine the alternative with the maximum weighted average. The higher the value of α, the more optimistic is the decision maker.

hypergeometric distribution A distribution of probabilities of the occurrence of *x* items in a sample of *n* when there are *A* of that same item in a population of *N*.

hypothesis A tentative explanation of a principle operating in nature.

hypothesis testing A process of testing hypotheses about parameters by setting up null and alternative hypotheses, gathering sample data, computing statistics from the samples, and using statistical techniques to reach conclusions about the hypotheses.

I

independent events Events such that the occurrence or non-occurrence of one has no effect on the occurrence of the others.

independent samples Two or more samples in which the selected items are related only by chance.

independent variable In regression analysis, the predictor variable.

index number A ratio, often expressed as a percentage, of a measure taken during one time frame to that same measure taken during another time frame, usually denoted as the base period.

indicator variable Another name for a dummy or qualitative variable; usually coded as 0 or 1 and represents whether or not a given item or person possesses a certain characteristic.

inferential statistics Statistics that have been gathered from a sample and used to reach conclusions about the population from which the sample was taken.

in-process quality control A quality control method in which product attributes are measured at various intervals throughout the manufacturing process.

interaction When the effects of one treatment in an experimental design vary according to the levels of treatment of the other effect(s).

interquartile range The range of values between the first and the third quartile.

intersection The portion of the population that contains elements that lie in both or all groups of interest.

interval estimate A range of values within which it is estimated with some confidence the population parameter lies.

Interval-level data Next to highest level of data. These data have all the properties of ordinal level data, but in addition, intervals between consecutive numbers have meaning.

irregular fluctuations Unexplained or error variation within time-series data.

Ishikawa diagram A tool developed by Kaoru Ishikawa as a way to display possible causes of a quality problem and the inter-relationships of the causes; also called a fishbone diagram or a cause-and-effect diagram.

J

joint probability The probability of the intersection occurring, or the probability of two or more events happening at once.

judgement sampling A non-random sampling technique in which items selected for the sample are chosen by the judgement of the researcher.

just-in-time inventory system An inventory system in which little or no extra raw materials or parts for production are stored.

K

Kruskal–Wallis test The non-parametric alternative to one-way analysis of variance; used to test whether three or more samples come from the same or different populations.

kurtosis The amount of peakedness of a distribution.

L

lambda (λ) Denotes the long-run average of a Poisson distribution.

Laspeyres price index A type of weighted aggregate price index in which the quantity values used in the calculations are from the base year.

lean manufacturing A quality-management philosophy that focuses on the reduction of wastes and the elimination of unnecessary steps in an operation or process.

least squares analysis The process by which a regression model is developed based on calculus techniques that attempt to produce a minimum sum of the squared error values.

leptokurtic Distributions that are high and thin.

level of significance The probability of committing a Type I error; also known as alpha.

levels The subcategories of the independent variable used by the researcher in the experimental design; also called classifications.

lower control limit (LCL) The bottom-end line of a control chart, usually situated approximately three standard deviations of the statistic below the centreline; data points below this line indicate quality control problems.

M

Mann–Whitney U test A non-parametric counterpart of the t test used to compare the means of two independent populations.

manufacturing quality A view of quality in which the emphasis is on the manufacturer's ability to target consistently the requirements for the product with little variability.

marginal probability A probability computed by dividing a subtotal of the population by the total of the population.

matched-pairs test A t test to test the differences in two related or matched samples; sometimes called the t test for related measures or the correlated t test.

maximax criterion An optimistic approach to decision making under uncertainty in which the decision alternative is chosen according to which alternative produces the maximum overall payoff of the maximum payoffs from each alternative.

maximin criterion A pessimistic approach to decision making under uncertainty in which the decision alternative is chosen according to which alternative produces the maximum overall payoff of the minimum payoffs from each alternative.

mean The long-run average of occurrences; also called the expected value.

mean absolute deviation (MAD) The average of the absolute values of the deviations around the mean for a set of numbers.

mean square error (MSE) The average of all errors squared of a forecast for a group of data.

measures of central tendency One type of measure that is used to yield information about the centre of a group of numbers.

measures of shape Tools that can be used to describe the shape of a distribution of data.

measures of variability Statistics that describe the spread or dispersion of a set of data.

median The middle value in an ordered array of numbers.

mesokurtic Distributions that are normal in shape – that is, not too high or too flat.

metric data Interval and ratio level data; also called quantitative data.

minimax regret A decision-making strategy in which the decision maker determines the lost opportunity for each decision alternative and selects the decision alternative with the minimum of lost opportunity or regret.

mn **counting rule** A rule used in probability to count the number of ways two operations can occur if the first operation has *m* possibilities and the second operation has *n* possibilities.

mode The most frequently occurring value in a set of data.

moving average When an average of data from previous time periods is used to forecast the value for ensuing time periods and this average is modified at each new time period by including more recent values not in the previous average and dropping out values from the more distant time periods that were in the average. It is continually updated at each new time period.

multicollinearity A problematic condition that occurs when two or more of the independent variables of a multiple regression model are highly correlated.

multimodal Data sets that contain more than two modes.

multiple comparisons Statistical techniques used to compare pairs of treatment means when the analysis of variance yields an overall significant difference in the treatment means.

multiple regression Regression analysis with one dependent variable and two or more independent variables or at least one non-linear independent variable.

mutually exclusive events Events such that the occurrence of one precludes the occurrence of the other.

N

naive forecasting models Simple models in which it is assumed that the more recent time periods of data represent the best predictions or forecasts for future outcomes.

nominal level data The lowest level of data measurement; used only to classify or categorize.

non-linear regression model Multiple regression models in which the models are non-linear, such as polynomial models, logarithmic models, and exponential models.

non-metric data Nominal and ordinal level data; also called qualitative data.

non-parametric statistics A class of statistical techniques that make few assumptions about the population and are particularly applicable to nominal and ordinal level data.

non-random sampling Sampling in which not every unit of the population has the same probability of being selected into the sample.

non-random sampling techniques Sampling techniques used to select elements from the population by any mechanism that does not involve a random selection process.

non-rejection region Any portion of a distribution that is not in the rejection region. If the observed statistic falls in this region, the decision is to fail to reject the null hypothesis.

non-sampling errors All errors other than sampling errors.

normal distribution A widely known and much-used continuous distribution that fits the measurements of many human characteristics and many machine-produced items.

null hypothesis The hypothesis that assumes the status quo (i.e., that the old theory, method, or standard is still true); the complement of the alternative hypothesis.

O

observed significance level Another name for the *p*-value method of testing hypotheses.

observed value A statistic computed from data gathered in an experiment that is used in the determination of whether or not to reject the null hypothesis.

ogive A cumulative frequency polygon; plotted by graphing a dot at each class endpoint for the cumulative or decumulative frequency value and connecting the dots.

one-tailed test A statistical test wherein the researcher is interested only in testing one side of the distribution.

one-way analysis of variance The process used to analyse a completely randomized experimental design. This process involves computing a ratio of the variance between treatment levels of the independent variable to the error variance. This ratio is an *F* value, which is then used to determine whether there are any significant differences between the means of the treatment levels.

operating-characteristic (OC) curve In hypothesis testing, a graph of Type II error probabilities for various possible values of an alternative hypothesis.

opportunity loss table A decision table constructed by subtracting all payoffs for a given state of nature from the maximum payoff for that state of nature and doing this for all states of nature; displays the lost opportunities or regret that would occur for a given decision alternative if that particular state of nature occurred.

Ordinal-level data Next highest level of data above nominal level data; can be used to order or rank items, objects, or people.

outliers Data points that lie apart from the rest of the points.

P

p **chart** A quality control chart for attribute compliance that graphs the proportion of sample items in non-compliance with specifications for multiple samples.

*p***-value method** A method of testing hypotheses in which there is no preset level of α. The probability of getting a test statistic at least as extreme as the observed test statistic is computed under the assumption that the null hypothesis is true. This probability is called the *p*-value, and it is the smallest value of α for which the null hypothesis can be rejected.

Paasche price index A type of weighted aggregate price index in which the quantity values used in the calculations are from the year of interest.

parameter A descriptive measure of the population.

parametric statistics A class of statistical techniques that contain assumptions about the population and that are used only with interval- and ratio-level data.

Pareto analysis A quantitative tallying of the number and types of defects that occur with a product or service, often recorded in a Pareto chart.

Pareto chart A vertical bar chart in which the number and types of defects for a product or service are graphed in order of magnitude from greatest to least.

partial regression coefficient The coefficient of an independent variable in a multiple regression model that represents the increase that will occur in the value of the dependent variable from a one-unit increase in the independent variable if all other variables are held constant.

payoff table A matrix that displays the decision alternatives, the states of nature, and the payoffs for a particular decision-making problem; also called a decision table.

payoffs The benefits or rewards that result from selecting a particular decision alternative.

Pearson product-moment correlation coefficient A measure of the linear correlation of two variables.

percentiles Measures of central tendency that divide a group of data into 100 parts.

pie chart A circular depiction of data where the area of the whole pie represents 100% of the data being studied and slices represent a percentage breakdown of the sublevels.

platykurtic Distributions that are flat and spread out.

point estimate An estimate of a population parameter constructed from a statistic taken from a sample.

Poisson distribution A discrete distribution that is constructed from the probability of occurrence of rare events over an interval; focuses only on the number of discrete occurrences over some interval or continuum.

poka-yoke Means 'mistake proofing' and uses devices, methods, or inspections in order to avoid machine error or simple human error.

population A collection of persons, objects, or items of interest.

post hoc After the experiment; pairwise comparisons made by the researcher *after* determining that there is a significant overall F value from ANOVA; also called 'a posteriori'.

power The probability of rejecting a false null hypothesis.

power curve A graph that plots the power values against various values of the alternative hypothesis.

prediction interval A range of values used in regression analysis to estimate a single value of y for a given value of x.

probabilistic model A model that includes an error term that allows for various values of output to occur for a given value of input.

probability matrix A two-dimensional table that displays the marginal and intersection probabilities of a given problem.

process A series of actions, changes, or functions that bring about a result.

product quality A view of quality in which quality is measurable in the product based on the fact that there are perceived differences in products and quality products possess more attributes.

proportionate stratified random sampling A type of stratified random sampling in which the proportions of the items selected for the sample from the strata reflect the proportions of the strata in the population.

Q

quadratic regression model A multiple regression model in which the predictors are a variable and the square of the variable.

qualitative variable Another name for a dummy or indicator variable; represents whether or not a given item or person possesses a certain characteristic and is usually coded as 0 or 1.

quality When a product delivers what is stipulated in its specifications.

quality circle A small group of workers consisting of supervisors and six to 10 employees who meet frequently and regularly to consider quality issues in their department or area of the business.

quality control The collection of strategies, techniques, and actions taken by an organization to ensure the production of quality products.

quartiles Measures of central tendency that divide a group of data into four subgroups or parts.

quota sampling A non-random sampling technique in which the population is stratified on some characteristic and then elements selected for the sample are chosen by non-random processes.

R

R chart A plot of sample ranges used in quality control.

R^2 The coefficient of multiple determination; a value that ranges from 0 to 1 and represents the proportion of the dependent variable in a multiple regression model that is accounted for by the independent variables.

random sampling Sampling in which every unit of the population has the same probability of being selected for the sample.

random variable A variable that contains the outcomes of a chance experiment.

randomized block design An experimental design in which there is one independent variable of interest and a second variable, known as a blocking variable, that is used to control for confounding or concomitant variables.

range The difference between the largest and the smallest values in a set of numbers.

Ratio-level data Highest level of data measurement; contains the same properties as interval-level data, with the additional property that zero has meaning and represents the absence of the phenomenon being measured.

rectangular distribution A relatively simple continuous distribution in which the same height is obtained over a range of values; also referred to as the uniform distribution.

re-engineering A radical approach to total quality management in which the core business processes of a company is redesigned.

regression analysis The process of constructing a mathematical model or function that can be used to predict or determine one variable by any other variable.

rejection region If a computed statistic lies in this portion of a distribution, the null hypothesis will be rejected.

related measures Another name for matched pairs or paired data in which measurements are taken from pairs of items or persons matched on some characteristic or from a before-and-after design and then separated into different samples.

relative frequency The proportion of the total frequencies that fall into any given class interval in a frequency distribution.

relative frequency of occurrence Assigning probability based on cumulated historical data.

repeated measures design A randomized block design in which each block level is an individual item or person, and that person or item is measured across all treatments.

research hypothesis A statement of what the researcher believes will be the outcome of an experiment or a study.

residual The difference between the actual y value and the y value predicted by the regression model; the error of the regression model in predicting each value of the dependent variable.

residual plot A type of graph in which the residuals for a particular regression model are plotted along with their associated values of x.

response plane A plane fit in a three-dimensional space and that represents the response surface defined by a multiple regression model with two independent first-order variables.

response surface The surface defined by a multiple regression model.

response variable The dependent variable in a multiple regression model; the variable that the researcher is trying to predict.

risk avoider A decision maker who avoids risk whenever possible and is willing to drop out of a game when given the chance even when the payoff is less than the expected monetary value.

risk taker A decision maker who enjoys taking risks and will not drop out of a game unless the payoff is more than the expected monetary value.

robust Describes a statistical technique that is relatively insensitive to minor violations in one or more of its underlying assumptions.

runs test A non-parametric test of randomness used to determine whether the order or sequence of observations in a sample is random.

S

sample A portion of the whole.

sample proportion The quotient of the frequency at which a given characteristic occurs in a sample and the number of items in the sample.

sample-size estimation An estimate of the size of sample necessary to fulfil the requirements of a particular level of confidence and to be within a specified amount of error.

sample space A complete roster or listing of all elementary events for an experiment.

sampling error Error that occurs when the sample is not representative of the population.

scatter plot (chart) A plot or graph of the pairs of data from a simple regression analysis.

search procedures Processes whereby more than one multiple regression model is developed for a given database, and the models are compared and sorted by different criteria, depending on the given procedure.

seasonal effects Patterns of data behaviour that occur in periods of time of less than one year, often measured by the month.

serial correlation A problem that arises in regression analysis when the error terms of a regression model are correlated due to time-series data; also called autocorrelation.

set notation The use of braces to group numbers that have some specified characteristic.

simple average The arithmetic mean or average for the values of a given number of time periods of data.

simple average model A forecasting averaging model in which the forecast for the next time period is the average of values for a given number of previous time periods.

simple index number A number determined by computing the ratio of a quantity, price, or cost for a particular year of interest to the quantity price or cost of a base year, expressed as a percentage.

simple random sampling The most elementary of the random sampling techniques; involves numbering each item in the population and using a list or roster of random numbers to select items for the sample.

simple regression Bivariate, linear regression; also known as bivariate regression.

Six Sigma A total quality management approach that measures the capability of a process to perform defect-free work, where a defect is defined as anything that results in customer dissatisfaction.

skewness The lack of symmetry of a distribution of values.

smoothing techniques Forecasting techniques that produce forecasts based on levelling out the irregular fluctuation effects in time-series data.

snowball sampling A non-random sampling technique in which survey subjects who fit a desired profile are selected based on referral from other survey respondents who also fit the desired profile.

Spearman's rank correlation A measure of the correlation of two variables; used when only ordinal level or ranked data are available.

standard deviation The square root of the variance.

standard error of the estimate (s_e) A standard deviation of the error of a regression model.

standard error of the mean The standard deviation of the distribution of sample means.

standard error of the proportion The standard deviation of the distribution of sample proportions.

standardized normal distribution z distribution; a distribution of z scores produced for values from a normal distribution with a mean of 0 and a standard deviation of 1.

states of nature The occurrences of nature that can happen after a decision has been made that can affect the outcome of the decision and over which the decision maker has little or no control.

stationary Time-series data that contain no trend, cyclical, or seasonal effects.

statistic A descriptive measure of a sample.

statistical hypothesis A formal hypothesis structure set up with a null and an alternative hypothesis to scientifically test research hypotheses.

statistics A science dealing with the collection, analysis, interpretation, and presentation of numerical data.

stem-and-leaf plot A plot of numbers constructed by separating each number into two groups, a stem and a leaf. The leftmost digits are the stems and the rightmost digits are the leaves.

stepwise regression A step-by-step multiple regression search procedure that begins by developing a regression model with a single predictor variable and adds and deletes predictors one step at a time, examining the fit of the model at each step until there are no more significant predictors remaining outside the model.

stratified random sampling A type of random sampling in which the population is divided into various non-overlapping strata and then items are randomly selected into the sample from each stratum.

subjective probability A probability assigned based on the intuition or reasoning of the person determining the probability.

substantive result Occurs when the outcome of a statistical study produces results that are important to the decision maker.

sum of squares of error (SSE) The sum of the residuals squared for a regression model.

sum of squares of x The sum of the squared deviations about the mean of a set of values.

systematic sampling A random sampling technique in which every kth item or person is selected from the population.

T

t distribution A distribution that describes the sample data when the standard deviation is unknown and the population is normally distributed.

t value The computed value of t used to reach statistical conclusions regarding the null hypothesis in small-sample analysis.

team building When a group of employees are organized as an entity to undertake management tasks and perform other functions such as organizing, developing, and overseeing projects.

time-series data Data gathered on a given characteristic over a period of time at regular intervals.

total quality management (TQM) A programme that occurs when all members of an organization are involved in improving quality; all goals and objectives of the organization come under the purview of quality control and are measured in quality terms.

transcendent quality A view of quality that implies that a product has an innate excellence, uncompromising standards, and high achievement.

treatment variable The independent variable of an experimental design that the researcher either controls or modifies.

trend Long-run general direction of a business climate over a period of several years.

Tukey–Kramer procedure A modification of the Tukey HSD multiple comparison procedure; used when there are unequal sample sizes.

Tukey's four-quadrant approach A graphical method using the four quadrants for determining which expressions of Tukey's ladder of transformations to use.

Tukey's honestly significant difference (HSD) test In analysis of variance, a technique used for pairwise a posteriori multiple comparisons to determine if there are significant differences between the means of any pair of treatment levels in an experimental design. This test requires equal sample sizes and uses a q value along with the mean square error in its computation.

Tukey's ladder of transformations A process used for determining ways to recode data in multiple regression analysis to achieve potential improvement in the predictability of the model.

two-stage sampling Cluster sampling done in two stages: a first round of samples is taken and then a second round is taken from within the first samples.

two-tailed test A statistical test wherein the researcher is interested in testing both sides of the distribution.

two-way analysis of variance (two-way ANOVA) The process used to statistically test the effects of variables in factorial designs with two independent variables.

Type I error An error committed by rejecting a true null hypothesis.

Type II error An error committed by failing to reject a false null hypothesis.

U

ungrouped data Raw data, or data that have not been summarized in any way.

uniform distribution A relatively simple continuous distribution in which the same height is obtained over a range of values; also called the rectangular distribution.

union A new set of elements formed by combining the elements of two or more other sets.

union probability The probability of one event occurring or the other event occurring or both occurring.

unweighted aggregate price index number The ratio of the sum of the prices of a market basket of items for a particular year to the sum of the prices of those same items in a base year, expressed as a percentage.

upper control limit (UCL) The top-end line of a control chart, usually situated approximately three standard deviations of the statistic above the centreline; data points above this line indicate quality-control problems.

user quality A view of quality in which the quality of the product is determined by the user.

utility The degree of pleasure or displeasure a decision maker has in being involved in the outcome selection process given the risks and opportunities available.

V

value quality A view of quality having to do with price and costs and whether the consumer got his or her money's worth.

variance The average of the squared deviations about the arithmetic mean for a set of numbers.

variance inflation factor A statistic computed using the R^2 value of a regression model developed by predicting one independent variable of a regression analysis by other independent variables; used to determine whether there is multicollinearity among the variables.

W

weighted aggregate price index number A price index computed by multiplying quantity weights and item prices and summing the products to determine a market basket's worth in a given year and then determining the ratio of the market basket's worth in the year of interest to the same value computed for a base year, expressed as a percentage.

weighted moving average A moving average in which different weights are applied to the data values from different time periods.

Wilcoxon matched-pairs signed rank test A non-parametric alternative to the t test for two related or dependent samples.

X

$\bar{x}$ chart A quality control chart for measurements that graphs the sample means computed for a series of small random samples over a period of time.

Z

z distribution A distribution of z scores; a normal distribution with a mean of 0 and a standard deviation of 1.

z score The number of standard deviations a value (x) is above or below the mean of a set of numbers when the data are normally distributed.